Classroom Companion: Economics

The Classroom Companion series in Economics includes undergraduate and graduate textbooks alike. It welcomes fundamental textbooks aimed at introducing students to the core concepts, empirical methods, theories and tools of the field, as well as advanced textbooks written for students at the Master and PhD level seeking a deeper understanding of economic theory, mathematical tools and quantitative methods.

Jörg Rothe
Editor

Economics and Computation

An Introduction to Algorithmic Game Theory,
Computational Social Choice,
and Fair Division

Second Edition

Illustrations by Irene Rothe

Editor
Jörg Rothe
MNF, Department of Computer Science
Heinrich-Heine-Universität Düsseldorf
Düsseldorf, Nordrhein-Westfalen, Germany

ISSN 2662-2882   ISSN 2662-2890 (electronic)
Classroom Companion: Economics
ISBN 978-3-031-60098-2   ISBN 978-3-031-60099-9 (eBook)
https://doi.org/10.1007/978-3-031-60099-9

© The Editor(s) (if applicable) and The Author(s), under exclusive license to Springer Nature Switzerland AG 2016, 2024
This work is subject to copyright. All rights are solely and exclusively licensed by the Publisher, whether the whole or part of the material is concerned, specifically the rights of translation, reprinting, reuse of illustrations, recitation, broadcasting, reproduction on microfilms or in any other physical way, and transmission or information storage and retrieval, electronic adaptation, computer software, or by similar or dissimilar methodology now known or hereafter developed.
The use of general descriptive names, registered names, trademarks, service marks, etc. in this publication does not imply, even in the absence of a specific statement, that such names are exempt from the relevant protective laws and regulations and therefore free for general use.
The publisher, the authors, and the editors are safe to assume that the advice and information in this book are believed to be true and accurate at the date of publication. Neither the publisher nor the authors or the editors give a warranty, expressed or implied, with respect to the material contained herein or for any errors or omissions that may have been made. The publisher remains neutral with regard to jurisdictional claims in published maps and institutional affiliations.

This Springer imprint is published by the registered company Springer Nature Switzerland AG
The registered company address is: Gewerbestrasse 11, 6330 Cham, Switzerland

If disposing of this product, please recycle the paper.

Foreword

Foreword by Michael Wooldridge for the Second Edition

The synthesis of ideas from economics and computer science has been one of the most important and influential developments in theoretical computer science and AI this century. This admirable book provides a comprehensive survey of the key ideas, results, and directions. It is a perfect graduate-level introduction to the area, written by many leaders of the field.

Michael Wooldridge
Hertford College, University of Oxford, UK
February 2024

Foreword by Matthew O. Jackson and Yoav Shoham for the First Edition

One of the most exciting, interesting, and important areas of interdisciplinary research over the past two decades has been at the juncture of computer science and economics, breathing new life into game-theoretic analyses of the many mechanisms and institutions that pervade our lives. It is an area that gives rise to fascinating intellectual problems, that are at the same time highly relevant to electronic commerce and other significant areas of life in the 21st century. In particular, the focus on complexity has not only forced important practical considerations to be taken into account in designing systems from elections to auctions, but has also provided new insights into why we see specific institutional features rather than more complex cousins that might be better from an unconstrained theoretical perspective.

The literature has rapidly advanced on this subject, and it is getting to the point where it is increasingly difficult to keep track of what is known and what is not, and what general insights are emerging. This volume fills a critical void. While there exist other, excellent publications covering some parts of this growing and sprawling literature, notably missing has been coverage of the areas of computational social choice and fair division, the focus of this volume. Its coverage is broad and encompassing: from voting systems, to judgment aggregation, to the allocation of indivisible goods, to the age-old problem of fair division viewed through a new lens. Moreover, it provides a very accessible introduction that should be required reading for anyone venturing into the area for the first time. We offer our congratulations to the editor and the authors for this impressive achievement.

Matthew O. Jackson and Yoav Shoham
Stanford University, Palo Alto, USA
May 2015

Preface

Our work on the second edition of this book has started in 2020. Its first edition, published in 2015, was a largely expanded successor of the German book *"Einführung in Computational Social Choice: Individuelle Strategien und kollektive Entscheidungen beim Spielen, Wählen und Teilen"* [822], published by Spektrum Akademischer Verlag in 2011. While this German book had been coauthored by only four of the current authors and presented its seven chapters on only 375 pages, I was proud and grateful to have found and persuaded six additional coauthors for the first edition of the English book, each an internationally renowned expert of his or her field, to contribute to this book. Now, for its second edition, I am again very proud and grateful to have found and persuaded three more internationally renowned experts as coauthors. Moreover, the second edition offers the new Chapter 6 on multiwinner voting. All "old" chapters have been updated and extended, and some of them (namely, Chapter 3 on cooperative game theory and Chapter 9 on fair division of indivisible goods) have been completely restructured and substantially extended.

Here is some information on each of the 13 authors and their chapters:

- Dorothea Baumeister[1] from Federal University of Applied Administrative Sciences in Brühl, Germany, has coauthored Chapter 4 on preference aggregation by voting, Chapter 6 on multiwinner voting, and Chapter 7 on judgment aggregation;
- Martin Bullinger[2] from TU München, Germany, and now at University of Oxford, UK, has coauthored Chapter 3 on cooperative game theory;
- Edith Elkind from University of Oxford, UK, also has coauthored Chapter 3 on cooperative game theory;

---

[1] Her work has been supported in part by DFG grants BA 6270/1-1 and RO 1202/15 1, an NRW grant for gender-sensitive universities supporting her as a junior professor for Computational Social Choice at Heinrich-Heine-Universität Düsseldorf, Germany, and by the project "Online Partizipation," both funded by the NRW Ministry for Innovation, Science, and Research.

[2] His work has been supported in part by DFG grant BR 2312/12-1.

- Gábor Erdélyi[3] from University of Canterbury in Christchurch, New Zealand, has coauthored Chapter 7 on judgment aggregation;
- Piotr Faliszewski[4] from AGH University of Science and Technology in Kraków, Poland, has coauthored Chapter 2 on noncooperative game theory and Chapter 6 on multiwinner voting;
- Ronald de Haan from University of Amsterdam, The Netherlands, has coauthored Chapter 7 on judgment aggregation;
- Edith Hemaspaandra[5] from Rochester Institute of Technology, USA, has coauthored Chapter 5 on the complexity of manipulative actions in single-peaked societies;
- Lane A. Hemaspaandra[6] from University of Rochester, USA, also has coauthored Chapter 5 on the complexity of manipulative actions in single-peaked societies;
- Jérôme Lang[7] from CNRS-LAMSADE, Université Paris-Dauphine, France, has coauthored Chapter 9 on fair division of indivisible goods;
- Claudia Lindner[8] from The University of Manchester, UK, has coauthored Chapter 8 on cake-cutting: fair division of divisible goods;
- Irene Rothe from Bonn-Rhein-Sieg University of Applied Sciences, Germany, has coauthored Chapter 2 on noncooperative game theory and has created numerous wonderful illustrations for all chapters in this book;

---

[3] His work has been supported in part by DFG grant ER-738/2-1, by "Förderverein des FB Wirtschaftswissenschaften, Wirtschaftsinformatik und Wirtschaftsrecht der Universität Siegen e.V.," and by the Short-term Scientific Mission program of COST Action IC1205 on Computational Social Choice.

[4] His work has been supported in part by the funds that the Polish Ministry of Science and Higher Education assigned to AGH University of Science and Technology and, in particular, by AGH University grant 11.11.230.124.

[5] Her work has been supported in part by NSF grants CCF-1101452 and DUE-1819546 and by ESF COST Action IC1205 on Computational Social Choice.

[6] His work has been supported in part by NSF grants CCF-0915792, CCF-1101479, CCF-2006496, and DUE-2135431 and by ESF COST Action IC1205 on Computational Social Choice.

[7] His work has been supported in part by the Alexander von Humboldt Foundation with a Humboldt Research Award (supporting many wonderful research visits to Düsseldorf, Germany) and the PRAIRIE 3IA Institute (ANR-19-P3IA-0001), by ANR Project CoCoRICo-CoDec, by the DAAD-PPP/PHC PROCOPE program entitled "Fair Division of Indivisible Goods: Incomplete Preferences, Communication Protocols and Computational Resistance to Strategic Behavior," and by ESF COST Action IC1205 on Computational Social Choice.

[8] Her work has been supported in part by a Sir Henry Dale Fellowship, jointly funded by the Wellcome Trust and the Royal Society of the United Kingdom (223267/Z/21/Z).

- Jörg Rothe[9] from Heinrich-Heine-Universität Düsseldorf, Germany, has edited this book, has written introductory Chapter 1, and has coauthored Chapters 2–9; and
- Piotr Skowron from University of Warsaw, Poland, has coauthored Chapter 6 on multiwinner voting.

The subject of this book, generally speaking, is *collective decision-making* in three areas, each having both an economical and a computational dimension. Accordingly, the book is divided into three parts:

Part I (Playing Successfully) is concerned with algorithmic game theory, where Chapter 2 introduces to noncooperative games and Chapter 3 to cooperative games, focusing on their computational aspects.

Part II (Voting and Judging) introduces to computational social choice. Chapter 4 is concerned with preference aggregation by voting, first providing some background from social choice theory and then focusing on the complexity of determining (possible and necessary) winners in elections and of manipulative actions to influence their outcomes. Chapter 5 sheds some light on the complexity of manipulative actions in single-peaked societies. Chapter 6 introduces to multiwinner voting where the goal is not to elect a single winner of an election but to find a *winning committee* of a certain size, and we again consider both axiomatic properties and computational aspects. Finally, Chapter 7 introduces to the field of judgment aggregation, again with a focus on the complexity of related problems.

Part III (Fair Division) deals with mechanisms of fair division among a number of agents, both for fairly dividing a divisible good (an area known as "cake-cutting") in Chapter 8 and for fairly dividing indivisible goods in Chapter 9, once more focusing on computational aspects.

These three parts are preceded by a brief introduction to playing, voting, and dividing in Chapter 1, which also gives some of the needed notions from computational complexity theory to be used throughout the book.

This book—which has grown from 612 pages in the first edition to 766 pages in the second—provides an accessible introduction to the areas mentioned above, which makes it a valuable source for teaching. Indeed, on a regular basis I have taught courses related to most of the single chapters of this book at my university since 2009, and so have the other authors at their universities. A noteworthy feature of this book is that its most important concepts and ideas are introduced not only in formal, technical terms but are

---

[9] His work has been supported in part by DFG grants RO-1202/14-1, RO-1202/15-1, and RO 1202/21-1, the DAAD-PPP/PHC PROCOPE program entitled "Fair Division of Indivisible Goods: Incomplete Preferences, Communication Protocols and Computational Resistance to Strategic Behavior," by the project "Online Partizipation" funded by the NRW Ministry for Innovation, Science, and Research, and by ESF COST Action IC1205 on Computational Social Choice.

also accompanied by numerous examples, usually told as a story from everyday life and featuring the same main characters throughout the book—have a look at the many gray boxes for the stories and at Figure 1.1 on page 3 for the book's main characters! The wonderful illustrations in all chapters have been created by Irene Rothe.

Care has been taken to unify notation and formalism throughout the book, and there are plenty of cross-references between the chapters to point the reader to identical or closely related notions in different contexts. Moreover, an extensive bibliography with 974 references (compared with the 625 references in the first edition of this book, a more than 50% increase) and a comprehensive index with 7,600 entries on more than 35 pages (to be compared with only 25 pages in the first edition) will be helpful for the reader. Note that authors are indexed even if their names are hidden in "et al." or in a plain reference without author names.

Regarding personal pronouns, referring to individual players, voters, candidates, judges, or agents by "she" alone or "he" alone would be inappropriate, and referring to them as "it" is simply wrong and ugly; therefore, we sometimes follow the approach of Chalkiadakis, Elkind, and Wooldridge [268] who promote an interleaved, (semi-)random usage of "she" and "he" and sometimes we write "she or he" and sometimes simply "they." This different usage also depends on the differing tastes of the single chapter authors.

On a personal note, I'm deeply indebted to many individuals for their help in proofreading the single chapters of this book. They have done a great job, and my collective thanks go to Dorothea Baumeister and Piotr Faliszewski (for proofreading some of the chapters they did not coauthor), and to my former PhD students Daniel Neugebauer, Nhan-Tam Nguyen, Anja Rey, and Lena Schend. All remaining typos, of course, are solely my responsibility.

I have been working on parts of this book during a number of research visits to Université Paris-Dauphine, Stanford University, Rochester Institute of Technology, and University of Rochester (in particular, I spent my research sabbaticals in 2013 and 2019 both at Stanford University in Palo Alto, California, and in Rochester, New York), and I'm deeply grateful to the hosts of these visits—Jérôme Lang, Yoav Shoham, Edith Hemaspaandra, and Lane A. Hemaspaandra—for their warm hospitality. Last but not least, I thank Matthew O. Jackson and Yoav Shoham from Stanford University and Michael Wooldridge from University of Oxford for reading an early draft of this book and for writing a foreword.

Jörg Rothe, Editor
Düsseldorf, Germany
March 2024

Contents

Foreword v

Preface ix

1 **Playing, Voting, and Dividing** 1
  J. Rothe
  1.1 Playing 3
    1.1.1 Noncooperative Game Theory 3
    1.1.2 Cooperative Game Theory 5
  1.2 Voting 5
    1.2.1 Preference Aggregation by Voting 6
    1.2.2 Manipulative Actions in Single-Peaked Societies 9
    1.2.3 Multiwinner Voting 9
    1.2.4 Judgment Aggregation 9
  1.3 Dividing 10
    1.3.1 Cake-cutting: Fair Division of Divisible Goods 10
    1.3.2 Fair Division of Indivisible Goods 11
    1.3.3 A Brief Digression to Single-Item Auctions 12
  1.4 Some Literature Pointers 17
  1.5 A Brief Digression to Computational Complexity 18
    1.5.1 Some Foundations of Complexity Theory 19
    1.5.2 The Satisfiability Problem of Propositional Logic 24
    1.5.3 A Brief Compendium of Complexity Classes 34

Part I Playing Successfully

2 **Noncooperative Game Theory** 43
  P. Faliszewski, I. Rothe, and J. Rothe
  2.1 Foundations 44
    2.1.1 Normal Form, Dominant Strategies, and Equilibria 45
    2.1.2 Further Two-Player Games 52

	2.2	Nash Equilibria in Mixed Strategies............................	62
		2.2.1 Definition and Application to Two-Player Games	62
		2.2.2 Existence of Nash Equilibria in Mixed Strategies	71
	2.3	Checkmate: Trees for Games with Perfect Information	84
		2.3.1 Sequential Two-Player Games	84
		2.3.2 Equilibria in Game Trees	97
	2.4	Full House: Games with Incomplete Information	103
		2.4.1 The Monty Hall Problem	104
		2.4.2 Analysis of a Simple Poker Variant	110
	2.5	How Hard Is It to Find a Nash Equilibrium?................	122
		2.5.1 Nash Equilibria in Zero-Sum Games	122
		2.5.2 Nash Equilibria in General Normal Form Games	125

3 Cooperative Game Theory................................. 139
M. Bullinger, E. Elkind, and J. Rothe

	3.1	Foundations	141
		3.1.1 Cooperative Games with Transferable Utility	141
		3.1.2 Special Subclasses of Cooperative Games	144
	3.2	Stability in Cooperative Games	149
		3.2.1 Imputations	150
		3.2.2 The Core of a Cooperative Game...................	150
		3.2.3 Further Stability Concepts in Cooperative Games	155
	3.3	Fairness in Cooperative Games	162
		3.3.1 The Shapley–Shubik Index and the Shapley Value	162
		3.3.2 The Banzhaf Indices	168
	3.4	Counting and Representing Cooperative Games	171
		3.4.1 A Universal Representation for Simple Games........	172
		3.4.2 Cooperative Games on Graphs	175
	3.5	Computational Complexity of Identifying Good Outcomes ...	178
		3.5.1 An Oracle-Based Approach: Core-Stable Outcomes in Convex and Simple Games	179
		3.5.2 Complexity of Problems for Weighted Voting Games ..	180
		3.5.3 Complexity of Problems for Induced Subgraph Games	194
	3.6	Hedonic Games	198
		3.6.1 Stability in Hedonic Games	199
		3.6.2 Representation Formalisms	203
		3.6.3 Complexity of Stability in Hedonic Games	207
		3.6.4 Dynamics in Hedonic Games.......................	222
		3.6.5 Other Important Concepts for Hedonic Games	225

Part II Voting and Judging

4 Preference Aggregation by Voting 233
D. Baumeister and J. Rothe
- 4.1 Some Basic Voting Systems 234
  - 4.1.1 Scoring Protocols 235
  - 4.1.2 Voting Systems Based on Pairwise Comparisons 237
  - 4.1.3 Approval Voting and Range Voting 249
  - 4.1.4 Voting Systems Proceeding in Stages 251
  - 4.1.5 Hybrid Voting Systems 258
  - 4.1.6 Overview of Some Fundamental Voting Systems 263
- 4.2 Properties of Voting Systems and Impossibility Theorems ... 264
  - 4.2.1 The Condorcet and the Majority Criterion 265
  - 4.2.2 Nondictatorship, Pareto Consistency, and Consistency 268
  - 4.2.3 Independence of Irrelevant Alternatives 272
  - 4.2.4 Resoluteness and Citizens' Sovereignty 273
  - 4.2.5 Strategy-Proofness and Independence of Clones 274
  - 4.2.6 Anonymity, Neutrality, and Monotonicity 276
  - 4.2.7 Homogeneity, Participation, and Twins Welcome 280
  - 4.2.8 Overview of Properties of Voting Systems 286
- 4.3 Complexity of Voting Problems 287
  - 4.3.1 Winner Determination 289
  - 4.3.2 Possible and Necessary Winners 297
  - 4.3.3 Manipulation 306
  - 4.3.4 Control 328
  - 4.3.5 Bribery 358

5 The Complexity of Manipulative Actions in Single-Peaked Societies 369
E. Hemaspaandra, L.A. Hemaspaandra, and J. Rothe
- 5.1 Single-Peaked Electorates 373
- 5.2 Control of Single-Peaked Electorates 376
- 5.3 Manipulation of Single-Peaked Electorates 386
- 5.4 Bribery of Single-Peaked Electorates 393
- 5.5 Do Nearly Single-Peaked Electorates Restore Intractability? 395
  - 5.5.1 K-Maverick-Single-Peakedness 397
  - 5.5.2 Swoon-Single-Peakedness 398

6 Multiwinner Voting 403
D. Baumeister, P. Faliszewski, J. Rothe, and P. Skowron
- 6.1 Introduction 403
- 6.2 Preferences as Rankings 404
  - 6.2.1 k-Borda and Individual Excellence 405
  - 6.2.2 The Chamberlin–Courant Rule and Diversity 407
  - 6.2.3 Committee Scoring Rules 410

		6.2.4	STV and Proportionality for Solid Coalitions	422

	6.3	Preferences as Approvals 426
		6.3.1	Approval Chamberlin–Courant and Diversity	428
		6.3.2	Extended Justified Representation and Proportional Approval Voting.	430
		6.3.3	The Phragmén Sequential Rule.	435
		6.3.4	Different Interpretations of Approvals	437
	6.4	Different Costs of Candidates and Score Voting	441
	6.5	Complexity of Computing Winning Committees	447
		6.5.1	The Problem of Winner Determination.	448
		6.5.2	Exact Agorithms	452
		6.5.3	Approximation Algorithms	456
		6.5.4	Other Approaches	459
	6.6	Further Topics Related to Committee Elections	459
		6.6.1	Robustness	460
		6.6.2	Manipulation, Control, and Bribery.	462

7   Judgment Aggregation 467
D. Baumeister, G. Erdélyi, R. de Haan, and J. Rothe

	7.1	Foundations 471
		7.1.1	Judgment Aggregation 471
		7.1.2	Other Frameworks for Judgment Aggregation 474
	7.2	Judgment Aggregation Procedures and Their Properties. 478
		7.2.1	Some Specific Judgment Aggregation Procedures 478
		7.2.2	Properties, Impossibility Results, and Characterizations 483
	7.3	Complexity of Judgment Aggregation Problems 486
		7.3.1	Winner Determination in Judgment Aggregation 487
		7.3.2	Safety of the Agenda. 488
		7.3.3	Manipulation in Judgment Aggregation 489
		7.3.4	Bribery in Judgment Aggregation 495
		7.3.5	Control in Judgment Aggregation 500
	7.4	Concluding Remarks. 504

Part III Fair Division

8   Cake-Cutting: Fair Division of Divisible Goods 507
C. Lindner and J. Rothe

	8.1	How to Have a Great Party with only a Single Cake 507
	8.2	Basics. 508
	8.3	Valuation Criteria 513
		8.3.1	Fairness. 513
		8.3.2	Efficiency 522
		8.3.3	Manipulability 523
		8.3.4	Runtime 527
	8.4	Cake-Cutting Protocols 528

		8.4.1	Two Envy-Free Protocols for Two Players 529
		8.4.2	Proportional Protocols for n Players 535
		8.4.3	Super-Proportional Protocols for n Players 557
		8.4.4	A Royal Wedding: Dividing into Unequal Shares 562
		8.4.5	Envy-Free Protocols for Three and More Players 564
		8.4.6	Oversalted Cream Cake: Dirty-Work Protocols 573
		8.4.7	Avoiding Crumbs: Minimizing the Number of Cuts . . . 577
		8.4.8	Degree of Guaranteed Envy-Freeness 598
		8.4.9	Overview of Some Cake-Cutting Protocols 602

9 Fair Division of Indivisible Goods 605
J. Lang and J. Rothe

9.1 Introduction 605
9.2 Definition and Classification of Allocation Problems 607
  9.2.1 Allocation Problems 607
  9.2.2 Classification of Allocation Problems 608
9.3 Preference Elicitation and Compact Representation 613
  9.3.1 Ordinal Preference Languages 615
  9.3.2 Cardinal Preference Languages 617
9.4 Pareto Efficiency and Envy-Freeness 622
  9.4.1 Pareto Efficiency 622
  9.4.2 Envy-Freeness 623
  9.4.3 Envy-Free and Pareto-Efficient Allocations: Existence
     and Computation 623
  9.4.4 Relaxing Pareto-Efficiency: Partial Assignments 624
9.5 Relaxations of Envy-Freeness 625
  9.5.1 Degrees of Envy 625
  9.5.2 Max-Min Fair Share and Min-Max Fair Share 626
  9.5.3 Envy-Freeness up to One Good and up to Any Good . 628
  9.5.4 Proportional Fair Share 629
  9.5.5 Pairwise Max-Min Fair Share and Group Max-Min
     Fair Share 631
  9.5.6 Epistemic Envy-Freeness 632
  9.5.7 Envy-Freeness up to k Hidden Goods 633
  9.5.8 Objective Envy-Freeness 633
  9.5.9 Discussion 634
9.6 Maximizing Social Welfare: The Santa Claus Problem 635
  9.6.1 Pure Egalitarian Social Welfare 636
  9.6.2 Maximum Leximin Social Welfare 640
  9.6.3 Maximum Nash Social Welfare 642
  9.6.4 Utilitarian Social Welfare 646
  9.6.5 Summary 647
9.7 Centralized Fair Division with Ordinal Preferences 649
9.8 Fair Division with Money, and Related Issues 653
  9.8.1 Money 653

| | | 9.8.2 One Divisible Good: The Adjusted Winner Procedure | 655 |

- 9.9 Decentralized Allocation Protocols 659
  - 9.9.1 The Descending Demand Protocols 660
  - 9.9.2 The Picking Sequences Protocols 663
  - 9.9.3 Contested Pile-Based Protocols: Undercut 665
  - 9.9.4 Protocols Based on Local Exchanges 667
- 9.10 Extensions and Variants 669
  - 9.10.1 Fair Division of Bads 669
  - 9.10.2 Constrained Fair Division 670
  - 9.10.3 Partial Knowledge 672
  - 9.10.4 Fair Division among Groups of Agents and Externalities 673
  - 9.10.5 Sharing Goods 674
  - 9.10.6 Online Fair Division 674
  - 9.10.7 Private Endowments and Asymmetric Agents 674
- 9.11 Further Issues 675
  - 9.11.1 Strategic Agents 675
  - 9.11.2 Matching 677
  - 9.11.3 Randomized Fair Division 678
  - 9.11.4 Fair Allocation of Public Goods 679

References 681

Index 731

List of Figures

1.1	The main characters in this book	3
1.2	Anna, Belle, and Chris are voting on which game to play	6
1.3	The Condorcet paradox	8
1.4	Division of a cake into three portions	10
1.5	A boolean circuit	27
1.6	Inclusions between some complexity classes	38
2.1	Relations between solution concepts for games in normal form	52
2.2	Inclusions among solution concepts for games in normal form	52
2.3	The chicken game	55
2.4	David and Edgar at the penalty shoot-out	57
2.5	Cycles of dominance relations in paper-rock-scissors	59
2.6	A convex and a nonconvex set	72
2.7	n-simplexes for $0 \leq n \leq 3$	74
2.8	Convex gain sets for pure and mixed strategy sets	75
2.9	A properly labeled simplicial subdivision of a 2-simplex	76
2.10	Walking through the 2-simplex $T_2 = \vec{x}_0 \vec{x}_1 \vec{x}_2$	78
2.11	All walks through the 2-simplex $T_2 = \vec{x}_0 \vec{x}_1 \vec{x}_2$	79
2.12	A course of the game in Tic-Tac-Toe	87
2.13	A part of the game tree for Tic-Tac-Toe	87
2.14	The game of geography based on English country names	92
2.15	A graph simulating a formula as a game of geography	95
2.16	Game tree for Edgar's campaign game in extensive form	100
2.17	Implementing backward induction through DFS	103
2.18	The Monty Hall problem	104
2.19	All cases of the Monty Hall problem	107
2.20	Getting von Neumann's simplified poker variant started	113
2.21	Von Neumann's simplified poker variant	115
2.22	Game tree for the simplified poker variant due to Binmore [150]	117
2.23	Belle bluffs with probability p so as to make David indifferent	118
2.24	David calls with probability q so as to make Belle indifferent	120

2.25 The structure of the subclasses of TFNP 132
2.26 Reducing NASH-EQUILIBRIUM to END-OF-THE-LINE 135

3.1 Four coalition structures 142
3.2 A weighted voting game.................................. 146
3.3 The cost of stability when sharing ice cream 157
3.4 A vector weighted voting game............................ 174
3.5 Scheme of reduction for ISG-IN-CORE 195
3.6 Overview of stability concepts for hedonic games 202
3.7 An aversion-to-enemies hedonic game represented as a graph.. 206
3.8 Illustration of the reduction for NP-hardness of Nash stability. 218

4.1 Preferences of Anna, Belle, and Chris 233
4.2 Preferences of Anna, Belle, Chris, David, and Edgar 238
4.3 Majority graph for the election in Table 4.2................. 240
4.4 Majority graph for the election in Table 4.3................. 243
4.5 Kemeny election from Table 4.3 as a weighted, directed graph. 248
4.6 A voting tree for the cup protocol 253
4.7 Cup protocol 254
4.8 Distinct schedules in the cup protocol...................... 255
4.9 Determining the Bucklin winners for the election in Table 4.3 . 257
4.10 Determining the fallback winners for the election in Table 4.3 . 260
4.11 Determining the SP-AV winners for the election in Table 4.3.. 262
4.12 Borda, Copeland, and Young scores........................ 271
4.13 Twin paradox in the cup protocol due to Moulin [715]........ 284
4.14 Trip preferences of Anna, Belle, and Chris 298
4.15 Relations between some variants of the possible winner, manipulation, and bribery problem 307
4.16 Reduction PARTITION $\leq_m^P$ Copeland-CCWM 317
4.17 Links between susceptibility results for various control types .. 340
4.18 Bucklin voting is susceptible to CCPV-TE and CCPV-TP 350

5.1 The annual charity Pumpkin Pie Taste-Off 370
5.2 Preferences regarding sweetness of pumpkin pie 372
5.3 Proving Theorem 5.1: input to SP-CCAV 378
5.4 Proving Theorem 5.1: two incomparable votes............... 379
5.5 Proving Theorem 5.1: dropping all votes not approving of p... 380
5.6 Proving Theorem 5.1: Anna suggests to not go step by step... 381
5.7 Proving Theorem 5.1: What if we don't go step by step? 382
5.8 Proving Theorem 5.1: Which votes can help in "smart greedy"? 383
5.9 Proving Theorem 5.1: drawing level with first dangerous rival . 384
5.10 A ruminating cow (left) and a maverick (right) 397
5.11 Beneficiaries of swooning 399

6.1 Visualization of the outcomes of the selected committee scoring rules rules for Euclidean preferences [374] 421

6.2	An approval ballot with three approvals for the eight books	427
6.3	Cost of utility in the proof of Theorem 6.8	446
6.4	An ILP for computing PAV	452
6.5	A MILP used by the FPT algorithm computing PAV	454
7.1	Three individual judgments	469
7.2	Doctrinal paradox	470
8.1	A cake as a metaphor of a heterogeneous resource	510
8.2	Belle and Edgar share a cake without envying each other	516
8.3	Anna, Belle, and Edgar share a cake: Envy raises its ugly head	518
8.4	Valuation criteria for divisions (implications)	521
8.5	Valuation criteria for divisions (inclusions)	522
8.6	Knife movement in a moving-knife protocol	528
8.7	A nonproportional cake-cutting protocol for three players	528
8.8	Cut & Choose protocol for two players	529
8.9	Box representations of two valuation functions	531
8.10	Cut & Choose protocol: Two examples	531
8.11	Cut & Choose is not Pareto-optimal	532
8.12	Austin's moving-knife protocol for two players	533
8.13	Austin's moving-knife protocol: p_2 calls "Stop!" at time t_0	534
8.14	Last Diminisher protocol for n players	535
8.15	Last Diminisher protocol for four players: 1st round	536
8.16	Last Diminisher protocol for four players: 2nd round	537
8.17	Last Diminisher protocol for four players: 3rd and final round	537
8.18	The Dubins–Spanier moving-knife protocol for n players	542
8.19	Lone Chooser protocol for n players	543
8.20	Cut Your Own Piece protocol: individual markings	548
8.21	Cut Your Own Piece protocol: all markings	549
8.22	Cut Your Own Piece protocol: assignment of the shares	549
8.23	Cut Your Own Piece protocol for n players	550
8.24	Divide & Conquer protocol for n players	553
8.25	Divide & Conquer protocol for four players: 1st round	554
8.26	Divide & Conquer protocol for four players: 2nd round	555
8.27	BJK Divide & Conquer protocol for n players	556
8.28	BJK Divide & Conquer protocol for four players: marking	559
8.29	BJK Divide & Conquer protocol for four players: cutting	560
8.30	BJK Divide & Conquer protocol for four players: 2nd round	561
8.31	Unequal shares: the lion's share	563
8.32	Selfridge–Conway protocol for three players	565
8.33	Selfridge–Conway protocol: Steps 1 and 2	568
8.34	Selfridge–Conway protocol: Step 3	568
8.35	Selfridge–Conway protocol: Step 4	569
8.36	Stromquist's moving-knife protocol for three players	570
8.37	Illustration of Stromquist's moving-knife protocol	571

8.38 The moving-knife protocol of Brams, Taylor, and Zwicker 573
8.39 Dirty-work Last Diminisher protocol: 1st round 575
8.40 Dirty-work Last Diminisher protocol: 2nd round............ 576
8.41 Dirty-work Last Diminisher protocol: 3rd round 577
8.42 Dirty-work Divide & Conquer protocol for n players 578
8.43 One cut suffices 581
8.44 The Quarter protocol for three players 591
8.45 The Quarter protocol of Even and Paz [416] for four players .. 594

9.1 George and Helena want to divide nine objects 607
9.2 Partial order over 2^R, where $R = \{c, \ell, s, p\}$, for Example 9.2 .. 616
9.3 Implications between various relaxations of envy-freeness 634
9.4 Santa Claus wants to make the most unhappy child happy.... 636
9.5 Division of nine objects by the adjusted winner procedure 658
9.6 Descending Demand Protocol: Preferences over bundles 662

List of Tables

1.1	Truth table for some boolean operations	25
1.2	Some deterministic upper bounds for k-SAT and SAT	28
1.3	Improving the deterministic upper bounds for 3-SAT	28
2.1	The prisoners' dilemma	46
2.2	The prisoners' dilemma without negative entries	47
2.3	The battle of the sexes	53
2.4	The chicken game	55
2.5	The penalty game	57
2.6	The paper-rock-scissors game	59
2.7	The penalty game: goalkeeper acting awkwardly on the left	66
2.8	Properties of some two-player games	70
2.9	Anna's gain (left) and Belle's gain (right)	74
2.10	Courses of two Nim games	88
2.11	When does there exist a winning strategy for Nim?	89
2.12	Edgar's campaign game in normal form	100
2.13	All cases of the Monty Hall problem	106
3.1	Allocation of seats in the *17th Deutscher Bundestag*	148
3.2	Some possible coalitions in the *17th Deutscher Bundestag*	149
4.1	Example of an election with some scoring protocols	236
4.2	Example of head-to-head contests	239
4.3	Example of an election without a Condorcet winner	242
4.4	Simpson scores of the candidates in the election in Table 4.3	245
4.5	Overview of some fundamental voting systems	264
4.6	Condorcet winner and loser in Borda and plurality	266
4.7	Counterexample for the consistency of some voting systems	270
4.8	Dodgson voting is not monotonic	279
4.9	Dodgson voting is not homogeneous	280
4.10	Overview of some properties of several voting systems	286

4.11	Election (C,V) constructed in the proof of Theorem 4.5 294
4.12	Complexity of POSSIBLE-WINNER and NECESSARY-WINNER .. 302
4.13	Complexity of POSSIBLE-WINNER-NEW-ALTERNATIVES 303
4.14	Complexity of POSSIBLE-WINNER-UNCERTAIN-WEIGHTS..... 305
4.15	Complexity of manipulation in various voting systems........ 313
4.16	Overview of the common control problems for voting system $\mathcal{E}$ 338
4.17	Overview of control complexity results for some voting systems 342
4.18	Complexity of control in Borda elections 346
4.19	Reduction X3C $\leq_{\mathrm{m}}^{\mathrm{P}}$ Bucklin-CCPV 352
4.20	Complexity of control by adding, deleting, and replacing either candidates or voters in various voting rules............ 357
4.21	Complexity of bribery problems for some voting systems 361
5.1	Control complexity for general, single-peaked, and nearly single-peaked electorates 401
7.1	Doctrinal paradox 470
7.2	Avoiding the doctrinal paradox........................... 480
7.3	Complexity of JA manipulation with $UPQR$ 496
7.4	Construction for the proof of Theorem 7.6 497
7.5	Complexity of JA bribery with the premise-based procedure .. 499
7.6	Example for control by bundling judges 502
8.1	Case 3 of Custer's procedure [319] 547
8.2	Envy-freeness in the Selfridge–Conway protocol 569
8.3	Minimal number of cuts for the Divide & Conquer protocol ... 586
8.4	Minimal numbers of cuts for some proportional protocols 587
8.5	Comparison of $D(n)$ and the best known upper bound on $P(n)$ 589
8.6	What do p_2, p_3, and p_4 think about the pieces in Case 2.2? ... 595
8.7	What do p_2, p_3, and p_4 think about the pieces in Case 2.2.1? . 595
8.8	What do p_2, p_3, and p_4 think about the pieces in Case 2.2.2? . 596
8.9	How many cuts guarantee how many players which share? 598
8.10	DGEF of selected finite bounded proportional protocols 601
8.11	Some cake-cutting protocols and their properties 603
9.1	Complexity of problems for (exact) social welfare optimization 647
9.2	Summary of (in)approximability of social welfare optimization 648
9.3	Adjusted winner procedure: George's and Helena's valuations . 656
9.4	Adjusted winner procedure: sorting the objects.............. 657

Chapter 1
Playing, Voting, and Dividing

Jörg Rothe

Playing, voting, and dividing are three everyday activities we all are familiar with. Having their personal chances of winning, their individual preferences, and their private valuations in mind, the players, voters, and dividers follow their individual strategies each. While everyone first and foremost is selfishly interested in his own advantage only, from the interplay of all actors' individual interests, strategies, and actions there will emerge a *collective* decision, an outcome of the game with winnings or losings for all players, an elected president ruling over all voters, or a division of the goods among all parties concerned. By the end of the day, there will be winners and losers.

However, it is just one thing to maximize one's individual profit in a game or in a division of goods, or to make one's favorite candidate win. It is quite another thing to look at this from a more global perspective: Is it possible to find mechanisms that increase the *social* or *societal* welfare and thus are beneficial to all and not only to single individuals? So as to help, for example, a whole family—both the parents and their children—choosing a consensual weekend trip destination by voting. Or, so as to help three siblings playing a parlor game to choose their strategies so optimally that none of them could do better by picking another strategy without making any other player be worse off at the same time. Or, so as to help them dividing a cake afterwards, whose single pieces are valued differently by everyone, in a way that no one will envy anyone else for their portion. Strategy-proofness of such mechanisms and procedures is another important goal. If someone tries to get an unfair advantage by choosing insincere strategies, is it possible to prevent that from happening by using a strategy-proof mechanism?

This book introduces to three emerging, interdisciplinary fields at the interface of economics and the political and social sciences on the one hand and (various fields of) computer science and mathematics on the other hand:

Algorithmic game theory will be handled in Chapters 2 and 3. Starting from classical game theory that was pioneered by von Neumann [734] (see also the early work by Borel [170] and the book by von Neumann and Morgenstern [736]), we will focus on algorithmic aspects of both noncooperative and cooperative games. Cooperative game theory—and especially the theory of hedonic games—has been much advanced in the past few years. Accordingly, Chapter 3 has been restructured and much extended in comparison with the first edition of this book, and Martin Bullinger has brought into play his great expertise in this field as an additional coauthor of this chapter.

Computational social choice, arising from classical social choice theory (see, e.g., the celebrated work of Nobel laureate Arrow [29]), is concerned with the computational aspects of voting. In Chapter 4, we will introduce a variety of voting systems and study their properties. Particular attention will be paid to the algorithmic feasibility and computational complexity of winner determination, the related concepts of possible and necessary winners, and of various ways of influencing the outcome of elections by manipulation, control, or bribery—three ways of strategic behavior that will also be handled in the next three chapters. For example, we will study the complexity of such manipulative attacks for so-called "single-peaked" electorates in Chapter 5. Chapter 6 will be concerned with multiwinner voting where the goal is to elect a winning committee rather than just a single winner. Much work has been done in this field during the past few years, and the winning committee of coauthors to present this progress consists of Dorothea Baumeister, Piotr Faliszewski, Piotr Skowron, and myself. In Chapter 7, we will introduce to the emerging field of judgment aggregation where individual judgment sets of possibly interconnected logical propositions (rather than individual preferences as in voting) are aggregated to come to a collective decision. Again, this field has been much advanced since the first edition of this book, and fortunately, Ronald de Haan accepted the judges' verdict who sentenced him to contribute to this chapter as another expert coauthor.

Fair division, finally, considers the problem of dividing goods among players who each can have quite different valuations of these goods. We will look at *"cake-cutting procedures"* in Chapter 8, aiming to divide one single, infinitely divisible good, the "cake." And, last but not least, we will study the issue of fairly allocating indivisible, nonshareable goods in Chapter 9, a field closely related to *"multiagent resource allocation"* and *"combinatorial auctions."* Concepts of fairness (such as envy-freeness) and social welfare optimization play a central role in these two chapters. Once more, much work has been done in fair division of indivisible goods since the first edition of this book; therefore, Chapter 9 has been restructured and much extended. It is only fair to say that Jérôme Lang was the coauthor of this chapter who had to bear the brunt of the chores when we were restructuring and extending it.

These three fields are closely interrelated, and we will highlight such cross connections throughout the book. In each of these areas, our particular focus is on the computational properties of the arising problems. Therefore, in Section 1.5, we will provide some background on algorithmics and complexity theory (and also on propositional logic) that will be useful in the subsequent chapters. Elementary basics of other mathematical fields (such as probability theory, topology, and graph theory) will be presented when they are needed, as tersely and informally as possible and with as many details as necessary. While all crucial concepts will be formally defined, most of them will be explained using examples and figures to make them easier to access. Moreover, many situations will be illustrated by short stories from everyday life, featuring Anna, Belle, Chris, David, Edgar (see Figure 1.1), and others.

(a) Anna (b) Belle (c) Chris (d) David

(e) Edgar

Fig. 1.1 The main characters in this book

1.1 Playing

Part I of this book is concerned with algorithmic game theory, covering both noncooperative and cooperative games.

1.1.1 Noncooperative Game Theory

Smith and Wesson, two bank robbers, have been arrested. Since there is only little evidence that incriminates them and would stand up in court, they are offered a deal:

- If one of them makes a confession, he will be free to go (on probation), provided the other one remains silent, and this other one will then have to serve his ten years alone;
- if both confess, they both will be sent to prison for four years;

- but if both stubbornly keep silent, they can be sentenced only to two years' imprisonment (not for bank robbery, due to lack of evidence; only for minor offenses such as possession of unregistered weapons).

Unfortunately, they cannot coordinate their actions. What would be best for Smith to do in order to get away with a prison sentence as short as possible, given that its length depends not only on his own but also on Wesson's action? And what would be a most clever decision for Wesson, whose prison sentence conversely depends on Smith's action as well? That is the famous *"prisoners' dilemma"*!

George and Helena want to spend their first anniversary together and want to have some fun. George suggests to watch an exciting soccer game with his wife. Helena, however, would rather like to go to a concert with her husband. These are quite different things to do. Would it be better for each of them to try to enforce their own wish or to give way to their partner's wish? If none of them gives in, they won't spend their anniversary together! Every couple is well familiar with this *"battle of the sexes."*

Situations like that, where several individuals interact when making their decisions and where their gains depend also on the other individuals' decisions, can be described by strategic games. About a century ago, Borel [170] and von Neumann [734] wrote the first mathematical treatises in game theory. And about twenty years later, the foundations of this theory as a standalone research area have been laid by von Neumann and Morgenstern in their groundbreaking work [736]. Evolving into a rich and central discipline within economics ever since, game theory has yielded many terrific insights and results, and a number of Nobel Prize winners in Economics, such as John Forbes Nash, Reinhard Selten, John Harsanyi, Lloyd S. Shapley, and Alvin E. Roth.

Basically, one distinguishes between cooperative and noncooperative games in this theory. The above examples are noncooperative games.

Noncooperative game theory, which we will be concerned with in Chapter 2, studies games where players face off against each other as lone, selfish fighters aiming to maximize their own gains. This category of games includes combinatorial games such as chess and go, but also card games such as poker, where the players have incomplete information about their opponents and where chance and the psychology of bluffing play an important role. However, not only board and gambling games can be expressed and studied in this theory, but all kinds of competitive situations (e.g., market strategies of companies or global strategies of states) can be modeled, too.

Some of the most central concepts in noncooperative games concern their stability, for example, with the intent to predict the outcome of such games. Is it possible for the individual players to choose their strategies so that they are all in equilibrium in the sense that no one has an incentive to deviate from the chosen strategy (provided all other players stick to their chosen strategies as well)? Historically, questions like this were of uttermost importance, for the entire human race, for instance during the cold war between the Eastern bloc and the NATO countries, in light of the arms race and the doctrine of nuclear deterrence. And such questions may be even more important in these days when many global players struggle in a multipolar world to prevail over their opponents. One of the questions we study for noncooperative games is: How hard is it to find such equilibrium strategies?

1.1.2 Cooperative Game Theory

In cooperative game theory, which will be introduced in Chapter 3, we are concerned with players forming coalitions and working together, seeking to achieve their common goals. It may be possible to increase the gains of single players by cooperation with others. Whether a player joins a coalition or deviates from it certainly depends on whether or not this player will benefit from this.

Stability concepts for cooperative games are of quite central importance. If there is an incentive for a player to deviate from the grand coalition (the set of all participating players), then the game is instable and breaks up into several smaller coalitions working on their own and sometimes even competing with each other. In this chapter, we will see various notions that capture the stability of cooperative games in different ways. Also, one can measure the influence—or power—of a single player in such a game in various ways. Roughly speaking, one fundamental such power index of a player is based on how often this player's membership in a coalition is decisive for its success. Stability concepts and power indices in cooperative games will also be studied in terms of their algorithmic and complexity-theoretic properties where we will focus on games that can be compactly represented, such as games on graphs and weighted voting games.

Finally, we will delve into hedonic games where players have preferences or utilities for the coalitions containing them. Again, we will study stability concepts and representation formalisms for them as well as their dynamics and will study the related problems in terms of their algorithmic solvability and computational complexity.

1.2 Voting

Part II of this book is concerned with preference aggregation by voting (both in general and when restricted to single-peaked societies), with multiwinner voting, and with judgment aggregation. In particular, we will focus on the social-choice-theoretic properties of single-winner and multiwinner voting systems and judgment aggregation procedures, and on the computational complexity of manipulative actions and strategic behavior.

1.2.1 Preference Aggregation by Voting

Anna, Belle, and Chris meet for a joint evening of gaming. First, however, they have to agree on which game to play. Up for election are chess, poker, and Yahtzee.

> "Let's play chess," Anna suggests. "And if that's not working for you, we might play Yahtzee. Poker I like the least."
>
> "Oh no!" grumbles Chris. "Chess is *to-tal-ly* boring, and it's three of us."
>
> "True, but we can just play a blitz chess tournament," replies Anna. "Then everyone will get to play."
>
> "But Yahtzee is much more fun!" Chris disagrees. "And if you don't like that, we should at least play poker."
>
> "I like poker the most," Belle pipes up now, "because I can bluff like hell, y'know. Yahtzee is what I like the least. If it's not poker, then we should play chess."
>
> "Listen up!" says Anna now. "If we want to play, we need to agree on a game. But since each of us has different preferences, we should perhaps simply vote on what we will play."

Figure 1.2 shows the preferences of Anna, Belle, and Chris over these three alternatives, ordered from left (the most preferred alternative) to right (the least preferred one).

Fig. 1.2 Anna, Belle, and Chris are voting on which game to play

Before they can vote, however, they first need to agree on a voting system, a rule that says how to determine a winner from their individual preferences. There are a multitude of voting systems (many of which will be introduced in

1.2 Voting

Section 4.1 starting on page 234). Democratic elections have been of central importance in human societies already since the roots of democracy have been planted in ancient Greece, and at least since the French Revolution and the Declaration of Independence of the United States of America. But, *which* rule should Anna, Belle, and Chris choose for their election?

"All right then," Belle agrees, "so let us vote like that: Each two of the games go in a head-on-head contest ... "

"I see," Chris interrupts. "Whatever game wins each of its pairwise comparisons, by which I mean it is preferred to each other game in at least two of our rankings, ... "

"... is the winner of the election and shall be played!" completes Anna.

"Exactly," says Belle. "Because then it must be better than every other game for us. After all, a majority of us prefer it to each other alternative."

The rule suggested by Belle was originally proposed by Marie Jean Antoine Nicolas de Caritat, the Marquis de Condorcet (1743–1794), a French philosopher, mathematician, and political scientist, in his essay from almost 240 years ago [300]. Condorcet's voting system is still very popular, since a Condorcet winner must be unique and there are good reasons indeed to consider a Condorcet winner the best alternative possible. However, Condorcet elections have their pitfalls as well.

"Great!" Anna bursts out. "Chess beats Yahtzee! Thanks to my and Belle's vote. Sorry, Chris, you can put away your dice box now and ... "

"Not so fast, please!" Belle interrupts her. "No frigging way are we playing chess! Poker beats chess due to my and Chris's vote, so ... as likely as not will we play poker." She thinks. "After all, if Yahtzee is beaten by chess and chess is beaten by poker, then poker must beat Yahtzee, too, must it not?"

"Not so fast, girls!" Chris pipes up now. "Screw poker! Because Yahtzee beats poker thanks to Anna's and my vote!"

The three look at each other, bewildered. "What the ... I mean, what *game* has won now?" they ask themselves in unison.

The winner is: ... *none* of the three games! This feature of the Condorcet voting system is known as the *Condorcet paradox*: A *Condorcet winner* (i.e., a candidate who beats every other candidate in pairwise comparison) does not always exist! Even though the individual rankings of the three voters are rational in the sense that there are no cycles, the societal ranking produced by the Condorcet rule is cyclic: Poker beats chess in their head-on-head contest, chess beats Yahtzee, and yet Yahtzee beats poker. This *Condorcet cycle* is depicted in Figure 1.3.

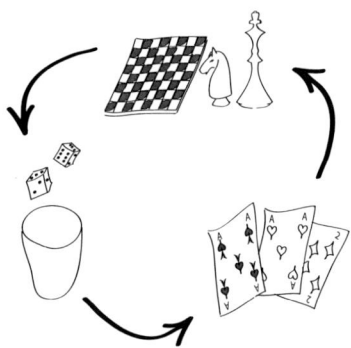

Fig. 1.3 The Condorcet paradox

Since the work of Condorcet [300], social choice theory has been concerned with collective decision-making by aggregating individual preferences via voting. Chapter 4 provides the foundations of this theory and introduces a variety of voting systems and their properties. In social choice theory, too, pathbreaking insights have been honored by Nobel Prizes in Economics, given to Kenneth Arrow and John Hicks for their pioneering contributions to general economic equilibrium theory and welfare theory (see, for example, Arrow's famous impossibility theorem [29], stated here as Theorem 4.1 on page 273) and to Amartya Sen for his contributions to welfare economics.

Not only the players in a game, but also the voters in an election can act strategically, by reporting a vote that they consider more useful for their goal than their true preferences and that they hope will—depending on the voting system and on the other voters' preferences—ensure their favorite candidate's victory. By the celebrated Gibbard–Satterthwaite theorem [835] (see Theorem 4.2 on page 275), no reasonable voting system is protected against this kind of manipulation. Again, we will focus on the algorithmic and complexity-theoretic aspects: How hard is it to know if one can set one's individual preferences strategically so as to successfully manipulate the election? Also other ways of influencing the outcome of elections, such as electoral control (e.g., by adding or deleting either candidates or voters) and bribery, will be introduced and thoroughly discussed in Chapter 4.

1.2 Voting

1.2.2 Manipulative Actions in Single-Peaked Societies

Relatedly, in Chapter 5 we will study the complexity of control, manipulation, and bribery problems when restricted to so-called *single-peaked electorates*. This notion has been introduced by Black [155, 156], and is now considered to be "*the* canonical setting for models of political institutions" [477, p. 336]. Single-peaked electorates can be used to model societies that are heavily focused on a single issue (such as taxes, public budgets, cutbacks in military expenditure, etc.) where political positions can be ordered on an axis (or, in a "left-right spectrum").

1.2.3 Multiwinner Voting

In Chapter 6, we will consider multiwinner voting rules, which aim at electing a winning committee and not just a single winner. If we are looking for a committee with k members, we might just take the "best" k candidates according to the collective ranking produced by some single-winner voting rule, thus short-listing them based on individual excellence. However, we might also have other goals. For instance, we might aim at a committee as diverse as possible or at a committee that represents each group of voters as well as possible.

In this chapter, we will give an overview of the most common multiwinner voting rules, discuss their properties in comparison, and then will focus on the complexity of computing winning committees exactly or approximately, and we will also briefly cover related topics such as robustness and strategic behavior in terms of manipulation, control, and bribery.

1.2.4 Judgment Aggregation

In Chapter 7, we will turn to a field called *"judgment aggregation,"* which is much younger than the related field of preference aggregation by voting. Unlike in voting, the individual judgments of experts (the *"judges"*) regarding propositions (that can be logically connected) are to be aggregated here. In judgment aggregation, too, very interesting paradoxical situations may occur, such as the so-called *doctrinal paradox* (relatedly, the *discursive dilemma* [620, 778]), which will be explained in more detail in this chapter.

Just as in elections, it is here possible to influence the outcome of a judgment aggregation procedure. The judges themselves might try to manipulate the outcome by reporting insincere individual judgments instead of their true ones. External actors might try to obtain their desired collective judgment set from the experts by bribing them. An external expert might also try to influence the outcome by controlling the structure of the used judgment aggregation procedure, e.g., by adding or deleting judges. The investigation of the algorithmic and complexity-theoretic properties of the related problems has been initiated by Endriss, Grandi, and Porello [394] and has opened a very active field of research, including manipulation, bribery, and control, which will also be surveyed in this chapter.

1.3 Dividing

Part III of this book is concerned with fair division of either divisible or indivisible goods.

1.3.1 Cake-cutting: Fair Division of Divisible Goods

"Ein Kompromiss," German economist and politician Ludwig Erhard (Chancellor of West Germany from 1963 until 1966) is quoted as saying, *"das ist die Kunst, einen Kuchen so zu teilen, dass jeder meint, er habe das größte Stück bekommen."* In English: "A compromise is the art of dividing a cake in such a way that everyone believes he has the biggest piece."

We will be concerned with the fair division of cake (or, *"cake-cutting"*) in Chapter 8. The "cake" is only a metaphor here that can be applied to any divisible resource or good.

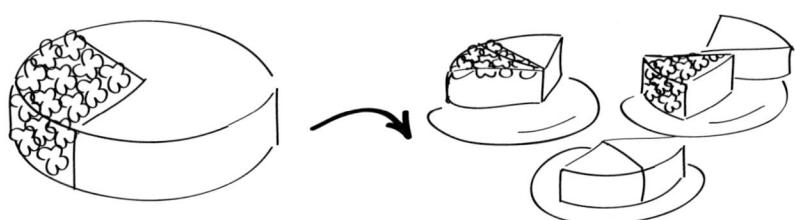

Fig. 1.4 Division of a cake into three portions

Figure 1.4 shows a division of a cake into three portions. But when can it be called a "fair" division? How many parents have failed miserably when trying to fairly divide a cake among their children? In the worst case, *all* the children believe to have received just the worst of all pieces and so feel *all* hard done by. Would that be easier perhaps if the children themselves were to cut the cake and divide it among them? Not very likely: Arguments and fights will start as soon as envy raises its ugly head. On a larger scale, how many mediators have failed miserably when trying to fairly divide debatable territories among opposing parties, be it in the Middle East, Central Africa, the Balkans, in East Europe, or in other parts of the world?

The purpose of cake-cutting protocols is to produce a fair division of the cake and to make all participating players as happy as possible. "Fairness" can be interpreted in various ways. It can mean, for instance, that no player receives a smaller share of the cake than is due to them with respect to their own individual valuation, i.e., everyone gets at least a proportional share. Or, it can also mean that there is no envy among the players. If possible, the protocol should even *guarantee* these properties. That is, proportionality or even envy-freeness of the allocation should be achieved independently of the players' individual valuations. Steinhaus [889] was one of the first to formulate the problem of fair division in cake-cutting as a most challenging and beautiful mathematical problem and to propose first solutions to it. In this chapter, we will survey much of the work that has been done in this field since his groundbreaking paper, focusing on valuation criteria such as various notions of fairness and introducing many concrete cake-cutting protocols.

1.3.2 Fair Division of Indivisible Goods

In Chapter 9, finally, we will be concerned with a problem that is closely related to cake-cutting, namely the problem of fair division of *indivisible, nonshareable* goods or resources. All participating agents (or, players) have individual, subjective (ordinal or cardinal) preferences over the single goods, or all bundles of goods. We will classify the resulting allocation problems and, distinguishing between ordinal preferences and cardinal utilities, will first focus on the problems of preference elicitation and compact representation. Turning then to (ordinal and cardinal) fairness criteria, we will first discuss the tension between efficiency and fairness (specifically, between Pareto efficiency and envy-freeness) and then a variety of relaxations of envy-freeness.

In addition to the individual utility each player can realize for herself, one is also interested in optimizing the *societal* utility, the "social welfare," which can be measured in various ways, including utilitarian, egalitarian, and Nash product social welfare. Finding an optimal allocation of all goods that maximizes social welfare is a difficult combinatorial problem whose algorithmic and complexity-theoretic properties will be surveyed.

Next, we will briefly turn to centralized fair division mechanisms for computing allocations under ordinal preferences. We will then describe fair division with money and the adjusted winner procedure as well as various decentralized allocation protocols, including the descending demand and the picking sequences protocols. Finally, we will briefly discuss a number of extensions and variants of the model in this chapter, including fair division of indivisible chores, and we will say a few words about further issues, such as strategy-proofness and matching.

As Chapter 9 focuses on *fair division*, its authors decided to not cover the related area of *(combinatorial) auctions* with its many applications, for example in e-commerce. Auctions will therefore be under-covered in this book with respect to their importance in the literature (see, e.,g., the textbook *"Combinatorial Auctions"* edited by Cramton, Shoham, and Steinberg [316] and also the books by Shoham and Leyton-Brown [864] and Wooldridge [950]). On the other hand, it would be unthinkable not to address auctions at all in a book on economics and computation. Therefore, in the following section we will briefly digress so as to informally introduce a number of common *single-item auctions* where the single objects are auctioned off to the bidders—as agents are called in the field of auctions—one by one (by contrast, in a *combinatorial auction*, the bidders make their bids for *bundles* of objects).

1.3.3 A Brief Digression to Single-Item Auctions

Every bidder knows his individual value for each of the single objects. However, he of course is not obliged to truthfully announce his actual values for the objects during the bidding phase, but he may bid quite different prices for the objects. The bidder's goal is to win each desired object for a price as low as possible. The auctioneer's goal, however, is to make as much profit as possible from auctioning off all items. In this sense, bidders and auctioneer are adversaries in this "auction game" and, in addition, the bidders compete with each other.

1.3.3.1 Classification

The bidders can make their bids

- either *openly*—in so-called *open-cry auctions*, i.e., all bidders know which bids have been made so far (possibly not by whom, though),
- or *concealed*—in so-called *sealed-bid auctions*, i.e., no bidder knows any of the previously made bids of other bidders.

Open-cry auctions can be further subdivided into

- *ascending-price auctions* where the bids go from lower to higher values, and
- *descending-price auctions* where the bids go from higher to lower values.

Winner determination is an important issue in an auction: Which bidder wins which object and what does she have to pay for it? One can distinguish, for example, between

- *first-price auctions* where a bidder wins an object by making the highest bid and has to pay this price, and

1.3 Dividing 13

- *second-price auctions* where, again, a bidder making the highest bid wins the corresponding object, but has to pay only the price of the second-highest bid.

We now introduce a number of "classical" types of single-item auctions. The above classification criteria can be combined with each other. For example, one thus obtains a *first-price, sealed-bid auction*: There is only a single round in which the bidders hand in an envelope containing their bid, and whoever has made the highest bid wins this object (if need be, ties are broken according to some preassigned rule). The winner pays the amount corresponding to his bid. This type of auction is often used, for example, in public auctions of real estate property.

What would be a good strategy for a bidder in such an auction? The highest bid wins, but the greater the difference is between the highest and the second-highest bid, the more money has been wasted by the winner. Making a bid just a tiny bit above the second-highest bid would have been enough to ensure victory, and would have saved a lot of money. Therefore, the best strategy is to make a bid that is lower than one's true value for this object. How much lower, though, depends on the circumstances—for example, it depends on what the unknown bids of the other bidders are, and on how desperately one desires to get just this object. There is no general solution to this problem; some risk is unavoidable. Auctions are like bets and how risk-loving a bidder is will influence his possible wins and losses.

1.3.3.2 English Auction

This is an ascending-price, first-price, open-cry auction. At the beginning, the auctioneer announces a minimum price (which may also be 0) for the object on the table and waits for higher bids. The bidders know the current bid, which must be exceeded by each new bid. If no bidder raises the current bid, the object goes to whoever made the last bid for the price of his final bid. If none of the bidders raises the initial minimum price, the auctioneer gets the object for this minimum price. Famous auction houses like *Sotheby's* (originally situated in London but now headquartered in New York City) and *Christie's* (with its main headquarters also located in London and New York City) conduct this type of English auction.

What would be a good strategy for a bidder in such an auction? The highest bid wins again, and the winner again wastes the more money, the greater the difference is between the highest and the second-highest bid. Therefore, it's in the interest of the bidder to keep this difference as small as possible. Since bids are made openly, it would be smart to raise the current bid only by a small amount. This approach, though, is based on the assumption that all bidders behave rationally. However, auctions are subject to psychological aspects, which are not that simple to be modeled game-theoretically. For example, it might be that two bidders work themselves up into fierce fighting, taking turns in outbidding each other again and again, and eventually the winner may have paid a lot more money than he would have paid if only he had raised—substantially!—the first bid just once, right at the beginning of the auction, in order to frustrate his rival. Would this strategy have worked better? Of course, that's also just a speculation.

Wooldridge [950] highlights another interesting feature of English (and other) auctions: If the true value of the object is unknown or uncertain, the *"winner's curse"* may occur, not only in English auctions, but particularly often in these. This refers to the tendency of the auction winner to overpay an object. For example, if there is an auction for a house that has not been thoroughly evaluated by experts, it is unclear if a bidder who wins this house in an auction should really be happy, or if he rather should be worried to have paid too much for it. Perhaps the other bidders stopped bidding only because they knew more than him about the constructional condition of the house.

1.3.3.3 Dutch Auction

This is a descending-price, first-price, open-cry auction (and is also called *clock auction*). At the beginning, the auctioneer announces an obviously too high price for the object on the table, which lies above the expected maximum price of the bidders. Then the auctioneer lowers the price step by step until there is a bidder who is willing to accept and pay the current price. This bidder receives the object for this price.

What would be a good strategy for a bidder in such an auction? It wouldn't make sense for a bidder to accept a price above her own, true value for this object. Once the auctioneer's offered price drops below this true value, the bidder's potential gain increases the longer she waits to accept a current price, but at the same time her risk also increases to lose the object to somebody else. As in the English auction, the *winner's curse* can occur here as well, namely if a bidder gets cold feet too early and accepts the current price, although it is only a tiny little bit lower than her true value for the object. This is why Dutch auctions tend to end pretty quickly.

1.3.3.4 Vickrey Auction

A *Vickrey auction*—named after its inventor William Vickrey, who in 1996 received the *Nobel Prize in Economics* jointly with James Mirrlees—is a second-price, sealed-bid auction. As in a first-price, sealed-bid auction, there is only one round where the bidders submit their bids in a sealed envelope. The winner is again a bidder with the highest bid, where a tie-breaking rule is applied to break ties, if there occur any. However, the winner does not have to pay the highest, but only the second-highest price. This may not seem intuitively sensible at first, but it does have a big advantage, as we will now see.

What would be a good strategy for a bidder in such an auction? The above-mentioned advantage of Vickrey auctions is that telling the truth—i.e., bidding their true value of the object—is the dominant bidder strategy (in the sense of Definition 2.2 on page 48). To understand why this is so, let's have a look at the following two cases, where we assume that the second-highest bid remains the same:

Case 1: The bidder bids more than her true value. Then it is more likely than when telling the truth for the bidder to get her bid accepted and win the object. If she really wins it, however, she may have to pay more than what she would have paid had she remained truthfully. In other words, by being dishonest the bidder only increases her chances to make a loss.

Case 2: The bidder bids less than her true value. Then it is less likely than when telling the truth for the bidder to get her bid accepted and win the object. However, even if she does win it, the price she has to pay has not been influenced by her bidding less than her true value: She still has to pay the second-highest price. That is, the bidder does not have an advantage in terms of gain maximization; all she has achieved by being dishonest is to lower her winning chances.

Since there is no advantage for the bidder to deviate from her true value in either way, it can be expected that she will make a truthful bid. Auctions similar to Vickrey auctions are used, for example, on the internet auction platform eBay. The *Vickrey–Clarke–Groves mechanism* generalizes Vickrey auctions from single-item to combinatorial auctions (see, e.g., the books by Cramton, Shoham, and Steinberg [316], Shoham and Leyton-Brown [864], and Wooldridge [950]).

1.3.3.5 American Auction

This is a special type of auction that is often used in charity events. In our classification scheme, it is an ascending-price, open-cry auction; however, in contrast with the English auction, the winner of an object does not pay the highest price, and does not pay the second-highest price either, but instead *each* bidder, immediately when making a new bid, has to pay the difference between his bid and the previous bid. The auctioneer may predetermine the allowed difference amounts between bids, and there is a variant in which the auctioneer also fixes the maximum duration of the auction, which is not known to the bidders. In this variant of the American auction, the last bid made prior to termination is accepted and wins.

What would be a good strategy for a bidder in such an auction? On the one hand, compared with, say, English auctions, American auctions increase the potential gain of the bidders, since a lucky bidder can win a valuable object with just one bid—the last one—for really little invested money (namely, merely the difference to the second-to-last bid). On the other hand, all bidders not winning the object come away empty-handed, that is, they lose all their invested money and might make a really big loss. In fact, the actual winner of an American auction is the auctioneer, who typically makes quite a lot more money than, e.g., in an English auction. One reason why American auctions are popular in charity events is that also the losers can better put up with their losses, knowing that their money has been gambled away for a good purpose. American auctions resemble gambling more than other auctions.

In an American auction with prefixed maximum duration, the winning decision is made only close to its end. Therefore, a good bidder strategy is to make bids only in the final phase. The problem, of course, is that the maximum length is unknown to the bidders, so they must guess when it is a good point in time to get in with their own bids. Moreover, when there are several objects to be auctioned off, one has to give consideration to which objects one would be most interested in bidding on. If one is too eager to bid on each object right from the start and then stops too early so as to not lose too much money, one is likely to maximize one's loss only.

Another variant of the American auction is known as *first-price, sealed-bid, all-pay auction*. In this variant, all bidders submit their bid in a sealed envelope, and the bidder with the highest bid wins (again making use of a tie-breaking rule if needed), but *all* bidders have to pay the price corresponding to their bid. In this way the auctioneer would carry his own gain to an extreme. However, he normally wouldn't find bidders willing to go into such an auction with their eyes open, since these rules are commonly considered to be rather unfair. That is why this quite special form of an auction is rather of theoretical interest; for example, the economic effect of lobbyism, party donations, and bribery can be modeled this way (cf. Section 4.3.5). Considering donators as bidders and parties as sellers of a "good" (namely political influence), the bidder with the highest donation wins in this model (assuming that this donation has helped the party to win the election) and increases his political influence. The other bidders come away empty-handed and increase their political influence not at all or only marginally, but their donations will still not be refunded (for details, see [5, pp. 83–84]).

1.3.3.6 Expected Revenue

The example of the American auction shows that an auctioneer has to be concerned about how to maximize the total revenue of an auction whenever possible. This concerns in particular choosing the right auction protocol, and different perspectives of the parties involved have to be taken into account. For example, American auctions are preferred from the point of view of an auctioneer, though they are unlikely to be accepted by the bidders.

Sandholm [831, p. 214] discusses this issue (see also [950, pp. 297–298]) and comes to the following result for the other types of auction considered above (first-price, sealed-bid auction; English auction; Dutch auction; and Vickrey auction). In particular, the answer to the question raised above very much depends on both the auctioneer's and the bidders' risk-taking propensity:

- For *risk-neutral bidders*, the auctioneer's expected revenue is provably identical for these four auction types (under certain simple assumptions). That is, the auctioneer can expect the same revenue on average, no matter which type of auction he chooses among these four.
- Maximizing the gain is not the most important issue for a *risk-loving bidder*, i.e., such a bidder would rather like to indeed receive an object, even if that means to pay something more for it. In this case, Dutch auctions and first-price, sealed-bid auctions provide the auctioneer with a higher expected revenue. That is so because risk-loving bidders may be inclined to increase their winning chances at the expense of their own profit by raising their bids a bit higher than risk-neutral bidders.
- By contrast, a *risk-averse auctioneer* would be better off on average when using English or Vickrey auctions.

These assertions are to be treated with caution, though. For example, the first statement critically depends on the bidders really having *private* values for the objects, known only to themselves.

1.4 Some Literature Pointers

In the single chapters about playing (Part I), voting and judging (Part II), and dividing (Part III), we will often focus on algorithmic aspects. In the central Part II, which with four chapters is more extensive than the other two parts, we are dealing with the problems of *computational social choice* (COMSOC). On the one hand, COMSOC applies methods of computer science (such as algorithm design, complexity analyses, etc.) to social choice mechanisms (such as voting systems or judgment aggregation procedures). On the other hand, concepts and ideas from social choice theory are integrated into computer science, for example, in the area of distributed artificial intelligence and, most notably, in the design of multiagent systems, networks, ranking algorithms, recommender systems, and others.

Since 2006, scientists from all over the world, who work in these areas ranging from economics and the political sciences to computer science, meet biennially at the *International Workshop on Computational Social Choice*, which so far took place at the universities of Amsterdam, The Netherlands (2006), Liverpool, UK (2008), Düsseldorf, Germany (2010), Kraków, Poland (2012), at Carnegie Mellon University in Pittsburgh, Pennsylvania, USA (2014), in Toulouse, France (2016), at RPI in Troy, New York, USA (2018), at the

Technion in Haifa, Israel, and—due to the pandemic—also online (2021), and at Ben-Gurion University of the Negev in Beersheba, Israel (2023). The (nonarchival) proceedings of these workshops have been edited by Endriss and Lang [398], Endriss and Goldberg [390], Conitzer and Rothe [306], Brandt and Faliszewski [216], Procaccia and Walsh [793], Grandi and Rosenschein [517], Elkind and Xia [385], Meir and Zwicker [701], and Pivato and Lev [782]. If you like reading this book, you will also enjoy reading these proceedings, to get deeper insights into the challenging research questions in this fascinating area.

Moreover, Brandt et al. edited the *"Handbook of Computational Social Choice"* [215] that was published by Cambridge University Press. More recently, Endriss [388] edited a book about trends in COMSOC published by AI Access Foundation. There are also many other, older textbooks on algorithmic game theory (e.g., the textbook *"Algorithmic Game Theory"* edited by Nisan et al. [755]), social choice theory and welfare economics (e.g., the two volumes of the *"Handbook of Social Choice and Welfare"* edited by Arrow, Sen, and Suzumura [30, 31]), and fair division (e.g., the books by Brams and Taylor [204] and Moulin [717]), and some more will be mentioned in the relevant chapters of this book.

The reader is also referred to the surveys by Chevaleyre et al. [284, 288], Conitzer [301], Daskalakis et al. [331], Faliszewski, Hemaspaandra, and Hemaspaandra [423], Faliszewski and Procaccia [438], Nguyen, Roos, and Rothe [746], Nguyen and Rothe [750], Rothe and Schend [824], and Rothe [820, 821], as well as to the book chapters by Brandt, Conitzer, and Endriss [214, 388], Faliszewski et al. [430], and Baumeister et al. [115]. They each cover some more specific topics in computational social choice, algorithmic game theory, and fair division.

But now, let's get ready to play, vote, and divide ...

Stop! Before we can actually start playing, voting, and dividing, a few more preliminaries are needed.

1.5 A Brief Digression to Computational Complexity

As mentioned earlier, we will often focus on the algorithmic and complexity-theoretic aspects of problems related to playing, voting, and dividing in this book. But, how exactly does one determine a problem's algorithmic or complexity-theoretic properties? To answer this question, we will now briefly digress, introducing to the foundations of complexity theory as terse and informally as possible and in as much detail as necessary. Crucial concepts of this theory, which will be important in almost all the following chapters, will be illustrated for SAT, the satisfiability problem of propositional logic. One can also skip this digression for now and, only when needed, come back later to look up one notion or another.

1.5.1 Some Foundations of Complexity Theory

For about half a century now, problems have been investigated and classified in terms of their computational complexity. Having developed from computability theory (embracing, in particular, recursive function theory) and the theory of formal languages and automata (see, e.g., the textbooks by Homer and Selman [556], Hopcroft, Motwani, and Ullman [557], and Rogers [815]), complexity theory is a subarea of theoretical computer science. Among other questions, computability theory studies which problems are algorithmically solvable and which are not; the latter are so-called *undecidable problems*, such as the so-called *"halting problem for Turing machines."* By contrast, complexity theory is concerned with algorithmically solvable (i.e., *"decidable"*) problems only, but specifically asks for the computational costs required to solve such problems. There are many useful textbooks and monographs on computational complexity, such as those by Bovet and Crescenzi [188], Hemaspaandra and Ogihara [551], Papadimitriou [764], Rothe [819], Wagner and Wechsung [926], Wechsung [937], and Wegener [938]. Algorithmic and complexity-theoretic concepts and methods for problems related to playing, voting, and dividing will be important in many chapters of this book.

1.5.1.1 Turing Machines and Complexity Measures

Complexity theory is to algorithmics as yin is to yang.[1] While the main objective of algorithmics is the design of algorithms that are as efficient as possible and thus provide as low *upper* (time) bounds as possible for solving a given problem, in complexity theory one tries to prove that certain problems cannot be solved efficiently at all, that is, those problems do not have any efficient algorithm whatsoever. In other words, the goal of complexity theory is to prove that solving these problems algorithmically requires as high a *lower* (time) bound as possible. And once the least upper and greatest lower bounds of a problem meet, its complexity analyst and algorithm designer (who often are just one and the same person) feels perfect harmony and enthusiastically calls this problem "classified."

Here, *computation time* (or, *runtime*) is a discrete complexity measure defined as the number of elementary computation steps an algorithm performs as a function of the size of the input. The input size depends on the chosen

---

[1] In Chinese philosophy, *yin and yang* denote opposite or contrary forces that are mutually dependent and interconnected: Opposites that give rise to each other by interrelating to one another and that cannot exist without each other.

encoding of problem instances. Usually, these are encoded in binary, i.e., as a string over the aphabet $\Sigma = \{0,1\}$, so that they can be processed by a computer executing the algorithm. Different encodings usually have only a negligible influence on the computation time of an algorithm; note that constant factors are commonly neglected in the runtime analysis of algorithms. Therefore, we will disregard coding details.[2] What is meant by an "elementary computation step" of an algorithm depends on the algorithm model at hand. In theoretical computer science, and particularly so in complexity theory, it is common to use the model of *Turing machine*, named after its inventor Alan Turing [918]. This is a very simple model, and yet it is universal: By Church's thesis [295], everything computable is computable on a Turing machine. Again, we omit formal details here and instead refer to the literature on computability and complexity theory given above. All algorithms in this book will be described only informally; whether they are implemented on an abstract algorithm model such as a Turing machine or in some concrete programming language does not matter for our purposes. We will distinguish, though, between various types of Turing machines or algorithms (where we confine ourselves to *"acceptors,"* suitable to solve decision problems only):

- In the computation of a *deterministic Turing machine* on any input, each computation step is uniquely determined, i.e., for each current *"configuration"* of the Turing machine (which is a complete *"instantaneous description"* of the computation at this point in time), there is a unique successor configuration until, eventually, an *end configuration* (or *halting configuration*) is reached that is either *accepting* or *rejecting*. Note that it is also possible, in principle, that a computation never terminates, i.e., it never reaches an end configuration. This is important in computability theory and is closely related to the halting problem's undecidability.[3]
- The computation of a *nondeterministic Turing machine* on any input is more general: In each computation step (before reaching an end configuration), it is possible to branch nondeterministically, i.e., at each point in time during the computation, the current configuration can have more than one successor configuration until, eventually, an accepting or rejecting end configuration is reached. Therefore, a nondeterministic compu-

---

[2] That, however, is not to say that we will disregard *representation details*. In fact, for many of the problems we will study it is utterly important to choose an appropriate, compact representation to be able to handle them algorithmically. This refers, for example, to Section 2.5.2 starting on page 125 about how to represent problems related to compute equilibria of (noncooperative) games, to Section 3.4 starting on page 171 about the representation of simple (cooperative) games, and to Section 9.3 starting on page 613 about preference elicitation for and compact representation of allocation problems for indivisible goods.

[3] Roughly speaking, the *halting problem (for Turing machines)* is the following: Given (an encoding of) a Turing machine and an input, does the machine on that input ever halt? That this famous problem is undecidable means that there is no algorithm that can solve this problem.

1.5 A Brief Digression to Computational Complexity

tation is not a deterministic sequence of configurations, but instead we have a *nondeterministic computation tree*

- whose root is the start configuration (which, in particular, encodes the input string),
- whose inner vertices are the configurations reachable from the start configuration before the computation has terminated, and
- whose leaves are the accepting and rejecting end configurations.

For an input to be accepted it is enough that there exists at least one accepting computation path in this computation tree. An input is rejected only if all paths of this tree lead to rejecting end configurations.

Again, note that the computation tree of a nondeterministic Turing machine running on some input can have infinite computation paths, in principle. However, since we will be concerned with *decidable* problems only and since for those it is possible to "clock" (deterministic and nondeterministic) Turing machines, we can safely ignore the possibility of infinite computations.

In addition to determinism and nondeterminism there are many other computational paradigms (for example, randomized algorithms), and in addition to computation time there are many other complexity measures (such as the *computation space*, or *memory*, required to solve the problem at hand). However, these will only rarely be considered here (namely, for PSPACE); we will mostly restrict ourselves to analyzing the computation time of (deterministic or nondeterministic) algorithms or to prove lower time bounds of problems.

1.5.1.2 The Complexity Classes P and NP

In complexity theory, one collects all those problems whose solutions require roughly the same computational cost with respect to some complexity measure (e.g., computation time) in so-called *complexity classes*, and the most central time classes are:

- P (*"deterministic polynomial time"*) and
- NP (*"nondeterministic polynomial time"*).

P (respectively, NP) is defined as the class of problems that can be solved by some deterministic (respectively, nondeterministic) Turing machine in polynomial time. Deterministic polynomial-time algorithms are thought of as being efficient, since a polynomial, such as $p(n) = n^2 + 13 \cdot n + 7$, typically grows relatively moderately, as opposed to the explosive growth of an exponential function, such as $e(n) = 2^n$. For very small input sizes n, it may be true that function e has rather benign values (for example, we have $e(n) = 2^n < n^2 + 13 \cdot n + 7 = p(n)$ for each $n < 8$), but already for slightly larger input sizes, the exponential function blows up compared with the polynomial (e.g., if $n = 30$, note that $e(30) = 1,073,741,824$ is considerably greater than $p(30) = 1,297$). For even larger input sizes (say, for $n = 100$), that are not uncommon in practice, exponential functions can reach nothing less than astronomically large values. For example, the number of all atoms in the visible universe (dark matter excluded) has been estimated to be roughly 10^{77}. Suppose that an algorithm having a runtime of, say, 10^n and destroying one atom per computation step, runs on an input of size 77. When this algorithm terminates, it will have wiped out the entire visible universe, including itself[4] (yes, this is a visual algorithm and so belongs to the *visible* universe). However, even for an algorithm that does not destroy atoms during its computation (which is the normal case for algorithms), whoever has started it to run on that input will have passed away a long, long time before the algorithm finally comes up with its answer.[5]

In contrast, *non*deterministic polynomial-time algorithms are viewed as not being efficient in general. If one would try to simulate an NP algorithm deterministically (which seemingly would amount to checking each path in the nondeterministic computation tree, one by one, systematically searching for an accepting end configuration), this deterministic algorithm would require exponential time. Namely, if the runtime of an NP algorithm running on inputs of size n is bounded by a polynomial $p(n)$ and if we assume that the NP algorithm, on any input, branches in each inner vertex of its computation tree into at most two successor vertices, then there can be up to $2^{p(n)}$ paths in the computation tree, and only after the last one of these paths has been checked without success, one can be sure that there really is no accepting computation path. Of course, this is not a proof that P is not equal to NP.

Indeed, the "P = NP?" question has been open since more than 40 years now. This is perhaps the most important open problem in all of theoretical computer science and it is one of the seven *millennium problems* whose solutions will each be rewarded with a prize money of one million US dollars by the Clay Mathematics Institute in Cambridge, Massachusetts. It is clear that $P \subseteq NP$; the challenging open question is whether or not this inclusion is strict. In 2002, Gasarch [492] conducted a "P =? NP poll" among complexity theoreticians where the overwhelming majority indicated by their vote that they believe that $P \neq NP$. However, no one succeeded so far with providing an actual *proof* of this widely believed inequality; and whenever a "proof" for it has spread in the past, it didn't take long to detect its flaws. The prize money of one million US dollars for a *correct* solution to this fascinating open problem is still out there waiting for you to pocket it!

1.5.1.3 Upper and Lower Bounds

In order to show an *upper bound* $t(n)$ on the (time) complexity of a problem, it is enough to find some *specific* algorithm solving this problem in time t,

---

[4] Which is why the assumption it would ever terminate is absurd.

[5] Like "42" as per Adams [2], although it is not quite clear in this case what the input was 7.5 million years before supercomputer Deep Thought came up with this answer.

1.5 A Brief Digression to Computational Complexity

i.e., it works for all inputs of size n in time at most $t(n)$. A P algorithm, for instance, shows that the corresponding problem can be solved in time $p(n)$ for some polynomial p and, therefore, is considered to be efficiently solvable. As mentioned earlier, when stating an upper bound we can neglect constant factors and also finitely many exceptions, as we are interested only in the *asymptotic rate of growth* of complexity functions.

The following notation describing the asymptotic rates of growth of functions (by orders of magnitude) are due to Bachmann [72] and Landau [638].

Definition 1.1 (asymptotic rates of growth). Let s and t be functions from $\mathbb{N}$ into $\mathbb{N}$, where $\mathbb{N}$ is the set of nonnegative integers.

1. $s \in \mathcal{O}(t)$ if and only if there exist a real constant $c > 0$ and an integer $n_0 \in \mathbb{N}$ such that $s(n) \leq c \cdot t(n)$ for all $n \in \mathbb{N}$ with $n \geq n_0$.
   If $s \in \mathcal{O}(t)$, we say that s *asymptotically does not grow faster than t*.
2. $s \in \Omega(t)$ if and only if $t \in \mathcal{O}(s)$.
   If $s \in \Omega(t)$, we say that s *asymptotically grows at least as fast as t*.
3. The class $\Theta(t) = \mathcal{O}(t) \cap \Omega(t)$ contains all functions that have the same *asymptotic rate of growth as t*.
4. $s \in o(t)$ if and only if for all real constants $c > 0$, there exists an integer $n_0 \in \mathbb{N}$ such that $s(n) < c \cdot t(n)$ for all $n \in \mathbb{N}$ with $n \geq n_0$.
   If $s \in o(t)$, we say that s *asymptotically grows strictly slower than t*.
5. $s \in \omega(t)$ if and only if $t \in o(s)$.
   If $s \in \omega(t)$, we say that s *asymptotically grows strictly faster than t*.

For example, $p \in \mathcal{O}(e)$ (and thus also $e \in \Omega(p)$) for the exponential function $e(n) = 2^n$ and the polynomial $p(n) = n^2 + 13 \cdot n + 7$ defined above. Since every exponential function asymptotically not only grows at least as fast as, but even strictly faster than each polynomial, we even have $p \in o(e)$ (and $e \in \omega(p)$). It is also clear that $\mathcal{O}(1)$ is the class of all constant functions. The crucial difference between the definitions of the $\mathcal{O}$ and o notation is the quantifier over the real positive constants c. While the existential quantifier before c in the definition of $\mathcal{O}(t)$ ensures that one may neglect arbitrarily large constant factors, the universal quantifier before c in the definition of $o(t)$ achieves that $c \cdot t(n)$ is greater than $s(n)$ even if the growth of $c \cdot t(n)$ is slowed down by an arbitrarily small constant factor c—that much faster grows t in comparison with s (see also, e.g., [819, 520] for further details, properties, and examples).

Upper bounds for the complexity of a problem are given in the $\mathcal{O}$ notation, and it is enough to find a suitable algorithm solving the problem within the time allowed by this upper bound. By contrast, a *lower bound* $t(n)$ on the (time) complexity of a problem means that *no* algorithm whatsoever can solve this problem in less time than allowed by $t(n)$. At least time $t(n)$ is required to solve this problem for inputs of size n (perhaps not for all of them, but at least for some inputs of this size), and no matter which algorithm of the considered algorithm type we choose. Again, we neglect constant factors and allow finitely many exceptions.

Obviously, both aspects—the upper and the lower time bounds for solving a problem algorithmically—are closely related, they are two sides of the same coin. If one succeeds in finding even *matching* (asymptotic) upper and lower bounds for some problem, one has determined its inherent complexity, at least with respect to the considered class of algorithms.[6] Unfortunately, that is not always possible—many important problems (related, in particular, to the topics of this book) still have a complexity gap between their currently known upper and lower bounds, giving rise to many fascinating open research questions.

1.5.2 The Satisfiability Problem of Propositional Logic

In this section, we will introduce SAT, the famous satisfiability problem of propositional logic, discuss its upper bounds and how to improve them, and then turn to how to prove lower bounds for computational problems such as SAT in general, provding some basic tools from complexity theory. Finally, we give some background on approximation theory.

1.5.2.1 Definitions

Consider, for example, the famous *satisfiability problem* of propositional logic, denoted by SATISFIABILITY (or, shorter, by SAT). Decision problems like that, whose question is to be answered by "yes" or "no," will be represented in the following form:

SATISFIABILITY (SAT)
Given: A boolean formula φ.
Question: Is there a satisfying truth assignment to the variables of φ, i.e., an assignment that makes φ evaluate to true?

Here, a *boolean* (or, *propositional*) *formula* consists of *atomic propositions* (also called the *variables of the formula*) connected by *boolean operations*, such as the following:

- *conjunction* (i.e., *AND* or, symbolically, $\wedge$),
- *disjunction* (i.e., *OR* or, symbolically, $\vee$),

---

[6] To avoid misunderstandings: Finding matching upper and lower bounds for a problem does not mean to find an algorithm with running time $\Theta(t)$ that solves the problem. That would merely show that the running time of *this particular* algorithm (which provides *some* upper bound only) has been analyzed well, so that no essential improvements are possible in the runtime analysis of this algorithm. By constrast, to show a lower bound matching this upper bound of $\Theta(t)$, one would have to show that *no* algorithm (of the considered type) for this problem has a runtime that asymptotically is better than that.

1.5 A Brief Digression to Computational Complexity

- *negation* (i.e., NOT or, symbolically, $\neg$),
- *implication* (i.e., $IF \ldots THEN \ldots$ or, symbolically, $\Longrightarrow$),
- *equivalence* (i.e., $\ldots IF\ AND\ ONLY\ IF \ldots$ or, symbolically, $\Longleftrightarrow$),
- etc.

Table 1.1 Truth table for some boolean operations

$x\ y$	$x \wedge y$	$x \vee y$	$\neg x$	$x \Longrightarrow y$	$x \Longleftrightarrow y$
0 0	0	0	1	1	1
0 1	0	1	1	1	0
1 0	0	1	0	0	0
1 1	1	1	0	1	1

The above-mentioned boolean operations can be defined via their truth tables (see Table 1.1), where the *truth values* (also called *boolean constants*) $TRUE$ and $FALSE$ are represented by 1 and 0. Every boolean formula φ with n variables represents a boolean function $f_\varphi : \{0,1\}^n \to \{0,1\}$. There are exactly two boolean functions of arity zero (namely, the two boolean constants), four 1-ary boolean functions (e.g., the identity and the negation), and 16 2-ary boolean functions (e.g., the functions corresponding to the boolean operations $\wedge$, $\vee$, $\Longrightarrow$, and $\Longleftrightarrow$ in Table 1.1). In general, there exist exactly 2^{2^n} n-ary boolean functions, since each of the 2^n possible truth assignments fixes two of them, namely the one with value 0 and the one with value 1 for the given assignment.

The truth value of a boolean formula (i.e., the value of the corresponding boolean function) can be determined from the truth values assigned to its variables. For example, the boolean formula

$$\varphi(x,y,z) = (x \wedge \neg y \wedge z) \vee (\neg x \wedge \neg y \wedge \neg z) \tag{1.1}$$

depends on three boolean variables, x, y, and z, and for the truth assignment $(1,1,1)$ to (x,y,z) it can be evaluated as follows:

$$\varphi(1,1,1) = (1 \wedge \neg 1 \wedge 1) \vee (\neg 1 \wedge \neg 1 \wedge \neg 1) = (1 \wedge 0 \wedge 1) \vee (0 \wedge 0 \wedge 0) = 0 \vee 0 = 0,$$

that is, φ is false under this assignment. Using the assignment $(1,0,1)$, however, we obtain

$$\varphi(1,0,1) = (1 \wedge \neg 0 \wedge 1) \vee (\neg 1 \wedge \neg 0 \wedge \neg 1) = (1 \wedge 1 \wedge 1) \vee (0 \wedge 1 \wedge 0) = 1 \vee 0 = 1,$$

so φ is true under this assignment. A boolean formula φ is *satisfiable* if there exists an assignment to its variables such that φ evalutes to true. The formula φ from (1.1) is satisfiable, since it has two satisfying assignments, $(0,0,0)$ and the already mentioned assignment $(1,0,1)$. None of the other six assignments satisfies φ.

A *literal* is a variable, such as x, or its negation, $\neg x$. An *implicant* is a conjunction of literals, such as $(x \wedge \neg y \wedge z)$, and a *clause* is a disjunction of literals, such as $(x \vee y \vee \neg z)$. Formulas like the one in (1.1) are in *disjunctive normal form* (DNF), i.e., they are disjunctions of implicants. Similarly, a formula is in *conjunctive normal form* (CNF) if it is a conjunction of clauses.

Every boolean formula can be transformed into an equivalent one in DNF, and also into an equivalent formula in CNF, where the new formulas can be exponentially larger than the given formula, though. Two formulas are said to be *equivalent* if they have the same truth value for all truth assignments to their variables. For example, one can easily check that

$$\varphi'(x,y,z) = (x \vee y \vee \neg z) \wedge (x \vee \neg y \vee z) \wedge (x \vee \neg y \vee \neg z) \wedge \\ (\neg x \vee y \vee z) \wedge (\neg x \vee \neg y \vee z) \wedge (\neg x \vee \neg y \vee \neg z) \qquad (1.2)$$

is a formula in CNF equivalent to φ from (1.1). If the number of literals in all clauses of a formula in CNF is bounded by a constant k, we say that the formula is in *k-CNF*. For example, the formula φ' from (1.2) is in 3-CNF. The restriction of the above-defined problem SAT to formulas in k-CNF is denoted by k-SAT.

1.5.2.2 Upper Bounds for SAT

How hard is SAT? Since formulas with n variables have 2^n truth assignments, the naive deterministic algorithm for SAT runs in time $\mathcal{O}(n^2 \cdot 2^n)$: On input φ (a given boolean formula with n variables), this algorithm tests for all possible truth assignments to the variables of φ, one after the other, whether they satisfy φ, and each single test can be done in quadratic time. If a satisfying assignment is found, the algorithm halts and accepts; but it can reject its input only after having tested all 2^n assignments unsuccessfully. This naive algorithm for SAT can be improved, as we will see below.

It is also known that certain restrictions of the problem SAT can be solved more efficiently. For example, if the given formula is in DNF, then it can be decided in polynomial time whether or not it is satisfiable. This is because in this case we can check the satisfiability of the implicants of the formula, one by one; as soon as a satisfiable one is found, the formula is accepted; otherwise, we can safely reject it as soon as the last unsatisfiable implicant has been checked.

If the given formula is in CNF, however, the problem seems to be much harder, namely as hard as the unrestricted problem SAT. Yet even in this case, an efficient algorithm is possible, provided there is no clause with more than two literals in the given formula: Jones, Lien, and Laaser [580] showed that 2-SAT is solvable in polynomial time, and that it even belongs to a complexity class that (presumably) is yet smaller than P, namely, to the class of problems solvable by a nondeterministic Turing machine in logarithmic

1.5 A Brief Digression to Computational Complexity

space (see, e.g., [819] for a detailed proof). This class is widely believed to be a proper subclass of P, just as P is widely believed to be a proper subclass of NP.

The runtime of the naive algorithm for SAT was given as a function of the number of variables of the given formula. Why? It would seem more natural to define the size of a formula as the number of *occurrences* of (positive or negated) variables (neglecting the encoding details with respect to parentheses, $\wedge$, $\vee$, and $\neg$ symbols, etc.), and each variable can occur very often in a formula. For example, x, y, and z each occur twice in the formula φ from (1.1), but they each occur six times in the equivalent formula φ' from (1.2). In general, every variable might occur exponentially often (in the number of variables). However, this is only due to the fact that the way we represented these formulas in (1.1) and (1.2) was more wasteful than needed. If one represents a boolean formula by a *boolean circuit* (as in Figure 1.5), then we have a compact representation of the given SAT instance whose size (under a reasonable encoding) is polynomial in the number of variables, and a polynomial blow-up is something we can easily cope with.

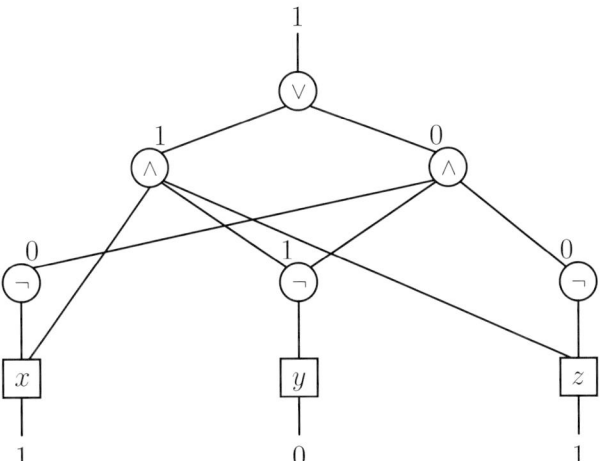

Fig. 1.5 A boolean circuit for the formula in (1.1) with satisfying assignment $(1, 0, 1)$

Alternatively, the runtimes of SAT and k-SAT algorithms are also given in the number of clauses or in relation to both parameters, the number of variables and the number of clauses.

As mentioned above, the naive SAT algorithm can be crucially improved. This fascinating, exceedingly important problem has been investigated in computer science for decades, in order to develop better and better SAT solvers. Not much different than in high-performance sports, new records are established for SAT again and again (see, e.g., the surveys by Schöning [842], Woeginger [942], and Riege und Rothe [811] for results on moderately exponential-time algorithm for SAT and other hard problems).

Table 1.2 Some deterministic upper bounds for k-SAT and SAT

Problem	Upper bound	Source
2-SAT	P	Jones, Lien, and Laaser [580]
3-SAT	1.4726^n	Brueggemann and Kern [239]
k-SAT, $k \geq 4$	$\left(2 - \frac{2}{(k+1)}\right)^n$	Dantsin at al. [324]
SAT	$2^{n\left(1 - \frac{1}{\log(2m)}\right)}$	Dantsin and Wolpert [325]

Table 1.2 lists some of these upper bounds for k-SAT and SAT, where n is the number of variables and m the number of clauses of the input formula and where polynomial factors are ignored.[7] Table 1.3 shows how the results for 3-SAT have evolved over time. All these upper bounds refer to *deterministic* algorithms (as opposed to *randomized* algorithms that may be more efficient, but can make errors) and to the *worst case*, i.e., when analyzing the runtimes of these algorithms, one assumes the "worst cases," the most difficult problem instances the algorithm must be able to handle.

Table 1.3 Improving the deterministic upper bounds for 3-SAT

Upper bound	Source
2^n	naive 3-SAT algorithm
1.6181^n	Monien and Speckenmeyer [712]
1.4970^n	Schiermeyer [838]
1.4963^n	Kullmann [628]
1.4802^n	Dantsin et al. [324]
1.4726^n	Brueggemann and Kern [239]

As one can see, all upper bounds listed in Tables 1.2 and 1.3 for these variants of the satisfiability problem are still exponential in the number of variables. Now, what is the point in improving an exponential-time algorithm running in time, say, 2^n to another one with a more moderate exponential runtime of c^n, where c is a constant with $1 < c < 2$? In practical applications, such an improved, more moderate exponential runtime can have a great impact, as one is then able to process significantly larger inputs in the same time. Note that the exponential growth hits only from a certain input size on; therefore, even exponential-time algorithms can be practicable for moderate input sizes below this threshold. Suppose, for instance, that within one hour we can

---

[7] It is common to neglect polynomial factors when analyzing exponential runtimes.

solve all inputs of size up to 30 by an algorithm running in time 2^n. If we are able to improve this algorithm so that our new algorithm runs in time, say, $\sqrt{2}^n \approx 1.4142^n$, then we will now be able to handle inputs up to size 60 on the same computer within one hour, since we have $\sqrt{2}^{60} = 2^{(1/2) \cdot 60} = 2^{30}$. In practice, this really can make a difference.

1.5.2.3 How to Prove Lower Bounds: Reducibility and Hardness

However, how can one prove a *lower* bound for SAT? Indeed, for (unrestricted) deterministic algorithms, no better lower bound than linear time is known for SAT to date (see, e.g., the work of Fortnow et al. [463]). But this lower bound is trivial, as linear time is needed just to read the input. Due to this difficulty to prove lower bounds of problems in the sense of the Ω or ω notation, one takes another approach: One compares the complexity of a given problem with that of other problems in a complexity class and tries to show that it is at least as hard to solve the given problem as it is to solve any of the other problems in the class. If one succeeds, the considered problem is *hard* for the entire complexity class, and also in this sense it is common to speak of a problem's *lower bound*: The related complexity class provides a lower bound for the problem at hand, since solving any problem in the class is no harder than solving this problem. If the latter belongs to this class in addition, it is said to be *complete* for it.

To compare two given problems in terms of their complexity, we now introduce the notion of reducibility on which the above notions of hardness and completeness are based. Intuitively, a reduction of a decision problem A to a decision problem B means that all instances of A can efficiently be transformed into instances of B such that the original instances are yes-instances of A if and only if their transformations are yes-instances of B. This notion of reducibility is just one among many, but perhaps the most important one, and it is called *polynomial-time many-one reducibility* because different instances of A can be mapped to one and the same instance of B by the transformation (see, e.g., the textbooks by Papadimitriou [764] and Rothe [819] for more details and for other reducibilities some of which will be presented in later chapters).

Definition 1.2 (reducibility, hardness, completeness, and closure).
An *alphabet* is a finite, nonempty set of characters (or symbols). Σ^* denotes the set of all strings over the alphabet Σ. A total function $f : \Sigma^* \to \Sigma^*$ is said to be *polynomial-time computable* if there is an algorithm that, given any string $x \in \Sigma^*$, computes the function value $f(x)$ in polynomial time. Let FP denote the class of all polynomial-time computable functions.

Let A and B be two given decision problems (for simplicity, they are assumed to be encoded over the same alphabet Σ, i.e., $A, B \subseteq \Sigma^*$). Let $\mathcal{C}$ be any complexity class.

1. We say that A is *(polynomial-time many-one) reducible to B* (denoted by $A \leq_m^P B$) if there is a function $f \in \mathrm{FP}$ such that for each $x \in \Sigma^*$, $x \in A \iff f(x) \in B$.
2. B is $\leq_m^P$-*hard for $\mathcal{C}$* (or, shorter, $\mathcal{C}$-*hard*) if $A \leq_m^P B$ for each set $A \in \mathcal{C}$.
3. B is $\leq_m^P$-*complete in $\mathcal{C}$* (or, shorter, $\mathcal{C}$-*complete*) if B is $\mathcal{C}$-hard and in $\mathcal{C}$.
4. $\mathcal{C}$ is *closed under the $\leq_m^P$-reducibility* (or, shorter, $\leq_m^P$-*closed*) if for any two problems A and B, it follows from $A \leq_m^P B$ and $B \in \mathcal{C}$ that A is in $\mathcal{C}$.

Cook [313] proved that SAT is NP-complete by encoding the computation of an arbitrary NP algorithm on any given input into a boolean formula that is satisfiable if and only if the algorithm accepts its input. This $\leq_m^P$-reduction of an arbitrary NP problem to SAT shows NP-hardness of, and thus a lower bound for, SAT. That SAT belongs to NP—which provides the corresponding upper bound for this problem—is easy to see: Given a boolean formula φ with n variables, an NP algorithm for SAT nondeterministically guesses a truth assignment for φ and then tests deterministically whether the guessed assignment satisfies the given formula. Here, the power of nondeterminism is exploited. In nondeterministic polynomial time, one can guess (and then deterministically verify) each of the 2^n possible assignments, where the corresponding computation tree has exactly 2^n paths whose length is polynomial in n. If there exists at least one accepting path in this tree (i.e., at least one satisfying assignment), the formula is accepted; otherwise, it is rejected.

SAT is the first natural problem whose NP-completeness could be proven; therefore, Cook's result [313] is a milestone in complexity theory. The notion of NP-comleteness has an immediate relation to the above-mentioned "P = NP?" problem, as the following lemma shows. In addition, we list a number of other fundamental properties of the $\leq_m^P$-reducibility and of the complexity classes P and NP that will be useful throughout the book.

Lemma 1.1. *1. P and NP are $\leq_m^P$-closed.*
2. If $A \leq_m^P B$ and B is in P, then A is in P.
3. If $A \leq_m^P B$ and A is NP-hard, then B is NP-hard.
4. P = NP if and only if SAT is in P.

Like most complexity classes, P and NP are closed under $\leq_m^P$-reductions according to the first statement in Lemma 1.1. This is a very useful property that can be applied in many proofs about these classes, such as in the proof of the fourth statement of this lemma, which says that the famous "P = NP?" problem can be solved simply by finding a (deterministic) polynomial-time algorithm for SAT. In this sense, SAT—just as every other NP-complete problem—represents the whole class NP. That is, one could take *any* other NP-complete problem to replace SAT in this equivalence. To see why this equivalence holds, just note that (a) if P = NP then the NP problem SAT is immediately in P, and (b) if SAT—or any other NP-complete problem—were in P, then all the rest of NP would trail behind due to the $\leq_m^P$-closure of P,

1.5 A Brief Digression to Computational Complexity

which immediately would imply equality of P and NP. Similarly simple proofs can be given for the remaining statements of this lemma that are immediate consequences of the definitions, too.

By the second statement of Lemma 1.1, upper bounds are inherited downward in terms of $\leq_m^P$, and by its third statement, lower (NP-hardness) bounds are inherited upward in terms of $\leq_m^P$. That is why the $\leq_m^P$-reducibility is a very useful tool:

- on the one hand, to prove new upper bounds (in particular, new P algorithms), and
- on the other hand, to prove new lower bounds (in particular, new NP-hardness results).

As regards the latter task, the problem SAT is particularly useful to show other problems NP-hard: Starting from a SAT instance, reductions to many other problems can easily be given. And even more suitable are certain restrictions of this problem. For example, the Cook reduction to SAT even provides a boolean formula in CNF. Thus, also the restriction of the satisfiability problem to formulas in CNF is NP-complete. Moreover, if one wants to start from that problem to show other problems NP-hard, it is sometimes very useful if the given boolean formula is not only in CNF, but is even a 3-CNF, i.e., if every clause of the formula has at most three literals. To this end, we first need to show that this restriction of SAT is NP-hard. That is, we need to give a $\leq_m^P$-reduction from, say, SAT restricted to CNF formulas to SAT restricted to 3-CNF formulas. And that is what we will do now to give a specific example of a $\leq_m^P$-reduction (see also, e.g., [483, 819]).

For any given formula φ in CNF, we want to construct an equivalent formula ψ in 3-CNF. To this end, it is enough to replace every clause φ with more than three literals, say $C = (\ell_1 \vee \ell_2 \vee \cdots \vee \ell_k)$ with $k \geq 4$ literals $\ell_1, \ell_2, \ldots, \ell_k$, by a new subformula C' that, first, is in 3-CNF and, second, is satisfiable if and only if C is satisfiable. This new subformula will have $k - 3$ new variables, $y_1, y_2, \ldots, y_{k-3}$, and will consist of $k - 2$ clauses of the following kind:

$$C' = (\ell_1 \vee \ell_2 \vee y_1) \wedge (\neg y_1 \vee \ell_3 \vee y_2) \wedge \cdots \wedge \\ (\neg y_{k-4} \vee \ell_{k-2} \vee y_{k-3}) \wedge (\neg y_{k-3} \vee \ell_{k-1} \vee \ell_k).$$

It is easy to see that if C is satisfiable, so is C', since a satisfying assignment for C can by extended to a satisfying assignment for C' as follows:

- If the clause $(\ell_1 \vee \ell_2 \vee y_1)$ is true under the satisfying assignment for C because ℓ_1 or ℓ_2 are satisfied already, then extend this assignment by setting all variables y_i to 0. Obviously, this ensures that every clause of C' is satisfied as well.
- If the clause $(\neg y_{k-3} \vee \ell_{k-1} \vee \ell_k)$ is true under the satisfying assignment for C because ℓ_{k-1} or ℓ_k are satisfied already, then extend this assignment

by setting all variables y_i to 1. Obviously, this ensures that every clause of C' is satisfied as well.
- Otherwise, there must exist another clause of C' (neither the first nor the last one) that is true under the satisfying assignment for C, since we know that at least one of the literals ℓ_j is satisfied. Let $(\neg y_{j-2} \vee \ell_j \vee y_{j-1})$ be the first such clause of C'. Now, extending the assignment for C by setting to 1 all variables y_i, $1 \leq i \leq j-2$, and by setting to 0 all the remaining variables, y_i with $j-1 \leq i \leq k-3$, obviously satisfies every clause of C'.

On the other hand, restricting any satisfying assignment for C' to the variables occurring in C (which correspond to the literals $\ell_1, \ell_2, \ldots, \ell_k$) provides a satisfying assignment for C. Therefore, if C' is satisfiable, so is C. It follows that the clause C is equivalent to the subformula C'. Since we introduce for each such clause with more than three literals a new (disjoint) set of new variables, the original formula φ is equivalent to the new formula ψ constructed in this way. It is also clear that the construction can be done in polynomial time. Hence, by the third statement of Lemma 1.1, even 3-SAT is NP-hard.

By the way, sometimes it is useful to start in a reduction from a formula that not only is in 3-CNF, but that even has the property that each clause has *exactly* three literals. Do you see how one would have to modify the above reduction in order to show that even that restriction of the satisfiability problem is NP-complete?

Garey and Johnson [483] collected hundreds of NP-complete problems from a variety of scientific fields already more than four decades ago; meanwhile, there are probably thousands, if not tens of thousands, of problems known to be NP-complete (see also Johnson's related column in the *Journal of Algorithms* [578]). For example, INDEPENDENT-SET is the problem to decide, given a graph G and a positive integer m, whether or not G has an independent set of size at least m.[8] It is easy to see that this problem is in NP. On the other hand, INDEPENDENT-SET can be seen to be NP-hard (and, therefore, NP-complete [483]) by a reduction from 3-SAT that, informally stated, works as follows. Given a boolean formula in 3-CNF with m clauses (each of which has exactly three literals), create a triangle for each clause (i.e., create three vertices—each being labeled with one of the literals of that clause—and connect any two of them by an edge), and create an edge between any two vertices of two distinct triangles whenever one is labeled with a literal and the other one with its negation. It is easy to see that this graph has an independent set of size m if and only if the given formula is satisfiable.

We will come across the satisfiability problem, SAT, and its variants in several places later in this book. On the one hand, we will apply its NP-completeness to determine the complexity of some of those problems we are

---

[8] An *independent set* of a graph is a subset I of its vertex set such that no two vertices in I are connected by an edge. The size of an independent set is the number of its vertices.

1.5 A Brief Digression to Computational Complexity   33

interested in here. On the other hand, in Chapter 7 we will be concerned with
boolean formulas in the context of judgment aggregation.

1.5.2.4 Some Background on Approximation Theory

If a decision problem is NP-hard, one cannot expect to find a deterministic
polynomial-time algorithm solving it exactly. However, one can try to find
efficient approximation algorithms for the optimization variants of this problem, i.e., polynomial-time algorithms that may not find an exact solution
to the problem but can approximate an optimal solution within a certain
factor. For example, an optimization variant of SAT is the problem MAX-SAT: Given a boolean formula in CNF, find an assignment of its variables
that satisfies as many as possible of its clauses. An optimization variant of
the problem INDEPENDENT-SET defined above is MAX-INDEPENDENT-SET:
Given a graph G, output the size of a maximum independent set of G.

Fix some α, $0 < \alpha < 1$. An α-*approximation algorithm* A for a maximization problem is a polynomial-time algorithm such that for each instance x,
A outputs a solution to the problem whose value is at least α times the optimal value for x. The value α is called the *approximation factor* (a.k.a. the
approximation ratio or *performance guarantee*) of an α-approximation algorithm. Note that α may also depend on the size of the given instance. For
minimization problems, the analogous notions are defined for $\alpha > 1$.

As an example, consider the restriction of INDEPENDENT-SET where the
degree of each vertex (i.e., the number of incident edges) is bounded by k; denote this problem by k-DEGREE-INDEPENDENT-SET. Note that for $k \geq 4$, this
problem is also NP-complete. However, there is a simple $\frac{k}{k+1}$-approximation
algorithm for MAX-k-DEGREE-INDEPENDENT-SET: Start with I set to the
empty set (which is trivially independent); while there are still vertices in
the given graph G, delete an arbitrary vertex v and all its neighbors from
G and add v to I. The resulting set I will obviously be independent. Since
in each while-loop, some vertex is added to I and at most $k+1$ vertices are
deleted from G (due to the bound on the vertex degree), I has size at least
the number of G's vertices divided by $k+1$, which is at least $1/(k+1)$ times
the size of a maximum independent set of G.

For the general MAX-INDEPENDENT-SET problem (without any given
bound on the vertex degree of the given graph), a quite similar algorithm, the
so-called *minimum-degree greedy heuristic*, provides a good approximation in
certain cases: Given a graph G, pick any vertex of minimum degree, delete
it and all its neighbors from G, and repeat this procedure until you are left
with an empty graph. Bodlaender, Thilikos, and Yamazaki [158] show that for
certain well-behaved classes of graphs (including trees, so-called split graphs,
and complete k-partite graphs, for any k), this simple heuristic indeed finds
the optimum, i.e., a maximum independent set.

We say an optimization problem has a *polynomial-time approximation scheme* (*PTAS*) if for each ε, $0 < \varepsilon < 1$, there is an α-approximation algorithm for the problem, where $\alpha = 1 - \varepsilon$ if it is a maximization problem, and $\alpha = 1 + \varepsilon$ if it is a minimization problem. A *fully polynomial-time approximation scheme* (*FPTAS*) is a PTAS whose running time is bounded by a polynomial of the input size and of $1/\varepsilon$. For example, it is known that MAX-3-DEGREE-INDEPENDENT-SET has no PTAS [134], unless P = NP (which, recall, is considered highly unlikely).

For more background on approximation theory, including techniques to show approximation results as well as techniques to prove inapproximability results (under certain complexity-theoretic hypotheses), we refer to the textbook by Vazirani [922] and the survey by Arora and Lund [28]. Approximation results will be mentioned in most of the following chapters. For concreteness, we in particular refer to the discussion on unary weights in weighted voting games and approximation of computing power indices in Section 3.5.2.2 starting on page 184; to the approximation algorithms for finding winning committees by certain multiwinner voting rules mentioned in Section 6.5.3 starting on page 456; and in fair division of indivisible goods, we refer to the approximations of envy-freeness and other fairness notions and of optimizing social welfare elaborated on in Sections 9.5 and 9.6 starting on pages 625 and 635, respectively.

We conclude the current chapter by giving a brief compendium of complexity classes and we will illustrate these classes by exhibiting suitable variants of SAT that are complete for them.

1.5.3 A Brief Compendium of Complexity Classes

We got to know the complexity classes P and NP already, see Section 1.5.1.2 starting on page 21. As we said there, these are the most central complexity classes with respect to the time complexity measure, and indeed, most of the problems we will come across in this book can be classified to either belong to P or be NP-complete. Moreover, we have already seen that SAT and 3-SAT, too, are NP-complete and that 2-SAT is in P.[9] However, as we will also see completeness results for classes other than NP later in this book, we now briefly describe these classes, along with "their" $\leq_m^P$-complete variants of SAT, and we will point the reader to the relevant chapters or sections for more details.

---

[9] We won't consider completeness for P here. Note that $\leq_m^P$-completeness in P is trivial, as $\leq_m^P$ is too coarse for P and smaller classes (see, e.g., [819, Lemma 3.37], which says that *every* P set distinct from $\emptyset$ and Σ^* is $\leq_m^P$-complete in P). There are also nontrivial completeness results for this class, but with respect to other reducibilities than $\leq_m^P$.

1.5.3.1 Polynomial Space

SAT is the satisfiability problem for propositional formulas with no quantifiers involved. Its quantified variant is the following:

QUANTIFIED-BOOLEAN-FORMULA (QBF)	
Given:	$H = (\exists X_1)(\forall X_2) \cdots (\mathfrak{Q} X_k)[\varphi(X_1, X_2, \ldots, X_k)]$, a quantified boolean formula, where the X_i are disjoint sets of variables, φ is a boolean formula without quantifiers, all variables occurring in φ are quantified, and $\mathfrak{Q} = \exists$ if k is odd, and $\mathfrak{Q} = \forall$ if k is even.
Question:	Does H evaluate to true, i.e., does there exist a truth assignment to the variables in X_1 such that for each truth assignment to the variables in X_2, there exists ... a truth assignment to the variables in X_k such that φ is true under this assignment?

Because of its quantifier structure QBF seems to be much harder a problem than SAT; in fact, it is PSPACE-complete, where PSPACE is the class of problems that can be solved by a Turing machine[10] in polynomial space; alternatively, PSPACE is the class of problems that can be solved by an *alternating Turing machine* in polynomial *time* (see the work of Chandra, Kozen, and Stockmeyer [273]).

We will come across PSPACE-complete problems in Section 2.3.1.4 on page 91, where we consider the hardness of detecting whether a player in a combinatorial two-player game with perfect information has a winning strategy. We will also see PSPACE-complete problems in the context of online manipulation (see Section 4.3.3.10 starting on page 321), online control (see Section 4.3.4.11 starting on page 354), and online bribery (see Section 4.3.5.5 starting on page 363) in sequential elections.

1.5.3.2 The Polynomial Hierarchy

Restricting QBF to a fixed number of i alternating quantifiers gives the problems QBF_i (starting with an existential quantifier) and their complements, $\overline{\text{QBF}_i}$ (starting with a universal quantifier). Note that QBF_1 is nothing other than SAT. The corresponding hierarchy of complexity classes has been introduced and studied by Meyer and Stockmeyer [706, 892]: the so-called *polynomial hierarchy*, $\text{PH} = \bigcup_{i \geq 0} \Sigma_i^p$. Its Σ_i^p levels are inductively defined by

$$\Sigma_0^p = \text{P}, \quad \Sigma_1^p = \text{NP}, \quad \text{and} \quad \Sigma_i^p = \text{NP}^{\Sigma_{i-1}^p}, \; i \geq 2,$$

where "$\text{NP}^{\Sigma_{i-1}^p}$" means that an NP "oracle Turing machine" accesses an "oracle" from Σ_{i-1}^p (see page 291 and [819, Def. 2.22] for more details). Since

---

[10] It does not matter whether this Turing machine is deterministic or nondeterministic; a consequence of Savitch's theorem [836] is that nondeterministic polynomial space coincides with deterministic polynomial space.

these classes are not likely to be closed under complementation, one has also studied their co-classes, the Π_i^p levels of the polynomial hierarchy, defined by $\Pi_i^p = \text{co}\Sigma_i^p = \{\overline{L} \mid L \in \Sigma_i^p\}$, $i \geq 0$.

In particular, $\Pi_0^p = \text{P}$ (because, as a deterministic class, P clearly is closed under complementation) and $\Pi_1^p = \text{coNP}$ is the class of complements of NP problems. Just as with the "P = NP?" question, it is open whether or not NP equals coNP, but it is widely believed that these two classes are distinct. It is also widely believed (but as yet unproven) that the polynomial hierarchy is strict, i.e., that all their levels differ from each other. Note that P = NP would imply NP = coNP, and even that PH = P. More generally, a collapse of any level of this hierarchy ($\Sigma_i^p = \Pi_i^p$, $i \geq 1$, or $\Sigma_i^p = \Sigma_{i+1}^p$, $i \geq 0$) would imply an upward collapse of the entire hierarchy to this level (PH $= \Sigma_i^p$).

It is known that for each $i > 0$, QBF$_i$ is Σ_i^p-complete and $\overline{\text{QBF}_i}$ is Π_i^p-complete. It is also known that the levels of the polyomial hierarchy can be characterized via alternating, polynomially length-bounded existential and universal quantifiers (where the Σ_i^p levels start with an existential and the Π_i^p levels with a universal quantifier), which nicely mirrors the structure of QBF$_i$ and $\overline{\text{QBF}_i}$ formulas, respectively; see the original work of Meyer and Stockmeyer [706, 892] or the textbook by Rothe [819, Section. 5.2] for a proof of this characterization.

Completeness for the levels of the polynomial hierarchy, specifically for Σ_{2k}^p, $k \geq 1$, will play a role in Section 3.4.2.3 on page 177 when we will be concerned with bribery in path disruption games [76, 807]; in Section 3.6 starting on page 198 when we will study stability problems in hedonic games; in, again, Sections 4.3.3.10, 4.3.4.11, and 4.3.5.5 starting on pages 321, 354, and 363, respectively, when we will consider online manipulation [543], online control [545, 544, 740], and online bribery [546] in sequential elections; in Section 7.3 starting on page 486 when we will study the complexity of problems related to judgment aggregation; and in Section 9.4.3 starting on page 623 when we will turn to the complexity of problems related to Pareto efficiency (see Definition 9.2 on page 622) and envy-freeness (see Definition 9.3 on page 623) in the allocation of indivisible goods.

The Δ_i^p levels of the polynomial hierarchy are $\Delta_0^p = \text{P}$ and $\Delta_i^p = \text{P}^{\Sigma_{i-1}^p}$ for $i \geq 1$; note that $\Delta_1^p = \text{P}^\text{P} = \text{P} = \Delta_0^p$. Once again related to online manipulation in sequential elections, we will specifically come across the class $\text{P}^\text{NP} = \Delta_2^p$ in Section 4.3.3.10 starting on page 321. In addition, we will also see completeness for the restriction $\text{P}^{\text{NP}[1]}$ of P^NP with only one oracle query allowed. Relatedly, the restriction of P^NP to logarithmically many oracle queries, denoted by $\text{P}^{\text{NP}[\log]}$, is known to be equivalent (due to results of Hemachandra [534] and Köbler, Schöning, and Wagner [616]) to *"parallel access to NP,"* denoted by $\text{P}_{\|}^\text{NP}$ and formally defined on page 291. $\text{P}_{\|}^\text{NP}$-completeness will be relevant to the winner problems for Dodgson, Young, and Kemeny elections in Section 4.3.1.3 starting on page 291. Similarly, in Section 7.3.1 starting on page 487 we will see $\text{P}_{\|}^\text{NP}$-completeness results for various problems in judgment aggregation. Another problem known to be complete for this class

1.5 A Brief Digression to Computational Complexity

is the problem of whether the minimum-degree greedy heuristic (described in Section 1.5.2.4 starting on page 33) can approximate the size of a maximum independent set of a given graph within a constant factor [547] (see the work of Hemaspaandra, Rothe, and Spakowski [548] for the analogous results regarding the problem MIN-VERTEX-COVER[11] and the so-called *edge deletion heuristic* and the *maximum-degree greedy heuristic*).

The following problem is a variant of the satisfiability problem that (by a result of Wagner [924]; see also [543] for some technical tweaks in the proof) is known to be complete in P^{NP}:

MAX-SATISFYING-ASSIGNMENT-EQUALITY (MAX-SAT-ASG$_=$)

Given: Two boolean formulas in CNF with exactly three literals per clause.
Question: Do they have the same maximal satisfying assignment?

And an example of a variant of the satisfiability problem that (again due to Wagner [924]) is known to be $P_{\|}^{NP}$-complete is the problem ODD-SAT: Given a list of boolean formulas, is it true that the number of satisfiable formulas in the list is odd?

1.5.3.3 DP: The Second Level of the Boolean Hierarchy over NP

There are also other hierarchies built upon NP, for example, the *boolean hierarchy over* NP (see, e.g., [248, 249, 819, 810]). In this book, we will only be interested in the second level of this hierarchy, the class DP, which was introduced by Papadimitriou and Yannakakis [766] and is defined as the class of differences of any two NP sets. An example of a variant of the satisfiability problem that is DP-complete is the problem SAT-UNSAT, which contains all pairs (φ, ψ) of boolean formulas such that φ is satisfiable and ψ is not.

We will come across further DP-complete problems that are related to the exact variant of the margin of victory problem in voting (see Section 4.3.5.4 starting on page 362) and to exactly maximizing social welfare in allocating indivisible goods (see Section 9.6.5 starting on page 647).

1.5.3.4 Probabilistic Polynomial Time

MAJORITY-SAT is the problem of whether a given boolean formula with n variables has at least 2^{n-1} satisfying assignments, i.e., at least half of the total number of assignments are required to be satisfying assignments. The complexity class that naturally corresponds to this problem has been introduced by Gill [501] as *probabilistic polynomial time*, abbreviated by PP. He

---

[11] MIN-VERTEX-COVER is defined as the problem: Given a graph G, output the size of a minimum vertex cover of G, where a *vertex cover of G* is a subset of the vertices of G that contains at least one of the two vertices of each edge in G.

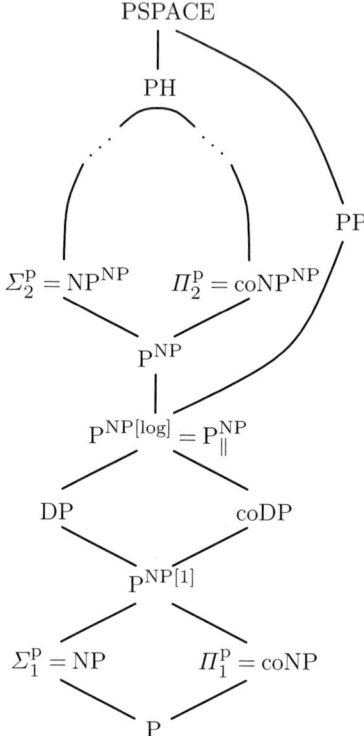

Fig. 1.6 Inclusions between some complexity classes

defines PP via *probabilistic (polynomial-time) Turing machines*, which can be viewed as nondeterministic (polynomial-time) Turing machines that accept their input if and only if at least half of their total number of computation paths accept. In Section 3.5.2.2 starting on page 184, we will see that certain problems related to "false-name manipulation" and to "control by deleting or adding players" in weighted voting games are PP-hard.

Overview

The landscape of the above-defined complexity classes is shown in Figure 1.6 as a Hasse diagram, i.e., every ascending line from some complexity class $\mathcal{C}$ to some other class $\mathcal{D}$ indicates an inclusion: $\mathcal{C} \subseteq \mathcal{D}$. Most of these inclusions follow immediately from the definitions; some require a nontrivial proof. For example, the inclusion $P_{\|}^{NP} \subseteq PP$ follows from the closure of PP under so-called parity reductions that was shown by Beigel, Hemachandra, and Wechsung [127]. Presumably, P^{NP} is not contained in PP, and PP and

1.5 A Brief Digression to Computational Complexity   39

PH are incomparable. None of the inclusions depicted in Figure 1.6 is known to be strict, though it is widely suspected that they are.

In later chapters, some further complexity classes will be introduced and discussed (e.g., those depicted in Figure 2.25 on page 132 or the parameterized complexity class W[2] occurring in Table 7.5 on page 499; see the textbooks [355, 753, 461] for background on parameterized complexity theory). However, as will be explained there, these classes are of a different type (those from Figure 2.25 are classes of functions and not classes of sets, while parameterized complexity classes contain *parameterized* problems that, unlike regular decision problems, specify one or more parameters in addition to the input) and, therefore, cannot directly be compared with those in Figure 1.6.

1.5.3.5 And Now, Finally, ...

... let's get ready to play, vote, and divide!

Part I
Playing Successfully

Chapter 2
Noncooperative Game Theory

Piotr Faliszewski · Irene Rothe · Jörg Rothe

Playing is something profoundly human,[1] and the ability to play is tightly tied to the intelligence of human beings,[2] to their capability of thinking foresightedly and strategically, of choosing a particularly profitable move among all possible moves, of anticipating possible response moves by their adversaries, and thus to their capability of maximizing their own profit. By playing a game we here mean, in general, an interaction under preassigned rules, amongst several players each interested in maximizing their gains and acting strategically to this end. Games are encountered everywhere, be it as a party game, a card game, a computer game, or a game of hazard, be it as an individual or team sport such as chess, foil fencing, soccer, or ice hockey, be it companies organizing their strategies in a market economy, or states and other global players deciding on their geopolitical strategies.

That all players at the same time aim at maximizing their individual gains constitutes the special fascination of playing and the grand intellectual challenge of game theory. The foundations of this theory have been laid in 1944 by von Neumann and Morgenstern in their groundbreaking work [736], see also the early papers by Borel [170] and von Neumann [734].

A quite fundamental classification of games is to distinguish cooperative games from noncooperative games. In this chapter we are concerned with noncooperative game theory. The players in a noncooperative game compete against each other: Every player is selfishly interested only in her own profit, and players do not take joint actions. By contrast, cooperative game theory is in particular concerned with forming coalitions of players that work to-

---

[1] Friedrich Schiller writes in his 15th letter of *The Aesthetic Education of Man* (1795): *"Der Mensch spielt nur, wo er in voller Bedeutung des Wortes Mensch ist, und er ist nur da ganz Mensch, wo er spielt"* ("Man only plays when he is in the full sense of the word a human being, and he is only entirely human when he plays").

[2] Christian Morgenstern says in *Gallows Hill*, one of *The Gallows Songs* (1905): *"Blödem Volke unverständlich treiben wir des Lebens Spiel"* ("Enigmatic for the masses, playfully with life we fool," as translated by Max Knight, University of California Press, 1964).

gether in order to reach their common goals. We will deal with this theory in Chapter 3.

To classify and analyze the tremendous variety of noncooperative games, we will turn to various aspects and criteria (or "solution concepts"), without making any claim of completeness. For example, one may distinguish games with perfect information (such as chess) from games with only incomplete information (including many card games). Further, some games are affected by randomness, while others are purely deterministic. There are games with one move made simultaneously by all players, and there are games with any number of moves made sequentially by the players taking turns. Of course, the order of moves that is determined by the rules of a game (e.g., simultaneity versus sequentiality) will have an impact on what information each of the players has in a certain situation of the game and thus affects their actions.

According to these criteria, concrete games can be classified into several categories. For example, chess is a sequential game for two players, not depending on randomness, and both chess players have always perfect information in each current game situation. Poker, however, is a sequential game for two or more players, where the hands of the players are (supposed to be) randomly dealt out and where no (honest) player has perfect knowledge of the hands of her adversaries, even though she might perhaps (hope to be able to) draw conclusions from their previous behavior in the game and so might have some guess as to whether they are bluffing or not (but certainly she can't be sure).

2.1 Foundations

Consider a set $P = \{1, 2, \ldots, n\}$ of players; occasionally, they will have names instead of numbers. Who these players are and how many of them there are depends on the game being played. Players can be individual persons as well as groups of individuals (as in a team sport), they can be computer programs, states (or their governments), companies, ethnic groups, organizations, etc. A game is defined by its *rules*, which describe how the game is to be played and what each player is allowed or not allowed to do in which situation. It must also be always clear what the single players can know about the current situation of the game, when a game is over, and who has won why and how much.

Some games feature a distinctly extensive set of rules; for example, the *Internationale Skatordnung* (the international laws of skat, a German card game) have 65 pages in A5 size (although many pages of this booklet do not contain actual rules, but other information related to skat that are worth knowing). However, to keep the analysis simple and comprehensible, we will here focus on relatively simple games that are still suitable to explain their crucial strategic or game-theoretic aspects.

2.1 Foundations

Unlike the rules of a game, the players' *strategies* constitute complete and precise policies of how to act in each possible situation they might encounter during the game. Depending on the current game situation, a player can have several options for how to proceed, and so can have a choice between alternative actions, which all must be rule-consistent, of course. If a player has no further alternatives to choose from, this often (although not always) means that the game is over and ends with this player's defeat (as, for example, in the case of a "checkmate" in chess).

2.1.1 Normal Form, Dominant Strategies, and Equilibria

We introduce the notion of normal form as well as the solution concepts of dominant strategies, Pareto optima, and equilibria in games by means of the following concrete game.

2.1.1.1 The Prisoners' Dilemma

A particularly simple, but well-known and beautiful example of a game is the following dilemma that two prisoners have gotten into.

Two most wanted criminals, in the underground milieu only known as "Smith & Wesson," have been arrested. They are being accused of a joint bank robbery, but incriminatory evidence is too thin to meet court standards, unfortunately. The detective superintendent responsible for this investigation interrogates both criminals separately, one after the other, and Smith is brought before him first. Despite intensive interrogation, Smith perseveringly remains silent, so the detective superintendent offers him a deal.

"You do know the maximum penalty for bank robbery, Smith," he points out, "ten years behind bars! However, if you confess that you and Wesson have mugged the bank together, then I can guarantee you that the judge will suspend your sentence on probation for good collaboration with the authorities, and Wesson has to serve his ten years alone, if he goes on to be stubborn."

Smith remains silent.

"Think about it by tomorrow," the detective superintendent adds. "I will now offer Wesson the same deal."

When Smith is led away, he turns around once more and asks: "What if Wesson makes a confession and incriminates me?"

"It depends," the detective superintendent replies. "If only he confesses and you do not, then his sentence will be suspended on probation and you'll go to jail for ten years. If you both confess, then you both will have to serve four years, despite your collaboration with us, because the other confession is less valuable for us, as the first one would have been enough, and that applies to each of you."

"And if none of us don't say nothing?"

"You mean if you both remain silent?" the detective superintendent asks. "Let me be honest with you. Incriminatory evidence is too thin to meet court standards, so we won't be able to make you serve the maximum penalty. If you both refuse to confess, we'll get you only for possession of unregistered weapons and for obstructing our police officers in the course of their duty—one of them is still in hospital. That would then make two years of prison for each of you."

Back in his solitary confinement cell, Smith is brooding over this offer all night long. Unfortunately, he cannot coordinate and come to an agreement with Wesson, who is sitting and brooding in his own solitary confinement cell next door. They both want to make their decision rationally and, being the ruthless criminals they are, they are selfishly interested in their own advantage only. Obviously, they are playing a strategic game, which leaves them both the choice between two strategies: either to make a confession or to remain silent. The outcome of the game, however, depends for each of the two players not only on his own, but also on the other player's strategy, and they both must make their move simultaneously, without knowing the other's decision.

Table 2.1 The prisoners' dilemma

		Wesson	
		Confession	Silence
Smith	Confession	$(-4, -4)$	$(0, -10)$
	Silence	$(-10, 0)$	$(-2, -2)$

Table 2.1 outlines the rules of the game as described by the detective superintendent. An entry $(-k, -\ell)$ means that Smith is sentenced to serve k years in prison and Wesson is sentenced to serve ℓ years in prison. The players thus maximize their gains if they get away with a sentence of as few years as possible. If one wants to avoid negative gains, one could scale all gains in equal measure, without causing the strategic aspects of the game to change. For example, dividing all entries from Table 2.1 by 2 and adding 5, one obtains the values in Table 2.2, which strategically are equivalent.

2.1 Foundations

Table 2.2 The prisoners' dilemma without negative entries

		Wesson	
		Confession	Silence
Smith	Confession	(3,3)	(5,0)
	Silence	(0,5)	(4,4)

2.1.1.2 Noncooperative Games in Normal Form

In general, for noncooperative games with any number of players, one considers the *normal form* (a.k.a. the *strategic form*), which is given above for the prisoners' dilemma. For more than two players, however, the gain vectors for all tuples of the players' strategies cannot be represented as a simple two-dimensional table as above. The normal form of games can be used to represent a multitude of games—albeit not the entirety of all games—in a uniform way and allows for a uniform analysis. The normal form, which is attributed to Borel [170] and von Neumann [734], is best suitable for representing one-move games where all players make their moves simultaneously and without knowing the other players' moves and where randomness is not involved. Sequential games, on the other hand, where the players move one after the other (for example, two players taking turns), can better be represented by the so-called *extended form* to be presented in Section 2.3.1. Games with one or more moves in which randomness plays a role are called *Bayesian games* and will be elaborated on in Section 2.4.2 below.

Definition 2.1 (normal form). A game with n players is in *normal form* if for all i, $1 \leq i \leq n$, it holds that:

1. Player i can choose from a (finite or infinite) set S_i of (pure) strategies (or actions). The *set of profiles of (pure) strategies (or actions)* of all n players is represented as the Cartesian product

$$\mathcal{S} = S_1 \times S_2 \times \cdots \times S_n.$$

2. Player i has gain function $g_i : \mathcal{S} \to \mathbb{R}$ that gives the player's *gain* $g_i(\vec{s})$ for the strategy profile $\vec{s} = (s_1, s_2, \ldots, s_n) \in \mathcal{S}$. Here, $\mathbb{R}$ denotes the set of real numbers and s_j, $1 \leq j \leq n$, is the strategy chosen by player j.

The term *pure strategies* in Definition 2.1 refers to the fact that each player chooses exactly one strategy. Alternatively, one can also consider *mixed strategies*, assuming a probability distribution on the set S_j of all possible strategies of each player j, $1 \leq j \leq n$. According to this distribution, every player chooses her strategy with a certain probability. We will further elaborate on this in Section 2.2.

2.1.1.3 Dominant Strategies

But how should Smith decide in his dilemma? If he could rely on Wesson remaining silent, it would be best for him, Smith, to confess the bank robbery and to incriminate Wesson, because a sentence on probabtion is better than two years of jail. However, he can of course not rely on Wesson to remain silent. On the other hand, even if Wesson confesses and incriminates him as well, it would be better for Smith to confess. He would then be imprisoned, to be sure, but serving only four years is better than ten. From Smith's point of view, making a confession is better than remaining silent for him, no matter which strategy is chosen by Wesson. In other words, making a confession is a dominant strategy for Smith. For symmetric reasons, making a confession is a dominant strategy for Wesson as well. This notion is formally defined as follows.

Definition 2.2 (dominant strategy). Let $\mathcal{S} = S_1 \times S_2 \times \cdots \times S_n$ be the set of strategy profiles of the n players in a noncooperative game in normal form and let g_i be the gain function of player i, $1 \leq i \leq n$. A strategy $s_i \in S_i$ of player i is said to be *dominant* (or *weakly dominant*) if

$$g_i(s_1, \ldots, s_{i-1}, s_i, s_{i+1}, \ldots, s_n) \geq g_i(s_1, \ldots, s_{i-1}, s'_i, s_{i+1}, \ldots, s_n) \qquad (2.1)$$

for all strategies $s'_i \in S_i$ and all strategies $s_j \in S_j$ with $1 \leq j \leq n$ and $j \neq i$.

If inequality (2.1) is strict for all $s'_i \in S_i$ with $s'_i \neq s_i$ and all $s_j \in S_j$ with $1 \leq j \leq n$ and $j \neq i$, then s_i is a *strictly dominant strategy* for player i.

Thus, the strategy of making a confession mentioned above is even strictly dominant for both Smith and Wesson. Having a (strictly) dominant strategy is of course very beneficial for a player: No matter what the other players do, this player has a best strategy independently of them.

Interestingly, from a *global* point of view, it would be better if *both* players in the prisoners' dilemma deviated from their strictly dominant strategies, i.e., if they remained silent. That is to say, if they were able to coordinate and come to an agreement that they both will remain silent, each of them would have to serve only two instead of four years in prison. Playing the dominant strategy and to confess, however, means to play it safe, because even if they would have agreed on jointly remaining silent, the risk that the other player would break his word and would confess out of selfishness is simply too high for both players. The bamboozled player would have to do his maximum time in that case, while the turncoat would be at large on probation.

As mentioned above, if both players had *cooperated* and agreed on jointly remaining silent, that would be beneficial for both of them and, from a global perspective, they would both prefer this outcome. Formally, this is expressed by the notion of *Pareto dominance* (relatedly, *Pareto optimality* or *Pareto efficiency*; cf. Definition 8.5 on page 522 in Section 8.3.2 and Definition 9.2 on page 622 in Section 9.4.1 for the same notions in other contexts), which is suitable to evaluate the chosen strategy profiles of all players globally.

2.1 Foundations

Definition 2.3 (Pareto dominance and Pareto optimality). Let $S = S_1 \times S_2 \times \cdots \times S_n$ be the set of strategy profiles of the n players in a noncooperative game in normal form. Let $\vec{s}$ and $\vec{t}$ be two strategy profiles from S.

1. We say $\vec{s}$ *weakly Pareto-dominates* $\vec{t}$ if for all i, $1 \leq i \leq n$:

$$g_i(\vec{s}) \geq g_i(\vec{t}). \tag{2.2}$$

2. If inequality (2.2) holds true for all i, $1 \leq i \leq n$, and is strict for at least one j, $1 \leq j \leq n$, we say $\vec{s}$ *Pareto-dominates* $\vec{t}$.
3. If inequality (2.2) is strict for all i, $1 \leq i \leq n$, we say $\vec{t}$ is *strongly Pareto-dominated* by $\vec{s}$.
4. We say $\vec{t}$ is *Pareto-optimal* if for all $\vec{s} \in S$ it holds that: If $\vec{t}$ is weakly Pareto-dominated by $\vec{s}$, then $g_i(\vec{s}) = g_i(\vec{t})$ for all i, $1 \leq i \leq n$. That is, $\vec{t}$ is Pareto-optimal if there is no $\vec{s} \in S$ that Pareto-dominates $\vec{t}$, i.e., if there is no $\vec{s}$ such that

   a. $g_i(\vec{s}) \geq g_i(\vec{t})$ for all i, $1 \leq i \leq n$, and
   b. $g_j(\vec{s}) > g_j(\vec{t})$ for at least one j, $1 \leq j \leq n$.

5. We say $\vec{t}$ is *weakly Pareto-optimal* if there is no $\vec{s} \in S$ that strongly Pareto-dominates $\vec{t}$.

Intuitively, Pareto optimality of a strategy profile $\vec{t} = (t_1, t_2, \ldots, t_n)$ means that no other strategy profile gives all players at least as much profit as $\vec{t}$ and in addition at least one player a strictly larger profit. A *Pareto optimum* exists if and only if no player can increase her gains without making another player getting off worse at the same time. On the other hand, $\vec{t}$ is weakly Pareto-optimal if no other strategy profile gives *all* players a strictly larger profit. Thus, every Pareto optimum is also a weak Pareto optimum; conversely, weak Pareto optima are not necessarily Pareto-optimal. It is easy to see that the strategy profile (Silence, Silence) is a Pareto optimum in the example of the prisoners' dilemma, since no other strategy profile is beneficial for one of the players without penalizing the other player at the same time. In addition, also the strategy profiles (Confession, Silence) and (Silence, Confession) are Pareto-optimal.

2.1.1.4 Nash Equilibria in Pure Strategies

So far we have got to know two distinct concepts that can help to predict the outcome of a game:

1. *dominant strategies*, such as the strategy profile (Confession, Confession) in the prisoners' dilemma,
2. *Pareto optima*, such as the profiles (Silence, Silence), (Confession, Silence), and (Silence, Confession) in the prisoners' dilemma.

A third concept that can help to predict the outcome of a game is the criterion of *stability of a solution*. Informally put, a solution (i.e., a profile of all players' strategies) is stable if no player has an incentive to deviate from her strategy in this profile, provided that also all other players choose their strategies according to this profile. Intuitively, this means that the strategies of this solution are in equilibrium.

There are various formal definitions of equilibria. The perhaps best known notion of equilibrium is due to Nash [730]. Just like John von Neumann, also John Forbes Nash is one of the great pioneers of game theory. Nash has won numerous prizes and awards and has been loaded with the highest academic honors for his superb insights and pathbreaking ideas, such as the 1978 *John von Neumann Theory Prize* for introducing the notion of equilibria in noncooperative games (which later was named after him) and the 1994 *Nobel Prize in Economics* (jointly with the game theoreticians Reinhard Selten and John Harsanyi). Nash's exciting and moving life—with all its brilliant moments and its downsides—is subject of the (unauthorized) biography by Nasar [727] and of the Hollywood motion picture *"A Beautiful Mind,"* USA 2001, distinguished by four Academy Awards, for example in the categories *Best Picture* (Brian Grazer and Ron Howard) and *Best Director* (Ron Howard), and nominated for Academy Awards in four additional categories.

The following definition of Nash equilibrium again refers to the players' pure strategies. The corresponding notion of Nash equilibrium in mixed strategies will be introduced later on in Definition 2.5.

Definition 2.4 (Nash equilibrium in pure strategies). Let $\mathcal{S} = S_1 \times S_2 \times \cdots \times S_n$ be the set of strategy profiles of the n players in a noncooperative game in normal form and let g_i be the gain function of player i, $1 \leq i \leq n$.

1. A strategy $s_i \in S_i$ of player i is said to be a *best response strategy to the profile* $\vec{s}_{-i} = (s_1, \ldots, s_{i-1}, s_{i+1}, \ldots, s_n) \in \mathcal{S}_{-i} = S_1 \times \cdots \times S_{i-1} \times S_{i+1} \times \cdots \times S_n$ *of the other players' strategies* if for all strategies $s'_i \in S_i$,

$$g_i(s_1, \ldots, s_{i-1}, s_i, s_{i+1}, \ldots, s_n) \geq g_i(s_1, \ldots, s_{i-1}, s'_i, s_{i+1}, \ldots, s_n). \quad (2.3)$$

2. If there is exactly one such strategy $s_i \in S_i$ of player i, then this is her *strictly best response strategy to the other players' strategy profile* $\vec{s}_{-i}$.
3. A strategy profile $\vec{s} = (s_1, s_2, \ldots, s_n) \in \mathcal{S}$ is in a *Nash equilibrium in pure strategies* if for all i, $1 \leq i \leq n$, $s_i \in S_i$ is a best response strategy of player i to the other players' strategy profile $\vec{s}_{-i}$.
4. If there is exactly one such strategy profile $\vec{s}$, then $\vec{s}$ is in a *strict Nash equilibrium in pure strategies*.

Since the inequalities (2.1) and (2.3) are identical, one might be tempted to think that a best response strategy is the same as a dominant strategy. However, there is a subtle, but decisive, distinction. In the definition of a best response strategy, (2.3) is true merely for all strategies $s'_i \in S_i$ of player i, while the context—the profile $\vec{s}_{-i} = (s_1, \ldots, s_{i-1}, s_{i+1}, \ldots, s_n)$ of the other players' strategies—is fixed. By contrast, in the definition of dominant strategy, (2.1) holds true for all strategies $s'_i \in S_i$ of player i *and for all of the other players' strategy profiles* $\vec{s}_{-i} = (s_1, \ldots, s_{i-1}, s_{i+1}, \ldots, s_n)$. Hence, this is a much sharper requirement: If a player has a dominant strategy, then this is also her best response strategy for *every* strategy profile of the other players. On the other hand, a strategy that is a best response for *some* strategy profile is not required to be a best response for all the other profiles.

2.1.1.5 Relations between Solution Concepts

There exists a Nash equilibrium in pure strategies if every player chooses a best response strategy to the strategies she expects her opponents to choose (and all opponents meet that expectation). Thus, no player has an incentive to deviate from her chosen best response strategy, and the solution is stable. It follows immediately from the argument in the previous paragraph that a profile of dominant strategies for all players, if it exists, is always in a Nash equilibrium in pure strategies; the converse, however, does not hold in general (a counterexample is the "battle of the sexes" to be introduced later on).

As we already know the dominant strategies of both players in the prisoners' dilemma (both make a confession), we also know that both making a confession is a Nash equilibrium in pure strategies: Neither Smith nor Wesson can improve his situation by "one-sided deviation" (that is, by deviating, assuming that the other player does not deviate from his strategy), and therefore they both can be expected (or predicted) to "stably keep to making a confession." This strategy profile even forms a strict Nash equilibrium, because there is no other one. This is due to the fact that both players even have *strictly* dominant strategies. That is generally true for every game in normal form: If all players have strictly dominant strategies, then these form a strict Nash equilibrium in pure strategies. Weakly dominant strategies, however, can occur in several Nash equilibria. Can you come up with a game in normal form that has no dominant strategies, even though there is a strict Nash equilibrium in pure strategies?

As we have seen, the (strict) Nash equilibrium (Confession, Confession) and the Pareto optimum (Silence, Silence) are distinct in the prisoners' dilemma. Thus, Nash equilibria and Pareto optima do not need to coincide in general. Figure 2.1 gives an overview of the relations between the solution concepts discussed so far. Here, an arrow $(A \to B)$ means an implication: If a strategy profile satisfies criterion A, then it also satisfies criterion B.

Figure 2.2 shows the set-theoretic inclusion relations between the solution concepts for noncooperative games in normal form introduced so far. In particular, both figures show that the property of Pareto optimality is incomparable with the existence of both (strictly) dominant strategies and (strict) Nash equilibria in pure strategies. That is, Pareto optimality of a so-

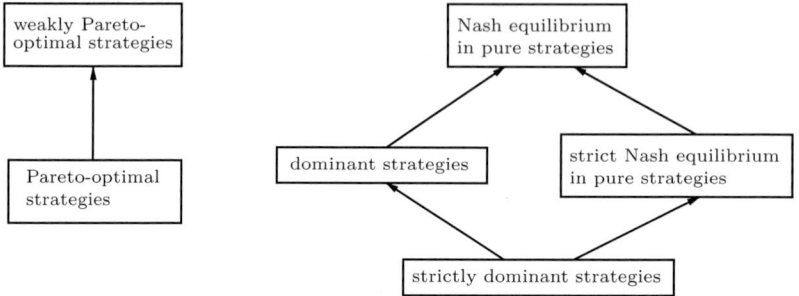

Fig. 2.1 Relations between solution concepts for noncooperative games in normal form

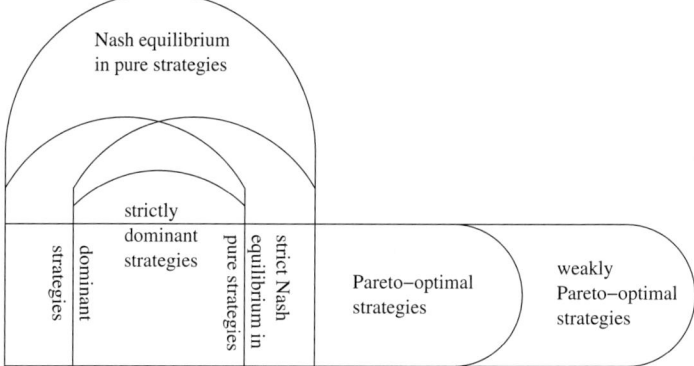

Fig. 2.2 Inclusions among solution concepts for noncooperative games in normal form

lution implies neither of these other properties for this solution in general, and conversely it is not implied by any of them.

2.1.2 Further Two-Player Games

In this section we analyze a number of further interesting two-player games in normal form with respect to the above-mentioned solution concepts. Just as the prisoners' dilemma, each of these games has its own strategic features and thus exemplifies a whole class of games with the same strategic features that differ only in the absolute values of the profit each of the two players can make.

2.1.2.1 The Battle of the Sexes

This game is commonly played by a couple, a woman and a man.

2.1 Foundations

To celebrate their first anniversary, George and Helena are going to go out together. Unsurprisingly, they make quite different suggestions.

"Let's just go to the stadium," George suggests. "By chance I got hold of two tickets for the soccer game tonight!" He proudly presents them to her.

"Too bad!" Helena replies disappointedly and lifts a pair of tickets as well. "I was going to surprise you with those! Lana del Rey is performing tonight, and so I figured ..."

"I'm really sorry!" George shouts. "Well, maybe you can sell them to somebody on the way to the stadium. But this is England against Germany, the classic! I surely can't miss that game!"

"Really not?" Helena snaps at him. "Then you just go to your classic! I'll certainly find someone who is interested in getting *your* Lana del Rey ticket!"

"I didn't mean it like this!" George wisely gives in. "That I want to spend this evening together with you, that is for sure, and for all I care we can also go to your concert. The main thing is we do something together! It's just that I would prefer going *with you together* to the soccer game ten times as much as going with you to the concert."

At this point, we better leave the two alone, for them to fight their "battle" undisturbedly and then to decide their controversy by coming to a mutual agreement. Instead, let us have a closer look at their game, where we assume that they have to make their moves simultaneously.

Table 2.3 The battle of the sexes

		Helena	
		Soccer	Concert
George	Soccer	**(10, 1)**	(0, 0)
	Concert	(0, 0)	**(1, 10)**

Table 2.3 shows the gain vectors of George and Helena in all four cases, where George's gain is the left entry and Helena's gain is the right entry. Obviously, spending the night at different events is for none of them beneficial: The strategy profiles (Soccer, Concert) and (Concert, Soccer) are both rewarded by (0, 0), since a separation from their partner kills all joy for both of them, even at their favorite events.

The strategy profiles (Soccer, Soccer) and (Concert, Concert) with the boldfaced gain vectors (10, 1) and (1, 10), however, form two Nash equilibria in pure strategies because if George and Helena decide upon either (Soccer, Soccer) or (Concert, Concert), they both have no incentive to deviate one-sidedly, as that would be punished by the loss of a positive gain. Even so, George wins ten times as much as Helena if they both go to the stadium, and conversely Helena's gains are ten times as high as George's gains if they both spend the evening in the concert hall. These two Nash equilibria in pure strategies are, unlike that in the prisoners' dilemma, not strict.

Are there dominant strategies? That choosing "Concert" is not a dominant strategy for George is immediately clear, since choosing "Soccer" instead of "Concert" increases his gains from 0 to 10, provided that Helena, only to please him, would also choose "Soccer." But choosing "Soccer" is not a dominant strategy for George either, since if Helena does not give in and firmly sticks to her wish to attend the concert, then George could increase his gains of 0, which insisting on "Soccer" would bring him, to 1 by making up his mind and going to the concert to please her. For symmetric reasons, Helena does not have a dominant strategy either. As this example shows, Nash equilibria in pure strategies can exist even if no player has a dominant strategy.

Unlike in the prisoners' dilemma, the Pareto optima coincide with the Nash equilibria in this game. The strategy profiles (Soccer, Concert) and (Concert, Soccer) with the gain vector (0,0) are not Pareto-optimal, since both players can be better off by switching to either (Soccer, Soccer) or (Concert, Concert) with the gain vectors of either (10,1) or (1,10), without the other player being worse off at the same time (on the contrary: also the other player will be better off that way). The strategy profile (Soccer, Soccer) with the gain vector (10,1), however, is Pareto-optimal, since George cannot improve at all in this case, and Helena can improve, to be sure, by switching to the profile (Concert, Concert) with the gain vector (1,10), but only at the cost of making George be worse off, and with the other two strategy profiles—(Soccer, Concert) and (Concert, Soccer)—she would even decrease her gains to 0. A symmetric argument shows that the strategy profile (Concert, Concert) is Pareto-optimal as well.

2.1.2.2 The Chicken Game

Unlike the battle of the sexes where two partners want to coordinate, the *chicken game* models a situation in which two enemies confront each other. Such situations occur with personal enmities as well as in politics, for example in the nuclear arms race of conflictive societal systems during the cold war. At the private level, the chicken game refers to a hazardous, possibly lethal, test-of-courage game that can be played in several variants and was featured

2.1 Foundations

in a number of movies already.[3] (The reader is strongly discouraged from playing this game.)

Fig. 2.3 The chicken game

To make sure nobody will get hurt or be harmed even worse in this book, we let David and Edgar, two ten year old boys, play the chicken game in their admittedly sexed up, but rather underperforming toy cars (see Figure 2.3). By the rules of the game, they "race" with a maximum speed of 5 miles per hour, approaching each other, and whoever cowardly weasels out of driving on or swerves is the chicken and has lost. However, since he at least has been wise and has survived, he gets one gummy bear as a consolation prize, while the heroic winner rakes in the top prize of three gummy bears. If they both are wise and swerve just in time, then each of them gets a prize of two gummy bears. However, if both are boldly driving on to the bitter end, they are declared "dead" (only in play) after the inevitable crash and win no gummy bears.

Table 2.4 The chicken game

		Edgar	
		Swerve	Drive on
David	Swerve	(2, 2)	(**1, 3**)
	Drive on	(**3, 1**)	(0, 0)

The gain vectors for these four possible outcomes of the game are shown in Table 2.4, where the left entry in a vector gives David's gain and the right entry gives Edgar's gain. Again, there are two Nash equilibria in pure strategies, namely if one player chooses "Drive on" and the other chooses "Swerve,"

---

[3] One variant of the chicken game is played by Jim (personated by James Dean in his second-to-last role before he was tragically killed in a car crash) and his adversary Buzz in *"Rebel Without a Cause,"* USA 1955. They both race in their cars toward a cliff with maximum speed. Whoever cowardly weasels out of driving on and hits the brakes or jumps out of the car is the chicken and has lost.

Another variant of this game is presented in John Waters's cult film *"Cry Baby,"* USA 1990. In this musical parody, Cry Baby (personated by Johnny Depp) and his adversary Baldwin approach one another with maximum speed, strapped onto the roofs of their cars. Whoever swerves is the chicken and has lost.

indicated by boldfaced gain vectors in the table. And as in the battle of the sexes, both Nash equilibria are Pareto-optimal as well. In addition, also the strategy profile (Swerve, Swerve) is Pareto-optimal, since with its gain vector $(2,2)$ no player can improve without making the other player be worse off at the same time. Does any one of the two players, or perhaps each of them due to symmetry, have a dominant strategy?

Chicken games are also referred to as *hawk-dove games* in the literature. The *hawk strategy* corresponds to "Drive on," i.e., a hawk is assumed to very aggressively fight for the resources with high risk and at all costs. The *dove strategy*, on the other hand, corresponds to "Swerve," i.e., a dove is assumed to back off when confronted with a hawk, as it does not fight for a resource if it is in danger of getting injured.

2.1.2.3 The Penalty Game

"Football is a simple game; 22 men chase a ball for 90 minutes, and at the end the Germans always win," Gary Lineker is quoted as saying when he explained the somewhat simplified rules, reduced to the essentials, of the game of soccer, after the English national team lost against the later champion Germany on penalties in the semifinals of the FIFA World Cup on July 4, 1990.[4] Of course, soccer is much more complicated than Lineker admitted in the moment of bitter disappointment; this game is way too diverse and subtle with respect to sporting, conditional, technical, and tactical aspects than being expressible as a noncooperative game in normal form. However, there is one situation in soccer matches that indeed can be represented as a noncooperative game in normal form: the penalty shoot-out.

The two players at a penalty shoot-out are the kicker and the goalkeeper, in Figure 2.4 impersonated by David and Edgar, who obviously have survived their chicken game unharmedly. Both would in fact have a choice between many actions; for example, the most popular actions of real penalty kickers are to kick the ball so that it hits either in the upper left or upper right corner, either in the lower left or lower right corner, or in the middle at the top or the bottom. There are even more actions, such as the less popular (and actually unintended), yet not so rare actions of hitting a post or the crossbar or lifting the ball above it. For simplification, we reduce this multitude of actually occurring strategies to only two for each player: The kicker can kick the ball to the left or to the right of the goal, and the goalkeeper can jump to the left or to the right to make a save. "Left" and "right" are here meant always from the goalkeeper's point of view, even if it is the kicker's turn. Abstracting from reality, we also assume that the goalkeeper is guaranteed to hold on to the ball when he jumps to the same side where the ball is being

---

[4] Lineker's explanation is somewhat imprecise in this case, since the particular match he frustratedly refers to actually took 120 minutes plus added time plus the time for the penalty shoot-out.

2.1 Foundations

Fig. 2.4 David as the kicker and Edgar as the goalkeeper at the penalty shoot-out

kicked, and that the kicker never misses the goal. That is, if he kicks to the left (or to the right) and if the goalkeeper also jumps to the left (or to the right), the goalkeeper has definitely thwarted a goal; but if the goalkeeper jumps to the wrong side, the kicker has definitely converted his penalty. (This explains why the penalty game is also known as *"matching pennies"*.)

Table 2.5 The penalty game

		Goalkeeper	
		Left	Right
Kicker	Left	$(-1, 1)$	$(1, -1)$
	Right	$(1, -1)$	$(-1, 1)$

Table 2.5 gives the gain vectors of all four cases, where the left entry refers to the kicker's gain and the right entry refers to the goalkeeper's gain. Unlike the two-player games introduced so far, the penalty game has *no* Nash equilibrium in pure strategies (not even if more than two strategies were allowed for each of the two players). In particular, neither (Left, Left) nor (Right, Right) is such a Nash equilibrium, since the kicker could increase his gain by one-sided deviation in either case. Similarly, neither (Left, Right) nor

(Right,Left) is such a Nash equilibrium, since the goalkeeper could increase his gain by one-sided deviation in each of these two cases.

Thus, the only reasonable strategy of the kicker is to *randomly* kick the ball to the left or to the right (and to not commit to one side at the outset), and the only reasonable strategy of the goalkeeper is to *randomly* jump to the left or to the right (and to not commit to one side at the outset).[5] Indeed, we will see later on that this game indeed possesses a Nash equilibrium in *mixed* strategies. Since there is no Nash equilibrium in pure strategies, there can be no profile containing a dominant strategy for both the kicker and the goalkeeper (recall Figures 2.1 and 2.2 from pages 52 and 52). It is easy to see that no player can have a dominant strategy at all. On the other hand, all four strategy profiles of this game are Pareto-optimal, since for each of them it is true that no player can be better off without making the other player be worse off at the same time. For example, for the profile (Left,Left), only the kicker can increase his gains from -1 to 1, but he achieves this only at the cost of decreasing the goalkeeper's gains from 1 to -1 at the same time.

2.1.2.4 The Paper-Rock-Scissors Game

In this well-known, popular game, two players—call them David and Edgar—count aloud until three together (or they say something like *"jan-ken-pon"*), and on three (or *"pon"*) they at the same time form with one of their hands a symbol: a flat hand for *"paper,"* a fist for *"rock,"* or two open fingers for *"scissors"* (see Figure 2.5(a)). These symbols stand for the players' three possible strategies, where the following rules apply:

- *Rock* defeats *scissors*, because the rock crushes the scissors.
- *Scissors* defeats *paper*, because the scissors cut a sheet of paper.
- *Paper* defeats *rock*, because the paper covers the rock.
- If both players form distinct symbols, then the player whose symbol defeats the other player's symbol receives one point, and the other player loses one point.
- If both players form the same symbol, then none of them wins; so nobody receives a point.

An extended variant of this game (see Figure 2.5(b)), known as *paper-rock-scissors-lizard-spock*, was invented by Sam Kass with Karen Bryla and has been mentioned, for example, in some episodes of the nerd sitcome *The Big Bang Theory* and in a book by Singh [868] about the mathematical secrets hidden in the episodes of *The Simpsons*, a famous animated sitcom created by Matt Groening. In this extended variant, two more symbols are added, *lizard*

---

[5] Abstracting from reality, we neglect that a goalkeeper with fast reactions might increase his chances by waiting to see where the ball is being kicked and to then jump to this side. We also neglect in our analysis the common tactics of goalkeepers to rattle the kicker and we similarly neglect the kickers' feints to rattle the goalkeeper.

2.1 Foundations

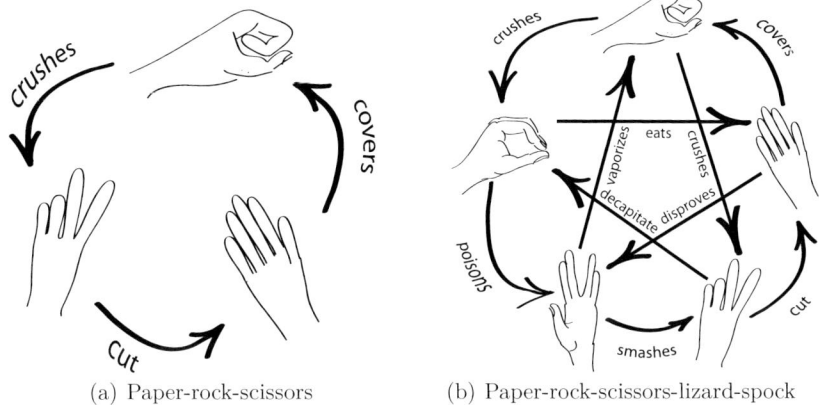

(a) Paper-rock-scissors   (b) Paper-rock-scissors-lizard-spock

Fig. 2.5 Cycles of dominance relations in two variants of paper-rock-scissors

and *spock*,[6] and each of these five symbols now defeats two other symbols and is defeated by the remaining two symbols.

In the following, however, we will discuss only the simpler variant of the game, *paper-rock-scissors*. Unlike in the games considered so far, each of the two players has a choice between three strategies here, which gives a total of nine strategy profiles and outcomes of the game, respectively. The above-mentioned dominance relations among the three strategies of each player form a cycle (see Figure 2.5(a)). The points the players can receive give their gains in the different outcomes of the game and are represented in Table 2.6, where the left entry is David's gain and the right entry is Edgar's gain.

Table 2.6 The paper-rock-scissors game

		Edgar		
		Rock	Scissors	Paper
David	Rock	(0,0)	(1,−1)	(−1,1)
	Scissors	(−1,1)	(0,0)	(1,−1)
	Paper	(1,−1)	(−1,1)	(0,0)

Again, there does not exist a Nash equilibrium in pure strategies in this game. Consider, for example, the strategy profile (Scissors, Rock), which gives Edgar his highest profit. However, if David thinks that Edgar will play the strategy *rock*, then he would benefit from one-sided deviation, because David maximizes his gain in this case by choosing *paper*. That is why (Scissors, Rock) is not a Nash equilibrium in pure strategies, and one can argue analogously in

---

[6] Mr. Spock is well known from the American science fiction TV series *Star Trek* created by Gene Roddenberry.

all other cases. We will see later on, however, that this game too has a Nash equilibrium in *mixed* strategies, namely, when each player chooses among his three strategies uniformly at random (see page 67).

Interestingly, human players in real life often tend to not pick their strategies according to the uniform distribution, but put more weight on one or the other strategy. The *World Rock Paper Scissors Society* found out that *rock* is the strategy preferred by most players, especially among boys (see [868]). Knowing that, Lisa Simpson—in an episode of *The Simpsons*—plays *paper* again and again to beat her brother Bart in this game [868]. This shows that, in addition to purely mathematical and rational aspects, gambling often has certain psychological aspects, such as evaluating or guessing your opponents' behavior. A game that highlights this point is considered next.

2.1.2.5 The Guessing Numbers Game

Any number of players can play this game. Each of them guesses a real number between 0 and 100 (including these two values). Take the average (i.e., the arithmetic mean) of the numbers chosen and let Z be exactly two thirds of this average. Whoever comes closest to this number Z wins. Thus, every player has infinitely many strategies to choose from, since there are infinitely (even uncountably) many real numbers in the interval $[0, 100]$.

Which number should one guess? A first thought is this: If all players were to choose the largest possible number, 100, then $Z_{\max} = 66.666\cdots$, and no Z can be larger than $Z_{\max}$. It would therefore be dull to choose a number greater than $Z_{\max}$. In other words, the strategy (or number) $Z_{\max}$ dominates all strategies (or numbers) exceeding it. These can thus be safely eliminated.

A second thought of a player might be this: How will my opponents behave? Suppose that all players randomly choose a number from $[0, 100]$ under the uniform distribution. The average would then be 50 and two thirds of this is $33.333\cdots$. But wait a minute! Actually, why should all players choose an arbitrary number from $[0, 100]$? If one assumes all players to behave *rationally*, they can be expected to already have eliminated all numbers above $Z_{\max} = 66.666\cdots$ as well; but then the average of the remaining numbers (assuming the uniform distribution) would be $Z_{\text{rat}} = 33.333\cdots$ and two thirds of that would be $22.222\cdots$.

This consideration outright takes us to a third thought: Every player behaves rationally and every player knows that everybody behaves rationally, and everybody knows that all players know that everybody behaves rationally, which again everybody is aware of, and so on. In other words, also the number $Z_{\text{rat}} = 33.333\cdots$ dominates all numbers exceeding it, which causes further numbers to be eliminated, and likewise $22.222\cdots$ dominates all greater numbers, which implies their elimination as well, and so on.

Why should one stop at any point with that argument? If you think this third thought consequently to the very end, only one number will remain that

2.1 Foundations

you should choose as your strategy: the zero. And indeed, the zero is the strict Nash equilibrium in pure strategies in this game. If all the players assume that all the other players, rationally, choose the zero, then so should they, since otherwise (i.e., if they would one-sidedly deviate from this strategy[7]) they would lose. If all players initially pay one dollar into a jackpot, which will be hit by the winner(s), and if all choose the zero and win, then they all will simply get their investment back.

Now, the fourth thought is interesting: Will all players in fact behave rationally? This game—or variants thereof—has frequently been played in public, often with several thousands of players. In these games it has never been the case that all players have chosen their strategies according to the Nash equilibrium. However, as soon as some players deviate from the Nash equilibrium, the other players are no longer guaranteed to win when playing their Nash equilibrium strategy. For example, four dull (or especially crafty?) players might dupe their rational, game-theoretically excellently trained adversary: While the latter one plays the zero of the Nash equilibrium, they might all play the 100 and thus win because their number is closest to the number

$$Z = \left(\frac{2}{3}\right) \cdot \left(\frac{100 + 100 + 100 + 100 + 0}{5}\right) = \left(\frac{2}{3}\right) \cdot \left(\frac{400}{5}\right) = \frac{800}{15} = 53.333\cdots.$$

If there are four players three of which choose the 100 and one of which chooses the zero, then $Z = (2/3) \cdot (300/4) = 50$ implies that all players win and fairly share the jackpot among themselves, that is, they simply get their investment back.

Nagel and Selten [723] (see also [853]) give an overview of such game-theoretic experiments. The resulting frequency distributions usually give average values around 22, which perhaps hints at the average player reaching the second of the thoughts mentioned above. Interestingly, the chosen numbers often accumulate especially around the values of 0, 22, and 33 and somewhat less eye-catching around the values 67 and 100. By the way, in one of these experiments, which was conducted with the readers of the *Financial Times*, a variant of this game was played that allowed the players only to choose one among the *integers* in the range [0, 100]. In this variant, there is no strict, but there are two Nash equilibria in pure strategies: 0 and 1.

These experimental results also point out that game theory alone, which assumes players to be rational, can only partially explain human behavior and allows to give only limited recommendations for action. For example, financial markets function quite similarly to the guessing numbers game: Analysts and brokers try to give a prediction for the future development of a stock, which depends not only on the fundamental data of the corresponding company, but also on the behavior of the other participants in the market. If some stock market broker relies only on the purely economical data of

---

[7] However, if more players deviate from this zero-strategy of the strict Nash equilibrium in pure strategies, then this deviation will perhaps not be punished.

a company under the assumption that all other brokers behave rationally like herself, then she might quickly lose a lot of money and ruin herself by speculations. The stock exchange is governed not only by the rules of the market and rational economic theory, including game theory. Markets are also ruled by psychological aspects and the behavior of market actors that may be extremely difficult to predict.

2.2 Nash Equilibria in Mixed Strategies

In all games introduced so far, all players had to choose exactly one strategy:

- Smith and Wesson had to either confess or remain silent in the prisoners' dilemma;
- George and Helena had to go either to the soccer match or the concert in the battle of the sexes;
- David and Edgar could only either swerve or go on driving in the chicken game;
- in the penalty game, the kicker and the goalkeeper had each to choose one side of the goal, left or right;
- in the paper-rock-scissors game, David and Edgar would form upon *pon* either paper, rock, or scissors with their hands; and
- each player had to choose exactly one number in the guessing numbers game.

All players play *pure* strategies in these games. However, if one such game is played several times in a row, the players might change their minds and choose different strategies. It would be pretty dull in certain games to always choose the same strategy. For example, a goalkeeper who always jumps to the left side will be very predictable; instead he should choose randomly, under the uniform distribution, where to jump, sometimes to the left, sometimes to the right. If the players make their decisions on which strategy to choose randomly under some probability distribution, we say they use a *mixed* strategy. In many games, especially so in those with mixed strategies, one does not win by intelligence only, one also has to be lucky.

2.2.1 Definition and Application to Two-Player Games

After defining the notion of Nash equilibria in mixed strategies, we will discuss it for the two-player games introduced in Section 2.1.1 starting on page 45 and in Section 2.1.2 starting on page 52. We will conclude this section by giving an overview of some properties of these two-player games.

2.2 Nash Equilibria in Mixed Strategies

2.2.1.1 Definition of Nash Equilibria in Mixed Strategies

As pointed out previously, games without a Nash equilibrium in pure strategies, such as the penalty and the paper-rock-scissors game, can well have a Nash equilibrium in mixed strategies. Formally, this concept is defined as follows.

Definition 2.5 (Nash equilibrium in mixed strategies). Let $\mathcal{S} = S_1 \times S_2 \times \cdots \times S_n$ be the set of strategy profiles of the n players in a noncooperative game in normal form and for each i, $1 \leq i \leq n$, let g_i be the gain function of player i. For simplicity, let us assume that all sets S_i are finite.[8]

1. A *mixed strategy for player i* is a probability distribution π_i on S_i, where $\pi_i(s_j)$ is the probability of the event that i chooses the strategy $s_j \in S_i$. Let Π_i be the set of all probability distributions on S_i (so $\pi_i \in \Pi_i$). Let $\Pi = \Pi_1 \times \Pi_2 \times \cdots \times \Pi_n$. For each player i, $1 \leq i \leq n$, let Π_{-i} denote the space $\Pi_1 \times \cdots \times \Pi_{i-1} \times \Pi_{i+1} \times \cdots \times \Pi_n$ of the other players' strategies.
2. The *expected gain of a mixed-strategy profile* $\boldsymbol{\pi} = (\pi_1, \pi_2, \ldots, \pi_n)$ *for player i* is

$$G_i(\boldsymbol{\pi}) = \sum_{\vec{s}=(s_1,\ldots,s_n) \in \mathcal{S}} g_i(\vec{s}) \prod_{j=1}^{n} \pi_j(s_j). \quad (2.4)$$

3. A mixed strategy $\pi_i \in \Pi_i$ is *player i's best response to the mixed-strategy profile* $\boldsymbol{\pi}_{-i} = (\pi_1, \ldots, \pi_{i-1}, \pi_{i+1}, \ldots, \pi_n) \in \Pi_{-i}$ of the other players if for all mixed strategies $\pi'_i \in \Pi_i$,

$$G_i(\pi_1, \ldots, \pi_{i-1}, \pi_i, \pi_{i+1}, \ldots, \pi_n) \geq G_i(\pi_1, \ldots, \pi_{i-1}, \pi'_i, \pi_{i+1}, \ldots, \pi_n). \quad (2.5)$$

4. A profile $\boldsymbol{\pi} = (\pi_1, \pi_2, \ldots, \pi_n)$ of mixed strategies is in a *Nash equilibrium in mixed strategies* if π_i is a best response to $\boldsymbol{\pi}_{-i}$ for all players i.

How does one mix one's pure strategies? This, of course, depends on how one evaluates the gains obtained by one's random choices. For example, a *risk-averse* player, aiming to be on the safe side, would prefer an outcome having a small positive gain with high probability, whereas a *risk-loving* player would favor an outcome with a large positive gain, even if the chances of getting it are low. In the above definition, we assume the players to be *risk-neutral*, i.e., they seek to simply maximize their expected gains. Intuitively, to compute the expected gain $G_i(\boldsymbol{\pi})$ for player i given $\boldsymbol{\pi}$, we first determine the probability of reaching each possible outcome, and we then calculate the average of the gains of the outcomes weighted by the probabilities of each outcome. For example, if $\vec{s}$ in (2.4) contains some pure strategy s_k that happens to be chosen with probability zero by player k (i.e., $\pi_k(s_k) = 0$), then $g_i(\vec{s}) \prod_{j=1}^{n} \pi_j(s_j) = 0$ for

---

[8] For suitable probability measures, the notions defined here can be extended to infinite strategy spaces.

each i, so $\vec{s}$ does not contribute anything to the expected gain of any player under $\boldsymbol{\pi}$.

By the above definition, a profile $\boldsymbol{\pi} = (\pi_1, \pi_2, \ldots, \pi_n)$ of mixed strategies is in a Nash equilibrium in mixed strategies if and only if no player i has a strategy $\pi'_i \in \Pi_i$ that would give her a higher expected profit than her mixed strategy π_i on S_i in response to the mixed strategies she expects the other players to choose. For each player, one-sided deviation from their mixed strategies would not, in expectation, be beneficial (and might even be punished), assuming that the other players stick to their mixed strategies of the Nash equilibrium. Consequently, for a Nash equilibrium in mixed strategies it must hold that every player is indifferent to each strategy she chooses with positive probabability in her mixed strategy. Also, the players' probability distributions in the profile of their mixed strategies are independent. (Compare this assumption with the notion of "correlated equilibrium," see, e.g., [755, Sections 1.3.6 and 2.7].)

In particular, the notion of a pure strategy is a special case of a mixed strategy where player i plays this concrete pure strategy, say s_j, with probability 1: $\pi_i(s_j) = 1$ for some $s_j \in S_i$ and $\pi_i(s_k) = 0$ for all $s_k \in S_i$ with $k \neq j$. For notational convenience, we identify pure strategies with their corresponding mixed strategies and denote both by s_i. For example, if $s_i \in \Pi_i$ occurs as the ith component of a mixed-strategy profile, this means that s_i is viewed as the mixed strategy of player i that uses the pure strategy s_i with probability 1.

Comparing pure and mixed strategies and the related notions of Nash equilibria, the following properties are easy to see. First, we need some notation. The *support of a mixed strategy* π_i *for player* i is the set $\{s_j \mid \pi_i(s_j) > 0\}$ of pure strategies properly involved in π_i. Thus, a pure strategy is the special case of a mixed strategy whose support is a singleton. As the other extreme, a strategy π_i can also be *fully mixed*, which means that it has full support, i.e., every pure strategy $s_j \in S_i$ occurs in it with nonzero probability.

Theorem 2.1. *1. Let* $\boldsymbol{\pi} = (\pi_1, \pi_2, \ldots, \pi_n)$ *be a profile of mixed strategies in a noncooperative game in normal form. A mixed strategy π_i is a best response to the mixed-strategy profile $\boldsymbol{\pi}_{-i}$ if and only if all pure strategies in its support are best responses to $\boldsymbol{\pi}_{-i}$.*
2. Every pure Nash equilibrium is also a mixed Nash equilibrium.

Let us informally explain why the first statement of Theorem 2.1 holds true, leaving the easy proof of the second statement to the reader. For a contradiction, suppose that a best response mixed strategy π_i contains in its support a pure strategy that itself is not a best response. Then player i's expected gain would be improved by decreasing the probability of the worst such pure strategy, say s_j, where the remaining nonzero probabilities $\pi_i(s_k)$, $k \neq j$, would have to be increased proportionally so that the probabilities in the resulting mixed strategy π'_i still add up to one. But now π'_i is a better response than π_i to the mixed-strategy profile $\boldsymbol{\pi}_{-i}$, contradict-

2.2 Nash Equilibria in Mixed Strategies

ing the assumption that π_i was a best response to $\boldsymbol{\pi}_{-i}$. The converse of the equivalence stated is immediate.

Let us explain this statement in slightly different terms. By Definition 2.5, a mixed-strategy profile $\boldsymbol{\pi}$ is a Nash equilibrium exactly if for each player i,

$$G_i(\boldsymbol{\pi}) = \max_{\pi'_i \in \Pi_i} G_i(\boldsymbol{\pi}_{-i}, \pi'_i), \tag{2.6}$$

where $G_i(\boldsymbol{\pi}_{-i}, \pi'_i)$ denotes $G_i(\pi_1, \ldots, \pi_{i-1}, \pi'_i, \pi_{i+1}, \ldots, \pi_n)$. As noted in the first statement of Theorem 2.1, however, it is enough to focus on the pure strategies in S_i. That is, it follows (essentially from the fact that the expected gains G_i are *multilinear functions*, i.e., G_i is linear in its ith component) that

$$\max_{\pi'_i \in \Pi_i} G_i(\boldsymbol{\pi}_{-i}, \pi'_i) = \max_{s_j \in \Pi_i} G_i(\boldsymbol{\pi}_{-i}, s_j), \tag{2.7}$$

which together with (2.6) implies the following necessary and sufficient condition for $\boldsymbol{\pi}$ to be a Nash equilibrium in mixed strategies:

$$G_i(\boldsymbol{\pi}) = \max_{s_j \in \Pi_i} G_i(\boldsymbol{\pi}_{-i}, s_j). \tag{2.8}$$

Consequently, for (2.8) to hold, $\pi_i(s_k) = 0$ whenever

$$G_i(\boldsymbol{\pi}_{-i}, s_k) < \max_{s_j \in \Pi_i} G_i(\boldsymbol{\pi}_{-i}, s_j).$$

In other words, a pure strategy does not occur in the support of some player i's mixed Nash equilibrium strategy π_i in $\boldsymbol{\pi}$ unless it is a best response to $\boldsymbol{\pi}_{-i}$.

By the second part of Theorem 2.1, every Nash equilibrium in pure strategies is also a Nash equilibrium in mixed strategies. In other words, if a particular game has Nash equilibria in pure strategies then it also has Nash equilibria in mixed strategies. It is tempting to say that, due to the first statement in Theorem 2.1, the reverse is true as well. Indeed, let $\boldsymbol{\pi} = (\pi_1, \ldots, \pi_n)$ be a profile of mixed strategies that are in Nash equilibrium in mixed strategies and let $s_j \in S_i$ be some pure strategy in the support of some (non-pure) mixed strategy π_i. Since, by Theorem 2.1, s_j is a best response to $\boldsymbol{\pi}_{-i}$, let $\boldsymbol{\pi}' = (\boldsymbol{\pi}_{-i}, s_j)$ be a new strategy profile. If we repeated this process sufficiently many times, we would get a profile of pure strategies that might not be a Nash equilibrium in pure strategies. The reason is that even though s_j is a best response to $\boldsymbol{\pi}_{-i}$, the mixed strategies in $\boldsymbol{\pi}_{-i}$ are not necessarily best responses to $\boldsymbol{\pi}'$. Indeed, as the following examples demonstrate, the existence of a Nash equilibrium in mixed strategies does not imply the existence of a Nash equilibrium in pure strategies. There can exist Nash equilibria in mixed strategies in addition to those in pure strategies, or even if Nash equilibria in pure strategies do not exist at all.

2.2.1.2 The Penalty Game

As already pointed out, the penalty game (see Table 2.5 on page 57) has no Nash equilibrium in pure strategies; however, it does have a Nash equilibrium in mixed strategies, namely if the kicker (denoted by K) and the goalkeeper (denoted by G) each choose the left side (denoted by L) or the right side (denoted by R) of the goal with probability $1/2$. These uniform distributions are written as follows:

$$\pi_K = (\pi_K(L), \pi_K(R)) = (1/2, 1/2);$$
$$\pi_G = (\pi_G(L), \pi_G(R)) = (1/2, 1/2).$$

It is obvious that this profile (π_K, π_G) of mixed strategies is a (unique) Nash equilibrium. For example, if the kicker would choose to deviate from his mixed equilibrium strategy one-sidedly, say by kicking the ball to the right side with a probability larger than $1/2$, then the goalkeeper would have a best response strategy punishing the kicker: He would jump to the right with probability 1. Thus, the kicker has no incentive to deviate from his mixed strategy $\pi_K = (1/2, 1/2)$ one-sidedly. A symmetric argument shows that also the goalkeeper would be ill-advised to one-sidedly deviate from his mixed strategy $\pi_G = (1/2, 1/2)$.

The probability distributions of a Nash equilibrium in mixed strategies depend on the values of the gain vectors of the given game in normal form. If the possible gains of the players in the single strategy profiles change, then so do the probabilities by which the players should choose their strategies to keep the solution of mixed strategies in equilibrium. Suppose, for example, that a goalkeeper will safely hold on to the ball when jumping to the right (whenever the ball is coming to the right side), but acts awkwardly when jumping to the left: He makes a save for only every second ball coming to this side of the goal and awkwardly lets the other shots slip through. For the strategy profile (Left,Left), the kicker and the goalkeeper thus have the same chances of success; all other gain vectors are as before. Table 2.7 shows the gain vectors in this modified game.

Table 2.7 The penalty game with a goalkeeper acting awkwardly on the left

		Goalkeeper	
		Left	Right
Kicker	Left	(0,0)	(1,−1)
	Right	(1,−1)	(−1,1)

For a mixed strategy profile to be in Nash equilibrium, the kicker has to find a mixed strategy π_K that makes the goalkeeper indifferent against each strategy in π_G he chooses with positive probability, and conversely the goalkeeper has to find a mixed strategy π_G that makes the kicker indifferent

2.2 Nash Equilibria in Mixed Strategies

against each strategy in π_K he chooses with positive probability. If the kicker chooses the left side, his gain is calculated depending on the mixed strategy π_G of the goalkeeper according to Table 2.7 as $0 \cdot \pi_G(L) + \pi_G(R) = \pi_G(R)$. If he chooses the right side, however, his gain is $\pi_G(L) - \pi_G(R)$. Thus, the kicker is made indifferent against a shot on goal to the left or to the right if the goalkeeper mixes his strategies such that

$$\pi_G(R) = \pi_G(L) - \pi_G(R).$$

Since π_G is a probability distribution, we in addition have

$$\pi_G(L) + \pi_G(R) = 1,$$

so the goalkeeper achieves the kicker's desired indifference by choosing $\pi_G(L) = 2/3$ and $\pi_G(R) = 1/3$. This can be interpreted as the goalkeeper trying to make up for this deficit on the left side by jumping there more often. He thus anticipates the fact that the kicker is more likely to try to catch him wrongfooted on his weak side.

Conversely, the goalkeeper's gain for a jump to the left is calculated depending on the kicker's mixed strategy π_K according to Table 2.7 as $0 \cdot \pi_K(L) - \pi_K(R) = -\pi_K(R)$. If he jumps to the right, however, his gain is $-\pi_K(L) + \pi_K(R)$. Thus, the goalkeeper is made indifferent against a jump to the left or to the right if the kicker mixes his strategies such that

$$-\pi_K(R) = -\pi_K(L) + \pi_K(R),$$

which together with

$$\pi_K(L) + \pi_K(R) = 1$$

gives the solution of $\pi_K(L) = 2/3$ and $\pi_K(R) = 1/3$ for the kicker as well. This mixed strategy reflects the above-mentioned fact that the kicker is more likely to challenge the goalkeeper's weak left side.

According to inequality (2.5) in Definition 2.5, one-sided deviation would be punished for any of the players, so this profile of mixed strategies, $(\pi_K, \pi_G) = ((2/3, 1/3), (2/3, 1/3))$, is a Nash equilibrium, and it is the only one.

2.2.1.3 The Paper-Rock-Scissors Game

This game (see Table 2.6 on page 59) has no Nash equilibrium in pure strategies either, as noted earlier. As in the penalty game, however, a (unique) Nash equilibrium in mixed strategies does exist. The difference is only in the number of strategies the players can choose from. Since each of the two players now can mix three pure strategies, this Nash equilibrium occurs exactly if each of the players chooses with a probability of one third either paper, rock, or scissors.

2.2.1.4 The Battle of the Sexes

Unlike the penalty and the paper-rock-scissors game, we have seen that the battle of the sexes (see Table 2.3 on page 53) did have two Nash equilibria in pure strategies. In addition, there is also a third Nash equilibrium, in mixed strategies. To determine this third Nash equilibrium, it is again enough for George to find a mixed strategy π_G that makes Helena indifferent against her two strategies, while conversely Helena mixes her pure strategies in a way that also George is indifferent against his possible actions.

If George chooses the soccer match (denoted by S) instead of the concert (denoted by C), his gain is calculated depending on Helena's mixed strategy π_H according to Table 2.3 as $10 \cdot \pi_H(S) + 0 \cdot \pi_H(C) = 10 \cdot \pi_H(S)$. If he chooses the concert, however, then he gains $0 \cdot \pi_H(S) + \pi_H(C) = \pi_H(C)$. To make him indifferent against these two actions, Helena must mix her strategies such that

$$10 \cdot \pi_H(S) = \pi_H(C),$$

and due to $\pi_H(S) + \pi_H(C) = 1$, we have $\pi_H(S) = 1/11$ and $\pi_H(C) = 10/11$. Since the gain vectors in Table 2.3 are symmetric for George and Helena, it follows that George's mixed strategy is analogously calculated to be $\pi_G(S) = 10/11$ and $\pi_G(C) = 1/11$. For this symmetric Nash equilibrium in mixed strategies,

$$(\pi_G, \pi_H) = ((10/11, 1/11), (1/11, 10/11)),$$

George and Helena would both stick to their own favorite ten times and give in only at the eleventh evening to finally fulfill their beloved one's desire. This, however, obviously causes trouble. In each round of the game, both have to commit themselves to one option, either the soccer game or the concert. If both are stubborn on ten out of eleven of their anniversaries and are gentle only once, they will spend only two of these special days together (assuming they choose different anniversaries to give in), and there is nothing in it for either of them. The reason for this lies in the intensity they each prefer their own favorite strategy over their partner's favorite strategy: Both valuate their own favorite ten times as much than their partner's! How would one have to change the gain vectors in Table 2.3 to obtain a Nash equilibrium in mixed strategies having the form

$$(\pi'_G, \pi'_H) = ((1/2, 1/2), (1/2, 1/2))?$$

This Nash equilibrium would enable them to take turns in following his or her desire, and their relationship would have been saved. As one can see, for a relation to work it is important that both partners are not too selfishly focused on their own preferences, but are open also for their partner's suggestions.

2.2.1.5 The Chicken Game

Also in the chicken game (see Table 2.4 on page 55), there exists an additional Nash equilibrium in mixed strategies that can be determined as above. Letting π_D denote David's and π_E denote Edgar's mixed strategy and letting S denote *"Swerve"* and T denote *"Drive on"*, we obtain

$$2 \cdot \pi_E(S) + \pi_E(T) = 3 \cdot \pi_E(S);$$
$$2 \cdot \pi_D(S) + \pi_D(T) = 3 \cdot \pi_D(S).$$

It follows that

$$(\pi_D, \pi_E) = ((\pi_D(S), \pi_D(T)), (\pi_E(S), \pi_E(T))) = ((1/2, 1/2), (1/2, 1/2))$$

is this Nash equilibrium in mixed strategies that exists in addition to the two Nash equilibria in pure strategies. Interpreting the three Nash equilibria in this game as recommendations for action, one could advice the players to do the following (and wish them good luck in evaluating their opponents well!):

1. If you expect your opponent to be a chicken, then you should definitely go all out and win heroically. This corresponds to one of the two Nash equilibria in pure strategies.
2. If you expect your opponent to be undaunted by death and risk it all, then you should be wise and swerve. Although you won't win, you will survive at least. This corresponds to the other one of the two Nash equilibria in pure strategies.
3. If you can't judge your opponent well and just have no idea of what he is up to do, then you should toss a coin and go all out with heads, but cautiously swerve with tails. Maybe you win that way; and if not, maybe you survive—good luck! This corresponds to the additional Nash equilibrium in mixed strategies.

2.2.1.6 The Prisoners' Dilemma

Does the prisoners' dilemma (see Table 2.1 on page 46) also have a further Nash equilibrium in mixed strategies, in addition to the (strict) Nash equilibrium in pure strategies, (Confession, Confession)? Let us try to apply the method described above. So we look for a mixed strategy for Smith that makes Wesson indifferent against his strategies—*"Confession"* (denoted by C) and *"Silence"* (denoted by S). And conversely, we look for a mixed strategy for Wesson that makes Smith indifferent against his strategies C und S. By the values in Table 2.1 (alternatively, one can also use the values from Table 2.2), for Smith to be indifferent Wesson's mixed strategy π_W must satisfy that:

$$4 \cdot \pi_W(C) = 10 \cdot \pi_W(C) + 2 \cdot \pi_W(S)$$
$$-3 \cdot \pi_W(C) = \pi_W(S).$$

However, substituting the value $-3 \cdot \pi_W(C)$ for $\pi_W(S)$ into

$$\pi_W(C) + \pi_W(S) = 1,$$

one obtains $\pi_W(C) = -(1/2)$, which as a negative number cannot be a probability. The same is true for Smith's mixed strategy when trying to make Wesson indifferent.

Hence, there is no additional Nash equilibrium in mixed strategies for the prisoners' dilemma. This is not surprising, considering that making a confession is a strictly dominant strategy for both Smith and Wesson, as we have seen in Section 2.1. Assuming that they both play rationally, they thus will never remain silent with positive probability. Strictly dominated strategies—such as to remain silent here—are never part of a Nash equilibrium.

2.2.1.7 Overview of Some Properties of Some Two-Player Games

Table 2.8 summarizes some of the properties of the two-player games in normal form as yet considered. The first two rows of this table answer the question of whether both players have a dominant or even a strictly dominant strategy in the respective game. The next two rows give the numbers of Nash equilibria (NE, for short) in pure or in mixed strategies. The last two rows of the table give the numbers of Pareto optima (PO, for short) and answer the question of whether the set of Pareto optima of a game coincides with the set of Nash equilibria in pure strategies in this game. If already the number of Pareto optima differs from that of Nash equilibria, the answer to this question is "no," of course, but even for equal numbers these sets need not be identical in general, as it happens to be the case in the battle of the sexes.

Table 2.8 Properties of some two-player games

	Prisoners' dilemma	Battle of the sexes	Chicken game	Penalty game	Paper-Rock-Scissors game
Dominant strategies?	yes	no	no	no	no
Strictly dominant strategies?	yes	no	no	no	no
Number of NE in pure str.	1	2	2	0	0
Number of NE in mixed str.	1	3	3	1	1
Number of PO	3	2	3	4	9
PO = NE?	no	yes	no	no	no

The notion of a mixed strategy (and, relatedly, of Nash equilibria in mixed strategies) raises some important questions. What does it mean to play a

2.2 Nash Equilibria in Mixed Strategies

mixed strategy? Do players really randomize when choosing their strategies, and, if so, why? There are (at least) four answers to these questions. First, players may randomize to *confuse* their opponents (such as in the penalty game and the paper-rock-scissors game). Second, players may randomize when they are *uncertain* about the other players' actions (such as in the battle of the sexes and the chicken game). Third, mixed strategies describe what might happen in *repeated play*,[9] i.e., we then look at the number or frequency of pure strategies in the limit (again, consider the penalty game played over and over again between the same kicker and goalkeeper). Fourth, mixed strategies describe *population dynamics*, i.e., picking some players at random from a population of players, each choosing a pure strategy (e.g., picking 100 out of 1000 freshly married couples, each playing the battle of the sexes on the occasion of their first anniversaries), a mixed strategy can be viewed simply as the probability of picking a player who will play one pure strategy or another.

2.2.2 Existence of Nash Equilibria in Mixed Strategies

As one can see in Table 2.8, all these games have a Nash equilibrium, if not one in pure strategies, then at least one in mixed strategies. That is not by coincidence. An outstanding result of Nash [729, 730] shows that under appropriate, fairly modest assumptions there always exists a Nash equilibrium in mixed strategies. In this section we will provide a proof of this celebrated result that will be stated in Theorem 2.3 on page 82. First, we need to introduce and discuss some notions and results from mathematical topology to be used in its proof.

2.2.2.1 Definition of Some Notions from Mathematical Topology

Let $\mathcal{S} = S_1 \times S_2 \times \cdots \times S_n$ be the set of strategy profiles of the n players in a noncooperative game in normal form. All sets S_i are here assumed to be finite. How can the notion of "strategy" (in pure and in mixed form) be made accessible to mathematical or, specifically, topological arguments? How can one prove statements applying to *all* finite games in normal form, to the chicken game with its very concrete strategies *Drive on* and *Swerve* as well as to the prisoners' dilemma with its concrete strategies *Confession* and *Silence*

---

[9] Jackson and Watts [573] introduce so-called *"social games"* where players not only choose their strategies but also their opponents, and they investigate equilibria in finitely repeated social games. In particular, if a group of players is not happy with their current opponents, they might join together to play a new equilibrium. This refines the standard notion of equilibrium in games.

and to the penalty game with its concrete strategies *Left* and *Right*? We obviously need some abstract notion of strategy.

To this end, Nash views pure strategies as the unit vectors in an appropriate real vector space; every strategy from S_i is thus in $\mathbb{R}^{m_i}$, where $m_i = \|S_i\|$ denotes the *cardinality of* S_i, i.e., the number of elements in S_i. Strategies can then be mixed using the common operations in vector spaces: Every mixed strategy is the linear combination of pure strategies, each weighted by a certain probability, and since a mixed strategy corresponds to a probability distribution on S_i, these probabilities sum up to 1.

Mathematically speaking, mixed strategies over S_i are the points of a simplex, which can be viewed as a convex subset of $\mathbb{R}^{m_i}$. Such a subset is said to be *convex* if the direct connection between any two points of this subset completely lies within this subset. Figure 2.6(a) shows a convex set; obviously, any two points in this set can be directly connected without leaving the set. The set displayed in Figure 2.6(b), however, is not convex; for example, the direct connection between the points $(1, 2)$ and $(3, 2)$ is not completely within this set.

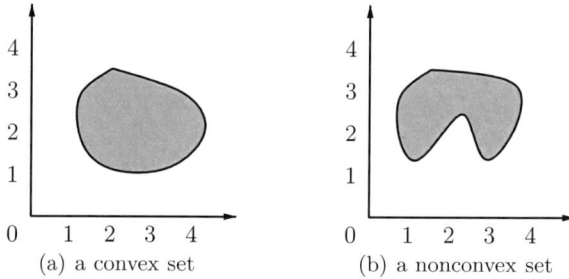

Fig. 2.6 A convex and a nonconvex set

Definition 2.6 (convexity, convex combination, affine independence).

1. A set $X \subseteq \mathbb{R}^m$ is *convex* if for all $\vec{x}, \vec{y} \in X$ and for all real numbers $\lambda \in [0, 1]$, $\lambda \cdot \vec{x} + (1 - \lambda) \cdot \vec{y} \in X$.
2. For vectors $\vec{x}_0, \vec{x}_1, \ldots, \vec{x}_n \in \mathbb{R}^m$ and nonnegative scalars $\lambda_0, \lambda_1, \ldots, \lambda_n$ satisfying $\sum_{i=0}^{n} \lambda_i = 1$, the vector

$$\sum_{i=0}^{n} \lambda_i \cdot \vec{x}_i$$

is a *convex combination of* $\vec{x}_0, \vec{x}_1, \ldots, \vec{x}_n$.
3. A finite set $\{\vec{x}_0, \vec{x}_1, \ldots, \vec{x}_n\}$ of vectors in $\mathbb{R}^m$ is said to be *affinely independent* if

2.2 Nash Equilibria in Mixed Strategies

$$\left(\sum_{i=0}^{n} \lambda_i \cdot \vec{x}_i = \vec{0} \text{ and } \sum_{i=0}^{n} \lambda_i = 0\right) \Rightarrow \lambda_0 = \lambda_1 = \cdots = \lambda_n = 0. \quad (2.9)$$

An equivalent condition for $\{\vec{x}_0, \vec{x}_1, \ldots, \vec{x}_n\}$ to be affinely independent is that the set $\{\vec{x}_1 - \vec{x}_0, \vec{x}_2 - \vec{x}_0, \ldots, \vec{x}_n - \vec{x}_0\}$ is linearly independent. That is, for no i is $\vec{x}_i - \vec{x}_0$ in the vector space spanned by

$$\vec{x}_1 - \vec{x}_0, \ldots, \vec{x}_{i-1} - \vec{x}_0, \vec{x}_{i+1} - \vec{x}_0, \ldots, \vec{x}_n - \vec{x}_0,$$

i.e., $\vec{x}_i - \vec{x}_0$ cannot be expressed as a linear combination of the above vectors. For example, the set $\{\vec{0}, \vec{u}_1, \ldots, \vec{u}_n\}$ is affinely independent, where $\vec{u}_i$ denotes the ith *unit vector*, i.e., the vector having a one in the ith position and zeros otherwise. The implication (2.9) says that the vector $\vec{0}$ can be combined only trivially in the vectors $\vec{x}_0, \vec{x}_1, \ldots, \vec{x}_n$. Intuitively, a set of points in a vector space is affinely independent if no three points lie on the same line, no four points lie in the same plane, etc.

We can now define the notion of a simplex formally. Intuitively, a simplex is a generalization of a two-dimensional triangle to other dimensions.

Definition 2.7 (simplex).

1. An *n-simplex* is the set of all convex combinations of the affinely independent set $\{\vec{x}_0, \vec{x}_1, \ldots, \vec{x}_n\}$ of vectors:

$$\vec{x}_0 \cdots \vec{x}_n = \left\{\sum_{i=0}^{n} \lambda_i \cdot \vec{x}_i \,\middle|\, \lambda_i \geq 0 \text{ for each } i, \, 0 \leq i \leq n, \text{ and } \sum_{i=0}^{n} \lambda_i = 1\right\}.$$

   a. Every $\vec{x}_i$ is a *vertex* of the n-simplex $\vec{x}_0 \cdots \vec{x}_n$.
   b. Every *k-simplex* $\vec{x}_{i_0} \cdots \vec{x}_{i_k}$, $i_0, \ldots i_k \in \{0, 1, \ldots, n\}$, is a *k-face* of the n-simplex $\vec{x}_0 \cdots \vec{x}_n$.

2. The *standard n-simplex* Δ_n is defined as

$$\Delta_n = \left\{\vec{y} = (y_0, y_1, \ldots, y_n) \in \mathbb{R}^{n+1} \,\middle|\, y_i \geq 0, 0 \leq i \leq n, \text{ and } \sum_{i=0}^{n} y_i = 1\right\}.$$

That is, $\Delta_n = \vec{u}_0 \cdots \vec{u}_n$, where $\vec{u}_i$ denotes the ith unit vector in $\mathbb{R}^{n+1}$.

Figure 2.7 shows a 0-simplex (i.e., a point in the Euclidean space), a 1-simplex (i.e., a line segment), a 2-simplex (i.e., a triangle), and a 3-simplex (i.e., a tetrahedron), where for each n, $1 \leq n \leq 3$, the n-simplex is obtained from the $(n-1)$-simplex by adding one point, $\vec{x}_n$, that is affinely independent from the previous points, $\vec{x}_0, \ldots, \vec{x}_{n-1}$. Note that the $(n-1)$-simplex $\vec{x}_0 \cdots \vec{x}_{n-1}$ is one $(n-1)$-face of the n-simplex $\vec{x}_0 \cdots \vec{x}_n$.

Mixed strategies over S_i are convex combinations of the vertices of the simplex representing the pure strategies in S_i. The players' gain functions map mixed-strategy profiles (whose components are points in these simplexes)

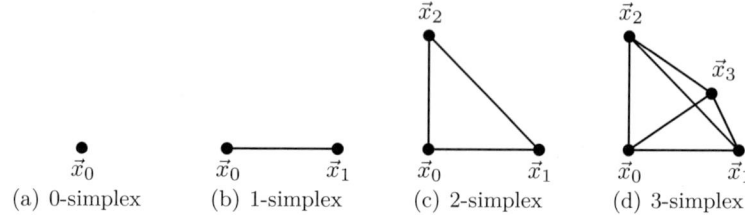

Fig. 2.7 n-simplexes for $0 \leq n \leq 3$

to real numbers, i.e., the gains of a player depend both on her own and the other players' strategies. For concreteness, suppose that Anna and Belle play a two-player noncooperative game in normal form with Anna having three pure strategies, a_1, a_2, and a_3, and Belle having two pure strategies, b_1 and b_2. Table 2.9 shows the gains of both players, depending on the six possible pure-strategy profiles.

Table 2.9 Anna's gain (left) and Belle's gain (right)

		Belle	
		Strategy b_1	Strategy b_2
	Strategy a_1	$(1,2)$	$(1,2)$
Anna	Strategy a_2	$(4,1)$	$(2,0)$
	Strategy a_3	$(3,0)$	$(4,3)$

The subset of $\mathbb{R}^2$ shown in Figure 2.8(a) represents Anna's gains with respect to her mixed strategies, depending on which of Belle's two pure strategies are chosen. Anna's pure strategies provide the three points $a_1 = (1,1)$, $a_2 = (4,2)$, and $a_3 = (3,4)$, the left entry of which is Anna's gain if Belle chooses the pure strategy b_1 and the right entry of which is Anna's gain if Belle chooses the pure strategy b_2, see Table 2.9. If Anna mixes any two of her pure strategies (so that both are chosen with positive probability and the third one is not chosen), then we obtain a point on the line connecting the corresponding points a_i and a_j, with $i, j \in \{1, 2, 3\}$ and $i \neq j$. If she mixes all three pure strategies with positive probability, then we obtain a point in the interior of the triangle shown in Figure 2.8(a). Obviously, the set of points representing Anna's gains for her mixed strategies is a 2-simplex.

The subset of $\mathbb{R}^3$ shown in Figure 2.8(b) represents Belle's gains with respect to her mixed strategies, depending on which of Anna's three pure strategies are chosen. Belle's pure strategies provide the two points $b_1 = (2,1,0)$ and $b_2 = (2,0,3)$, the ith entry of which is Belle's gain if Anna chooses the pure strategy a_i, $i \in \{1,2,3\}$, see Table 2.9. If Belle mixes her two pure strategies, then we obtain a point on the line connecting the points b_1 and b_2.

2.2 Nash Equilibria in Mixed Strategies

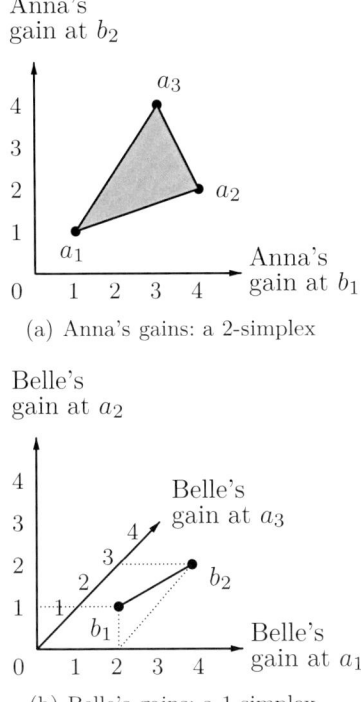

Fig. 2.8 Convex gain sets for pure and mixed strategy sets

Obviously, the set of points representing Belle's gains for her mixed strategies is a 1-simplex.

To prove Nash's theorem, we will now show Brouwer's fixed point theorem, which uses Sperner's lemma. To state this lemma, we need some more definitions.

Definition 2.8 (simplicial subdivision).

1. A *simplicial subdivision* of an n-simplex $T = \vec{x}_0 \cdots \vec{x}_n$ is a finite set of n-subsimplexes $\{T_i \mid 1 \leq i \leq k\}$ such that

   a. $\bigcup_{T_i \in T} T_i = T$ (i.e., all these n-subsimplexes cover the same space as T), and
   b. for each $T_i, T_j \in T$, $i \neq j$, $T_i \cap T_j$ is either empty or equal to a common face (i.e., any two distinct n-subsimplexes can overlap only on a complete k-face of both, $k < n$).

2. Let $T = \vec{x}_0 \cdots \vec{x}_n$ be a simplicial subdivided n-simplex, and let V denote the set of all distinct vertices of all the n-subsimplexes. For a point $\vec{y} \in T$,

$\vec{y} = \sum_{i=0}^{n} \lambda_i \cdot \vec{x}_i$, let $\sigma(\vec{y}) = \{i \mid \lambda_i > 0\}$ be the set of vertices "involved" in $\vec{y}$.

A function $\mathcal{L}: V \to \{0, 1, \ldots, n\}$ is a *proper labeling of a subdivision of T* if $\mathcal{L}(\vec{v}) \in \sigma(\vec{v})$ for each $\vec{v} \in V$.

3. An n-subsimplex T_i of T is *completely labeled by a proper labeling $\mathcal{L}$* if $\mathcal{L}$ takes on all the values $0, 1, \ldots, n$ on its set of vertices, i.e., if

$$\{\mathcal{L}(\vec{v}) \mid \vec{v} \text{ is a vertex of } T_i\} = \{0, 1, \ldots, n\}.$$

Note that if $\vec{y}$ is a mixed strategy in which each pure strategy $\vec{x}_i$ occurs with probability λ_i, then $\sigma(\vec{y})$ is essentially nothing other than the support of $\vec{y}$. A proper labeling maps the vertices of all n-subsimplexes of T to the subscripts of the vertices $\vec{x}_0, \vec{x}_1, \ldots, \vec{x}_n$ of T such that each vertex $\vec{v}$ of an n-subsimplex of T can be labeled only by the subscript i of some $\vec{x}_i$ that really is involved in $\vec{v}$.

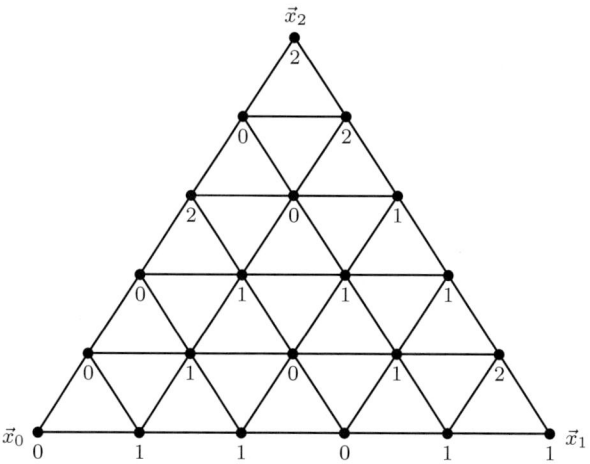

Fig. 2.9 A properly labeled simplicial subdivision of a 2-simplex

Figure 2.9 shows a properly labeled simplicial subdivision of a triangle (i.e., of the 2-simplex $T = \vec{x}_0 \vec{x}_1 \vec{x}_2$). The 25 triangles (i.e., the 25 2-subsimplexes of T) in this subdivision overlap only at their edges (i.e., their 1-faces) or vertices (i.e., their 0-faces), or they are completely disjoint. Since the labeling is proper, the six vertices on each of the three 1-faces of T are labeled only by the subscripts of those $\vec{x}_i$ involved: All six vertices on $\vec{x}_0 \vec{x}_1$ have labels 0 and 1 only, all six vertices on $\vec{x}_0 \vec{x}_2$ have labels 0 and 2 only, and all six vertices on $\vec{x}_1 \vec{x}_2$ have labels 1 and 2 only. All other vertices can have any of the labels 0, 1, and 2 under a proper labeling (but happen here to have labels 0 and 1 only). There are three completely labeled 2-subsimplexes of T, an odd number. That is not by coincidence, as Sperner's lemma shows.

2.2.2.2 Sperner's Lemma and Brouwer's Fixed Point Theorem

Lemma 2.1 (Sperner's lemma). *Let $T_n = \vec{x}_0 \cdots \vec{x}_n$ be a simplicially subdivided n-simplex and let $\mathcal{L}$ be a proper labeling of the given subdivision of T_n. There are an odd number of n-subsimplexes that are completely labeled by $\mathcal{L}$ in this subdivision of T_n.*

Proof. The proof is by induction on n. The base case, $n = 0$, holds trivially. Indeed, the only simplicial subdivision of $T_0 = \vec{x}_0$ is $\{\vec{x}_0\}$, which can be labeled only by $\mathcal{L}(\vec{x}_0) = 0$, a proper labeling, so there is exactly one completely labeled 0-subsimplex of T_0, T_0 itself.

Now suppose the claim holds true for $n-1$. We show that it then also holds for n. Note that the given simplicial subdivision of the n-simplex $T_n = \vec{x}_0 \cdots \vec{x}_n$ induces a simplicial subdivision of its $(n-1)$-face $T_{n-1} = \vec{x}_0 \cdots \vec{x}_{n-1}$, which is an $(n-1)$-simplex. Furthermore, the labeling function $\mathcal{L}$ restricted to T_{n-1} is still proper. By induction hypothesis, there are an odd number of $(n-1)$-subsimplexes in T_{n-1} with labels $0, 1, \ldots, n-1$. Indeed, looking at our example in Figure 2.9, the 1-simplex $T_1 = \vec{x}_0 \vec{x}_1$ has three 1-subsimplexes with labels 0 and 1 (and two 1-subsimplexes with labels 1 only).

We now describe certain walks on T_n some of which will end in a completely labeled n-subsimplex of T_n. The first type of walk we consider starts from T_{n-1}:

1. Start from any $(n-1)$-subsimplex in T_{n-1} with labels $0, 1, \ldots, n-1$. Call this $(n-1)$-subsimplex T'_{n-1}.
2. There is a unique n-subsimplex of T_n with $(n-1)$-face T'_{n-1}. Call this n-subsimplex T'_n. Walk into T'_n. Note that T'_n has the same vertices as T'_{n-1}, plus one additional vertex, say $\vec{z}$. Distinguish the following two cases.

    a. If $\mathcal{L}(\vec{z}) = n$, we have found a completely labeled n-subsimplex of T_n, namely T'_n, and the walk ends.
    b. If $\mathcal{L}(\vec{z}) = j \neq n$, the $n+1$ vertices of T'_n have the labels $0, 1, \ldots, n-1$, so label j occurs twice and all other of these labels once.
    We claim that in this case, T'_n has exactly one additional $(n-1)$-face, $T''_{n-1} \neq T'_{n-1}$, which is an $(n-1)$-subsimplex with labels $0, 1, \ldots, n-1$. But this follows immediately from the fact that every $(n-1)$-face of T'_n has all vertices of T'_n except one. Since only label j occurs twice, an $(n-1)$-face of T'_n has the labels $0, 1, \ldots, n-1$ if and only if one of the two vertices labeled by j is missing in it. T'_{n-1} is one such $(n-1)$-face of T'_n, so there must be exactly another one, T''_{n-1}. Continue the walk via T''_{n-1}. Again, distinguish the following two cases.

      i. If T''_{n-1} belongs to an $(n-1)$-face of T_n, the walk ends.
      ii. Otherwise, walk into the unique n-subsimplex of T_n having $(n-1)$-face T''_{n-1} with labels $0, 1, \ldots, n-1$. Call this n-subsimplex T''_n

and proceed as in the beginning of step 2, with T_n'' and T_{n-1}'' playing the roles of T_n' and T_{n-1}', respectively.

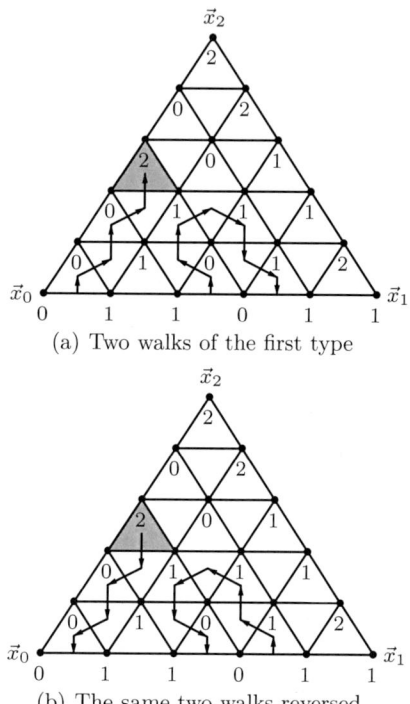

Fig. 2.10 Walking through the 2-simplex $T_2 = \vec{x}_0\vec{x}_1\vec{x}_2$

The second type of walk we consider does not start from an $(n-1)$-subsimplex of T_{n-1} but from any completely labeled n-subsimplex of T_n, but otherwise follows the same rules, so only step 1 is skipped.

Walks of both types are uniquely and completely determined by their starting points. These starting points are either on $(n-1)$-subsimplexes of T_{n-1} with labels $0, 1, \ldots, n-1$ (for walks of the first type) or on completely labeled n-subsimplexes of T_n (for walks of the second type).[10] The walks end either in completely labeled n-subsimplexes of T_n (step 2(a); see the left walk in Figure 2.10(a) for an example), or in $(n-1)$-subsimplexes of T_n's $(n-1)$-face T_{n-1} (step 2(b)i; see the right walk in Figure 2.10(a) for an example). They cannot end at another $(n-1)$-face of T_n because $\mathcal{L}$ is a proper labeling.

Each such walk (of either type) can be reversed by essentially the same rules, leading from the former end point to the former starting point; Fig-

---

[10] That is, any two walks (of either type) with the same starting point must be identical. We refer to this property as *uniqueness of walks*.

2.2 Nash Equilibria in Mixed Strategies

ure 2.10(b) shows the two walks of Figure 2.10(a) reversed. This implies that if a walk starts from an $(n-1)$-subsimplex T'_{n-1} on T_{n-1} and ends in an $(n-1)$-subsimplex T''_{n-1} on T_{n-1}, then $T'_{n-1} \neq T''_{n-1}$, for otherwise we could reverse this walk and would have two distinct walks with the same starting point, contradicting the uniqueness of walks.

Since the number of $(n-1)$-subsimplexes with labels $0, 1, \ldots, n-1$ on T_{n-1} is odd by the induction hypothesis, there are an odd number of walks starting from T_{n-1} and ending in a completely labeled n-subsimplex of T_n.[11] All these walks must end in *distinct* completely labeled n-subsimplexes of T_n, since otherwise they could be reversed, leading to distinct walks with the same starting point, again contradicting the uniqueness of walks.

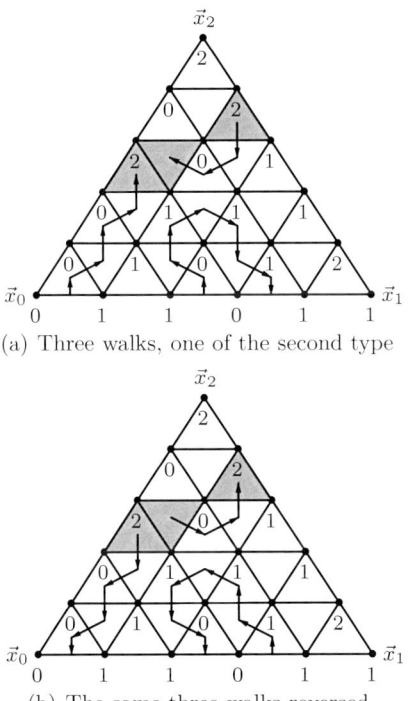

(a) Three walks, one of the second type

(b) The same three walks reversed

Fig. 2.11 All walks through the 2-simplex $T_2 = \vec{x}_0 \vec{x}_1 \vec{x}_2$

Not all completely labeled n-subsimplexes of T_n can be reached by walks of the first type (i.e., by walks starting from an $(n-1)$-subsimplex of T_{n-1}). However, such n-subsimplexes of T_n are connected by walks of the second type. That is, all such completely labeled n-subsimplexes of T_n form pairs

---

[11] All other walks starting from an $(n-1)$-subsimplex of T_{n-1} end in another $(n-1)$-subsimplex of T_{n-1}, as argued above. These walks thus form pairs.

again, since (again by the reversal argument) they can neither be the starting points of walks leading to T_{n-1} (or any other $(n-1)$-face of T_n), nor the starting points of walks that return to themselves (forming cycles). Figure 2.11 shows one such pair of 2-subsimplexes of T_2 in our example and one connecting walk in Figure 2.11(a) and its reversal in Figure 2.11(b).

Summing up, we have shown that there are an odd number of completely labeled n-subsimplexes of T_n. ❑

In addition to being convex sets, the simplexes representing the players' strategy sets are also compact sets. A subset T of $\mathbb{R}^m$ is said to be *compact* if it is closed and bounded. Intuitively, T is *closed* if for each point $\vec{x}$ outside T there is an $\varepsilon > 0$ such that every point in the ε-ball around $\vec{x}$ is not in T either. T is said to be *bounded* if there is a ball of finite radius containing T. An important property of compact sets to be used later on is that every infinite sequence in such sets has a convergent subsequence. It is immediately clear that the standard m-simplex Δ_m is compact. We also need the following definition for the proof of Brouwer's fixed point theorem.

Definition 2.9 (centroid). The *centroid of an n-simplex $\vec{x}_0 \cdots \vec{x}_n$* is the "average" of its vertices:

$$\frac{1}{n+1}\sum_{i=0}^{n}\vec{x}_i.$$

For example, the 2-simplex representing Anna's gains for her mixed strategies in Figure 2.8(a) has the centroid $(1/3) \cdot ((1,1)+(4,2)+(3,4)) = (1/3) \cdot (8,7) = (8/3, 7/3)$, and the 1-simplex representing Belle's gains for her mixed strategies in Figure 2.8(b) has the centroid $(1/2) \cdot ((2,1,0)+(2,0,3)) = (1/2) \cdot (4,1,3) = (2, 1/2, 3/2)$.

Finally, the expected gain functions G_i, $1 \le i \le n$, mapping each strategy profile $\boldsymbol{\pi} = (\pi_1, \pi_2, \ldots, \pi_n) \in \Pi$ to a real number (the expected gain of player i under $\boldsymbol{\pi}$), must be *continuous* for Brouwer's fixed point theorem to be applicable. Informally speaking, continuity means that if there are only very small changes in the profiles of mixed strategies, then also the corresponding gains change only very little, i.e., there are no "jumps" (technically speaking, no points of discontinuity) in these gain functions.

Theorem 2.2 (Brouwer's fixed point theorem). *Every continuous function $f : \Delta_m \to \Delta_m$ has a fixed point, i.e., there exists some $\vec{z} \in \Delta_m$ such that*

$$f(\vec{z}) = \vec{z}.$$

Proof. The proof proceeds in two parts. First, we construct a simplicial subdivision with a proper labeling function $\mathcal{L}$ for Δ_m so that Sperner's lemma can be applied, yielding at least one completely labeled m-subsimplex in this subdivision. Second, making such subdivisions finer and finer, we show that this m-subsimplex contracts to a fixed point of f.

2.2 Nash Equilibria in Mixed Strategies

For the first part, fix an $\varepsilon > 0$. Subdivide Δ_m simplicially such that the Euclidean distance[12] between any two points in the same m-subsimplex of this subdivision is at most ε.[13] Now define a labeling function $\mathcal{L} : V \to \{0, 1, \ldots, m\}$ as follows. For each vertex $\vec{v} \in V$ of the m-subsimplexes in this subdivision, we choose a label $\mathcal{L}(\vec{v})$ from the set $\sigma(\vec{v}) \cap \{i \mid f_i(\vec{v}) \le v_i\}$, where $\vec{v} = (v_0, v_1, \ldots, v_m)$ and $f(\vec{v}) = (f_0(\vec{v}), f_1(\vec{v}), \ldots, f_m(\vec{v}))$ are points in Δ_m and, recall, $\sigma(\vec{v}) = \{i \mid v_i > 0\}$ for $\vec{v} = \sum_{i=0}^{m} v_i \cdot \vec{u}_i$, since $\vec{u}_i$ is the ith unit vector in $\mathbb{R}^{m+1}$. That is, $\mathcal{L}(\vec{v}) = i$ means that $v_i > 0$ and $f_i(\vec{v}) \le v_i$.

We have to show that this labeling function is well-defined, i.e., that $\sigma(\vec{v}) \cap \{i \mid f_i(\vec{v}) \le v_i\} \ne \emptyset$. Intuitively, this is true because $\vec{v}$ and $f(\vec{v})$ are points in Δ_m, so their components each add up to one by definition of Δ_m. Thus, there exists an i such that $f_i(\vec{v}) \le v_i$, and this holds true even when restricted to $\sigma(\vec{v})$, so $v_i > 0$.

Formally, for a contradiction suppose that $\sigma(\vec{v}) \cap \{i \mid f_i(\vec{v}) \le v_i\} = \emptyset$. Since $\vec{v}$ is a point in Δ_m (i.e., $\sum_{i=0}^{m} v_i = 1$) and $v_j > 0$ exactly if $j \in \sigma(\vec{v})$, we have

$$\sum_{j \in \sigma(\vec{v})} v_j = \sum_{i=0}^{m} v_i = 1.$$

From our assumption we know that $f_j(\vec{v}) > v_j$ for each $j \in \sigma(\vec{v})$, which implies

$$\sum_{j \in \sigma(\vec{v})} f_j(\vec{v}) > \sum_{j \in \sigma(\vec{v})} v_j = 1. \tag{2.10}$$

However, since $f(\vec{v})$ is a point in Δ_m as well, we have

$$\sum_{j \in \sigma(\vec{v})} f_j(\vec{v}) \le \sum_{i=0}^{m} f_i(\vec{v}) = 1,$$

contradicting (2.10). Thus $\mathcal{L}$ is well-defined. By construction, $\mathcal{L}(\vec{v}) \in \sigma(\vec{v})$ for each $\vec{v} \in V$. Thus $\mathcal{L}$ is also proper. By Sperner's lemma (see Lemma 2.1), in this simplicial subdivision of Δ_m there exists at least one m-subsimplex T_m^ε that depends on ε and is completely labeled by $\mathcal{L}$.

In the second part of the proof, we will show that when ε goes to zero, the resulting m-subsimplex $T_m^\varepsilon = \vec{t}_0 \cdots \vec{t}_m$ contracts to a fixed point of f. T_m^ε is completely labeled; without loss of generality, we may assume that $\mathcal{L}(\vec{t}_i) = i$. (Otherwise, we simply rename the labels accordingly.) Furthermore, by construction of $\mathcal{L}$, we have

$$f_i(\vec{t}_i) \le (\vec{t}_i)_i \tag{2.11}$$

---

[12] The *Euclidean distance* between any two points $\vec{x} = (x_1, \ldots, x_m)$ and $\vec{y} = (y_1, \ldots, y_m)$ in $\mathbb{R}^m$ is defined by $\sqrt{(x_1 - y_1)^2 + \cdots + (x_m - y_m)^2}$.
[13] We here assume that it is always possible to find such a simplicial subdivision of Δ_m, regardless of the dimension m, which is true, but not trivial to show.

for each i, $0 \leq i \leq m$, where $(\vec{t}_i)_i$ denotes the ith component of $\vec{t}_i$.

For ε going to zero, we consider the infinite sequence of centroids in these completely labeled m-subsimplexes T_m^ε. Since Δ_m is compact, there exists a convergent subsequence with limit $\vec{z}$. The vertices of these m-subsimplexes T_m^ε then move toward $\vec{z}$ with ε going to zero, that is, $\vec{t}_i \xrightarrow[\varepsilon \to 0]{} \vec{z}$ for each i, $0 \leq i \leq m$. Since f is continuous, it follows from (2.11) that

$$f_i(\vec{z}) \leq \vec{z}_i$$

for each i, $0 \leq i \leq m$. This implies that $f(\vec{z}) = \vec{z}$, as desired, since otherwise, by the same argument as used in the first part of this proof, we would have

$$1 = \sum_{i=0}^m f_i(\vec{z}) < \sum_{i=0}^m \vec{z}_i = 1,$$

a contradiction. □

In the proof of Nash's theorem, we will need a slightly more general fixed point theorem than Brouwer's because the set $\Pi = \Pi_1 \times \Pi_2 \times \cdots \times \Pi_n$ of the players' mixed-strategy profiles is in fact a Cartesian product of simplexes, each simplex representing the set of mixed strategies of one player. A Cartesian product of simplexes is called a *simplotope*.

Corollary 2.1 (Brouwer's fixed point theorem, for simplotopes).
Let $K = \prod_{j=1}^k \Delta_{m_j}$ be a simplotope. Every continuous function $f : K \to K$ has a fixed point.

We omit the proof of Corollary 2.1. Now we are ready to prove Nash's theorem.

2.2.2.3 Nash's Theorem

Theorem 2.3 (Nash [729, 730]). *For each noncooperative game in normal form with a finite number of players each having a finite set of strategies, there exists a Nash equilibrium in mixed strategies.*

Proof. Let $\boldsymbol{\pi} = (\pi_1, \pi_2, \ldots, \pi_n) \in \Pi$ be a profile of mixed strategies with the expected gain functions $G_i(\boldsymbol{\pi})$ (see Definition 2.5 on page 63). Let $\mathcal{S} = S_1 \times S_2 \times \cdots \times S_n$ be the underlying set of pure strategy profiles, where each S_i is finite. For each pure strategy s_j of each player i, let $G_i(\boldsymbol{\pi}_{-i}, s_j)$ be i's gain when switching one-sidedly from π_i to s_j. Define the functions

$$\varphi_{ij}(\boldsymbol{\pi}) = \max(0, G_i(\boldsymbol{\pi}_{-i}, s_j) - G_i(\boldsymbol{\pi}))$$

2.2 Nash Equilibria in Mixed Strategies

for each i and j with $1 \leq i \leq n$ and $1 \leq j \leq \|S_i\|$. Since the expected gain functions are continuous, so is each function φ_{ij}.

Now, define the function $f : \Pi \to \Pi$ by $f(\boldsymbol{\pi}) = \boldsymbol{\pi}' = (\pi'_1, \pi'_2, \ldots, \pi'_n)$, where the modifications π'_i of π_i are defined by

$$\pi'_i(s_j) = \frac{\pi_i(s_j) + \varphi_{ij}(\boldsymbol{\pi})}{\sum_{s_k \in S_i}(\pi_i(s_k) + \varphi_{ik}(\boldsymbol{\pi}))} = \frac{\pi_i(s_j) + \varphi_{ij}(\boldsymbol{\pi})}{1 + \sum_{s_k \in S_i} \varphi_{ik}(\boldsymbol{\pi})}. \quad (2.12)$$

Intuitively, $\boldsymbol{\pi}'$ puts more probability weight π'_i on those pure strategies of each player i that are "better" responses to the other players' mixed strategies $\boldsymbol{\pi}'_{-i}$.

Since every function φ_{ij} is continuous, so is f. Since Π, as a simplotope, is convex and compact and since $f : \Pi \to \Pi$ is continuous, f has at least one fixed point by Corollary 2.1. It remains to show that $\boldsymbol{\pi}$ is a Nash equilibrium in mixed strategies if and only if $f(\boldsymbol{\pi}) = \boldsymbol{\pi}$.

From left to right, if $\boldsymbol{\pi}$ is a Nash equilibrium in mixed strategies, we have $\varphi_{ij}(\boldsymbol{\pi}) = 0$ for all i and j. Hence, $f(\boldsymbol{\pi}) = \boldsymbol{\pi}' = \boldsymbol{\pi}$, so $\boldsymbol{\pi}$ is a fixed point.

From right to left, suppose $f(\boldsymbol{\pi}) = \boldsymbol{\pi}$. Consider player i. Since G_i is linear in its ith component, there exists at least one pure strategy s_j in the support of π_i (i.e., $\pi_i(s_j) > 0$) such that

$$G_i(\boldsymbol{\pi}_{-i}, s_j) \leq G_i(\boldsymbol{\pi}).$$

(In other words, by linearity of G_i in its ith component, we see that the situation where for each pure strategy s_k (in the support of π_i) it holds that $G_i(\boldsymbol{\pi}_{-i}, s_k) > G_i(\boldsymbol{\pi})$ is impossible.) By definition of φ_{ij}, it follows that $\varphi_{ij}(\boldsymbol{\pi}_{-i}, s_j) = 0$. Since $f(\boldsymbol{\pi}) = \boldsymbol{\pi}$, this enforces that

$$\pi'_i(s_j) = \pi_i(s_j).$$

That is, the enumerator in (2.12) simplifies to $\pi_i(s_j)$ (due to $\varphi_{ij}(\boldsymbol{\pi}_{-i}, s_j) = 0$) and it is positive because s_j is in the support of π_i. This implies, by simple arithmetic, that the denominator in (2.12) must be one. Consequently, $\sum_{s_k \in S_i} \varphi_{ik}(\boldsymbol{\pi}) = 0$. This holds true for each player i and, in effect, for all i and k, we have $\varphi_{ik}(\boldsymbol{\pi}) = 0$. That is, no player i can increase her gain by moving one-sidedly from her mixed strategy π_i to some pure strategy. However, since we know that

$$\max_{\pi'_i \in \Pi_i} G_i(\boldsymbol{\pi}_{-1}, \pi'_i) = \max_{s_j \in \Pi_i} G_i(\boldsymbol{\pi}_{-1}, s_j)$$

(see Equation (2.7) on page 65), this means that $\boldsymbol{\pi}$ is a Nash equilibrium in mixed strategies. ⊔⊓

Since finite sets cannot be convex, the existence of a Nash equilibrium in *pure* strategies cannot be guaranteed by the proof of Theorem 2.3.

Nash's original proof for Theorem 2.3, stated in [729], uses the fixed point theorem of Kakutani instead of Brouwer's. Kakutani's fixed point theorem [587] generalizes that of Brouwer from fixed points for mappings to fixed points for correspondences. Nash writes [730, p. 288]: *"The proof given here is a considerable improvement over that earlier version and is based directly on the Brouwer theorem."* Nasar [727] reports on page 94 that the great John von Neumann, when Nash was going to mention his result to him, gave him short shrift by saying: *"That's trivial, you know. That's just a fixed point theorem."*

2.3 Checkmate: Trees for Games with Perfect Information

In all previously considered games, all players made only *one move* and they did so *simultaneously*. They each had to choose their strategy without knowing how their opponents would act. However, many games are played *sequentially* and have *multiple moves* that all the players can observe. We now turn to this type of games.

2.3.1 Sequential Two-Player Games

The rules of a sequential game determine the number of players, the order in which they make their moves, and the particular actions that constitute these moves (e.g., "draw a circle on the board" or "move a piece from one position to the other"). Further, all players (in particular, the player whose turn it is currently) always have perfect information about all the previous moves. (For sequential games, "moves" are used synonymously with "strategies.")

Chess, checkers, nine men's morris, and *go* are examples of two-player games with perfect information in which the two players, say Black and White, take turns in drawing. In each situation of such a game, the player whose turn it is to draw, seeks to find an answer to the question: *"Does there exist a move such that for each response move of my opponent, there exists a move for me such that . . . I will win?"* To "win" often means that by the rules of the game no further move is possible for the opponent,[14] for example, when the winner announces the end of a chess match by: "Checkmate!"

The normal form of a game introduced in Definition 2.1 on page 47 can be extended so that this sequentiality—the succession of moves—can be expressed. Such *games in extensive form* not only give a complete description

---

[14] There are also games in which a player who cannot move in the current situation simply skips. One may "to skip" also interpret as a move in the game, though.

2.3 Checkmate: Trees for Games with Perfect Information   85

of the single players' strategies, they also determine by their rules whose turn it is next, what the players know in which situation of the game, and which strategies then are available to them. A formal definition of the notion of game in extensive form would be rather extensive, which is why it is omitted here. For an important class of such games, namely the *(finite) games with perfect information*, a representation by game trees is appropriate. Trees are special graphs (namely, connected, acyclic graphs), and we assume familiarity with the common graph-theoretical notions and notation (such as *root* and *leaf of a tree* and *(inner) vertex* and *edge of a graph*, etc.). The foundations of graph theory can be found, e.g., in the textbook by Diestel [342]; algorithmic aspects of problems on graphs are treated, e.g., by Gurski et al. [520].

2.3.1.1 Game Trees

Game trees can be described as follows:

- The *root* (a distinguished vertex of the tree) represents the start situation. For chess and checkers, for example, this is the basic arrangement of chess or checkers pieces, for nine men's morris and go it is the empty board, etc.
- Every *leaf* (i.e., every vertex with exactly one incident edge) represents a final situation.[15] One can get into such a situation from exactly one preceding situation closer to the root, but there is no subsequent situation—the game is over. Distinct leaves may represent the same final situation. For example, one and the same checkmate situation can be reached via quite a number of different paths (i.e., courses of the game) from the root. Every leaf is labeled by the player whose turn it is to draw in this situation, but who has no choice of a further move anymore. Every leaf also lists the gains for all players resulting for them by the game ending in this leaf. If only *win*, *draw*, or *loss* is at stake in a game, such as in chess, for example, then the players' gain functions map, e.g., to only the values 1, 1/2, or 0. More generally, these function may also map into $\mathbb{R}$.
- Every *inner vertex* (i.e., every vertex that is not a leaf) represents the game situation reached by following the course of the game along the path from the root to this vertex. Every inner vertex is a *decision vertex*, labeled by the player whose turn it is to draw in this situation. In every game situation, all players know the complete course of the game so far, from the root up to the current situation. The children of each inner vertex represent the subsequent situations resulting from the move chosen

---
[15] Although in undirected trees also "leafs" might be distinguished as the root in principle, for game trees it is reasonable to require that the root be an inner vertex, thus representing the start and not a final situation. This implies that the root has at least two children, so the starting player has a choice between at least two moves. This is not a restriction, since if this player had only one possible starting move, one could omit this move and declare the situation resulting from it to be the starting situation instead.

by the current player; each edge in the game tree is thus labeled by the chosen strategy.

The game trees of finite games are finite and, thus, subject to computer-aided analysis. For example, using a computer program, Allis [13] has shown that there is a winning strategy for the starting player in Go-Moku (Go-Moku, a.k.a. *five in a row*, is a version of Tic-Tac-Toe played on a larger board and where the players need five of their pieces in a row for a victory; see the next section for a discussion of Tic-Tac-Toe). Due to great progress in algorithmic analysis of game trees, currently computers can play many games as well, or even better, than most humans. For example, currently even programs running on smartphones can reach the level of a chess grandmaster. On the other hand, there still are games, such as go, where human masters beat even the best computer programs. Human players can oppose the machine's enormous computing power by their experience, cleverness, and intuition, and this often is enough to be successful. To make the next move in a reasonable amount of time, a player—human or machine—can always afford only a limited foresightful horizon of a few moves.

As a first example of a two-player sequential game, let us have a look at a popular, but very simple combinatorial two-player game with perfect information that can be played using pencil and paper, where initially the "playing board" is drawn onto the sheet of paper.

2.3.1.2 Tic-Tac-Toe

There are (at least) two variants of this game: Tic-Tac-Toe can be played on a (3×3)-board or on a (4×4)-board. Initially, all nine (respectively, 16) fields of the board are empty. The two players, X and O, take turns in entering their symbol into one as yet empty field of the board. (In a different variant of the game that we will not go into here, the board stands upright and the gaming pieces can be tossed in from above only.) The goal of the game is to place one's symbol (X or O) in the three (respectively, four) board fields of one row, or column, or diagonal. If none of the two players succeeds, the game ends in a draw as soon as all fields of the board are filled. Figure 2.12 shows a course of the game of Tic-Tac-Toe that X wins, where not all 16 moves of the game are displayed.

Figure 2.13 shows some part of the game tree for Tic-Tac-Toe, which even for this simple game is rather large. The first two levels of this tree are displayed, i.e., the root and its children, and in addition three leaves with their respective predecessors. Each vertex is labeled with the player whose turn it is now, and for the sake of clarity, only some of the edges are labeled with the corresponding strategy: (i,j) for $1 \leq i,j \leq 3$ means that this player puts her symbol in this move in the ith column and the jth row.

One can figure out that there are a total of 765 different game situations reachable from the start situation in Tic-Tac-Toe (this is the size of the state

2.3 Checkmate: Trees for Games with Perfect Information   87

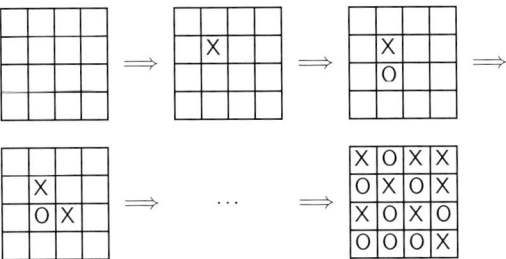

Fig. 2.12 A course of the game in Tic-Tac-Toe

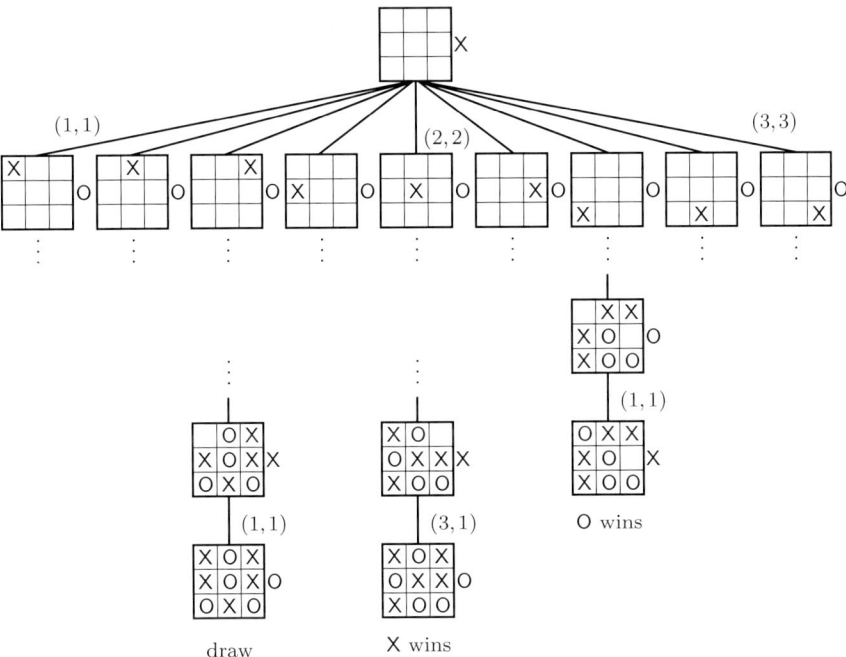

Fig. 2.13 A part of the game tree for Tic-Tac-Toe

space), and modulo rotations and mirrorings of the board there are 26,830 different courses of the game possible (this is the number of leaves in the game tree). That the number of possible courses of the game is so much larger than the state space is due to the fact that one and the same game situation can occur in many matches. For example, the final situation ▦ from Figure 2.13, in which X wins, can result from ▦ (as in this figure) and as well from ▦ or ▦.

It is easy to see that if one is focused when playing Tic-Tac-Toe and does not make a mistake, a draw is always possible.

2.3.1.3 Nim

Another game for which one can easily find a winning strategy is Nim. This game is so old that it is not clear today who invented it and when and where it has been played for the first time. It is said that the roots of this game date back to ancient China; in Europe it was mentioned for the first time in the 16th century. The name of Nim is due to Bouton [175], a US-American mathematician from Harvard University, who analyzed it mathematically. Even though the actual origin of the name of Nim is still unsettled, it is assumed (as Bouton did his PhD in Leipzig, Germany) that it may come from the German verb *"nimm"* (to take), which in the form *"nim"* is also an obsolete English verb of the same meaning. It is also noteworthy that if one turns **NIM** upside down, one obtains **WIN**.

Nim is played by two players, who take turns in removing things from different heaps. In each move, they have to remove at least one object, and they may take only a certain number of objects, which all must be from the same heap. For simplicity, we assume that David and Edgar play Nim with only one heap, say with a heap of 16 balls. In our variant of the game the rules are specified so that each of the two players in one move must take at least one ball but may take also two or three balls. Thus each of the players has a choice between three strategies in each move. The player taking the last ball loses, and the other player wins.

David and Edgar play twice. David starts in their first match, and Edgar in their second. How many balls should they take when it is their turn? Table 2.10(a) shows the course of the first game and Table 2.10(b) shows the course of the second.

Table 2.10 Courses of two Nim games

(a) David starts and Edgar wins

Number of balls	16	15	13	10	9	7	5	4	1	0
David takes		1		3		2		1	1	
Edgar takes			2		1		2			3 win

(b) Edgar starts and wins

Number of balls	16	13	11	9	6	5	3	1	0
David takes			2		3		2	1	
Edgar takes	3		2			1		2	win

2.3 Checkmate: Trees for Games with Perfect Information

Edgar has won twice. What did he do better than David? Is there a winning strategy that leads to success with certainty? To explore this question, it is useful to have a look at the end of the game and then to unroll the game from the back end forward.

In both games, when David made the last move, only one ball was left. By the rules of the game, this move must lead to defeat necessarily. David's bad luck was that he had to move at this point, when only one ball was left. To win, one thus has to try to push one's opponent into this position. Edgar succeeded in doing so, because there were still four balls in the second-to-last move of the game in Table 2.10(a) and there were still three balls in the second-to-last move of the game in Table 2.10(b). He would also have won with two remaining balls in this move, since by the rules he may take one, two, or three balls. That is, a player has a winning strategy that leaves the opponent no chance whenever there are two, three, or four balls left.

But what when there are still five balls? By the rules one has to take at least one and no more than three balls. Hence, with five balls you are in a situation where you have no choice but to allow your opponent to come into the winning situation described above that forces your defeat when the opponent plays well. It follows that the situation to have to move with five balls left is as bad as to have to move with only one ball left. However, if there are still six, seven, or eight balls, then you can force your opponent to be in the bad situation with five balls left. This argumentation chain goes on like this. In Table 2.11, a • indicates for which number of balls a player has a winning strategy when it is her turn to move.

Table 2.11 When does there exist a winning strategy for Nim?

Number of balls	1	2	3	4	5	6	7	8	9	10	11	12	13	14	15	16
Winning strategy		•	•	•		•	•	•		•	•	•		•	•	•

Of course, this is also the case when one starts with more than 16 balls. Depending on which player makes the first move and how many balls there are initially, there exists a winning strategy for exactly one of the players. In the example shown in Table 2.10(a), David, who was allowed to make the first move, actually would have had a winning strategy to enforce his victory, but already in his first move he screwed it up and gave his advantage away to Edgar. Edgar, however, did a better job in the match of Table 2.10(b) and converted the advantage of his winning strategy into victory in cold blood. This example also shows that it may matter which player is going to move first and that it is not always an advantage to be the first player to move, for example not if one starts with a heap of 13 balls and thus, up to the very end, can never come into a winning position against an unrelentingly playing opponent.

Another example that it can be disadvantageous to move first is a "sequentialized" variant of the penalty game. If the goalkeeper would move first, for example by jumping to the left, then the kicker would be able to simply shoot to the other side of the goal and win, and conversely the goalkeeper would make a save and win with certainty if he knew where the kicker would shoot. Of course, this sequentialized variant of the penalty game would never be played, as it is trivial and thus boring.

The algorithm implementing the above-described winning strategy is a very simple application of the principle of *dynamic programming*. This principle is attributed to the US-American mathematician Richard Bellman, who applied it to optimization problems in control theory in the 1940s, and later also to optimization problems in game theory. Bellman is said to be the father of dynamic programming, even though this principle itself was common even earlier in physics under a different name. Dynamic programming can be used to solve optimization problems when their instances can be divided into smaller optimization problem instances of the same kind such that solutions to these smaller problem instances recursively yield a solution to the original problem instance. For example, the question of whether a Nim player has a winning strategy for n balls can be reduced to the question of whether she has a winning strategy for $n-1$, $n-2$, and $n-3$ balls. Eventually, one has divided the original problem instance into smaller instances that are solvable immediately; in our example, it would be enough to find a solution for one, two, and three balls. Now, one obtains a solution to the original problem instance by going backward from one recursion level to the next, each time using the solutions to the smaller problem instances to solve the larger problem instance.

This algorithm based on dynamic programming for the winning strategy of Nim runs in pseudo-polynomial time, i.e., the time this algorithm needs to find the solution is proportional to the number n of balls in the given problem instance. Since it is common to represent numbers in binary (not in unary) and since the size of the binary representation of a number n is logarithmic in n, such a pseudo-polynomial-time algorithm is actually not efficient. It would be better if the problem could be solved in proper polynomial time, i.e., in a time proportional to the size of the representation of the number n of given balls, which is $\log n$. The algorithm described above and a detailed analysis of its run time can be found in the survey by Könemann [619], for example. However, it is actually not necessary to go through the full recursion depth, which makes the algorithm's running time so bad. As one sees in Table 2.11, it would be enough to determine for the given number n of balls whether it is in the remainder class 1 mod 4, that is, in the set $\{1, 5, 9, \ldots\}$, and this can be done efficiently indeed. A player has a winning strategy if and only if the answer to this question is "no" and it is this player's turn to move. If one plays Nim not in the simplified variant presented here but with several heaps and other, more complicated rules for removing objects, the solution is less trivial.

2.3.1.4 Geography and the Hardness of Finding Winning Strategies

In the two games we have seen so far, the problem of deciding if a given player has a winning strategy was, in one way or another, computationally easy. Specifically, for Tic-Tac-Toe the whole game tree was small enough for an algorithm to simply consider every possible play. For Nim, we have found an efficient algorithm that exploited the structure of the game. Indeed, through our dynamic programming algorithm we have shown that it is possible to prune the Nim game tree down to a single (though, possibly exponentially long) path. Then we have shown how to exploit the properties of this pruned tree to obtain a polynomial-time algorithm.

However, intuitively, there is a large qualitative difference between the elegant algorithm for Nim and the brute-force game-tree search for Tic-Tac-Toe. In the former case we use clever mathematical analysis to derive an algorithm that exploits the properties of the game, wheras in the latter case the only information that we need about the game is the game tree, with leaves labeled as either the X player winning, or the O player winning, or a tie. The only reason why we may possibly consider the brute-force algorithm for Tic-Tac-Toe as an efficient one is that the game tree is finite and, technically speaking, the search algorithm requires only a constant amount of time to analyze it. For this reason, when considering the computational complexity of games, we usually cannot directly study their standard versions that we play in real life, but rather we have to introduce their generalizations. For example, we may have to allow arbitrary board sizes and arbitrary starting positions (for Tic-Tac-Toe this would mean allowing boards of arbitrary size with some symbols already placed; intuitively, the symbols on the board would correspond to a situation in the middle of the game). Sometimes we might even have to modify the rules of the game, for example, to account for the larger boards. We will illustrate the process of generalizing a game and of proving the hardness of finding its winning strategies through the children's game of geography.

The "real-world" game of geography is typically played by two children, say David and Edgar, who want to see which one of them knows world's geography better. The boys agree on a particular group of geographical names (for example, names of cities, names of countries, etc.), and then one of them gives the first name. From then on, the boys alternate in giving names that must satisfy the following requirement: A name given in a particular round must start with the same letter as the final letter of the name from the previous round. Whoever ends up in a position where he cannot produce a name satisfying this condition (and that was not used before) loses. The boys have just started, so let us have a look at their play.

> Due to certain negotiations comprehensible only to two little boys (and, fortunately, not involving any physical violence), David and Edgar have agreed that they would use country names and that Edgar would give the initial name.
>
> After a moment's thought, Edgar looks at David in a menacing way and bursts out with "GERMANY," hoping that there are no countries starting with "Y."
>
> David is hestitating for a moment, but all of a sudden his eyes brighten and he shouts, "YEMEN!," the only possible response to this challenge.
>
> However, Edgar only smiles and counters, "NORWAY!"
>
> David's outrageous answer (omitted here for the sake of grace) signifies his deep dissatisfaction with Edgar's seemingly low moral standards and the fact that he allowed Edgar to start in the first place, but eventually he is forced to stop swearing and to accept defeat.

The above example, which is depicted in Figure 2.14, shows an easy winning strategy for the game of geography based on country names. There is only one country starting with "Y" (YEMEN) and so whoever begins should simply give a name that ends with "Y" (but not NORWAY, since this is the only country that starts with "N" and ends with "Y"), wait for the only possible response (or, celebrate victory already if there is no response at all) and, then, riposte "NORWAY." Thus, in this basic form, the game is very easy. However, the reason for this is the particular form of the dictionary of English country names. But what if we used a different dictionary?

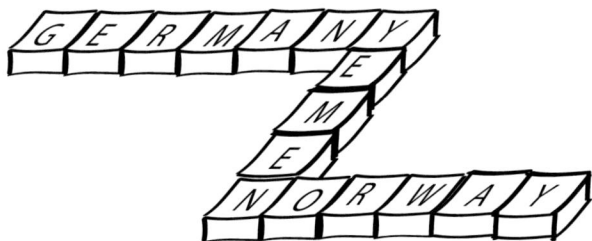

Fig. 2.14 The game of geography based on country names from the English dictionary

To analyze the game of geography with arbitrary dictionaries, it is most convenient to represent it as a directed graph with a designated starting vertex s. Now, the game proceeds as follows:

1. Initially, there is gaming piece on vertex s.
2. In the following rounds, the two players alternately move the piece using the following rules:

2.3 Checkmate: Trees for Games with Perfect Information

  a. The piece can only move along the directed edges, in the direction of the edge, and
  b. the piece can never move to a vertex on which it stood previously.

3. The player who cannot make a legal move loses, and the other one wins.

We can encode the standard version of the game, based on country names, using this graph approach as follows: The vertices in the graph are the names of the countries and the special starting vertex s; we have directed edges from s to every other vertex and we have an edge directed from country A to country B if and only if the last letter of the name of A is the same as the first letter of the name of B.

Now, we can represent the task of finding a winning strategy for this generalized game of geography as the following decision problem.

GEOGRAPHY

Given: A directed graph $G = (V, E)$ with a distinguished vertex s.

Question: Is there a winning strategy for the first player in the game of geography based on G with starting point s?

From the point of view of computational complexity theory, GEOGRAPHY is a very difficult problem. The main reason for this difficulty comes from the fact that to decide if the first player has a winning strategy, we have to consider a statement of the following form:

Does there *exist* a move for the first player such that *for all* possible moves of the second player, there *exists* a move for the first player such that *for all* possible moves of the second player ..., it holds that the second player cannot make a legal move.

In effect, given a propositional formula F over boolean variables $x_1, \ldots, x_n$ (recall these notions from Chapter 1), it is possible to form an instance (G, s) of GEOGRAPHY where the first player has a winning strategy if and only if the following quantified boolean formula is true:

$$H = (\exists x_1)(\forall x_2)(\exists x_3) \cdots (Q_n x_n)[F(x_1, \ldots, x_n)], \qquad (2.13)$$

where Q_n stands for $\exists$ if n is odd, and for $\forall$ if n is even, and F is a boolean formula without quantifiers. (We will soon show why this equivalence holds.)

However, the problem of deciding if such a formula is true, known as QBF (the satisfiability problem for quantified boolean formulas; again, recall the definition of this problem from Chapter 1 on page 35), is complete for the complexity class PSPACE, the class of problems that can be solved on a Turing machine using a polynomial amount of space. In other words, the PSPACE-complete problem QBF reduces to GEOGRAPHY. By an argument analogous to Lemma 1.1 on page 30, this means that GEOGRAPHY is itself PSPACE-hard and, since a careful reader can easily verify that GEOGRAPHY also belongs to PSPACE,[16] we have that GEOGRAPHY is PSPACE-complete. Indeed, PSPACE is a natural "habitat" for the decision problems of determining whether there exist winning strategies for interesting generalizations of sequential games. For example, determining whether there exists a winning strategy starting from a given position for a generalization of Tic-Tac-Toe (and, in effect, of Go-Moku) is PSPACE-complete as well [799].

We will now show that we indeed can transform instances of QBF into instances of GEOGRAPHY in polynomial time such that these instances are equivalent in terms of membership in their problem (i.e., they both are either yes-instances or no-instances), thus providing a reduction from QBF to GEOGRAPHY. We will do so through an example, based on the following quantified boolean formula:

$$H = (\exists x_1)(\forall x_2)(\exists x_3)(\forall x_4)\left[(\neg x_1 \vee x_2) \wedge (x_2 \vee x_3 \vee \neg x_4) \wedge (x_1 \vee \neg x_3)\right],$$

i.e., the boolean formula F without quantifiers from (2.13) here takes the form $F(x_1,\ldots,x_4) = (\neg x_1 \vee x_2) \wedge (x_2 \vee x_3 \vee \neg x_4) \wedge (x_1 \vee \neg x_3)$. This formula is in the so-called conjunctive normal form (CNF; again, recall the definition from Chapter 1), which means that it is a conjunction of clauses, where each clause is a disjunction of variables and their negations. In particular, F has three clauses, $C_1 = (\neg x_1 \vee x_2)$, $C_2 = (x_2 \vee x_3 \vee \neg x_4)$, and $C_3 = (x_1 \vee \neg x_3)$. Our construction will depend on the given formula being in conjunctive normal form, but since QBF is PSPACE-complete even in this restricted case, our argument will still be valid. Following our example, it should be clear how to encode any other instance of QBF as an instance of GEOGRAPHY (provided the input formula is in conjunctive normal form).

Figure 2.15 presents the graph G for the game of geography such that the first player has a winning strategy if and only if the quantified boolean formula H is true. Let us explain the idea behind this graph. Our game will proceed in two rounds. In the first round, the *value-selection* round, each of the two players (to whom we will now refer as the ∃-player and the ∀-player) picks the values for their variables in the following way:

1. The ∃-player starts and picks the value for variable x_1. To do so, she moves the piece from the starting vertex $s = x_1$ either to the vertex $x_1 = \mathsf{T}$ or to the vertex $x_1 = \mathsf{F}$. However, there is a trick here. If she moves the piece to $x_1 = \mathsf{T}$, this means that, in fact, she wants x_1 to have the value *false*. If she moves the piece to $x_1 = \mathsf{F}$, then she intends x_1 to have the value *true*. The reason for this strange behavior is that each player wants to keep the vertices that represent the values of "their" variables available for the *challenge* round (to be described soon).

---

[16] To see the containment of GEOGRAPHY in PSPACE, briefly put, it is possible to consider the whole game tree while looking at only polynomially many vertices at a time; we point the reader to the next section, to the discussion of the complexity of backward induction, for more details.

2.3 Checkmate: Trees for Games with Perfect Information

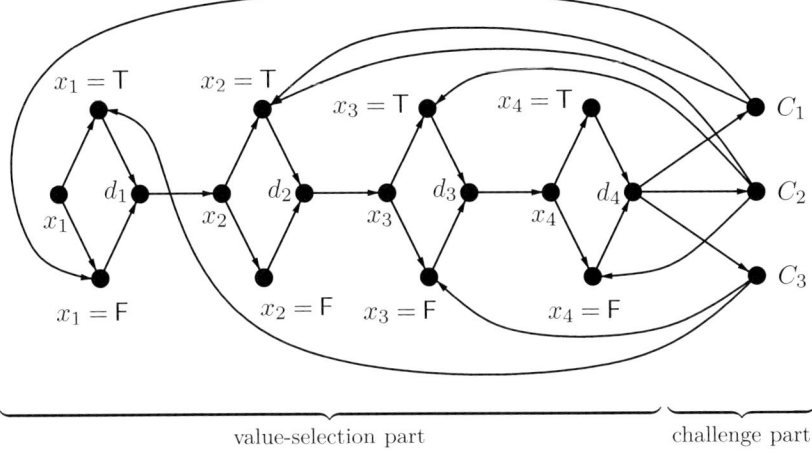

Fig. 2.15 The graph G used to simulate formula H as a game of geography

2. After the $\exists$-player has made her move, the next two moves are completely forced:

    a. First, the $\forall$-player has to move the piece to vertex d_1, and
    b. then the $\exists$-player has to move the piece to vertex x_2.

3. Then, the players are in a situation analogous to the one in the beginning of the game, but now it's the $\forall$-player's turn to pick the value of x_2.
4. Then, again, there are two forced moves,
5. the $\exists$-player picks the value of x_3, and,
6. after the two forced moves,
7. the $\forall$-player picks the value of x_4.
8. Finally, after a single forced move, we end up in vertex d_4 where it is the $\forall$-player's turn to move.

Note that, by this time, the $\exists$-player picked the values of x_1 and x_3 (the existentially quantified variables), and the $\forall$-player picked the values of x_2 and x_4 (the universally quantified variables).

At this point, the *challenge* round starts. It consists of only two moves. The gaming piece is standing on vertex d_4 and it is the $\forall$-player's turn to move. He can move the piece to one of the vertices C_1, C_2, and C_3 (which correspond to the clauses in the formula) in order to challenge the $\exists$-player. Intuitively, the $\forall$-player wants to prove that in the preceding (value-selection) round the two players collectively chose values for the variables $x_1, \ldots, x_4$ that make the quantified boolean formula H evaluate to *false*. Therefore, he wants to pick a clause that witnesses this fact (that is, a clause for which all the unnegated variables have the value *false* and for which all the negated variables have the value *true*). Let us say that the $\forall$-player has picked clause C_1. Now, the $\exists$-player has a chance to counter the $\forall$-player's challenge by showing that, after all, the suggested clause in fact is true. So, if at this point the piece is on C_1 then the ∃-player has three possibilities:

1. If in the value-selection round she did not put the piece on the vertex $x_1 = \mathsf{F}$ (which means she selected the value *false* for x_1; hence, $C_1 = (\neg x_1 \vee x_2)$ is *true*), then she can now move the piece there and force the ∀-player to lose.
2. If in the value-selection round the ∀-player did not place the piece on the vertex $x_2 = \mathsf{T}$ (which means that the ∀-player set the value of x_2 to *true*, making C_1 *true* as well), then she can move the piece to $x_2 = \mathsf{T}$ and force the ∀-player to lose.
3. If neither of the previous two options are available, then the ∃-player cannot make a move and loses.

By this description, we see that the ∃-player has a winning strategy if and only if there is a value of x_1 such that, for every value of x_2, there exists a value of x_3 such that, for every value of x_4 and for each clause C in $\{C_1, C_2, C_3\}$, clause C evaluates to *true*. However, it is clear that this is true if and only if the quantified boolean formula H is true.

The game of geography captures the essential hardness of determining whether there exist winning strategies for a designated player in two-player sequential games, that is, the necessity to evaluate propositional formulas with alternating ∃ and ∀ quantifiers. Many other generalizations of well-known games have the same property. For example, appropriately defined problems, corresponding to the existence of winning strategies, are PSPACE-complete for go [655], chess [893], go-moku [799], and many other games. Interestingly, solving Sokoban puzzles, a popular single-player game, is PSPACE-complete as well [352, 318].

PSPACE is a class believed to be significantly harder than NP. This reflects the fact that a winning strategy in a game is a much more complicated mathematical object than, for example, a satisfying assignment of truth values to propositional variables. Thus, finding a winning strategy for one player in a two-player sequential game (and, for that matter, even finding out if one exists) is much harder than, for example, finding a satisfying assignment to the variables of a given boolean formula or, more generally, an accepting computation of a nondeterministic polynomial-time Turing machine (and even finding out if one exists). It is quite easy to convince someone that a given propositional formula F over variables $x_1, \ldots, x_n$ is true by simply showing the particular values of the variables witnessing this fact. For example, setting all variables to *true* in our formula $F(x_1, \ldots, x_4) = (\neg x_1 \vee x_2) \wedge (x_2 \vee x_3 \vee \neg x_4) \wedge (x_1 \vee \neg x_3)$ witnesses its satisfiability. However, it seems much more difficult to convince the same person that the quantified boolean formula $H = (\exists x_1)(\forall x_2) \cdots (Q_n x_n)[F(x_1, \ldots, x_n)]$ is true, where Q_n stands for ∃ if n is odd, and for ∀ if n is even.

2.3 Checkmate: Trees for Games with Perfect Information

2.3.2 Equilibria in Game Trees

The gains of the players in a game with perfect information represented by a game tree depend on which leaf the game ends in, i.e., it depends on the course of the game. The notion of equilibrium can also be applied to game trees, to make a prediction about the course of the game, i.e., about which strategy will most likely be chosen by the corresponding (rational) player in each vertex on the path from the root of the tree to the leaf where it ends.

Applying, for example, the notion of Nash equilibrium in pure strategies (see Definition 2.4 on page 50) to game trees, one requires for each inner vertex of the tree that the corresponding player chooses a "best response strategy" to the other players' strategies. In the Nim game (see page 88), for example, such a best response strategy consists in pushing the opponent—if at all possible—into a position from which she cannot win, that is, one tries to take as many balls as needed to ensure that the number of remaining balls is in the remainder class 1 mod 4. However, this best response is possible only if one oneself is in a position that allows to make such a move, i.e., in a position marked by • in Table 2.11 on page 89. In the two courses of the game given in Tables 2.10(a) and 2.10(b) on page 88, Edgar was for each of his moves in such a lucky position and so was able to counter each of David's moves of taking, say, i balls by the best response strategy of taking $4-i$ balls to remain in a winning position. Do you see why this expression, $4-i$, does not apply to Edgar's first move in the match of Table 2.10(a)?

However, Nash equilibria do not always make reasonable predictions about the course of a sequential game in a game tree. In particular, Nash equilibria can contain threats that become implausible when players move in sequence rather than simultaneously. For example, we have seen that the chicken game in normal form (see Table 2.4 on page 55) possesses two Nash equilibria in pure strategies: (Drive on, Swerve) with the gain vector $(3,1)$ and (Swerve, Drive on) with the gain vector $(1,3)$. However, sequentializing this game by letting David move (i.e., drive) first, his threat to keep on driving undauntedly by death will no longer intimidate Edgar. Since he is going to move (or drive) second, he can calmly wait and see what David really does in his first move and can then react optimally.

2.3.2.1 Edgar's Sequential Campaign Game

Sure enough, one would never play the chicken game in sequential form, since it then—similarly to the sequentialized penalty game mentioned earlier—would be trivial. Therefore, consider yet another example of a game, played by Anna and Belle who are among the best students in their school.

To supplement their small pocket money, Anna and Belle usually give less talented students extra lessons. There are ten such students whose parents are willing to pay for private tutoring, and currently seven of them are tutored by Anna and the remaining three by Belle. As she isn't quite happy with this tough competition, Belle complains to her little brother.

"Cheer up, sis!" Edgar replies. "I might be able to help you. That's what brothers are there for, aren't they?"

"What would you do?"

"OK, it's easy! What you need is someone to help you enhance your reputation. It's like when you are selling a product and want to boost up the sales, you may want to run an ad on TV. Of course, I won't go on air with a spot praising your teaching abilities, but I will go around in the neighborhood and in school and will tell everybody that I owe my smartness and wisdom all to you and your tutoring."

"Thank you so much, Edgar, that is so generous of you!"

"Well, actually, y'know, it is a lot of work and time, and you'll make more money then. So I wouldn't offer this for free, it would cost you ... what you earn by teaching four students."

"But I currently have only three!"

"Yes, I know, but it will be eight after my campaign, I guarantee you that! So you will still make more money than now. Plus, you'll have the joy of knowing that there are only two students left for Anna. So, what do you think? Do you want me to run a campaign for you?"

"How much time do I have to make this decision?"

"Let's say one week from now."

The next day, Edgar visits Anna. "Have you heard the news?" he asks. "Big sis is thinking to hire me as her campaign manager. She wants to give extra lessons to more students." And he tells her everything.

"No kidding!" Anna is nonplussed. "What can I do about it?"

"OK, it's easy! What you need is someone to help you enhance your reputation. It's like when you are selling a product and want to boost up the sales, you may want to run an ad on TV. Of course, I won't go on air with a spot praising your teaching abilities, but I will go around in the neighborhood and in school and will tell everybody that I owe my smartness and wisdom all to you and your tutoring."

"Thank you so much, Edgar, that is so generous of you!"

"Well, actually, y'know, it is a lot of work and time. So I wouldn't offer this for free, it would cost you ... what you earn by teaching four students."

"But I currently have seven! This leaves only three students' payments for me."

2.3 Checkmate: Trees for Games with Perfect Information

"Yes, I know, but if I run this campaign only for you (Belle is not sure about it yet), you will have eight students afterwards, I guarantee you that! So you may make a bit less money than now, but you'll have the joy of knowing that there are only two students left for Belle. On the other hand, if Belle wants me to run the campaign for her and you do not, *she* will have eight students afterwards and you'll drop down to two! Keep that in mind!"

"And what if both I and Belle hire you to run a campaign?"

"I guess you'd then split the market equally: five for her, five for you. So, what do you think? Do you want me to run a campaign for you?"

"How much time do I have to make this decision? Can I wait for Belle to decide first?"

"Of course not. I'd need you to tell me by tomorrow."

Unlike the chicken or the penalty game, this game is played sequentially: Edgar has requested that Anna makes her decision prior to Belle.[17] This gives Belle the presumed advantage to adequately react on Anna's decision. We will soon see, however, that the advantage actually lies with Anna, who moves first.

Since our goal is to show that Nash equilibria in normal form games may become implausible in the tree of the corresponding sequential game, it will be instructive to first consider the above game in *nonsequential* (i.e., in normal) form, shown in Table 2.12. In the normal form variant of this game, Anna and Belle compete for a total profit of 10 and are now considering whether to ask Edgar to run a campaign for them that would cause costs of 4 for each of them. Both must make their move simultaneously. Without Edgar's campaign, Anna currently has a profit of 7 and Belle consequently has a profit of 3. If only one of them asks Edgar for help, she can expect a "market share" of 8 (so Anna would gain another student, while Belle would even gain 5 new students to teach), although she would also have to shoulder the campaign costs. Thus we obtain a gain vector of $(4,2)$ if Anna alone asks Edgar, and the gain vector $(2,4)$ if Belle alone asks him. If they both ask him for help, they each get to tutor 5 students, and after deducting the campaign costs we have a gain vector of $(1,1)$.

The analysis of this game in normal form is simplified by the fact that Anna has a strictly dominant strategy: no campaign. Belle, however, does not have a dominant strategy, since her decision depends on the strategy chosen by Anna: If Anna does not ask Edgar to run a campaign for her, it would be beneficial for Belle to do so; otherwise, it would be better for her not

---

[17] Jackson [570, p. 311 ff] gives an example where two companies consider to run an advertisement campaign each, in order to increase their current market share. As in our example, it is quite plausible that this game can be played sequentially. Suppose, for example, the advertising agency sets a deadline for both companies by which they have to decide, and the deadline for company A is prior to that for company B.

Table 2.12 Edgar's campaign game in normal form

		Belle	
		No campaign	Campaign
Anna	No campaign	(7,3)	(**2,4**)
	Campaign	(4,2)	(1,1)

to ask him. It is easy to see that the boldfaced gain vector $(2,4)$ in Table 2.12 belongs to a profile of pure strategies in a strict Nash equilibrium: None of the girls has an incentive to deviate from her strategy according to $(2,4)$, provided that also the other one sticks to her strategy according to $(2,4)$.

Now, sequentializing this game, where Anna makes the first move, we obtain the game tree displayed in Figure 2.16. (Unlike for example in the Tic-Tac-Toe game tree, no "situations" of the game have to be represented here, which is why this game tree formally differs somewhat from that in Figure 2.13 on page 87.) The gain vectors in the leaves correspond to the choices of the two players according to Table 2.12, where the gain of Anna is the left entry and the gain of Belle is the right entry.

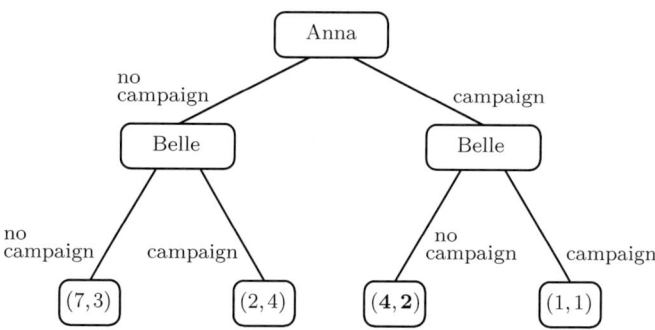

Fig. 2.16 Game tree for Edgar's campaign game in extensive form

2.3.2.2 Nash Equilibria in Edgar's Sequential Campaign Game

There are two Nash equilibria in pure strategies for the game tree in Figure 2.16. The first one is that occurring also in the normal form game: If Anna chooses to not ask Edgar to run a campaign for her, the best response strategy of Belle is to do so, and we end up in the leaf labeled by $(2,4)$. The second Nash equilibrium is now added: If Anna chooses to ask Edgar, the best response strategy of Belle is to not ask him, and we end up in the leaf labeled by $(4,2)$.

2.3 Checkmate: Trees for Games with Perfect Information

However, the first Nash equilibrium, which leads to the leaf labeled by (2,4) and which corresponds to the strict Nash equilibrium in the normal form game, is *implausible* for the game in extensive form, since Anna knows the strategies of Belle in the perfect information model and thus can predict that choosing not to run a campaign would lower her own profit from 4 to 2, because Belle(moving second) would go with running a campaign of her own to increase her profit from 3 to 4. Therefore, Anna would have an incentive to deviate from choosing to not run a campaign—unlike in the normal form game where both players move simultaneously, no campaign would no longer be a dominant strategy for Anna. In other words, since the game is played sequentially, Anna can take advantage of moving first, based on the following strategic considerations. Starting from the gains of both players at the leaves of the game tree, she can go backward through the tree and check which decisions in the vertices above the leaf level are made, and then iterate this process. This method is known as *backward induction* and works in this example as follows:

1. First, consider the left subtree of the game tree with leaves (7,3) and (2,4). Since Belle is to make a decision in the root of this subtree, Anna can predict that Belle will choose to ask Edgar to run a campaign for her, since Belle's profit is larger in the gain vector (2,4) than it is in (7,3).
2. Consider now the right subtree of the game tree with leaves (4,2) and (1,1). Again, Belle is to make her decision in the root of this subtree, so Anna can predict that Belle will here choose to not run a campaign, since Belle's profit is larger in the gain vector (4,2) than it is in (1,1).
3. From these considerations and predictions Anna concludes that she can maximize her own profit by making the right choice between the two subtrees, namely by choosing the right subtree (the campaign), since her own profit is larger in the gain vector (4,2) than it is in (2,4).

The result of this backward induction thus is that Anna runs a campaign, but Belle does not, and their gains are (4,2). Interestingly, this result differs from the Nash equilibrium in pure strategies for the corresponding game in normal form. Also, it is somewhat disappointing for both girls that their gains were higher *before* the game than *after* it: Had they never cared about a campaign, they'd be better off now. The actual winner of this game is Edgar, who was not even a player in it.

2.3.2.3 Subgame-Perfect Equilibria

Selten [852] developed an appropriate solution concept that formally expresses the idea of backward induction and is known as *subgame-perfect equilibrium*. A subgame of a finite game with tree T is represented by a subtree T' of T whose root is an arbitrary decision vertex in the original game tree T. For simplicity, we identify (sub)games with the (sub)trees representing them.

A subgame-perfect equilibrium in a finite game tree T is given by the respective strategies being in equilibrium in every subtree T' of T (even if only one more move is to be made in T'). That is, no matter from which vertex in T we start, the strategy chosen by the corresponding player in this vertex is always a best response to the other strategies in the equilibrium profile.

Subgame-perfect equilibria refine the notion of Nash equilibrium in pure strategies for game trees, since every subgame-perfect equilibrium is such a Nash equilibrium as well. The converse does not hold in general, as we have seen in the example of the advertisement campaign game: One of the two Nash equilibria—$(4, 2)$—is also a subgame-perfect equilibrium, whereas the other—$(2, 4)$—is not. The purpose of subgame-perfect equilibria is just to exclude implausible Nash equilibria in game trees.

Unfortunately, in general, finding subgame-perfect Nash equilibria may be quite difficult from a computational point of view: This notion generalizes that of a winning strategy in a two-player sequential game and, thus, can be PSPACE-hard. However, if our game satisfies several simple conditions, then backward induction can be implemented in PSPACE. These simple conditions, in essence, boil down to requiring that:

- the evaluation of the game needs at most polynomial space (that is, the depth of the game tree is polynomially bounded),
- the description of each game situation is of at most polynomial size, and
- given a game situation, there are at most polynomially many possible moves (and we can compute all these moves in polynomial space).

If a game satisfies these conditions, then we can implement backward induction as a simple depth-first search (DFS) of the game tree. The key observation here is that we never have to keep the whole game tree in memory. Instead, it suffices to only keep the path from the root to the currently analyzed game situation. More formally, it suffices to keep in memory each vertex on the path from the currently analyzed game situation to the root and, for each game situation on this path, which moves from this game situation were already considered and what the so-far computed outcome of the subgame starting from this game situation is. We depict this process in Figure 2.17. We encourage the reader to formally describe the depth-first search algorithm for backward induction and to give a formal proof of its time and space complexity.

The notion of subgame-perfect Nash equilibrium can also be carried over to more general games in extensive form, such as games with imperfect information. In such games, one in addition defines the notion of *information set* for each player as the subset of decision vertices in the game tree such that every vertex in this subset belongs to some player and when the game reaches this subset, this player is not able to differentiate between the vertices in this subset (i.e., when making her move, she may not know exactly which vertex in the information set the game has reached).[18] In the definition of subgame-

---

[18] In games with perfect information, all information sets consist of a single vertex only.

2.4 Full House: Games with Incomplete Information 103

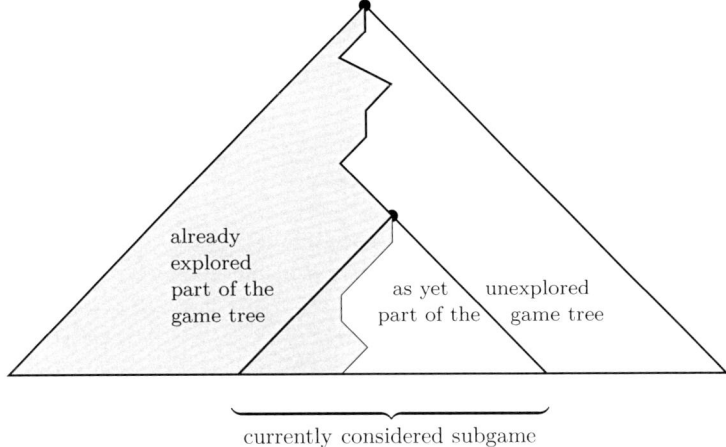

Fig. 2.17 Implementing backward induction through DFS

perfect equilibrium for games with imperfect information, it is additionally required that in its subgames the corresponding information sets must not be separated.

Many other notions of equilibrium have been considered in the literature. For instance, building on previous work of Piccione and Rubinstein [779] and Battigalli [105], Marple and Shoham [685] generalize the traditional equilibrium concepts for finite games in extensive form so as to apply also to more general classes of games, such as games with imperfect recall.

2.4 Full House: Games with Incomplete Information

As opposed to games with perfect information[19] (in which every player has always complete knowledge of the course of the game thus far and in particular about her current position in the game tree), information is withheld from the players in *games with incomplete information*. In many card games, for example, the players don't let other players have a look at their hands. In particular, in such games players typically do not know where exactly in the game tree they are in a certain stage of the game, and thus they may be uncertain about which strategy to choose in the current game situation. As mentioned above, this can be formalized by the notion of information set for all players. However, we will continue to refrain from excessive formalism and will rather illustrate central concepts using suitably chosen examples. To

---

[19] We will elaborate on the distinction between games with perfect information and games with complete information later on in Section 2.4.2.

this end, we will introduce the Monty Hall problem in this section and will also analyze a simplified variant of poker.

2.4.1 The Monty Hall Problem

A delightful game that has been argued about very controversially and that alone provides enough material to fill an entire book (by von Randow [798], available only in German) is the *Monty Hall problem*, also known as the *Monty Hall dilemma*. Monty Hall was the host of the US-American TV game show *"Let's make a deal,"* which made the Monty Hall problem so popular in the United States. Hall hosted this show from 1963 through 1986 and later again from 1990 through 1991. Adaptations of the show have been broadcast in other countries as well, for example in Germany (there known under the title *"Geh aufs Ganze!"*).

(a) Choosing a door.   (b) Changing one's choice or not?

Fig. 2.18 The Monty Hall problem

A player is shown three closed doors behind which there are prizes. Behind one door there is an expensive car, and behind each of the other two doors there is a goat. Sure enough, the player wants to get the car and not a bleating goat. The player is now asked to choose one of the doors, see Figure 2.18(a). Then the host opens one of the two other doors, making sure that a goat is behind the opened door. In more detail, if the player has chosen the door with the car, the host opens—randomly and with the same probability—one of the two goat doors; but if the player has chosen a door with a goat, the host opens the other goat door. Now the player is offered to think her choice over: She can either stay with her first choice or change her mind, see Figure 2.18(b). The question is whether changing the choice of the door will pay off or not.

2.4 Full House: Games with Incomplete Information

The fierce, public debate about the right answer to this question was triggered by the US-American journalist Marilyn vos Savant,[20] who in 1990 presented her solution of the puzzle to the readers of her column *Ask Marilyn* in the September issue of the *Parade Magazine*. The correct solution she presented evoked an unexpectedly large echo of controversial opinions, sometimes bordering on insults.[21]

Among the about 10,000 letters sent in, there were close to 1,000 from readers with a PhD, many of them being mathematics professors, and some of their taunting comments and convincingly presented (and yet wrong) assertions, of which Footnote 21 gives a few examples, are collected in the article *"Behind Monty Hall's Doors: Puzzle, Debate and Answer?"* by John Tierney in the *New York Times* of July 21, 1991, see also [798].

2.4.1.1 Intuitive Solutions to the Monty Hall Problem

The solution to the Monty Hall problem proposed by vos Savant is as follows: It does pay off for the player to change her originally chosen door, since doing so increases her chances to win the car from one third to two thirds! Intuitively, this is because the host reveals some additional information by opening a goat door: The main prize can certainly not be behind the open door, as there a goat is bleating. In some sense, the chances of success carry over from this door to the two other doors, but surprisingly enough, not equally to both other doors—as some may have guessed (see again Footnote 21)—but exclusively to the door the player originally did not choose. That is why changing her mind will increase her winning chances.

But what is the reason for this? Roughly speaking, one explanation is that, after one door has been opened, a player who changes her originally chosen door has *learned* something, whereas a player who does not change her mind but stubbornly stays with her first decision has not learned anything. The latter player does not take advantage from the host opening one door, so it was just a waste of time for the host to do so.

Another intuitive explanation was originally given by vos Savant: Suppose there is a million of doors only one of which hides the main prize, and the player chooses the first one. The host, who knows what is behind which door

---

[20] Marilyn vos Savant established an impressive record already at the age of ten: With an IQ of 228 she is, according to the *Guinness Book of World Records*, the person with the highest intelligence quotient as yet measured.

[21] *"You blew it! [...] Please help by confessing your error and, in the future, being more careful."* (Prof. Robert Sachs, George Mason University in Fairfax, Virginia) – *"You are utterly incorrect. [...] How many irate mathematicians are needed to get you to change your mind?"* (Prof. E. Ray Bobo, Georgetown University, Washington D.C.) – *"Our math department had a good, self-righteous laugh at your expense."* (Prof. Mary Jane Still, Palm Beach Junior College, Florida) – *"Maybe women look at math problems differently than men."* (Don Edwards, Sunriver, Oregon) – *"You are the goat!"* (Glenn Calkins, Western State College, Colorado) – etc.

and who certainly does not want to give away the right door, opens all doors except the first one and that with number 777,777. A rational player would now of course switch from the first door to the one with number 777,777, would she not?

Due to the extraordinarily fiercy criticism of her argumentation (see, once more, Footnote 21), vos Savant provided yet another explanation that considers every possible case systematically (see Table 2.13). We here assume that behind the doors 1 and 2 there is a goat, and the car is behind door 3. We repeat the game six times in a thought experiment, without keeping track of the game outcomes, since otherwise we would still recall in the second run of the game behind which door we found the car in the first game. For example, we may assume that we let six different players play these six runs of the game and that the players do not exchange any information.

In the first three runs of the game, the player chooses the first, the second, and the third door subsequently, and after the host has opened one goat door, the player stays with her first choice. In the other three runs of the game, the player chooses again the first, the second, and the third door subsequently, but after the host has opened one goat door, the player now changes her mind and goes with the other closed door.

Table 2.13 All cases of the Monty Hall problem

	Door 1: Goat	Door 2: Goat	Door 3: Car	Strategy	Gain
Game 1	chosen door			stay	goat
Game 2		chosen door		stay	goat
Game 3			chosen door	stay	car
Game 4	chosen door			change	car
Game 5		chosen door		change	car
Game 6			chosen door	change	goat

As one can see, the car is won with a changing strategy in two out of three cases, but with the staying strategy only in one out of three cases. This case distinction is also represented by the tree in Figure 2.19. Every leaf of the tree represents exactly one of the game runs in Table 2.13, namely, the games 1, 4, 2, 5, 3, and 6 from left to right.

2.4.1.2 Solution of the Monty Hall Problem Using the Law of Total Probability

For those who are still not completely convinced by these more or less intuitive arguments, we now provide a formal mathematical solution to the Monty Hall problem in addition. In particular, we will use the law of total probability and the related formula of Bayes to explain why opening one door not hiding

2.4 Full House: Games with Incomplete Information

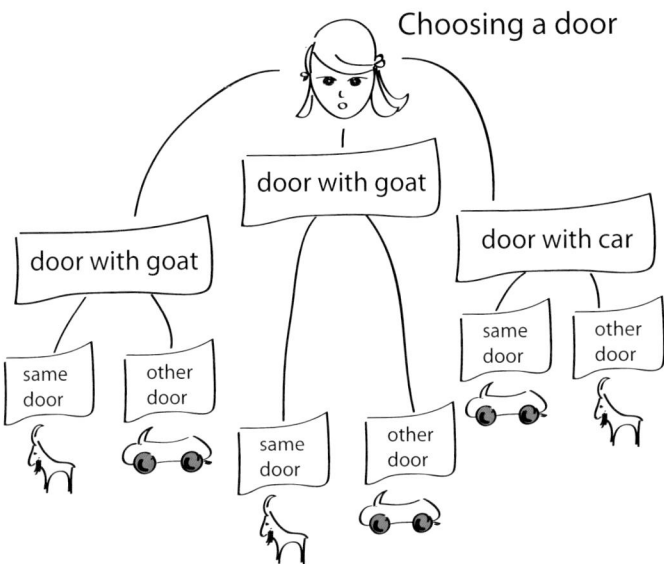

Fig. 2.19 All cases of the Monty Hall problem

the car does not convey the success chances from this door equally to both other doors, but only to the one other door that originally was not chosen by the player, i.e., we formally explain why the changing strategy makes sense.

First, to explain the formula of Bayes, we need some basic notions from probability theory. A *(finite) probability space* is given by a finite set $\mathcal{E} = \{e_1, e_2, \ldots, e_k\}$ of *atomic events*, where event e_i occurs with the probability $w_i = P(e_i)$ and we have $\sum_{i=1}^{k} w_i = 1$. Such a correspondence between probabilities w_i and atomic events e_i specifies a *probability distribution*. If all atomic events occur with the same probability ($w_i = 1/k$ for each i, $1 \leq i \leq k$), we have the *uniform distribution*.

This can be extended to every subset $E \subseteq \mathcal{E}$, which represents an *event*, as follows:

$$P(E) = \sum_{e_i \in E} w_i.$$

$P(E)$ denotes the *(total) probability* for event E to occur. For example, for the uniform distribution on $\mathcal{E}$, $P(E) = \|E\|/k$ simply gives the frequency that among all possible atomic events some event from E occurs.

For all events $A, B \subseteq \mathcal{E}$, we have the following basic properties:

1. $0 \leq P(A) \leq 1$, where $P(\emptyset) = 0$ and $P(\mathcal{E}) = 1$.
2. $P(\overline{A}) = 1 - P(A)$, where $\overline{A} = \mathcal{E} - A$ is the event complementary to A.
3. $P(A \cup B) = P(A) + P(B) - P(A \cap B)$.

Now we define the notion of conditional probability.

Definition 2.10 (conditional probability). Let A and B be events with $\mathrm{P}(B) > 0$.

1. The *conditional probability for event A to occur under the condition of event B occurring* is defined as

$$\mathrm{P}(A|B) = \frac{\mathrm{P}(A \cap B)}{\mathrm{P}(B)}.$$

2. A and B are said to be *(stochastically) independent* if

$$\mathrm{P}(A \cap B) = \mathrm{P}(A) \cdot \mathrm{P}(B).$$

For two independent events, whether or not one occurs does not depend on whether or not the other occurs: A and B are independent if and only if $\mathrm{P}(A|B) = \mathrm{P}(A)$ or $\mathrm{P}(A|B) = \mathrm{P}(A|\overline{B})$ (equivalently, $\mathrm{P}(B|A) = \mathrm{P}(B)$ or $\mathrm{P}(B|A) = \mathrm{P}(B|\overline{A})$). The formula of Bayes states how to calculate with conditional probabilities.

Theorem 2.4 (Bayes).

1. If A and B are two events with $\mathrm{P}(A) > 0$ and $\mathrm{P}(B) > 0$, then

$$\mathrm{P}(A) \cdot \mathrm{P}(B|A) = \mathrm{P}(B) \cdot \mathrm{P}(A|B).$$

2. If A and $B_1, B_2, \ldots, B_\ell$ are events with $\mathrm{P}(A) > 0$ and $\mathrm{P}(B_i) > 0$, $1 \leq i \leq \ell$, where $\bigcup_{i=1}^{\ell} B_i = \mathcal{E}$ is a partition of the finite probability space $\mathcal{E}$ in disjoint events, then for all i, $1 \leq i \leq \ell$,

$$\mathrm{P}(B_i|A) = \frac{\mathrm{P}(A|B_i) \cdot \mathrm{P}(B_i)}{\sum_{j=1}^{\ell} \mathrm{P}(A|B_j) \cdot \mathrm{P}(B_j)} = \frac{\mathrm{P}(A|B_i) \cdot \mathrm{P}(B_i)}{\mathrm{P}(A)}. \quad (2.14)$$

The first statement of Theorem 2.4 is the special case of its second statement for $\ell = 1$. The proof of this theorem follows immediately from the definition of conditional probability. The equality

$$\mathrm{P}(A) = \sum_{j=1}^{\ell} \mathrm{P}(B_j \cap A) = \sum_{j=1}^{\ell} \mathrm{P}(A|B_j) \cdot \mathrm{P}(B_j) \quad (2.15)$$

used in (2.14) is also denoted as the *law of total probability*.

Now, to apply this formula to the Monty Hall problem, we calculate the probabilities for the player to win the car (denoted by $\mathrm{P}(A)$) if she either chooses the staying or the changing strategy. Letting B_1 denote the event that she has initially chosen a goat door and letting B_2 denote the event that she has initially chosen the door hiding the car, we obtain $\mathrm{P}(B_1) = 2/3$ and $\mathrm{P}(B_2) = 1/3$.

If the player now goes with the strategy to stay with her initial choice, then $\mathrm{P}(A)$ is obviously equal to the probability that she initially has chosen

2.4 Full House: Games with Incomplete Information

the door with the car, so $P(A) = P(B_2) = 1/3$. This also follows from the law of total probability, since the conditional probabilities for winning the car under these two conditions, either B_1 or B_2, of having initially chosen either a wrong or the right door, are due to the staying strategy of the player either $P(A|B_1) = 0$ or $P(A|B_2) = 1$, which according to (2.15) implies:

$$P(A) = P(A|B_1) \cdot P(B_1) + P(A|B_2) \cdot P(B_2) = 0 \cdot \frac{2}{3} + 1 \cdot \frac{1}{3} = \frac{1}{3}.$$

That the host will give a hint in addition by opening one door is completely irrelevant for such a player who is resistant to learning and stubbornly stays with her initial choice.

If the player chooses to go with the changing strategy, however, the conditional probabilities for winning the car under these two conditions, either B_1 or B_2, amount to either $P(A|B_1) = 1$ or $P(A|B_2) = 0$. Hence, by the law of total probability according to (2.15), we now obtain that this player has a total success probability of

$$P(A) = P(A|B_1) \cdot P(B_1) + P(A|B_2) \cdot P(B_2) = 1 \cdot \frac{2}{3} + 0 \cdot \frac{1}{3} = \frac{2}{3}.$$

Alternatively to the above approach, one can also apply the formula of Bayes (i.e., the first statement of Theorem 2.4) directly to come to the same conclusion (see, e.g., [809]). Do you see how?

Actually, why are we concerned with the Monty Hall problem in this chapter about noncooperative game theory? After all, there is only *one* player here, since the role of the host is not that of an adversary, but rather that of an advisor giving hints. Is the Monty Hall problem a game at all, and if so, against whom is the player playing?

The Monty Hall problem can conceivably be seen as a game, not only because it is played in a TV game show, but most notably because there is a player who wants to maximize her gains by choosing the right strategy. One might say that this player gambles "against nature," which means she plays against chance. By *games against nature* one expresses the endeavors of human beings to defeat merciless randomness.

In some sense, one might also say that the player is playing against the host, since sometimes the host does not content with the passive role of the doorkeeper and door opener, but seeks to actively influence the strategy chosen by the player and thus the outcome of the game. For example, in the above-mentioned *New York Times* article about host Monty Hall, it is reported that he occasionally side-stepped the rules of the game:

On the first trial, the contestant picked Door 1.
"That's too bad," Mr. Hall said, opening Door 1. "You've won a goat."

> "But you didn't open another door yet or give me a chance to switch."
> "Where does it say I have to let you switch every time? I'm the master of the show. Here, try it again."

Or Hall offered money to the players to make them change their chosen strategy:

> *On the second trial, the contestant again picked Door 1. Mr. Hall opened Door 3, revealing a goat. The contestant was about to switch to Door 2 when Mr. Hall pulled out a roll of bills.*
> "You're sure you want Door No. 2?" he asked. "Before I show you what's behind that door, I will give you $3,000 in cash not to switch to it."
> "I'll switch to it."
> "Three thousand dollars," Mr. Hall repeated, shifting into his famous cadence. "Cash. Cash money. It could be a car, but it could be a goat. Four thousand."
> "I'll try the door."
> "Forty-five hundred. Forty-seven. Forty-eight. My last offer: Five thousand dollars."
> "Let's open the door."
> "You just ended up with a goat," he said, opening the door.

As one can see, games are not only subject to chance, but also to the psychology of the players involved, in particular in games with incomplete information where opponents can bluff. The game where bluffing and other psychological feints are part of the strategy repertoire is poker.

2.4.2 Analysis of a Simple Poker Variant

In a chapter about noncooperative game theory one cannot avoid elaborating on poker. Both Borel [170] and von Neumann [734], the fathers of game theory (see also [171, 736]), were fascinated by this game that motivated them to explore the laws of gambling by mathematical models.

Poker is different from all previously considered games. As in all games with incomplete information, there is uncertainty in poker, yet not only about which hands your opponents hold but also about which types of players they are. Are they cunning bluffers or wary chickens? The answer to this question determines which strategy a player should choose, and thus also her gains. Since it is common to not play a single game of poker but several in a row, one usually gets to know one's opponents and their gambling tactics better and

2.4 Full House: Games with Incomplete Information

better during a long poker night. Conversely, one also gives away information about one's own way of gambling, which might help the other players to better assess them. A nervous blinking or a twitching with the eyebrows when bluffing has caused many a financial ruin!

When studying the Monty Hall problem, we have already seen that calculating with conditional probabilities can be useful in a learning process. The formula of Bayes stated in Theorem 2.4 shows how to mathematically express learning by observing. When playing poker, we constantly observe our opponents and try to learn from their behavior. That is why poker is a so-called *Bayesian game*. In such a game, players make assumptions about the behavior of their opponents and intuitively determine the probabilities of their future actions. Do the other players favor certain moves in the game? Who of them would hazard everything with a relatively good hand, who would play it safe? Does this poker face give away any information? For which players is a bluff easy to detect?

Based on these observations, the players constantly adapt their own strategies to their opponents' strategies and their assumptions about them. To poker successfully and to maximize one's gains, the ability to well assess one's opponents is important, perhaps even more important than having a good hand. And if one is a great bluffer, it is not at all necessary to have a lucky hand. All that matters is to be able to bluff well enough. Crucially, in poker it is important what everyone knows (or believes to know) about her opponents. Similarly to the "third thought" in the guessing numbers game on page 60, one may imagine a sequence of considerations of the following kind:

- I think my opponent will bluff.
- But because she thinks that I know she will bluff, she doesn't bluff.
- But as I think that she knows this, I actually don't believe she will bluff.
- But because she in turn thinks that I think that she thinks that ...
- And so on out to infinity.

Harsanyi developed a method for how a player can deal with such an infinite chain of thoughts about her opponents. He proposes to interrupt this chain of thoughts by the assumption that all players know the roles—or types—of their opponents after a certain number of moves in the game. They are then *common knowledge*. Knowledge, belief, information—these notions are central to games such as poker. There are also relations to philosophy, epistemology, and logic, which study questions such as: What does it mean for a group of individuals (or, in the context of game theory, of players) to have common knowledge or to share generally accepted beliefs, and what consequences result from the interaction of individuals (e.g., in a game), based on this shared knowledge, for society?

The notions of *complete information* and *perfect information* are similar, but not identical. In games with *complete information*, every player has complete knowledge of the structure of the game (and, in particular, knows at each point of the game whose turn it is to move and which strategies are then available to this player) and of the gain functions of all players, yet the players do not necessarily know which actions the other players have performed so far or are performing now. For example, in the one-move, two-player games from Sections 2.1.1 and 2.1.2 (the prisoners' dilemma, the battle of the sexes, the chicken game, etc.) each player knows exactly which strategies are available to her opponent, but they do not know which strategy her opponent has chosen in the moment of their simultaneously made move. That is why these games are considered as games with complete, yet imperfect information.

However, in games with *perfect information* (in chess or Tic-Tac-Toe, for example), every player has not only complete knowledge of the game structure and the gain functions of all players, but knows also in each of her moves which actions the other players have performed so far. That is, in each of her moves she knows her exact position in the game tree and can choose her next action based on this perfect information.

Bayesian games like poker are games with *incomplete information*. For poker, this not only means that the players don't know the hands of their opponents (which is a basic precondition in this game because one cannot play poker with uncovered hands). To begin with, this concealed information requires the players to estimate the probabilities of their opponents' expected moves. But the special challenge in poker is (and that is crucial in every Bayesian game) that the players initially don't even know which *types of players* are sitting opposite them at the table. The type of a player (e.g., whether she is a *risk-averse*, a *risk-neutral*, or a *risk-loving* player) determines the probabilities by which she chooses a strategy among those available, and thus which gains can be expected. A game has incomplete information even if only one player is uncertain about the type of one other player (i.e., about the probability distribution under which this opponent chooses her strategies) and thus also about the corresponding gain function.

Harsanyi proposes a transformation to obtain games with complete, even though imperfect, information from games with incomplete information by letting "nature"—the goddess of destiny or, to say it with a bit less pathos, letting randomness—come into play as an additional player and by letting the gains of the other players depend on the unknown random moves of nature (see also the *"games against nature"* in the Monty Hall problem). By this method players can learn during a game. They observe the actions of their opponents and, according to the formula of Bayes from Theorem 2.4, can suitably adapt their initial belief about the types of their opponents, where "belief" here is simply viewed as a probability distribution over the possible player types.

Coming back to poker, with intelligence alone one cannot achieve anything against bad luck. For example, one cannot influence one's own hand in poker or those of one's opponents. Therefore, one can lose miserably, even if one is a professional poker player. It's just a lottery! In a game subject to randomness, winning can never be taken for granted. To maximize one's gain here means

2.4 Full House: Games with Incomplete Information

to outsmart and defeat both randomness and the other players as well as possible. Whoever has played poker at one time or another knows that every once in a while one has to bluff (i.e., to pretend one had a better—or worse— hand than one actually has), or otherwise one will lose more money than needed.

Von Neumann considered the following simplified variant of poker (see [736, Chapter 19]) that nonetheless displays the typical features of this game.[22]

2.4.2.1 Von Neumann's Simplified Poker Variant

Two players, Belle and David, randomly draw a card from an infinite deck of cards, each card showing an arbitrary real number between 0 and 1. Since both players distrust each other, they have asked Belle's little brother Edgar to first shuffle the deck (which, given it is an *infinite* deck, is just the type of task big sisters ask their little brothers to do), and we assume that each card can be drawn independently and with the same probability. Before the game, both players pay one dollar into the jackpot (see Figure 2.20).

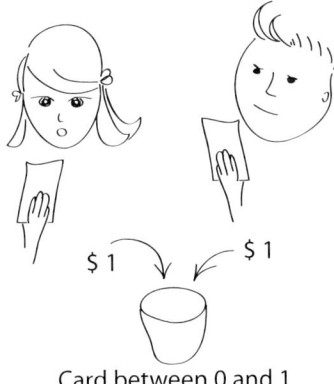

Fig. 2.20 Getting von Neumann's simplified poker variant started

Now the betting stage starts. Let us say that Belle moves first. She can:

1. either *pass on*, that is, she pays no further money into the jackpot,
2. or *raise*, which means that she pays another dollar into the jackpot.

Then it is David's turn. His actions depend on Belle's strategy:

1. Suppose that Belle has passed on. Then David has no choice and *must* call, which in this case means that he does not have to pay money into

---

[22] Borel [170, 171], whom Fréchet [465] referred to as the initiator of the theory of psychological games, studied a similar variant of poker, which differs from von Neumann's: Belle, with a bad hand, will simply fold before it comes to the showdown (cf. Figure 2.21).

the jackpot, as Belle hasn't paid anything either. Now it comes to the *showdown*: Both players show their cards and whoever has the card with the greater number wins the jackpot.

2. Suppose that Belle has raised. Then David does have a choice. He can:

   a. either *fold* (since he has a bad hand and therefore gives up)—in this case Belle wins the jackpot without her or David needing to show their cards,

   b. or *call*—in this case he pays the same amount as Belle (another dollar) into the jackpot, and it comes to the *showdown* also in this case: The greater card wins.

Figure 2.21 shows the course of the game in this simplified poker variant. In comparison with real poker (in any of its many variants), the possibilities of the players are severely restricted in von Neumann's simplified poker. In particular, both players have at most one option to choose among one of two alternative strategies before the game ends, whereas in most real poker variants all players have many options; there are more cards available to them, they can draw new cards or put cards down, etc., until they eventually, with some luck, end up with a good poker hand—such as a *Full House*, a *Flush*, a *Straight Flush*, a *Poker*, a *Three of a Kind*, a *Two Pair*, etc.—or do not. Also, they have many ways of making bets, they can raise several times, and of course they can win also with a bad hand when they are good bluffers.

These severe restrictions do not hurt, though. The analysis of the game can be simplified this way, while the crucial details and characteristics of the game are still captured.

2.4.2.2 Analyzing a Simplified Variant of von Neumann's Poker

We will not analyze von Neumann's original simplified variant of poker, however, but rather an even simpler variant due to Binmore [150]. In his poker variant, Belle and David draw their cards not from an infinite deck of cards, but they can choose among exactly three cards with the values 1, 2, and 3. Except for this change, the rules of the game are identical. In particular, the higher card wins the *showdown*.

What course of the game can be predicted? To analyze this game, an equilibrium concept is useful again. Equilibria in Bayesian games are called *Bayesian Nash equilibria*. In normal form games, a Nash equilibrium in pure or mixed strategies (see Definition 2.4 on page 50 and Definition 2.5 on page 63) is a strategy profile such that every strategy in this profile is a best response to the other strategies in the profile, i.e., all strategies in such a profile guarantee each player a highest possible gain, provided the other players stick to their strategies from the equilibrium profile.

In a Bayesian game, the influence of randomness supervenes; here, rational players seek to maximize their *expected* gain, depending on their beliefs about

2.4 Full House: Games with Incomplete Information

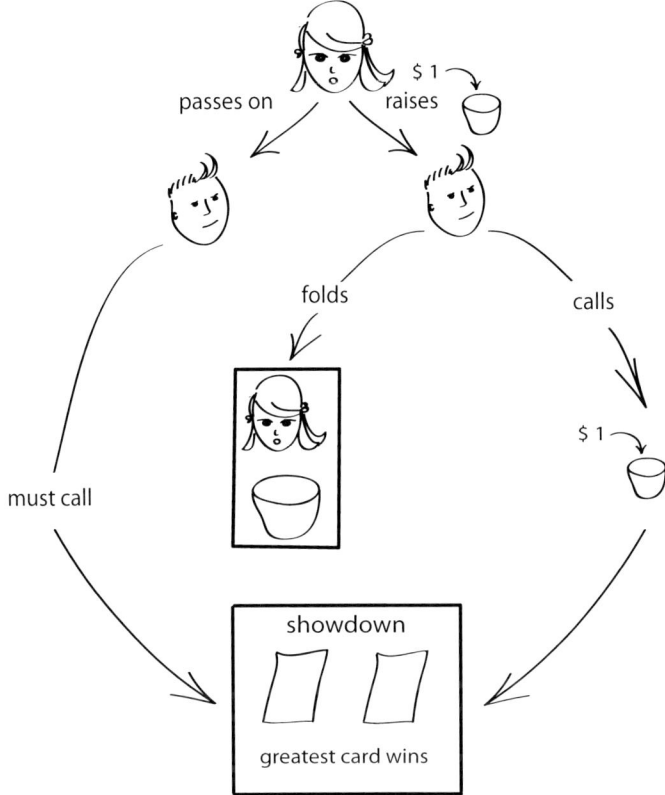

Fig. 2.21 Von Neumann's simplified poker variant

the other players. Recall that "belief" is viewed as a probability distribution over the possible player types. More concretely, one views the gain of a player as a *discrete random variable* X that, depending on the strategies of all players, can take on any of the values $x_1, x_2, \ldots, x_k$.[23] The *expectation of X* is defined as

$$\mathrm{E}(X) = \sum_{i=1}^{k} x_i \cdot \mathrm{P}(X = x_i), \tag{2.16}$$

where $\mathrm{P}(X = x_i)$ is the probability that X takes on the value x_i.

---

[23] A *random variable* is a function mapping atomic events to real numbers. Here, the atomic events are the courses of the game depending on the strategies chosen by the players, which are mapped to the corresponding gains of the players. We restrict ourselves to finitely many discrete values for X; the notion of expectation can also be extended to countably many discrete values and to real random variables with a probability density function.

We now define the notion of Bayesian Nash equilibrium for the type of risk-neutral player. The corresponding definitions for risk-averse or risk-loving players are even more general, and we do not elaborate on them here.

Definition 2.11 (Bayesian Nash equilibrium). In a Bayesian game, a *Bayesian Nash equilibrium for risk-neutral players* is defined to be a strategy profile that maximizes the expected gains of all players depending on what each player believes about the strategies chosen by the other players.

Turning again to the poker game of Belle and David, we assume that they both are risk-neutral players and that they both, via a Bayesian learning process, have come to believe that the other player is risk-neutral as well. The three cards (1, 2, and 3) are now shuffled by Edgar, and we obtain the following six possibilities for arranging the cards in the deck:

$$\begin{array}{cccccc} 1 & 1 & 2 & 2 & 3 & 3 \\ 2 & 3 & 1 & 3 & 1 & 2 \\ 3 & 2 & 3 & 1 & 2 & 1 \end{array}$$

Every arrangement occurs with a probability of 1/6. Belle obtains the card on top and David the card in the middle. We may assume that David—in case he has a choice because Belle raises—will fold with the 1 on his hand, since he would only lose more money by calling. It is also clear that he will call whenever he wins for sure, i.e., whenever he has the 3 on his hand. Therefore, we may further assume that Belle passes on (and does not raise) if she has the card 2 on her hand. If she would raise, she would lose more than necessary in case David has the 3 (as he will call then), but she couldn't win more than she would win by passing on in case David has the 1 (as he will fold then). Of course, Belle raises with the 3 on her hand. These strategies, which can be taken for granted, are shown by double lines in Figure 2.22.

This figure shows a game tree as described on page 85 ff., with the difference that here some vertices are labeled by probabilities, since it is a Bayesian game. The fields enclosed by dashed lines represent the information sets of the players; e.g., David at the bottom of Figure 2.22 sees his own card 2, but he does not know in which decision vertex of this information set he actually is, i.e., he does not know whether Belle has drawn the card 1 (left vertex) or the card 3 (right vertex).

In the square leaves of the game tree, Belle's gain is given in the upper left corner and David's gain is given in the lower right corner of the square. Since the gains in each leaf add up to 0, this is a so-called *zero-sum game*.

By the above discussion, only the following two cases are still unclear:

1. Should Belle bluff if she has drawn the card 1?
2. And should David call if he has drawn the card 2?

Consider this first from Belle's point of view, who has to decide how to act when drawing the 1. David is always dependent on her first move:

- Either he has a choice (namely, if she raises, i.e., bluffs), or

2.4 Full House: Games with Incomplete Information

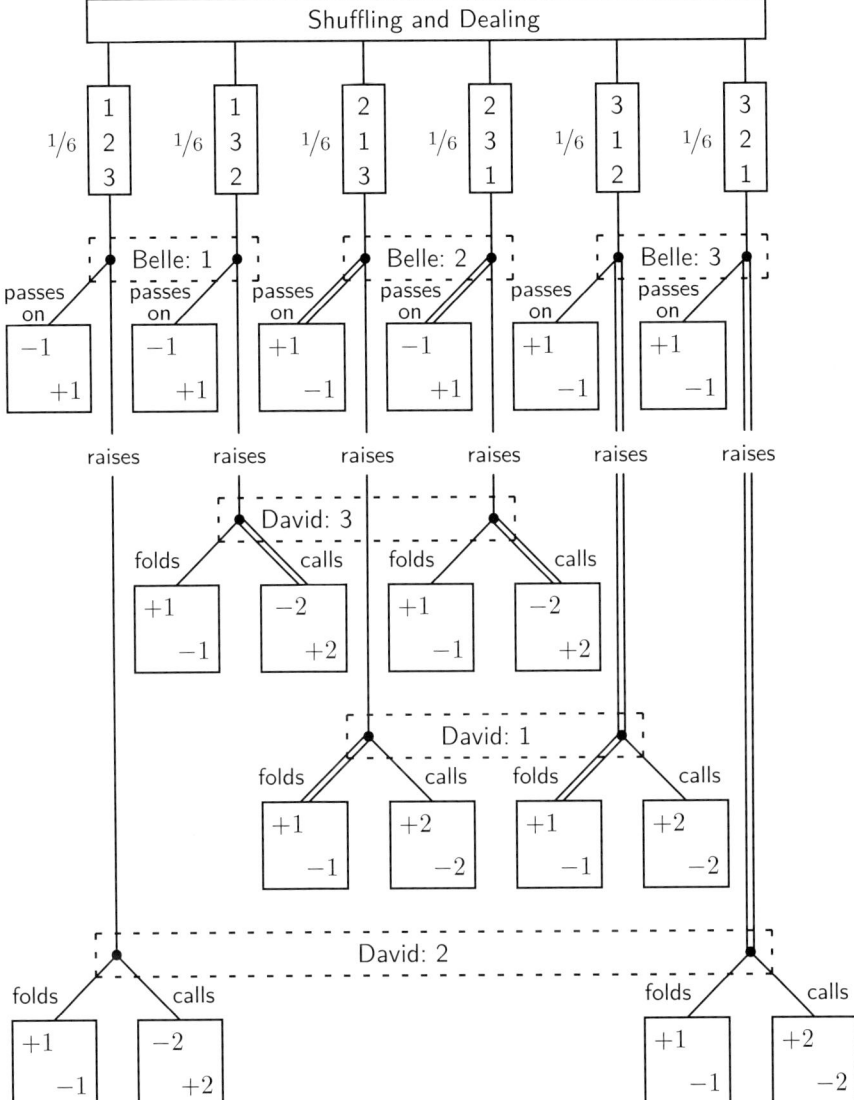

Fig. 2.22 Game tree for the simplified poker variant due to Binmore [150]

- he is forced to call (namely, if she passes on).

If David has a choice, his decision as to whether he calls or folds is made randomly if he holds the 2. The question now is "how randomly," i.e., what are the probabilities for these two events to occur?

Let p be the probability that Belle bluffs (i.e., raises), provided she has drawn the card 1. Both know that the probability for Belle to draw the card 3 and for David to subsequently draw the card 2 is exactly $1/6$, and that the probability for Belle to draw the card 1 and for David to subsequently draw the 2 is $1/6$ as well.

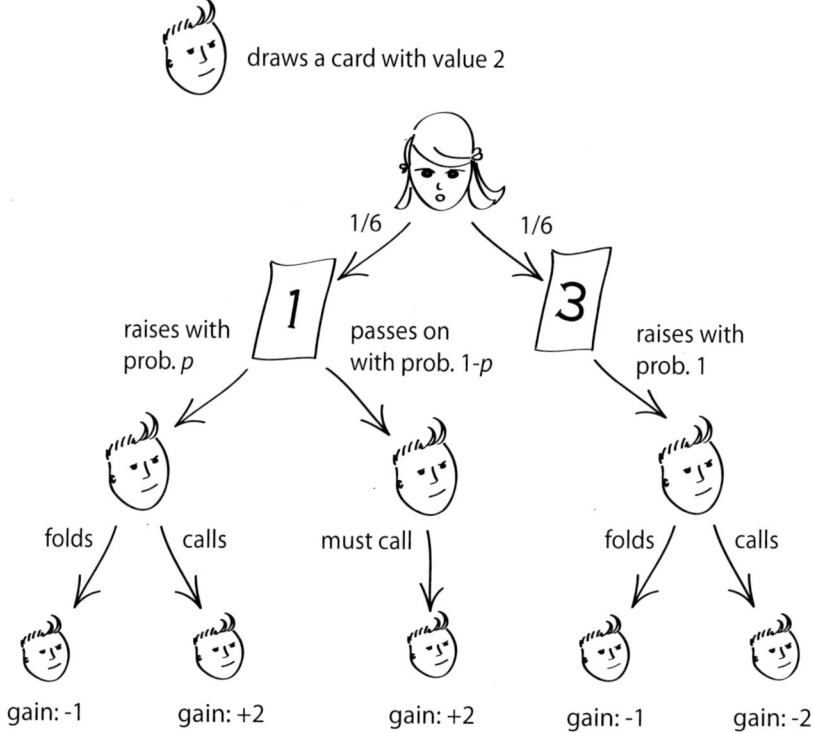

Fig. 2.23 Belle bluffs with probability p so as to make David indifferent

If David holds the 2, he must mix his strategies. Therefore, Belle tries to choose her bluffing probability p so that David is indifferent with respect to his strategies. Since Belle has the 1 against David's 2 with probability $1/6$ and raises with probability p, and since she has the 3 against David's 2 with probability $1/6$ and raises with probability 1 (see Figure 2.23), the left decision vertex in David's information set 2 in Figure 2.22 is reached with probability $(1/6)p$, but the right one with probability $1/6$. Thus, the corresponding conditional probabilities are $p/(p+1)$ and $1/(p+1)$. Why?

Denote by A_ℓ and A_r, respectively, the event that the left and the right decision vertex in David's information set 2 in Figure 2.22 is reached, and denote by B the condition that David holds the 2 on his hand and Belle raises. We have already determined the probabilities $P(A_\ell) = (1/6)p$ and $P(A_r) = 1/6$. Of course, we have $P(B|A_\ell) = P(B|A_r) = 1$, since if the left or right decision

2.4 Full House: Games with Incomplete Information

vertex in his information set 2 has been reached, we can be certain that David holds the 2 and Belle has raised. We want to determine the conditional probabilities $P(A_\ell | B)$ and $P(A_r | B)$. By the Bayesian formula from Theorem 2.4 on page 108, we have that:

$$P(A_\ell | B) = \frac{P(A_\ell) \cdot P(B | A_\ell)}{P(B)} = \frac{P(A_\ell)}{P(B)}, \tag{2.17}$$

$$P(A_r | B) = \frac{P(A_r) \cdot P(B | A_r)}{P(B)} = \frac{P(A_r)}{P(B)}. \tag{2.18}$$

But how can we determine the probability $P(B)$ in (2.17) and (2.18)? The direct calculation of this probability can be avoided by the following little trick. Setting $b = 1/P(B)$, we can write (2.17) and (2.18) as follows:

$$P(A_\ell | B) = b \cdot P(A_\ell), \tag{2.19}$$
$$P(A_r | B) = b \cdot P(A_r). \tag{2.20}$$

However, if the condition B is satisfied, either one of the two decision vertices in David's information set 2 must be reached. Hence, $P(A_\ell | B) + P(A_r | B) = 1$. Now, adding (2.19) and (2.20), we obtain

$$P(A_\ell | B) + P(A_r | B) = 1 = b(P(A_\ell) + P(A_r))$$

and thus $b = 1/(P(A_\ell) + P(A_r))$. Substituting this value for b in (2.19) and (2.20), we obtain the conditional probabilities given above:

$$P(A_\ell | B) = b \cdot P(A_\ell) = \frac{P(A_\ell)}{P(A_\ell) + P(A_r)} = \frac{(1/6)p}{(1/6)p + 1/6} = \frac{p}{p+1},$$

$$P(A_r | B) = b \cdot P(A_r) = \frac{P(A_r)}{P(A_\ell) + P(A_r)} = \frac{1/6}{(1/6)p + 1/6} = \frac{1}{p+1}.$$

At the bottom of Figure 2.23, we see David's possible gains. If David has a choice (since Belle raises), he will be made indifferent if his gains (or losses) balance out when folding (which is -1 in either case) with his gains (or losses) when calling (which is $+2$ or -2, depending on whether Belle bluffs or not), that is, he is indifferent whenever we have

$$-1 = \frac{2p}{p+1} - \frac{2}{p+1},$$

which implies a bluffing probability of $p = 1/3$ for Belle in case she holds the 1 on her hand.

Conversely, let q be the probability for David to call if he has drawn the 2. If Belle holds the 1 on her hand, she has to mix her strategies (to pass on or to raise) as well. Therefore, David tries to choose q so that she is indifferent with respect to her strategies. This is shown in Figure 2.24 that lists Belle's possible gains at the bottom.

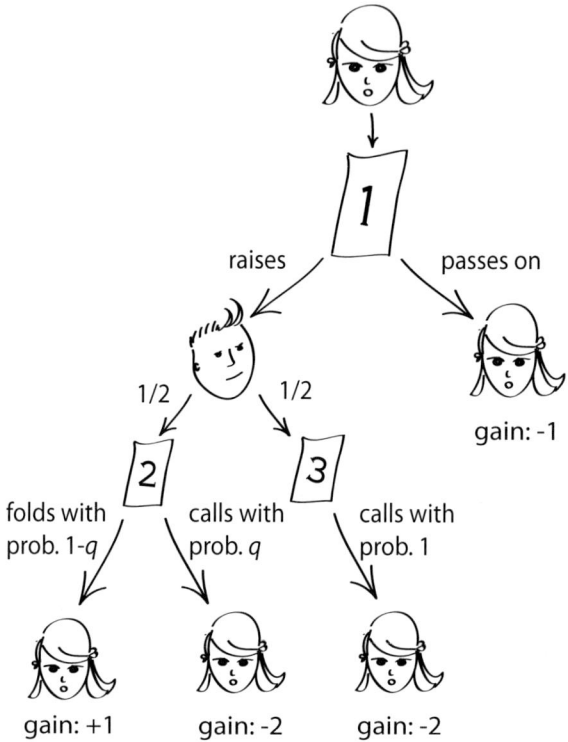

Fig. 2.24 David calls with probability q so as to make Belle indifferent

For Belle with the 1 on her hand to be indifferent against her strategies, David has to choose q so that Belle's gains (or losses) when passing on (which obviously is -1) balance out with her gains (or losses) when raising (which is $+1$ or -2, depending on whether David folds or calls). To fold here means that David falls for her bluff, and to call means he does not. If he has drawn the 3, he will certainly not fall for her bluff and will call instead. With the 2 on his hand, however, David must choose q so that

$$-1 = (1/2)((1-q) - 2q) + (1/2)(-2)$$

holds to make Belle indifferent against her two strategies in her information set 1 in Figure 2.22, where she does not know for sure whether she is in the left or right decision vertex. From this equation it follows that David should call with a probability of $q = 1/3$, in case he has drawn the 2.

2.4 Full House: Games with Incomplete Information

Thus, we have a Bayesian Nash equilibrium for this simplified poker variant if Belle bluffs with probability $1/3$ whenever she holds the 1, and if David calls with probability $1/3$ whenever he holds the 2 and she raises. In principle, this result is meaningful for the more general variants of this game as well, such as for von Neumann's simplified poker game or the more common poker variants with *Full House*, *Flush*, etc. For example, beginners often make the mistake to bluff only with a relatively good hand, most likely because they do not dare to do so with a really bad hand. As we have seen, however, Belle does bluff with her worst hand ever possible, the card 1. Moreover, it is important for poker to bluff frequently enough, or otherwise one will stay way below one's expected gains. Using the formula (2.16), you can easily determine the expected gains of Belle and David in the above Bayesian Nash equilibrium.

On the other hand, one should not exaggerate bluffing. If Belle were to bluff every time she draws the 1, this might take its toll eventually, since David observes her and learns from her way to play. This is true not only in poker, but also in many other situations in life where one might increase one's profit by cheating. For example, when a buyer purchases a car, she does not know necessarily whether she looks at a good or a bad car and whether it is worth its price. Buying a car is a bit like poker. However, if a car dealer sells always bad cars for excessive prices, this will get around eventually and becomes common knowledge in the city. A car dealer who bluffs too often might soon have to close his shop, since the buyers have learned and he has exhausted his credit of trust with them.

It takes not only logical thinking to be successful in poker, but psychological empathy and intuition as well. However, to *really* have success in this game, a poker player needs to be able to well disguise her player type. Not only the *poker face* matters here, it is also important to be unpredictable. This helps to frustrate one's opponents when they are trying to learn something from one's own way of playing.

Also other games, such as the two-player games in normal form discussed in Sections 2.1.1 and 2.1.2, can be viewed as Bayesian games. For example, if one would play the prisoners' dilemma from Section 2.1.1 many times (i.e., finitely often, without knowing exactly how often) and if this were common knowledge of both players, then according to Harsanyi a Bayesian Nash equilibrium would emerge with the move *Silence* for both players, since Smith and Wesson would now learn from the behavior of their partners. Interestingly, this Bayesian Nash equilibrium differs from the regular Nash equilibrium in pure strategies, which, recall, occurs when both players make a confession (see Section 2.1.1), and in addition it is Pareto-optimal.

2.5 How Hard Is It to Find a Nash Equilibrium?

Even though the existence of a Nash equilibrium in pure strategies cannot be guaranteed, we know from Theorem 2.3 that a Nash equilibrium in *mixed* strategies always exists. This outstanding result due to Nash [729, 730] shows the universality of such equilibria and establishes the central importance of this solution concept in game theory.

When the strategies of all players are in equilibrium, none of them has an incentive to deviate from her (mixed) equilibrium strategy as long as the other players do not deviate from their (mixed) equilibrium strategies either. Hence, one may assume that during the course of a game an equilibrium will eventually emerge. However, how useful is such a prediction if the game is over before one was able to compute it? How long does it take to find a strategy profile in equilibrium?

This is the main focus of *algorithmic game theory* (see, e.g., the surveys by Papadimitriou [765] and Shoham [863]): Rather than being satisfied with merely an existence result for an important solution concept such as Nash equilibrium in mixed strategies, one seeks to determine the complexity of finding it. Some of the foundations of computational complexity theory have already been laid in Section 1.5. For determining the complexity of finding Nash equilibria, we will now need some more.

2.5.1 Nash Equilibria in Zero-Sum Games

Of course, it would be wonderful if Nash equilibria could always be computed efficiently. Unfortunately, for most games we know only algorithms either with an unknown running time (because it is very difficult to analyze them in these cases) or with an exponential running time in the worst case. Merely for some special cases, it was possible to design efficient algorithms for computing Nash equilibria. For example, the result that Nash equilibria in mixed strategies can be computed in polynomial time for two-player zero-sum games is possible due to linear programming methods, as von Neumann [734] realized. In zero-sum games, the gains of all players add up to zero for each strategy profile; examples are the simplified poker variant due to von Neumann that was mentioned in the previous section, and also the penalty game and the paper-rock-scissors game from Section 2.1.2.

Linear programming plays an important role in optimization. That linear programs can actually be solved in polynomial time was shown only near the end of the 1970s by Hačijan [525], whose algorithm is based on the *ellipsoid method*. His work was a milestone in linear programming, since it had been unclear before whether linear programs are solvable in polynomial time. Also later developed procedures such as the *interior point methods*, which too run in polynomial time, have been inspired by his work. In practice, the

2.5 How Hard Is It to Find a Nash Equilibrium?

older *simplex methods* (see, e.g., [326, 327]) are still in use. They do require exponential time for certain, particularly constructed linear programs, but run more efficiently than, e.g., the algorithm by Hačijan [525] for many other linear programs.

By linear programming, linear objective functions can be optimized, such as those occurring in the context of gain maximization in games, where the solutions are subject to certain constraints that are given in the form of linear equations or inequalities. Formally, a linear program is of the following form:

$$
\begin{aligned}
\text{Maximize} \quad & \vec{c}^T \vec{x} \\
\text{under the constraints} \quad & A\vec{x} \leq \vec{b} \\
\text{and} \quad & \vec{x} \geq \vec{0}.
\end{aligned}
$$

Here, $\vec{x}$, $\vec{b}$, and $\vec{c}$ are vectors and A is a matrix of matching dimensions, and $\vec{c}^T$ denotes the transposed vector for $\vec{c}$, i.e., $\vec{c}^T$ is the row vector corresponding to the column vector $\vec{c}$. The coefficients in the vectors $\vec{b}$ and $\vec{c}$ and in the matrix A are known, and we want to find the solution vector $\vec{x}$. If we are looking for integer solutions (i.e., for a solution vector $\vec{x}$ with entries $x_i \in \mathbb{Z}$, where $\mathbb{Z}$ denotes the set of integers), then we say this is an *integer linear program*. Unlike for linear programs, the decision problem corresponding to integer linear programming is NP-complete (see, e.g., [483]), except when the number of variables is bounded by a constant—in this case, Lenstra [651] showed that this problem can be solved in polynomial time.

By the way, von Neumann also considered the connection between linear programming and so-called matrix games, and he suggested the first interior point method ever, even though in terms of efficiency it was no match for the simplex method due to Dantzig, see [327].

The following result, known as the minimax theorem, is due to von Neumann [734] and applies to two-player zero-sum games.

Theorem 2.5 (von Neumann [734]). *In every zero-sum game with two players who have gain functions g_1 and g_2 and can both choose among a finite number of strategies, where $\mathcal{A}$ is the set of mixed strategies for player 1 and $\mathcal{B}$ is the set of mixed strategies for player 2, there exists a value v^* such that the following hold:*

1. *For player 1, there exists a mixed strategy $\alpha^* \in \mathcal{A}$ such that*

$$\max_{\alpha \in \mathcal{A}} \min_{\beta \in \mathcal{B}} g_1(\alpha, \beta) = \min_{\beta \in \mathcal{B}} g_1(\alpha^*, \beta) = v^*.$$

2. *For player 2, there exists a mixed strategy $\beta^* \in \mathcal{B}$ such that*

$$\min_{\beta \in \mathcal{B}} \max_{\alpha \in \mathcal{A}} g_1(\alpha, \beta) = \max_{\alpha \in \mathcal{A}} g_1(\alpha, \beta^*) = v^*.$$

Proof. This theorem was first published in 1928 by von Neumann [734]. However, here we will give a different proof, based on the fact that a Nash equilibrium in mixed strategies exists for every noncooperative game in normal form.

We now proceed with the proof. Let $(\alpha^*, \beta^*) \in \mathcal{A} \times \mathcal{B}$ be a Nash equilibrium in mixed strategies for our game. We know that such an equilibrium exists by Theorem 2.3.[24] We set $v^* = g_1(\alpha^*, \beta^*)$ and we claim that

$$\max_{\alpha \in \mathcal{A}} \min_{\beta \in \mathcal{B}} g_1(\alpha, \beta) \geq \min_{\beta \in \mathcal{B}} g_1(\alpha^*, \beta) \geq v^*. \quad (2.21)$$

The first inequality holds because on the left-hand side we have the maximum over all possible values of α, and on the right-hand side we use one particular value, α^*. The second inequality holds because (α^*, β^*) is a Nash equilibrium in mixed strategies and, thus, β^* is a best response for α^*. By definition of a best response (see Definition 2.4 on page 50), it must hold that for each mixed strategy $\beta \in \mathcal{B}$ we have $g_2(\alpha^*, \beta) \leq g_2(\alpha^*, \beta^*)$. Since our game is a zero-sum game, $g_2(\alpha^*, \beta) \leq g_2(\alpha^*, \beta^*)$ implies $g_1(\alpha^*, \beta) \geq g_1(\alpha^*, \beta^*) = v^*$.

However, by applying similar reasoning, we can also show that

$$\max_{\alpha \in \mathcal{A}} \min_{\beta \in \mathcal{B}} g_1(\alpha, \beta) \leq \max_{\alpha \in \mathcal{A}} g_1(\alpha, \beta^*) \leq v^*. \quad (2.22)$$

The first inequality holds because on the left-hand side the maximum operator has to find a mixed strategy α that does well against every possible mixed strategy β, whereas on the right-hand side it only has to do well against one fixed strategy, β^*. The second inequality holds because α^* is a best response to β^* and, thus, for every mixed strategy $\alpha \in \mathcal{A}$ we have that $g_1(\alpha, \beta^*) \leq g_1(\alpha^*, \beta^*) = v^*$.

By combining inequalities (2.21) and (2.22), we obtain the first part of the theorem:

$$\max_{\alpha \in \mathcal{A}} \min_{\beta \in \mathcal{B}} g_1(\alpha, \beta) = \min_{\beta \in \mathcal{B}} g_1(\alpha^*, \beta) = v^*.$$

The reader should verify that we can apply analogous reasoning to the case of player 2 and obtain the second part as well:

$$\min_{\beta \in \mathcal{B}} \max_{\alpha \in \mathcal{A}} g_1(\alpha, \beta) = \max_{\alpha \in \mathcal{A}} g_1(\alpha, \beta^*) = v^*.$$

This completes the proof. ❑

Briefly put, the above theorem says that player 1 is guaranteed a gain of at least v^* when using her mixed strategy α^*, independent of the mixed strategy chosen by player 2. Conversely, player 2 can make sure by choosing her mixed strategy β^* that player 1's gain does not exceed v^*. Since we have a zero-sum game, it follows that $g_1(\boldsymbol{\pi}) = -g_2(\boldsymbol{\pi})$ for each strategy profile $\boldsymbol{\pi}$; hence, the loss of player 2 can never be worse than $-v^*$. This result is called the minimax theorem, since each of the two players minimizes the maximum gain of the

---

[24] Interestingly, Nash, who first proved this result, was born in 1928.

other player—and thus, again due to the zero-sum property, maximizes her own gain. Since this minimax strategy coincides with the Nash equilibrium in mixed strategies for zero-sum games, and since it can be expressed as a linear program, such Nash equilibria can be computed in polynomial time in zero-sum games with two players. The minimax principle from Theorem 2.5 can be extended to sequential games as well.

2.5.2 Nash Equilibria in General Normal Form Games

In general normal form games that are not restricted to two players, Nash equilibria appear to be not as efficiently computable. This conjecture has been shown by Daskalakis, Goldberg, and Papadimitriou [330] (see also [331] for a concise summary of the main ideas, and [332] for all proof details). Specifically, they have shown the following result (to be stated as Theorem 2.7 on page 134):

> The problem of (approximately) computing Nash equilibria in mixed strategies is PPAD-complete.

In the remainder of this section we will explain why we have to consider the complexity of *approximately* computing Nash equilibria, why it requires a new complexity class, PPAD, and what this class really is. For readers not so much interested in the technical details, in essence, PPAD is a complexity class that captures a certain family of problems that do not seem to be polynomial-time solvable, but which appear to be easier than NP-complete problems. For readers interested in the details, you are warmly invited to read the rest of this section, but be warned that it may be a bit more technical than the preceding discussions. Cling on tight!

2.5.2.1 Issues to Deal with

Unlike the decision problems for discrete structures introduced in Section 1.5, such as SAT, we are here concerned with

- a functional search problem (i.e., rather than just finding an answer to a yes/no query, we here have to actually *find* a Nash equilibrium in mixed strategies)
- that is defined over continuous structures (for example, the gain functions of the players may have irrational values, which cannot exactly be handled by a computer but can only be approximated).

The latter problem can be sidestepped by searching not for an exact Nash equilibrium $\boldsymbol{\pi}$, but for merely an ε-approximation $\boldsymbol{\pi}_\varepsilon$ of $\boldsymbol{\pi}$, i.e., the expected gain of each player in the strategy profile $\boldsymbol{\pi}_\varepsilon$ is by at most the factor ε better than the expected gain of each player in the strategy profile π. Such an ε-approximation of a Nash equilibrium π thus allows the players to deviate from their mixed equilibrium strategies so as to improve their gains, but by no more than the factor $\varepsilon > 0$, which can be chosen arbitrarily small. This approach is justified by the fact that computing an ε-approximation of a Nash equilibrium is no harder than computing this Nash equilibrium itself (since even an ε-error is allowed); hence, a hardness result for the problem of computing ε-approximations of Nash equilibria can be transferred to the problem of computing Nash equilibria (cf. Statement 3 in Lemma 1.1 on page 30).

The next problem in the proof of Daskalakis, Goldberg, and Papadimitriou [330] is that the methods of classical complexity theory, the foundations of which have been sketched in Section 1.5, are actually not applicable here. By Theorem 2.3, there *always* exists a Nash equilibrium in mixed strategies; for the corresponding decision problem, the answer for a given problem instance will thus always be "yes," which makes the decision problem trivial. By constrast, common NP-hard decision problems such as SAT are typically hard because telling "yes"-instances apart from "no"-instances is difficult. How can one show computational hardness of such functional search problems that have a solution for all instances? To be sure, there does exist a theory of complexity of functions (see, e.g., the survey by Selman [851]), but this theory is not well applicable here.

To illustrate the approach to circumventing this difficulty in characterizing the complexity of computing Nash equilibria, let us consider the following simple problem, a modfication of the standard SUBSET-SUM problem.

EQUAL-SUBSETS

Given: A sequence $a_1, \ldots, a_n$ of positive integers such that $\sum_{i=1}^{n} a_i < 2^n - 2$.
Find: Nonempty, disjoint subsets I, J of $\{1, \ldots, n\}$, such that

$$\sum_{i \in I} a_i = \sum_{j \in J} a_j.$$

First, let us note that, indeed, there always exists a solution for EQUAL-SUBSETS. The reason for this is that

1. there are exactly $2^n - 1$ nonempty subsets I of $\{1 \ldots, n\}$, and
2. for each nonempty subset I of $\{1, \ldots, n\}$, the value $\sum_{i \in I} a_i$ has one out of $2^n - 3$ values (these values come from the set $\{1, 2, \ldots, 2^n - 3\}$).

Thus, by the pigeonhole principle, there must be two distinct subsets, $I', J' \subset \{1, \ldots n\}$, such that $\sum_{i \in I'} a_i = \sum_{j \in J'} a_j$. If we let $I = I' \setminus (I' \cap J')$ and $J = J' \setminus (I' \cap J')$, then we get our two disjoint sets that have the same sums. (It must be the case that I and J are nonempty because I' and J' are distinct and the numbers $a_1, \ldots, a_n$ are positive.)

2.5 How Hard Is It to Find a Nash Equilibrium?

If we would have defined EQUAL-SUBSETS as a decision problem where we ask about the existence of the sets I and J then, of course, each problem instance would have trivially the answer "yes." Still, computing the two sets that witness this "yes" answer intuitively seems to be difficult. The reason for this is that the proof that there always is a "yes" answer relies on the pigeonhole principle, which is inherently *nonconstructive*. The pigeonhole principle says that if we have some m objects and each of them has one of $m-1$ "features" (for EQUAL-SUBSETS these features are the sums of the subsequences of $a_1, \ldots, a_n$), then there must be at least two objects that have the same "feature." However, the pigeonhole principle gives no hint whatsoever as to how to find these two objects.

2.5.2.2 Four Types of Nonconstructive Proof Steps

To capture the complexity of such problems based on the pigeonhole principle, Papadimitriou [763] introduced a new complexity class, PPP (the name stands for *Polynomial Pigeonhole Principle*). Interestingly, it turns out that computing a Nash equilibrium in mixed strategies is, in fact, a problem in PPP. However, PPAD captures the complexity of this problem more precisely.

Papadimitriou defined the class PPAD in the same paper in which he also introduced the class PPP. Indeed, Papadimitriou showed that the problem of computing a Nash equilibrium in normal form games is in PPAD, but the proof of its PPAD-completeness appeared well over a decade later, in the paper by Daskalakis, Goldberg, and Papadimitriou [330]).

As for the case of PPP, the idea behind PPAD is that for total search problems, that is, for problems for which there always exist solutions, there must exist a mathematical proof of this fact (such as the proof of Theorem 2.3 about the existence of Nash equilibria), and if the problem is not solvable in polynomial time, then there must be a nonconstructive step in that proof. For the case of EQUAL-SUBSETS, this nonconstructive step is applying the pigeonhole principle, and for computing Nash equilibria it is using Brouwer's fixed point theorem (recall Corollary 2.1 on page 82). Papadimitriou [763] identifies four types of such nonconstructive proof steps, and he suggests a distinction of such problems according to the following four complexity classes (including the classes PPP and PPAD that we have briefly discussed already):

1. PPP, for "**P**olynomial **P**igeonhole **P**rinciple." For each problem in PPP, the nonconstructive proof step can be described by the pigeonhole argument:

   Every function mapping n elements to $n-1$ elements has a collision, i.e., $f(i) = f(j)$ for $i \neq j$.

2. PLS, for "**P**olynomial **L**ocal **S**earch." For each problem in PLS, the nonconstructive proof step can be described by the following argument:

Every directed acyclic graph has a sink, i.e., a vertex without any outgoing edges.

3. PPA, for "**P**olynomial **P**arity **A**rgument for graphs." For each problem in PPA, the nonconstructive proof step can be described by the following parity argument:

   If an undirected graph has a vertex of odd degree, then it has at least one other such vertex.

   Here, the *degree of a vertex* in an undirected graph is the number of edges incident with this vertex.

4. PPAD, for "**P**olynomial **P**arity **A**rgument for **D**irected graphs." For each problem in PPAD, the nonconstructive proof step can be described by the following parity argument:

   If a directed graph has an unbalanced vertex (i.e., a vertex with distinct indegree and outdegree), then it has at least one other such vertex.

   Here, the *indegree of a vertex* in a directed graph is the number of incoming edges and its *outdegree* is the number of outgoing edges.

At first glance, it might be surprising that the above complexity classes contain problems that are not polynomial-time solvable. For example, for the case of PPAD, given a directed graph G with an unbalanced vertex v it is very easy to find some other vertex with that property. It suffices to start from v and move along outgoing edges as yet unvisited (possibly visiting the same vertices several times). Such a walk would, eventually, lead to an unbalanced vertex $v' \neq v$ (we leave the detailed proof of this fact as an exercise to the reader). So, where does the hardness of PPAD really come from? The point is that the graphs in the problems from PPAD (and in other problems from the above-defined classes) can be encoded in a *succinct* way, so that a graph described by n bits of input can have $\mathcal{O}(2^n)$ vertices. For such cases, the simple walk algorithm described above would require exponential time.

2.5.2.3 The Polynomial Pigeonhole Principle

Let us now define the above complexity classes formally, starting with the class PPP. We first define the following problem:

PIGEONHOLE-FUNCTION

Given: A function f, $f \colon \{0,1\}^n \to \{0,1\}^n$, expressed as an algorithm computing f in time linear in the length of the encoding of the algorithm.

Find: An input $x \in \{0,1\}^n$ such that $f(x) = 0^n$, or two distinct inputs $x, x' \in \{0,1\}^n$ such that $f(x) = f(x')$.

2.5 How Hard Is It to Find a Nash Equilibrium?

We make two observations. First, the function f in this problem can essentially be any polynomial-time computable function. This is so because every reasonable way of encoding the function (either as a Turing machine transducer, or in languages resembling the C or Java programming languages) allows to extend the algorithm for computing f by a polynomial factor. Suppose we have an algorithm that computes f in time $\mathcal{O}(n^k)$ for some fixed k, but the encoding of the algorithm of f is very short (perhaps even of constant size). We can extend this encoding of the algorithm by including dead code (that is, instructions that are never executed) such that after this process the algorithm has an encoding of size $\mathcal{O}(n^k)$. Our second observation is that there always is a solution for PIGEONHOLE-FUNCTION. If there is no input x such that $f(x) = 0^n$ then, by the pigeonhole principle, there must be two distinct inputs, x and x', such that $f(x) = f(x')$.

We define PPP as the class of all problems that can be reduced to PIGEONHOLE-FUNCTION using the following notion of reducibility for search problems (see [969]): If $F, G: \Sigma^* \to \Sigma^*$ are two total functions, then F is *functionally (many-one-)reducible in polynomial time to G* if there exist functions $r, s \in \mathrm{FP}$ such that
$$F(x) = s(G(r(x)))$$
for all $x \in \Sigma^*$. That is,

1. the efficiently computable function r transforms a given instance x of search problem F into an equivalent instance of search problem G, and
2. the efficiently computable function s transforms a solution of this instance $r(x)$ of search problem G back into an equivalent solution of the given instance x of search problem F.

It might be surprising at first that we define a complexity class not through an appropriate machine representation (as is typical for P, NP, PSPACE, and many other classes), but through a particular problem, namely as the class of all problems reducing to it. However, P, NP, PSPACE, etc. can be defined in this way as well. For example, we can define NP to be the class of all problems that can be $\leq_m^P$-reduced to SAT. A convenient feature of this approach is that a thus-defined complexity class clearly has at least one complete problem (the one used in the definition).

We can now formally prove that EQUAL-SUBSETS is in PPP: Given a sequence $a_1, \ldots, a_n$ of positive integers such that $\sum_{i=1}^n a_i < 2^n - 2$, we define function f such that $f(0^n) = 1^n$ and for each $(x_1, \ldots, x_n) \in \{0, 1\}^n$ we have $f(x_1, \ldots, x_n) = \sum_{i=1}^n a_i x_i$ (or, formally, f's output is the vector $(y_1, \ldots, y_n) \in \{0, 1\}^n$ encoding the number $\sum_{i=1}^n a_i x_i$ in binary; note that this vector is never 1^n). Clearly, f never outputs 0^n and, thus, there are two distinct inputs $x, x' \subset \{0, 1\}^n$ such that $f(x) = f(x')$. These inputs x and x' clearly define the two disjoint sets that we seek in the EQUAL-SUBSETS problem (recall the discussion on how each instance of EQUAL-SUBSETS always has a solution to see why we can obtain nonempty, disjoint sets with the same sums back from the inputs x, x').

2.5.2.4 Polynomial Local Search

Using the above approach, we can define the class PLS as the class of problems that reduce to the following problem, which corresponds to finding a local optimum of a function:

LOCAL-OPTIMUM

Given: Two functions, $f\colon \{0,1\}^n \to \{0,1\}^n$ and $p\colon \{0,1\}^n \to \{0,\ldots,2^n-1\}$, expressed as algorithms computing these functions in time linear in the lengths of their encodings.

Find: An input x such that $p(f(x)) \geq p(x)$.

Since for each argument, function p takes one of finitely many values, it is clear that there always exists a solution x such that $p(f(x)) \geq p(x)$. Intuitively, LOCAL-OPTIMUM captures the idea of a heuristic local search: We want to find a local minimum of the function p and, given a potential solution x, we use function f to provide a candidate for a better one. We consider the sequence $x, f(x), f(f(x)), \ldots$, until we finally find an input x', $x' = f(\cdots f(x)\cdots)$, such that $p(f(x')) \geq p(x')$. Then, x' is a local minimum of p with respect to the local search function f. Local search approaches are successfully used for multiple tasks, including, for example, finding solutions for the traveling salesperson problem.[25] There is also a variant of this class, called CLS, which captures *continuous local search*, that is, local search where both underlying functions are continuous [333]. We will soon see why this class is quite interesting as well.

2.5.2.5 The Polynomial Parity Argument for Graphs

To define PPA, we need a convention regarding the representation of a certain family of undirected graphs. Let f be some function that given input x, $x \in \{0,1\}^n$, outputs a set of zero, one, or two elements from $\{0,1\}^n$. This function defines an undirected graph $U(f)$ as follows: Each $x \in \{0,1\}^n$ is a vertex and there is an edge connecting the vertices x and x' exactly if $x' \in f(x)$ and $x \in f(x')$. Note that each vertex in $U(f)$ has degree at most two. We define PPA as the class of problems that reduce to the following one:

ODD-DEGREE-VERTEX

Given: A function $f\colon \{0,1\}^n \to \{\emptyset\} \cup \{0,1\}^n \cup \{0,1\}^{2n}$, represented by an algorithm computable in time linear in the length of its encoding.

Find: If vertex 0^n has odd degree in $U(f)$, then find another vertex of odd degree in $U(f)$.

---

[25] We omit a detailed discussion here, but interested readers can find a comprehensive treatment of local search, including its theoretical aspects, in a survey by Aarts and Lenstra [1]; PLS was first defined and formally studied by Johnson, Papadimitriou, and Yannakakis [579].

2.5 How Hard Is It to Find a Nash Equilibrium?

2.5.2.6 The Polynomial Parity Argument for Directed Graphs

We can represent directed graphs with indegrees and outdegrees at most one in a very similar way. Let s and p be two functions from $\{0,1\}^n$ to $\{0,1\}^n$. We define the graph $G(s,p)$ to have vertex set $\{0,1\}^n$ and to have the following edges: For each $x, x' \in \{0,1\}^n$, there is a directed edge from x to x' if and only if $s(x) = x'$ and $p(x') = x$. Intuitively, s is the *successor* function, i.e., $s(x)$ is the vertex we can move to from vertex x (provided that $p(s(x)) = x$), and p is the *predecessor* function, i.e., $p(x)$ is the vertex from which we could reach x (provided that $s(p(x)) = x$). Based on this representation of graphs with indegrees and outdegrees bounded by one, we define the following problem:

END-OF-THE-LINE

Given: Two functions, $s\colon \{0,1\}^n \to \{0,1\}^n$ and $p\colon \{0,1\}^n \to \{0,1\}^n$, represented by algorithms with linear running times with respect to the lengths of their encodings.

Find: If in graph $G(s,p)$ vertex 0^n has indegree zero, then find a vertex x with either the outdegree equal to zero or the indegree equal to zero (in the latter case, this x must be different from 0^n).

PPAD is defined as the class of problems that reduce to END-OF-THE-LINE. Note that, naturally, there always is a solution for END-OF-THE-LINE. It suffices to start from vertex 0^n and proceed along the edges until finally we will reach a vertex with outdegree zero. The reason for that is that, since each vertex has indegree and outdegree at most one, if the current vertex has outdegree one, then we move to a not-yet-visited vertex. Thus, since there is only a finite number of vertices, we eventually have to reach a vertex with outdegree zero. In essence, this is simply an application of the pigeonhole principle.

"Just a second!" a careful reader might say, "wasn't the pigeonhole principle the idea behind the class PPP?" And, of course, the reader would be right. It turns out that the four (or, including CLS, five) complexity classes we have just discussed are tightly related. However, before we discuss these relations further, let us briefly describe just one more complexity class: TFNP is the class of all *total* NP *functions*, that is, of all the search problems for which there *always* is a solution (of size polynomially bounded in the size of the input) that can be found by a nondeterministic Turing machine in polynomial time.

2.5.2.7 Relations Among These Complexity Classes

Naturally, each of the classes PPP, PLS, PPA, PPAD, and CLS is a subclass of TFNP. In each of the problems defining these classes, a nondeterministic Turing machine can simply guess the desired, guaranteed to exist, solution and then verify that it indeed has the desired properties. Given this observation, we can state the following theorem.

Theorem 2.6. *1.* CLS $\subseteq$ PPAD $\subseteq$ PPA $\subseteq$ TFNP,
2. CLS $\subseteq$ PLS $\subseteq$ TFNP, *and*
3. PPAD $\subseteq$ PPP $\subseteq$ TFNP.

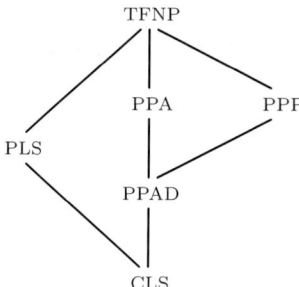

Fig. 2.25 The structure of the subclasses of TFNP

The complexity class inclusions indicated by Theorem 2.6 are depicted in Figure 2.25 as a Hasse diagram, i.e., an ascending line from complexity class $\mathcal{C}$ to some other class $\mathcal{D}$ means that we know that $\mathcal{C} \subseteq \mathcal{D}$. However, we do not know whether these inclusions are strict. For example, if P = NP, then all the classes would be equal, but if P $\neq$ NP then they are free to differ (and we strongly suspect that they do).

It is interesting to analyze what Theorem 2.6 means from the point of view of the class PPAD, which we have said to capture the complexity of computing Nash equilibria (even though we would not at all be surprised if the reader, having now seen the definition, would still have some doubts!—we will soon shed some more light on why this utterly nontrivial result indeed holds). First, since we have PPAD $\subseteq$ PPA$\cap$PPP (as PPAD is contained both in PPA and in PPP, it certainly is contained in their intersection as well), we know that PPAD problems, such as computing Nash equilibria, can easily be solved both by applying the pigeonhole principle and by considering the parity argument on graphs. On the other hand, since PPAD $\subseteq$ CLS is *not* known to hold, it seems that continuous local search heuristics on continuous functions might not be sufficient for capturing PPAD (and, in particular, for computing Nash equilibria).

We do not give a complete, formal proof of Theorem 2.6 (in particular, because for that we would need a careful definition of CLS), but it is instructive to consider two complexity-class inclusions, PPAD $\subseteq$ PPA and PPAD $\subseteq$ PPP. The former one is particularly interesting, because, in essence, PPA is a variant of PPAD for undirected graphs. The typical intuition is that directed graphs are more general and, thus, should allow one to express more complicated problems. On the intuitive level, however, one can understand PPAD as a more restrictive variant of PPA because in PPAD (and, more specifically, in the END-OF-THE-LINE problem) the directed edges are used to limit and guide our search from the initial vertex (for which we used the encoding 0^n) toward the sink (the vertex without outgoing edges). On the other hand, in PPA (or, more specifically, in the ODD-DEGREE-VERTEX problem) we also have to move along the edges of the graph, but without these guiding directions.

On slightly more formal grounds, to see that PPAD ⊆ PPA, it suffices to note that END-OF-THE-LINE reduces to ODD-DEGREE-VERTEX: Change all the edges in the graph described by functions s and p in a given instance of END-OF-THE-LINE to be undirected and encoded by the function f, as formally required by ODD-DEGREE-VERTEX. Now, if a vertex in the thus-transformed graph has an odd degree (which means that it has degree one) then it must either be the initial vertex 0^n, or a vertex that in the original graph had either outdegree zero or indegree zero. (Strictly speaking, this argument disregards vertices with *both* indegree and outdegree zero, i.e., isolated vertices; we encourage the reader to come up with a technical trick that works around this problem.)

For the inclusion of PPAD in PPP, we give an informal description of the reduction from END-OF-THE-LINE to PIGEONHOLE-FUNCTION. Using the functions s and p from the definition of END-OF-THE-LINE, we define the function f from PIGEONHOLE-FUNCTION as follows: If for vertex $x \in \{0,1\}^n$ there is a successor vertex, then we define $f(x) = s(x)$. Otherwise, we set $f(x) = x$. In the graph defined by functions s and p, we are guaranteed to have a sink, i.e., a vertex $y \in \{0,1\}^n$ with outdegre zero. Thus, if y has a predecessor then $f(p(y)) = f(y) = y$ and so we have a solution to the constructed instance of END-OF-THE-LINE. Otherwise, if there is no sink that has a predecessor, the vertex 0^n itself has no successor and, thus, $f(0^n) = 0^n$, which is the other type of solution that PIGEONHOLE-FUNCTION accepts. In either case, we reduce the process of finding a solution to END-OF-THE-LINE to the process of finding a solution to PIGEONHOLE-FUNCTION. Moreover, given a solution for the constructed instance of PIGEONHOLE-FUNCTION, it is easy to reconstruct the solution to the original instance of END-OF-THE-LINE.

2.5.2.8 Complexity of Computing Mixed Nash Equilibria

Now that we have developed some understanding of the class PPAD and of its complexity-class neighbors, let us shift the focus back to computing Nash equilibria. First, let us formally express the problem of computing a Nash equilibrium as a search problem.

NASH-EQUILIBRIUM
Given: A cooperative game in normal form, discretely represented in terms of an ε-approximation for an arbitrarily small given positive constant ε.   **Find:** An ε-approximation of a Nash equilibrium in mixed strategies.

The result of Daskalakis, Goldberg, and Papadimitriou [330] can then be expressed as follows.

Theorem 2.7 (Daskalakis, Goldberg, & Papadimitriou [330]).
NASH-EQUILIBRIUM *is* PPAD-*complete*.

The crucial steps of the proof of Theorem 2.7 can be sketched as follows. First, we have to show that the problem NASH-EQUILIBRIUM belongs to PPAD, and second, that it is PPAD-hard. The first claim can be shown by reducing NASH-EQUILIBRIUM to END-OF-THE-LINE; the second claim by conversely reducing END-OF-THE-LINE to NASH-EQUILIBRIUM. These reductions are far from being simple and we will not give their formal proofs. However, it is very instructive to give some intuition for them. In particular, why would such a natural problem as NASH-EQUILIBRIUM reduce to such a seemingly unnatural problem as END-OF-THE-LINE?

The intuitive answer to the above question lays in our proof of Sperner's lemma (recall Lemma 2.1 on page 77). Let us recall the statement of the lemma (we encourage the reader to recall the notation regarding its proof):

> Let $T_n = \vec{x}_0 \cdots \vec{x}_n$ be a simplicially subdivided n-simplex and let $\mathcal{L}$ be a proper labeling of the given subdivision of T_n. There are an odd number of n-subsimplexes that are completely labeled by $\mathcal{L}$ in this subdivision of T_n.

Intuitively, the proof that there is a Nash equilibrium in mixed strategies for every game in normal form (stated as Theorem 2.3 on page 82) relied on the fact that, in the limit, the completely labeled n-subsimplexes correspond to Nash equilibria. Strictly speaking, we would need Brouwer's fixed point theorem (recall Corollary 2.1 on page 82) in order to consider all the players at the same time, but it is easier to deal with the simplification here; the main idea and the relation between games and the END-OF-THE-LINE problem will be more transparent this way.

Now, let us recall the idea behind the proof of Sperner's lemma. The proof proceeded by constructing walks over the n-subsimplexes and used the fact that on one of the faces of the simplex, there must be an n-subsimplex T'_n such that

1. the face that T'_n shares with the main simplex is labeled with $0, 1, \ldots, n-1$, and
2. the remaining vertex v is labeled either with n (in which case we have found a completely labeled n-subsimplex), or with a label from set $\{0, \ldots, n-1\}$.

2.5 How Hard Is It to Find a Nash Equilibrium?

In the latter case, there is a unique face that includes v and is labeled with $0, \ldots, n-1$. We cross over this face to the next n-subsimplex and repeat the process. Sperner's lemma guarantees that there is a starting n-subsimplex that, through this process, leads to finding a completely labeled n-subsimplex. However, note that the proof of Sperner's lemma is not constructive! It says that there is a good starting n-subsimplex that will lead to the completely labeled n-subsimplex, but some of the promising starting subsimplexes can, just as well, lead us "through the simplex" and then "out of it." (For example, recall the walk on the right-hand side in Figure 2.10(a) on page 78.) Still, these walks that we have first defined in the proof of Sperner's lemma and which we have recalled now are, in essence, the "indegree/outdegree bounded by one" graphs necessary for the END-OF-THE-LINE problem (where we take the subsimplexes as the vertices of the graph). From a high-level point of view, the only missing piece of the puzzle is finding a good starting point.

The solution to the problem of picking the starting point is to make the "in and out" walks work for us, instead of against us. We do so by building their traversal into the problem. We extend the space available to our walks by attaching a number of additional n-simplexes at the face through which we used to enter. In Figure 2.26 we show an example of doing so in two dimensions (note, however, that in this figure our "subsimplexes," or rather "triangles," have three very flexible edges and we have squeezed them a bit).

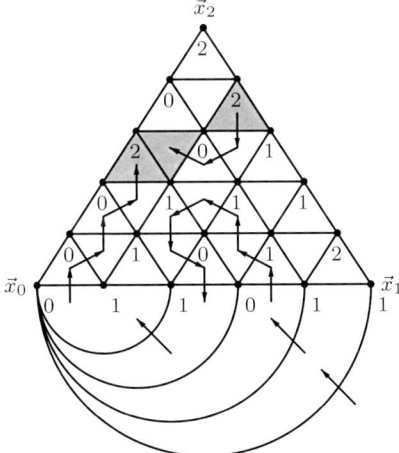

Fig. 2.26 Reducing NASH-EQUILIBRIUM to END-OF-THE-LINE by Sperner's lemma

Now, the starting point is simply the outermost added subsimplex. Following the walk from this point (using exactly the same rules as before, see page 77), we will eventually reach the completely labeled n-subsimplex, possibly winding in and out of the main simplex several times, but eventually reaching our goal.

This gives the outline of how to use (the computational variant of) Sperner's lemma to reduce NASH-EQUILIBRIUM to the END-OF-THE-LINE problem. To be techincally correct, we would have to deal with all the issues of actually encoding the simplex and the underlying graph in the format required by END-OF-THE-LINE. We omit these details and recommend reading the original proof by Daskalakis, Goldberg, and Papadimitriou [330] instead.

Thus we know that NASH-EQUILIBRIUM is in PPAD. Unfortunately, the proof in the other direction—that establishes PPAD-hardness of this problem via reducing END-OF-THE-LINE to NASH-EQUILIBRIUM—is technically much more involved, and we do not outline it here. However, it is important to understand the meaning of the result of Daskalakis, Goldberg, and Papadimitriou [330], stated in Theorem 2.7, and we hope that our discussion gave the reader appropriate intuition. Theorem 2.7 is a milestone in algorithmic game theory. It builds on preceding results that focus on the idea underlying the problem END-OF-THE-LINE as well as, for example, on the algorithm of Scarf [837] and the algorithm of Lemke and Howson [650].

Daskalakis, Goldberg, and Papadimitriou [330] actually show more than what is stated in Theorem 2.7, namely, they show that the problem NASH-EQUILIBRIUM is PPAD-complete even if restricted to normal form games with three players. This result could be improved later on. An outstanding and perhaps even somewhat surprising result is due to Chen and Deng [283], who showed that this problem is PPAD-complete even if restricted to normal form games with *two* players. The reason this result is surprising is that very often things qualitatively change when going from two to more players. For example, as we have seen on page 122, Nash equilibria in mixed strategies can be computed in polynomial time for two-player zero-sum games, but this is not known to hold for more than two players.[26] To give another example, Jackson and Srivastava [572] have shown that Nash equilibria exist in two-player finite games if and only if there exist sets of strategies that are undominated relative to each other in these games, yet for three or more players they gave a counterexample disproving this equivalence in the general case.

Many more algorithmic and complexity-theoretic results on the complexity of computing Nash equilibria in various classes of games have been obtained later on. Without going into details here, we mention as examples the work of Brandt, Fischer, and Holzer [218, 219], Conitzer and Sandholm [307, 309],

---

[26] In fact, even in a given two-player *general-sum* game in normal form, computing all the Nash equilibria requires worst-case time that is exponential in the number of pure strategies of the players. While the problem of finding a Nash equilibrium in a two-player general-sum game cannot be formulated as a linear program, it can at least be expressed as a so-called *"linear complementary problem" (LCP)*. In general-sum normal form games with more than two players, this problem can no longer be represented even as an LCP, but only as a *"nonlinear complementary problem,"* and those are hopelessly impractical to solve exactly. For a detailed discussion of how to approximate them by sequences of LCPs and of other methods, we refer to the textbook by Shoham and Leyton-Brown [864].

Elkind et al. [380], Porter, Nudelman, and Shoham [785], and Sandholm, Gilpin, and Conitzer [832].

Chapter 3
Cooperative Game Theory

Martin Bullinger · Edith Elkind · Jörg Rothe✉

This chapter provides an introduction to cooperative game theory, which complements noncooperative game theory (Chapter 2). Both of these fields study strategic aspects of cooperation and competition among the players and have been considered since the early days of game theory by von Neumann and Morgenstern [736]. In noncooperative game theory, players are assumed to choose their actions individually, selfishly seeking to realize their own goals and to maximize their own profit. While this does not mean that players are necessarily adversarial to each other (for example, they may prefer the dove strategy over the hawk strategy in the chicken game), they do not coordinate their actions and cannot make payments to each other.

In contrast, in cooperative games players work together by forming groups, also called *coalitions*, and take *joint* actions so as to realize their goals. However, in general the players are not assumed to be altruistic: They only join a coalition if this helps them increase their individual profit. Indeed, strategic reasoning plays an important role in cooperative games, as the players have to divide the fruits of their collective labor.

The following example shows how, by forming coalitions and working together, players can increase the profit of each coalition member.

Sunday is the market day. Anna, Belle, Chris, David, and Edgar have plundered their parents' pantries, and each of them wants to sell their haul for $\$1$ on the market.

"I've got one pound of sour cherries," Anna says. "An easily earned buck!"

"Or we could team up," Belle suggests. "I've got a pound of quark and half a pound of sugar. When we offer four portions of sweet cherry quark for $\$1$ each, we'll get $\$4$, so each of us gets $\$2$, twice as much as what you or me can get alone!"

> "Or you let me, too, join the game," proposes Chris. "I've got a quarter gallon of milk, so we could mix a yummy, sweet cherry-quark milkshake!"
>
> "Hey, man, think!" shouts David. "From your quarter gallon of milk, you can get no more than two milkshakes, and you can sell them for no more than $2 each. That's $4 all in all, but because there are three of you, after dividing the gains that makes $4/3 for each of you, which is less than $2. It's the economy, stupid! The girls wouldn't let you join their game!"
>
> "You're right!" admits disenchanted Chris. "They, too, can count."
>
> "Cheer up, dude!" David carries on. "I snarfed half a pound of flour and two awesome eggs. With your milk we can make pancakes. I guess four are in there, each good for $1, makes $4 in total, or $2 for each of us. We don't need the girls! Let's bring off this coup together!"
>
> Before Chris can answer, Belle turns to her little brother: "Edgar, what do you have to offer, anyway?"
>
> "I've carried off a pack of baking soda, and vanilla pudding powder, and half a piece of butter," Edgar itemizes. "Well, that's what has been left. With the milk we could make vanilla pudding, but it appears to me that making pancakes with David would be more profitable for Chris."
>
> "Hey!" Anna bursts out joyfully. "If we throw together all our ingredients, we can make an absolutely delicious cherry-quark cake. We can easily sell it for $15, which, divided by five, gives a clear profit of $3 for each of us!"
>
> "Great idea!" agrees Belle. "And if we then decide to rather have it ourselves, we simply divide it among us."
>
> "Awesome!" says Edgar. "Meanwhile, I'll try to figure out how to divide the cake as fairly as possible, to make sure there will be no envy."

While Edgar sneaks off to Chapter 8 starting on page 507 to take a sneak peek at fair division of cakes, or simply cake-cutting, in the present chapter we are concerned with cooperative game theory. In Section 3.1, we describe its foundations; in particular, we will talk about how coalitions form and how the players divide the payoffs from cooperation. We discuss two approaches to payoff division—one based on stability considerations (Section 3.2 starting on page 149) and another one based on fairness concerns (Section 3.3 starting on page 162), captured by the notions of the core and the Shapley[1] value, respectively. Further, we will discuss representation issues for cooperative games in Section 3.4 starting on page 171, and will then investigate the

---

[1] Lloyd S. Shapley was awarded the 2012 *Nobel Prize in Economics*, jointly with Alvin E. Roth, for the theory of stable allocations and the practice of market design. We will meet Shapley many times in this chapter.

complexity of problems associated with representing cooperative games and computing good outcomes of succinctly represented games in Section 3.5 starting on page 178. Finally, we will study a particularly interesting class of cooperative games, the hedonic games, in Section 3.6 starting on page 198.

3.1 Foundations

In this section we lay the foundations of cooperative game theory and introduce our notation. In particular, we will introduce the class of cooperative games with transferable utility and some of its important subclasses.

3.1.1 Cooperative Games with Transferable Utility

We start by describing cooperative games with transferable utility.

3.1.1.1 Coalition Structures, Characteristic Functions, and Payoff Vectors

As in Chapter 2, let $P = \{1, 2, \ldots, n\}$ be a set of players. Sometimes we refer to players by names rather than numbers. Let 2^P denote the *power set of* P, i.e., the set of subsets of P. For example, Figure 3.1 shows the five players—Anna, Belle, Chris, David, and Edgar—from the game given above and the four coalition structures mentioned there.

The game in our example is a *cooperative game with transferable utility*. Such a game is given by a pair $G = (P, v)$, where $P = \{1, 2, \ldots, n\}$ is the set of players and $v : 2^P \to \mathbb{R}^+$ is the *characteristic function* (sometimes also referred to as the *coalitional function*), which for each subset (or *coalition*) $C \subseteq P$ of players indicates the utility (or gain) $v(C)$ that these players attain by working together. Here $\mathbb{R}^+$ denotes the set of nonnegative real numbers. (Occasionally, we will also consider games whose characteristic function can take negative values.) It is common to assume (and we will do so unless stated otherwise) that the characteristic function v of such a game $G = (P, v)$ has the following properties:

1. *Normalization*: $v(\emptyset) = 0$.
2. *Monotonicity*: $v(C) \leq v(D)$ for all coalitions C and D with $C \subseteq D$.

The characteristic function of the game from Figure 3.1 appears to be monotonic—at least insofar as one can tell from the gains of the coalitions shown there. For example, for the coalitions $C_1 = \{\text{Anna}\}$, $C_2 = \{\text{Anna}, \text{Belle}\}$, $C_3 = \{\text{Anna}, \text{Belle}, \text{Chris}\}$, and $C_4 = \{\text{Anna}, \text{Belle}, \text{Chris}, \text{David}, \text{Edgar}\}$, which

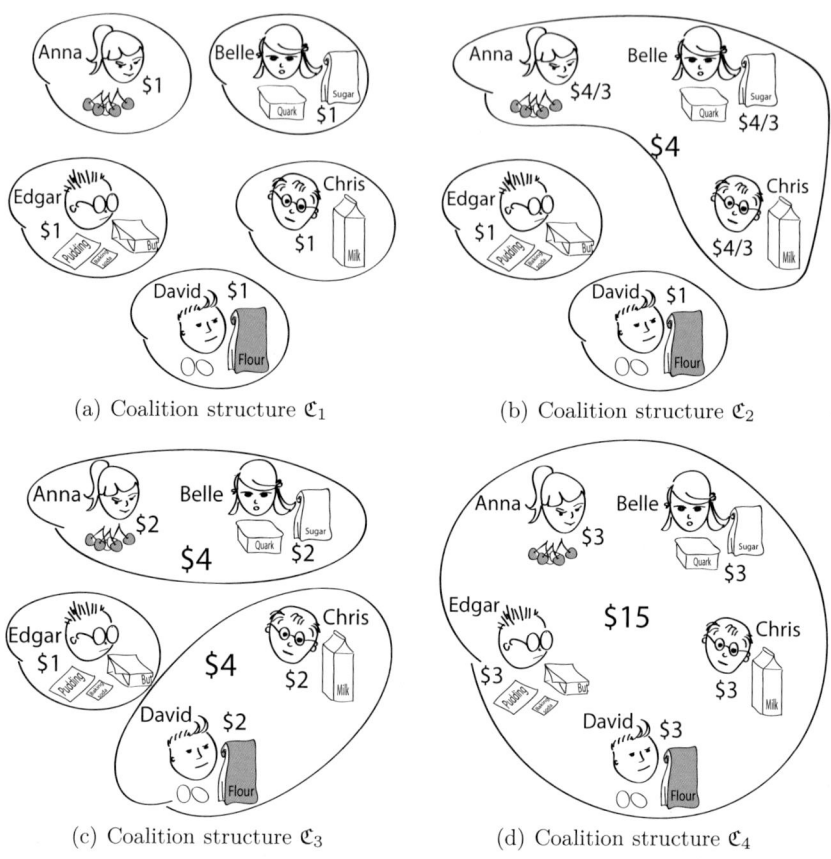

Fig. 3.1 Four coalition structures

are ordered in the ascending inclusion chain $C_1 \subseteq C_2 \subseteq C_3 \subseteq C_4$, we have the (in)equalities $v(C_1) = 1 \leq 4 = v(C_2)$, $v(C_2) = 4 = v(C_3)$, and $v(C_3) = 4 \leq 15 = v(C_4)$.

Figure 3.1 only gives the values of the characteristic function for some of the possible coalitions. For $n = 5$ players, one would need to specify the values for a total of $2^5 = 32$ coalitions. In general, cooperative games require a representation whose size is exponential in the number of players, which is asking for trouble regarding algorithmic feasibility. However, there are ways to handle this problem, and some of them will be discussed later on.

A *coalition structure* for a cooperative game $G = (P, v)$ with transferable utility is a partition $\mathfrak{C} = \{C_1, C_2, \ldots, C_k\}$ of P into pairwise disjoint coalitions, i.e., $\bigcup_{i=1}^{k} C_i = P$ and $C_i \cap C_j = \emptyset$ for $i \neq j$. The simplest coalition structure consists of only one coalition, the so-called *grand coalition*, embracing all players. Figure 3.1(d) shows the grand coalition in our example. The other extreme can be seen in Figure 3.1(a): A coalition structure made up of $n = 5$

3.1 Foundations

coalitions each containing only a single player. Here, all players act on their own. Formally, the four coalition structures from Figure 3.1 are represented as follows:

1. Figure 3.1(a): $\mathfrak{C}_1 = \{\{\text{Anna}\}, \{\text{Belle}\}, \{\text{Chris}\}, \{\text{David}\}, \{\text{Edgar}\}\}$,
2. Figure 3.1(b): $\mathfrak{C}_2 = \{\{\text{Anna, Belle, Chris}\}, \{\text{David}\}, \{\text{Edgar}\}\}$,
3. Figure 3.1(c): $\mathfrak{C}_3 = \{\{\text{Anna, Belle}\}, \{\text{Chris, David}\}, \{\text{Edgar}\}\}$,
4. Figure 3.1(d): $\mathfrak{C}_4 = \{\{\text{Anna, Belle, Chris, David, Edgar}\}\}$.

For each coalition C, the value $v(C)$ merely indicates the joint gains of the players in C. One would then have to determine how these gains are to be divided amongst the players. An *outcome* of a cooperative game $G = (P, v)$ with transferable utility is given by a pair $(\mathfrak{C}, \vec{a})$, where $\mathfrak{C}$ is a coalition structure and $\vec{a} = (a_1, a_2, \ldots, a_n) \in \mathbb{R}^n$ is a *payoff vector* such that

$$a_i \geq 0 \text{ for each } i \in P \quad \text{and} \quad \sum_{i \in C} a_i = v(C) \text{ for each coalition } C \in \mathfrak{C}.$$

That is, every player $i \in P$ receives the nonnegative amount a_i, and, for each coalition C, the profit $v(C)$ earned jointly by the members of C is completely distributed within this coalition (this is referred to as the *efficiency* condition). The phrase "transferable utility" refers to the fact that players in a coalition can make monetary transfers to each other, which allows them to share the value of their coalition in any way they like. In the four coalition structures from Figures 3.1(a) through 3.1(d), the profit $v(C)$ made by each coalition C in the corresponding coalition structure is simply distributed uniformly among the players, so that the amount each player receives is simply the value of their coalition, divided by the number of players in it:

1. Figure 3.1(a) shows the outcome $(\mathfrak{C}_1, \vec{a}_1)$ with $\vec{a}_1 = (1, 1, 1, 1, 1)$,
2. Figure 3.1(b) shows the outcome $(\mathfrak{C}_2, \vec{a}_2)$ with $\vec{a}_2 = (4/3, 4/3, 4/3, 1, 1)$,
3. Figure 3.1(c) shows the outcome $(\mathfrak{C}_3, \vec{a}_3)$ with $\vec{a}_3 = (2, 2, 2, 2, 1)$, and
4. Figure 3.1(d) shows the outcome $(\mathfrak{C}_4, \vec{a}_4)$ with $\vec{a}_4 = (3, 3, 3, 3, 3)$.

It is also feasible, however, for some players in a coalition to receive a relatively larger share while some other players receive a relatively smaller share of the gains, depending, for instance, on their contribution to success. The only requirement is that the gains of a coalition are completely distributed among its members. For example, while $\vec{a}_2 = (4/3, 4/3, 4/3, 1, 1)$ is a valid payoff vector for the coalition structure $\mathfrak{C}_2$ from Figure 3.1(b), the payoff vector $\vec{a}_2' = (1/3, 4/3, 4/3, 2, 1)$ does not match $\mathfrak{C}_2$, since transfers between distinct coalitions are not allowed, i.e., $(\mathfrak{C}_2, \vec{a}_2')$ is not a valid outcome of this game.

There are also *cooperative games with nontransferable utility*. Think, for example, of n huskies that are supposed to drag several sledges from a research ship to a research station in Antarctica. Every husky has a different owner. Every sledge is dragged by a pack of huskies that tackle their task jointly. Depending on how such a pack (or "coalition") of huskies is assembled, they can solve their task more or less successfully. However, every husky will be rewarded by its owner only, for example by getting more or less food, depending on how fast this husky's sledge has reached its destination. That means that gains are not transferable within a coalition.

In what follows, up to and including Section 3.5, we will be concerned with cooperative games with transferable utility only, and therefore we will omit the phrase "with transferable utility" when speaking of a cooperative game.

3.1.1.2 Superadditivity

A cooperative game $G = (P, v)$ is said to be *superadditive* if for any two disjoint coalitions C and D we have

$$v(C \cup D) \geq v(C) + v(D). \tag{3.1}$$

In a superadditive game one may expect the grand coalition to form, since any two coalitions can merge without loss: In the worst case, the gains of the two merging coalitions would only add up, and it is quite possible that the total gains are even higher due to synergy effects. For this reason, we may identify outcomes of superadditive cooperative games with payoff vectors for the grand coalition, and not consider more complicated coalition structures. In the rest of the chapter, we will focus on superadditive games unless specified otherwise.

For instance, the game G whose characteristic function is defined by $v(C) = \|C\|^2$ is superadditive because for any two disjoint coalitions C and D, we have

$$v(C \cup D) = \|C \cup D\|^2 = (\|C\| + \|D\|)^2 \geq \|C\|^2 + \|D\|^2 = v(C) + v(D).$$

In contrast, the game from Figure 3.1 is not superadditive, since

$$v(\{\text{Anna}, \text{Belle}\} \cup \{\text{Chris}\}) = v(\{\text{Anna}, \text{Belle}, \text{Chris}\})$$
$$= 4$$
$$< 5 = 4 + 1 = v(\{\text{Anna}, \text{Belle}\}) + v(\{\text{Chris}\}),$$

contradicting (3.1).

3.1.2 Special Subclasses of Cooperative Games

We will now discuss two prominent subclasses of cooperative games, namely, convex games and simple games. Briefly, convex games describe settings with increasing returns to cooperation, and simple games capture scenarios where each coalition either wins or loses.

3.1.2.1 Convex Games

Convex games form a subclass of superadditive games, and were introduced by Shapley [858].

A function $s: 2^P \to \mathbb{R}^+$ is said to be *supermodular* if for any two subsets $C, D \subseteq P$, we have

$$s(C \cup D) + s(C \cap D) \geq s(C) + s(D).$$

One can show (see, e.g., [357]) that this condition is equivalent to

$$s(C \cup \{i\}) - s(C) \leq s(D \cup \{i\}) - s(D) \qquad (3.2)$$

for all subsets $C, D \subseteq P$ with $C \subseteq D \subseteq P$ and all $i \in P \setminus D$. A cooperative game $G = (P, v)$ is said to be *convex* if its characteristic function v is supermodular. Both sides of the inequality in Equation (3.2) are a *marginal contribution of player i to the gains of an existing coalition*. That is why this inequality is also called the *increasing marginal return* condition (the term "nondecreasing marginal return condition" would actually be more correct).

A large cocktail party is in full swing: Well-dressed men and women move around, forming small groups. Some of the people know each other, but not even the hosts know everyone. Each partygoer's enjoyment is determined by the composition of her current group and is equal to the number of people she knows in this group (not counting herself). We can think of this setting as a coalitional game, where the players are the partygoers, and the characteristic function v associates each group (coalition) C with the total enjoyment of its members. Clearly, $v(C)$ equals the number of ordered pairs $i, j \in C$ such that $i \neq j$ and i and j know each other.

To see that this game is convex, consider two coalitions C and D such that $C \subseteq D$ and a player $i \in P \setminus D$. Let F_i be the set of people i knows. Then $v(C \cup \{i\}) - v(C) = 2\|F_i \cap C\|$, $v(D \cup \{i\}) - v(D) = 2\|F_i \cap D\|$. Since C is a subset of D, we have $\|F_i \cap C\| \leq \|F_i \cap D\|$, which proves our claim.

We remark that this game belongs to the class of *induced subgraph games*, which will be discussed in Section 3.4.2.1; however, in general, induced subgraph games are not necessarily convex.

Convex games have a number of advantageous properties. First, it follows immediately from the definition that every convex game is superadditive; the converse implication does not hold in general. Moreover, every subgame of a convex game is convex, where the *subgame of a cooperative game* $G = (P, v)$ for a nonempty subset $T \subseteq P$ is defined to be the pair $G_T = (T, v_T)$ with $v_T(C) = v(C)$ for all coalitions $C \subseteq T$. Further, in the next sections we will see that convex games always admit outcomes that are simultaneously fair and stable.

3.1.2.2 Simple Games

Another important class of cooperative games are games whose characteristic function only takes two values, namely, 0 and 1, indicating whether the respective coalition is successful ($v(C) = 1$) or not ($v(C) = 0$). Such games are known as simple games, and were already studied by von Neumann and Morgenstern [736] and Shapley [857]. The book by Taylor and Zwicker [910] is a comprehensive study of simple games.

Formally, a cooperative game $G = (P, v)$ is said to be *simple* if

1. $v(C) \in \{0, 1\}$ for each coalition $C \subseteq P$,
2. if $C \subseteq D$ and $v(C) = 1$ then $v(D) = 1$, and
3. $v(\emptyset) = 0$ and $v(P) = 1$.

In a simple game a coalition C *wins* if $v(C) = 1$, and it *loses* if $v(C) = 0$. The second property expresses monotonicity of v. The third property excludes the trivial games where all coalitions win or all coalitions lose.

A particularly well-studied class of simple games is that of *weighted voting games*. In these games, each player is associated with a weight, and there is a threshold q (the *quota*) that determines whether a coalition is winning or losing; namely, a coalition *wins* if its total weight is at least q, and it *loses* otherwise. These games are sometimes referred to as *weighted majority games* (if the quota is half the total weight of the grand coalition) or *weighted threshold games*. The following example describes a game in this class.

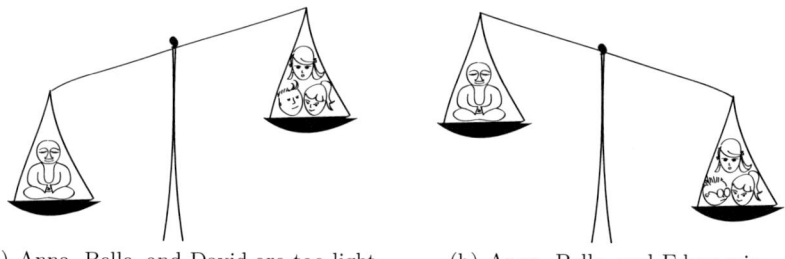

(a) Anna, Belle, and David are too light   (b) Anna, Belle, and Edgar win

Fig. 3.2 A weighted voting game

Anna, Belle, David, and Edgar (who just in time returned from Chapter 8; unfortunately, Chris is currently in custody—the reason will be

3.1 Foundations

revealed only in Chapter 7) make an incredible discovery on a walk in the woods: They find a treasure!

First they can't believe their eyes. But when they approach, they recognize beyond any doubt a great golden statue formed like a buddha, resting in a hollow in the forest ground. Edgar's suggestion to call some adults for help is outright rejected because they of course wish to unearth this treasure on their own and keep it for themselves. They quickly go and get a trolley and tools and then they build a lifting device that looks like a pair of scales, which they intend to use to lift the buddha from its hollow and to load it onto their trolley. They roll the buddha on one of the two scales and one by one they take turns, jumping onto the other scale to lift the buddha, but they all fail—the buddha is simply too heavy. Apparently, it truly is made from solid gold.

"How about," asks Belle, "if the four of us jump on together?"

"No," David replies, and all of a sudden greed twinkles in his eyes. "Actually, I'd say the buddha belongs to whoever is able to lift it, and if all four of us lift it together, I'm gonna have to share it with three others. Why don't we try first if two or three of us succeed?"

But no party of two is successful either. The buddha immovably sits on its scale in the hollow. Next, David tries it jointly with Anna and Belle who are some years older and accordingly heavier than the two boys. But the buddha still doesn't move (see Figure 3.2(a)).

"O.K.!" David turns to Edgar, "then you too may join in."

"Not before you get off!" Edgar retorts unrelentingly.

"Why me?" outraged David asks. "Why don't Anna or your sister get off in the first place?"

"That's bullshit, David!" Anna explains to him. "I'm heavier than Edgar and so is Belle. If we weren't able to lift the buddha with you, what sense would it make for him to take my place or hers? But you are about the same weight as him, I would say, perhaps Edgar might even be a bit heavier than you."

David can't help but concede, and the two boys swap their places. Finally, the buddha is rising (see Figure 3.2(b)) and they can load it onto their trolley.

"We win!" Edgar shouts out with joy and, turning to David, he adds, "well, bad luck for you!"

"By his own rules," Belle says, "David indeed has bad luck and has lost. However, if he would have followed my suggestion, he would have won with us. That is why I'd say the buddha should belong to all of us—including Chris, although he couldn't be here at all."

All agree and David is relieved. "Speaking of Chris," he then says, "isn't his court hearing scheduled for this afternoon? I think we should hurry up."

We will turn to Chris's court hearing only in Chapter 7. Here, however, we are concerned with weighted voting games—a class of simple games that admit a succinct representation.

A *weighted voting game* is a simple cooperative game $G = (P, v)$ given by the weights $w_1, w_2, \ldots, w_n \in \mathbb{R}^+$ of the n players and a quota $q \in \mathbb{R}^+$ with $0 < q \leq \sum_{i=1}^{n} w_i$. We write $G = (w_1, w_2, \ldots, w_n; q)$. A coalition $C \subseteq P$ *wins* (i.e., $v(C) = 1$) if $\sum_{i \in C} w_i \geq q$, otherwise C *loses* (i.e., $v(C) = 0$). The example above can be described as a weighted voting game: If Anna weighs 108 pounds, Belle 106 pounds, David 62 pounds, Edgar 66 pounds, and the buddha weighs 277 pounds, then the corresponding game is $G = (108, 106, 62, 66; 277)$. Besides the grand coalition $P = \{$Anna, Belle, David, Edgar$\}$, which puts $\sum_{i \in P} w_i = 342$ pounds on the scale, the only winning coalition is $E = \{$Anna, Belle, Edgar$\}$ (see Figure 3.2(b)), exceeding the quota of 277 pounds with $\sum_{i \in E} w_i = 280$ pounds. All other coalitions lose, including the coalition $D = \{$Anna, Belle, David$\}$ from Figure 3.2(a), since $\sum_{i \in D} w_i = 276 < 277$.

The name "weighted voting game" refers to the fact that such games can model decision processes in the political bodies of representative democracies (such as the US House of Representatives, the British House of Commons, and *Deutscher Bundestag*, the Lower House of the German Parliament) as well as in the United Nations Security Council and in the governing body of the International Monetary Fund.

Table 3.1 Allocation of seats in the *17th Deutscher Bundestag* (as of May 30, 2011)

Party	CDU/CSU	SPD	FDP	Die Linke	Bündnis 90/Die Grünen
Number of seats	237	146	93	76	68

For example, Table 3.1 shows how the 620 seats in the *17th Deutscher Bundestag* are allocated. The number of seats of a party can be taken as its weight. Abstracting from the details of the actually rather complicated, somewhat bureaucratic legislative procedure, let us assume that a simple majority suffices to pass a bill. This gives a quota of $q = 311$. If we also ignore the possibility of abstentions and, furthermore, assume that the parliament members belonging to the same party always vote in unison (i.e., there are no dissenters), we can represent the *Bundestag* as the weighted voting game

$$G = (237, 146, 93, 76, 68; 311).$$

3.2 Stability in Cooperative Games

Altogether there are $2^5 = 32$ possible coalitions C of parties in the *17th Deutscher Bundestag*. For some of these, Table 3.2 shows which of them win in this weighted voting game (i.e., $v(C) = 1$) and which of them lose (i.e., $v(C) = 0$).

Table 3.2 Some theoretically possible coalitions in the *17th Deutscher Bundestag*

Coalition C	Weight of C	$v(C)$
{CDU/CSU, SPD}	383	1
{CDU/CSU, FDP}	330	1
{CDU/CSU, Die Linke}	313	1
{CDU/CSU, Bündnis 90/Die Grünen}	305	0
{SPD, FDP}	239	0
{SPD, Die Linke}	222	0
{SPD, Bündnis 90/Die Grünen}	214	0
{SPD, FDP, Die Linke}	315	1
{SPD, FDP, Bündnis 90/Die Grünen}	307	0
{SPD, Die Linke, Bündnis 90/Die Grünen}	290	0
{SPD, FDP, Die Linke, Bündnis 90/Die Grünen}	383	1
{FDP, Die Linke, Bündnis 90/Die Grünen}	237	0

The reader may wonder if every simple game is, in fact, a weighted voting game; we will revisit this question in Section 3.4.

3.2 Stability in Cooperative Games

Not all outcomes of cooperative games are equally appealing. In this section, we discuss an approach to identify desirable outcomes that is based on stability considerations: An outcome is viewed as good as long as it is resistant to deviations by the players. Then, in the next section, we investigate a very different approach, which focuses on fairness.

Stability plays an important role in both cooperative and noncooperative game theory. In noncooperative games, it is usually associated with the notion of Nash equilibrium (a vector of strategies such that no player has an incentive to deviate from her strategy, provided that the other players do not deviate from their strategies either, see Definition 2.4 on page 50). In contrast, when defining stability in cooperative games, we usually consider deviations by *coalitions* of players.

We focus on superadditive games, and hence we assume that the only acceptable outcomes are the ones where the grand coalition forms. Now, the way in which the value of the grand coalition is distributed among the players is crucial for the stability of the grand coalition. Indeed, since every player in a cooperative game is primarily interested in maximizing her own profit, a group of players may break off from the grand coalition if it is profitable for each of them to do so. If that happens, the game is unstable and the grand coalition breaks up into several smaller coalitions.

3.2.1 Imputations

In this section, we will introduce the notion of imputations, which capture a basic idea of stability. They will motivate us to subsequently consider more general deviations, which then leads to the definition of the core.

Players have an incentive to join the grand coalition in a cooperative game $G = (P, v)$ if the profit $v(P)$ of the grand coalition can be distributed among the individual players according to a payoff vector $\vec{a} = (a_1, a_2, \ldots, a_n) \in \mathbb{R}^n$ in a way that satisfies the *individual rationality* condition:

$$a_i \geq v(\{i\}) \qquad \text{for all } i \in P.$$

That is, no player can make more profit alone than as a member of the grand coalition. We collect all such payoff vectors, which are called the *imputations for G*, in the following set:

$$\mathcal{I}(G) = \left\{ (a_1, a_2, \ldots, a_n) \in \mathbb{R}^n \;\middle|\; \sum_{i=1}^{n} a_i = v(P) \text{ and } a_i \geq v(\{i\}) \text{ for all } i \in P \right\}.$$

The set $\mathcal{I}(G)$ is nonempty exactly if $v(P) \geq \sum_{i \in P} v(\{i\})$.

3.2.2 The Core of a Cooperative Game

Imputations only pose weak constraints to a payoff vector. We can make further assumptions on them to obtain a stronger concept of stability. The *core of G* is defined as follows:

$$\mathrm{Core}(G) = \left\{ (a_1, a_2, \ldots, a_n) \in \mathcal{I}(G) \;\middle|\; \sum_{i \in C} a_i \geq v(C) \text{ for all coalitions } C \subseteq P \right\}.$$

If the profit of the grand coalition is distributed among the players according to an imputation in the core of G, no group of players has an incentive to break off from the grand coalition, as it is impossible to ensure that every member of the deviating group profits from the deviation. In this sense the game is stable.

The following example is based on an example by Osborne and Rubinstein [762, page 259].

3.2 Stability in Cooperative Games

Anna, Belle, and their friends go to a gaming night to play chess. In total, there are $n \geq 3$ players who want to play. The success of the gaming night depends on how many players can actually play, i.e., on how many player pairs can be formed. Every pair of players that can be formed receives one success point. That is, the characteristic function of this game $G = (P, v)$ is given by

$$v(C) = \begin{cases} \|C\|/2 & \text{if } \|C\| \text{ is even} \\ (\|C\|-1)/2 & \text{if } \|C\| \text{ is odd} \end{cases}$$

for each coalition $C \subseteq P$.

If $n \geq 4$ is even, we have $(1/2, \ldots, 1/2) \in Core(G)$: It suffices to consider deviations by pairs of players (do you see why?), and any two players (whether or not they are currently matched to play with each other) jointly receive one point under this imputation, so they cannot do better by deviating. In fact, it can be shown that $(1/2, \ldots, 1/2)$ is the only imputation in the core of G, i.e., $Core(G) = \{(1/2, \ldots, 1/2)\}$; we leave the proof of this fact as an exercise to the reader.

However, if $n \geq 3$ is odd, one player remains without a partner. This implies that the core of G is empty in this case. Indeed, suppose that the core of G is not empty for, say, $n = 3$ players, and hence contains a vector $\vec{a} = (a_1, a_2, a_3)$. Then, since $a_1 + a_2 + a_3 = v(\{1,2,3\}) = 1$, at least one of the values a_i is positive, say $a_1 > 0$. Thus $a_2 + a_3 < 1$. However, since $v(\{2,3\}) = 1$, we have a contradiction to the assumption that $\vec{a}$ is in $Core(G)$. Hence, $Core(G)$ is empty for $n = 3$ (and, by a similar argument, for any odd n).

Cooperative games whose core is empty are unstable. When does a game have a nonempty core? Notably, the core is defined by a series of linear constraints and it is therefore natural to approach its computation by linear programming. Indeed, for superadditive games $G = (P, v)$, this question boils down to checking whether the following linear program (see also page 122) with variables $a_1, a_2, \ldots, a_n \in \mathbb{R}$ and $2^n + n$ constraints has a feasible solution:

$$\begin{aligned} a_i &\geq 0 & \text{for all } i \in P \\ \sum_{i \in P} a_i &= v(P) \\ \sum_{i \in C} a_i &\geq v(C) & \text{for all } C \subset P. \end{aligned}$$

Note that the first set of constraints is redundant. However, we prefer this presentation because it is helpful to illustrate relationships with relaxations of the core that we define later. Moreover, for some classes of cooperative games, it is possible to solve this linear program efficiently (i.e., in time polynomial in the number n of players), even though the number of constraints in it is exponential in n.

We will now discuss the existence of core-stable outcomes in the two classes of cooperative games introduced earlier in this chapter, namely, convex games and simple games.

3.2.2.1 The Core of a Convex Game

Intuitively, inequality (3.2) in the definition of a convex game implies that in a convex game a player's incentive to join a coalition increases as the coalition gets larger. This makes the grand coalition particularly attractive. Therefore, it is not too surprising that in a convex game the players always have a way of dividing up the profits of the grand coalition so as to ensure stability.

Theorem 3.1 (Shapley [858]). *Every convex game has a nonempty core.*

Proof. Given a convex game $G = (P, v)$ with $P = \{1, 2, \ldots, n\}$, we construct a payoff vector $\vec{a} = (a_1, a_2, \ldots, a_n) \in \mathbb{R}^n$ and show that $\vec{a}$ is in the core of G. Set $D_0 = \emptyset$, and define $D_i = \{1, \ldots, i\}$ and $a_i = v(D_i) - v(D_{i-1})$ for $1 \leq i \leq n$. We have

$$a_1 = v(\{1\}), \quad a_2 = v(\{1,2\}) - v(\{1\}), \quad \ldots, \quad a_n = v(P) - v(P \smallsetminus \{n\}),$$

i.e., every player gets her *marginal contribution* to the coalition of her predecessors, which is the amount by which a player increases the profit of a coalition by joining it. It follows that $\vec{a}$ is individually rational: For each $i \in P$, inequality (3.2) with $C = \emptyset$ and $D = D_{i-1}$ implies:

$$\begin{aligned} a_i = v(D_i) - v(D_{i-1}) &= v(D_{i-1} \cup \{i\}) - v(D_{i-1}) \\ &\geq v(\emptyset \cup \{i\}) - v(\emptyset) = v(\{i\}). \end{aligned} \quad (3.3)$$

Moreover, $\vec{a}$ is efficient, since

$$\sum_{i=1}^n a_i = \sum_{i=1}^n (v(D_i) - v(D_{i-1})) = v(D_n) - v(D_0) = v(P),$$

which implies $\vec{a} \in \mathcal{I}(G)$. It remains to show that $\vec{a}$ is in $Core(G)$. On the one hand, for each coalition $C = \{i, j, \ldots, k\}$ with $1 \leq i < j < \cdots < k \leq n$, we have

$$v(C) = v(\{i\}) - v(\emptyset) + v(\{i,j\}) - v(\{i\}) + \cdots + v(C) - v(C \smallsetminus \{k\}).$$

On the other hand, for each term of this expression, we have:

$$\begin{aligned} v(\{i\}) - v(\emptyset) &\leq v(D_i) - v(D_{i-1}) = a_i \\ v(\{i,j\}) - v(\{i\}) &\leq v(D_j) - v(D_{j-1}) = a_j \\ &\vdots \\ v(C) - v(C \smallsetminus \{k\}) &\leq v(D_k) - v(D_{k-1}) = a_k. \end{aligned}$$

3.2 Stability in Cooperative Games

The first of these inequalities is just (3.3), and the others can be derived by further applications of (3.2). Hence, $v(C) \leq a_i + a_j + \cdots + a_k$. Thus, the members of coalition C have no incentive to leave the grand coalition when they get paid according to $\vec{a}$. Since C is an arbitrary coalition, this is true for all coalitions. It follows that $\vec{a}$ is in the core of G. This completes the proof. ❑

3.2.2.2 The Core of a Simple Game

In simple games, it is conventional to assume that players form the grand coalition, even if this is not a social welfare-maximizing outcome (see, e.g., Section 9.6 starting on page 635 for various measures of social welfare). Following the literature, we will make this assumption throughout this section. In particular, when discussing the existence of core-stable outcomes, we assume that only outcomes where players form the grand coalition can be in the core.

An important notion in simple games is that of a veto player: A player $i \in P$ in a simple game $G = (P, v)$ is said to be a *veto player* if $v(C) = 0$ for each coalition $C \subseteq P \smallsetminus \{i\}$ (by monotonicity, this is equivalent to $v(P \smallsetminus \{i\}) = 0$). That is, a veto player is necessary (but not always sufficient) to form a winning coalition. Note that a simple game may have several veto players: For instance, in the game $G = (P, v)$ given by

$$v(C) = \begin{cases} 1 & \text{if } C = P \\ 0 & \text{otherwise,} \end{cases}$$

every player is a veto player. Further, in the weighted voting game $G = (108, 106, 62, 66; 277)$ from Figure 3.2, in which Anna, Belle, David, and Edgar want to jointly lift a treasure, each of Anna, Belle, and Edgar is a veto player.

On the other hand, some simple games have no veto players. Consider, for instance, a three-player game where a coalition wins if and only if it consists of two or more players: In this game, for every player there is a winning coalition that does not contain him. As another example, there is no veto player in the *Bundestag* game, as no party in Table 3.1 (where each of CDU/CSU and Bündnis 90/Die Grünen counts as just *one* party or, respectively, player) is indispensable for victory. For example, all parties except the heaviest, CDU/CSU, have a joint total weight of $383 > 311$. Assuming that all these parties form a special-purpose alliance against CDU/CSU, they can pass a bill or shoot it down against the will of CDU/CSU. A central point in forming the government, where CDU/CSU coalesced with FDP in 2011, was the fact that this coalition with a joint weight of $330 > 311$ was capable of making decisions alone, against the opposition parties.

The following theorem, which seems to be folklore, establishes that the presence of veto players is crucial for core stability. (We use the terms *core stability* or *core-stable outcome* synonymously with the statements that *the core of a game is nonempty* or that *it contains a payoff vector*.)

Theorem 3.2. *A simple game has a nonempty core if and only if it has a veto player.*

Proof. Suppose that $i \in P$ is a veto player in a simple game $G = (P, v)$. Define a payoff vector $\vec{a} = (a_1, a_2, \ldots, a_n)$ by setting $a_i = 1$ and $a_j = 0$ for all $j \neq i$, $1 \leq j \leq n$. Then $\vec{a}$ is in $\mathcal{I}(G)$, since $a_k \geq v(\{k\})$ for all $k \in P$ and $\sum_{k \in P} a_k = 1 = v(P)$. For an arbitrary coalition $C \subseteq P$, we have $\sum_{k \in C} a_k = 1 \geq v(C)$ if $i \in C$, and $\sum_{k \in C} a_k = 0 = v(C)$ if $i \notin C$. Thus, if the players are paid according to $\vec{a}$, no coalition has an incentive to leave the grand coalition. It follows that $\vec{a}$ is in the core of G, so $Core(G) \neq \emptyset$.

Conversely, suppose now that no player in G is a veto player. Let $\vec{a} = (a_1, a_2, \ldots, a_n)$ be an arbitrary imputation in $\mathcal{I}(G)$. Since $\sum_{k \in P} a_k = v(P) = 1$, there exists an i with $a_i > 0$, $1 \leq i \leq n$. Since in particular i is not a veto player in G, we have $v(P \setminus \{i\}) = 1$. However, $\sum_{k \in P \setminus \{i\}} a_k = 1 - a_i < 1$. Thus, if the players in $P \setminus \{i\}$ are paid according to $\vec{a}$, they do have an incentive to leave the grand coalition, or, in other words, to throw i out. Therefore, $\vec{a}$ is not in the core of G. Since $\vec{a}$ is an arbitrary imputation, this is true for all imputations in $\mathcal{I}(G)$, and thus $Core(G) = \emptyset$. ❑

By a similar argument, one can show that for each simple game G with a nonempty core the following holds: An imputation $\vec{a} = (a_1, a_2, \ldots, a_n)$ is in $Core(G)$ if and only if $a_i = 0$ for each player i that is not a veto player.

Can you construct a specific simple game with a veto player, which therefore, by Theorem 3.2, has a nonempty core? Can you also construct a simple game *without* a veto player? By Theorem 3.2, the core of this game would be empty, which, by Theorem 3.1, would imply that it is not convex. And can you come up with a game that has a nonempty core, but is nevertheless not convex?

Our definition of the core assumes that the players form the grand coalition. Another approach is to allow the players to form arbitrary coalition structures, with the payoff of each coalition shared among its members, but nevertheless require that no group of players can benefit by splitting off from their respective coalitions so as to form a new coalition of their own. For superadditive games, focusing on the grand coalition is without loss of generality: It can be verified that a superadditive game has a nonempty core under the modified definition if and only if it has a nonempty core under the original definition.

However, if the game is not superadditive, this is not the case, and in particular Theorem 3.2 is no longer true if more complicated coalition structures are permitted. Indeed, consider a four-player simple game where a coalition wins if it contains at least two players, and loses otherwise. This

3.2 Stability in Cooperative Games

game is not superadditive: If the set of players is $P = \{1,2,3,4\}$, we have $v(\{1,2\}) + v(\{3,4\}) = 1 + 1 > v(P)$ (one can think of this game as a non-superadditive version of the chess game on page 150). It has no veto player, as any three players form a winning coalition. However, it admits a stable outcome where the players form two size-two coalitions and the payoff of each coalition is shared equally among its members (so that the payoff vector is $(1/2, 1/2, 1/2, 1/2)$). Importantly, in this outcome the grand coalition does not form: This makes sense, as, by splitting into two coalitions, the players increase their total payoff.

3.2.3 Further Stability Concepts in Cooperative Games

Besides the core there are a number of further stability concepts some of which we will consider now.

3.2.3.1 The ϵ-Core and the Least Core of a Cooperative Game

Since the core of a cooperative game G can be empty, Shapley and Shubik [861] introduced the notion of *(strong) ϵ-core of G*, which for $\epsilon \in \mathbb{R}$ is defined by

$$\epsilon\text{-}Core(G) = \left\{ (a_1, a_2, \ldots, a_n) \in \mathcal{PV}(G) \;\middle|\; \sum_{i \in C} a_i \geq v(C) - \epsilon \text{ for all } C \subset P \right\},$$

where $\mathcal{PV}(G)$ denotes the *set of all payoff vectors for P*.

This notion can be interpreted as follows: If $\epsilon > 0$, ϵ-$Core(G)$ consists of payoff vectors that are stable as long as each coalition $C \subset P$ has to pay a penalty of ϵ in order to leave the grand coalition. If $\epsilon < 0$, deviating from the grand coalition is made easier for C, as it receives a bonus of $-\epsilon$. For $\epsilon = 0$, the ϵ-core of G coincides with the core of G, i.e., 0-$Core(G) = Core(G)$. Thus, the ϵ-core generalizes the core. Clearly, even if the core of a cooperative game G is empty, G has a nonempty ϵ-core for large enough ϵ. On the other hand, even if G has a nonempty core, its ϵ-core is empty for small enough (indeed, negative) values of ϵ.

Maschler, Peleg, and Shapley [686] extended this idea by introducing the *least core of G* as the intersection of all nonempty ϵ-cores of G. Alternatively, the least core of G can be defined as its $\tilde{\epsilon}$-core, where $\tilde{\epsilon}$ is chosen so that the $\tilde{\epsilon}$-core of G is nonempty, but the ϵ-core of G is empty for all values $\epsilon < \tilde{\epsilon}$; it can be shown that the value of $\tilde{\epsilon}$ is well-defined. By definition, the least core of a cooperative game is never empty. Intuitively, the least core consists of the most stable outcomes of the game.

The quantity $\tilde{\epsilon}$ is called the *value of the least core*. It can be computed by a linear program similar to the one for the core (see page 151):

$$\begin{aligned}
\min \quad & \epsilon \\
& a_i \geq 0 && \text{for all } i \in P \\
& \sum_{i \in P} a_i = v(P) \\
& \sum_{i \in C} a_i \geq v(C) - \epsilon && \text{for all } C \subset P.
\end{aligned}$$

3.2.3.2 The Cost of Stability

The concept of cost of stability has been proposed much more recently than the stability notions discussed in the previous sections. The intuition behind this concept is as follows. If a game is not stable, i.e., has an empty core, it is possible to stabilize it by offering payments to the players, provided they stick together. That is, an external authority can offer a *subsidy* to the grand coalition, conditional on players not deviating. How much would this stabilization cost? And how hard is it to determine the required subsidy?

An ice cream truck has come to the neighborhood where Anna, Belle, Chris (who luckily has been released on bail), David, and Edgar live. Anna got a chocolate chip cone, and Edgar selected an ice cream sandwich. However, Belle, Chris, and David decided they can get better value for money by buying a whole tub of ice cream and dividing it among themselves. There are two types of ice cream tubs for sale: a small one (24 oz) for $ 8 and a large one (30 oz) for $ 10. Belle has $ 5, and Chris and David have $ 4 each, so they can afford the large tub, and the question is how to divide it among themselves.

"Let us share the ice cream equally, so that each of us gets 10 oz," proposes Belle.

"We can," Chris responds. "But then David and I can get more ice cream by abandoning you and going for the small tub, which we can split equally."

"In fact, something like this is going to happen no matter how we share the ice cream," muses David. "Indeed, at least one of us will get 10 oz or more, which means that there is at most 20 oz left for others, and they are better off getting the small tub."

Chris's mother overhears this conversation. "I do not want you to fight, and I've got some ice cream in the freezer," she says. "How about you buy the big tub, I supplement it with 12 oz of stracciatella ice cream from the freezer, and you share all this ice cream equally?"

"Works for me!" rejoices Belle. "This way any two of us get 28 oz of ice cream, so there is no reason to consider splitting off."

3.2 Stability in Cooperative Games

"In fact, this is a bit too generous," observes David. "10 oz extra would do the trick, or maybe even less—all that matters is that any two of us get at least 24 oz of ice cream."

Figure 3.3 depicts the situation described above.

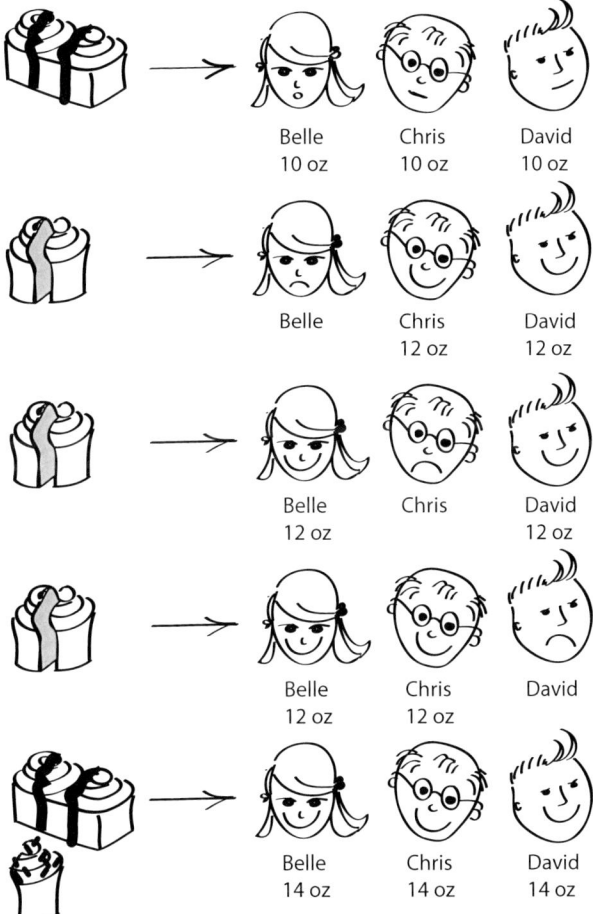

Fig. 3.3 The cost of stability when sharing ice cream

Meir, Bachrach, and Rosenschein [699] give another motivating example, where a significant amount of money is at stake:[2]

---

[2] In their example, the goal is to minimize shared costs rather than to maximize shared gains of the grand coalition, i.e., it refers to an "expense-sharing game" rather than to a "surplus-sharing game."

Suppose three private hospitals in the same community want to purchase an X-ray machine. The standard X-ray machine costs $5 million, but can comply with the requirements of only two hospitals. A more advanced machine, which is capable of fulfilling the needs of all three hospitals, costs $9 million. If all three hospitals join their forces and buy the more advanced, though more expensive, X-ray machine, this will cost them less than buying two standard X-ray machines. However, the three hospital managers cannot settle on how to distribute the cost for this more expensive machine among themselves: If each hospital pays one third of the $9 million, every pair of hospitals (which together had to pay $6 million then) would be better off by leaving the grand coalition and purchasing the $5 million machine for themselves. To resolve this issue, the municipal council proposes to subsidize the more advanced X-ray machine by offering a supplemental payment of $3 million, so each hospital only needs to add $2 million. Then every pair of hospitals only needs to pays $4 million, and so no longer has an incentive to break off.

But wait! Isn't this an incredible waste of tax money? Indeed, the positive effect of stabilizing the grand coalition could have been achieved at a much lower cost: A subsidy of $1.5 million towards the purchase of the more expensive machine would have been enough to ensure that no pair of hospital managers has an incentive to leave the grand coalition. In that case, each individual hospital would have to contribute $2.5 million, which is $5 million for each pair of hospitals, i.e., exactly the amount they would have to pay for the cheaper, less efficient X-ray machine. Since the municipal council is interested in keeping its expenditure as low as possible, that is the right way to go.

Thus, the question motivated by our examples is: What is the *minimum* external supplemental payment needed to stabilize the grand coalition, and how hard is it to determine this payment? This is the question studied by Bachrach et al. [74] (see also this paper's predecessor [75]).[3]

Suppose an external party (like the municipal council in the second example above and Chris's mother in the first example) is interested in stabilizing the grand coalition in a cooperative game with an empty core, and is willing to pay for that. The payment is provided only if the players do not deviate from the grand coalition. This amount plus the actual gains of the grand

---

[3] In noncooperative game theory (see Chapter 2 starting on page 43), the somewhat related notion of *"cost of implementation"* has been studied by Monderer and Tennenholtz [710]. In their model, an interested external party seeks to influence the outcome of a noncooperative game by committing to nonnegative monetary transfers for the different strategy profiles that may be chosen by the players. In a follow-up work, Anderson, Shoham, and Altman [26] introduce and investigate *"internal implementation"* by modeling this interested party explicitly as a player in the game.

3.2 Stability in Cooperative Games

coalition will then be distributed among the players to ensure stability. The *cost of stability for G* is defined to be the smallest supplemental payment that stabilizes G. Formally, for a game $G = (P, v)$ and a supplemental payment of $\Delta \geq 0$, the *adjusted game* $G_\Delta = (P, v')$ is given by $v'(C) = v(C)$ for $C \neq P$ and $v'(P) = v(P) + \Delta$, and the *cost of stability of G* is defined as

$$CoS(G) = \inf\{\Delta \mid \Delta \geq 0 \text{ and } Core(G_\Delta) \neq \emptyset\}.$$

Just as the least core, the cost of stability can be computed by solving a linear program that has a constraint for each coalition:

$$\begin{aligned}
\min \ & \Delta \\
\Delta \ & \geq 0 \\
a_i \ & \geq 0 & \text{for all } i \in P \\
\sum_{i \in P} a_i \ & = v(P) + \Delta \\
\sum_{i \in C} a_i \ & \geq v(C) & \text{for all } C \subset P.
\end{aligned}$$

Note that we have to impose the constraint $\Delta \geq 0$: Without this constraint, if the game has a nonempty core, the value of this linear program may be negative, which would correspond to imposing a fine on the grand coalition.

What is the relationship between the value of the least core and the cost of stability? Both of these quantities are (strictly) positive if and only if the core is empty. However, they capture two very different approaches to dealing with coalitional instability: The least core corresponds to punishing undesirable behavior (i.e., making deviations more costly), whereas the cost of stability corresponds to encouraging desirable behavior by rewarding it (i.e., by making staying in the grand coalition more attractive for the players).

For instance, let $n \geq 2$ and consider the following two n-player games, $G_1 = (P, v_1)$ and $G_2 = (P, v_2)$. In G_1, a coalition has value $n-1$ if it contains player 1 or player 2 (or both of them), and 0 otherwise. In G_2, the value of every nonempty coalition is 1; the empty coalition has value 0. (Note that these games are not superadditive; however, we still seek to stabilize the grand coalition.) We have $CoS(G_1) = CoS(G_2) = n-1$. Indeed, the cheapest way to stabilize G_1 is to pay $n-1$ to each of the players 1 and 2 and 0 to every other player, and this requires a subsidy of $n-1$, whereas for G_2 all players are symmetric, so it is optimal to pay 1 to each of them, which, again, requires a subsidy of $n-1$. On the other hand, the value of the least core of G_1 is $\frac{n-1}{2}$ (it is optimal to split the value of the grand coalition evenly between players 1 and 2, in which case each of them receives $\frac{n-1}{2}$, and can gain as much by deviating), whereas for G_2 it is $\frac{n-1}{n}$ (once the value of the grand coalition is split evenly, each player's "deficit" is $1 - \frac{1}{n} = \frac{n-1}{n}$). Thus, two games may have the same cost of stability, but very different values of the least core (and, by rescaling the characteristic function of these games, we can get two games with the same value of the least core, but very different cost of stability).

The game G_2 described in the previous paragraph illustrates that the ratio between the cost of stability and the value of the grand coalition can be as large as n. However, Bachrach et al. [74] show that if a cooperative game $G = (P, v)$ with n players is superadditive, then $CoS(G) \leq (\sqrt{n} - 1)v(P)$, and this bound is asymptotically tight.

3.2.3.3 Stable Sets in a Cooperative Game

The next stability concept for cooperative games that we are going to discuss is historically the first one: It dates back to the classic book of von Neumann and Morgenstern [736]. Let $G = (P, v)$ be a cooperative game with more than two players, and let $\vec{a} = (a_1, a_2, \ldots, a_n)$ and $\vec{b} = (b_1, b_2, \ldots, b_n)$ be two imputations in $\mathcal{I}(G)$. We say that $\vec{a}$ *dominates* $\vec{b}$ *via a coalition* $C \neq \emptyset$ if $a_i > b_i$ for all $i \in C$ and $\sum_{i \in C} a_i \leq v(C)$; we write $\vec{a} \succ_C \vec{b}$ if this is the case. To interpret this condition, note that if $a_i > b_i$ for all $i \in C$, each player in C (strictly) prefers the payoffs to be distributed according to $\vec{a}$ rather than according to $\vec{b}$. Together with the condition $\sum_{i \in C} a_i \leq v(C)$, this means that the players in C can plausibly threaten to leave the grand coalition: If they do so, they have enough payoff to share according to $\vec{a}$, which all of them prefer to $\vec{b}$. We say that $\vec{a}$ *dominates* $\vec{b}$ (and write $\vec{a} \succ \vec{b}$) if $\vec{a} \succ_C \vec{b}$ for some nonempty set $C \subseteq P$. We remark that the dominance relation is not necessarily antisymmetric: It may be the case that $\vec{a} \succ_{C_1} \vec{b}$ and $\vec{b} \succ_{C_2} \vec{a}$, as long as C_1 and C_2 are disjoint.

A *stable set of* G is a subset $S \subseteq \mathcal{I}(G)$ satisfying the following two conditions:[4]

1. *Internal stability*: No vector $\vec{a} \in S$ is dominated by a vector $\vec{b} \in S$.
2. *External stability*: For all $\vec{b} \in \mathcal{I}(G) \setminus S$, there is a vector $\vec{a} \in S$ such that $\vec{a} \succ \vec{b}$.

The stability of such a set S in a cooperative game can be explained as follows: Due to internal stability of S, there is no reason to remove a payoff vector from S; on the other hand, due to external stability of S, there is no reason to add another payoff vector to S. This is interpreted by von Neumann and Morgenstern [736] by saying that a stable set can be seen as a list of "acceptable behaviors" in a society. No behavior within this list is strictly superior to another behavior in the list, but for each unacceptable behavior there is an acceptable behavior that is preferable.

Stable sets exist in some, yet not in all cooperative games [671]. Moreover, if they exist, they are usually not unique [672], and appear to be hard to find;

---

[4] The combination of internal and external stability due to a dominance relation is a common theme in many other contexts in economic theory. For instance, it is the basis for *Shapley's saddle* and *CURB sets* in noncooperative game theory [103, 856], the *minimal covering set* in voting [365, 217, 110], or *stable semantics* in argumentation theory [363].

3.2 Stability in Cooperative Games

in fact, it is not even clear if this problem is algorithmically decidable (see the work of Jain and Vohra [575] for a discussion of this and related issues). These are some of the reasons why stable sets have not received as much attention in the literature as, e.g., the core.

How is the core of a game related to its collection of stable sets? Suppose that $\vec{a}$ is an imputation in the core of G. Then it cannot be dominated by any other imputation. Indeed, suppose for the sake of contradiction that it is dominated by some imputation $\vec{b}$, i.e., there exists a coalition C such that $b_i > a_i$ for all $i \in C$ and $\sum_{i \in C} b_i \leq v(C)$. Then we have $\sum_{i \in C} a_i < \sum_{i \in C} b_i \leq v(C)$, a contradiction with $\vec{a}$ being in the core of G. This implies that $\vec{a}$ belongs to every stable set of G. That is, if the core of G is nonempty, it is contained in all stable sets of G.

However, the core itself is not necessarily a stable set: While the argument above implies that it satisfies the condition of internal stability, it may fail external stability. A simple example is a game $\hat{G} = (P, v)$ with $P = \{1, 2, 3\}$ and

$$v(C) = \begin{cases} 1 & \text{if } 1 \in C \text{ and } \|C\| \geq 2 \\ 0 & \text{otherwise.} \end{cases}$$

As $\hat{G}$ is a simple game and player 1 is the only veto player in this game, by Theorem 3.2 the only imputation in the core of this game is $\vec{a} = (1, 0, 0)$. Now, consider the imputation $\vec{b} = (0, 1/2, 1/2)$. It is easy to see that $\vec{a}$ does not dominate $\vec{b}$: The only coalition C such that $a_i > b_i$ for all $i \in C$ is $\{1\}$, and

$$\sum_{i \in \{1\}} a_i = a_1 = 1 > 0 = v(\{1\}).$$

In fact, it is not hard to see that if the core of a game G is a stable set, then it is the only stable set for G: This follows from the observation that no stable set can be a strict subset of another stable set. We further remark that every convex game has a unique stable set, which coincides with its core.

What are the stable sets in the game $\hat{G}$ defined above? One example is $S_{12} = \{(x, 1-x, 0) \mid x \in [0, 1]\}$. To see that this set satisfies the internal stability condition, consider a pair of imputations $(x, 1-x, 0), (y, 1-y, 0) \in S_{12}$. If $(x, 1-x, 0) \succ (y, 1-y, 0)$, it has to be the case that $x > y$ and $1 - x > 1 - y$, which is impossible. To verify that S_{12} satisfies external stability, note that an arbitrary imputation (x, y, z) with $x + y + z = 1$ and $z > 0$ is dominated by $(x + \frac{z}{2}, y + \frac{z}{2}, 0) \in S_{12}$ via $C = \{1, 2\}$. Similarly, $S_{13} = \{(x, 0, 1-x) \mid x \in [0, 1]\}$ is also a stable set in $\hat{G}$. We leave it to the reader to determine whether this game has other stable sets.

For an overview of other stability concepts, such as the "kernel" or the "nucleolus" of a cooperative game, we refer the reader to the textbook by Osborne and Rubinstein [762].

3.3 Fairness in Cooperative Games

We have discussed how to divide the payoff of the grand coalition in a stable manner. However, a somewhat different issue is how to divide it *fairly*. Consider, for instance, a two-player game where the value of each of the singleton coalitions is 5, but the value of the grand coalition is 20. Then each of the payoff vectors (5,15), (15,5), and (10,10) is in the core of this game. Intuitively, however, the first two payoff vectors are unfair: Even though the two players appear to be indistinguishable, they receive different rewards. Thus, core stability does not capture our intuition of what it means for an outcome to be fair.

Building on this example, we may want to say that an outcome is *fair* if it rewards all players according to their contribution, or influence. However, we still need to explain how to measure a player's influence. We now introduce several methods for this task. In the context of simple games, such methods are collectively known as *power indices*.

3.3.1 The Shapley–Shubik Index and the Shapley Value

Consider the weighted voting game $G = (108, 106, 62, 66; 277)$ from Figure 3.2. How can we measure the players' power in this game? It is common to answer this question by using the value division scheme introduced by Shapley [856] and known as the *Shapley value*. This scheme was defined for general cooperative games. Later on, its restriction to *simple* cooperative games was proposed by Shapley and Shubik [860] as a measure of players' influence in such games (see also the work of Dubey and Shapley [859] and Roth [818]). We start with this restricted notion, and then move on to general cooperative games.

3.3.1.1 Definitions and an Example: Simple Games

Let $G = (P, v)$ be a simple game. A player $i \in P$ is said to be *pivotal for a coalition* $C \subseteq P \setminus \{i\}$ if $C \cup \{i\}$ wins, but C does not. We define the *marginal contribution of player i to the gains of coalition C in G* as

$$d_G(C, i) = v(C \cup \{i\}) - v(C); \qquad (3.4)$$

we have $d_G(C, i) = 1$ if i is a pivotal player for C, and $d_G(C, i) = 0$ otherwise. We say that a player i is a *null player* if i is not pivotal for *any* coalition, i.e., if $d_G(C, i) = 0$ for each coalition $C \subseteq P \setminus \{i\}$.

Given a *permutation* π of P (i.e., a bijective mapping $P \to \{1, \ldots, n\}$, which defines an ordering over the players), we say that a player i is *pivotal for* π

3.3 Fairness in Cooperative Games

if i is pivotal for the coalition $\{j \mid \pi(j) < \pi(i)\}$ of her predecessors in π. A player's *marginal contribution to π* is denoted by $d_G(\pi,i)$: It is defined to be 1 if i is pivotal for π, and 0 otherwise. Observe that for each permutation π there exists exactly one player who is pivotal for it. We denote the *set of all permutations of P* by Π_P. Naturally, the size of Π_P is given by the *factorial function* $n! = 1 \cdot 2 \cdot \cdots \cdot n$ with $0! = 1$.

The *raw Shapley–Shubik index of a player i in G* is defined by

$$\text{Shapley-Shubik}^*(G,i) = \sum_{\pi \in \Pi_P} d_G(\pi,i). \tag{3.5}$$

That is, it simply counts how many permutations i is pivotal for. For a weighted voting game, this works as follows. We choose a permutation of players and invite them to join the coalition one by one, in the order specified by the permutation, until the total weight of the coalition meets or exceeds the quota (so that the coalition becomes winning). We then award a point to the last player joining, as this player has been pivotal for the victory of the coalition. We repeat this process for all possible permutations of the players, awarding a point to one of the players each time. The number of points accumulated by a player at the end of this process is this player's raw Shapley–Shubik index.

By summing expression (3.5) over all players $i \in P$, we obtain

$$\sum_{i \in P}\sum_{\pi \in \Pi_P} d_G(\pi,i) = \sum_{\pi \in \Pi_P}\sum_{i \in P} d_G(\pi,i) = \sum_{\pi \in \Pi_P} 1 = n!, \tag{3.6}$$

where the first equality is obtained by changing the order of summation, and the second equality follows from our observation that for each permutation there exists exactly one player that is pivotal for it. Thus, the sum of the players' raw Shapley–Shubik indices in an n-player game is always $n!$.

It is convenient as well as computationally advantageous to express the raw Shapley–Shubik index Shapley-Shubik$^*(G,i)$ in terms of $d_G(C,i)$ rather than $d_G(\pi,i)$. To this end, observe that one coalition of size k that i is pivotal for corresponds to $k!(n-k-1)!$ permutations that i is pivotal for: If i is pivotal for C, it is pivotal for all permutations obtained by placing the elements of C in the first $\|C\|$ positions (in an arbitrary order), followed by i, followed by the elements of $P \setminus (C \cup \{i\})$ (in an arbitrary order). Thus, we can rewrite expression (3.5) as follows:

$$\text{Shapley-Shubik}^*(G,i) = \sum_{C \subseteq P \setminus \{i\}} \|C\|! \cdot (n - \|C\| - 1)! \cdot d_G(C,i). \tag{3.7}$$

Observe that expression (3.5) has $n!$ terms, while expression (3.7) has only 2^{n-1} terms. The factorial function grows much faster than the exponential function: For instance, for $n = 6$ we have $6! = 720$, while $2^{6-1} = 32$. This means that, if we are to compute a player's raw Shapley–Shubik index directly from the definition, it is more practical to use expression (3.7). Of course, even computing 2^{n-1} terms is not feasible for large values of n, and later in the chapter we will discuss other approaches to computing the players' Shapley–Shubik indices.

The raw Shapley–Shubik index is then normalized by setting

$$\text{Shapley-Shubik}(G,i) = \frac{1}{n!} \cdot \text{Shapley-Shubik}^*(G,i);$$

the quantity Shapley-Shubik(G,i) is known as the *Shapley–Shubik index of i in G*. By normalizing the raw Shapley–Shubik index, we ensure that Shapley-Shubik(G,i) is always between 0 and 1, and that all players' indices sum up to 1. Indeed, we have

$$\sum_{i \in P} \text{Shapley-Shubik}(G,i) = \sum_{i \in P} \frac{1}{n!} \cdot \text{Shapley-Shubik}^*(G,i) = \frac{n!}{n!} = 1.$$

In other words, this index gives the probability that i is pivotal for a randomly chosen permutation of P under the uniform distribution. It is 0 if and only if i is a null player, since this means that i is not pivotal for any coalition.

When is i's Shapley–Shubik index equal to 1? It may be tempting to conjecture that this is the case whenever i is a veto player. However, this is not necessarily true: A veto player is necessary but not sufficient to form a winning coalition, so it may be the case that $v(C) = v(C \cup \{i\}) = 0$ for some coalition C even if i is a veto player. In fact, since the Shapley–Shubik indices of all players in the game add up to 1, a player's Shapley–Shubik index is 1 if and only if all other players are null players.

Let us determine the Shapley–Shubik index of, say, Anna in the weighted voting game $G = (108, 106, 62, 66; 277)$ from Figure 3.2. We have already seen that there are only two winning coalitions in G: $E = \{\text{Anna, Belle, Edgar}\}$ and the grand coalition $P = \{\text{Anna, Belle, David, Edgar}\}$. Consequently, Anna is pivotal for $E \smallsetminus \{\text{Anna}\} = \{\text{Belle, Edgar}\}$ and $P \smallsetminus \{\text{Anna}\} = \{\text{Belle, David, Edgar}\}$, i.e., we have

$$d_G(E \smallsetminus \{\text{Anna}\}, \text{Anna}) = 1 \quad \text{and} \quad d_G(P \smallsetminus \{\text{Anna}\}, \text{Anna}) = 1.$$

Only these terms contribute to the sum in (3.7); all other terms vanish because $d_G(C, \text{Anna}) = 0$ for all other coalitions $C \subseteq P \smallsetminus \{\text{Anna}\}$.

It follows that

Shapley-Shubik$^*(G, \text{Anna})$
$= \|E \smallsetminus \{\text{Anna}\}\|! \cdot (4 - \|E \smallsetminus \{\text{Anna}\}\| - 1)! \cdot d_G(E \smallsetminus \{\text{Anna}\}, \text{Anna}) +$
$\quad \|P \smallsetminus \{\text{Anna}\}\|! \cdot (4 - \|P \smallsetminus \{\text{Anna}\}\| - 1)! \cdot d_G(P \smallsetminus \{\text{Anna}\}, \text{Anna})$
$= 2! \cdot (4 - 2 - 1)! \cdot 1 + 3! \cdot (4 - 3 - 1)! \cdot 1$
$= 2 + 6 = 8.$

3.3 Fairness in Cooperative Games

Normalizing Anna's raw Shapley–Shubik index, we obtain her Shapley–Shubik index in G:

$$\text{Shapley-Shubik}(G, \text{Anna}) = \frac{1}{4!} \cdot \text{Shapley-Shubik}^*(G, \text{Anna})$$
$$= \frac{1}{24} \cdot 8 = \frac{1}{3}.$$

A similar argument shows that Edgar and Belle have the same Shapley–Shubik index in G, whereas David, the null player, has no influence in G at all:

$$\text{Shapley-Shubik}(G, \text{Edgar}) = \text{Shapley-Shubik}(G, \text{Belle}) = 1/3,$$
$$\text{Shapley-Shubik}(G, \text{David}) = 0.$$

3.3.1.2 Beyond Simple Games: The Shapley Value

The Shapley–Shubik index is defined for simple games, but a similar approach can be used to define a power measure for general cooperative games, which is known as the *Shapley value*. This measure was originally proposed by Shapley [856]. To define it formally, we need to extend the notion of marginal contribution to general games: Given a cooperative game $G = (P, v)$ and a permutation π of the players in P, we set

$$d_G(\pi, i) = v(\{j \mid \pi(j) \leq \pi(i)\}) - v(\{j \mid \pi(j) < \pi(i)\}).$$

We define player i's *raw Shapley value* in game G (denoted by $\text{Shapley}^*(G,i)$) using Equation (3.5); i's *Shapley value* $\text{Shapley}(G,i)$ in G is then obtained by normalization:

$$\text{Shapley}(G,i) = \frac{1}{n!} \cdot \text{Shapley}^*(G,i).$$

The reader can verify that if G is a simple game, the definition of marginal contribution given above coincides with the definition of $d_G(\pi, i)$ given on page 163; thus, a player's Shapley value in a simple game is equal to her Shapley–Shubik index.

Just like the Shapley–Shubik index, the Shapley value can be defined in terms of a player's marginal contribution to coalitions (rather than permutations) of players, i.e., via Equation (3.7), where for an arbitrary cooperative game $G = (P, v)$ and a player $i \in P$ we set $d_G(C, i) = v(C \cup \{i\}) - v(C)$. This alternative definition can be useful for computational purposes.

3.3.1.3 Properties and Axiomatic Characterization

We have argued that for every simple game $G = (P, v)$, the Shapley–Shubik index of each player is nonnegative and the players' indices sum up to 1, which is also the value of the grand coalition. This is also the case for general cooperative games. Specifically, we have

1. Shapley$(G, i) \geq 0$ for all $i \in P$, and
2. it holds that

$$\sum_{i=1}^{n} \text{Shapley}(G,i) = \frac{1}{n!} \sum_{i=1}^{n} \text{Shapley}^*(G,i) = \frac{1}{n!} \sum_{i=1}^{n} \sum_{\pi \in \Pi_P} d_G(\pi, i)$$

$$= \frac{1}{n!} \sum_{\pi \in \Pi_P} \sum_{i=1}^{n} (v(\{j \mid \pi(j) \leq i\}) - v(\{j \mid \pi(j) < i\}))$$

$$= \frac{1}{n!} \sum_{\pi \in \Pi_P} v(P) = v(P).$$

Hence, the vector

$$(\text{Shapley}(G, 1), \text{Shapley}(G, 2), \ldots, \text{Shapley}(G, n))$$

is a valid payoff vector for P.

An important advantage of this payoff division scheme is that it is always well-defined, i.e., it provides a concrete suggestion for how to share the payoff of the grand coalition; in contrast, the core of G may be empty. In addition, it has a number of desirable properties.

First, by construction, it assigns no payoff to null players.

Further, it guarantees equal treatment to players who contribute to all coalitions in the same way. To make this claim formal, we need a notion of symmetry: Two players i and j in a cooperative game $G = (P, v)$ are said to be *symmetric* if $v(C \cup \{i\}) = v(C \cup \{j\})$ for all coalitions $C \subseteq P \setminus \{i, j\}$. Now, it is not hard to show that if i and j are symmetric players in G, we have Shapley$(G, i) = $ Shapley(G, j).

We remark that, in weighted voting games, two players of the same weight are symmetric, but the converse is not necessarily true: For instance, in the weighted voting game $G = (108, 106, 62, 66; 277)$ from Figure 3.2, Anna, Belle, and Edgar are pairwise symmetric despite having very different weights. This illustrates that the players' Shapley values are a better measure of their contribution to success than their weights.

Another important property of the Shapley value is its *additivity*, defined as follows. The sum of two games $G_1 = (P, v_1)$ and $G_2 = (P, v_2)$ with the same set of players is defined to be the game $G_+ = G_1 + G_2 = (P, v_+)$, whose characteristic function is given by $v_+(C) = v_1(C) + v_2(C)$ for all coalitions $C \subseteq P$. Then, for all players $i \in P$, it holds that

3.3 Fairness in Cooperative Games

$$\text{Shapley}(G_+, i) = \text{Shapley}(G_1, i) + \text{Shapley}(G_2, i).$$

Interestingly, the four properties we just listed provide an axiomatic characterization of the Shapley value. To state this characterization, we say that a *value* for cooperative games is a mapping φ that for any cooperative game $G = (P, v)$ and every player $i \in P$ defines a real number $\varphi(G, i)$. It turns out that the Shapley value is the unique value that satisfies a short list of desirable properties, or axioms.

Theorem 3.3 (Shapley [856]). *The Shapley value is the only value φ for cooperative games that satisfies the following four properties:*

1. Null player: *For every game $G = (P, v)$, it holds that if $i \in P$ is a null player then $\varphi(G, i) = 0$.*
2. Efficiency: *For every game $G = (P, v)$, it holds that*

$$\sum_{i \in P} \varphi(G, i) = v(P).$$

3. Symmetry: *For every game $G = (P, v)$, it holds that if i and j are symmetric players then*

$$\varphi(G, i) = \varphi(G, j).$$

4. Additivity: *For every pair of games (G_1, G_2) such that $G_1 = (P, v_1)$ and $G_2 = (P, v_2)$, and for every $i \in P$, it holds that*

$$\varphi(G_1 + G_2, i) = \varphi(G_1, i) + \varphi(G_2, i).$$

A remarkable feature of Theorem 3.3 is that none of the axioms used to characterize the Shapley value refers to the notion of marginal contribution, yet they uniquely determine a payoff distribution scheme that is defined in terms of marginal contributions.

If a game $G = (P, v)$ is superadditive, then the payoff vector given by the Shapley value is an imputation, i.e., it satisfies the individual rationality condition: $\text{Shapley}(G, i) \geq v(\{i\})$ for all $i \in P$. This is because for each player $i \in P$, by superadditivity, we have

$$d_G(C, i) = v(C \cup \{i\}) - v(C) \geq v(C) + v(\{i\}) - v(C) = v(\{i\}).$$

In general, however, it may happen that $\text{Shapley}(G, i) < v(\{i\})$.

Finally, consider the converse of the null player property for the Shapley value: "*If $\text{Shapley}(G, i) = 0$ then i is a null player.*" Note that this converse is true if the cooperative game G is monotonic, but is not always true in nonmonotonic games.

3.3.1.4 Shapley Value versus Core

What is the relationship between the Shapley value and the core? One may expect that the payoff vector (Shapley$(G, 1), \ldots,$ Shapley(G, n)) is in the core of G whenever $Core(G)$ is nonempty. However, in general this is not the case. The easiest way to see this is to focus on simple games: In such games, a player receives a positive payoff in a core imputation only if she is a veto player, whereas her Shapley value is positive as long as she is not a null player. Thus any simple game that has some veto players (so that its core is nonempty) as well as some players that are neither veto players nor null players provides a counterexample to our conjecture. For concreteness, consider the three-player simple game $\hat{G}$ discussed on page 161, in which a coalition wins if and only if it contains player 1 and at least one of the other two players. Clearly, player 1 is the only veto player in this game, so its core is given by $Core(\hat{G}) = \{(1, 0, 0)\}$. On the other hand, players 2 and 3 are not null players, so their Shapley value is (strictly) positive. (The reader can verify that Shapley$(\hat{G}, 1) = 2/3$ and Shapley$(\hat{G}, 2) =$ Shapley$(\hat{G}, 3) = 1/6$.) This example suggests that (at least in simple games) the Shapley value provides a more nuanced measure of a player's contribution than the core.

However, for convex games, the Shapley value is always in the core. Indeed, there is an obvious similarity between the definition of the Shapley value and the method we used to construct an outcome in the core of a convex game in the proof of Theorem 3.1 on page 152: The only difference is that the latter pays the players according to their marginal contributions with respect to a *specific* permutation of the players, while the former averages over *all* player permutations (Equation (3.5)). This observation, coupled with the fact that the core of any cooperative game is a convex set (i.e., for every $\alpha \in [0, 1]$, it holds that if $\vec{x}$ and $\vec{y}$ are payoff vectors in the core of a game G, then their *convex combination* $\alpha\vec{x} + (1 - \alpha)\vec{y}$ is also in the core of G; recall Definition 2.6 from page 72), implies that for every convex game G, it holds that (Shapley$(G, 1), \ldots,$ Shapley(G, n)) is in the core of G. We leave it as an exercise for the reader to work out a complete proof of this claim.

3.3.2 The Banzhaf Indices

The Shapley value and the Shapley–Shubik index are defined by averaging a player's marginal contribution over all possible permutations of players. While this definition can be equivalently expressed as a weighted sum over all coalitions (Equation (3.7)), the weight of each coalition depends on its size. Alternatively, one can average over all coalitions, assigning the same weight to each coalition. This idea was first proposed by Penrose [775] and subsequently rediscovered by Banzhaf [86] in the context of simple games. The resulting power index is sometimes referred to as the *Penrose–Banzhaf*

3.3 Fairness in Cooperative Games

index, or simply the *Banzhaf index*. For this index, all that matters is the number of coalitions each player is pivotal for; in particular, unlike for the Shapley–Shubik index, the size of these coalitions is not important.

Formally, for a simple game $G = (P, v)$, the *raw Banzhaf index of a player i in G* is defined by

$$\text{Banzhaf}^*(G, i) = \sum_{C \subseteq P \setminus \{i\}} d_G(C, i), \tag{3.8}$$

where the marginal contribution $d_G(C, i)$ of player i to coalition C is defined as in (3.4), i.e., $d_G(C, i) = 1$ if i is pivotal for C, and $d_G(C, i) = 0$ otherwise. The value $\text{Banzhaf}^*(G, i)$ thus gives the number of coalitions that i is pivotal for.

Since we are interested in players' *relative* rather than *absolute* power, it again makes sense to normalize. We will now discuss two ways of doing so, and compare their advantages and disadvantages.

Banzhaf [86] proposed to normalize the raw Banzhaf index by setting

$$\overline{\text{Banzhaf}}(G, i) = \frac{\text{Banzhaf}^*(G, i)}{\sum_{j \in P} \text{Banzhaf}^*(G, j)}, \tag{3.9}$$

obtaining the *(normalized) Banzhaf index of i in G*.

Dubey and Shapley [359] proposed a different normalization method, which computes the *fraction* of coalitions in $P \setminus \{i\}$ that i is pivotal for. More formally, their *probabilistic Banzhaf index of i in G* is defined by

$$\text{Banzhaf}(G, i) = \frac{\text{Banzhaf}^*(G, i)}{2^{n-1}}.$$

Consider once again the weighted voting game $G = (108, 106, 62, 66; 277)$ from Figure 3.2, in which Anna, Belle, David, and Edgar try to lift a golden buddha. We have seen that in this game Anna, Belle, and Edgar are pivotal for two coalitions each, but David is not pivotal for any coalition. Hence, the raw Banzhaf indices of these four players are:

$\text{Banzhaf}^*(G, \text{Anna}) = \text{Banzhaf}^*(G, \text{Belle}) = \text{Banzhaf}^*(G, \text{Edgar}) = 2,$
$\text{Banzhaf}^*(G, \text{David}) = 0.$

Thus, their normalized Banzhaf indices are the same as their Shapley–Shubik indices:

$\overline{\text{Banzhaf}}(G, \text{Anna}) = \overline{\text{Banzhaf}}(G, \text{Belle}) = \overline{\text{Banzhaf}}(G, \text{Edgar}) = 1/3,$
$\overline{\text{Banzhaf}}(G, \text{David}) = 0.$

In contrast, their probabilistic Banzhaf indices are different:

Banzhaf(G, Anna) = Banzhaf(G, Belle) = Banzhaf(G, Edgar) = $1/4$,
Banzhaf(G, David) = 0.

Dubey and Shapley [359] analyzed the different variants of the Banzhaf index with respect to a number of properties. In addition to the properties that are used to characterize the Shapley value (recall Theorem 3.3), they considered the *valuation property* (as Peleg and Sudhölter [774] call it). A power index $\mathbb{PI}$ has this property if for any pair of simple games $G_1 = (P, v_1)$ and $G_2 = (P, v_2)$ over the same set of players P, it holds that

$$\mathbb{PI}(G_\vee, i) + \mathbb{PI}(G_\wedge, i) = \mathbb{PI}(G_1, i) + \mathbb{PI}(G_2, i),$$

where the games $G_\vee = (P, v_1 \vee v_2)$ and $G_\wedge = (P, v_1 \wedge v_2)$ are defined as

$$(v_1 \vee v_2)(C) = \max(v_1(C), v_2(C)) \quad \text{and}$$
$$(v_1 \wedge v_2)(C) = \min(v_1(C), v_2(C))$$

for all coalitions $C \subseteq P$.

Dubey and Shapley [359, Theorem 1 on page 104] showed that the normalized Banzhaf index, as defined in (3.9) above, does not satisfy the valuation property, and conclude: *"This may be taken as an initial sign of trouble with the normalization [of the normalized Banzhaf index]"* [359, Footnote 21]. Furthermore, the normalized Banzhaf index is not additive.

In contrast, the probabilistic Banzhaf index satisfies the null player, symmetry, additivity, and valuation axioms, and has other desirable properties (which we do not describe here). Moreover, Dubey and Shapley stated that the probabilistic Banzhaf index is *"better behaved when analyzing convergence"* [359, page 116], and other authors (e.g., Bachrach and Rosenschein [77], Felsenthal and Machover [448, 449], and Rey and Rothe [802, 805]) make similar comments.

However, an important disadvantage of the probabilistic Banzhaf index is that the vector (Banzhaf(G, 1), Banzhaf(G, 2), ..., Banzhaf(G, n)) may fail to be efficient, i.e., it may happen that the players' indices do not sum up to the value of the grand coalition P. An easy way to see this is to note that the probabilistic Banzhaf index satisfies the other three properties stated in Theorem 3.3, and only the Shapley value fulfills all four of them; an explicit example is provided by the weighted voting game $G = (108, 106, 62, 66; 277)$ analyzed earlier in this section.

In contrast, the normalized Banzhaf index is efficient, i.e., we have

$$\overline{\text{Banzhaf}}(G, 1) + \overline{\text{Banzhaf}}(G, 2) + \cdots + \overline{\text{Banzhaf}}(G, n) \;=\; v(P).$$

Thus, for simple games the normalized Banzhaf index provides a way to share the value of the grand coalition, whereas the probabilistic Banzhaf index does not. It also satisifies the null player axiom and the symmetry axiom. Further

3.4 Counting and Representing Cooperative Games

arguments in favor of the normalized Banzhaf index are given by Aziz et al. [39].

3.4 Counting and Representing Cooperative Games

Note that even simple games appear to be costly to represent: Naively, for each of the 2^n coalitions $C \subseteq P$ we need to specify the value $v(C)$. Of course, if there are very few winning coalitions, it suffices to list them explicitly, with the understanding that all remaining coalitions are losing. In fact, the monotonicity property means that it suffices to list the *minimal* winning coalitions only, i.e., create a list that contains a coalition C if and only if $v(C) = 1$, but $v(C') = 0$ for all $C' \subsetneq C$. Indeed, given such a list $C_1, \ldots, C_k$ and a coalition C, we can compute the value of C by checking whether $C_i \subseteq C$ for some $i \in \{1, \ldots, k\}$: We have $v(C) = 1$ if and only if the answer is positive. Similarly, we can fully specify a game by listing all *maximal* losing coalitions—this is useful if almost all coalitions are winning.

However, there are simple games where both the number of minimal winning coalitions and the number of maximal losing coalitions are exponential in n. Consider, for instance, an n-player game, where $n = 2k$ is an even number, such that all coalitions of size at most $k-1$ as well as exactly half of the coalitions of size k have value 0, whereas all other coalitions (i.e., all coalitions of size at least $k+1$ and the remaining half of the coalitions of size k) have value 1. It is easy to see that this game has the monotonicity property, so it is a simple game. On the other hand, it has at least $\frac{1}{2} \cdot \binom{n}{k}$ minimal winning coalitions (i.e., the winning coalitions of size k) and at least $\frac{1}{2} \cdot \binom{n}{k}$ maximal losing coalitions (i.e., the losing coalitions of size k).

The construction above establishes another interesting fact about simple games: There are simply too many of them for a succinct representation scheme to exist! Indeed, the construction in the previous paragraph produces a simple game no matter which $\frac{1}{2} \cdot \binom{n}{k}$ k-player coalitions have value 1. This means that there are at least $\binom{N}{N/2}$ distinct simple games, where $N = \binom{n}{k}$, i.e., the number of n-player simple games is *doubly exponential* in n. Now, suppose for the sake of contradiction that there is a representation language that can encode every n-player simple game using $p(n)$ bits, where p is some fixed polynomial. Then each n-player simple game corresponds to a binary string of length at most $p(n)$, and there are exactly $2^{p(n)}$ such strings. On the other hand, we just argued that the number of n-player simple games grows faster than $2^{p(n)}$, for any fixed polynomial p. So we cannot represent all simple games in this language, a contradiction.

Thus, there does not exist a succinct language for encoding all simple games (and, *a forteriori*, there is no succinct language for encoding all cooperative games). Hence, if we would like to develop a representation formalism for cooperative games, we have two options.

First, we can design formalisms that are universal, i.e., can capture every coalitional game (perhaps within a certain class of games, such as simple games) but are not necessarily succinct. In contrast to the standard representation, which encodes every cooperative n-player game using 2^n numbers, such formalisms may be able to represent many games succinctly, but some games will require exponentially many bits. Indeed, representing a simple game by the list of (minimal) winning coalitions is an example of this approach. In Section 3.4.1, we will describe another formalism of this type, which can capture all simple games, but is only succinct for games that are, in some sense, similar to weighted voting games. Another notable universal formalism, which was proposed by Ieong and Shoham [566] is *"marginal contribution nets"*; we will not discuss it here, but refer the interested reader to the book of Chalkiadakis, Elkind, and Wooldridge [268].

Second, we can develop representations that are not universal, but are tailored for a specific class of cooperative games and offer a succinct encoding of every game in this class. Indeed, we have already seen one such formalism, namely, weighted voting games. In Section 3.4.2, we provide further examples of this approach, focusing on games that can be described by weighted graphs.

3.4.1 A Universal Representation for Simple Games

The goal of this section is to describe a representation formalism that is universally expressive for simple games (i.e., can represent any simple game), but is not always succinct.

A natural starting point when looking for such a formalism is to ask whether each simple game can be represented as a weighted voting game. That is, given a simple game $G = (P, v)$ with n players, can we come up with n weights $w_1, \ldots, w_n$ and a quota q so that for each coalition $C \subseteq P$, it holds that $v(C) = 1$ if and only if $\sum_{i \in C} w_i \geq q$? Intuitively, it seems that this should not be the case: A weighted voting game can be represented by $n+1$ numbers, and we have just argued that there is no way to succinctly encode all simple games. However, one has to be careful here: Perhaps every simple game can be represented as a weighted voting game with *huge* weights, e.g., for an n-player game, we may have to use weights that are doubly exponential in n (and, therefore, their binary representation has $\Omega(2^n)$ bits)? This would be compatible with our lower bound on the number of simple games. It can be shown, however, that for every weighted voting game, there is an equivalent weighted voting game (i.e., one with the same set of winning coalitions) with integer weights that do not exceed $\mathcal{O}(n^n)$ [721]. This means that our original intuition is correct and there are simple games that are not weighted voting games.

3.4 Counting and Representing Cooperative Games

However, this argument is not constructive, i.e., it does not give us an example of a simple game that is not a weighted voting game. Is it possible to explicitly construct such a game?

Anna, Belle, David, and Edgar would like to participate in a local talent show. The program they have in mind requires at least one singer and one dancer. Anna and David can sing, and Belle and Edgar can dance, but none of them can both sing and dance. Thus, for instance, {Anna, Belle} is a valid team, but {Anna, David} is not. Is there a way to describe this situation as a weighted voting game?

Suppose there are weights w_A, w_B, w_D, and w_E and a quota q such that the weight of any valid team is at least q, but the weight of any nonvalid team is less than q. Since {Anna, Belle} and {David, Edgar} are valid teams, we have

$$w_A + w_B \geq q, \quad w_D + w_E \geq q \text{ and, hence, } w_A + w_B + w_D + w_E \geq 2q.$$

On the other hand, since the teams {Anna, David} and {Belle, Edgar} are not valid, we obtain

$$w_A + w_D < q, \quad w_B + w_E < q \text{ and, hence, } w_A + w_B + w_D + w_E < 2q.$$

a contradiction. Thus, there is no way to represent this scenario as a weighted voting game.

In a sense, the simple game described above (see also Figure 3.4) is a combination of two weighted voting games: The first game corresponds to singing and has weights $w_A^{sing} = w_D^{sing} = 1$ and $w_B^{sing} = w_E^{sing} = 0$ and quota $q^{sing} = 1$, while the second game corresponds to dancing and has weights $w_A^{dance} = w_D^{dance} = 0$ and $w_B^{dance} = w_E^{dance} = 1$ and quota $q^{dance} = 1$. A coalition of players is winning (corresponds to a valid team) if and only if it wins in *both* of these games, i.e., has at least one singer *and* at least one dancer. The games that are obtained by combining weighted voting games in this manner are called *vector weighted voting games*, and the minimum number of weighted voting games needed to represent a given simple game G is called the *dimension of G*.

It turns out that every simple game has a finite dimension, i.e., every simple game can be obtained by combining a finite number of weighted voting games. Isn't this a contradiction to our counting argument? Well, it would be if we claimed that the dimension of every simple game is polynomial in the number of players. However, the construction that we will now describe represents each simple game as a combination of possibly an *exponential* number of weighted voting games (this argument is implicitly given in the constructive proof by Taylor and Zwicker [910, Theorem 1.7.2] for the result that every simple game has a dimension; see also [909, 472]). A mathematically inclined

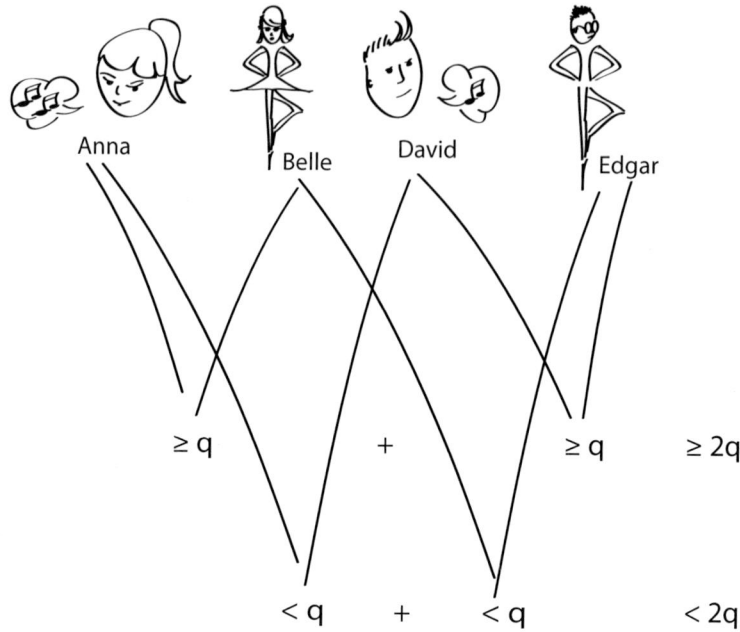

Fig. 3.4 A vector weighted voting game expressing coalition formation in a talent show

reader may observe that our argument is similar to Cantor's diagonalization argument; however, unlike Cantor, we deal with finite objects.

Theorem 3.4 (Taylor and Zwicker [909, 910]). *Every n-player simple game has dimension at most 2^n.*

Proof. Consider a simple game $G = (P, v)$ with n players. Let $\mathcal{L} = (L_1, \ldots, L_k)$ be the list of all maximal losing coalitions in G; note that $k < 2^n$. For each $L \in \mathcal{L}$, we construct a weighted voting game $G^L = (w_1^L, \ldots, w_n^L; q^L)$ given by $w_i^L = 1$ if $i \notin L$ and $w_i^L = 0$ otherwise, and $q^L = 1$. Now, let G' be the vector weighted voting game given by the combination of all such games. We claim that a coalition is winning in G if and only if it is winning in G'.

Indeed, let C be a winning coalition in G. Then by monotonicity it cannot be fully contained in a losing coalition, so for each $L \in \mathcal{L}$, there exists a player $i \in C \setminus L$, which means that the total weight of C in G^L is at least 1, i.e., C is winning in G^L. As this holds for every $L \in \mathcal{L}$, C is a winning coalition in G'. On the other hand, suppose that C is a losing coalition in G. Then it is contained in some maximal losing coalition $L \in \mathcal{L}$, and therefore its weight in G^L is 0. This means that C is a losing coalition in G'. □

Notably, the upper bound from the preceding theorem is asymptotically tight. There exist simple games that have an exponentially large dimension [471, 910].

3.4 Counting and Representing Cooperative Games

In the definition of vector weighted voting games, the underlying weighted voting games are combined *conjunctively*: To win in the overall game, a coalition must win in *each* of the component games. Alternatively, we can combine weighted voting games disjunctively, i.e., say that a coalition wins in the combined game if it wins in *at least one* of the component games [470]. It is not hard to see that an analogue of Theorem 3.4 holds for this setting: Each simple game can be represented as a disjunctive combination of at most 2^n weighted voting games. We leave the proof of this statement to the reader. (Hint: Consider minimal winning coalitions.)

One can go one step further and combine weighted voting games using arbitrary boolean connectives. For instance, given $2k$ weighted voting games $G_1, G'_1, \ldots, G_k, G'_k$, we can define a game G where a coalition is winning if it wins both in G_1 and in G'_1, or both in G_2 and in G'_2, etc.; this corresponds to the boolean expression $(G_1 \wedge G'_1) \vee \cdots \vee (G_k \wedge G'_k)$. Such games are called *boolean weighted voting games*; they have been defined and studied by Faliszewski, Elkind, and Wooldridge [419]. The main advantage of this richer formalism, as compared to vector weighted voting games, is that there are simple games that can be represented very compactly as boolean weighted voting games, but have a huge dimension; e.g., Kurz and Napel [631] show that the simple game corresponding to the EU Council under the Lisbon treaty can be represented as a boolean combination of just three games, but its dimension is at least 7. Kober and Weltge [615] improve this lower bound to 8. On the other hand, once we have described a simple game as a vector weighted voting game, we can use essentially the same representation to describe this game as a boolean vector weighted voting game, so the additional expressivity comes at no cost.

3.4.2 Cooperative Games on Graphs

Our goal in this section is to present several examples of formalisms that can compactly encode interesting classes of games but are not universal. In each of our examples, every game in the class can be fully described by a weighted graph: Players may correspond to vertices or edges, and the value of a coalition is determined by the properties of the corresponding subgraph. This encoding is succinct as long as we require that all edge weights are rational numbers that can be represented using $\text{poly}(\nu, \mu)$ bits, where ν and μ are, respectively, the number of vertices and the number of edges of the underlying graph. Throughout most of this section, we assume that a game is defined by a graph $\mathcal{G} = (\mathcal{V}, \mathcal{E})$ with $\|\mathcal{V}\| = \nu$ and $\|\mathcal{E}\| = \mu$, and a weight function $w : \mathcal{E} \to \mathbb{Q}$ (note that the edge weights need not be positive). Note that we require the edge weights to be rational numbers: Allowing for irrational weights is problematic, as it is not clear how to represent them.

We emphasize that the weighted graph-based representations described in the remainder of this section are not fully expressive: For each of the formalisms that we present, there are cooperative games that cannot be encoded in this formalism. We provide an explicit example for the first class of games that we consider, namely, induced subgraph games; the readers are invited to construct their own examples for other representations discussed in this section.

3.4.2.1 Induced Subgraph Games

Induced subgraph games were defined by Deng and Papadimitriou [338]. In these games, the players are vertices of an undirected graph, and the value of a coalition is the total weight of the edges of the corresponding induced subgraph. Formally, a graph $\mathcal{G} = (\mathcal{V}, \mathcal{E})$ with a weight function $w : \mathcal{E} \to \mathbb{Q}$ defines a game $G = (\mathcal{V}, \mathcal{E}, w)$ with the set of players $P = \mathcal{V}$ and the characteristic function $v : 2^{\mathcal{V}} \to \mathbb{Q}$ given, for all $C \subseteq P$, by

$$v(C) = \sum_{i,j \in C, \{i,j\} \in \mathcal{E}} w(\{i,j\}).$$

The graph $\mathcal{G}$ may contain self-loops, so that singleton coalitions may have nonzero value.

An interesting special case of these games is when the weight of each edge is either 0 or 1. This setting admits a simple interpretation: Each edge indicates a friendship relation between players, a player's utility from a coalition is the number of friends she has in that coalition, and the value of a coalition is one half of the total utility of all players in it. This intuition extends to the general model: A player may have both friends and enemies,[5] and, moreover, each relationship has a certain intensity, indicated by the edge weight.

However, this representation formalism is not universal: There are cooperative games (and even simple games) that cannot be represented as induced subgraph games. Indeed, consider the three-player simple game where the grand coalition is winning, and all smaller coalitions are losing. Suppose that this game can be represented as an induced subgraph game. Then, as every coalition of size at most two is losing, the weight of every edge would have to be 0, a contradiction with the grand coalition having value 1.

3.4.2.2 Network Flow Games

A *network flow game* is defined by a directed graph $\mathcal{G} = (\mathcal{V}, \mathcal{E})$, an edge weight function $w : \mathcal{E} \to \mathbb{Q}$, and two special vertices—a *source* $s \in \mathcal{V}$ and a *sink* $t \in \mathcal{V}$.

---

[5] This is different, though, from the class of friend- and enemy-oriented hedonic games to be described in Section 3.6 starting on page 198.

3.4 Counting and Representing Cooperative Games

In contrast with induced subgraph games, players are associated with edges of the graph, and the value of a coalition $C \subseteq \mathcal{E}$ is the size of s–t flow it can carry, assuming that the capacity of an edge e is given by $w(e)$. These games were introduced by Kalai and Zemel [588, 589].[6]

An important feature of these games is that they are *totally balanced*: If $G = (P, v)$ is a network flow game then G has a nonempty core, and, moreover, each subgame of G (i.e., each game $G' = (P', v')$ with $P' \subseteq P$ and $v'(C) = v(C)$ for each $C \subseteq P'$) also has a nonempty core. To see why G has a nonempty core, consider an *s-t-cut* in $\mathcal{G} = (\mathcal{V}, \mathcal{E})$, i.e., a partition $(S, \mathcal{V} \setminus S)$ of $\mathcal{V}$ such that if we delete the edges between S and $\mathcal{V} \setminus S$ from $\mathcal{G}$, then the resulting graph contains no path from s to t. This cut can be associated with the coalition that consists of the edges between S and $\mathcal{V} \setminus S$. Pick a minimum-capacity cut in $\mathcal{G}$, and let C be the associated coalition. By the max-flow/min-cut theorem, the capacity of a minimum cut equals the size of a maximum flow in $\mathcal{G}$, i.e., $\sum_{e \in C} w(e) = v(\mathcal{E})$. Thus, if we pay each player in C an amount that equals the capacity of his edge (and 0 to the other players), we obtain an imputation for our game; let us denote this imputation by $\vec{a}$. Now, consider an arbitrary coalition D. For $C' = D \cap C$, all the s–t flow carried by D has to pass through C', and hence the value of D does not exceed $\sum_{e \in C'} w(e)$, so D cannot profitably deviate from $\vec{a}$. As this holds for any D, $\vec{a}$ is in the core of G. Further, by the same argument, each subgame of G has a nonempty core, as it, too, is a network flow game.

It follows immediately that network flow games are not a universal representation formalism, as some cooperative games have no core-stable outcomes (see the examples given in Section 3.2 starting on page 149).

We can generalize network flow games by allowing each player to control more than one edge. The reader can check that the games in this broader class are totally balanced, too. In fact, for this class of games the converse is true as well: Every totally balanced game can be represented as a generalized network flow game [589].

3.4.2.3 Path Disruption Games

Similarly to network flow games, path disruption games, which were introduced by Bachrach and Porat [76], are also defined on networks. However, in these games the players' goal is opposite to that in network flow games: Instead of connecting the two terminal vertices, the players aim to disconnect them. Further, the players are associated with the vertices rather than the edges of the network.

---

[6] A quite different connection between networks and cooperative games has been studied by Jackson and van den Nouweland [571], who investigate network formation among individuals and, in particular, show that a "strongly stable" network exists if and only if the core of a derived cooperative game is nonempty.

Such games model situations where an intruder seeks to transfer data in a computer network from a source computer to a target computer, and a security system has to prevent this. To achieve its goal, the security system would have to choose certain vertices in the network and intercept the intruder's messages at these vertices. The security system wins if it can ensure that all paths from the source to the target computer can be blocked in this way.

Another example of a setting modeled by path disruption games, due to Jain et al. [576],[7] is inspired by the terrorist attack in Mumbai, India, of 2008: The attackers entered the city at certain locations and tried to reach certain destinations to hit. Since then, to prevent such attacks, the Mumbai police forces have set up a number of inspection checkpoints at various locations in the city's road network in order to make sure that no terrorist finds a way from an entrance point to a presumed target point.

Formally, a path disruption game is modeled by a graph $\mathcal{G} = (\mathcal{V}, \mathcal{E})$ and two special vertices—a *source* $s \in \mathcal{V}$ and a *sink* $t \in \mathcal{V}$. The players are associated with the vertices of the graph. A coalition $C \subseteq \mathcal{V}$ *wins* (has value 1) if, after we remove the vertices in C from the graph, s and t become disconnected; otherwise, it *loses* (has value 0).

Bachrach and Porat [76] analyzed the complexity of computing several solution concepts for this class of games. More recently, Marple, Rey, and Rothe [684, 803, 804] investigated the complexity of bribery problems for path disruption games.[8] Bribery has been studied in the context of voting (see Section 4.3.5 starting on page 358) and, in particular, for single-peaked electorates in Section 5.4 starting on page 393, as well as in judgment aggregation (see Section 7.3.4 starting on page 495).

3.5 Computational Complexity of Identifying Good Outcomes

From a computer scientist's perspective, a key question to ask about solution concepts in cooperative game theory is whether they can be computed efficiently. In this context, "efficiently" is typically understood to mean "in time polynomial in the number of players." However, we have already observed

---

[7] It is interesting to note that such a situation can also be modeled as a *"zero-sum security game on graphs"* [576], i.e., using the framework of noncooperative game theory (see Chapter 2 starting on page 43). In operations research, related problems have been studied under the name *"network interdiction"* (see, e.g., [935, 880]). For further reading on noncooperative games on networks, we point the reader to the work of Galeotti et al. [480] and to the book chapter by Jackson and Zenou [574].

[8] In particular, they have shown that bribery in path disruption games with vertex costs is NP-complete for the case of a single adversary (or intruder) [803], and is complete for $\Sigma_2^p = \text{NP}^{\text{NP}}$, the second level of the polynomial hierarchy [706, 892], for the case of multiple adversaries [684, 807].

3.5 Computational Complexity of Identifying Good Outcomes

that there is no representation formalism that can describe every cooperative game so that the size of the encoding of each game is polynomial in the number of players, n. This appears to be highly problematic: If we describe a game in a nonsuccinct manner, e.g., by listing the values of all coalitions, or, in case of simple games, by encoding it as a vector weighted voting game of exponentially large dimension, then no algorithm that runs in time polynomial in n can hope to read the full description of the game.

There are various ways to (at least partially) circumvent this issue.

First, it turns out that for some classes of games and some solution concepts, knowing the values of all coalitions is not necessary: One can compute "good" outcomes (or decide whether "good" outcomes exist) by looking at the values of a few carefully selected coalitions. In this case, instead of an explicit representation of the input game, it suffices to have a subroutine that, given a coalition, can compute its value in time polynomial in n. This approach is sometimes referred to as "oracle-based": One's goal is to design an algorithm for computing a "good" outcome, assuming one has access to an *oracle* that can be queried about coalitional values and answers each query in one unit of time.

Alternatively, one can focus on a specific representation, and develop algorithms for computing "good" outcomes that run in time polynomial in the input size. This approach results in algorithms that run in time polynomial in n whenever the input game is represented using polynomially many (in n) bits. The representation itself can be universally expressive (in which case the runtime of the algorithms will be polynomial in n only for the games that admit a succinct representation within that formalism), or it can be tailored to a specific class of games, which can be represented succinctly (in the latter case, the runtime of the algorithms will be polynomial in n for all games that can be represented in that formalism).

In this section, we explore both approaches. First, we present oracle-based algorithms for computing core-stable outcomes in convex games and in simple games, and then we focus on two of the succinct representation formalisms described earlier in this chapter—namely, weighted voting games and induced subgraph games—and provide algorithms and complexity results for key solution concepts under these representations.

3.5.1 An Oracle-Based Approach: Core-Stable Outcomes in Convex and Simple Games

The reader may have observed that, to find a core-stable outcome of a convex game with a set of players $P = [n] = \{1, \ldots, n\}$, we did not have to look at all coalitions: It was sufficient to know the values of coalitions $[i] = \{1, \ldots, i\}$ for each $i \in P$. Similarly, to decide if a simple game (P, v) admits a stable outcome,[9] it was sufficient to determine if the game has any veto players (by Theorem 3.2 on page 154), and this question boils down to computing the value of each coalition of the form $P \smallsetminus \{i\}$, where $i \in P$: i is a veto player if and only if $v(P \smallsetminus \{i\}) = 0$. Moreover, having identified the veto players, we can also construct a core-stable outcome if one exists (by sharing the value of the grand coalition among the veto players), or decide whether a given outcome is core-stable (by checking that none of the nonveto players receives a positive payoff). We summarize these observations in the following corollary.

Corollary 3.1. *Suppose that (P, v) is a cooperative game, and we have access to an algorithm that, given a coalition $C \subseteq P$, computes $v(C)$ in time polynomial in $n = \|P\|$. Then, if (P, v) is convex, we can find an outcome in the core of (P, v) in polynomial time. Moreover, if (P, v) is simple, we can*

(a) decide whether the core of (P, v) is empty, and if not, find an outcome in the core of (P, v) in polynomial time;

(b) given an outcome for (P, v), decide whether it is in the core of (P, v) in polynomial time.

Unfortunately, this approach does not extend to fairness-based solution concepts such as the Shapley value; indeed, in the next section, we will see that computing players' Shapley values in weighted voting games is computationally demanding.

3.5.2 Complexity of Problems for Weighted Voting Games

Recall that weighted voting games with n players can be represented by $n+1$ numbers, i.e., the weights of the players and the quota. This implicitly tells us which of the altogether 2^n possible coalitions of players are winning and which ones are losing. As before, we assume that these $n+1$ numbers and other inputs to our computational problems (e.g., ϵ in problems concerning the ϵ-core) are rational numbers.

We will now discuss the computational complexity of various problems related to solution concepts for weighted voting games. We first consider stability-related solution concepts, and then discuss power indices, such as the Shapley–Shubik index and the Banzhaf index.

---

[9] Recall that, when considering stability in simple games, we assume that players form the grand coalition, even if this is not a social welfare-maximizing outcome.

3.5.2.1 Complexity of Stability-Related Solution Concepts for Weighted Voting Games

In this section, we are concerned with the complexity of problems regarding stability-related solution concepts for weighted voting games, such as the core, the ϵ-core, the least core, and the cost of stability.

Core

One of the most basic questions concerning a cooperative game is whether this game admits a stable outcome, i.e., whether its core is nonempty. For weighted voting games (which we abbreviate as "WVG" in the following problem names), this question corresponds to the following decision problem.

WVG-EMPTY-CORE
Given: A weighted voting game $G = (w_1, \ldots, w_n; q)$.
Question: Does it hold that $Core(G) = \emptyset$?

Alternatively, one might consider the complementary problem by asking whether the core of a given weighted voting game is *non*empty.

Just knowing that a stable outcome exists is usually not enough: Typically, we want to construct such an outcome, or verify that a specific outcome is stable. These computational problems can be formalized as follows.

WVG-CONSTRUCT-CORE
Given: A weighted voting game $G = (w_1, \ldots, w_n; q)$.
Task: Construct an imputation $\vec{a}$ in the core of G.

WVG-IN-CORE
Given: A weighted voting game $G = (w_1, \ldots, w_n; q)$ and an imputation $\vec{a}$.
Question: Is $\vec{a}$ in the core of G?

Now, we have established (Corollary 3.1) that in simple games we can decide in polynomial time whether the core is nonempty and whether a given outcome is in the core, as well as to construct an outcome in the core as long as we have access to an oracle that, given a coalition, can return its value in time polynomial in the number of players. Weighted voting games are simple games, and for a weighted voting game such an oracle is straightforward to implement: To determine the value of a given coalition, we compute its total weight and compare it with the quota. It follows that for weighted voting games, the core-related problems we have formulated can be decided by an algorithm that runs in time polynomial in n. We thus have the following corollary, called a folklore result by Elkind et al. [381].

Corollary 3.2. *The problems WVG-EMPTY-CORE, WVG-CONSTRUCT-CORE, and WVG-IN-CORE can be solved in polynomial time.*

Least Core and ϵ-Core

In addition, Elkind et al. [381] investigate the complexity of least-core-related problems for weighted voting games. We will now briefly overview their results. They consider the following computational problems.

WVG-LEAST-CORE
Given: A weighted voting game $G = (w_1, w_2, \ldots, w_n; q)$ and $\epsilon \geq 0$.
Question: Does it hold that $\epsilon\text{-}Core(G) \neq \emptyset$?

WVG-CONSTRUCT-LEAST-CORE
Given: A weighted voting game $G = (w_1, \ldots, w_n; q)$.
Task: Construct an imputation $\vec{a}$ in the least core of G.

WVG-IN-LEAST-CORE
Given: A weighted voting game $G = (w_1, \ldots, w_n; q)$ and an imputation $\vec{a}$.
Question: Is $\vec{a}$ in the least core of G?

We remark that the first of these problems is called WVG-LEAST-CORE, since the smallest value of $\epsilon \geq 0$ for which (G, ϵ) is a yes-instance of this problem is the value of the least core of G.

Elkind et al. [381] show that none of the problems WVG-LEAST-CORE, WVG-CONSTRUCT-LEAST-CORE, and WVG-IN-LEAST-CORE can be solved in polynomial time, unless P = NP. In particular, they demonstrate coNP-hardness[10] of WVG-LEAST-CORE and NP-hardness of WVG-IN-LEAST-CORE by a $\leq_m^P$-reduction (see Definition 1.2 on page 29) from the following NP-complete problem (see, e.g., [483]):

PARTITION
Given: A nonempty sequence $(k_1, k_2, \ldots, k_n)$ of positive integers such that $\sum_{i=1}^{n} k_i$ is even.
Question: Is there a subset $A \subseteq \{1, 2, \ldots, n\}$ with $\sum_{i \in A} k_i = \sum_{i \in \{1,2,\ldots,n\} \setminus A} k_i$?

Given an instance $(k_1, k_2, \ldots, k_n)$ of PARTITION with $\sum_{i=1}^{n} k_i = 2K$, construct the following weighted voting game with $n+1$ players:

$$G = (k_1, k_2, \ldots, k_n, K; K).$$

The following lemma states some useful properties of this construction (the easy proof is left as an exercise).

Lemma 3.1 (Elkind et al. [381]).

---

[10] The class coNP contains all problems whose complements are in NP, and similarly to the "P = NP?" question described in Section 1.5 starting on page 18, it is an open problem whether or not NP $\neq$ coNP is true.

3.5 Computational Complexity of Identifying Good Outcomes

1. If $(k_1, k_2, \ldots, k_n)$ is a yes-instance of PARTITION then the value of the least core of G is $2/3$ and the imputation $(k_1/3K, \ldots, k_n/3K, 1/3)$ is in the least core of G. Moreover, for any imputation $(p_1, \ldots, p_{n+1})$ in the least core of G, we have $p_{n+1} = 1/3$.
2. If $(k_1, k_2, \ldots, k_n)$ is a no-instance of PARTITION then the value of the least core of G is at most $2/3 - 1/6K$.

Lemma 3.1 says that if $(k_1, \ldots, k_n)$ is a yes-instance of PARTITION then $(G, 2/3 - 1/6K)$ is a no-instance of WVG-LEAST-CORE, and vice versa. By Lemma 1.1 on page 30, it follows that WVG-LEAST-CORE is coNP-hard. Further, the second part of Lemma 3.1 implies that if we start with a no-instance of PARTITION, then the imputation $(k_1/3K, \ldots, k_n/3K, 1/3)$ is not in the least core: Indeed, under this imputation player $n+1$ is only paid $1/3$, even though $\{n+1\}$ is a winning coalition, whereas in the least core each winning coalition is paid at least $1/3 + 1/6K$. Combining this observation with the first part of Lemma 3.1, we conclude that $(k_1/3K, \ldots, k_n/3K, 1/3)$ is in the least core if and only if we start with a yes-instance of PARTITION, and therefore WVG-IN-LEAST-CORE is NP-hard.

This argument also shows that WVG-CONSTRUCT-LEAST-CORE is not solvable in polynomial time, unless P = NP: If one could construct an imputation $\vec{a} = (a_1, a_2, \ldots, a_{n+1})$ in the least core of G in polynomial time, then one could also solve the NP-complete problem PARTITION in polynomial time, simply by looking at a_{n+1}: $(k_1, k_2, \ldots, k_n)$ is a yes-instance of PARTITION if and only if $a_{n+1} = 1/3$.

Cost of Stability

We will now consider computational questions related to the cost of stability (recall the notions defined in Section 3.2 starting on page 156). Bachrach et al. [74] define the following two problems:

WVG-SUPER-IMPUTATION-STABILITY
Given: A weighted voting game $G = (w_1, \ldots, w_n; q)$, a parameter $\Delta \geq 0$, and an imputation $\vec{a} = (a_1, a_2, \ldots, a_n)$ in the adjusted game G_Δ.
Question: Is $\vec{a}$ in the core of G_Δ?

WVG-COST-OF-STABILITY
Given: A weighted voting game $G = (w_1, \ldots, w_n; q)$ and a parameter $\Delta \geq 0$.
Question: Is it true that $CoS(G) \leq \Delta$ (i.e., is it true that $Core(G_\Delta) \neq \emptyset$)?

By reductions from PARTITION, Bachrach et al. [74] show that these two problems are coNP-complete, provided that the weights and the quota of the given weighted voting game are represented in binary.

3.5.2.2 Complexity of Power Indices in Weighted Voting Games

We now consider various problems related to power indices in weighted voting games.

Computing and Comparing Power Indices

In contrast with the core and the least core, the Shapley–Shubik index and the Banzhaf index are single-valued solution concepts. Therefore, the most natural computational question associated with these notions is simply computing the index of each player. Thus, for a power index $\mathbb{PI}$ (such as any of the variants of the Shapley–Shubik index and the Banzhaf index defined earlier in this chapter), we define the following problem.

WVG-$\mathbb{PI}$

Given: A weighted voting game $G = (w_1, \ldots, w_n; q)$.
Task: Construct an imputation $\vec{a}$ such that $a_i = \mathbb{PI}(G, i)$.

Note that WVG-$\mathbb{PI}$ is a functional (rather than a decision) problem, so it is captured by complexity classes other than P and NP (as these by definition contain only decision problems).

Deng and Papadimitriou [338] showed that WVG-Shapley-Shubik is complete for the class #P under a suitable notion of reduction. This complexity class, introduced by Valiant [921], can be viewed as the counting version of the class NP. It consists of functions that give the number of solutions to instances of an NP problem. For instance, this class contains the problem of computing the number of satisfying truth assignments for a given boolean formula; the corresponding NP-complete problem is SAT. If one could efficiently compute the number $f(x)$ of solutions to a given instance x of an NP problem X, then one could efficiently decide X itself as well: x is a yes-instance if and only if $f(x) > 0$. Prasad and Kelly [787] (implicitly) proved that the problem WVG-Banzhaf is #P-complete under another suitable notion of reduction. Matsui and Matsui [688] showed NP-completeness of certain decision problems concerning WVG-Shapley-Shubik und WVG-Banzhaf.

A related question is deciding whether some player is a null player: Indeed, a player is a null player in a weighted voting game if and only if her Shapley–Shubik index (or, equivalently, her Banzhaf index) is 0.

WVG-Null

Given: A weighted voting game $G = (w_1, \ldots, w_n; q)$ and a player i in G.
Question: Is i a null player in G?

This problem is obviously in coNP: To show that a player i is *not* a null player in a given weighted voting game $G = (P, v)$, it suffices to exhibit a coalition $C \subseteq P \setminus \{i\}$ such that i is pivotal for C (i.e., $v(C) = 0$,

3.5 Computational Complexity of Identifying Good Outcomes

yet $v(C \cup \{i\}) = 1$), which is an NP question. Moreover, a simple reduction from PARTITION to the complement of WVG-NULL shows that WVG-NULL is coNP-complete. Indeed, given an instance $(k_1, \ldots, k_n)$ of PARTITION with $\sum_{i=1}^{n} k_i = 2K$, consider the weighted voting game $(k_1, \ldots, k_n, 1; K+1)$. Player $n+1$ (with weight 1) is *not* a null player in this game if and only if there is some coalition $C \subseteq P \smallsetminus \{n+1\}$ that $n+1$ is pivotal for (which is the case exactly if there exists a coalition $C \subseteq P \smallsetminus \{n+1\}$ of total weight exactly K), which is equivalent to $(k_1, \ldots, k_n)$ being a yes-instance of PARTITION.

Another related problem is to compare a player's power in two given games. This question can be formalized as follows (as before, $\mathbb{PI}$ is an arbitrary power index).

WVG-$\mathbb{PI}$-POWER-COMPARE

Given: Two weighted voting games, G and G', and a player i occurring in both games.
Question: Is it true that $\mathbb{PI}(G, i) > \mathbb{PI}(G', i)$?

Faliszewski and Hemaspaandra [435] showed that this problem is PP-complete for both the Shapley–Shubik and the probabilistic Banzhaf power index. The complexity class PP, which was introduced by Gill [501], stands for *probabilistic polynomial time*, and is widely believed to be even larger than NP.[11]

Beneficial Merging and Beneficial Splitting of Players

If power indices are used to distribute the payoffs, the players may want to manipulate them. Perhaps the most natural form of manipulation in the context of weighted voting games involves a single player splitting into multiple players, or multiple players merging into one.

Anna and Chris each buy a ticket in a raffle. There are red, green, and blue tickets worth 50, 75, and 100 points, respectively. Once they have accumulated tickets worth a total of 150 points, they can trade them for a box of cookies. Anna's ticket is green, i.e., worth 75 points, whereas Chris has received a blue ticket worth 100 points. Happily, they want to trade their tickets for cookies. They agree to share the cookies according to their Shapley–Shubik index in the weighted voting game induced by their contributions and the threshold for receiving cookies, that is, $G = (75, 100; 150)$, which yields Shapley-Shubik$(G, \text{Anna}) = $ Shapley-Shubik$(G, \text{Chris}) = 1/2$.

---

[11] As with the famous "P = NP?" question, whether or not NP = PP (or even P = PP) is an open problem in complexity theory. It is easy to see that $P \subseteq NP \subseteq PP$, and it is considered highly unlikely that any one of these two inclusions is in fact an equality.

On the way to the raffle booth, Anna sees Belle and stops for a quick chitchat with her, and when Chris arrives alone at the raffle booth, he meets David who is waiting there for Belle and Edgar to play a little weighted voting game with them that involves raffle tickets and a box of cookies to share. David shows Chris his two red tickets, each worth 50 points.

"I have a great idea!" Chris shouts out very excitedly. "How about if we trade our tickets: I give you my blue ticket in exchange for your two red tickets!"

"OK, why not?" says David, who doesn't care.

When Anna eventually arrives, Chris says: "Guess who just stopped by? My old buddy Christopher! He says he wants to play with us, y'know, for the cookies. He was in a hurry and is gone already, but he left his ticket here, so he has a red ticket and I have a red ticket and your ticket is still green. Let's play!"

But now, Anna and two copies of Chris (namely, the real Chris and his fake-identity buddy Christopher) participate in the new weighted voting game $G' = (75, 50, 50; 150)$, and the latter two players' Shapley–Shubik indices sum up to ²/₃, as Shapley-Shubik$(G',$ Anna$) =$ Shapley-Shubik$(G',$ Chris$) =$ Shapley-Shubik$(G',$ Christopher$) = $ ¹/₃.

Chris grabs two thirds of the cookies, greedily squinting his eyes, and says: "Let me bring him his share."

Given a weighted voting game $G = (w_1, \ldots, w_n; q)$, a player i with weight w_i can split into two players i_1 and i_2 with weights w_{i_1} and w_{i_2} satisfying $w_{i_1} + w_{i_2} = w_i$ (specifically, $50 + 50 = 100$ in the above example). More generally, i can split into k players $i_1, \ldots, i_k$ with weights $w_{i_1}, \ldots, w_{i_k}$ such that $\sum_{\ell=1}^{k} w_{i_\ell} = w_i$. To avoid double subscripts, let us rename the weights of the k players i has split into by $u_1, \ldots, u_k$ such that $\sum_{\ell=1}^{k} u_\ell = w_i$. We then obtain the new weighted voting game $G' = (w_1, \ldots, w_{i-1}, u_1, \ldots, u_k, w_{i+1}, \ldots, w_n; q)$ that results from G by splitting i into k players with these weights.

If the sum of the Shapley–Shubik indices of the players i has split into in the resulting game G' exceeds the Shapley–Shubik index of player i in the original game G, then this manipulation is *successful* (or *beneficial*) with respect to the Shapley–Shubik index. Beneficial splitting of (the weight of) players is also referred to as *false-name manipulation*. Formally, for a given power index $\mathbb{PI}$, we define the following computational problem:

3.5 Computational Complexity of Identifying Good Outcomes

WVG-PI-Beneficial-Split

Given: A weighted voting game $G = (w_1, \ldots, w_n; q)$, a player $i \in \{1, \ldots, n\}$, and an integer $k > 2$.

Question: Is it possible to split i into k new players with positive integer weights $u_1, \ldots, u_k$ satisfying $\sum_{\ell=1}^{k} u_\ell = w_i$ so that

$$\sum_{\ell=1}^{k} \mathbb{PI}(G', i-1+\ell) > \mathbb{PI}(G, i),$$

where $G' = (w_1, \ldots, w_{i-1}, u_1, \ldots, u_k, w_{i+1}, \ldots, w_n; q)$?

WVG-PI-Beneficial-Split was originally defined by Bachrach and Elkind [73], who showed that it is NP-hard for the Shapley–Shubik power index, even if a player is only allowed to split her weight between two new identities (i.e., for $k \geq 2$). NP-hardness of the beneficial splitting problem is also known for the two Banzhaf power indices, as shown in the works of Aziz et al. [39, 65] and Rey and Rothe [801, 802]. In particular, Rey and Rothe proved that WVG-Banzhaf-Beneficial-Split is in P whenever $k = 2$, but is NP-hard when splitting into more than two false identities is allowed. Subsequently, they provided improved lower bounds for the Shapley–Shubik and the probabilistic Banzhaf power index by proving PP-hardness of WVG-Shapley-Shubik-Beneficial-Split for $k \geq 2$, and of WVG-Banzhaf-Beneficial-Split for $k \geq 3$ [805]. They also observed that for both power indices, if the number of false identities is given (in unary) in the problem instance, the best known upper bound is NP^{PP}, the class of problems solvable by an NP oracle machine with access to a PP oracle.[12]

Conversely, multiple players may want to merge into a single player in order to increase their overall power. For example, two players of weights 5 and 3, respectively, can merge to form one new player of weight 8. Given a weighted voting game $G = (w_1, \ldots, w_n; q)$ and a coalition $C \subseteq \{1, \ldots, n\}$ with $\|C\| = k$, we define the weighted voting game

$$G_{\&C} = \left(\sum_{i \in C} w_i, w_{j_1}, \ldots, w_{j_{n-k}}; q\right)$$

with the set of players $\{j_{\&C}, j_1, \ldots, j_{n-k}\} = \{j_{\&C}\} \cup (\{1, \ldots, n\} \setminus C)$. In this game, the players from C have merged into a single player, $j_{\&C}$, whose weight is equal to the sum of the weights of the players in C.

We can now define the computational problem that captures the notion of beneficial merging for a given power index $\mathbb{PI}$:

---

[12] By Toda's result that, in terms of polynomial-time Turing reductions, PP is at least as hard as any problem in the polynomial hierarchy [915], NP^{PP} contains the entire polynomial hierarchy: $\text{PH} \subseteq \text{P}^{\text{PP}} \subseteq \text{NP}^{\text{PP}}$. On the other hand, PH and PP are widely believed to be "incomparable" (i.e., neither $\text{PH} \subseteq \text{PP}$ nor $\text{PP} \subseteq \text{PH}$ is known); the largest part of the polynomial hierarchy known to be contained in PP is the class $\text{P}_{\|}^{\text{NP}}$ (see [127]), which will be considered more closely in Section 4.3.1 starting on page 289.

WVG-ℙI-Beneficial-Merge
Given: A weighted voting game $G = (w_1, \ldots, w_n; q)$ and a nonempty coalition $C \subseteq \{1, \ldots, n\}$.   **Question:** Is it true that $\mathbb{PI}(G_{\&C}, j_{\&C}) > \sum_{i \in C} \mathbb{PI}(G, i)$?

For the normalized Banzhaf and the Shapley–Shubik index, Aziz et al. [39][13] showed that WVG-$\overline{\text{Banzhaf}}$-Beneficial-Merge and WVG-Shapley-Shubik-Beneficial-Merge are both NP-hard. Subsequently, Rey and Rothe [801, 802] proved NP-hardness of the beneficial merging problem for the probabilistic Banzhaf index as well.

Regarding upper bounds, Faliszewski and Hemaspaandra [435] showed that WVG-Shapley-Shubik-Beneficial-Merge is in PP and conjectured this problem to be complete for this class. Rey and Rothe [801, 802] observed that their proof technique can also be applied to the probabilistic Banzhaf index, so WVG-Banzhaf-Beneficial-Merge is in PP as well.[14] Raising two of the above-mentioned NP-hardness results to matching upper bounds and confirming the conjecture of Faliszewski and Hemaspaandra [435], Rey and Rothe [805] subsequently showed that, in fact, WVG-Banzhaf-Beneficial-Merge is PP-complete, even if only three players of equal weight merge, and that WVG-Shapley-Shubik-Beneficial-Merge is PP-complete, even if only two players of equal weight merge.

Whether or not beneficial merging or splitting is PP-hard for the normalized Banzhaf index remains an open question. However, the decisions whether to split or to merge become easy in *unanimous weighted voting games*, i.e., in games where the total weight of all players equals the quota, so that only the grand coalition can reach the quota and win. Aziz et al. [39] showed that in such games, no coalition of players can increase their total payoff by merging, with respect to either the Shapley–Shubik or the normalized Banzhaf index. This result generalizes to games in which each player is a veto player. In contrast, splitting is always beneficial in unanimous weighted voting games, with respect to either of these two indices. Unlike for the normalized Banzhaf index, Rey and Rothe [805] showed that for the probabilistic Banzhaf index in unanimous weighted voting games, splitting a player is always neutral (i.e., neither beneficial nor disadvantageous) or even disadvantageous, whereas merging players is neutral for coalitions of size two, but is beneficial for larger coalitions.[15]

---

[13] This paper is a beneficial merge of two earlier conference papers by Bachrach and Elkind [73] and Aziz and Paterson [65].

[14] However, this technique does not seem to be capable of establishing PP membership of this problem for the normalized Banzhaf index.

[15] This may look somewhat counterintuitive at first glance. If merging two players in a weighted voting game (unanimous or not) is never beneficial in terms of the probabilistic Banzhaf index, why doesn't this outright imply that merging three or more players is never beneficial in terms of that index either? As noted by Rey and Rothe [805], the reason is that merging two players can change a third player's probabilistic Banzhaf

Control by Deleting or Adding Players

After Chris is gone with his treasure of two thirds of all cookies, Anna becomes a bit suspicious of him. She had never met his buddy Christopher, nor heard anything about him, and she somehow doubts he really exists. Shortly decided, she pursues Chris, and eventually catches him (with all the cookies gone).

"How about, Chris, if we play this weighted voting game again?"

"Sure, I'd like to," answers Chris, wiping some chocolate crumbs off his face. They return to the raffle booth, this time aiming for a box of fine candies that is worth 300 points.

"That's pretty expensive, though!" Chris is worried.

"Then let's ask Belle to join, she is still here," Anna suggests and asks her right away. Belle agrees. The girls buy a blue and a red ticket each, so both contribute a total of 150 points. Chris takes a blue ticket for himself, which is worth 100 points, and (as he made good experiences with beneficially splitting his point weight on page 185) a red ticket "for my buddy Christopher," worth 50 points. They look at their weighted voting game $G = (150, 150, 100, 50; 300)$. A quick calculation reveals that Shapley-Shubik$(G, \text{Anna}) = $ Shapley-Shubik$(G, \text{Belle}) = 1/3$ and Shapley-Shubik$(G, \text{Chris}) = $ Shapley-Shubik$(G, \text{Christopher}) = 1/6$.

"OK, that's not too bad!" says Chris. He grabs one third of the candies and adds: "Let me bring Christopher his share."

"Wait a minute," Anna replies. "I don't believe there is someone called Christopher! You just made him up, right? So either you present him to us now to show he really exists, or he must be deleted from the game with his weight of 50 points, and we have to recalculate!"

She has him cornered, so Chris admits everything. After removing fake Christopher's weight of 50 points from the game, their new weighted voting game is $G' = (150, 150, 100; 300)$, and a quick recalculation gives that Shapley-Shubik$(G', \text{Anna}) = $ Shapley-Shubik$(G', \text{Belle}) = 1/2$ and Shapley-Shubik$(G', \text{Chris}) = 0$.

"Oops, so sorry, Chris!" says Anna full of true regret. "It seems you are a null player now, so I will share the candies fifty-fifty with Belle."

What Anna has exerted here is known as *control by deleting players* and has been introduced by Rey and Rothe [806], inspired by electoral control actions in voting, such as control by adding or deleting either candidates or voters—see Section 4.3.4 starting on page 328; see also Section 5.2 starting

---

index in this new game. Therefore, the probabilistic Banzhaf index of, say, a size-three merger can differ from that of a player resulting from first merging two players and then merging this size-two merger with a third player.

on page 376 for control of single-peaked elections, and Section 7.3.5 starting on page 500 for control in the context of judgment aggregation.

When players are deleted from a weighted voting game with the goal of increasing the power index $\mathbb{PI}$ of a given player (for instance, by deleting Christopher in the example above, Anna's and Belle's Shapley–Shubik indices have been increased from $1/3$ to $1/2$), we can define the computational problem that captures this notion as follows:

WVG-Control-by-Deleting-Players-to-Increase-$\mathbb{PI}$

Given: A weighted voting game $G = (w_1, \ldots, w_n; q)$, a distinguished player $i \in \{1, \ldots, n\}$, and a positive integer $k < n$.

Question: Can at most k players $M \subseteq \{1, \ldots, n\} \smallsetminus \{i\}$, $\|M\| \leq k$, be deleted from G such that for the new game $\mathcal{G}_{\smallsetminus M} = (w_{j_1}, \ldots, w_{j_{n-\|M\|}}; q)$, where $\{j_1, \ldots, j_{n-\|M\|}\} = \{1, \ldots, n\} \smallsetminus M$, it holds that $\mathbb{PI}(G_{\smallsetminus M}, i) > \mathbb{PI}(G, i)$?

Rey and Rothe [806] also introduce the problems corresponding to control by deleting players with the goals of *decreasing* a given player's power index $\mathbb{PI}$ (WVG-Control-by-Deleting-Players-to-Decrease-$\mathbb{PI}$) and of *maintaining* it (WVG-Control-by-Deleting-Players-to-Maintain-$\mathbb{PI}$) by asking instead whether $\mathbb{PI}(G_{\smallsetminus M}, i) < \mathbb{PI}(G, i)$ and $\mathbb{PI}(G_{\smallsetminus M}, i) = \mathbb{PI}(G, i)$, respectively.[16] Further, they consider the corresponding problems capturing control by *adding players to* a given weighted voting game.

Rey and Rothe [806] and, later on, Kaczmarek and Rothe [584] have studied the complexity of these problems. While for control by adding players, Rey and Rothe [806] establish PP-hardness for the various goals via the techniques that were previously used to prove PP-hardness for the problems of beneficial merging and splitting players [805], the known complexity results are more diverse for control by deleting players. In particular, while WVG-Control-by-Deleting-Players-to-Increase-Shapley-Shubik is NP-hard and WVG-Control-by-Deleting-Players-to-Maintain-Banzhaf and WVG-Control-by-Deleting-Players-to-Maintain-Shapley-Shubik are coNP-hard [806], Kaczmarek and Rothe [584] show that WVG-Control-by-Deleting-Players-to-Increase-Banzhaf is DP-hard, WVG-Control-by-Deleting-Players-to-Decrease-Banzhaf is $P_{\|}^{NP}$-hard, and WVG-Control-by-Deleting-Players-to-Decrease-Shapley-Shubik is NP-hard; the reader is referred to Section 1.5.3 starting on page 34 for the definitions of these complexity classes.

As for beneficial splitting players, the best known upper bound for all these problems is NP^{PP} (recall Footnote 12 on page 187). While it is still an open question whether these lower bounds can be raised so as to meet this upper bound in many cases, Kaczmarek and Rothe [585] have recently shown

---

[16] In addition, they also study these problems for the goals of *nonincreasing* and of *nondecreasing* a given player's power index. For these two goals and for the goal of maintaining a power index, it is also required that at least one player *must* be deleted, to avoid that the problems become trivial.

3.5 Computational Complexity of Identifying Good Outcomes

NP^{PP}-completeness for control by adding players to change or maintain the probabilistic Banzhaf index in weighted voting games.

A difference between beneficial splitting or merging players and control by deleting or adding players is that in the former case the total weight of the grand coalition remains preserved in the new game, whereas in the latter case the total weight of the grand coalition is changed in the new game—it decreases by deleting players and increases by adding them. But when the total weight of the grand coalition has been changed by such a control action, is it then fair to use in the new game the same quota as in the original one?

"No, that's totally unfair!" Edgar complains, thoroughly annoyed. "I reckon, deleting players could mean that *all* coalitions of players, even the grand coalition, become losing because they no longer reach the quota ..."

Kaczmarek and Rothe [584] also introduce modified control problems where, for some real parameter $r \in [0,1]$,

- the quota of the original weighted voting game G with players P is given by $q = r \cdot \sum_{i \in P} w_i$ and
- the quota of the new weighted voting game G' with players P' (resulting by either deleting players from P or adding them to P) is analogously specified by $q' = r \cdot \sum_{i \in P'} w_i$.

For these modified control problems with changing quota, they show

- for the goals of increasing or decreasing the given player's power index,
  - NP-hardness for control by deleting players when using the Shapley–Shubik index;
  - DP-hardness for control by deleting players when using the probabilistic Banzhaf index; and
  - PP-hardness for control by adding players when using either of the Shapley–Shubik or the probabilistic Banzhaf index; and
- for the goal of maintaining the given player's power index, coNP-hardness for control by either deleting or adding players when using either of the Shapley–Shubik or the probabilistic Banzhaf index.

The best known upper bound for all these modified control problems again is NP^{PP}, and the question of whether one can find matching upper and lower bounds for them remains open.

Manipulating the Quota

Another form of manipulation in weighted voting games is changing the quota. The complexity of computational problems related to this form of manipulation, for the Shapley–Shubik index and the normalized Banzhaf index, was investigated by Zuckerman et al. [972] and subsequently by Zick, Skopalik, and Elkind [971]. In particular, Zuckerman et al. [972] describe a polynomial-time algorithm for determining whether a given player's voting power can be reduced to 0 by this type of manipulation, i.e., whether a player can be made a null player by changing the quota. This result is remarkable, as we have argued on page 185 that *checking* whether a player is a null player is coNP-complete. On the other hand, Zick et al. [971] show that a player's Shapley–Shubik index is maximized by setting the quota to this player's weight. In contrast, they prove that it is coNP-hard to decide whether a given value of the quota maximizes or minimizes a given player's Shapley–Shubik index, and Zuckerman et al. [972] demonstrate that determining which of two given values of the quota is better for a particular player is PP-complete; the latter result holds both for the Shapley–Shubik index and for the probabilistic Banzhaf index and strengthens the previously mentioned result of Faliszewski and Hemaspaandra [435] that the problem of comparing the power index of a given player in two given weighted voting games (which generalizes the quota manipulation problem) is PP-complete.

Unary Weights and Approximation

So far, we have implicitly assumed that, when encoding a weighted voting game with n players, we represent the weights and the quota in binary notation. In contrast, if these $n+1$ numbers are given in unary (or if the weights are polynomially bounded in n), then many questions concerning weighted voting games can be answered in polynomial time. In particular, Matsui and Matsui [688] observe that the Shapley value can be computed by a dynamic program similar to the one for the classic KNAPSACK problem (see, e.g., [600]); this approach extends to other power indices we consider. Moreover, quota manipulation problems can be solved by considering all possible quota values, and, similarly, for small values of k, weight-splitting problems can be solved by considering all possible splits. The efficient algorithms for core-related questions are based on encoding the input as an exponential-size linear program and observing that a separation oracle for this linear program needs to solve an instance of KNAPSACK.

Moreover, for weights and quota given in binary, both the value of the least core and the cost of stability can be computed approximately in polynomial time, with any given degree of precision. More specifically, both of these problems admit fully polynomial-time approximation schemes (FPTAS) (see, respectively, the work of Elkind et al. [381] and Bachrach et al. [74]).

3.5 Computational Complexity of Identifying Good Outcomes

Graph-Restricted Weighted Voting Games

As mentioned in Section 3.1.2.2 starting on page 146, weighted voting games can be used to model decision-making in the political bodies of representative democracies. Sometimes, however, not all players are willing or able to form a joint coalition; there may be ideological discrepancies or personal aversions among the players that prevent them from working together. On the other hand, such ideological or personal kloofs can occasionally be bridged via other players who are connected with the antagonists. A stunning example is the 36th Israeli government where extreme left-wing and right-wing parties formed a government coalition in 2021 and took turns in providing the prime minister. Similarly, it often makes sense to consider weighted voting games together with communication structures that enable some of the players to communicate with each other, whereas other players cannot communicate.

Taking such topological structures into account, Myerson [722] introduced *graph-restricted cooperative games* that are played on undirected graphs whose vertices are the players, and the graph structure describes which players can form a coalition. Based on his model, Napel, Nohn, and Alonso-Meijide [725] define *graph-restricted weighted voting games* as pairs $(G, \mathcal{H})$, where $G = (w_1, \ldots, w_n; q)$ is a weighted voting game and $\mathcal{H}$ an undirected graph whose vertices are the players of G. A coalition C in such a game *wins* if C has a connected part C' with $\sum_{i \in C'} w_i \geq q$; otherwise, C *loses*. Skibski et al. [870] study the *Shapley–Shubik index* and the *probabilistic Banzhaf index* in graph-restricted weighted voting games and show that they are #P-complete under a suitable notion of reducibility due to Valiant [921]. They complement this hardness result by designing a pseudo-polynomial-time algorithm for graphs with bounded treewidth.

Kaczmarek, Rothe, and Talmon [586, 583] define control by either *deleting edges from* or *adding edges to* the graph in a given graph-restricted weighted voting game and study the corresponding problems in terms of their computational complexity. For both deleting and adding edges and for both the Shapley–Shubik and the probabilistic Banzhaf index, they show PP-hardness for the goals of increasing and decreasing a given player's power and coNP-hardness for the goal of maintaining a given player's power in such games. Once more, NP^{PP} is the best known upper bound for all these control problems, and the question of whether one can find matching upper and lower bounds for them is left open.

This concludes a our treatment of weighted voting games and their computational aspects. For further details, we refer to the book chapter by Chalkiadakis and Wooldridge [269] and to the book by Chalkiadakis, Elkind, and Wooldridge [268]. Now, we turn to another class of games that are played on graphs and consider the complexity of the associated problems.

3.5.3 Complexity of Problems for Induced Subgraph Games

Deng and Papadimitriou [338] systematically explored the computational complexity of many solution concepts for induced subgraph games. To present their results, we first have to give formal definitions of the associated computational problems. We start with stability-related solution concepts.

Just as for weighted voting games, the key questions are whether the core is empty, whether a given outcome is in the core, and how to construct an outcome in the core. For induced subgraph games (which we abbreviate as "ISG"), these questions correspond to the following decision problems.

ISG-EMPTY-CORE

Given: An induced subgraph game $G = (\mathcal{V}, \mathcal{E}, \omega)$.
Question: Does it hold that $Core(G) = \emptyset$?

ISG-CONSTRUCT-CORE

Given: An induced subgraph game $G = (\mathcal{V}, \mathcal{E}, \omega)$.
Task: Construct an imputation $\vec{a}$ in the core of G.

ISG-IN-CORE

Given: An induced subgraph game $G = (\mathcal{V}, \mathcal{E}, \omega)$ and an imputation $\vec{a}$.
Question: Is $\vec{a}$ in the core of G?

We now provide an overview of the complexity results for ISG-EMPTY-CORE, ISG-CONSTRUCT-CORE, and ISG-IN-CORE, first for graphs with nonnegative edge weights, and then for general graphs.

Observe that when all edge weights are nonnegative, the resulting game is convex: If C and D are two sets of players such that $C \subset D$, then all friends of a player $i \in P \setminus D$ in the set C are still present in D. Thus, every induced subgraph game with nonnegative edge weights has a nonempty core, i.e., every instance of ISG-EMPTY-CORE with nonnegative weights is a no-instance. Moreover, the convexity property implies that ISG-CONSTRUCT-CORE with nonnegative weights is in P as well: This follows from Corollary 3.1, as it is straightforward to compute the value of a given coalition in polynomial time. Further, Deng and Papadimitriou [338] show that ISG-IN-CORE is also in P, by reducing this problem to finding a network flow in an auxiliary graph. We will now describe their construction. An illustration is provided in Figure 3.5.

Given an induced subgraph game $G = (\mathcal{V}, \mathcal{E}, \omega)$ and an imputation $\vec{a} = (a_1, \ldots, a_{\|\mathcal{V}\|})$ for this game, we build a network with $\|\mathcal{V}\| + \|\mathcal{E}\| + 2$ nodes: a source s, a sink t, a node u_i for every vertex $i \in \mathcal{V}$, and a node $w_{i,j}$ for every edge $\{i,j\} \in \mathcal{E}$. The source s is connected to each node $w_{i,j}$ by an arc of capacity $\omega(\{i,j\})$, and each node u_i is connected to the sink t by an arc of capacity a_i. Also, there are infinite-capacity arcs between $w_{i,j}$ and u_i

3.5 Computational Complexity of Identifying Good Outcomes

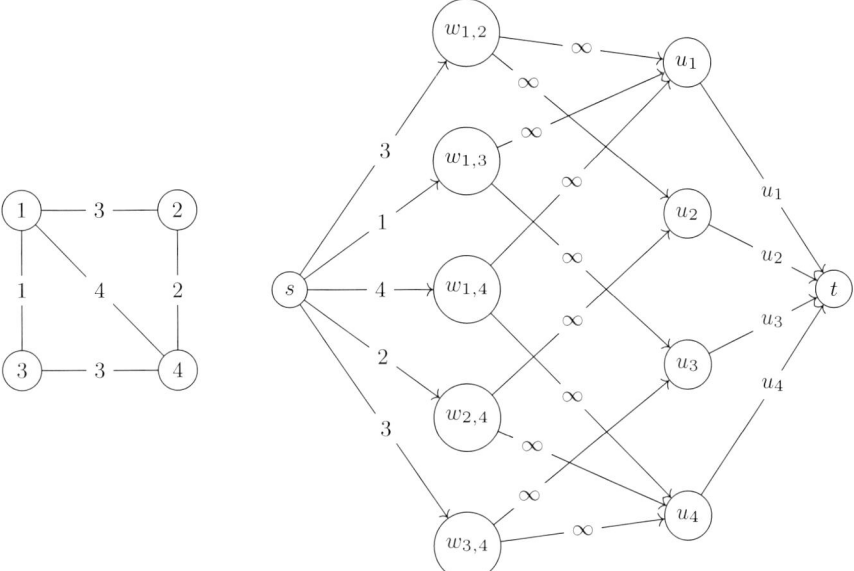

Fig. 3.5 Scheme of reduction for ISG-IN-CORE. The underlying weighted graph is on the left and the associated network on the right.

and between $w_{i,j}$ and u_j for every edge $\{i,j\} \in \mathcal{E}$. Set $U = \{u_i \mid i \in \mathcal{V}\}$ and $W = \{w_{i,j} \mid \{i,j\} \in \mathcal{E}\}$.

Obviously, the size of a source-sink flow in this network is at most $\sum_{\{i,j\} \in \mathcal{E}} w(\{i,j\}) = \sum_{i \in \mathcal{V}} a_i$. Now, suppose that there is indeed a flow of this size. We will argue that in this case $\vec{a}$ is in the core of G. Note that this flow saturates all arcs leaving s and all arcs entering t. Consider a coalition of players $C \subseteq \mathcal{V}$ and the corresponding set of nodes $U_C = \{u_i \mid i \in C\}$. Let W_C be the set of nodes that correspond to the edges between the players in C, i.e., $W_C = \{w_{i,j} \mid i,j \in C\}$. The entire flow that passes through W_C has to enter U_C, so it cannot exceed $\sum_{i \in C} a_i$. But the amount of flow that enters W_C is exactly $\sum_{i,j \in C} w(\{i,j\}) = v(C)$, where v is the characteristic function of our game. Thus, the payment that coalition C receives under $\vec{a}$ is at least its value, i.e., C does not want to deviate. As this holds for an arbitrary coalition C, it follows that $\vec{a}$ is in the core.

Conversely, suppose that there is no flow through this network of size $\sum_{\{i,j\} \in \mathcal{E}} w(\{i,j\})$. Then, by the classic max-flow/min-cut theorem (see, e.g., the book by Ahuja, Magnanti, and Orlin [3]), this network admits a cut of size (strictly) less than $\sum_{\{i,j\} \in \mathcal{E}} w(\{i,j\})$. Obviously, this cut cannot involve any of the infinite-capacity arcs. Consider the set of arcs from U to t that have been cut, and let C be the corresponding coalition, i.e., $C = \{i \mid \text{the arc from } u_i \text{ to } t \text{ has been cut}\}$. We can assume, without loss of generality, that C contains all nodes u_i such that $a_i = 0$: Adding all such nodes to C does not affect the capacity of the cut. Now, to separate s and t, we also have to cut all arcs from s to nodes $w_{i,j}$ such that $i \notin C$ or $j \notin C$. Thus, we have

$$\sum_{i \in C} a_i + \sum_{\{i,j\} \not\subseteq C} \omega(\{i,j\}) < \sum_{\{i,j\} \in \mathcal{E}} \omega(\{i,j\}).$$

Rearranging the terms, we obtain

$$\sum_{i \in C} a_i < \sum_{i,j \in C} \omega(\{i,j\}) = v(C),$$

i.e., C has a profitable deviation under $\vec{a}$.

Since a maximum flow can be computed in polynomial time, it follows that for nonnegative weights, ISG-IN-CORE admits an efficient algorithm.

We can summarize these observations as follows.

Theorem 3.5 (Deng and Papadimitriou [338]). *For induced subgraph games with nonnegative edge weights, the problems ISG-EMPTY-CORE, ISG-CONSTRUCT-CORE, and ISG-IN-CORE are solvable in polynomial time.*

Deng and Papadimitriou [338] also consider the general model, which allows for negative edge weights. They assume that players are required to form the grand coalition, i.e., they identify the core with the set of outcomes where the grand coalition forms and no group of players can benefit from deviating from the grand coalition.[17] Under this assumption, they prove that the core is nonempty if and only if the graph has no negative cut, i.e., there does not exist a set of players $C \subseteq \mathcal{V}$ such that $\sum_{i \in C, j \in \mathcal{V} \setminus C} \omega(\{i,j\}) < 0$, and checking whether a negative cut exists is NP-complete.

It is straightforward to see that a negative cut precludes stability: Assuming that $(C, \mathcal{V} \setminus C)$ is a negative cut, we have

$$v(C) + v(\mathcal{V} \setminus C) = v(\mathcal{V}) - \sum_{i \in C, j \in \mathcal{V} \setminus C} \omega(\{i,j\}) > v(\mathcal{V}),$$

whereas for any imputation $\vec{a}$ we have

$$\sum_{i \in C} a_i + \sum_{i \in \mathcal{V} \setminus C} a_i = v(\mathcal{V}).$$

Thus, it has to be the case that $v(C) > \sum_{i \in C} a_i$ or $v(\mathcal{V} \setminus C) > \sum_{i \in \mathcal{V} \setminus C} a_i$ (or both), i.e., $\vec{a}$ is not in the core.

---

[17] We remark that if weights may be negative, this assumption is not without loss of generality: Consider a two-player induced subgraph game with no self-loops and a single negative-weight edge connecting the players; in this game, the coalition structure consisting of two singleton coalitions is core-stable, but there is no stable outcome in which the grand coalition forms.

3.5 Computational Complexity of Identifying Good Outcomes

To prove the converse direction, consider an imputation $\vec{a}^*$ that splits the value of each edge evenly between its endpoints, i.e.,

$$a_i^* = \frac{1}{2} \sum_{\{i,j\} \in \mathcal{E}} w(\{i,j\})$$

for each $i \in \mathcal{V}$. Under this imputation, the payoff of a coalition C is the total weight of its internal edges plus half of the weight of the cut $(C, \mathcal{V} \setminus C)$. If this coalition deviates, its new payoff is just the total weight of its internal edges. Thus, if all cuts are nonnegative, no coalition has a reason to deviate under $\vec{a}^*$.

This argument also shows that ISG-IN-CORE is computationally hard: To decide whether the imputation $\vec{a}^*$ described in the previous paragraph is in the core, we have to decide whether the graph has a negative cut. Further, an algorithm for ISG-CONSTRUCT-CORE would enable us to decide whether the core is nonempty. Hence, we obtain the following hardness results.

Theorem 3.6 (Deng and Papadimitriou [338]). *For induced subgraph games, the decision problems ISG-EMPTY-CORE and ISG-IN-CORE are NP-complete and coNP-complete, respectively, and the functional problem ISG-CONSTRUCT-CORE is not solvable in polynomial time unless $P = NP$.*

In contrast, it is easy to compute the players' Shapley values in induced subgraph games: In fact, the Shapley values are given by the imputation $\vec{a}^*$ constructed above. To show this, we decompose our game into $\mu = \|\mathcal{E}\|$ "elementary" games (one for each edge), so that in the game $G_{i,j}$ a coalition has value $w(\{i,j\})$ if it contains i and j, and zero otherwise. We then use the properties of the Shapley value listed in Theorem 3.3: The null player property implies that in $G_{i,j}$ the Shapley value of each player other than i and j is 0, symmetry means that in $G_{i,j}$ players i and j have the same Shapley value, and hence the efficiency property implies that the Shapley values of players i and j in $G_{i,j}$ are given by $w(\{i,j\})/2$. Finally, additivity means that a player's Shapley value is obtained by summing over all elementary games. Thus, we obtain the following result.

Theorem 3.7 (Deng and Papadimitriou [338]). *The Shapley value of player $i \in \mathcal{V}$ in an induced subgraph game $G = (\mathcal{V}, \mathcal{E}, w)$ is given by*

$$\text{Shapley}(G, i) = \frac{1}{2} \sum_{\{i,j\} \in \mathcal{E}} w(\{i,j\}).$$

Deng and Papadimitriou [338] also analyze the complexity of other solution concepts for induced subgraph games, including the ϵ-core and the stable set as well as further solution concepts that we did not include formally, such as the nucleolus, the kernel, and the bargaining set; for instance, they show that the Shapley value coincides with the nucleolus.

3.6 Hedonic Games

So far in this chapter, we focused on cooperative games with transferable utility, i.e., games where the payoff to a coalition can be distributed arbitrarily among its members. However, there are settings where groups can benefit from cooperation, but the payoffs accrue to individual group members and cannot be transferred. For instance, a group of countries can benefit from a trade agreement, but such agreements typically do not involve direct monetary transfers among the countries.

Do you still remember the attempt by Anna, Belle, Chris, David, and Edgar to get rich quick by making a cherry-quark cake and selling it at the market? (If not, just look it up on page 139 at the beginning of this chapter.) Unfortunately, their plan failed: It turns out that only adults are allowed to rent a market stall. They thus have to reconsider their options.

"Frankly, I do not like cake all that much," confesses Belle.

"But that's the best we can do if we all work together!" counters David.

"I know...," muses Belle. "Still, I think I like milkshake better, and I can make it easily with Anna and Chris."

Anna agrees: "Baking a cake is too much work, and a cherry milkshake tastes great!"

Chris, too, is happy with this idea: "If we cannot make money anyway, I, too, prefer milkshake over cake."

"But how about pancakes?" David chimes in. "Chris, are you sure you prefer milkshake over pancakes?"

"Ugh, maybe pancakes are an even better option," agrees Chris. "So perhaps I prefer working with you to working with Anna and Belle after all."

In the example above, Belle and Chris were reasoning about their preferences over food items rather than monetary payoffs. In fact, Chris's concluding argument suggests that such preferences can be reframed as preferences over coalitions: Chris prefers the coalition {Chris, David} over {Anna, Belle, Chris}, as he prefers pancakes over milkshake, and he prefers {Anna, Belle, Chris} over the grand coalition, as he prefers milkshake over cherry-quark cake.

In their early work, Drèze and Greenberg [356] start to break away from the assumption of transferable utility among players. They coin the term *hedonic games* for games in which players express preferences over coalitions they are part of. Hedonic games in the form they are typically researched today (and presented in this section) have been introduced by Banerjee, Konishi, and Sönmez [84] and, independently, by Bogomolnaia and Jackson [162].

3.6 Hedonic Games

To define hedonic games formally, we first introduce additional notation and terminology. A relation $\succeq$ on a set S is *reflexive* if $x \succeq x$ for all $x \in S$; it is *transitive* if for all $x, y, z \in S$, it holds that $x \succeq y$, $y \succeq z$ implies $x \succeq z$; it is *total* if for all $x, y \in S$, it holds that $x \succeq y$ or $y \succeq x$; and it is *antisymmetric* if for all $x, y \in S$, it holds that $x \succeq y$, $y \succeq x$ implies $x = y$. A *weak order* on S is a transitive and total (thus, in particular, a reflexive) relation on S.

Given a set of players $P = \{1, 2, \ldots, n\}$ and an $i \in P$, let 2^{P_i} be the set of coalitions over P containing i. Player i's *weak preferences over* 2^{P_i} are expressed by a weak order $\succeq_i$ on 2^{P_i}; if $C \succeq_i D$ for some $C, D \in 2^{P_i}$, we say that i *weakly prefers* C over D. Note that $\succeq_i$ need not be antisymmetric, i.e., it may be the case that C and D are distinct coalitions in 2^{P_i}, yet player i's preferences satisfy both $C \succeq_i D$ and $D \succeq_i C$; in this case, we say that i is *indifferent* between C and D and write $C \sim_i D$. On the other hand, if $C \succeq_i D$ but not $D \succeq_i C$, we write $C \succ_i D$ and say that i *(strictly) prefers* C to D.

We are now ready to define the key concept of this section. A *hedonic game* is given by a pair $(P, \succeq)$, where P is a set of players and $\succeq = (\succeq_1, \succeq_2, \ldots, \succeq_n)$ is a *preference profile*, where for each $i \in P$ the relation $\succeq_i$ expresses player i's weak preferences over the coalitions in 2^{P_i}.

Recall that, in cooperative games with transferable utility, an outcome of a game is a partition of the players into pairwise disjoint coalitions together with a payoff vector, which distributes the value of each coalition among its members. As a consequence, when comparing two outcomes, players need to consider their payoffs in both outcomes. In contrast, in a hedonic game an outcome is fully specified by a coalition structure alone, and players compare outcomes based on what coalition they end up in.

An *outcome of a hedonic game* $(P, \succeq)$ is a *coalition structure*, i.e., a partition $\mathfrak{C} = \{C_1, C_2, \ldots, C_k\}$ of P into pairwise disjoint coalitions, with $\bigcup_{i=1}^{k} C_i = P$ and $C_i \cap C_j = \emptyset$ for $i \neq j$. The coalition in $\mathfrak{C}$ that contains player i is denoted by $\mathfrak{C}(i)$. A player i *weakly prefers* an outcome $\mathfrak{C}$ to an outcome $\mathfrak{C}'$ if i weakly prefers $\mathfrak{C}(i)$ over $\mathfrak{C}'(i)$.

Our next task is to define what it means for an outcome of a hedonic game to be viewed as desirable. Unlike for games with transferable utility, there is no commonly accepted general notion of fairness for hedonic games;[18] on the other hand, there is a rich and varied collection of stability notions.

3.6.1 Stability in Hedonic Games

In this section, we assume that we are given a fixed hedonic game $(P, \succeq)$. Just as in games with transferable utility, the minimum stability requirement is

---

[18] However, we point the interested reader to the work of Kerkmann, Nguyen, and Rothe [606], who propose various notions of *local fairness* in hedonic games that are inspired by the corresponding notions from fair division of indivisible goods, a topic that will be handled in Chapter 9 starting on page 605.

that of individual rationality: We assume that each player can always deviate by forming a singleton coalition, so an outcome is clearly unstable if some of the players prefer being alone to their current coalition. More formally, an outcome $\mathfrak{C}$ is said to be *individually rational* if for each $i \in P$, it holds that $\mathfrak{C}(i) \succeq_i \{i\}$. Relatedly, we call a coalition C *individually rational for player* $i \in C$ if $C \succeq_i \{i\}$, i.e., if i weakly prefers being in C to being alone.

More broadly, players may deviate from the current outcome by moving from their current coalition to a different one. A *single-player deviation* performed by player i transforms a partition $\mathfrak{C}$ into a partition $\mathfrak{C}'$ where $\mathfrak{C}(i) \neq \mathfrak{C}'(i)$ and, for all players $j \neq i$, it holds that $\mathfrak{C}(j) \setminus \{i\} = \mathfrak{C}'(j) \setminus \{i\}$. We write $\mathfrak{C} \xrightarrow{i} \mathfrak{C}'$ to denote a single-player deviation performed by player i transforming a partition $\mathfrak{C}$ into a partition $\mathfrak{C}'$. Our first stability notion based on single-player deviations is Nash stability, the analogue of Nash equilibrium in noncooperative game theory (recall Definitions 2.4 and 2.5 on pages 50 and 63, respectively).

A *Nash deviation by player* i is a single-player deviation $\mathfrak{C} \xrightarrow{i} \mathfrak{C}'$ making player i better off, i.e., $\mathfrak{C}'(i) \succ_i \mathfrak{C}(i)$. A coalition structure that allows for no Nash deviation is said to be *Nash-stable*.

Note that the definition of a single-player deviation allows the deviator to form a new coalition, i.e., a deviation to an empty coalition. Hence, we ensure that Nash-stable outcomes are individually rational by definition.

The definition of Nash stability allows individual players to join and abandon coalitions at will. In particular, a player i is allowed to join a coalition C even if some of the current members of C object. Our next definition gives all players in the target coalition a veto power over new joiners.

An *individual deviation by player* i is a Nash deviation $\mathfrak{C} \xrightarrow{i} \mathfrak{C}'$ such that $\mathfrak{C}(j) \cup \{i\} \succeq_j \mathfrak{C}(j)$ for all $j \in \mathfrak{C}'(i)$. A coalition structure is said to be *individually stable* if it allows for no individual deviation. In other words, $\mathfrak{C}$ is individually stable if for each $i \in P$ and each $C \in \mathfrak{C} \cup \{\emptyset\}$, it holds that

(1) $\mathfrak{C}(i) \succeq_i C \cup \{i\}$ or
(2) $C \succ_j C \cup \{i\}$ for some $j \in C$.

Conditions (1) and (2) in the definition of individual stability capture two reasons for why i may be unable to deviate by joining C: Condition (1) says that i finds her current coalition at least as attractive as $C \cup \{i\}$, and condition (2) says that one of the current members of C objects to i joining C.

The notion of individual stability stipulates that players may prohibit other players from joining them. Of course, it may also be the case that players are contractually bound to their current coalitions and therefore may be unable to leave them if one of their current coalition partners objects. Depending on whether the consent to join another coalition is still required, this leads to two further notions of stability.

A *contractual deviation by player* i is a Nash deviation $\mathfrak{C} \xrightarrow{i} \mathfrak{C}'$ such that $\mathfrak{C}(j) \setminus \{i\} \succeq_j \mathfrak{C}(j)$ for all $j \in \mathfrak{C}(i) \setminus \{i\}$. A single-player deviation that is both

3.6 Hedonic Games

an individual and a contractual deviation is called a *contractual individual deviation*. A partition is said to be *contractually Nash-stable* if it allows for no contractual deviation, and it is said to be *contractually individually stable* if it allows for no contractual individual deviation.

We have already observed that if an outcome is Nash-stable, then it is individually rational. A similar argument shows that every individually stable outcome is individually rational, too: If an outcome violates individual rationality, i.e., some player prefers to deviate by forming a singleton coalition, this deviation also constitutes a violation of individual stability. However, a contractually individually stable outcome is not necessarily individually rational, as illustrated by our next example.

Anna has been charged with looking after Edgar for a few hours after school. She is not in a mood to play with him, as she would rather go to her room and read a book, i.e., she prefers the singleton coalition {Anna} to the grand coalition {Anna,Edgar}.

Edgar, however, does not want her to leave, so he prefers the grand coalition {Anna,Edgar} over the singleton coalition {Edgar}. Under these preferences, the grand coalition {Anna,Edgar} is contractually individually stable, even though it is not individually rational.

Moreover, it is easy to see that every Nash-stable outcome is individually stable and contractually Nash-stable, and the weakest concept of those defined above is contractual individual stability; generally, if we place additional restrictions on what constitutes a permissible deviation, we expand the set of stable outcomes. A taxonomy of stability concepts for hedonic games is presented in Figure 3.6, which already includes some stability concepts based on group deviations that we introduce next. An arrow $(A \to B)$ between two stability concepts A and B means an implication: Every outcome of a hedonic game that satisfies A also satisfies B.

So far, we have focused on notions of stability that guard against deviations by individual players. However, just as in games with transferable utility, it is important to consider group deviations as well.

An outcome $\mathfrak{C}$ of a hedonic game $(P, \succeq)$ is *core-stable* if for each nonempty coalition of players $C \subseteq P$, there exists a player $i \in C$ such that $\mathfrak{C}(i) \succeq_i C$. In words, an outcome is *core-stable* as long as for each potential deviation involving a group of players C, at least one of the players in C does not benefit from deviating. That is, this player prevents C from *blocking* this outcome $\mathfrak{C}$; C would be blocking the current coalition structure if deviating to C would make every member of C happier. In other words, *core stability* of $\mathfrak{C}$ means that there exists no nonempty coalition that blocks $\mathfrak{C}$.

A more permissive notion of group deviation already allows deviations that make only one of the deviating players happier, while not making any of the deviators less happy. An outcome $\mathfrak{C}$ is *strictly core-stable* if for each coalition

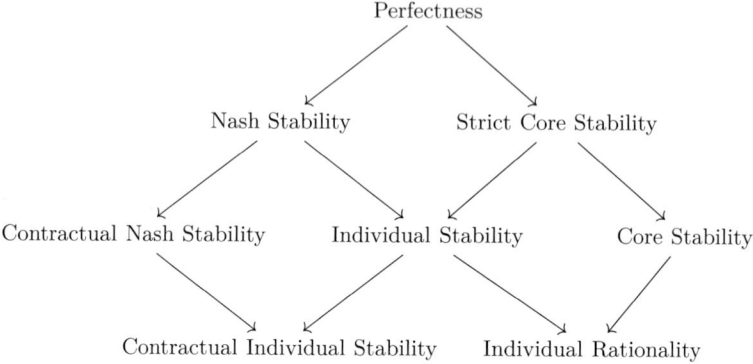

Fig. 3.6 Overview of stability concepts for hedonic games

of players $C \subseteq P$, it holds that $\mathfrak{C}(i) \succeq_i C$ for all $i \in C$, or there exists a player $i \in C$ such that $\mathfrak{C}(i) \succ_i C$. In other words, *strict core stability of* $\mathfrak{C}$ means that there exists no coalition that *weakly blocks* this outcome $\mathfrak{C}$, i.e., no coalition C whose members weakly prefer deviating to C while at least one member of C would even (strictly) prefer to be in C over being in the current coalition.

If an outcome is strictly core-stable, it is also core-stable, but the converse is not true. Also, an individual deviation makes the deviator better off while none of the joined players is worse off. Hence, the new coalition of the deviator violates strict core stability. Hence, strict core stability also implies individual stability.

> Consider again the outcome in which all the children form the grand coalition and make the cherry-quark cake.
>
> Anna sighs: "I actually like the cake a bit more than I like quark, but quark is much easier to make. So I guess I am indifferent between making quark with Belle and baking the cake with everybody."
>
> Belle responds enthusiastically: "If it's all the same to you, let's make quark—I like it better, because it's healthier!"

In the example above, we can conclude that the grand coalition is not strictly core-stable, as the deviation by Anna and Belle makes both of them weakly better off and one of them (Belle) even strictly better off. However, this example does not establish that the grand coalition is not core-stable.

Of course, if no player's ranking of possible coalitions contains indifferences, then strict core stability is equivalent to core stability, as any deviation that weakly benefits all players must strictly benefit them. It is also worth contemplating on why we do not make a distinction between *core* and *strict core* in the transferable-utility setting. Intuitively, the reason is that the payoff is earned by the entire coalition in games with transferable utility; so if the value of this coalition exceeds the sum of the individual players' payoffs in the outcome prior to deviation, it can always be shared so that each of the players obtains a positive surplus.

Finally, we define the purest and strongest of all our stability concepts: *perfectness*. An outcome $\mathfrak{C}$ of a hedonic game $(P, \succeq)$ is *perfect* if $\mathfrak{C}(i) \succeq_i C$ for each player $i \in P$ and for each coalition $C \in 2^{P_i}$. That is, in a perfect coalition structure, all players end up in one of their most preferred coalitions. It is clear that perfectness implies both Nash stability and strict core stability (but not vice versa), and thus also all other stability concepts we have defined.

3.6.2 Representation Formalisms

Just as for games with transferable utility, it is important to consider how to encode hedonic games, as the preferences of each of the n players correspond to a weak order over 2^{n-1} coalitions. In what follows, we will discuss several classes of hedonic games that admit succinct encodings. We focus on two prominent ways to encode preferences, merely providing some pointers to other interesting representation formalisms.

3.6.2.1 Anonymous Hedonic Games

Consider a player that is indifferent among all potential coalition partners. The preferences of such a player over coalitions are uniquely determined by the coalition sizes. Games where all players' preferences have this structure are known as anonymous games [162].

A hedonic game $(P, \succeq)$ is said to be *anonymous* if for every player $i \in P$ and every pair of coalitions $C, C' \in 2^{P_i}$ with $\|C\| = \|C'\|$, it holds that $C \sim_i C'$.

In an anonymous hedonic game $(P, \succeq)$, the players' preferences over coalitions that contain them can be fully described by specifying their preferences over coalition *sizes*, i.e., numbers in $\{1, \ldots, n\}$, where $n = \|P\|$. Indeed, given a player $i \in P$, construct a weak order $\rhd_i$ over $\{1, \ldots, n\}$ by setting $x \rhd_i y$ if and only if $C \succeq_i C'$ for some pair of coalitions $C, C' \in 2^{P_i}$ with $\|C\| = x$ and $\|C'\| = y$; note that $\rhd_i$ is well-defined exactly because the underlying game is anonymous.

In the context of the so-called *"group activity selection problem,"* Darmann et al. [328] introduce a variant of the anonymous encoding where every player equally likes some sizes of coalitions, but equally dislikes the remaining ones. For example, if the goal is to set up pairs of players to compete in the first round of a chess tournament, the players would like size-two coalitions but would dislike coalitions of any other size; the same task in a soccer tournament would make 11 the favored coalition size, whereas all other coalition sizes would not be favorable, and so on. Such games are called *black-and-white anonymous hedonic games*.

3.6.2.2 Additively Separable Hedonic Games

One of the most prominent and natural classes of hedonic games are additively separable hedonic games [162]. In these games, individual players possess cardinal valuation functions for the other players and determine their preferences by comparing an aggregated value of a coalition.

More formally, a hedonic game $(P, \succeq)$ is called an *additively separable hedonic game* if for every player $i \in P$, there exists a function $u_i \colon P \to \mathbb{Q}$ such that for every pair of coalitions $C, C' \in 2^{P_i}$, it holds that

$$C \succeq_i C' \iff \sum_{j \in C} u_i(j) \geq \sum_{j \in C'} u_i(j).^{19}$$

Since additively separable hedonic games are fully specified by their valuation functions, we also represent them by the pair (P, u). Given a coalition $C \in 2^{P_i}$ or a coalition structure $\mathfrak{C}$, we also write $u_i(C) = \sum_{j \in C} u_i(j)$ and $u_i(\mathfrak{C}) = u_i(\mathfrak{C}(i))$.

Dimitrov et al. [347] introduced related classes of games based on the consideration of friends and enemies. Given a hedonic game $(P, \succeq)$, they assume that every player $i \in P$ has a *set of friends*, $F_i \subseteq P$ (including i herself), and a *set of enemies*, $E_i = P \setminus F_i$. Note that friendship is not necessarily a symmetric relation.

$(P, \succeq)$ is called an *appreciation-of-friends hedonic game* if for every pair of coalitions $C, C' \in 2^{P_i}$, it holds that $C \succeq_i C'$ if and only if

1. $\|C \cap F_i\| > \|C' \cap F_i\|$, or
2. $\|C \cap F_i\| = \|C' \cap F_i\|$ and $\|C \cap E_i\| \leq \|C' \cap E_i\|$.

Appreciation-of-friends hedonic games are also known as *friend-oriented hedonic games*.

Similarly, $(P, \succeq)$ is called an *aversion-to-enemies hedonic game* if for every pair of coalitions $C, C' \in 2^{P_i}$, it holds that $C \succeq_i C'$ if and only if

1. $\|C \cap E_i\| < \|C' \cap E_i\|$, or
2. $\|C \cap E_i\| = \|C' \cap E_i\|$ and $\|C \cap F_i\| \geq \|C' \cap F_i\|$.

Aversion-to-enemies hedonic games are also known as *enemy-oriented hedonic games*.

In other words, in these games, players try to simultaneously maximize the number of their friends while minimizing the number of their enemies. While the priority is on the former objective in appreciation-of-friends hedonic

---

[19] Additive separability is a frequently considered property of preferences, for instance, in the context of fair division, see Definition 9.1 on page 613.

3.6 Hedonic Games

games, the latter objective is more important in aversion-to-enemies hedonic games. An appreciation-of-friends or aversion-to-enemies hedonic game can be represented by an additively separable hedonic game where the valuation functions of the players have only the range $\{-1, n\}$ or $\{-n, 1\}$, respectively. Recall that n is the number of players.

Anna (also abbreviated by a) was teased by Chris (c), David (d), and Edgar (e) and definitely doesn't want to see them today. Instead, she would like to spend the afternoon with Belle (b). Her preferences can be expressed as follows (where we omit set parentheses and separating commas for better readability):

$$ab \succ_a a \succ_a abc \sim_a abd \sim_a abe \succ_a ac \sim_a ad \sim_a ae \succ_a$$
$$abcd \sim_a abce \sim_a abde \succ_a acd \sim_a ace \sim_a ade \succ_a abcde \succ_a acde.$$

To express her preferences on the coalitions that include her, Anna explicitly lists $2^4 = 16$ coalitions above. However, she can express her preferences much more succinctly by dividing the other children into friends and enemies as $F_{\text{Anna}} = \{\text{Anna, Belle}\}$ and $E_{\text{Anna}} = \{\text{Chris, David, Edgar}\}$.

Figure 3.7 shows the undirected graph representing the associated aversion-to-enemies hedonic game where the two girls, Anna and Belle, are mutual friends, and so are the three boys, Chris, David, and Edgar, and in addition the siblings Belle and Edgar like each other. It is easily checked that Anna's enemy-oriented preferences are represented correctly. Now, consider the coalition structure $\mathfrak{C} = \{\{\text{Anna, Belle}\}, \{\text{Chris, David, Edgar}\}\}$ that is displayed by circles in Figure 3.7. $\mathfrak{C}$ is strictly core-stable, as can be seen as follows. Anna, Chris, and David each are in their unique most preferred coalition in $\mathfrak{C}$. Therefore, they cannot belong to any coalition that weakly blocks $\mathfrak{C}$. Hence, we only have to check the coalitions $\{\text{Belle, Edgar}\}$, $\{\text{Belle}\}$, and $\{\text{Edgar}\}$: The coalition containing Belle and Edgar does not weakly block $\mathfrak{C}$ because Edgar prefers his coalition $\mathfrak{C}(\text{Edgar}) = \{\text{Chris, David, Edgar}\}$ to $\{\text{Belle, Edgar}\}$ under enemy-oriented preferences: Both coalitions have the same number of Edgar's enemies—namely, none! Kind-hearted Edgar is enemies with Anna alone (he is a bit jealous of her, for her being BFF with his sister), but Anna belongs to neither of these two coalitions. However, $\mathfrak{C}(\text{Edgar})$ contains more of his friends. Moreover, the singleton coalitions are not weakly blocking because $\mathfrak{C}$ is individually rational.

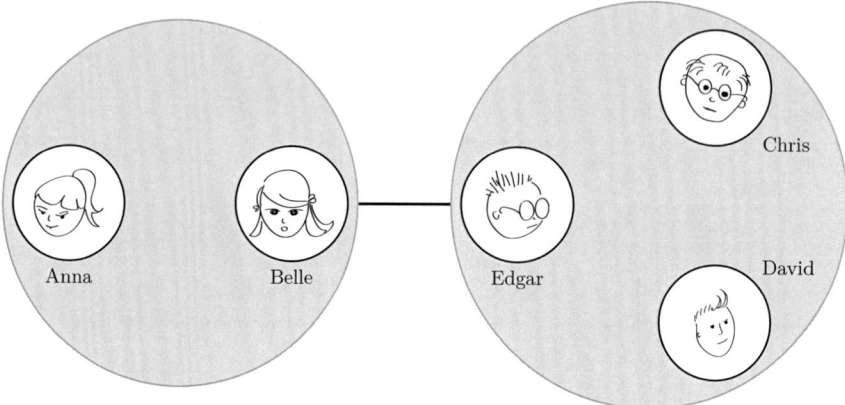

Fig. 3.7 An aversion-to-enemies hedonic game represented as an undirected graph (see Footnote 27 on page 221), a.k.a. a *network of friends*. Displayed edges indicate friends, missing edges enemies. The highlighted coalition structure where the girls and boys each form a coalition is strictly core-stable.

3.6.2.3 Further Classes of Hedonic games

While we will focus on anonymous hedonic games and additively separable hedonic games in the remainder of this chapter, we want to briefly discuss some other representation formalisms that have been proposed.

First, as for cooperative games with transferable utility, there also exist universal representation formalisms. Similar to marginal contribution nets by Ieong and Shoham [566], Elkind and Wooldridge [384] have introduced *hedonic coalition nets*, a rule-based representation for hedonic games that is universally expressive and at least as compact as other representation schemes for hedonic games.

Ballester [83] has proposed a representation where every player provides a complete *list of individually rational coalitions*. While this representation is not able to express preferences over coalitions that are not individually rational, it is, without loss of generality, suitable for the consideration of some solution concepts. For instance, every core-stable coalition structure is individually rational. Otherwise, it would be blocked by some singleton coalition.

While a representation by either of hedonic coalition nets and lists of individually rational coalitions needs exponential space in the worst case, it yields a succinct representation for large classes of games that are rich enough to allow for computational hardness results [83, 211, 384].

A second prominent idea is to consider cardinal valuation functions like for additively separable hedonic game, but to use another preference aggregation mechanism. For instance, Aziz et al. [44] introduce *fractional hedonic games*, where the valuation of a coalition is the average value assigned to the members of this coalition, with a player's own value always being zero. A variant of these games, so-called *modified fractional hedonic games*, were introduced by Olsen [761]. In these games, the average is taken without considering the players themselves when determining their valuations of a coalition containing them.

Another way to aggregate preferences apart from taking sums or averages is to consider the maximum or minimum valuation in a coalition. Since these games do not really need the consideration of single cardinal values for other players, they are usually represented by ordinal preference rankings over other players, which is also referred to as the *singleton encoding of hedonic games* [263]. Cechlárová and Romero-Medina [263] (see also the work of Cechlárová and Hajduková [261, 262]) consider the *optimistic* and *pessimistic* preference extension, where players aggregate preferences according to the maximally and minimally preferred player, respectively.

Kerkmann et al. [604] propose another representation for hedonic games where each player subdivides the other players into *friends*, *enemies*, and *neutral players*, with friends and enemies being ordinally ranked according to a weak preference order. Assuming that these preferences are monotonic (respectively, antimonotonic) with respect to the addition of friends (respectively, enemies), they propose a *bipolar extension of the responsive extension principle* (this principle will be considered and discussed in more detail in the chapter on fair division of indivisible goods, see Section 9.3.1.1 starting on page 615). They apply this principle to derive the (partial) preferences of players over coalitions. For some of the stability concepts introduced in Section 3.6.1 starting on page 199, they characterize coalition structures that necessarily or possibly satisfy them, and they study the corresponding computational problems in terms of their complexity—a topic we will turn to in the next section.

Relatedly, Ohta et al. [760] investigate the impact of neutrals on the core stability of hedonic games with friends, enemies, and neutral players. More recently, Chen et al. [281] also study the (parameterized) complexity of problems related to stability notions in hedonic games with friends, enemies, and neutral players, in particular disproving some conjectures by Woeginger [943].

But now, let us turn to the computational complexity of existence and verification for various stability notions in anonymous and in additively separable hedonic games.

3.6.3 Complexity of Stability in Hedonic Games

In this section, we want to discuss the question of computing stable outcomes in hedonic games and to have a closer look at the computational complexity of problems related to stability concepts for compactly represented hedonic games. We first define these problems and then focus on anonymous hedonic games and additively separable hedonic games.

3.6.3.1 Existence and Verification for Stability in Hedonic Games

> Anna and Belle were having an argument about whether or not to hang out with the boys. Belle wanted to play not only with Anna but wanted to let also her little brother Edgar join in. Anna, however, was staunchly opposed to him joining; it's the girls' eve, after all. Now, Anna is very angry and just wants to be alone. But when Belle realizes that Anna is her most important friend, she wants to approach Anna again, to be with her and spend time together anyway, even if without Edgar.

The previous example describes a situation where one player wants to be in a certain coalition while the other player doesn't want to. This can be modeled as a hedonic game with $P = \{\text{Anna}, \text{Belle}\}$ and $\{\text{Anna}\} \succ_{\text{Anna}} \{\text{Anna}, \text{Belle}\}$, whereas $\{\text{Anna}, \text{Belle}\} \succ_{\text{Belle}} \{\text{Belle}\}$. In this game, there does not exist a Nash-stable outcome.

Such *"run-and-chase"* games are a problem to stability in many classes of hedonic games. For instance, this occurs in anonymous hedonic games if Anna prefers a coalition size of 1 over 2, whereas Belle prefers a coalition size of 2 over 1. The same phenomenon happens in an additively separable game where $u_{\text{Anna}}(\text{Belle}) = -1$, whereas $u_{\text{Belle}}(\text{Anna}) = 1$. Hence, one important computational question is whether we can decide about the existence of a stable outcome in a given game. Formally, we define the following problem.

NASH-STABILITY-EXISTENCE (NSE)

Given: A hedonic game that is compactly represented.
Question: Does this game possess a Nash-stable coalition structure?

Relatedly, when we are given a (succinctly represented) hedonic game and a coalition structure, we may want to verify whether or not it satisfies some notion of stability. For Nash stability, we define the following problem.

NASH-STABILITY-VERIFICATION (NSV)

Given: A hedonic game that is compactly represented and a coalition structure $\mathfrak{C}$ for it.
Question: Is $\mathfrak{C}$ a Nash-stable coalition structure for this game?

Similarly, INDIVIDUAL-STABILITY-EXISTENCE (ISE) defines the existence problem and INDIVIDUAL-STABILITY-VERIFICATION (ISV) the verification problem for individual stability.

3.6 Hedonic Games

The same two computational problems can, of course, also be defined for any other solution concept. For example, regarding core stability, we define the existence and verification problems as follows:

CORE-STABILITY-EXISTENCE (CSE)

Given: A hedonic game that is compactly represented.
Question: Does there exist a core-stable coalition structure for this game?

CORE-STABILITY-VERIFICATION (CSV)

Given: A hedonic game $(P, \succeq)$ that is compactly represented and a coalition structure $\mathfrak{C}$ for it.
Question: Does there exist a coalition $C \subseteq P$ that blocks $\mathfrak{C}$?

The verification problem is expressed by a (polynomially length-bounded) existential quantifier, followed by a property (namely, whether some coalition blocks the given coalition structure) that can be checked in polynomial time, provided that preference relations can be evaluated in polynomial time for our input representation:

$$((P, \succeq), \mathfrak{C}) \in \text{CSV} \iff (\exists C \subseteq P)[C \text{ blocks } \mathfrak{C}].$$

Furthermore, the existence problem is expressed by a (polynomially length-bounded) existential quantifier, followed by a (polynomially length-bounded) universal quantifier, followed by a property (namely, whether some coalition structure is not blocked by any coalition) that can be checked in polynomial time, again assuming that preference relations can be evaluated efficiently:

$$(P, \succeq) \in \text{CSE} \iff (\exists \mathfrak{C} \text{ for } (P, \succeq))(\forall C \subseteq P)[C \text{ does not block } \mathfrak{C}].$$

Thus, by the well-known characterization of the classes of the polynomial hierarchy via alternating (polynomially length-bounded) existential and universal quantifiers (recall Section 1.5.3.2 starting on page 35), the verification problem CSV is in NP and the existence problem CSE is in $\Sigma_2^p = \text{NP}^{\text{NP}}$.

Replacing "*core-stable*" by "*strictly core-stable*" and "*blocks*" by "*weakly blocks*" in the problem definitions and characterizations above, we obtain the strict variants of the verification and the existence problem, and the same upper bounds on their complexity: STRICT-CORE-STABILITY-VERIFICATION (SCSV) is in NP and STRICT-CORE-STABILITY-EXISTENCE (SCSE) is in $\Sigma_2^p = \text{NP}^{\text{NP}}$.

"However we divide ourselves up," says Edgar, "I would always prefer to be in a group where everyone is friends with each other. If any two in my group were enemies, I would really hate joining them!"

"Not surprising," diagnoses his sister Belle. "Edgar, the purist! You are a perfectionist of harmony, right?"

"But," Anna chimes in, "doesn't this mean that we have to put up with losing some solutions? Look at the aversion-to-enemies hedonic game in Figure 3.7 on page 206. In the coalition structure $\mathfrak{C} = \{\{\text{Anna}, \text{Belle}\}, \{\text{Chris}, \text{David}, \text{Edgar}\}\}$ shown there, the three boys are together. However, even though Belle is enemies with both Chris and David, Edgar would prefer to be in the larger group that also contains his sister: $\{\text{Belle}, \text{Chris}, \text{David}, \text{Edgar}\}$ has as few of Edgar's enemies as $\{\text{Chris}, \text{David}, \text{Edgar}\}$ (namely, none) but more friends of his."

"I don't care!" stubborn Edgar replies. "I insist that only cliques of friends should be together!"

As already mentioned, both friend- and enemy-oriented hedonic games have a purely graph-theoretical interpretation. Relatedly, we can define the following notions (see, e.g., [943, 808]). Given an undirected graph G, a partition $\mathfrak{C}$ of the vertex set of G into cliques is said to be *wonderfully stable* if every vertex of G ends up in a largest clique among those containing it: $\|\mathfrak{C}(v)\| = \omega_G(v)$ for each vertex v of G, where $\omega_G(v)$ denotes the *clique number of v in G*, i.e., the size of a largest clique in G containing v.

"Hey, folks!" Edgar yells excitedly. "Look again at the coalition structure $\mathfrak{C} = \{\{\text{Anna}, \text{Belle}\}, \{\text{Chris}, \text{David}, \text{Edgar}\}\}$ that is displayed by circles in Figure 3.7 on page 206! We already knew that it is is strictly core-stable under aversion-to-enemies preferences, right? But $\mathfrak{C}$ is even *wonderfully* stable! Isn't that wonderful?"

"Really, is it?" asks David. "Why?"

"Yep, I see," says Belle and explains: "It's quite easy, David: Each of us is in their largest clique! And if we all have aversion-to-enemies preferences, this means that we all are together with no enemies at all and with as many friends as possible, that is, each of us is in their *most preferred* clique. Perfect!"

"That's right!" agrees Anna and scratches her head. "But what would happen if I leave the game?" she wonders. "Would the coalition structure $\mathfrak{C}' = \{\{\text{Belle}\}, \{\text{Chris}, \text{David}, \text{Edgar}\}\}$ still be strictly core-stable or even wonderfully stable under aversion-to-enemies preferences in this new game?"

"Good question!" replies Edgar. "Let me think." After a while he says: "I'd say $\mathfrak{C}'$ should still be strictly core-stable, essentially by the same reasoning as before: There is simply no weakly blocking coalition for $\mathfrak{C}'$."

"That's right, Edgar," adds his sister. "Unfortunately, however, $\mathfrak{C}'$ is not wonderfully stable. And I'm afraid I am the one who makes it fail to be wonderfully stable because when I'm alone, I'm not in my largest clique, $\{\text{Belle}, \text{Edgar}\}$."

3.6 Hedonic Games   211

"OK, but what about another coalition structure?" Chris is curious. "If Belle is happy only when she is together with Edgar, then let's consider $\mathfrak{C}'' = \{\{\text{Belle}, \text{Edgar}\}, \{\text{Chris}, \text{David}\}\}$. Huh? That's wonderfully stable, isn't it?"

"Nope!" Edgar disagrees. "Think! When I leave you and David to join Belle, no one of us three is in his largest clique anymore. Who needs the girls? Only together we three are happy! So $\mathfrak{C}''$ is not wonderfully stable because both {Belle, Edgar} and {Chris, David} contains some dude who is not in his *largest* clique."

"Even worse!" adds Belle regretfully. "When Anna leaves the game, there does not exist *any* wonderfully stable coalition structure anymore!"

The existence problem for wonderful stability in graphs (respectively, in appreciation-of-friends or aversion-to-enemies hedonic games) is defined as follows:

WONDERFUL-STABILITY-EXISTENCE (WSE)

Given: An undirected graph (e.g., a network of friends for an appreciation-of-friends or aversion-to-enemies hedonic game).

Question: Does there exist a wonderfully stable partition of its vertex set into cliques?

In the verification problem WONDERFUL-STABILITY-VERIFICATION (WSV), we are given an undirected G graph and a partition $\mathfrak{C}$ of its vertex set into cliques, and we ask whether $\|\mathfrak{C}(v)\| < \omega_G(v)$ for some vertex v of G.

We now survey some of the known results on the computational complexity of the problems defined above, focusing on anonymous hedonic games and additively separable hedonic games. We will see that the complexity of these problems critically depends on the representation used.

3.6.3.2 Anonymous Hedonic Games

Our first result concerns computing individually stable outcomes in anonymous hedonic games. We will present an algorithm by Bogomolnaia and Jackson [162], where we impose a structural property for the preferences.[20] Let $(P, \succeq)$ be an anonymous hedonic game with n players. $(P, \succeq)$ is said to be *single-peaked* if for every player $i \in P$, there exists a number $p_i \in \{1, \ldots, n\}$, called i's *peak*, such that for all $C, D \in 2^{P_i}$ with $\|C\| < \|D\| \leq p_i$ or $\|C\| >$

---

[20] Bogomolnaia and Jackson [162] present this algorithm for a larger class of hedonic games called *hedonic games with ordered characteristics*. We present a simplified version of the algorithm for the important special case of anonymous hedonic games.

$\|D\| \geq p_i$, it holds that $D \succ_i C$ [162].[21] More informally, players' preferences in an anonymous hedonic game are *single-peaked* if each player can order the coalition sizes and specify a most preferred coalition size such that the preferences over coalition sizes decay towards both sides of the most preferred coalition size on the axis of coalition sizes.[22]

Interestingly, without single-peakedness, the existence of individually stable outcomes is not guaranteed. However, examples that show this fact are not easy to construct. Bogomolnaia and Jackson [162] provide a game with 63 players having this property. A smaller game with 15 players was found by Brandt, Bullinger, and Wilczynski [213].

Algorithm 1 ISE for single-peaked anonymous hedonic games

Input: Single-peaked anonymous hedonic games where $p_1 \geq p_2 \geq \cdots \geq p_n$
Output: Individually stable coalition structure

    INITIALIZATION
    $k \leftarrow 1$ ▷ *Number of coalition*
    $C_1 \leftarrow \emptyset$ ▷ *First coalition*
    $\mathfrak{C} \leftarrow \{C_1\}$ ▷ *Final coalition structure*
    $\mathfrak{O} \leftarrow \emptyset$ ▷ *Open coalitions*
    MAIN CONSTRUCTION
    for all $i = 1, \ldots, n$ **do**
      if $p_i \geq \|C_k\| + 1$ **then**
        $C_k \leftarrow C_k \cup \{i\}$ ▷ *PHASE I*
      else   ▷ $p_i \leq \|C_k\|$. *PHASE II*
        while $p_i \leq \|C_k\|$ & $\exists j \in C_k, O \in \mathfrak{O}$ with $\|O\| + 1 \succ_j \|C_k\|$ **do**
          $O' \leftarrow \mathrm{argmax}_{\succ_j}\{\|O\| + 1 \,|\, O \in \mathfrak{O}\}$
          $\mathfrak{O} \leftarrow \mathfrak{O} \setminus \{O'\}, \mathfrak{C} \leftarrow (\mathfrak{C} \setminus \{O'\}) \cup \{O' \cup \{j\}\}, C_k \leftarrow C_k \setminus \{j\}$
        if $p_i \geq \|C_k\| + 1$ **then**
          $C_k \leftarrow C_k \cup \{i\}$
        else
          if $\|C_k\| + 1 \succ_j \|C_k\|$ for all $j \in C_k$ **then**
            $\mathfrak{O} \leftarrow \mathfrak{O} \cup \{C_k\}$
          $k \leftarrow k + 1, C_k \leftarrow \{i\}$
          $\mathfrak{C} \leftarrow \mathfrak{C} \cup \{C_k\}$
    while $\exists j \in C_k, O \in \mathfrak{O}$ with $\|O\| + 1 \succ_j \|C_k\|$ **do**
      $O' \leftarrow \mathrm{argmax}_{\succ_j}\{\|O\| + 1 \,|\, O \in \mathfrak{O}\}$
      $\mathfrak{O} \leftarrow \mathfrak{O} \setminus \{O'\}, \mathfrak{C} \leftarrow (\mathfrak{C} \setminus \{O'\}) \cup \{O' \cup \{j\}\}, C_k \leftarrow C_k \setminus \{j\}$
    return $\mathfrak{C}$

Now, we describe Algorithm 1. Consider a single-peaked anonymous hedonic game $(P, \succeq)$. Assume that $P = \{1, \ldots, n\}$ and that the players are ordered by nonincreasing order of their peaks, i.e., $p_1 \geq p_2 \geq \cdots \geq p_n$. The algorithm

---

[21] Brandt, Bullinger, and Wilczynski [213] also investigate anonymous hedonic games under weaker notions of single-peakedness. The definition used here additionally implies a *strict* decay of the preferences away from the peak.

[22] In Chapter 5 starting on page 369, we will be concerned with single-peaked preferences in voting.

3.6 Hedonic Games

iteratively builds a coalition structure $\mathfrak{C}$ that eventually will be individually stable once the last player has been added. While doing so, it maintains a set $\mathfrak{O}$ of *open* coalitions, defined as coalitions that would allow an additional player to join. We start by creating a coalition C_1 containing player 1.

Now, we try to add players in two phases. As long as the next player's peak is at least as large as the coalition size after she has been added, we simply add her to the coalition. This is referred to as *PHASE I*. Once this is not possible anymore, we do not know if it is good to add a further player. This player might be much worse off than in a smaller coalition (even worse than in a singleton coalition), so we have to postpone such a decision. We essentially end up with a coalition that is as large as possible, provided that its size is no larger than the peak of any of its members.

Once we have reached this stage, we investigate whether some of the members of this coalition would actually prefer to be in a larger coalition, in which case we move them to a best such coalition. This is the start of *PHASE II*. We move players up until the coalition is small enough to admit a new player according to this player's peak. We continue to move players up and add new players until this is no longer possible. At this stage, we start a new coalition. After the last player has been added to a coalition, we try to move up players for a final time.

Theorem 3.8 (Bogomolnaia and Jackson [162]). *Algorithm 1 computes an individually stable coalition structure in single-peaked anonymous hedonic games in polynomial time.*

Proof. We use the notation developed in the description of Algorithm 1. In particular, let $\mathfrak{C}$ denote the coalition structure produced by Algorithm 1. We will perform an induction on the number of created coalitions, that is, on the value k at the end of the algorithm. To this end, we will simultaneously prove the following two claims:

1. The coalition structure $\mathfrak{C}$ is individually stable.
2. It holds that $|C_k| \le |C|$ for all $C \in \mathfrak{C} \setminus \{C_k\}$.

The first claim is our main claim and establishes the assertion. The second claim is an auxiliary claim that captures a bound on the size of the last created coalition, and is helpful to prove the main claim.

If $k = 1$, that is, $\mathfrak{C} = \{P\}$, then, by design of the algorithm, every player is added to the first coalition in Phase I and therefore the peak of each player is n. Hence, by single-peakedness, for every player $i \in \{1, \ldots, n\}$, it holds that $n \succ_i 1$, and no player has an incentive to form a singleton coalition. Thus, the output is individually stable. Also, the auxiliary claim is trivially true.

Now, assume that $\mathfrak{C}$ contains $k > 1$ coalitions. Assume that ℓ is the player that initiated forming C_k in the final part of the for-loop of Algorithm 1. Consider the coalition structure

$$\mathfrak{C}' = \{C \cap \{1, \ldots, \ell-1\} \mid C \in \mathfrak{C} \setminus \{C_k\}\}.$$

Note that $\mathfrak{C}'$ not only differs from $\mathfrak{C}$ by removing the coalition C_k but also by removing all players that have moved up, i.e., that have been added to the existing coalitions $C_1,\ldots,C_{k-1}$ during the creation of C_k. Then $\mathfrak{C}'$ is an output of Algorithm 1 for the instance restricted to the player set $P' = \{1,\ldots,\ell-1\}$.[23] In the restricted instance, the preferences over coalition sizes are only expressed with respect to feasible coalition sizes. This can in principal affect the peaks of players but only if they are at least ℓ. Hence, by single-peakedness, such players have a reduced preference over coalition sizes of $1 \prec \cdots \prec \ell-1$ and do not affect the execution of the algorithm.

Before we deal with the individual stability of $\mathfrak{C}$, we prove the auxiliary claim. Assume that C'_{k-1} is the last coalition created in the execution of Algorithm 1 for P'. Note that C'_{k-1} can differ from C_{k-1} because some player might have moved to C'_{k-1} during the creation of C_k. Now, we observe that player ℓ was not added to C'_{k-1} and, therefore, $\|C'_{k-1}\| \geq p_\ell$. Moreover, since player n was added to C_k, it holds that $p_n \geq \|C_k\|$. By the ordering of the players according to peaks, it follows that $\|C'_{k-1}\| \geq p_\ell \geq p_n \geq \|C_k\|$. The auxiliary claim follows by induction from the auxiliary claim for $\mathfrak{C}'$ because the coalitions in $\mathfrak{C}'$ can only have increased in size during the creation of C_k.

We are finally ready to prove the individual stability of $\mathfrak{C}$. First, note that the only coalitions that are not necessarily vetoed to be joined by an existing member are C_k and the coalitions in the set of open coalitions $\mathfrak{O}$. In particular, if some player has moved up (i.e., has deviated) in Phase II, then this player will veto any other player to join her new coalition. This legitimates the removal of the joined coalition from the set of open coalitions. Indeed, the size of the coalition joined by the deviating player is at least as large as the peak of the player who started the abandoned coalition, and therefore, by the ordering of the players according to peaks, also at least as large as the deviating player's peak. Hence, this player will not agree to further enlarge her new coalition. We now make a case distinction showing that no player can perform an individual deviation.

First, consider a player $i \in \{1,\ldots,\ell-1\}$. Note that an open coalition with respect to $\mathfrak{C}$ was already open with respect to $\mathfrak{C}'$, and therefore was not altered. Hence, since $\mathfrak{C}'$ is individually stable, i cannot perform a deviation to join such a coalition if $\mathfrak{C}(i) = \mathfrak{C}'(i)$. However, if $\mathfrak{C}(i) \neq \mathfrak{C}'(i)$, i.e., if some player has moved up to join $\mathfrak{C}'(i)$, then player i is simply in a better coalition, and a deviation towards an open coalition is not beneficial either. We still have to consider a potential deviation to join C_k. If player i never moved up, then she is in a coalition that is at most as large as the size of her peak. Hence, by the auxiliary claim and single-peakedness, joining C_k is not beneficial for player i. Moreover, if i moved up at some point, then she has left a coalition smaller than p_i. Afterwards, this coalition can only increase again until reaching the

---

[23] In principal, Algorithm 1 allows for a different output if the tie-breaking for ordering players according to peaks or for selecting coalitions when moving them up is performed differently, but $\mathfrak{C}'$ is certainly *some* output of Algorithm 1 for the same tie-breaking as in its original execution.

size when i abandoned it. If this happens because another new player joins, then, because players moved up from this coalition already, the size of this coalition is exactly at the peak of the last new player, and no other player is allowed to join it. Moreover, if it reaches the size again in Phase II, where some player moves up to it, then, again, this coalition is not open anymore and cannot be joined. Hence, no deviation of i to a coalition of at most the size of the final size of the abandoned coalition is beneficial. Thus, since the size of C_k is bounded by this size, it is not beneficial for i to join C_k.

Second, consider a player $i \in \{\ell, \ldots, n\}$. Assume first that $i \in C_k$. Then, by the final while-loop of Algorithm 1, player i has no incentive to join any open coalition. In addition, since i joined the coalition C_k, it holds that $p_i \geq p_n \geq \|C_k\|$. Hence, by single-peakedness, i has no incentive to perform a deviation towards a singleton coalition. Finally, assume that $i \notin C_k$, i.e., that i has moved up during the creation of C_k. Then i has moved to a best open coalition and cannot move to another open coalition. Moreover, moving back to C_k, even if new players join, is no improvement because C_k can only reach the size again at which i performed her deviation to move up, but, as we have discussed above, if it reaches this size, then no player is allowed to join.

This concludes the proof of the individual stability of $\mathfrak{C}$. Finally, it is easy to check that the algorithm runs in polynomial time. ❑

Of course, not all stability notions yield problems that can be easily solved in anonymous hedonic games. For example, Darmann et al. [328] show that it is NP-complete to decide whether or not a given black-and-white anonymous hedonic game (defined in Section 3.6.2.1 on page 203), represented as a graph, has a wonderfully stable partition, i.e., WSE is NP-complete for these games.

3.6.3.3 Additively Separable Hedonic Games

For additively separable hedonic games, we first consider computational aspects of stability concepts based on single-player deviations and then turn to group stability concepts and wonderful stability.

Stability Concepts Based on Single-Player Deviations

Our next result concerns the existence problem for Nash stability in additively separable hedonic games. As we have discussed earlier, the run-and-chase phenomenon can happen in such games and therefore, Nash-stable outcomes are not guaranteed to exist. However, the reason for this nonexistence is the asymmetry of the utility functions. Indeed, Nash-stable (and, therefore, individually stable, etc.) outcomes exist whenever such games are symmetric. An additively separable hedonic game (P, u) is called *symmetric* if $u_i(j) = u_j(i)$ for each pair of players $i, j \in P$.

Theorem 3.9 (Bogomolnaia and Jackson [162]). *Every symmetric additively separable hedonic game contains a Nash-stable outcome.*

Proof. Let (P, u) be a symmetric additively separable hedonic game. Given a coalition structure $\mathfrak{C}$, let $\Phi(\mathfrak{C}) = \sum_{i \in P} u_i(\mathfrak{C})$. Consider a coalition structure $\mathfrak{C}^* \in \arg\max_{\mathfrak{C}} \Phi(\mathfrak{C})$. We claim that $\mathfrak{C}^*$ is Nash-stable. Let $i \in P$, and let $\mathfrak{C}'$ be a coalition structure such that $\mathfrak{C}^* \xrightarrow{i} \mathfrak{C}'$ is a single-player deviation. Then

$$\Phi(\mathfrak{C}^*) - \Phi(\mathfrak{C}') = u_i(\mathfrak{C}^*) - u_i(\mathfrak{C}') + \sum_{j \in \mathfrak{C}^*(i) \setminus \{i\}} u_j(i) - \sum_{j \in \mathfrak{C}'(i) \setminus \{i\}} u_j(i)$$

$$= u_i(\mathfrak{C}^*) - u_i(\mathfrak{C}') + \sum_{j \in \mathfrak{C}^*(i) \setminus \{i\}} u_i(j) - \sum_{j \in \mathfrak{C}'(i) \setminus \{i\}} u_i(j)$$

$$= 2(u_i(\mathfrak{C}') - u_i(\mathfrak{C}^*)).$$

In the first inequality, we split the difference into the change for player i, the value that players in $\mathfrak{C}^*(i)$ had in $\mathfrak{C}^*$ compared to $\mathfrak{C}'$, and the value that players in $\mathfrak{C}'(i)$ gained in $\mathfrak{C}'$ compared to $\mathfrak{C}^*$. The second inequality uses symmetry.

Hence, maximality of $\mathfrak{C}^*$ with regard to Φ implies that i cannot improve her utility by joining $\mathfrak{C}'(i) \setminus \{i\}$. Since the single-player deviation was chosen arbitrarily, it follows that $\mathfrak{C}^*$ is Nash-stable. ❏

The approach in the previous proof is also called the *potential function technique*. Φ, the *potential function*, assigns a value to every outcome, and this value is strictly increasing with every deviation that is beneficial for the deviator. Hence, maximizing this value yields a Nash-stable outcome. This technique is quite universal and has been applied in similar contexts as well [148, 246].

If we do not have symmetric additively separable hedonic games, we may not only fail to have Nash-stable outcomes, but we may also face computational boundaries. Sung and Dimitrov [902] show that NSE is NP-complete in nonsymmetric additively separable hedonic games. Extending their technique, Brandt, Bullinger, and Tappe [212] show that this hardness persists even if the value range of the valuation functions in the involved games is severely restricted.

Theorem 3.10 (Brandt, Bullinger, and Tappe [212]). *Let x and y be two rational numbers with $x \geq y$. NSE is NP-complete for additively separable hedonic games with values restricted to $\{-x, y\}$.*

Proof. Let x and y be two rational numbers with $x \geq y$, and consider the class of additively separable hedonic games with utility values restricted to $\{-x, y\}$. Membership of NSE in NP is easy to see for such games. To show NP-hardness of NSE, we provide a reduction from the NP-complete problem EXACT-COVER-BY-THREE-SETS (X3C) (see, e.g., [595, 483]). An instance

3.6 Hedonic Games

of X3C consists of a pair (R, S), where R is a ground set together with a family S of three-element subsets of R, and we ask whether there exists an *exact cover* of R in S, i.e., a subfamily $S' \subseteq S$ that partitions R. Given an instance (R, S) of X3C, for every $r \in R$, we define $S_r = \{s \in S \mid r \in s\}$, i.e., S_r comprises the elements of S containing r, and $n_r = \|S_r\|$.

Now, from a given instance (R, S) of X3C, we produce an additively separable hedonic game (P, u) such that (R, S) has an exact cover if and only if (P, u) has a Nash-stable outcome. The reduction is illustrated in Figure 3.8. Define the player set as

$$P = \{c\} \cup \bigcup_{s \in S} A^s \cup \bigcup_{r \in R} \{b_i^r \mid 1 \leq i \leq n_r - 1\}, \quad \text{where}$$

$$A^s = \{a_{r_1}^s, a_{r_2}^s, a_{r_3}^s, a^s\} \quad \text{for } s = \{r_1, r_2, r_3\} \in S.$$

That is, the player set consists of copies of the elements in R corresponding to how often they occur in the sets of S minus 1, copies of the three members of each set in S together with one specific player for each such set, and an auxiliary player c. Now, define the following valuations u:

- For each $s \in S$ and for each a, a' with $a \neq a' \in A^s$: $u_a(a') = y$.
- For each $r \in R$, for each $s \in S_r$, and for each i, $1 \leq i \leq n_r - 1$:

$$u_{a_r^s}(b_i^r) = u_{b_i^r}(a_r^s) = u_{b_i^r}(c) = y.$$

- All other valuations are $-x$.

Obviously, the reduction can be computed in polynomial time. It remains to prove that (R, S) has an exact cover if and only if (P, u) has a Nash-stable outcome.

Suppose (R, S) has an exact cover $S' \subseteq S$. We construct a Nash-stable outcome $\mathfrak{C}$ as follows:

- We have coalitions in $\mathfrak{C}$ corresponding to the exact cover, i.e., for each $s \in S$, we have $A^s \in \mathfrak{C} \iff s \in S'$.
- For each $r \in R$, this leaves exactly $n_r - 1$ sets $s \in S_r$ such that $A^s \notin \mathfrak{C}$. Enumerate (by renaming) these sets as $s_1, \ldots, s_{n_r-1}$, and for each i, $1 \leq i \leq n_r - 1$, define the coalition $\{a_r^{s_i}, b_i^r\}$.
- All players a^s with $A^s \notin \mathfrak{C}$ are in a singleton: $\mathfrak{C}(a^s) = \{a^s\}$.
- Player c is also in a singleton: $\mathfrak{C}(c) = \{c\}$.

The fact that this coalition structure is Nash-stable follows from a straightforward case analysis.

Conversely, suppose now that (P, u) contains a Nash-stable outcome $\mathfrak{C}$. We show that then there must be an exact cover $S' \subseteq S$ of R. We make the following observations:

1. Player c must be in a singleton coalition, for otherwise c would deviate to form such a coalition.

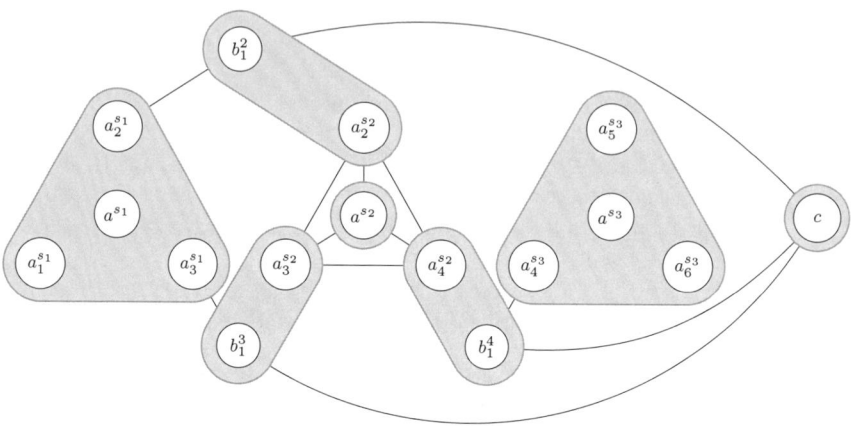

Fig. 3.8 Illustration of the reduction from the proof of Theorem 3.10 for the yes-instance $(\{1,\ldots,6\},\{s_1,s_2,s_3\})$ of X3C with $s_1 = \{1,2,3\}$, $s_2 = \{2,3,4\}$, and $s_3 = \{4,5,6\}$. Drawn edges have weight y, omitted edges have weight $-x$. The highlighted partition corresponds to the exact cover $\{s_1, s_3\}$.

2. Players b_i^r must have utility $u_{b_i^r}(\mathfrak{C}) \geq y$, for otherwise they would join $\{c\}$.
3. Coalitions of players a^s satisfy $\mathfrak{C}(a^s) \cap A^{s'} = \emptyset$ for $s' \neq s$. Indeed, for the sake of contradiction, suppose that there exists a player $a \in \mathfrak{C}(a^s) \cap A^{s'}$. Consider the sets

$$A = \{i \in \mathfrak{C}(a^s) \,|\, u_a(i) = y\} \quad \text{and} \quad A' = \{i \in \mathfrak{C}(a^s) \,|\, u_{a^s}(i) = y\}.$$

We have $A \cap A' = \emptyset$. If $\|A\| \leq \|A'\|$, then a has an incentive to deviate to a singleton, as a dislikes all players from A' as well as a^s. Similarly, if $\|A'\| \leq \|A\|$, then a^s has an incentive to form a singleton coalition, as a^s dislikes all players from A as well as a.

4. Using Observation 3, we must have $\mathfrak{C}(a^s) \neq \mathfrak{C}(b_i^r)$, as otherwise $u_{b_i^r}(\mathfrak{C}) \leq 0$, contradicting Observation 2. Hence, we have $\mathfrak{C}(a^s) \subseteq A^s$ for all $s \in S$.
5. Now, consider a player b_i^r. Define the sets

$$A = \{a_r^s \,|\, s \in S_r\} \quad \text{and} \quad B = \{b_j^r \,|\, 1 \leq j \leq n_r - 1\}.$$

By Observation 2, we have $\|A \cap \mathfrak{C}(b_i^r)\| \geq \|\mathfrak{C}(b_i^r) \setminus A\|$. We show that we even have $\|A \cap \mathfrak{C}(b_i^r)\| = \|\mathfrak{C}(b_i^r) \setminus A\|$. Indeed, for a contradiction, suppose that $\|A \cap \mathfrak{C}(b_i^r)\| > \|\mathfrak{C}(b_i^r) \setminus A\|$. Then each player $a_r^s \in A \cap \mathfrak{C}(b_i^r)$ has utility $u_{a_r^s}(\mathfrak{C}) \leq 0$ and would, by Observation 4, rather deviate to $\mathfrak{C}(a^s)$.
Moreover, we show that we must have $\mathfrak{C}(b_i^r) \setminus A \subseteq B$. Again, suppose for the sake of contradiction that this is not true. Then there are two cases. In the first case, there is a player $b_j^{r'} \in \mathfrak{C}(b_i^r) \setminus A$ with $r \neq r'$. This player

3.6 Hedonic Games

dislikes all players in A, and so would rather deviate to join $\{c\}$. In the second case, there is a player $a_{r'}^s \in \mathfrak{C}(b_i^r) \smallsetminus A$ with $r \neq r'$. This player dislikes all but one players from A as well as b_i^r, so would rather deviate to join $\mathfrak{C}(a^s)$.

Observation 5 shows that coalitions of players b_i^r are of the form $A \cup B$, where $A \subseteq \{a_r^s \,|\, s \in S_r\}$, $B \subseteq \{b_j^r \,|\, 1 \leq j \leq n_r - 1\}$, and $\|A\| = \|B\|$. This leaves for each $r \in R$ exactly one player a_r^s that is not in such a coalition. For these players, we have $\mathfrak{C}(a_r^s) = A^s$, yielding an exact cover $S' = \{s \in S \,|\, A^s \in \mathfrak{C}\}$. This completes the proof. ❑

Brandt, Bullinger, and Tappe [212] also prove versions of the preceding theorem for appreciation-of-friends and aversion-to-enemies hedonic games.

Similar hardness results for additively separable hedonic games can also be obtained for individual stability and contractual Nash stability [243, 902]. However, contractually individually stable outcomes always exist for such games and can be computed in polynomial time [48].

Moreover, in contrast to Nash-stable outcomes, individually stable and contractually Nash-stable outcomes always exist and can be computed efficiently in appreciation-of-friends and aversion-to-enemies hedonic games [43, 212, 347]. We will prove an even stronger statement for the case of individual stability (see upcoming Theorem 3.11 on page 223).

Notably, for all stability concepts based on single-player deviations, it is at least clear how to efficiently verify stable outcomes: One can simply try all possible deviations (at most n) for all players. In particular, NSV and ISV can thus be solved in polynomial time.

Core Stability, Strict Core Stability, Perfectness, and Wonderful Stability

In contrast to stability concepts based on single-player deviations, for group stability it is not always clear how to approach the verification problem. Also, concepts like the core seem to yield problems that possibly are even harder than NP. Specifically, Sung and Dimitrov [901] have shown that it is NP-complete to verify whether a given coalition structure in a given additively separable hedonic game admits a blocking coalition, i.e., is not core-stable. In other words, CSV is NP-complete in this representation. On the other hand, Sung and Dimitrov [902] only provide an NP-hardness lower bound for the corresponding existence problem, CSE. Aziz, Brandt, and Seedig [48] strengthen this result by showing NP-hardness of CSE even if the players' additively separable preferences are *symmetric*, i.e., even if $u_i(j) = u_j(i)$ for all players i and j. The complexity of this problem has eventually been settled by Woeginger [944]: For additively separable preferences, CSE is complete in $\Sigma_2^p = \text{NP}^{\text{NP}}$ (recall Section 1.5.3 starting on page 34 for the definition of this complexity class in the second level of the polynomial hierarchy).

Regarding *strictly* core-stable coalition structures for additively separable preferences, Sung and Dimitrov [901] establish NP-completeness of the verification problem, SCSV, while they again provide only an NP-hardness lower bound for the existence problem [902], SCSE, leaving the precise complexity of the latter problem open. Indeed, Woeginger [943] suspects that this problem may be Σ_2^p-complete as well.

For perfectness, however, both the existence and the verification problem can be solved in polynomial time for additively separable preferences and perfect coalition structures can be computed efficiently in such hedonic games [48].

Once again, one can restrict the utilities and investigate whether core stability then becomes more tractable. So far, we have seen that, in general, the verification problem has an NP and the existence problem has a Σ_2^p upper bound, and for the representations and stability notions considered above, the existence problem tends to be at least as hard as the verification problem. However, hardness of verification does not necessarily give hardness of recognizing existence: We will now see cases where the existence problem is easy, whereas verification can be hard.

Dimitrov et al. [347] show that there always exists a strictly core-stable coalition structure for appreciation-of-friends hedonic games and a core-stable coalition structure for aversion-to-enemies hedonic games. That is, SCSE trivially is in P for appreciation-of-friends hedonic games and CSE trivially is in P for aversion-to-enemies hedonic games (as every syntactically valid input is a yes-instance). However, the task of actually *finding* such a coalition structure is quite different in these two settings: It is easy for appreciation-of-friends hedonic games, yet hard for aversion-to-enemies hedonic games.

In more detail, in appreciation-of-friends hedonic games, every coalition in a core-stable coalition structure induces a strongly connected component of the associated (directed) graph,[24] which are easy to determine (see, e.g., the book by Cormen et al. [315]). This does not imply, though, that the verification problem for core stability would be in P in appreciation-of-friends hedonic games. In fact, there exist friend-oriented hedonic games with a core-stable coalition structure where every coalition is a proper subset of a strongly connected component.[25] The computational complexity of CSV in appreciation-of-friends hedonic games was open for more than a decade. Refuting a conjecture by Woeginger [943] (who suspected this problem to

---

[24] A *strongly connected component of a graph* is a subset S of its vertex set satisfying that for any two vertices $u, v \in S$, there is a path from u to v.

[25] The following example is due to Woeginger [943]: Suppose Anna and Belle are mutual friends, and so are David and Edgar, and Anna views David as a friend (but not vice versa), and Edgar views Belle as a friend (but not vice versa). Then the coalition structure {{Anna, Belle}, {David, Edgar}} is core-stable under friend-oriented preferences, but all four players form a strongly connected component of the corresponding directed graph.

3.6 Hedonic Games

be in P), Chen et al. [281] prove that CSV in fact is NP-complete in such games.[26]

In contrast, in aversion-to-enemies hedonic games, every coalition in a core-stable coalition structure induces a clique in the associated undirected graph:[27] Indeed, since a player would rather stay alone than being in a coalition with an enemy, every core-stable coalition structure must place enemies in distinct coalitions. In particular, a maximum clique must be contained in a core-stable coalition structure. Otherwise, by forming the coalition induced by this clique, all these players would improve. Hence, finding core-stable coalition structures in aversion-to-enemies hedonic games is sufficient to determine maximum cliques. However, determining the maximum cliques of an undirected graph is an NP-hard problem (see, e.g., the book by Garey and Johnson [483]). Therefore, it is also hard to find core-stable coalition structures in aversion-to-enemies hedonic games.

On the other hand, as mentioned previously, CSE is in P for aversion-to-enemies hedonic games. But what about the *strict* core stability existence problem, SCSE, in aversion-to-enemies hedonic games? Rey et al. [808] show that SCSE is DP-hard, where DP is the second level of the boolean hierarchy over NP (recall Section 1.5.3.3 starting on page 37). The precise computational complexity of this problem is still an open question. Sung and Dimitrov [901] show that the verification problem for core stability, CSV, is NP-complete in the enemy-oriented setting, and the same is true for strict core stability: SCSV is NP-complete for such preferences as well [901, 943].

Finally, turning to wonderful stability, recall that the existence problem for wonderfully stable partitions of the vertex set of a graph into cliques, WSE, is NP-complete for black-and-white anonymous hedonic games [328]. Considering only black-and-white anonymous preferences, however, is a severe restriction. For aversion-to-enemies hedonic games, Rey et al. [808] show that the verification problem for wonderfully stable partitions, WSV, is NP-complete (using the proof technique by which Sung and Dimitrov [901] showed NP-completeness of CSV in such games) and that the existence problem, WSE, is DP-hard. The best known upper bound for the latter problem, though, is $P_{\|}^{NP}$, as has been shown by Woeginger [943]. (Recall the definition of $P_{\|}^{NP}$, "parallel access to NP," from Section 1.5.3.2 starting on page 35.) Closing the complexity gap between DP-hardness of WSE for aversion-to-enemies hedonic games and its membership in $P_{\|}^{NP}$ remains an interesting open problem.

---

[26] They actually state coNP-completeness for the complement of CSV in these games.

[27] As observed by Woeginger [943], only symmetric friendship relations (i.e., $i \in F_j$ if and only if $j \in F_i$) matter in aversion-to-enemies hedonic games when we are concerned with the concept of core stability from Section 3.6.1 starting on page 199. This is why aversion-to-enemies hedonic games can be represented by *undirected* graphs as in Figure 3.7 on page 206.

3.6.4 Dynamics in Hedonic Games

So far, we have discussed the search for suitable outcomes in hedonic games in a static way: We assume that we are given a fixed game, and we wonder whether some coalition structure satisfies a specific solution concept. In this section, we add a dynamic view point on stability in hedonic games.

We formalize this idea as follows. Assume that we are given a hedonic game $(P, \succeq)$. A sequence $(\mathfrak{C}_k)_{k \geq 0}^{K}$ of coalition structures for $(P, \succeq)$ is called an *execution of the IS dynamics* (IS for *individual stability*) if, for every k with $1 \leq k \leq K$, there exists a player $d_k \in P$ such that $\mathfrak{C}_{k-1} \xrightarrow{d_k} \mathfrak{C}_k$ is an individual deviation. The partition $\mathfrak{C}_0$ is then called the *starting partition*. Similarly, we define *(an execution of the) NS dynamics* as the dynamics based on Nash deviations. Analogously, dynamics based on other stability concepts can be defined.

In the definition of a dynamics, we allow for the case $K = \infty$, which indicates an infinite execution of the dynamics. We want to investigate the convergence of dynamics. We say that a dynamics for some stability concept *possibly converges from starting partition* $\mathfrak{C}_0$ if there exists an execution of this dynamics with starting partition $\mathfrak{C}_0$ that terminates with an outcome satisfying this stability concept. Analogously, we say that a dynamics for some stability concept *necessarily converges from starting partition* $\mathfrak{C}_0$ if every execution of this dynamics with starting partition $\mathfrak{C}_0$ terminates with an outcome satisfying this stability concept. We say that a dynamics *possibly* (or, *necessarily*) *converges for some stability concept* if it possibly (or, necessarily) converges for this stability concept from every starting partition. Apart from convergence, another interesting question regards the running time of a dynamics, i.e., the number of deviations until convergence.

We have already touched upon dynamics during the analysis of the complexity of stability in Section 3.6.3 starting on page 207. First, Algorithm 1 on page 212 can be interpreted as a specific execution of the IS dynamics where we start from the coalition structure consisting only of singleton coalitions. Then players join a coalition in PHASE I only if they thus move towards their peaks, and in PHASE II only if they can improve by moving to an open coalition. Clearly, both of these steps are individual deviations. Hence, Theorem 3.8 on page 213 even establishes possible convergence of the IS dynamics in single-peaked anonymous hedonic games when starting from the coalition structure consisting only of singleton coalitions. Moreover, in this execution, every player performs at most two deviations, hence a linear number of deviations with respect to the number of players suffices for possible convergence. In addition, Brandt, Bullinger, and Wilczynski [213] even prove necessary convergence of the IS dynamics in anonymous hedonic games, independent from the starting partition.

Second, the proof of Theorem 3.9 on page 216 implies that the NS dynamics converges in additively separable hedonic games. However, a careful

3.6 Hedonic Games

reader may have realized that we did not discuss the computational complexity of computing a Nash-stable outcome in symmetric additively separable hedonic games when stating Theorem 3.9. Indeed, this is a delicate issue because NSE is trivial, but the proof of existence of Nash-stable outcomes in such games does not give rise to an efficient algorithm. In fact, Gairing and Savani [478] show that computing a Nash-stable or individually stable outcome in symmetric additively separable hedonic games is PLS-complete, i.e., complete for the complexity class associated with local search algorithms.[28] The IS dynamics may even take an exponential number of rounds. Such an example is explicitly provided by Brandt, Bullinger, and Tappe [212].

In contrast, we have seen examples of nonconvergence of dynamics. Indeed, run-and-chase games are, for instance, possible in both the framework of anonymous and of additively separable hedonic games. Hence, both classes contain games, where any kind of convergence is impossible, independent of the starting partition.

As our next result, we will show necessary convergence of the IS dynamics in appreciation-of-friends and aversion-to-enemies hedonic games. The proof follows from a lemma that we state without proof. Interestingly, this lemma gives a purely combinatorial insight that can be applied in various ways to prove the convergence of different types of dynamics [212].

Lemma 3.2 (Brandt, Bullinger, and Tappe [212]). *Consider a sequence of k successive single-player deviations*

$$\mathfrak{C}_0 \xrightarrow{i_1} \mathfrak{C}_1 \xrightarrow{i_2} \ldots \xrightarrow{i_k} \mathfrak{C}_k.$$

The following bounds hold:

$$-\frac{n(n-1)}{2} \leq \sum_{j \in [k]} \|\mathfrak{C}_j(i_j)\| - \|\mathfrak{C}_{j-1}(i_j)\| \leq \frac{n(n-1)}{2}.$$

The next result by Brandt, Bullinger, and Tappe [212] considers an even more general subclass of additively separable hedonic games (than appreciation-of-friends or aversion-to-enemies hedonic games).

Theorem 3.11 (Brandt, Bullinger, and Tappe [212]). *The IS dynamics is guaranteed to converge in additively separable hedonic games where valuations attain at most one nonnegative number.*

Proof. Let (P, u) be an additively separable hedonic game with n players. If there are no nonnegative valuations, then all individual deviations are singleton formations, so the IS dynamics necessarily converges after at most n deviations. Hence, suppose that there is exactly one nonnegative utility value $x \geq 0$. If there are no negative valuations, then in case $x = 0$ we terminate

---

[28] Recall this complexity class from page 127 where it was discussed with respect to computing Nash equilibria in general normal form games.

immediately, and in case $x > 0$ the grand coalition will form after at most n^2 deviations. The latter holds because every deviation increases the number of pairs of players that are part of the same coalition. Therefore, we will now assume that in addition to the single nonnegative utility value x, there is at least one negative utility value, and we denote the largest absolute value of all negative utility values by y. Further, define

$$\Delta = \min\{u_i(C) - u_i(C') \mid i \in P, C, C' \in 2^{P_i}, u_i(C) > u_i(C')\}.$$

Intuitively, $\Delta > 0$ is the minimum improvement any player is guaranteed from an NS deviation. Finally, consider the potential function Φ defined by the utilitarian social welfare of a partition as $\Phi(\mathfrak{C}) = \sum_{i \in N} u_i(\mathfrak{C})$. (See, e.g., Section 9.6 starting on page 635 for various measures of social welfare.)

Let us investigate how this potential changes for a single individual deviation $\mathfrak{C} \xrightarrow{i} \mathfrak{C}'$.

$$\Phi(\mathfrak{C}') - \Phi(\mathfrak{C}) = \underbrace{u_i(\mathfrak{C}') - u_i(\mathfrak{C})}_{\text{deviator } i}$$

$$+ \underbrace{\sum_{j \in \mathfrak{C}'(i) \smallsetminus \{i\}} u_j(\mathfrak{C}') - u_j(\mathfrak{C})}_{\text{welcoming coalition } \mathfrak{C}'(i)} + \underbrace{\sum_{j \in \mathfrak{C}(i) \smallsetminus \{i\}} u_j(\mathfrak{C}') - u_j(\mathfrak{C})}_{\text{abandoned coalition } \mathfrak{C}(i)}$$

$$= u_i(\mathfrak{C}') - u_i(\mathfrak{C}) + \sum_{j \in \mathfrak{C}'(i) \smallsetminus \{i\}} u_j(i) - \sum_{j \in \mathfrak{C}(i) \smallsetminus \{i\}} u_j(i)$$

$$= u_i(\mathfrak{C}') - u_i(\mathfrak{C}) + x\left(\|\mathfrak{C}'(i)\| - 1\right) - \sum_{j \in \mathfrak{C}(i) \smallsetminus \{i\}} u_j(i)$$

$$\geq \Delta + x\left(\|\mathfrak{C}'(i)\| - 1\right) - x\left(\|\mathfrak{C}(i)\| - 1\right)$$

$$= \Delta + x\left(\|\mathfrak{C}'(i)\| - \|\mathfrak{C}(i)\|\right).$$

The third equality comes from the fact that i performs an individual deviation, so all players $j \in \mathfrak{C}'(i) \smallsetminus \{i\}$ must accept i, which means they must have $u_j(i) = x$. Now, let $\mathfrak{C}_0$ be any initial partition and consider any sequence of k successive individual deviations

$$\mathfrak{C}_0 \xrightarrow{i_1} \mathfrak{C}_1 \xrightarrow{i_2} \ldots \xrightarrow{i_k} \mathfrak{C}_k.$$

Telescoping and term-wise application of the above inequality yields

$$\Phi(\mathfrak{C}_k) - \Phi(\mathfrak{C}_0) = \sum_{j \in [k]} \Phi(\mathfrak{C}_j) - \Phi(\mathfrak{C}_{j-1})$$
$$\geq \sum_{j \in [k]} \Delta + x \left(\|\mathfrak{C}_j(i_j)\| - \|\mathfrak{C}_{j-1}(i_j)\| \right)$$
$$= k\Delta + x \sum_{j \in [k]} \|\mathfrak{C}_j(i_j)\| - \|\mathfrak{C}_{j-1}(i_j)\|.$$

We recognize the sum from Lemma 3.2 and obtain

$$\Phi(\mathfrak{C}_k) - \Phi(\mathfrak{C}_0) \geq k\Delta - x \frac{n(n-1)}{2}.$$

As only the right hand side depends on k, the sequence must be finite. To be precise, we can bound the potentials of the initial and final partitions by

$$\Phi(\mathfrak{C}_0) \geq -n(n-1)y \quad \text{and} \quad \Phi(\mathfrak{C}_k) \leq n(n-1)x.$$

Substituting in these bounds and rearranging for k gives

$$k \leq \frac{(2y + 3x)n(n-1)}{2\Delta}. \tag{3.10}$$

This completes the proof. ∎

Note that Equation (3.10) directly implies polynomial running time of dynamics in appreciation-of-friends and aversion-to-enemies hedonic games. Moreover, Lemma 3.2 can also be applied to obtain a convergence result for deviations based on contractual Nash stability.

We conclude the section by quickly discussing an interesting extension of dynamics. Boehmer, Bullinger, and Kerkmann [161] enhance the dynamic model of coalition formation with the consideration of *dynamic valuation functions*. This offers the players to react to deviations by altering their utilities. For instance, a player might be resentful of a player who abandons her. This allows to obtain new convergence results that were not possible in a static utility model. Variable utility models are very rich and could be an avenue for new research on hedonic games.

3.6.5 Other Important Concepts for Hedonic Games

We conclude our coverage of hedonic games by giving a brief outlook on important concepts that we decided not to cover in more depth.

3.6.5.1 Altruistic Hedonic Games

Up to now, we have assumed that players in hedonic games take only their *own* preferences or utilities into account when considering which coalitions to join. However, in many real-life situations, especially in a context of cooperation, people do not always behave purely selfishly but may also consider their friends' preferences and utilities—for example, various models of altruism in both noncooperative and cooperative game theory have been surveyed by Rothe [821]. In particular, focusing on the appreciation-of-friends model of representation, Kerkmann et al. [605, 745] introduce and study *altruistic hedonic games* where players take into account not only their own but also their friends' preferences, i.e., the preferences of their neighbors in a network of friends. Depending on the order in which players look at their own and their friends' appreciation-of-friends preferences or utilities when comparing two coalitions they may want to join, Kerkmann et al. [605, 745] distinguish three degrees of altruism:

- For *selfish-first preferences*, players first look at their own preferences, and only in case of indifference they consider the average of their friends' utilities in both coalitions to make a decision as to which coalition they (weakly) prefer.
- For *equal-treatment preferences*, players treat their own and their friends' utilities for both coalitions equally when taking the average to make this decision.
- And, finally, for *altruistic-treatment preferences*, players first look at their friends' preferences, and only in case of indifference they make this decision based on their own utilities for these two coalitions.

Instead of taking averages of utilities, Kerkmann et al. [605] also introduce and study *minimum-based altruistic hedonic games* where players only view those friends who are worst off in the coalition at hand.[29] They study both the axiomatic properties of altruistic and minimum-based altruistic hedonic games and the common stability concepts for them and investigate the corresponding existence and verification problems in terms of their computational complexity.

Kerkmann, Cramer, and Rothe [603, 607] extend the model of Kerkmann et al. [605, 745] to coalition formation beyond hedonic games, i.e., *all* the friends of a player matter (be they inside or outside of this player's coalition) when determining the players' preferences for any of the three degrees of altruism.

---

[29] Interestingly, minimum-based altruistic hedonic games with equal-treatment preferences are equivalent to the *"loyal"* variant of symmetric appreciation-of-friends hedonic games introduced and studied by Bullinger and Kober [244].

3.6.5.2 Solution Concepts Beyond Stability

The absence of incentives to perform deviations may not be the only desideratum for an outcome of a hedonic game. Often, some sort of optimality is also required. A fundamental notion of optimality, which has already been presented in the early research on hedonic games by Drèze and Greenberg [356], is *Pareto optimality*.[30] Informally speaking, a coalition structure is *Pareto-optimal* if there is no other coalition structure that is more preferred by some player and no less preferred by any other player. In other words, if a coalition structure is Pareto-optimal, but some player could be better off in another coalition structure, then some other player must be worse off in this alternative coalition structure. It can be shown that the algorithm from Theorem 3.8 on page 213 even yields outcomes that satisfy a weak version of Pareto optimality in anonymous hedonic games [162]. For additively separable hedonic games, Pareto optimality on its own is usually computationally easy to achieve but Pareto-optimal outcomes can have other significant drawbacks [48, 242, 379]. Aziz, Brandt, and Harrenstein [47] study Pareto optimality in a variety of other classes of hedonic games.

A strengthening of Pareto optimality for games with a cardinal preference representation is social welfare optimality (see, e.g., Section 9.6 starting on page 635 for various measures of social welfare in the context of fair division of indivisible goods), which for instance requires that an outcome maximizes the so-called *utilitarian social welfare*, defined as the sum of all players' values for a coalition structure [48]. The paper by Aziz, Brandt, and Seedig [48] is a standard reference for computational considerations regarding many solution concepts in additively separable hedonic games and defines further solution concepts, which, for instance, also incorporate the idea of fairness.

When Anna discloses to the boys that she and Belle have decided on page 202 that they actually prefer to make cherry quark together instead of baking a cherry-quark cake with them, Edgar is outraged.

"You know how much I was looking forward to baking the cake and how much I love cherry-quark cake!" he rants at his sister.

"Let me remind you," Belle hits back at him, "how glad you've been on page 210 to discover that $\mathfrak{C} = \{\{\text{Anna}, \text{Belle}\}, \{\text{Chris}, \text{David}, \text{Edgar}\}\}$ is a wonderfully stable coalition structure."

---

[30] Pareto optimality is also considered in other chapters of this book—see Definition 2.3 on page 49 for Pareto optimality in the context of noncooperative game theory; Section 4.2.2.2 starting on page 268 for the related notion of Pareto consistency in the context of voting; Section 8.3.2 starting on page 522 for Pareto optimality in the context of fair division of divisible goods (cake-cutting); and last but not least Section 9.4 starting on page 622 for Pareto optimality in the context of fair division of indivisible goods.

> "Indeed, I was glad about that, but back then I didn't realize that this would mean to have *no cake*!" Edgar snaps back. "I insist on the cake and the grand coalition: $\mathfrak{G} = \{\{\text{Anna}, \text{Belle}, \text{Chris}, \text{David}, \text{Edgar}\}\}$!"
>
> "OK, calm down!" Chris tries to mediate in this dispute. "How about if we all just take a deep breath ... and then take a vote?"
>
> "You want us to vote?" David is stunned. "Like holding an election? I thought the chapter on voting does not start before page 233?"
>
> "Voting on which of $\mathfrak{C}$ and $\mathfrak{G}$ a majority of us prefer is a great idea!" Anna weighs in. "I prefer $\mathfrak{C}$ to $\mathfrak{G}$."
>
> After all children have cast their votes, it turns out that the grand coalition wins with the votes of the boys.
>
> "Yeah! 3 to 2 for $\mathfrak{G}$!" Edgar crows. "Let's bake the cake!"

A coalition structure $\mathfrak{C}$ for a hedonic game is said to be *popular* if there exists no other coalition structure for this game in which more players are better off than worse off. That is, if the players would take a vote on $\mathfrak{C}$ and every other coalition structure $\mathfrak{D} \neq \mathfrak{C}$, a weak majority of them would prefer $\mathfrak{C}$ to $\mathfrak{D}$ in this pairwise comparison. This is akin to the notion of *weak Condorcet winner* in voting, to be discussed in Section 4.1.2.1 starting on page 239. Similarly, $\mathfrak{C}$ is called *strongly popular* (sometimes called *strictly popular*) if $\mathfrak{C}$ is preferred to every other coalition structure $\mathfrak{D} \neq \mathfrak{C}$ by a (strict) majority of players. As before, this is akin to the notion of *Condorcet winner* in voting; again, see Section 4.1.2.1.

Popularity and strong popularity have been thoroughly studied by Brandt and Bullinger [211], who also study variations thereof, such as *mixed popularity*, which is particularly fascinating because existence is guaranteed by the minimax theorem, stated in the chapter on noncooperative game theory as Theorem 2.5 on page 123. Previously, popularity and strong popularity have been studied in additively separable hedonic games by Aziz, Brandt, and Seedig [48], and by Kerkmann et al. [604] in their model of hedonic games with friends, enemies, and neutral players.

For altruistic hedonic games, the verification problem for popularity and strong popularity has been investigated by Nguyen et al. [745], Wiechers and Rothe [940], and Kerkmann and Rothe [608]. In both the average-based and the minimum-based model of altruistic preferences and for all three degrees of altruism, verifying either popular or strongly popular coalition structures has been shown to be coNP-complete. The complexity of the existence problem for popularity and strong popularity in these games, however, remains unresolved.

3.6.5.3 Quantifying Social Welfare Loss

Social welfare optimality, despite being computationally demanding, gives a natural benchmark for the value of a coalition that the entirety of players can jointly achieve. It is therefore natural to compare the utilitarian social welfare of outcomes satisfying certain solution concepts with the maximally achievable utilitarian social welfare, in the spirit of the *price of anarchy* [621]. By considering solution concepts for hedonic games, this yields various *price* concepts.

Bilò, Monaco, and Moscardelli [148] consider the *price of a worst Nash-stable outcome* in additively separable hedonic games. Since these games are very rich, this price can be very large, even if the utility structure is very restricted. They find bounds independent of the value range under the assumption that the sizes of coalitions are bounded. Bilò et al. [147] study the *price of a best Nash-stable outcome* in fractional hedonic games. Finally, Elkind, Fanelli, and Flammini [379] propose the *price of Pareto optimality* and study it in typical cardinal classes of hedonic games.

3.6.5.4 Online Hedonic Games

A feature of many realistic coalition formation scenarios is that the composition of the game is not static. In other words, players might join and leave the market while coalitions are formed.[31] Motivated by this idea, Flammini et al. [460] propose an *online variant of hedonic games*. This extends the prominent work on *online matching* problems [596] by the consideration of coalitions of size larger than two (an excellent survey of this research is due to Huang and Tröbst [564]).

Flammini et al. [460] find that the optimal algorithm for additively separable hedonic games is a greedy algorithm that always assigns players to the coalition where they yield the largest gain in the moment when they arrive. However, this result makes the strong assumption of an adaptive adversary. If the players arrive in a random order or if the algorithm has the power to dissolve coalitions under certain circumstances, then algorithms that conduct waiting periods outperform the greedy algorithm [245].

This concludes the chapter on cooperative game theory, and we will now turn to the next major subject of this book, preference aggregation, which in particular is central to voting.

---

[31] This idea is also prevalent in other areas and will be covered in other chapters of this book, such as in the chapter on voting (see, e.g., Section 4.3.3.10 starting on page 321 where a model of online manipulation in sequential elections is explained, with voters arriving over time [543]; also online control and online bribery are described) or in the chapter on fair division of indivisible goods (see Section 9.10.6 starting on page 674).

Part II
Voting and Judging

Chapter 4
Preference Aggregation by Voting

Dorothea Baumeister · Jörg Rothe

Anna, Belle, and Chris want to spend the afternoon together. They consider to either play miniature golf, or go on a bicycle tour, or go for a swim. However, they cannot come to an agreement, as not all of them have the same preferences.

"I would love to go on a bicycle tour," says Anna. "But if that's not possible, I would rather play miniature golf than go for a swim."

"Hey!" Chris bursts out delightedly. "That's just how I see it."

"Oh no!" replies Belle. "Just yesterday I was going on a bicycle tour and my knees still hurt terribly, so I'd rather play miniature golf or go for a swim. And it has been such a long time since I've played miniature golf: That's my favorite!"

Fig. 4.1 Preferences of Anna, Belle, and Chris

© The Author(s), under exclusive license to Springer Nature Switzerland AG 2024
J. Rothe (ed.), *Economics and Computation*, Classroom Companion: Economics,
https://doi.org/10.1007/978-3-031-60099-9_4

Figure 4.1 shows the preferences of Anna, Belle, and Chris. What alternative should they go for to make sure everyone is (more or less) happy? Anna and Chris think they should go for the bicycle tour because it is ranked on top of the three preferences twice, a (strict[1]) majority! Belle, however, dissents from their opinion and argues that they should go for miniature golf, as this alternative is the only one that is never ranked last. To make their decision, Anna, Belle, and Chris need to choose a voting system, a rule that says how to aggregate their individual preferences on (or rankings of) the available alternatives so as to determine the winner of the election.

4.1 Some Basic Voting Systems

Formally, an *election* is given by a pair (C, V), where C is a set of *candidates* (or, *alternatives*) and V is a list of *votes* (or, *ballots*). All voters are represented by their votes, which express their preferences over the candidates in C.[2] How often the same vote occurs, of course, must be taken into account when determining the winner(s) of an election. A list V of votes over candidate set C is also called a *preference profile*; for clarity, we sometimes write (C, V).

A *voting system* is a rule that defines how to determine the winner(s) of a given election. In some settings, we will be asking whether a given candidate is a winner (but it is fine if there are other winners, too), and in some cases we will be asking whether a given candidate is the one and only winner (we will sometimes call such a candidate a "unique winner"). As we'll discuss further in Section 4.3.1 (see, in particular, page 290), problems related to the first goal—making the candidate a winner—are said to be in the *nonunique-winner model*, and problems related to the second goal—making the candidate a unique winner—are said to be in the *unique-winner model*. In either of the models, in some cases it is possible that no candidate emerges victorious, in which case the election has no winner at all; recall the example related to the Condorcet paradox in Section 1.2 (see Figures 1.2 and 1.3 on pages 6 and 8).

- Let C be a set of candidates. Formally, a *voting system* can be described by a so-called *social choice correspondence*, f, that maps each preference profile V over C to a subset of C.
- A *social choice function* maps any given preference profile to a single winner.
- A *social welfare function* describes not only how to select a winner or set of winners by a voting system, but even returns a complete ranking of (or, preference list over) the candidates.

---

[1] By default, when we write "majority" we usually mean a "strict majority," though occasionally we may explicitly write "strict majority" to emphasize this.

[2] V is a *list*, not a set, of votes because distinct voters may have the same preferences, just like Anna and Chris in Figure 4.1.

4.1 Some Basic Voting Systems

How the votes are represented depends on the voting system. Many systems require preference lists that rank the candidates in decreasing order according to the degree of preference. That is, the top position in a (linear) preference order of a voter is taken by the candidate this voter prefers the most, and the least preferred candidate is in the last position. Belle's vote in Figure 4.1, for example, is given in the form

$$\text{miniature golf} > \text{swim} > \text{bicycle tour}.$$

Miniature golf is Belle's most preferred alternative and the bicycle tour her least preferred. As in this figure, we sometimes omit the relation symbol ">" and simply write, e.g., $a\ d\ b\ c$ instead of $a > d > b > c$.

Mathematically speaking, a preference is expressed by a (strict) *linear order* on the set C of candidates, i.e., the underlying preference relation $>$ is

1. *connected* (for each two distinct candidates c and d in C, $c > d$ or $d > c$),
2. *transitive* (for each three candidates c, d, and e in C, $c > d$ and $d > e$ imply $c > e$), and
3. *asymmetric* (for each two candidates c and d in C, $c > d$ implies that $d > c$ does not hold).[3]

For most of the voting systems to be introduced in the following, votes are linear orders on the set of candidates. When a voting systems requires votes of a different form, we will explicitly point that out.

4.1.1 Scoring Protocols

Anna and Chris have a different criterion than Belle for determining the winners of an election. Both criteria correspond to voting systems that belong to the important class of so-called scoring protocols. In a *scoring protocol* (a.k.a. *scoring rule*), a scoring vector determines how many points a candidate scores by taking a certain position in a vote. A scoring vector for m candidates is of the form

$$\boldsymbol{\alpha} = (\alpha_1, \alpha_2, \ldots, \alpha_m),$$

where the α_i are nonnegative integers satisfying $\alpha_1 \geq \alpha_2 \geq \cdots \geq \alpha_m$. A candidate taking the ith position in the preference list of a voter receives α_i points from this voter. Whoever scores the most points wins; it is possible that there are several winners.

---

[3] Asymmetry of $>$ in particular implies irreflexivity of $>$. A relation R on C is said to be *reflexive* if cRc for all $c \in C$, and R is said to be *irreflexive* if cRc for no $c \in C$. In the social choice literature (see, e.g., the work of Arrow [29]), voters are occasionally also allowed to be *indifferent* about the candidates; in this case the preference relation is not required to be asymmetric.

4.1.1.1 Plurality Voting

The suggestion of Anna and Chris that, in the example from the beginning of this chapter, the bicycle tour should be elected as the winner is best met by the plurality rule, one of the simplest scoring rules. In *plurality voting*, only the top alternative of each vote scores a point, all other candidates come away empty-handed. The scoring vector for *plurality* thus is

$$\alpha = (1, 0, \ldots, 0).$$

Table 4.1 shows how to determine the plurality winner (and the winners in two other scoring protocols, veto and Borda) of the election from this example, where the three alternatives are abbreviated by their initial letters. The bicycle tour with two points is ahead of miniature golf and swimming, which get one point and zero points, respectively.

Table 4.1 Example of an election with the scoring protocols plurality, veto, and Borda. Key: B = bicycle tour, M = miniature golf, and S = swimming.

	Points								
	Plurality			Veto			Borda		
Preference profile	B	M	S	B	M	S	B	M	S
Anna: $B > M > S$	1	0	0	1	1	0	2	1	0
Belle: $M > S > B$	0	1	0	0	1	1	0	2	1
Chris: $B > M > S$	1	0	0	1	1	0	2	1	0
Sum of points:	2	1	0	2	3	1	4	4	1

4.1.1.2 Veto (Antiplurality)

On the other hand, Belle's suggestion from our introductory example—that miniature golf should win because it never is in the last position—in some sense corresponds to the scoring protocol *veto* (a.k.a. *antiplurality*), with the scoring vector

$$\alpha = (1, \ldots, 1, 0).$$

In this system, all candidates receive one point, except for the candidate ranked last, who comes away empty-handed. Table 4.1 shows that miniature golf wins with three points according to veto, since the bicycle tour gets only two points and swimming only one point.

4.1.1.3 k-Approval and k-Veto

A scoring protocol remotely related to approval voting, a voting rule to be introduced in Section 4.1.3, is *k-approval*. For a fixed $k \leq m$ (where m is the number of candidates), k-approval has the scoring vector

$$\boldsymbol{\alpha} = (\underbrace{1,\ldots,1}_{k},0,\ldots,0).$$

That is, the voters here cast complete preference lists over the candidates, and the first k candidates in each vote get one point each. Obviously, 1-approval is nothing other than plurality, and $(m-1)$-approval for m candidates is nothing other than veto. More generally, k-veto is defined to be $(m-k)$-approval.

4.1.1.4 Borda Count

Another well-known scoring protocol is *Borda Count*, due to Borda [169]. When there are m candidates, the candidate in the top position of a vote gets $m-1$ points from this voter, the candidate in the second position gets $m-2$ points, and so on until the candidate in the last position gets zero points. The scoring vector for Borda with m candidates thus is

$$\boldsymbol{\alpha} = (m-1, m-2, \ldots, 0).$$

Table 4.1 shows that regarding the decision about the activity Anna, Belle, and Chris will engage in in the afternoon, both the bicycle tour and miniature golf with four points each are the Borda winners, whereas swimming gets only one point. The Borda system and its variants are widely used in real-world elections, for example, in political elections in Slovenia, and the winner of the Eurovision Song Contest is determined by a Borda-like scoring scheme.

4.1.2 Voting Systems Based on Pairwise Comparisons

Anna, Belle, and Chris eventually opt for the bicycle tour. On the way back they come across David and Edgar, and they decide to do something together with them. The boys suggest to have either some ice cream or a barbecue. Anna and Belle, however, would rather like to either watch a movie or have a hamburger. To make a joint decision, they all let the others know their preferences (see Figure 4.2).

Chris comments, "If one compares having ice cream with having a barbecue, more of us are in favor of having ice cream, namely, Anna, Belle, and me."

"Great!" Anna agrees. "So we can cancel the barbecue already. But how about our suggestions? Belle and I think a hearty hamburger is much more yummy than ice cream, and watching a movie would be better than having ice cream, too."

"True, but we are three and you are only two," replies Chris. "Having ice cream also wins the pairwise comparison with having a hamburger or watching a movie. No doubt, having ice cream wins altogether!"

Although Anna and Belle aren't totally happy with this decision (and even less so than David and Edgar), Chris has been able to convince them, and they all head for the closest ice cream parlor.

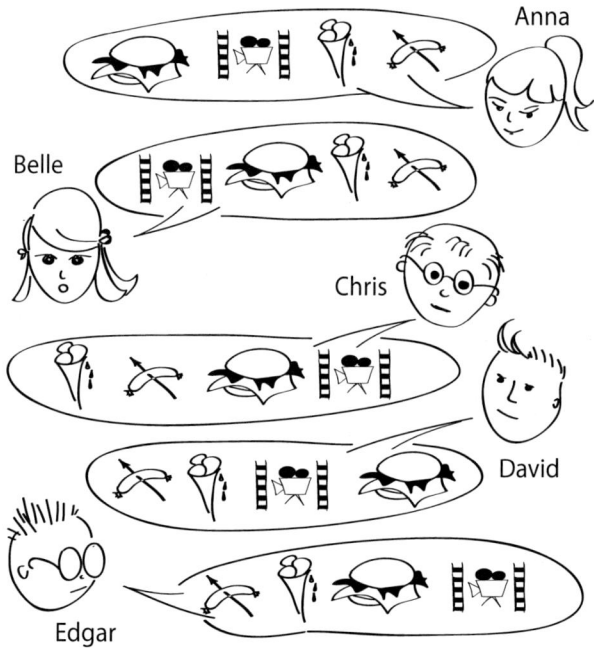

Fig. 4.2 Preferences of Anna, Belle, Chris, David, and Edgar

Unlike in scoring protocols, candidates here do not receive points immediately from the voters according to their positions in their votes, but instead all candidates are compared with each other in head-to-head contests. In such a pairwise comparison, it is tested who of these two candidates is preferred

4.1 Some Basic Voting Systems

(i.e., is ahead of the other in the voters' preference lists) by a majority of voters. If there is an even number of voters, ties may occur.

All head-to-head contests for the preferences from Figure 4.2 are given in Table 4.2. Again, the alternatives are represented by their initial letters; for example, the abbreviation $B?M$ stands for the head-to-head contest between barbecue and movie.

Table 4.2 Example of head-to-head contests: Who wins a majority of votes? Key: B = barbecue, H = hamburger, I = ice cream, and M = movie.

	Pairwise comparison					
Preference profile	$B?H$	$B?I$	$B?M$	$H?I$	$H?M$	$I?M$
Anna: $H > M > I > B$	H	I	M	H	H	M
Belle: $M > H > I > B$	H	I	M	H	M	M
Chris: $I > B > H > M$	B	I	B	I	H	I
David: $B > I > M > H$	B	B	B	I	M	I
Edgar: $B > I > H > M$	B	B	B	I	H	I
Winner of the comparison:	B	I	B	I	H	I

The following voting systems are based on such pairwise comparisons, yet they all determine their winners in different ways.

4.1.2.1 Condorcet Voting

Chris has explained the important notion of Condorcet winner, which has been mentioned already in Section 1.2. This notion and the corresponding voting system are due to the Marquis de Condorcet [300]. A *Condorcet winner* (respectively, a *weak Condorcet winner*) is a candidate who is preferred to each other candidate in a pairwise comparison by more than (respectively, by at least) half of the voters. Clearly, a Condorcet winner is also a weak Condorcet winner, but not every weak Condorcet winner is a Condorcet winner.

As shown in Table 4.2, ice cream wins all pairwise comparisons with all other alternatives by a majority and therefore is the Condorcet winner of this election. Figure 4.3 shows the *majority graph* corresponding to the pairwise comparisons from Table 4.2. The vertices of this graph are the alternatives, and there is an edge directed from X to Y if and only if X wins the head-to-head contest against Y by a majority of votes.

As in this figure, the edges may be labeled by the exact result of each comparison: If an edge from X to Y is labeled by $i:j$ with $i > j$, then i voters prefer X to Y and only j voters prefer Y to X. If there is a tie $i:i$ for an even number of voters, one may indicate this either by two edges in opposite directions between the related vertices, or by one undirected edge between

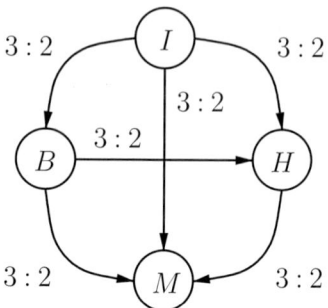

Fig. 4.3 Majority graph for the election in Table 4.2

them. In our example, every edge is directed and they each coincidentally happen to be labeled by the same relation of 3 : 2.

As already mentioned, I (having ice cream) is the Condorcet winner of the election from Table 4.2. Alternative M (watching a movie), however, is the *Condorcet loser*, since M is defeated by every other alternative in their head-to-head contest with a majority. Analogously to the notion of weak Condorcet winner, a *weak Condorcet loser* is preferred to no other candidate by more than one half of the voters and so never reaches a majority in any head-to-head contest.

As we have seen in Figures 1.2 and 1.3 of Section 1.2, a Condorcet winner is not guaranteed to exist, though, and the same is true for weak Condorcet winners, Condorcet losers, and weak Condorcet losers. If there is no Condorcet winner, we have a *Condorcet top cycle* expressing the Condorcet paradox, i.e., the alternatives in this cycle defeat all alternatives outside of the cycle in their pairwise comparison, yet none of them defeats all other alternatives in the cycle. Since all alternatives in Figure 1.3 on page 8 belong to the top cycle, one simply refers to it as a *Condorcet cycle*. On the other hand, Condorcet winners and Condorcet losers are unique whenever they exist. This is in general not the case for weak Condorcet winners and weak Condorcet losers.

A voting system *respects the (weak) Condorcet winner* if it elects the (weak) Condorcet winner whenever one exists (which is just the Condorcet criterion that is to be considered on page 265). A voting system respecting the (weak) Condorcet winner is also called *Condorcet-consistent* (or *weakCondorcet-consistent*). In the following, we introduce various voting systems that each respect the Condorcet winner (respectively, the weak Condorcet winner), but—unlike Condorcet's system—always have one winner or more, even if there is no Condorcet winner (respectively, no weak Condorcet winner). In each of these systems, winners are as "close" as possible to become a (weak) Condorcet winner with respect to a suitable distance function. As in scoring protocols or the Condorcet system, the voters rank the candidates by decreasing preference in each of these voting systems, i.e., votes are linear rankings.

4.1.2.2 The Family of Llull and Copeland Voting Systems

A voting system that is based on pairwise comparisons and respects the Condorcet winner is due to Copeland [314]. To determine the points of the candidates, they all enter a head-to-head contest with everyone else. If one of the two candidates in such a contest is preferred by a majority of voters, he receives one point and the other one no point. If none of them has a majority (i.e., if they tie, which is possible only for an even number of voters), then they both receive half a point. Adding up the points for a candidate gives this candidate's *Copeland score*. Whoever has the most points (i.e., the highest Copeland score) is a *Copeland winner*.

Already in the 13th century, the philosopher, poet, and missionary Ramon Llull proposed a voting system that also is based on pairwise comparisons and respects the Condorcet winner. Llull's systems and the above-mentioned Copeland system are very much alike. The only crucial difference is that ties in the pairwise comparisons (which may occur if there is an even number of voters) are rewarded with one point, not half a point, in Llull's system.[4]

But why shouldn't one reward such a tie also with other points than $1/2$ or 1? In the literature, "Copeland voting" sometimes also refers to a rule that gives zero points to both candidates if they tie in a head-to-head contest. Faliszewski et al. [429] therefore propose an entire family of voting systems: For each rational number α between 0 and 1, *Copeland$^\alpha$* denotes the same system as that described above, except that a tie in a pairwise comparison is rewarded with α points. Thus, the $C^\alpha Score(c)$ of any candidate c is the number of c's victories plus α times the number of ties in c's pairwise comparisons, and all candidates c with a highest $C^\alpha Score(c)$ win.

In this notation, Copeland$^{1/2}$ is the common Copeland system, and Llull's system is Copeland^1. If there is an odd number of voters, no tie is possible in any majority comparison; the systems Copeland$^\alpha$ are then identical for all α.

The German champion in the Bundesliga (or for that matter the Premiere League's champion, or etc.) is not elected, of course, but is determined by the results of soccer matches that each reward the two contestants with points: A victory gives three points, a tie one point for both, and a defeat zero points. Considering this as a "Copeland$^\alpha$ election" and scaling these points into the range $[0, 1]$, we get the value of $\alpha = 1/3$. There are nine fixtures per match day, and after 34 match days the "Copeland$^{1/3}$ winner" celebrates the German championship.

---

[4] In his work *Artifitium Electionis Personarum*, Llull in fact described several voting systems, and since he was proposing them for the election of an abbot in a friary or an abbess in a nunnery, or for electing a bishop or even the pope, the candidates are identical with the voters in his systems. Such details will be neglected here, though, and "Llull's system" will instead refer to the voting system described in the text. By the way, Llull's systems could never really prevail during his lifetime and later sank into oblivion, see the investigation of Hägele and Pukelsheim [526]. For example, since centuries the conclave, where cardinals meet for electing the pope, is held using a variant of plurality that requires a two-third majority of votes of the participating cardinals.

If there is a Condorcet winner in an election with m candidates, she is (independently of α) the only one to have the maximum Copeland$^\alpha$ score of $m-1$, and thus is the (unique) *Copeland$^\alpha$ winner* for all α.

Table 4.3 Example of an election without a Condorcet winner

	Pairwise comparison					
Preference profile	A?B	A?C	A?D	B?C	B?D	C?D
$A > D > C > B$	A	A	A	C	D	D
$C > D > B > A$	B	C	D	C	D	C
$C > D > B > A$	B	C	D	C	D	C
$B > D > A > C$	B	A	D	B	B	D
$A > C > D > B$	A	A	A	C	D	C
$A > C > B > D$	A	A	A	C	B	C
Winner of the comparison:	?	A	?	C	D	C

Table 4.3 gives an example of an election with pairwise comparisons in which no Condorcet winner exists and the number of voters is even. Here, it may happen that when two candidates, such as A and B, are compared, none of them wins the head-to-head contest (indicated by a question mark in the corresponding columns of the table). Evaluating this election for an arbitrary rational value $\alpha \in [0,1]$, we obtain the following points of the candidates:

$$C^\alpha Score(A) = 1 + 2\alpha,$$
$$C^\alpha Score(B) = \alpha,$$
$$C^\alpha Score(C) = 2,$$
$$C^\alpha Score(D) = 1 + \alpha.$$

Thus, the Copeland^0 winner for the preference profile from Table 4.3 is C with 2 points, whereas a Copeland^1 election is won by A with 3 points, and A and C both are Copeland$^{1/2}$ winners, each with 2 points. The majority graph of this election is depicted in Figure 4.4. As one can see, A is not a Condorcet winner here, but is a weak Condorcet winner, and the only one at that.

4.1.2.3 Dodgson Voting

Another voting system whose winners are determined by pairwise comparison and that respects the Condorcet winner, yet always has (at least) one winner, has been attributed to Dodgson [349], who (as the author of chil-

4.1 Some Basic Voting Systems

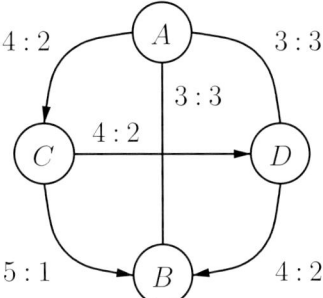

Fig. 4.4 Majority graph for the election in Table 4.3

dren's books such as *"Alice's Adventures in Wonderland"* and *"Through the Looking-Glass"*) may be better known by his pen name, Lewis Carroll.[5]

Here again, the winner should be as close as possible to become a Condorcet winner. To this end, this voting system considers swapping adjacent candidates in the votes, which again are linear rankings. The *Dodgson score* of any given candidate c (denoted by $DScore(c)$) is the minimum number of such swaps in the votes required to make c a Condorcet winner. In *Dodgson's voting system*, all candidates with the lowest Dodgson score win.

If there exists a Condorcet winner, no swaps are needed at all. He thus is the only one having a Dodgson score of 0 and so is the Dodgson winner. In the example from Table 4.3, we have $DScore(A) = 2$. Note that A defeats B and D in their head-to-head contests by two swaps in, say, the second vote:

$$C > D > B \overset{\frown}{>} A \quad \rightsquigarrow \quad C > D \overset{\frown}{>} A > B \quad \rightsquigarrow \quad C > A > D > B$$

and then is the Condorcet winner. This shows that $DScore(A) \leq 2$. However, we also have $DScore(A) \geq 2$ and thus equality, since with only one swap it is not possible for A to become a Condorcet winner. C's Dodgson score is 2 as well. To become a Condorcet winner, C must win the pairwise comparison against A and thus must move up in two more votes to be ahead of A (because if C were only in one more vote ahead of A, they would tie with $3:3$ and there would be no strict majority supporting C). For example, C can be turned into a Condorcet winner by one swap in each of the last two votes:

---

[5] Since Dodgson's system, as we will see in Section 4.2, has a number of undesirable properties, it is only fair to add that it has never been proposed as a voting system by Dodgson himself. The manuscript by Dodgson [349] is headed *"not yet published,"* and Fishburn [454] states on page 474: *"Since Dodgson's function has serious defects, it may be a bit unfair to label [Dodgson's rule] with his name in view of the fact that the idea of counting inversions was cautiously proposed as a part of a more complex procedure."* Tideman [914] adds: *"Dodgson did not actually propose the rule that has been given his name. Rather, he used it implicitly to criticize other rules."* Brandt [208] discusses these historic remarks and investigates the properties of Dodgson's system in more detail.

$$A \overset{\frown}{>} C > D > B \rightsquigarrow C > A > D > B,$$
$$A \overset{\frown}{>} C > B > D \rightsquigarrow C > A > B > D.$$

The Dodgson scores of B and D, however, are greater than 2. Do you see which swaps in the original election from Table 4.3 would be needed to turn either of B and D into a Condorcet winner? Can you show that the numbers of these swaps are both minimal, i.e., can you determine the Dodgson scores of B and D exactly?

This shows that A and C are the Dodgson winners of this election. Obviously, determining the winners in this system can be rather difficult, due to the many possibilities of swaps of adjacent candidates in the voters' preference lists, and this first impression is not deceiving, as we will see in Section 4.3.1 (see also the paper by Hemaspaandra, Hemaspaandra, and Rothe [537]).

Also the analogue of Dodgson's system that is based on the notion of *weak* Condorcet winner has been studied in the literature (see, e.g., the papers by McCabe-Dansted, Pritchard, and Slinko [695] and Brandt et al. [210]).

4.1.2.4 Simpson Voting (Maximin)

Maximin voting has been proposed by Simpson [867] and works as follows. For any two distinct candidates, c and d, in a given election, let $N(c,d)$ be the number of voters who prefer c to d. These numbers can be read off directly from the edge labelings $i:j$ of the majority graph. For example, in the election from Table 4.3 whose majority graph is depicted in Figure 4.4, we have:

$$N(A,B) = N(B,A) = N(A,D) = N(D,A) = 3, \quad N(A,C) = 4, \quad N(C,A) = 2,$$

and so on. The *Simpson score* of a candidate c is defined as

$$SScore(c) = \min_{d \neq c} N(c,d).$$

Simpson winners of the election are the candidates with maximum Simpson score. Since the maximum is taken over the minima of all values $N(c,d)$ with $d \neq c$, this score is also called the *maximin score of c*, and that is why this voting system is also called *maximin*. That is, a Simpson winner comes off best against her worst rival.

Whenever there is a Condorcet winner, he is the only one to have the maximum Simpson score, since he defeats even his worst rival in pairwise comparison. Hence, Condorcet winners are always Simpson winners as well.

Table 4.4 shows the Simpson scores of the candidates in the election from Table 4.3. The four middle columns have the values $N(c,d)$ for $c,d \in \{A,B,C,D\}$, $c \neq d$, where the minima of each row are boldfaced. These minima are the candidates' Simpson scores and are collected in the right column,

4.1 Some Basic Voting Systems

Table 4.4 Simpson scores of the candidates in the election in Table 4.3

	A	B	C	D	Simpson score
A	×	3	4	**3**	3
B	3	×	**1**	2	1
C	2	5	×	4	2
D	3	4	**2**	×	2

where the maximum of the Simpson scores is boldfaced as well. Thus, candidate A is the Simpson winner of this election with $SScore(A) = 3$.

4.1.2.5 Young Voting

In the voting system of Young [968], the Young score of each candidate is determined. As in Dodgson voting, all candidates with the lowest Young score win. This score tells us how close a candidate is to becoming a weak Condorcet winner. The analogous notion based on the notion of Condorcet winner has also been investigated in the literature (see, e.g., the paper by Rothe, Spakowski, and Vogel [825]), and we will call the corresponding system *StrongYoung voting*. Unlike in Dodgson's system, Young and StrongYoung voting are not based on swaps of adjacent candidates, but instead on deleting votes. The *Young score* (respectively, the *StrongYoung score*) of a candidate c, denoted by $YScore(c)$ (respectively, by $SYScore(c)$) is the minimum number of votes that must be deleted to make c a Condorcet winner (respectively, a weak Condorcet winner).

Alternatively to the here defined Young and StrongYoung scores, one might as well define the corresponding scores of a candidate c in an election (C,V) to be the size of a largest sublist $V' \subseteq V$ such that c is a (weak) Condorcet winner in (C,V'). Let us denote these scores by $\overline{Y}Score(c)$ and $\overline{SY}Score(c)$. The associated scores are dually related:

$$\overline{Y}Score(c) = \|V\| - YScore(c) \quad \text{and} \quad \overline{SY}Score(c) = \|V\| - SYScore(c).$$

That is, computing $YScore(c)$ or $SYScore(c)$ corresponds to a minimization problem and computing $\overline{Y}Score(c)$ or $\overline{SY}Score(c)$ corresponds to a maximization problem. For determining the Young or StrongYoung winners, it does not matter which of the two notions one chooses. And also the associated problems have the same complexity (see Section 4.3.1 for details). However, as Betzler, Guo, and Niedermeier [139] note, from the perspective of parameterized complexity, computing $YScore(c)$ or $SYScore(c)$ may be viewed to be more natural than computing $\overline{Y}Score(c)$ or $\overline{SY}Score(c)$, simply because the parameter "number of votes to be deleted" in the minimization problem can be expected to have smaller values than the parameter "number of votes in a largest sublist V'" in the maximization problem.

Whenever there exists a Condorcet winner, no votes need to be deleted for him to become a StrongYoung winner, of course. That is, a Condorcet winner is the only one to have the minimum StrongYoung score of 0. StrongYoung voting thus respects the Condorcet winner, just as the systems of Copeland, Llull, Dodgson, and Simpson do. Further, every weak Condorcet winner has the minimum Young score of 0 and therefore is a Young winner.

In the election from Table 4.3 (see Figure 4.4 for the corresponding majority graph), candidate A already is the one and only weak Condorcet winner and so has a Young score of 0 and is the Young winner as well. Both C and D can be turned into weak Condorcet winners by two deletions of votes for each (for example, by deleting $A > D > C > B$ and $B > D > A > C$ for C to draw level with A while still defeating B and D, and by deleting $A > C > D > B$ and $A > C > B > D$ for D to draw level with C while still defeating B and now also defeating A), and fewer deletions are not enough for them. Thus, $YScore(C) = YScore(D) = 2$. On the other hand, for B to become a weak Condorcet winner, four votes need to be deleted (e.g., $A > D > C > B$, $C > D > B > A$, $C > D > B > A$, and $A > C > D > B$), and no three votes would be enough. Therefore, $YScore(B) = 4$.

Being tied with B and D, A is not a Condorcet winner, but can be made one by deleting, e.g., the second vote, $C > D > B > A$, and at least one deletion is necessary. Thus, $SYScore(A) = 1$. For all other candidates, however, at least two votes need to be deleted to turn them into Condorcet winners. It follows that A is also the one and only StrongYoung winner of this election.

Since for an election (C,V) there are $2^{\|V\|}$ sublists of the votes in V in total that might be deleted, and since one would have to check all these possibilities in the worst case in order to determine the Young or StrongYoung scores of a candidate in C, winner determination is a hard problem for Young and StrongYoung elections as well. For more details about the complexity of these problems, see Section 4.3.1 and the work of Rothe, Spakowski, and Vogel [825], Hemaspaandra, Hemaspaandra, and Rothe [542], and Brandt et al. [210].

4.1.2.6 Kemeny Voting

This voting system has been introduced by Kemeny [602] and has been further specified by Levenglick [653]. The original definition of this voting system is not restricted to strict linear orders, but explicitly allows voters to be indifferent about the candidates. A preference of the form $a > b = c > d$, for example, means that a is the favorite, b and c are equally liked, both less than a and more than d, who is the least preferred candidate. Again, a Condorcet winner, if there exists one, is also a Kemeny winner.

Kemeny winners are determined in several steps. Let (C,V) be an election, where indifferences in the preferences are allowed. For any two potential preference lists P and Q over C (which are not necessarily contained in V),

4.1 Some Basic Voting Systems

define the distance functions $d_{P,Q}$ that give the distance between any two candidates $c, d \in C$ as follows:

$$d_{P,Q}(c,d) = \begin{cases} 0 & \text{if } P \text{ and } Q \text{ agree regarding the order of } c \text{ and } d \\ 2 & \text{if } P \text{ and } Q \text{ strictly disagree regarding the order of } c \text{ and } d \\ 1 & \text{otherwise.} \end{cases}$$

Thus, the distance between c and d with respect to the preference lists P and Q is defined by this function to be:

- 0 if either of $c > d$, $d > c$, and $c = d$ holds true in both preference lists;
- 2 if $c > d$ holds true in one preference list and $d > c$ in the other;
- 1 if we have $c = d$ in one preference list and either $c > d$ or $d > c$ in the other.

By means of the functions $d_{P,Q}$, the distance of any two preference lists P and Q can now be determined as follows:

$$dist(P,Q) = \sum_{\{c,d\} \subseteq C} d_{P,Q}(c,d),$$

i.e., $dist(P,Q)$ is the sum of the values $d_{P,Q}(c,d)$ over all unordered pairs $\{c,d\} \subseteq C$. In the next step, the *Kemeny score* of a preference list P over C is determined (note that this is *not* the Kemeny score of a candidate from C):

$$KScore_{(C,V)}(P) = \sum_{Q \in V} dist(P,Q).$$

Kemeny winners of the given election are then the candidates on top of the preference lists having the smallest Kemeny score. Each of these preference lists is said to be a *Kemeny consensus*. Determining the winners of Kemeny elections is fairly complicated, and again this first impression is not deceiving (see Section 4.3.1 and the paper by Hemaspaandra, Spakowski, and Vogel [550]).

Conitzer, Davenport, and Kalagnanam [302] define the Kemeny rule slightly differently than above; to avoid confusion, we will refer to their definition as the *CDK-Kemeny rule*. Let (C,V) be an election with only strict rankings allowed. For any two candidates $c, d \in C$, any ranking R over C, and any vote P in V, let $\delta_{P,R}(c,d) = 1$ if R and V agree on the relative ranking of c and d (either $c > d$ in both, or $d > c$ in both), and let $\delta_{P,R}(c,d) = 0$ if R and V disagree on the relative ranking of c and d. The total number of pairwise agreements, $\sum_{c,d \in C} \delta_{P,R}(c,d)$, then describes the agreement of R with P, and a *CDK-Kemeny ranking* R maximizes the sum of the agreements with V, $\sum_{P \in V} \sum_{c,d \in C} \delta_{P,R}(c,d)$. Conitzer, Davenport, and Kalagnanam [302] reinterpret this maximization problem of finding a CDK-Kemeny ranking as a minimization problem on edge-weighted, directed graphs. They construct a graph similar to the majority graph from a given election as follows. The

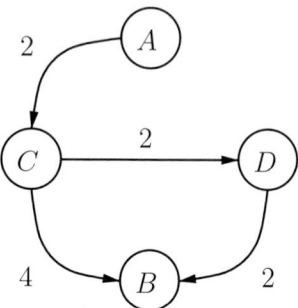

Fig. 4.5 Kemeny election from Table 4.3 as a weighted, directed graph

vertices again correspond to the candidates, and there is a directed edge from vertex c to vertex d if and only if c defeats d in their head-to-head contest. The weight of such an edge is given by the number of votes preferring c to d minus the number of votes preferring d to c, i.e., $N(c,d) - N(d,c)$. If there is a tie between any two candidates, the corresponding edge is omitted in this graph. Figure 4.5 shows this graph for the election from Table 4.3.

Suppose there is an edge from c to d in this graph. If a ranking R agrees with this edge by preferring c to d, there are $N(c,d)$ pairwise agreements, but if R disagrees with this edge by preferring d to c, there are only $N(d,c)$ (i.e., $N(c,d) - N(d,c)$ fewer) pairwise agreements. That is, by disagreeing with the orientation of an edge, R loses exactly as many pairwise agreements as indicated by the weight of this edge. Therefore, finding a CDK-Kemeny ranking boils down to the problem of finding a ranking that minimizes the total weight of the edges it disagrees with. In our example from Table 4.3 and the corresponding graph in Figure 4.5, $A > C > D > B$ will do and is therefore a CDK-Kemeny ranking.

4.1.2.7 Ranked Pairs, Schwartz Voting, Schulze's Rule, etc.

There are many other voting systems that are based on pairwise comparisons and respect the Condorcet winner (see the paper by Fishburn [454] for an early overview). We mention the following three important examples:

- *Ranked pairs* is a rule proposed by Tideman [914] in 1987: In each step, consider a new pair (c,d) of candidates with the highest $N(c,d)$ value (using some tie-breaking mechanism in the case of ties), and fix the order $c > d$ unless this contradicts already previously fixed orderings of candidate pairs (so as to satisfy transitivity). Choose a candidate at the top of the resulting ranking (again, possibly making use of the tie-breaking mechanism chosen).
- Schwartz [844, 845] proposed a voting system (also known as *top cycle*) where the winner set consists of the maximal elements of the asymmetric

part of the transitive closure of the majority relation (i.e., of the *Condorcet top cycle* of the election).
- In *Schulze's rule* [843] (also known, among other names, as *Schwartz sequential dropping*), we first construct the *weighted majority graph*, call it $G_{(C,V)}$, from a given election (C,V), where the candidates in C correspond to the vertices in $G_{(C,V)}$, and there is an edge (c,d) of weight $N(c,d) - N(d,c)$ between any two distinct vertices $c,d \in C$. Define the *strength of a path from c to d in* $G_{(C,V)}$ to be the smallest weight of any edge on this path. For each pair (c,d) of candidates, let $P(c,d)$ denote the strength of a *strongest* path from c to d (i.e., of a path with the greatest minimum edge weight among all paths from c to d). Every candidate $c \in C$ satisfying $P(c,d) \geq P(d,c)$ for each $d \in C \smallsetminus \{c\}$ is a *Schulze winner* of election (C,V). Observe that a candidate $c \in C$ is the unique Schulze winner of (C,V) exactly if $P(c,d) > P(d,c)$ for each $d \in C \smallsetminus \{c\}$.

Regarding Schwartz voting, let us just mention that a Condorcet winner (if there is one) is the one and only Schwartz winner, and a unique Schwartz winner is at least a weak Condorcet winner.

And regarding Schulze's rule, we mention that determining the winners in Schulze's system is a bit complicated, as it involves nontrivial—though still efficient—graph algorithms. However, the users of this system (currently including political parties such as the Pirate Parties of Sweden, Germany, Australia, and many other countries as well as numerous other organizations such as *The Debian Project*, the television channel MTV, etc.) will be partially compensated for this slight disadvantage by the large number of beneficial properties of this voting system (see Table 4.10 on page 286 in Section 4.2.8).

4.1.3 Approval Voting and Range Voting

We now turn to a different type of voting system, approval and (normalized) range voting, which do not ask the voters to rank the candidates but only to either approve or disapprove of them, possibly expressing different intensities of approval.

4.1.3.1 Approval Voting

The previously described voting systems expect each voter to provide a complete preference list over the candidates, i.e., a linear order. In *approval voting*, however, which has been introduced by Brams and Fishburn [191], each voter approves of some candidates and disapproves of the others. A vote thus consists of an unordered (possibly empty, possibly complete) list of candidates approved by this voter.

Approval votes are commonly represented by so-called *approval vectors*: For a universally fixed order of the candidates in $C = \{c_1, c_2, \ldots, c_m\}$, this is a vector in $\{0,1\}^m$, which for $1 \leq i \leq m$ has a 1 in the ith position if this voter approves of candidate c_i, and a 0 otherwise. Approval vectors are not to be mistaken for the scoring vectors introduced in Section 4.1.1. For example, $(1,0,1)$ would be a valid approval vector for three candidates, but would not be a valid scoring vector because its components are not nonincreasing: 1, the last component, is greater than 0, the second-to-last component.

Every voter may approve of any number of candidates. Every candidate scores one point for each approval. The *approval score* of a candidate c (denoted by $AScore(c)$) is the sum of points c scores in total. Whoever has the highest approval score is an *approval winner*.

Suppose the decision from our introductory example (as to whether Anna, Belle, and Chris go on a bicycle tour, go for a swim, or play miniature golf) is to be made by approval voting. Then, instead of using preference lists as in Figure 4.1, they need to either approve or disapprove of these activities. Assuming that all three alternatives are ordered alphabetically—bicycle tour, miniature golf, swim—and Anna, Belle, and Chris have the approval vectors $(1,1,0)$, $(0,1,0)$, and $(1,0,0)$, then the bicycle tour and miniature golf are the approval winners with two points each, whereas swimming comes away empty-handed.

Recall the scoring protocol k-approval, which has been introduced in Section 4.1.1. One difference between k-approval and approval voting is that the voters are required to cast linear orders in the former, yet approval vectors in the latter system. Another difference is that the number of approved candidates, which in k-approval is the same for each voter (namely, k), may be different for the voters in approval voting.

4.1.3.2 Range Voting and Normalized Range Voting

Range voting (or, RV) is a natural way to generalize approval voting by allowing the voters to put more weight than only weight 0 (for disapproval) or weight 1 (for approval) on some candidates. That is, range voting works just as approval voting, except with k-range voting vectors chosen from $\{0, 1, \ldots, k\}^m$.

Normalized range voting (or, NRV) is a variant of range voting that alters the votes so that the potential impact of each voter is maximized (see, e.g., the paper by Menton [703]). Formally, when the voters have cast their k-range voting vectors, each candidate's RV score from each vote is normalized to be a rational number in the range $[0, k]$ as follows. Let $v_a \in \{0, 1, \ldots, k\}$ be the RV score some candidate a gets from a vote v and let, respectively, $v_{\min}$ and $v_{\max}$ be v's minimum and maximum RV scores among all candidates. In NRV, a scores

$$\frac{k(v_a - v_{\min})}{v_{\max} - v_{\min}}$$

points from v, where votes with $v_{\min} = v_{\max}$ will not be taken into account (which is well justified because these voters are indifferent about all candidates). Finally, a's NRV score is the sum of the points a scores from all votes, and whoever has the greatest NRV score wins under normalized range voting.

4.1.4 Voting Systems Proceeding in Stages

In all previously described voting systems, the winners are determined directly from the votes in one step. However, there are also voting systems that proceed in several stages (or, synonymously, steps or rounds), where in each stage possibly only a subset of the candidates or only certain parts of the voters' preference lists are considered and the next stage depends on the result of this consideration.[6] Some of the most important voting systems proceeding in stages will now be introduced. In particular, plurality with run-off and single transferable vote are prominent voting rules that in each round eliminate the "weakest" candidates until only the winner remains.[7] Bucklin voting, on the other hand, is a voting system that, in each stage, considers only the candidates ranked up to a certain position in the votes until it finds a winner, i.e., a candidate supported by a majority of votes up to this position.

4.1.4.1 Plurality with Run-Off and Veto with Run-Off

The winners of plurality with run-off elections are determined in two rounds. After the voters have cast complete preference lists over all candidates, it is checked in the first round how often each candidate is in the top position. The two candidates with the most top positions take part in the run-off, which comprises the second round.[8] In the run-off, all candidates except these two are canceled in the votes. Whoever now has the most top positions is the plurality with run-off winner. If a tie occurs in any of the two rounds, it will

---

[6] With respect to the Kemeny system we mentioned that winners are determined "in several steps"; this merely means that we were going to explain the rather complicated procedure of determining Kemeny winners in a more comprehensible way. It does not mean, however, that Kemeny itself would proceed in several stages. A similar comment applies to other earlier described voting systems such as Dodgson and Schulze voting.

[7] If (C, V) is an election and $C' \subseteq C$ is a subset of the candidates, we will write (C', V) to denote the election with candidates C' and the votes in V restricted to C'.

[8] A run-off is not necessary if some candidate has an absolute majority already in the first round.

be broken by a tie-breaking rule, making sure that only two candidates go to the run-off and only one of them wins it.⁹

If the winner of the election from Table 4.3 on page 242 is to be determined by plurality with run-off, then candidate A receives three points in the first round, C gets two points, B only one point, and D zero points. Hence, A faces C in the run-off, and the two other candidates, B and D, are deleted. In the run-off, the votes of the original election are restricted to $\{A,C\}$ (see Footnote 7):

$$A > C$$
$$C > A$$
$$C > A$$
$$A > C$$
$$A > C$$
$$A > C$$

and since C wins two and loses four head-to-head contests against A, A is the plurality with run-off winner.

A variant of this voting system is used for electing the president of France. The crucial differences are that the voters

- cast their votes not as complete preference lists of all candidates, but vote only for their most preferred candidate, and
- cast their votes anew whenever no candidate has received an absolute majority in the first round.

The voting system *veto with run-off* is defined analogously.

4.1.4.2 Single Transferable Vote (STV)

This voting system (a.k.a. *instant-runoff voting* and *alternative vote*) again requires complete linear orderings of the candidates as votes. The number of rounds is not fixed beforehand, though, but of course there are at most as many rounds as there are candidates. In each round, every candidate in a vote's top position receives one point (just as in plurality). A candidate who scores a point from more than half of the voters wins immediately, and no further rounds are performed. If there is no such candidate in the current round, however, the candidate with the least score is deleted from all votes. Whenever a candidate in top position is deleted from some vote, this point is transferred to the candidate in the second position, who now moves up to the top position. This procedure is repeated until a winner emerges. This voting system, too, makes use of a tie-breaking rule in case of ties, to make sure that

---

⁹ A tie-breaking rule might work alphabetically, for instance. Of course, this rule is rather unfair, since it favors, e.g., a candidate named Aaron and is disadvantageous for, e.g., Züleyha. It would be fairer to choose the tie-breaking winner by lot, so everyone has the same chance of winning.

4.1 Some Basic Voting Systems

at most one candidate is deleted in every round and that there is only one winner in the end.

For the election from Table 4.3 on page 242, we have the following rounds:

1. In the first round, candidate A receives three points, C two points, B one point, and D zero points. Therefore, D is deleted; for example, the fourth vote $(B > D > A > C)$ thus becomes $B > A > C$.
2. Since D was never in the top position, the other candidates' points remain unchanged, and now candidate B is deleted in the second round. In the fourth vote $(B > A > C)$, candidate A moves up to the top position $(A > C)$ and gains one more point.
3. With four points now, A has more than half of the votes in the third round and thus wins the election.

This voting systems (or variants thereof) is applied, for example, in elections in Australia and New Zealand.

4.1.4.3 The Cup Protocol

The winner of the *UEFA Champions League* is not elected, of course, but is determined by the results of soccer matches. The procedure used, however, is somewhat akin to the cup protocol: Whoever wins the cup in the end had to prevail over every opponent encountered in the knockout phase comprising four knockout rounds: the round of sixteen, the quarter-finals, the semi-finals, and finally the final.

In the *cup protocol*, which is also sometimes called *voting tree*, the structure of an election is given by a balanced binary tree whose leaves are labeled with the candidates. Every inner vertex of a *binary tree* has two children, and it is *balanced* if the distances between each leaf and the root differ by at most one. Figure 4.6 shows an example for four candidates. If the number of candidates were not a power of two, the leaf level of the tree would be incomplete, but the tree would still be balanced.

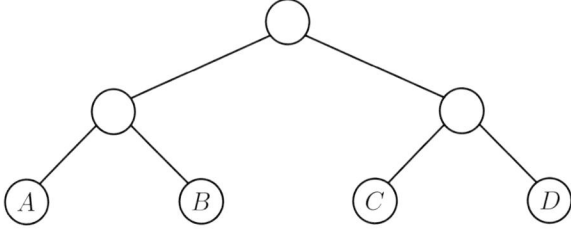

Fig. 4.6 A voting tree for the cup protocol

The labeling of the leaves in the tree from Figure 4.6 corresponds to the draws of the fixtures in the semi-finals of a knockout round, and is called the

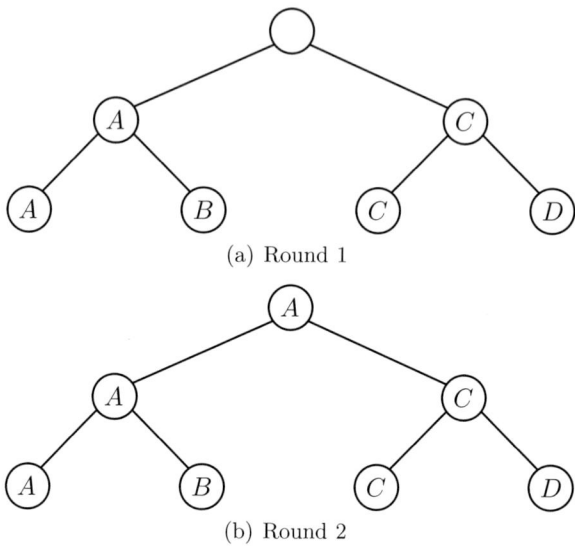

Fig. 4.7 Cup protocol for the election from Table 4.3 using the tree from Figure 4.6

schedule of the election. Working our way from the leaf level up to the root, in each round we decide the head-to-head contests (see also Section 4.1.2) for each current pair of siblings in the tree (i.e., for those pairs of candidates that the two children of one and the same inner vertices are labeled with), except that the winner of such a comparison is not determined by the result of a soccer match here, but by the voters.

The votes are again complete preference lists. The winner of a pairwise comparison—i.e., the candidate who is preferred by a majority of voters—moves on to the next round, which is indicated by copying the label from the child to the corresponding parental vertex. If there is a tie in a pairwise comparison, we again apply a tie-breaking rule. Whenever there is an odd number of candidates initially, one candidate is not matched by an opponent for the pairwise comparison in the first round and thus is given a bye to move on to the next round unchallenged. In this manner one proceeds round by round from the leaf level toward the root. The winner of the election is the candidate the root is labeled with eventually.

For the election from Table 4.3 and applying the (somewhat arbitrary) tie-breaking rule by which the alphabetically lesser candidate wins a tie, we obtain the course of election shown in Figure 4.7. In the first round (see Figure 4.7(a)), A wins due to the tie $(3:3)$ by the tie-breaking rule against B, while C prevails with $4:2$ votes against D. A and C move on to the second and final round (see Figure 4.7(b)), which A wins by $4:2$ votes against C.

Of course, the outcome of an election with the cup protocol depends not only on the voters' preference lists, but also on the specified tie-breaking rule

4.1 Some Basic Voting Systems

and on the schedule (note that the fixtures on the leaf level influence the pairings in later rounds as well). The tie-breaking rule we agreed upon above makes candidate A as strong as a Condorcet winner in the election from Table 4.3, since A defeats C by a majority of votes and B and D by means of the tie-breaking rule. Obviously, the cup protocol respects the Condorcet winner, because a Condorcet winner (if there exists one) wins every pairwise comparison and cannot be stopped on its marching-through from the leaf to the root. Therefore, we now apply a different tie-breaking rule: A tie is won by the alphabetically greater candidate. According to this rule, A is defeated both by B and by D in the election from Table 4.3. With this tie-breaking rule we obtain the course of election shown in Figure 4.8 for two distinct schedules: A versus B and C versus D makes C the winner in Figure 4.8(a), and A versus C and B versus D makes D the winner in Figure 4.8(b).

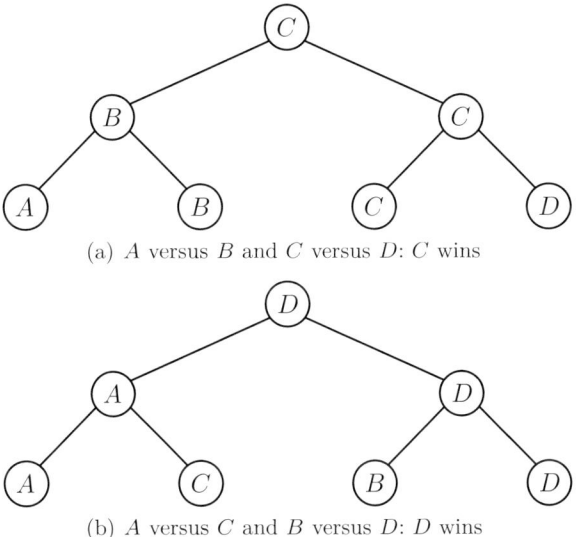

Fig. 4.8 Distinct schedules in the cup protocol for the election from Table 4.3

As C wins in the voting tree from Figure 4.8(a) and D wins in that from Figure 4.8(b), the chosen schedule obviously plays an important role. Can you find a schedule (i.e., a labeling of the leaves with candidates that gives the fixtures of candidate pairs in the first round) such that either A or B wins, assuming we use the same tie-breaking rule? If so, which? If not, why not?

4.1.4.4 Bucklin Voting

This voting system is named after the US-American James W. Bucklin from Grand Junction, Colorado. Since it was used there (between 1909 and 1922) for the first time in political elections, it is also known as the *Grand Junction system*.

For an election (C,V), let $maj(V) = \lfloor \|V\|/2 \rfloor + 1$ denote the threshold value required for a majority.[10] Again, the votes in V are complete preference lists. To determine the winners in the Bucklin system, we proceed in stages (or, levels), where the number of stages varies, but can be at most $m = \|C\|$. In stage $i \leq m$, we consider only the first i positions in the votes:

1. In the first stage, we check if there is a candidate who is in the first position of a majority (i.e., of at least $maj(V)$) of the votes:

   a. If so, this candidate is the (unique) Bucklin winner, and we are done.
   b. Otherwise, go to the next stage.

2. In the second stage, we check if there are candidates who reach a majority in the first two positions of the votes:

   a. If so, those among these candidates having the most votes in the first two positions are the Bucklin winners, and we are done.
   b. Otherwise, go to the next stage.

3. Etc.

In this way, a *Bucklin score* is assigned to each candidate $c \in C$ (denoted by $BScore(c)$) and the *Bucklin winners* are determined as follows:

- The *Bucklin score of c in stage i* (denoted by $BScore^i(c)$) is the number of votes having c among their top i positions.
- The *Bucklin score of c* is the smallest stage i with $BScore^i(c) \geq maj(V)$, i.e., $BScore(c) = \min\{i \,|\, BScore^i(c) \geq maj(V)\}$.
- Among all candidates with the least Bucklin score, say k, those candidates c having the greatest value of $BScore^k(c)$ are the Bucklin winners of (C,V).

If there exists a Bucklin winner already in the first stage, there can be no other Bucklin winner. Bucklin winners from the second stage on, however, need not be unique. Clearly, in the last stage $m = \|C\|$, every candidate occurs in the first m positions of each voter because there are no more positions. Therefore, each candidate's Bucklin score in stage m equals the number of voters, which implies that every candidate would have a majority in this final possible stage. Thus, there always exists at least one Bucklin winner.

Let us again have a look at the election from Table 4.3 on page 242. Figure 4.9 shows how to determine this election's Bucklin winner. In Figure 4.9(a), which presents the first stage, the candidates in the votes' top

---

[10] For a real number x, let $\lfloor x \rfloor$ denote the greatest integer not exceeding x.

4.1 Some Basic Voting Systems

$$\begin{array}{c}
\mathbf{A} > D > C > B \\
\mathbf{C} > D > B > A \\
\mathbf{C} > D > B > A \\
\mathbf{B} > D > A > C \\
\mathbf{A} > C > D > B \\
\mathbf{A} > C > B > D
\end{array}$$

(a) Stage 1

$$\begin{array}{c}
\mathbf{A} > \mathbf{D} > C > B \\
\mathbf{C} > \mathbf{D} > B > A \\
\mathbf{C} > \mathbf{D} > B > A \\
\mathbf{B} > \mathbf{D} > A > C \\
\mathbf{A} > \mathbf{C} > D > B \\
\mathbf{A} > \mathbf{C} > B > D
\end{array}$$

(b) Stage 2

	A	B	C	D
$BScore^1(\cdot)$	3	1	2	0
$BScore^2(\cdot)$	3	1	**4**	**4**

(c) Bucklin scores in stages 1 and 2

Fig. 4.9 Determining the Bucklin winners for the election from Table 4.3

positions are boldfaced. In this first stage, candidate A has the highest score with $BScore^1(A) = 3$ (see Figure 4.9(c)), but because of $maj(V) = 4$ that is not enough for victory. Therefore, the election enters the second stage, shown in Figure 4.9(b), where the candidates in the first two positions are in boldface. A and B gain no more votes in this stage, but C raises his score by two and D massively catches up by gaining four points. C and D reach the critical majority threshold with $BScore^2(C) = BScore^2(D) = 4 = maj(V)$ in this stage and thus win the election, with $BScore(C) = BScore(D) = 2$, and no further stage is needed, as the winners are definite already. A and B, on the other hand, would have reached the majority threshold only in the third stage, with $BScore^3(A) = BScore^3(B) = 4$, so we have $BScore(A) = BScore(B) = 3$.

4.1.4.5 Nanson and Baldwin Voting

Nanson [724] proposed a voting system that based on the Borda system proceeds in stages. At each stage, the candidates whose Borda score is at most the average Borda score of all candidates still participating at this stage are eliminated unless they all have the same Borda score, in which case they all win.

A number of variations of Nanson's rules have also been proposed. For example, Baldwin [82] suggests to successively eliminate a candidate with the *lowest* Borda score at each stage (among those candidates still participating at this stage, and using a tie-breaking mechanism if needed) until only one candidate remains. Schwartz's variant of Nanson voting at each stage eliminates a candidate whose Borda score is *less* than the average Borda score of all candidates. Relatedly, see also the work of Fishburn [454].

4.1.5 Hybrid Voting Systems

So far, only "pure" voting systems have been introduced, each of which is based on another basic idea to aggregate individual preferences. In this section, we will present some *hybrid voting systems*, which combine such pure voting systems and, often, also inherit their advantages.

4.1.5.1 Black Voting

As we know, a Condorcet winner defeats all other candidates in pairwise comparison. This is a very strong argument to support the view that electing the Condorcet winner is utmost reasonable, as there can be no better choice. However, as we also know (recall Figure 1.3 on page 8), the problem with the Condorcet system is that a Condorcet winner does not always exist.

One can make a good case for the Borda system (see page 237) as well, and here we know that there always is at least one Borda winner.

Which of these two systems, Borda and Condorcet, is the better one? This issue has been the subject of great controversy, carried on fiercely in part between the Chevalier de Borda and the Marquis de Condorcet at the French Academy of Sciences[11] in the 18th century and ever since, and the pros and cons of both systems have been carefully discussed and weighed against each other (see, e.g., the survey by Saari [827]).

Black [156] realized that one can combine the advantages of one of the two systems with those of the other, thus getting rid of the crucial disadvantage of the Condorcet system. The *voting system of Black* proceeds in two stages (i.e., in principle we might have listed it in Section 4.1.4 as well, as a voting system proceeding in stages), where the order of the two stages is important:

1. A Condorcet winner (whenever there exists one) also is the winner of Black's voting system.
2. Otherwise, all Borda winners are the Black winners.

It is clear that Black's system, on the one hand, respects the Condorcet winner, yet on the other hand, it always has at least one winner. Thus it combines the advantages of both systems, Borda and Condorcet, and avoids the most crucial disadvantage of Condorcet voting. Moreover, Black winners

---

[11] The French Academy of Sciences eventually adopted Borda's method (see Section 4.1.1) to elect its members for about two decades, before Napoleon Bonaparte insisted that his own election method be used when he became the president of this academy in 1801.

4.1.5.2 Fallback Voting

Brams and Sanver [202] proposed a hybrid voting system that combines Bucklin with approval voting. That requires, to begin with, that the voters express their preferences both as linear orders and approval or disapproval of the candidates, where only the approved candidates need to be ranked in a linear order. All other candidates need not be ranked, as they do not get any points anyway and therefore are irrelevant for determining the winners. We will represent votes in fallback voting as follows:

1. There is an "approval line" to the left of which there are the approved candidates and to the right of which there are the disapproved candidates, and
2. the approved candidates are ranked in a linear order from left to right.

Figure 4.10 on page 260 shows this representation of votes for the election from Table 4.3 on page 242, supplemented by (arbitrarily chosen) approval lines of the voters. The irrelevant disapproved candidates are shown in gray to the right of the approval line as an unordered set. Alternatively, one might also drop them entirely.

To determine the winners, one first tries—by applying the Bucklin system (though, restricted to the approved candidates only)—to find those candidates having a majority of the votes, again "falling back" stage by stage as long as one does not succeed. As soon as one does succeed, all Bucklin winners are the fallback winners, and the process of winner determination is over. However, unlike in Bucklin voting, it here may happen that no candidate at all reaches a majority of the votes at any stage, due to disapprovals. If so, one applies the approval system in the second phase, i.e., the candidates with the highest approval score win. In the extreme case that every voter refuses to approve of any candidate, no candidate would win during the first phase according to Bucklin's system, but every candidate would be an approval winner (with an approval score of 0) in the second phase and therefore would be a fallback winner as well.

In more formal detail, the *fallback winners* of an election (C,V) are determined as follows:

1. First, try to find candidates having a majority of the votes by applying the Bucklin system:

    a. In the first stage, we check if there is a candidate who is in the first position of a majority (i.e., of at least $maj(V)$) of the votes:

i. If so, this candidate is the (unique) fallback winner, and we are done.
   ii. Otherwise, go to the next stage.
  b. In the second stage, we check if there are candidates who reach a majority in the first two positions of the votes:
   i. If so, those among these candidates having the most votes in the first two positions are the fallback winners, and we are done.
   ii. Otherwise, go to the next stage.
  c. Etc.
2. If we do not succeed in determining one or more fallback winners, those candidates with the highest approval score are the fallback winners.

In this way, a *fallback score* is assigned to each candidate $c \in C$ (denoted by $FScore(c)$) and the *fallback winners* are determined as follows:

- The *fallback score of c in stage i* (denoted by $FScore^i(c)$) is the number of votes having c among their top i positions.
- If there exists an i with $FScore^i(c) \geq maj(V)$ for some candidate c, then the *fallback score of c* is the smallest such i. In this case, among all candidates with the least fallback score, say k, those candidates c having the greatest value of $FScore^k(c)$ are the fallback winners of (C,V).
- Otherwise (i.e., if there is no i with $FScore^i(c) \geq maj(V)$ for any candidate c), we set the *fallback score of c* to ∞, and in that case all candidates with the highest approval score are the fallback winners.

A |	$\{B,C,D\}$
C $>$ **D** $>$ **B** |	$\{A\}$
|	$\{A,B,C,D\}$
B $>$ **D** |	$\{A,C\}$
A |	$\{B,C,D\}$
A $>$ **C** $>$ **B** |	$\{D\}$

(a) Stage 1

A |	$\{B,C,D\}$
C $>$ **D** $>$ **B** |	$\{A\}$
|	$\{A,B,C,D\}$
B $>$ **D** |	$\{A,C\}$
A |	$\{B,C,D\}$
A $>$ **C** $>$ **B** |	$\{D\}$

(b) Stage 2

A |	$\{B,C,D\}$
C $>$ **D** $>$ **B** |	$\{A\}$
|	$\{A,B,C,D\}$
B $>$ **D** |	$\{A,C\}$
A |	$\{B,C,D\}$
A $>$ **C** $>$ **B** |	$\{D\}$

(c) Stage 3

	A B C D
$FScore^1(\cdot)$	3 1 1 0
$FScore^2(\cdot)$	3 1 2 2
$FScore^3(\cdot)$	3 3 2 2
$FScore(\cdot)$	**3 3 2 2**

(d) Fallback scores in stages 1 through 3: Since no one reaches a majority, the winners are determined by approval score

Fig. 4.10 Determining the fallback winners for the election from Table 4.3

4.1 Some Basic Voting Systems

For illustration, we again look at the election from Table 4.3, supplemented by (arbitrarily chosen) approval lines of the voters, where one of the voters approves of no candidate at all. Figures 4.10(a) through 4.10(c) show how to determine the fallback scores of all candidates in stages 1 through 3. Obviously, no candidate reaches a majority of at least $maj(V) = 4$ votes. Thus, the winners, A and B, are determined by approval score (see Figure 4.10(d)). By the way, A and B happen to be exactly the candidates who did not win in the Bucklin election shown in Figure 4.9 on page 257 (which was quite similar, except that there were no disapprovals), while the Bucklin winners C and D do not win here.

The special case of a fallback election where every voter approves of all candidates is nothing other than a Bucklin election, and the fallback and Bucklin winners then coincide. If you think that determining the winners in these two systems were as difficult as, e.g., in the Dodgson, Young, or Kemeny systems, then you are wrong. Although determining the winners of a Bucklin or fallback election indeed is a bit more complicated than, for example, doing so in plurality, Borda, or veto elections, the difference is not crucial. This slight disadvantage may be compensated for by the beneficial properties regarding electoral control that Bucklin and fallback voting possess, as we will see in Section 4.3.4.

4.1.5.3 Sincere-Strategy Preference-Based Approval Voting

Another hybrid system due to Brams and Sanver [201] combines again the approval system with preference-based voting. As in the fallback system, every voter divides the candidate set into two subsets, one consisting of the approved candidates (to the left of the approval line) and the other consisting of the disapproved candidates (to the right of the approval line). In addition, every voter linearly orders the candidates according to his preference. Unlike in fallback voting, however, he also ranks the disapproved candidates.

For a given election (C, V), we denote by $S_v \subseteq C$ the *approval strategy* of each voter $v \in V$, i.e., S_v contains those candidates v approves of. Such a voter's approval strategy, however, has to "match" his preference list. For example, a vote preferring a disapproved candidate to an approved candidate would be inconsistent. One might also call it "insincere." In the voting system *sincere-strategy preference-based approval voting* (or, *SP-AV*), only sincere votes are allowed. They are represented just as a regular preference list, supplemented by an approval line to separate the approved from the disapproved candidates. Further, for the sake of conciseness, we will omit the symbol ">." A vote of the form

$$A \quad D \quad C \mid B$$

thus means that this voter approves of A, D, and C, yet not of B, and that he ranks the candidates by $A > D > C > B$. The approval strategy of this voter is sincere. An insincere approval strategy for the same preference list would, e.g., approve of only A and C, yet not of B and D, which would require three lines separating approved from disapproved candidates. Since we exclude insincere approval strategies, however, it is possible to represent all SP-AV votes with a single approval line. The winner of a given preference profile consisting of such sincere-strategy votes only is determined just as in approval voting: All candidates with highest approval score (denoted by $AScore(\cdot)$) win.

Formally, the notion of sincerity is defined as follows. The approval strategy of a vote given in the above format is said to be *sincere* if the following holds: Whenever a candidate c is approved by some voter, then this voter also approves of all candidates he prefers to c. Thus, there cannot be any "gaps" in a sincere approval strategy with respect to the underlying preference list.

Moreover, Brams and Sanver [201] require that the sincere approval strategies must not be dominated in a game-theoretic sense (see Definition 2.2 in Section 2.1.1 on page 45). Such approval strategies are said to be *admissible*. A voter with an admissible approval strategy must approve of her most preferred candidate and must not approve of her least preferred candidate. In particular, this means that admissible approval strategies have to be nontrivial: For each voter $v \in V$ in an SP-AV election (C, V), we have $\emptyset \neq S_v \neq C$, i.e., no voter is allowed to approve of all candidates, nor to disapprove of them all. For an election in this format, all candidates with the highest approval score win.

For example, the voters' approval strategies given for the fallback election in Figure 4.10 on page 260 would not be allowed for an SP-AV election because one voter disapproves of all candidates. By contrast, the election given in Figure 4.11(a) would be a possible SP-AV election, since it combines the same preference profile with a profile of sincere, admissible approval strategies. The candidates to the right of the voters' approval lines, who thus are all disapproved, are shown in gray. Figure 4.11(b) gives the approval scores of all candidates. Thus, D wins this SP-AV election.

$A\;D\;\mid\;C\;B$
$C\;D\;B\;\mid\;A$
$C\;D\;\mid\;B\;A$
$B\;D\;\mid\;A\;C$
$A\;\mid\;C\;D\;B$
$A\;\mid\;C\;B\;D$

(a) SP-AV election

	A	B	C	D
$AScore(\cdot)$	3	2	2	4

(b) SP-AV winner D

Fig. 4.11 Determining the SP-AV winners for the election from Table 4.3 with sincere, admissible approval strategies

4.1 Some Basic Voting Systems

Brams and Sanver [201] then introduce the notion of *critical profile of approval strategies* in order to analyze the *"SP-AV outcome"*—the set of all potential SP-AV winners for a given preference profile with a matching profile of sincere, admissible approval strategies. It turns out that the SP-AV outcome includes the winners under scoring protocols (such as Borda) and also the winners under STV, Bucklin, and voting systems respecting the Condorcet winner. Without going into detail here, the advantage of this combination of preference profiles with a matching profile of sincere, admissible approval strategies is that it allows an analysis from a game-theoretic perspective. In contrast to the voting systems mentioned above, SP-AV guarantees electing the Condorcet winner as a (strict) Nash equilibrium (see Definition 2.4 in Section 2.1.1) in which the voters use sincere, admissible approval strategies. On the other hand, SP-AV can also elect the Condorcet loser—sometimes in the strict Nash equilibrium as well. For further details, we refer the reader to the work of Brams and Sanver [201].

As we will see in Section 4.3.4, SP-AV has beneficial properties also in other regards: Erdélyi, Nowak, and Rothe [412] show that a slightly modified variant of the voting system SP-AV resists many types of electoral control.

4.1.6 Overview of Some Fundamental Voting Systems

Many central voting systems have been introduced in this section, all being based on quite different ideas. In principle, they can be classified both in terms of their structural organization and in terms of the method by which they assign points to the candidates so as to determine the winners. Regarding structural organization, there are voting systems proceeding

- in one stage or
- in several stages.

Regarding the method used to assign points to the candidates, there are

- voting systems based on the positions of the candidates in the votes,
- voting systems based on pairwise comparisons, and
- approval-based voting systems.

Table 4.5 gives a systematic overview.

The hybrid systems introduced in Section 4.1.5 combine these classification features and therefore occur in several columns of this table. For example, the Black system is based on both Borda and Condorcet voting and thus is based both on the candidates' positions in the votes and on pairwise comparisons. Fallback voting and SP-AV combine position-based systems with approval voting. The hybrid systems Black and fallback proceed in several stages, where the Black system combines two systems proceeding in one stage each and the fallback system combines a one-stage system with a system proceeding in several stages; by contrast, SP-AV is to be viewed as a one-stage

Table 4.5 Overview of some fundamental voting systems

	Voting systems		
	position-based	pairwise comparisons	approval-based
one stage	scoring protocols (plurality, veto, Borda, etc.) SP-AV	Condorcet Copeland$^\alpha$ (Copeland, Llull, etc.) Dodgson Kemeny Tideman (ranked pairs) Simpson (maximin) Schulze Schwartz (top cycle) Young	approval NRV range voting SP-AV
several stages	Baldwin Black Bucklin fallback Nanson plurality with run-off STV	Black cup protocol	fallback

voting system. Note, however, that in some cases—such as ranked pairs, which here is classified as a one-stage system, but might perhaps also viewed to proceed in several stages—the distinction between one and several stages is not quite clear.

4.2 Properties of Voting Systems and Impossibility Theorems

All voting systems presented so far aggregate individual preference lists or approval vectors of the voters to a social consensus so as to determine one (or more) winner(s)—or, for some systems, sometimes no winner at all. Depending on the voting system used, the outcome of an election can be quite diverse; there can even be complementary winner sets for the same election. Which one is the "right" one, i.e., which voting system "best" reflects the preferences and/or approvals of the voters? This question does not have one answer only, but many, and they are partially conflicting because what the "right" social choice is depends on the properties one has in mind and cares most about. In this section, we will formulate a number of criteria that express such desirable properties of voting systems. At the end of this section (see Table 4.10 on page 286), an overview summarizes—at least for some of the properties to be introduced in this section and for some of the voting

4.2 Properties of Voting Systems and Impossibility Theorems

systems from Section 4.1—which criteria are satisfied (or are not satisfied) by which voting systems. In this discussion, we disregard tie-breaking issues.

A voting system simply is a rule for how to determine the winners of an election. Whether this rule is "good" or "democratic" and whether the winners adequately represent the voters' preferences is a different matter. For example, from a purely formal point of view, even a dictatorship is a voting system, although a degenerate one. In a dictatorship, the outcome of an election depends on the dictator's vote only, and not on the votes of any of the other voters. For the overwhelming majority of voters, this is a really bad voting system, since their votes do not have any influence whatsoever on the outcome of the election. Clearly, a "fair" voting system should satisfy certain reasonable criteria (it should not be a dictatorship, for example), to provide an acceptable and comprehensible outcome.

Besides single properties and criteria, we will also consider combinations of desirable properties in this section. In particular, we will present some impossibility results showing that there is no voting system that satisfies certain combinations of criteria simultaneously. Many of the criteria stated here apply only to voting systems in which the votes are complete rankings of the candidates. These criteria have to be appropriately adapted to also apply to approval voting and its variants.

4.2.1 The Condorcet and the Majority Criterion

Two of the most central properties of voting systems are the Condorcet and the majority criterion. We start with the former.

4.2.1.1 The Condorcet Criterion

The notion of Condorcet winner—a candidate that beats every other candidate in their head-to-head contest—has already been introduced in Section 4.1. A voting system satisfies the *Condorcet criterion* if it elects the Condorcet winner whenever one exists. We will also say that such a voting system *respects the Condorcet winner*.

Condorcet voting trivially satisfies this criterion. As mentioned before, the voting systems of Copeland and Llull (as well as the whole family of Copeland$^\alpha$ systems, for each rational $\alpha \in [0,1]$) and the voting systems of Dodgson, Simpson, Young, Kemeny, Black, Baldwin, Nanson, Schulze, Schwartz, ranked pairs, and the cup protocol each satisfy the Condorcet criterion, too. Scoring protocols such as plurality, veto, or Borda, however, do not satisfy it. Consider, for example, the election in Table 4.6. The first column, under #, shows how often the vote given in the second column occurs, so there is a total of 19 voters ranking these three candidates. Since A wins the pairwise comparison with each of B and C (defeating both by $10:9$), A is the Condorcet winner of this election. However, evaluating the same election according to the Borda rule, candidate A is with 20 points only the runner-up behind candidate B with 21 points; thus, the Borda winner of this election is B and not the Condorcet winner A.

Table 4.6 Condorcet winner and loser in Borda and plurality

#	Vote	Pairwise comparison			Borda			Plurality		
		$A?B$	$A?C$	$B?C$	A	B	C	A	B	C
6	$A>B>C$	A	A	B	12	6	0	6	0	0
4	$B>A>C$	B	A	B	4	8	0	0	4	0
2	$B>C>A$	B	C	B	0	4	2	0	2	0
4	$C>A>B$	A	C	C	4	0	8	0	0	4
3	$C>B>A$	B	C	C	0	3	6	0	0	3
Result:		A	A	B	20	21	16	6	6	7

Already in Section 4.1.2 the notion of *Condorcet loser* of an election has been defined—analogously to that of Condorcet winner—as the candidate that loses against all other candidates in their head-to-head contest. Just as Condorcet winners, Condorcet losers may not exist, but whenever they do, they are unique.

The *Borda paradox* refers to a situation where a candidate turns out to be a plurality winner (i.e., a candidate with the most top positions in the votes), even though he is a Condorcet loser at the same time. The example in Table 4.6 shows that candidate C—despite losing his head-to-head contests against both A and B and, therefore, being the Condorcet loser—wins the plurality election with 7 points ahead of A and B. Even though the name of this paradox might be mistaken to point to Borda's voting system, the corresponding effect in fact does never occur there, since a Condorcet loser can never be a Borda winner. Rather, the name of this paradox comes from Borda being the first to discover this defect of the plurality rule (his original example has three candidates and 21 voters), which motivated him to propose his own voting system, Borda count, as an alternative.

It is easy to see that none of plurality with run-off, STV, Bucklin, and fallback voting satisfy the Condorcet criterion. Can you come up with a counterexample for each of these voting systems? For approval voting, the notion of Condorcet winner is actually not defined (but recall the related discussion of SP-AV and Condorcet winners at the end of Section 4.1.5 that is due to Brams and Sanver [201]).

The Condorcet criterion is based on the fact that the pairwise comparisons must be won with a majority. In the *weak Condorcet criterion*, however, for a candidate to win such a comparison it is enough to be supported by at least half of the votes, and in the case of a tie both candidates win their

4.2.1.2 The Majority Criterion

Like the Condorcet criterion, the majority criterion is a condition that concerns the winner determination of a voting system. A voting system satisfies the *majority criterion* if a candidate who is placed on top in more than half of the votes always is a winner of the election. A voting system satisfies the *simple majority criterion* if a candidate who is placed on top in at least half of the votes always wins.

For example, the majority criterion is satisfied by the very simple voting system under which a candidate with a majority of top positions wins (and whenever such a candidate exists, she is unique). As with the Condorcet criterion, however, the problem is that such a winner does not always exist (for example, if there are three candidates each being ranked first in exactly one third of the votes).

Plurality avoids this problem by replacing the winning criterion of *absolute majority* (on which the majority criterion is based) by a *relative majority*. A plurality winner has the most top spots in the votes, and at least one such candidate always exists. Obviously, plurality satisfies the majority criterion, since if there is a candidate with an absolute majority of top spots, she also has—in relation to all other candidates—the *most* top spots and, hence, is the one and only plurality winner.

In Bucklin voting, the problem that no candidate with an absolute majority of top spots in the votes may exist is circumvented differently than in plurality. In contrast to plurality, Bucklin relies on the notion of absolute majority expressed by the threshold

$$maj(V) = \lfloor \|V\|/2 \rfloor + 1$$

for an election (C,V). If there is no Bucklin winner on the first level, however, one looks for a candidate with an absolute majority up to the second position, then up to the third position, and so on. On some level up to the $\|C\|$th level, the Bucklin winner(s) must be definite. Obviously, since a candidate with an absolute majority of top spots will immediately be the level 1 Bucklin winner, the majority criterion is also satisfied by Bucklin voting.

By contrast, Borda voting is an example for a voting system that does not satisfy the majority criterion. Suppose there are four candidates, A, B, C, and D, and three voters, two of which have cast the vote $A > B > C > D$ and the remaining one has cast the vote $B > C > D > A$. Since A is placed on top in more than half of the votes, she has an absolute majority. In Borda voting, however, B scores $2 \cdot 2 + 3 = 7$ points, A only $2 \cdot 3 = 6$ points, C only $2 \cdot 1 + 2 = 4$ points, and D as few as one point. Thus, B and not A is the Borda winner.

4.2.2 Nondictatorship, Pareto Consistency, and Consistency

Nondictatorship and Pareto consistency are properties that every natural voting system should fulfill. Indeed, all the voting systems defined in Section 4.1 starting on page 234 are nondictatorial and Pareto-consistent. However, we will see (in Theorem 4.1 on page 273) that when these two properties are satisfied at the same time by a preference-based voting system for at least three candidates, then it cannot be independent of irrelevant alternatives, a property to be defined in Section 4.2.3 starting on page 272. This impossibility result is due to Arrow [29] and is one of the most celebrated results in social choice theory.

On the other hand, consistency is a bit less natural a property and is satisfied by far fewer of the commonly used voting systems.

4.2.2.1 Nondictatorship

This property has already been mentioned briefly at the beginning of this section. Intuitively, it says that the outcome of an election must not depend solely on a single voter. Formally, a voting system (in the sense of a social welfare function) f is *dictatorial* if there is a voter v (called the *dictator*) such that for every election $P = (C, V)$ with v's vote being contained in V, the outcome $f(P)$ coincides with v's preference list, regardless of the remaining votes in V. (This notion can directly be transferred to social choice correspondences and functions, respectively.)

The property of *nondictatorship* requires from a voting system that there be no dictator. This is a basic democratic requirement, which is obviously satisfied by all the voting systems presented here.

4.2.2.2 Pareto Consistency

The property of Pareto consistency is also an intuitively desirable property. It is closely related to the game-theoretic notion of Pareto optimality from Definition 2.3 on page 49, but it is defined in a different context: Instead of gain functions of players, we are dealing with the voters' preferences here. A voting system (in the sense of a social welfare function) f satisfies the property

4.2 Properties of Voting Systems and Impossibility Theorems

of *Pareto consistency* if whenever in a preference profile P all voters prefer a candidate c to a candidate d, c is also preferred to d in the outcome $f(P)$.

Like nondictatorship, this requirement is an utterly natural requirement. It would be close to impossible to convince an electorate that, say, Philipp should win the election against Angela, in spite of all voters considering her better than him. The property of Pareto consistency is also satisfied by all the voting systems presented here.

4.2.2.3 Consistency

In 1990, the first parliamentary elections for the Bundestag after the German reunification took place. It was not only of high interest who would win the election as a whole, but also how the voters in East Germany and West Germany would vote. In this election, it turned out that the partial East and West German results were essentially matching with the all-German result: They were *consistent* in the sense that the same party (or, as we here say, the same alternative) who would have won alone in East and West Germany, respectively, was also the nationwide winner and thus in charge of appointing the chancellor of reunified Germany.

A voting system is said to be *consistent* if whenever some candidate wins in all the subelections that result from partitioning the voters into two (or more) disjoint groups, then this candidate also wins the election as a whole, where we assume that all voters cast the same ballot there as in the subelections. Consistency has been called *separability* by Smith [879] and *convexity* by Woodall [948]. Violation of consistency is also known as *Simpson's paradox*.

This intuitively highly attractive and reasonable property is satisfied by a number of voting systems; for instance, by the plurality rule. Consider a plurality election (C, V) and divide V into two disjoint groups, V_1 and V_2 with $V_1 \cap V_2 = \emptyset$ and $V_1 \cup V_2 = V$. Suppose a candidate $c \in C$ wins both subelections, (C, V_1) and (C, V_2). Then, in both subelections, c has at least as many points (i.e., top positions in the ballots) as every other candidate in C. Since the scores of the candidates in election (C, V) are obtained by adding their scores from both subelections, c has at least as many points as every other candidate also in this combined election, and thus wins it.

But yet, by far not all voting systems have this nice property; quite a few are inconsistent. Consider the election in Table 4.7 with the candidates A, B, and C, and the list V containing 25 votes. The first column of the table shows a partition of the voters into two disjoint groups, $V = V_1 \cup V_2$ with $\|V_1\| = 16$ and $\|V_2\| = 9$. There are five types of votes in V, and the table's second column gives the number of votes of each type, and the third column gives the corresponding vote itself.

Table 4.7 Counterexample for the consistency of some voting systems

Partition of V	#	Vote
V_1	7	$A > B > C$
	5	$B > C > A$
	4	$C > A > B$
V_2	5	$A > C > B$
	4	$C > B > A$

For which voting systems does this election with the given partition show that consistency is not satisfied? The pairwise comparisons and the Borda, Copeland,[12] and Young scores for

- the entire election $(\{A,B,C\},V)$ from Table 4.7 are shown in Figure 4.12(a),
- the first subelection $(\{A,B,C\},V_1)$ in Figure 4.12(b), and
- the second subelection $(\{A,B,C\},V_2)$ in Figure 4.12(c).

The boldfaced scores indicate the winners of the corresponding election.

The example from Table 4.7 does not show inconsistency of Borda. That is not a surprise, though: No other example whatsoever would show this either, since Borda is consistent, which can be seen similarly to the argument for plurality. As an exercise, can you either show or disprove that all scoring protocols are consistent?

Now consider the pairwise comparisons. For example, the $16:9$ in the row of candidate A and the column of candidate B in Figure 4.12(a) means that 16 voters prefer A to B and only 9 voters prefer B to A, which makes A the winner of this pairwise comparison. Note that C is the Condorcet winner of the election as a whole and that A is the Condorcet winner of the second subelection. Therefore, C and A both have the maximal Copeland score of 2 and the minimal Young score of 0 in the respective elections.[13] Although the first subelection, $(\{A,B,C\},V_1)$, has no Condorcet winner, we see that A is both a Copeland winner and a Young winner of this subelection as well. A thus wins both subelections under these two voting systems, yet C (and not A) is the Copeland winner and also the Young winner of the election as a whole. Hence, neither Copeland nor Young is consistent.

In Black voting (defined on page 258), one first tries to determine the Condorcet winner, and if this fails (because there doesn't exist a Condorcet winner in the given election), one identifies the Borda winners and declares them to be Black winners. For example, there is no Condorcet winner in the

---

[12] Since there are no ties in the pairwise comparisons, the Copeland$^\alpha$ scores in fact are independent of α; therefore, what we observe here for Copeland (i.e., Copeland$^{1/2}$) applies to Copeland$^\alpha$ for all values of α.

[13] Since the Young scores of the other candidates in these elections are irrelevant for our counterexample, we will not give them explicitly. As an exercise: Can you determine them? And what about the StrongYoung scores in these elections?

4.2 Properties of Voting Systems and Impossibility Theorems

Pairwise comparison				Scores		
	A	B	C	Borda	Copeland	Young
A	–	16 : 9	12 : 13	**28**	1	–
B	9 : 16	–	12 : 13	21	0	–
C	13 : 12	13 : 12	–	26	2	0

(a) Election $(\{A, B, C\}, V)$

Pairwise comparison				Scores		
	A	B	C	Borda	Copeland	Young
A	–	11 : 5	7 : 9	**18**	1	2
B	5 : 11	–	12 : 4	17	1	6
C	9 : 7	4 : 12	–	13	1	8

(b) First subelection $(\{A, B, C\}, V_1)$

Pairwise comparison				Scores		
	A	B	C	Borda	Copeland	Young
A	–	5 : 4	5 : 4	10	2	0
B	4 : 5	–	0 : 9	4	0	–
C	4 : 5	9 : 0	–	13	1	–

(c) Second subelection $(\{A, B, C\}, V_2)$

Fig. 4.12 Borda, Copeland, and Young scores in the election from Table 4.7 and both its subelections

first subelection from Table 4.7, $(\{A, B, C\}, V_1)$, and the Borda (hence also the Black) winner is A. Again, A wins both subelections under Black (the first one right away by being the Condorcet winner), but not the election as a whole, which is won by the Condorcet (thus also the Black) winner C. This shows that Black voting, too, is inconsistent.

For Condorcet voting itself the election from Table 4.7 is not appropriate, to show inconsistency, since there is no Condorcet winner in the first subelection. But this property can easily be proven by other counterexamples—can you find one?

The voting systems by Condorcet, Black, Copeland, and Young all satisfy the Condorcet criterion, but are inconsistent. Are there any systems at all that respect the Condorcet winner and nevertheless are consistent?

Yes, but not many! Among "neutral" voting systems (a very natural requirement to be defined in Section 4.2.6.2), Kemeny voting is the *only* system that is consistent and satisfies the Condorcet criterion.

The election from Table 4.7 is also suitable to prove inconsistency of plurality with run-off (defined on page 251) and STV (defined on page 252). Do you see how? By the way, these two systems are identical when restricted to three candidates.

4.2.3 Independence of Irrelevant Alternatives

This property may not be as self-evident as the former properties. Intuitively, independence of irrelevant alternatives means that the preference between any two alternatives in the outcome of an election as a "societal consensus," must depend only on the individual preferences of the voters between these two alternatives. All other alternatives are irrelevant and should not play any role in determining society's preference between these two.

Suppose the voters have cast their votes over the three candidates A, B, and C, and A is ahead of C in the aggregated ranking that is output by the voting system, which satisfies the property of *independence of irrelevant alternatives* (*IIA*). But then, suddenly, candidate D announces to run for office as well, which is why the election has to be repeated with appropriately modified individual preference lists of the voters, i.e., every voter inserts D according to his preference into his vote, leaving everything else unchanged. For the relative ranking of A and C, candidate D thus is irrelevant! When applying the same voting system to the modified preference profile, independence of irrelevant alternatives guarantees that A is still ahead of C in the new aggregated ranking of society. Instead of adding D, one could also make other changes in the original preference lists of the voters (e.g., suppose candidate B has passed away shortly after the election, which is why it has to be repeated without her), as long as the original order between A and C remains the same. Furthermore, for independence of irrelevant alternatives to hold, this must apply not only to A and C, but to any pair of candidates.

The notion of *independence of irrelevant alternatives* goes back to Arrow [29], who defined this property (as in the example above) for social welfare functions that return a full societal preference list over the candidates. Taylor [907, 908] adapted this definition to social choice correspondences that return a set of winners of an election in the following way: If a candidate C is a winner of the election and another candidate D is not, then in a new election where the relative rankings of C and D remain unchanged, it must not be the case that D wins this new election. That means that by modifying the preference profile (but leaving the relative rankings of C and D unchanged), it is possible that C no longer wins (e.g., because another candidate now performs much better than C), yet it is not allowed that D becomes a winner. That C remains a winner cannot be taken for granted even in the presence of independence of irrelevant alternatives, but a victory of D is excluded.

The following small anecdote, attributed to the philosopher Sidney Morgenbesser by Poundstone [786, p. 50], illustrates a violation of this kind of independence of irrelevant alternatives.

Morgenbesser, ordering dessert, is told by the waitress that he can choose between apple pie and blueberry pie. He orders the apple pie. Shortly thereafter, the waitress comes back and says that cherry pie is

4.2 Properties of Voting Systems and Impossibility Theorems

also an option; Morgenbesser says, "In that case I'll have the blueberry pie."

Even if one may be tempted to think, intuitively, that a reasonable voting system ought to satisfy the three properties of nondictatorship, Pareto consistency, and independence of irrelevant alternatives, the following impossibility result by Arrow [29], shows that this is not possible—at least not for preference-based voting systems with more than two candidates.

Theorem 4.1 (Arrow [29]). *If there are at least three candidates, there is no preference-based voting system that satisfies the following three properties at the same time:*

1. *nondictatorship,*
2. *Pareto consistency, and*
3. *independence of irrelevant alternatives.*

Originally, Arrow [29] formulated his impossibility result in a slightly different formal context. The formulation given here goes back to Taylor [907, 908]. Due to this fundamental result, Kenneth Arrow may be seen as one of the founders of modern social choice theory; he was honored by a *Nobel Prize in Economics* in 1972, jointly with John Hicks, for their pioneering contributions to general economic equilibrium theory and welfare theory.

Since all preference-based voting systems introduced in Section 4.1 are nondictatorial and Pareto-consistent, it follows from Theorem 4.1 that they cannot be independent of irrelevant alternatives. Theorem 4.1 does not apply to approval voting, which is not preference-based, and indeed it is easy to see that this system is not only nondictatorial and Pareto-consistent, but also independent of irrelevant alternatives (with these properties appropriately adapted so as to apply to approval voting).

Besides the property of independence of irrelevant alternatives as described here, there are other definitions of similar properties, sometimes under the same name. In addition, there are several weaker forms of this criterion, since some social choice theoreticians think that the definition used here may be too restrictive.

4.2.4 Resoluteness and Citizens' Sovereignty

We now turn to resoluteness and citizens' sovereignty.

4.2.4.1 Resoluteness

This property describes the difference between social choice functions and correspondences. A voting system is *resolute* if it always chooses a single candidate as the winner. A voting system that can return more than one winner or no winner at all, however, is not resolute.

For instance, the cup protocol, plurality with run-off, and STV are resolute due to the built-in tie-breaking rule, whereas, e.g., approval, Borda, Copeland, Black, Dodgson, Simpson, Young, Schulze, and Kemeny are irresolute. By applying a tie-breaking rule, these systems can be made resolute as well (even though this might affect some of their other properties).

4.2.4.2 Citizens' Sovereignty

A voter may expect from a voting system that, based on the voters' preferences only, in principle every candidate can be made a winner of an election. It is a fundamental democratic dictate that the voters, and only the voters, ought to decide on victory and defeat in an election. This dictate is named *citizens' sovereignty*. A voting system f, in the sense of a social choice function (or correspondence), satisfies this property if for every candidate c, there is at least one preference profile P such that $f(P) = c$ (or $c \in f(P)$). This is nothing other than to require *surjectivity* of a social choice function (or correspondence), i.e., to require that f be an onto mapping: For each element c in the codomain of f, there is a corresponding element in the domain of f that is mapped to c. All voting systems considered in Section 4.1 satisfy this property.

4.2.5 Strategy-Proofness and Independence of Clones

The next two properties, strategy-proofness and independence of clones, are particularly interesting and important. We start with the former, which again gives rise to a famous impossibility result (stated here as Theorem 4.2).

4.2.5.1 Strategy-Proofness

Intuitively, a voting system is *strategy-proof* if no voter can benefit from reporting an untruthful vote. If a voting system is not strategy-proof, it is said to be *manipulable*. In practice, this means that for a voter with full knowledge of the other voters' preferences, it is possible to achieve a better election outcome for herself by reporting a vote that does not reflect her sincere preferences over the candidates.

4.2 Properties of Voting Systems and Impossibility Theorems

This kind of manipulation is also called *strategic voting* and can occur, for instance, in Borda voting, as the following example shows.

Example 4.1. Suppose there are four candidates, A, B, C, and D, and three voters. The sincere preference of the manipulator is $D > C > B > A$, but she knows that the other two (nonmanipulative) voters have the preference $A > B > C > D$. Her most favorite candidate D has no chance to win, but at least she can try to prevent her most despised candidate, A, from winning. Specifically, it would be reasonable for her to cast the insincere vote $B > D > C > A$ instead of her sincere vote $D > C > B > A$: With the sincere vote, A would have 6 points and so would be the winner of the election ahead of runner-up B having only 5 points; with the insincere (or, strategic) vote $B > D > C > A$, however, candidate B (having now 7 points) would win the election ahead of A (again, with 6 points). Thus, by reporting an insincere vote the strategic voter is able to influence the outcome of the election to her own advantage.

The following impossibility result concerning strategy-proofness of voting systems has been shown independently by Gibbard [500] and Satterthwaite [835].

Theorem 4.2 (Gibbard [500] and Satterthwaite [835]). *If there are at least three candidates, there is no preference-based voting system that satisfies the following four properties at the same time:*

1. *nondictatorship,*
2. *resoluteness,*
3. *citizens' sovereignty, and*
4. *strategy-proofness.*

Since most of the common voting systems that always have a winner are neither a dictatorship nor exclude any candidate from winning right at the outset, this means that all these voting systems are manipulable. A disillusioning—and, at the same time, a fascinating—result!

The restriction to resolute voting systems is not essential here: Duggan and Schwartz [362] generalize the Gibbard–Satterthwaite theorem from social choice functions to social choice correspondences. Possible ways to circumvent—or at least hinder—manipulation, will be considered in more detail in Section 4.3 when we are concerned with the computational complexity of the manipulation problem.

In an interesting work, Brandt and Geist [220] apply SAT solving techniques to find strategy-proof social choice functions.

4.2.5.2 Independence of Clones

In addition to independence of irrelevant alternatives, another independence criterion is independence of clones. A *clone* of a candidate c is a candidate that is right next to c in every voter's preference list. Tideman [914], who introduces this criterion, very nicely describes how the outcome of an election can be influenced by introducing clones:

> When I was 12 years old, I was nominated for the election of the treasurer of my class. One girl named Michelle was also nominated. Being a treasurer was very attractive for me, so after a quick calculation I also nominated Michelle's best friend, Charlotte. In the following election I got 13 votes, whereas Michelle got 12 and Charlotte 11, hence I became the treasurer.

In this example, Tideman has introduced a clone of his rival Michelle by nominating Charlotte, who as Michelle's best friend is likely to be very similar to her—or, at least, likewise popular among their classmates. He could win the election only because the votes of Michelle's supporters have now been split between two candidates; otherwise, he would have clearly lost to Michelle, who apparently was favored over him by many.

A voting system satisfies *independence of clones* if it is never possible to turn a candidate, who originally does not win the election, into a winner by introducing clones into the election. Since the election for the treasurer in Tideman's class was held under the plurality rule, this example shows that this voting system violates independence of clones. As one can easily see, this equally applies to, e.g., Borda, Simpson, and Bucklin voting.

That also Dodgson voting is not independent of clones was shown by Brandt [208]; his counterexample has three original candidates, one clone, and 12 voters. Examples of voting systems that fulfill the criterion of independence of clones are approval voting (in which a *clone* of a candidate c is a candidate who is approved of by exactly the same voters that approve of c) and Schulze's rule.

4.2.6 Anonymity, Neutrality, and Monotonicity

We now turn to three fundamental properties that should be satisfied by every natural voting system. While the voting systems defined in Section 4.1 starting on page 234 indeed do satisfy the first two properties, anonymity and neutrality, there are natural voting systems that are not monotonic, which may be seen as a paradoxical situation. We again will state an impossibility

4.2 Properties of Voting Systems and Impossibility Theorems

result, which involves a strong variant of monotonicity and nondictatorship, resoluteness, and citizens' sovereignty (see Theorem 4.3 on page 279).

4.2.6.1 Anonymity

This is again a fundamental property: In an election, it should not play a role which voter reports which vote; only the votes themselves should decide on victory and defeat. A voting system satisfies the property of *anonymity* if the same candidates win under every permutation of the votes. An anonymous voting system hence does not consider the order in which the votes come in.

All voting systems presented here are anonymous. It follows immediately from the definition that an anonymous voting system must not have a dictator (except for the case of only one voter—but, in this trivial case, whether or not nondictatorship or anonymity hold is irrelevant anyway).

4.2.6.2 Neutrality

The likewise fundamental property of *neutrality* is satisfied if all candidates are treated equal. This means that if any two candidates swap their positions in every vote, then they also swap their positions in the outcome of the election. All voting systems considered here are neutral.

4.2.6.3 Monotonicity

Everyone knows what the property of monotonicity means for a function $f : \mathbb{R} \to \mathbb{R}$. For example, f is *monotonically increasing* if $f(x) \geq f(y)$ follows from $x \geq y$ for all $x, y \in \mathbb{R}$. And if even $f(x) > f(y)$ for all $x, y \in \mathbb{R}$ with $x > y$, then f is *strictly monotonically increasing*. But, what does monotonicity mean for a voting system?

As mentioned in Section 4.1, voting systems may be seen as, respectively, social choice functions, social choice correspondences, and social welfare functions. Social choice functions, for example, do not map real numbers to real numbers, but instead they map preference profiles to candidates. On $\mathbb{R}$, there is a simple order relation, so one can easily say whether or not one real number is greater than another one. Now, however, we want to compare *preference profiles*—the arguments of, respectively, social choice functions, social choice correspondences, and social welfare functions—with each other, and that with a focus on single candidates and based on the underlying preference relations of the voters. Intuitively, a preference profile is the better for a designated candidate, the higher she is positioned in the voters' individual preference lists. We can also compare candidates—the function values of social choice functions or correspondences—with each other; intuitively, a winner is better off than a loser. For social welfare functions, whose function values are rankings of all candidates, a candidate is the better off, the higher his place in the ranking is.

Obviously, there are many possibilities for how to formalize the above-mentioned intuitive notions. Accordingly, quite a number of different notions of monotonicity are common for voting systems. We focus on the following. A voting system (seen as a social choice function or correspondence) f is *monotonic* if for every preference profile P and for every candidate w, we have the following:

1. If w wins the election (i.e., if $f(P) = w$ or $w \in f(P)$, depending on whether f is a social choice function or correspondence) and
2. if a new profile P' is generated from P by improving the position of w in some of the votes, leaving everything else—i.e., the relative rankings of all other candidates—unchanged,

then w also wins the new election with P' (i.e., $f(P') = w$ or $w \in f(P')$).

In this sense, a voting system violates monotonicity if a winner of an election becomes a loser by improving her position in some of the votes, while everything else remains unchanged. If a voting system violates monotonicity, one also refers to this as the *winner-turns-loser paradox*.

A more demanding requirement is *strong monotonicity* of a voting system. This property is satisfied if a winner w of an election still wins whenever he is placed in a better position in the votes than before, but now it is no longer required that everything else remains unchanged; instead, the voters may change the preferences over the remaining candidates arbitrarily, as long as all candidates that originally were ranked behind w are still behind w in the new election. The latter requirement is necessary, since otherwise one could not actually say that w has improved in the new preference profile. Obviously, every strongly monotonic voting system is also monotonic.

Applied to social welfare functions, these notions are defined analogously, except that the same hypothesis now implies that, compared with her position in the original societal ranking, w keeps or improves her position in the new societal ranking, which results from applying the voting system to the modified preference profile.

Obviously, monotonicity is satisfied by all scoring protocols. The winning criterion here is the maximal score. But since the score of a candidate w either remains the same or will increase whenever she is placed higher in some votes and is not placed lower in any vote, and since the scores of all other candidates cannot increase due to this change, w cannot turn from a winner into a loser.

The example by Fishburn [454] in Table 4.8, however, shows that the intuitively very reasonable and desirable property of monotonicity is not satisfied by Dodgson voting. The numbers in the first column, under #, indicate how often each vote has been cast. The Dodgson winner in the election with the original votes (in the second column of the table) is candidate A. Why? In

4.2 Properties of Voting Systems and Impossibility Theorems

Table 4.8 Dodgson voting is not monotonic [454]

#	Original votes	Modified votes
15	$C > A > D > B$	$C > A > D > B$
9	$B > D > C > A$	$B > D > C > A$
9	$A > B > D > C$	$A > B > D > C$
5	$A > C > B > D$	$A > C > B > D$
5	$B > A > C > D$	$A > B > C > D$
	Dodgson winner: A	Dodgson winner: C
	(3 swaps)	(2 swaps)

order to become a Condorcet winner, A has to be preferred to each of the other candidates by a majority of votes (i.e., by at least 22 votes, since there are 43 voters in total). For candidates B and D, this is the case in the original election already, but to beat also C in their pairwise comparison, A has to be placed ahead of C in three more votes. This is possible, for example, by a swap in three of the first votes: $C \overset{\frown}{>} A > D > B \leadsto A > C > D > B$. To become a Condorcet winner, candidate C lacks two votes in which he has to be placed ahead of B in the original preference profile. However, at least four swaps are needed for that, since in the original votes C either is already placed ahead of B or he is not placed directly behind B. The other candidates, B and D, need even more swaps to become Condorcet winners, which shows, as claimed, that A is the Dodgson winner of the original election.

A swap of A and B in each of the five last votes are the only changes that are made in the votes. In the resulting modified preference profile (shown in the rightmost column of the table), the previous Dodgson winner A has therefore improved her position. The situation for A remains unchanged, since she still needs three swaps so as to become a Condorcet winner. However, since in the modified preference profile candidate C is now placed directly behind B in five votes, C can pass candidate B with one swap each in two of the last five votes, which makes him the one and only Dodgson winner of the modified election. Hence, by improving A's position in some of the votes, leaving everything else unchanged, A has lost her status as a Dodgson winner, which shows that Dodgson voting is not monotonic.

Similarly to the impossibility theorems by Arrow [29] (see Theorem 4.1 on page 273) and by Gibbard [500] and Satterthwaite [835] (see Theorem 4.2 on page 275), Muller and Satterthwaite [718] show that it is impossible to combine the strong monotonicity criterion with other reasonable criteria. Their impossibility theorem will also be stated without proof here.

Theorem 4.3 (Muller and Satterthwaite [718]). *If there are at least three candidates, there is no preference-based voting system that satisfies the following four properties at the same time:*

1. *nondictatorship,*
2. *resoluteness,*

3. *citizens' sovereignty*, and
4. *strong monotonicity*.

4.2.7 Homogeneity, Participation, and Twins Welcome

The last three properties we will describe are homogeneity, the participation criterion, and the twins-welcome criterion. We start with homogeneity of voting systems.

4.2.7.1 Homogeneity

Suppose one conducts an election and determines the winner. Afterwards, the same election will be conducted again, but now every voter reports the same vote as in the first election *twice*. Then one would expect that the same candidate as in the first election wins, and no one else, would one not? Suppose that the same election is conducted a third time, and every voter reports the same vote as before *three times* now. Then, again, the same candidate as before should emerge victorious, shouldn't she?

A voting system is said to be *homogeneous* if multiplying each vote by the same factor leads to the same result as the original election. A bit more formally, a voting system f is *homogeneous* if $f(P) = f(P^{(k)})$ for every preference profile P and every number $k \geq 1$, where $P^{(k)} = (C, V^{(k)})$ is built from the preference profile $P = (C,V)$ by repeating every vote from V so as to appear exactly k times in $V^{(k)}$.

Table 4.9 Dodgson voting is not homogeneous [208]

Vote	Original number	Multiplied number
$D > C > A > B$	2	6
$B > C > A > D$	2	6
$C > A > B > D$	2	6
$D > B > C > A$	2	6
$A > B > C > D$	2	6
$A > D > B > C$	1	3
$D > A > B > C$	1	3
	Dodgson winner: A (3 swaps)	Dodgson winner: D (6 swaps)

Intuitively, every reasonable voting system should satisfy this property. However, Fishburn [454] shows that, just like monotonicity, Dodgson voting fails to satisfy homogeneity. The counterexample by Fishburn [454] has eight candidates and seven voters and actually refers to the variant of Dodgson

4.2 Properties of Voting Systems and Impossibility Theorems

voting that relies on the *weak Condorcet criterion* (where, recall, a candidate is already a weak Condorcet winner if she does not lose any pairwise comparison). Table 4.9 shows the counterexample given by Brandt [208], which refers to the common Dodgson voting procedure and makes do with only four candidates, even though it has 12 voters in the original election. For A to become a Condorcet winner, A must be preferred to each of the other candidates by at least seven voters. In this election, A already wins the pairwise comparisons with all candidates except C. To also beat C, A still needs three swaps, e.g., in the two votes of the first group and in one vote of the second group:

- one each in the first two votes: $D > C \overset{\frown}{>} A > B \rightsquigarrow D > A > C > B$ and
- one in the third vote: $B > C \overset{\frown}{>} A > D \rightsquigarrow B > A > C > D$.

All other candidates need more swaps to become Condorcet winners. Hence, with $DScore(A) = 3$, A is the Dodgson winner.

If now every vote is multiplied by three, we obtain the modified numbers of votes in the rightmost column of the table. Altogether, there are now 36 voters and a candidate needs the support of at least 19 votes so as to win a pairwise comparison against another candidate. With this modified number of votes, candidate A now needs seven swaps to beat candidate D and become a Condorcet winner. On the other hand, though D loses all pairwise comparisons in the tripled election, he can

- beat A via four swaps, one each in four of the votes from the second group: $B > C > A \overset{\frown}{>} D \rightsquigarrow B > C > D > A$, and
- beat also B and C via one swap in one vote of the third group and one vote in the fifth group:
  - $C > A > B \overset{\frown}{>} D \rightsquigarrow C > A > D > B$ and
  - $A > B > C \overset{\frown}{>} D \rightsquigarrow A > B > D > C$.

Altogether, D thus needs only six swaps to become a Condorcet winner. Since B and C each need at least eight swaps, D is the Dodgson winner in the tripled election. And since the tripled election is won by a different candidate than the original Dodgson winner, this example shows that Dodgson voting violates the homogeneity criterion.

Fishburn [454] proposes a method to define a homogeneous variant of an originally inhomogeneous voting system, such as Dodgson voting. One gets a different voting system, but it still relies on the same idea as Dodgson voting (namely, swapping candidates in the votes so as to make a given candidate the Condorcet winner) and it is homogeneous. Rothe, Spakowski, and Vogel [825] show that—in contrast to the original Dodgson system (see Section 4.3.1)—the winners in this homogeneous variant of Dodgson voting can be determined efficiently. This relies on the method of linear programming (see, e.g., Section 2.5.1), which has also been used by Young [968] to show that winner determination in a homogeneous variant of the Young System is efficient. The original Young voting is also inhomogeneous [454].

4.2.7.2 Participation Criterion

Imagine it is election day and nobody goes to the polls! Although such an extreme situation rarely occurs, the low electoral participation that can be observed in many democratic countries and that is often explained by "political cynicism" is still a bit alarming. In the 2009 parliamentary elections for the German Bundestag, with $70,8\%$ the lowest electoral participation since World War II was registered, and only a slight increase (to $71,5\%$) has been observed in 2013. Almost every third eligible voter abstained from voting! The (purely fictional) "Party of the Non-Voters" (PNV) would be a serious danger for the established parties and might, in a suitable coalition, even provide the Federal Chancellor ... if only they'd run for office and would cast their ballots.

In other countries it is even worse. For example, in the United States of America, the home country of democracy where the right to vote was hard-won more than two centuries ago in the bloody battles of the Revolutionary War,[14] participation in the presidential and parliamentary elections usually is only barely above 50%—almost every second eligible voter abstains from voting there. The contrary extreme, by the way, could be observed in the dictatorships and quasi-dictatorships of the Third Reich of Nazi Germany and of the Eastern bloc countries during the Cold War: For example, the 1936 and 1939 parliamentary elections for the Reichstag had an electoral participation of more than 99% and likewise the 1963, 1967, 1981, and 1986 parliamentary elections for the Volkskammer of the GDR (in the other years, electoral participation dropped down a bit, but still was above 98%). Obviously, this has nothing to do with democracy either!

Also the right to abstain from voting should be granted to the voters of a free democracy, unlike, e.g., in the GDR (and even though there are definitely democratic countries like Australia or Austria where an obligation to vote partially existed or still exists). Nevertheless, also nonvoters may have preferences over the nominated candidates, even though they may see their "favorite candidate" only as the least evil. By not casting their votes, they obviously refuse to support this least evil, and they accept that somebody even worse will be elected.

One may be tempted to suspect that, for every reasonable voting system, a nonvoter weakens his "favorite candidate" by abstaining from voting. But this suspicion is wrong! For certain voting systems, the so-called *no-show*

---

[14] Until about the middle of the 19th century, though, this right—guaranteed by the constitution of 1787—was reserved for white men of the privileged middle and upper class of society in the USA.

4.2 Properties of Voting Systems and Impossibility Theorems

paradox occurs, which has been introduced by Brams and Fishburn [192], see also the work of Moulin and Woodall [715, 949].

There are several variants of this paradox in the literature. One strong variant says that sometimes only by abstaining from voting a (non)voter can make sure that her favorite candidate wins. In another strong variant, a (non)voter prevents a victory of her most despised candidate only by abstaining from voting (equivalently, casting her vote will cause that her most despised candidate wins). We will use the general variant of the no-show paradox where it is advantageous for a (non)voter to abstain from voting, since this makes a candidate win that she prefers to a candidate who would have won if only she had participated in the election.

A paradox usually occurs if a corresponding (namely, a desirable) property is violated for some election; this is the case here, too. Voting systems for which the no-show paradox does not occur are said to satisfy the *participation criterion*. In its general variant it says that casting an additional vote (i.e., the participation of an additional voter in an election) must not make a candidate win the election who in this additional vote is ranked behind an original winner. In other words, casting a new vote must do no harm for this voter, since if it does, it would have been better for her to stay at home and abstain from voting.

The participation criterion also reminds of other criteria in various ways. Similarly to the monotonicity criterion, violation of the participation criterion describes a situation in which a candidate is worse off, even though the performed action in fact appears to be advantageous for her: either turning a winner into a loser by improving the position of this candidate in some votes (if monotonicity is violated), or making another candidate win by introducing an additional vote in which this new winner actually is ranked worse than a current winner (if the participation criterion is violated). Similarly to manipulation (which violates strategy-proofness), voters may also try to influence the outcome of an election by abstaining from voting. In both cases, they behave differently than expected: To make their favorite candidate win, they either cast an insincere—or strategic—vote (manipulation), or they do not vote at all (no-show paradox).

Furthermore, the participation criterion is closely related to the twins-welcome criterion (equivalently, the no-show paradox is closely related to the twin paradox) that will be presented below. Moulin [715] also shows the following theorem, which establishes a connection between the participation and the Condorcet criterion.

Theorem 4.4 (Moulin [715]).

1. *For elections with at most three candidates, there are voting systems that satisfy the participation and Condorcet criterion at the same time.*
2. *For elections with at least four candidates (and at least 25 voters), there is no voting system that satisfies the participation and the Condorcet criterion at the same time.*

Examples of voting systems that satisfy the (general) participation criterion are all scoring protocols (plurality, veto, Borda, and so on), as well as the simple majority rule by which a candidate with an absolute majority wins (see also Table 4.10 on page 286). Voting systems violating the participation criterion (i.e., voting systems where the no-show paradox occurs) include STV, plurality with run-off, the cup protocol (see [192, 715]), and Dodgson voting (an example showing the latter will be given in Section 4.2.7.3).

4.2.7.3 Twins-Welcome Criterion

Voter	Vote
Anna	$C > B > A$
Belle	$A > B > C$
Chris	$A > B > C$
David	$C > A > B$
Edgar	$B > C > A$
Felix	$B > C > A$

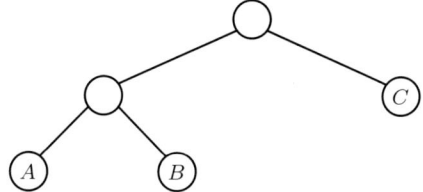

(a) Preference profile   (b) Voting tree and schedule for the cup protocol

Fig. 4.13 Twin paradox in the cup protocol due to Moulin [715]

Figure 4.13(a) shows an election with the three candidates A, B, and C and the six voters Anna, Belle, Chris, David, Edgar, and Felix, and Figure 4.13(b) shows the tree and schedule for which the cup protocol is to be executed. Possible ties will be broken in alphabetical order, i.e., the tie-breaking rule favors the alphabetically smaller candidate.

In the first round, candidates A and B compete against each other, and since they both are preferred by three voters, the tie-breaking rule makes A win. In the second round, A competes against C, and C beats A by $4:2$ and hence wins the election. This makes especially the first voter happy, Anna, since C is her preferred candidate. But then, as always a bit delayed, her twin sister Alice arrives and requests that the election be repeated, since she also wants to cast her vote. A *twin* of a voter is another voter with the same preference list, i.e., Alice has the same preference list as Anna: $C > B > A$.

"Even better!" Anna converses with herself, "if Alice, who as my twin votes exactly like me, takes part in the election, then our common favorite candidate C must win as well!"

But, what happens? Now B wins in the first round against A with $4:3$ and passes on to the second round where she faces C. Indeed, C is preferred to B by Anna, Alice, and David, but the remaining four voters disagree, which ensures B's victory!

Such a situation has been named *twin paradox* by Moulin [715]. If this can never happen for a voting system, it satisfies the *twins-welcome criterion*, and this example given by Moulin [715] shows that the cup protocol does not satisfy this criterion. Similarly to the no-show paradox, introducing a twin of a voter into the election has in fact harmed their common favorite candidate.

Also the no-show paradox itself can be observed in this example: If Alice would not have shown up, candidate C, which is preferred to B by Alice, would have won the election. The victory of B has been made possible by Alice's imprudent participation. This connection between these two paradoxes always holds: If the twin paradox occurs in some voting system, then so does the no-show paradox. In other words, if a voting system satisfies the participation criterion, then twins are always welcome in an election held under this voting system, because they never do harm and may even be useful.

Accordingly, the first part of Theorem 4.4 trivially holds for the twins-welcome criterion, too. But also the second part of this theorem can be transferred to the twins-welcome criterion, and thus provides a slightly stronger impossibility result regarding the Condorcet criterion.

Corollary 4.1 (Moulin [715]). *For elections (C,V) with at least four candidates (and $25 + \frac{\|C\|(\|C\|-1)}{2}$ or more voters), there is no voting system that satisfies the twins-welcome and the Condorcet criterion at the same time.*

Since the twins-welcome criterion is a special case of the participation criterion, it has similarities with the same properties (such as monotonicity and strategy-proofness) that are akin to the participation criterion. Moreover, in some sense the twins-welcome criterion is for voters what independence of clones is for candidates: Instead of clones of original candidates, twins of original voters are added to an election.

That also Dodgson voting violates the twins-welcome criterion—and thus also the participation criterion—can be seen by the following example. In the original election from Table 4.8 on page 279 (that was used there to show that Dodgson voting is not monotonic), A is the Dodgson winner. Now suppose that four twins of the voters in the second-to-last group, each with the same preference $A > C > B > D$, decide to also take part in the election. Then candidate C is the Condorcet winner of the modified election, and hence its Dodgson winner, too. Thus, the added votes prevent the victory of A, although A is in the first position of these votes.

4.2.8 Overview of Properties of Voting Systems

Now we have got to know about numerous voting systems and a long list of properties of voting systems. Which voting system satisfies which property?

Table 4.10 Overview of some properties of several voting systems

	anonymity	neutrality	nondictatorship	Pareto consistency	citizens' sovereignty	majority criterion	Condorcet criterion	monotonicity	homogeneity	IIA	independence of clones	strategy-proofness	participation criterion	consistency
plurality	✓	✓	✓	✓	✓	✓	✗	✓	✓	✗	✗	✗	✓	✓
Borda	✓	✓	✓	✓	✓	✗	✗	✓	✓	✗	✗	✗	✓	✓
Copeland	✓	✓	✓	✓	✓	✓	✓	✓	✓	✗	✗	✗	✗	✗
Dodgson	✓	✓	✓	✓	✓	✓	✓	✗	✗	✗	✗	✗	✗	✗
Young	✓	✓	✓	✓	✓	✓	✓	✓	✗	✗	✗	✗	✗	✗
Bucklin	✓	✓	✓	✓	✓	✓	✗	✓	✓	✗	✗	✗	✗	✗
STV	✓	✓	✓	✓	✓	✓	✗	✗	✓	✗	✓	✗	✗	✗
Schulze	✓	✓	✓	✓	✓	✓	✓	✓	✓	✗	✓	✗	✗	✗

Table 4.10 gives an overview for some selected properties and some preference-based voting systems. Note that it has to be taken into account that winner determination may depend on a tie-breaking rule (such as in STV). Whether or not a property is satisfied for such a system should not be an artifact of the tie-breaking rule, but rather should reflect a typical feature of the voting system. For instance, some properties (monotonicity, homogeneity, and so on) require a transformation of an election into another election. In that case, violation of such a property should not merely be due to the fact that the tie-breaking rule favors a different candidate in the modified election than in the original election. For example, one could easily construct an election that shows "inhomogeneity" of STV (which, in fact, is homogeneous for "well-behaved" tie-breaking rules), by using the following very artificial tie-breaking rule: *"If there are at most ten voters then A is preferred to B; if there are at least eleven voters, however, B is preferred to A."* Without formalizing this further, we exclude such "irrational" rules.

Tie-handling in elections is a challenging issue that has been discussed in detail in the literature. Properties of different tie-breaking rules and models have been studied, and also the extent to which properties of voting systems depend on tie-breaking mechanisms (see, e.g., the work of Obraztsova, Elkind, and Hazon [757]). It may also be appropriate to not implement a specific tie-breaking rule, but to randomly choose one from several "reasonable" tie-breaking rules. Or one may use a *randomized tie-breaking rule* that determines the winner in case of a tie between the involved candidates each time by chance.

4.3 Complexity of Voting Problems

In the previous two sections, we were concerned with the foundations of social choice theory, introduced some of the most important voting systems, and considered their properties and some impossibility theorems. The present section is devoted to the algorithmic aspects of voting systems, that is, we now turn to *computational social choice*. As mentioned in Chapter 1, this field is situated at the interface of political science, social choice theory, and economics on the one hand and computer science at the other hand.

Why would a computer scientist be interested in voting systems at all? The interest of researchers working in computational social choice is not limited to evaluating merely political elections, which perhaps would not be an overly ambitious task from an algorithmic perspective.[15] However, the goals and the motivation of computational social choice go far beyond political elections. By interlacing two fields, social choice theory and computer science, that have evolved completely independently of each other until not long ago, and that both are strongly affected by mathematical methods and their formal rigor (thus speaking, to a certain extent, "the same language"), computational social choice enables a transfer of knowledge in two directions:

- **From social choice theory to computer science:** Mechanisms of social choice theory such as voting systems (as well as fair division procedures, game-theoretic approaches, etc.) can be applied in computer science, e.g., in network design or in the design of ranking algorithms. For example, Dwork et al. [367] developed a method for aggregating website rankings that is based on the Kemeny system and can be used by meta-search engines that aggregate the website rankings created by several individual search engines to produce a consensus ranking.[16] Most notably, social choice mechanisms can be integrated into systems of artificial intelligence. To name just a few examples, this has been done in:
  - multiagent systems where independent "selfish" software agents may have different individual preferences over certain alternatives and use

---

[15] Certainly, political elections are interesting and important for everyone living in a democracy, including computer scientists, and indeed some important and fascinating work has been done by political scientists and economists who use algorithmic tools to evaluate political elections, predict their outcome, etc. A good example of such work is the *"Handbook on Approval Voting"* edited by Laslier and Sanver [644].

[16] Here, websites are the alternatives and individual search engines are the voters, and unlike in political elections, the alternatives by far outnumber the voters in this scenario.

preference aggregation by voting to come to a consensus (see, e.g., the textbooks by Shoham and Leyton-Brown [864] and Wooldridge [950]),
- the design of so-called *recommender systems* that may be used, for instance, to select movies according to certain criteria (see, e.g., the paper by Ghosh et al. [499]), and
- many other tasks of artificial intelligence, such as multiagent planning (see, e.g., the work of Ephrati and Rosenschein [401, 402, 403]).

- **From computer science to social choice theory:** Methods of computer science (such as algorithm design, complexity analyses, etc.) can be applied to social choice mechanisms. This is mainly necessary because such mechanisms are used not only in political elections but also in automated systems, as mentioned above. For example,
  - ranking algorithms for meta-search engines in the internet and other applications (see, e.g., the work of Betzler, Bredereck, and Niedermeier [136]) as well as recommender systems [499, 670] should work as efficiently as possible,
  - "selfish" software agents in multiagent systems who may try to influence the outcome of an election strategically should be prevented or at least hindered from doing so via computational barriers, and
  - another reason to apply state-of-the-art algorithmic techniques to social choice mechanisms is that with the rise of the internet many social and economic activities take place in highly complex environments (think, e.g., of social and economic networks, exquisitely presented in the book by Jackson [570]) and require huge amounts of data to be processed with feasible expenditure.

In this section, we will focus on the algorithmic and complexity-theoretic properties of voting systems considered in the previous two sections. By the Gibbard–Satterthwaite theorem (see Theorem 4.2 on page 275), we know that in principle all reasonable voting systems are prone to manipulation by strategic voters. Is it possible to somehow circumvent this disenchanting result? Is it possible to prevent strategic voters, who seek to influence the outcome of an election to their advantage, from being successful in their attempts?

Bartholdi, Tovey, and Trick [100] (see also the work of Bartholdi and Orlin [99]) had the path-breaking insight that the complexity of manipulation problems can provide some protection against manipulation attacks. Even though manipulation is possible in principle for all reasonable systems (recall Theorem 4.2 on page 275), it might be that the effort to compute a successful manipulation action is too high to actually be performed. With that insight, the foundation stone was laid for investigating problems of social choice theory in terms of their computational complexity. The complexity of manipulation problems, in particular, will be presented in Section 4.3.3.

4.3 Complexity of Voting Problems

Also other problems related to voting have been considered from a complexity-theoretic perspective. Bartholdi, Tovey, and Trick [101] initiated the study of the complexity of winner determination, which we will turn to in Section 4.3.1. In Section 4.3.2, we will consider the complexity of determining possible and necessary winners when the voters' preferences are only partially known; Konczak and Lang [618] were the first to study these problems. In addition to manipulation, Bartholdi, Tovey, and Trick [102] also considered other ways of influencing the outcome of elections, namely so-called electoral control, which we will focus on in Section 4.3.4. In Section 4.3.5, finally, we will address bribery in elections, another way of influencing their outcomes that has been introduced by Faliszewski, Hemaspaandra, and Hemaspaandra [422].

Other properties of voting systems have been considered algorithmically and complexity-theoretically as well, such as independency of clones (see the work of Elkind, Faliszewski, and Slinko [377]), and while we will mention some of them in passing, we will not go into much detail there.

4.3.1 Winner Determination

A desirable property of voting systems is that their winners can be determined efficiently. Since we will be concerned with the computational complexity of the corresponding problems in the following, it is advisable to recall the basics of complexity theory from Section 1.5 starting on page 18. As mentioned there, it is common to distinguish efficiently solvable from intractable problems. The former belong to the complexity class P, which contains all problems solvable by a deterministic algorithm in polynomial time. Among the intractable problems, which presumably are not efficiently solvable, are the NP-complete problems (see Definition 1.2 on page 29). Most of the problems that we will encounter in the rest of this chapter are either in P or NP-complete. For a problem to belong to the classes P or NP, it needs to be defined formally as a decision problem. The problem of winner determination for a given voting system $\mathcal{E}$ has first been considered by Bartholdi, Tovey, and Trick [101] and is formally defined as follows:

$\mathcal{E}$-WINNER

Given: An election (C,V) and a distinguished candidate $c \in C$.
Question: Is c a winner of (C,V) according to voting system $\mathcal{E}$?

Neither the number of candidates nor the number of voters is bounded in the definition of this problem. That means that the complexity-theoretic results to be presented in the following only apply to this general case. If the number of candidates or the number of voters (or both) is bounded by a constant, this problem may become efficiently solvable.

On a related note, voting problems have also been studied in terms of their so-called *parameterized complexity*. Despite its NP-hardness, these problems may be *fixed-parameter tractable*, for natural parameters such as the number of voters or the number of candidates in a given election. It may also happen, however, that such a problem is hard to solve, in terms of parameterized complexity, even for bounded problem parameters. Parameterized complexity theory has been introduced by Downey and Fellows [355] (see also the books by Niedermeier [753] and Flum and Grohe [461]). Parameterized complexity results for voting problems (which will not be presented here) can be found, e.g., in the original research papers by Betzler et al. [142, 138, 139, 136], Binkele-Raible et al. [149], Erdélyi et al. [406, 404], and Faliszewski et al. [429] and in the survey papers by Betzler et al. [135] and Lindner and Rothe [658].

4.3.1.1 Unique versus Nonunique Winners

In the WINNER problem defined above we ask whether the distinguished candidate is a (not necessarily a unique) winner. As already mentioned in Section 4.1 on page 234, this is often referred to as the *nonunique-winner model* (or as the *co-winner model*, or simply the *winner model*). Analogously, one can formulate the WINNER problem also in the *unique-winner model* where, depending on the voting system used, we may need to specify a tie-breaking rule in the problem definition that tells us how to break ties between two or more candidates whenever they occur. Of the two models, the nonunique-winner model is the more elegant and attractive, as it avoids blurring issues of winnership with issues of tie-breaking. Alternatively, one may ask whether the distinguished candidate is the only winner, no matter how ties are broken. Another possibility is to use the *parallel-universes tie-breaking* model due to Conitzer, Rognlie, and Xia [305]. In this model, we ask whether there is at least one possibility to break all occurring ties in a given election such that the distinguished candidate wins. This makes sense in particular for voting systems proceeding in stages, such as STV for which this model has been proposed in the first place.

4.3.1.2 Scoring Protocols, Copeland, and Other Voting Systems

For all scoring protocols (see Section 4.1.1), it is easy to determine the winners. It is enough to add up the points each candidate gets according to her position in each vote and then to compare the scores of all candidates. This obviously is possible in time polynomial in the input size, in fact even in linear time. Therefore, $\mathcal{E}$-WINNER is in P for all scoring protocols.

For the family of Copeland$^\alpha$ systems (see Section 4.1.2), it is a bit more complicated to determine the winners. The candidates' scores result from pairwise comparisons: Everyone takes part in a head-to-head contest with everyone else. Obviously, this is possible in quadratic time, so again we have that the problem Copeland$^\alpha$-WINNER is in P for each rational α in $[0,1]$.

4.3 Complexity of Voting Problems 291

Also in the voting systems proceeding in stages that have been described in Section 4.1.4, the WINNER problem can be solved efficiently, since the number of stages (or rounds) is always polynomial in the input size and also each of the single stages/rounds requires only polynomial time. This is only true, though, if the possibly applied tie-breaking rule (e.g., for STV) requires no more than polynomial time as well.

Similar arguments can be given for most of the voting systems presented in Section 4.1. Usually, the WINNER problem can be solved efficiently, and that is a nice property. However, there are also exceptions, and we are going to elaborate on those now.

4.3.1.3 Dodgson, Young, and Kemeny Voting

It was mentioned already in Section 4.1 that, intuitively, determining the winners for the voting systems of Dodgson [349] (see page 242), Young [968] (see page 245), and Kemeny [602] (see page 246), is harder than, e.g., for scoring protocols. Bartholdi, Tovey, and Trick [101] were the first to back this intuition by formal arguments: They show that Dodgson-WINNER and Kemeny-WINNER are NP-hard problems. Hence, they cannot belong to P unless P = NP (which is considered highly unlikely). Their results show that these two problems are at least as hard to solve as any other NP problem. The question of whether they themselves might be contained in the class NP, though, was left open by Bartholdi, Tovey, and Trick [101].

This open question has later been solved for Dodgson elections by Hemaspaandra, Hemaspaandra, and Rothe [537] and for Kemeny elections by Hemaspaandra, Spakowski, and Vogel [550]. They show that both Dodgson-WINNER and Kemeny-WINNER are complete for the complexity class $P_{\|}^{NP}$. Rothe, Spakowski, and Vogel [825] show the corresponding result for the problem StrongYoung-WINNER, which is complete for $P_{\|}^{NP}$ as well; below, we will present the analogous result for Young-WINNER. Finally, let us mention that the *Slater rule* [874], another voting system that we have not discussed so far, can be viewed as an unweighted variant—and thus as a special case—of the Kemeny rule. NP-hardness of determining Slater winners was shown by Hudry [565], and Lampis [637] optimally improved his result by establishing $P_{\|}^{NP}$-completeness of this problem as well.

To better comprehend these results, another brief detour to complexity theory is in order. As already mentioned in Section 1.5, NP is the class of decision problems solvable by a nondeterministic algorithm (or, a nondeterministic Turing machine) in polynomial time. The problems in the class $P_{\|}^{NP}$, however, can be decided in polynomial time by a deterministic Turing machine that can access a so-called NP *oracle*. That is, the P algorithm can ask queries to its NP oracle, e.g., to the NP problem SAT. In this case, the oracle queries would be boolean formulas and the oracle SAT would give the answer of whether or not the queried formulas are satisfiable. Since the running time of the P algorithm is polynomial, only polynomially many (in the input size) queries may be asked in total, and the size of each query is polynomial in the input size as well. For problems in the class $P_{\|}^{NP}$, the oracle access is restricted, though: All queries must be asked *in parallel*, i.e., no query may depend on the answer to a previously asked query. Instead, a list of all oracle queries $q_1, q_2, \ldots, q_m$ (in our example, the q_i are boolean formulas) is computed by the P algorithm first, and the NP oracle (in our example, SAT) answers by providing—within just one computation step, regardless of the computing cost to obtain these answers (that is why it is called an "oracle")—a length-m list of answers (in our example, the ith entry in this list is "yes" if $q_i \in \text{SAT}$, and "no" if $q_i \notin \text{SAT}$). Obviously, NP is contained in $P_{\|}^{NP}$, since just one query to the oracle is enough to decide the input. However, whether the equality $NP = P_{\|}^{NP}$ or the inequality $NP \neq P_{\|}^{NP}$ holds again is an open question (see also the survey by Hemaspaandra, Hemaspaandra, and Rothe [538]).

Because of this restricted oracle access, the class $P_{\|}^{NP}$ is called *"parallel access to NP."* A number of characterizations of this complexity class, and further information regarding its properties, can be found in the papers by Papadimitriou and Zachos [767] (who introduced it), Wagner [924, 925], Hemachandra [534], and Köbler, Schöning, and Wagner [616], and in the textbook by Rothe [819].

Hardness of a problem for the class $P_{\|}^{NP}$ can, of course, be directly shown by means of a reduction from another problem that is already known to be $P_{\|}^{NP}$-hard. Alternatively, one can also apply a technique due to Wagner [924], which requires a reduction with certain special properties from an NP-complete problem. This technique has been used, e.g., by Hemaspaandra, Hemaspaandra, and Rothe [537] for showing $P_{\|}^{NP}$-hardness of Dodgson-WINNER,[17] which may be seen as the first *natural* problem shown to be $P_{\|}^{NP}$-complete.[18]

Rothe, Spakowski, and Vogel [825] prove the same result for the problem StrongYoung-WINNER, which is $P_{\|}^{NP}$-complete as well. We now demonstrate how their proof works by showing the corresponding result for the closely related problem Young-WINNER (which has been explicitly noted by Brandt et al. [210]). Recall from the definition of Young's system in Section 4.1.2 that the Young score of a candidate c in an election (C, V) can be defined as the size of a largest sublist $V' \subseteq V$ such that c is a weak Condorcet winner

---

[17] For a rough overview and an easily accessible presentation of the crucial ideas of this relatively involved proof, see the book chapter by Faliszewski et al. [430].

[18] The previously known $P_{\|}^{NP}$-complete problems are usually relatively artificially constructed. Consider, for instance, the problem MAX-SAT-ASG$_=$ defined on page 37 in Section 1.5.3, which asks whether two given boolean formulas in 3-CNF have the same maximal satisfying assignment; $P_{\|}^{NP}$-completeness of this problem has been shown by Wagner [924] (see also [543] for some technical details).

4.3 Complexity of Voting Problems

in (C,V'), and that we denote this score by $\overline{Y}Score_{(C,V)}(c)$ or sometimes, shorter, $\overline{Y}Score(c)$. (We will use this score instead of its dual score, $YScore(c)$, which is based on the minimum number of votes required to be deleted for c to become a weak Condorcet winner.)

On our way to proving that Young-WINNER is $P_{\parallel}^{NP}$-complete, we will need the following decision problem that compares the Young scores of two given candidates in the same election.

Young-RANKING

Given: An election (C,V) and two distinguished candidates $c,d \in C$.
Question: Is it true that $\overline{Y}Score_{(C,V)}(c) \geq \overline{Y}Score_{(C,V)}(d)$?

We will first show that Young-RANKING is $P_{\parallel}^{NP}$-complete.[19] Our proof is based on a reduction starting from the following problem that compares the maximum size of two set packings and has been shown to be $P_{\parallel}^{NP}$-complete by Rothe, Spakowski, and Vogel [825].

MAXIMUM-SET-PACKING-COMPARE (MSPC)

Given: Two finite, nonempty sets B_1 and B_2 and two collections $\mathcal{S}_1$ and $\mathcal{S}_2$ of sets such that $\mathcal{S}_i$ contains only finite, nonempty subsets of B_i, $i \in \{1,2\}$.
Question: Is it true that $\kappa(\mathcal{S}_1) \geq \kappa(\mathcal{S}_2)$, where $\kappa(\mathcal{S}_i)$ denotes the maximum number of pairwise disjoint sets in $\mathcal{S}_i$, $i \in \{1,2\}$?

Theorem 4.5. *Young-RANKING is $P_{\parallel}^{NP}$-complete.*

Proof. We first show that Young-RANKING is in $P_{\parallel}^{NP}$. To see this, consider the following problem:

Young-SCORE

Given: An election (C,V), a designated candidate $c \in C$, and an integer $k > 0$.
Question: Is it true that $\overline{Y}Score_{(C,V)}(c) \geq k$?

Young-SCORE is easily seen to be in NP by nondeterministically guessing all sublists $V' \subseteq V$ having size at least k, accepting on such a nondeterministic computation path if c is a weak Condorcet winner in (C,V'), and otherwise rejecting on this path. (Exercise: Prove that Young-SCORE is NP-complete.) Now, for the ranking problem with given candidates c and d in election (C,V), we use the NP oracle Young-SCORE to ask in parallel all plausible values of $\overline{Y}Score_{(C,V)}(c)$ and $\overline{Y}Score_{(C,V)}(d)$ in order to determine $\overline{Y}Score_{(C,V)}(c)$ and $\overline{Y}Score_{(C,V)}(d)$ exactly, and accept if and only if $\overline{Y}Score_{(C,V)}(c) \geq \overline{Y}Score_{(C,V)}(d)$. This works since the highest possible Young score of c and d in (C,V') is polynomially bounded in the input size.

---

[19] Similarly, Dodgson-WINNER and Kemeny-WINNER have been shown $P_{\parallel}^{NP}$-complete via first proving Dodgson-RANKING and Kemeny-RANKING $P_{\parallel}^{NP}$-complete [537, 550].

$P_{\|}^{NP}$-hardness of Young-RANKING is shown by a polynomial-time many-one reduction from MSPC that slightly modifies the corresponding reduction from MSPC to StrongYoung-RANKING given in [825].[20] Let $(B_1, B_2, \mathcal{S}_1, \mathcal{S}_2)$ be a given instance of MSPC, where $B_1 = \{x_1, x_2, \ldots, x_m\}$ and $B_2 = \{y_1, y_2, \ldots, y_n\}$ and where $\mathcal{S}_i$ is a collection of subsets of B_i for $i \in \{1, 2\}$. Without loss of generality, we may assume that $\kappa(\mathcal{S}_i) > 2$. Define an election (C, V) with candidate set $C = \{a, b, c, d\} \cup B_1 \cup B_2$, where c and d are the distinguished candidates, and let V consist of the $2(\|\mathcal{S}_1\| + \|\mathcal{S}_2\|)$ votes, subdivided into six groups, that are shown in Table 4.11. Here, assuming a fixed ordering of all candidates in C, for any subset $D \subseteq C$ with $D = \{d_1, d_2, \ldots, d_\ell\}$ according to this ordering, we let "D" in a vote from V be a shorthand for "$d_1 > d_2 > \cdots > d_\ell$," and for $E \subseteq B_1$ and $F \subseteq B_2$ we let $\overline{E} = B_1 \smallsetminus E$ and $\overline{F} = B_2 \smallsetminus F$.

Table 4.11 Election (C, V) constructed in the proof of Theorem 4.5

Group		#	Vote
(1)	For each $E \in \mathcal{S}_1$	1	$v_E = E > a > c > \overline{E} > B_2 > b > d$
(2)		1	$c > B_1 > a > B_2 > b > d$
(3)		$\|\mathcal{S}_1\| - 1$	$B_1 > c > a > B_2 > b > d$
(4)	For each $F \in \mathcal{S}_2$	1	$v_F = F > b > d > \overline{F} > B_1 > a > c$
(5)		1	$d > B_2 > b > B_1 > a > c$
(6)		$\|\mathcal{S}_2\| - 1$	$B_2 > d > b > B_1 > a > c$

The only difference to the original reduction from MSPC to StrongYoung-RANKING given in [825] is that the second and the fifth group now each consist of only one vote instead of two votes. The proof of correctness essentially follows the lines of the original proof of correctness in [825], but as we now argue about Young (which is based on weak Condorcet winners) instead of StrongYoung (which is based on Condorcet winners), the proof differs in various minor details. We start by proving that:

$$\overline{Y}\mathit{Score}_{(C,V)}(c) = 2 \cdot \kappa(\mathcal{S}_1); \qquad (4.1)$$

$$\overline{Y}\mathit{Score}_{(C,V)}(d) = 2 \cdot \kappa(\mathcal{S}_2). \qquad (4.2)$$

To prove Equation (4.1), let $E_1, E_2, \ldots, E_{\kappa(\mathcal{S}_1)} \in \mathcal{S}_1$ be a maximum number of pairwise disjoint subsets of B_1. Consider the sublist $\widehat{V} \subseteq V$ that consists of the following votes:

---

[20] By contrast, in the $P_{\|}^{NP}$-completeness proofs for Dodgson-RANKING and Kemeny-RANKING [537, 550], not only the upper bounds are shown via the NP oracles Dodgson-SCORE and Kemeny-SCORE, but also the proofs of the $P_{\|}^{NP}$-hardness lower bounds of the ranking problems are based on first showing NP-hardness of the corresponding scoring problems via suitable reductions satisfying certain special properties so as to make Wagner's technique [924] applicable.

4.3 Complexity of Voting Problems

- the vote v_{E_i} from group (1) corresponding to the set E_i, $1 \leq i \leq \kappa(\mathcal{S}_1)$,
- the vote from group (2), and
- $\kappa(\mathcal{S}_1) - 1$ of the votes from group (3).

Since c is preferred to each other candidate by at least $\kappa(\mathcal{S}_1) = \|\widehat{V}\|/2$ of the votes in $\widehat{V}$, c is a weak Condorcet winner in $(C, \widehat{V})$. Thus

$$\overline{Y}Score_{(C,V)}(c) \geq \|\widehat{V}\| = 2 \cdot \kappa(\mathcal{S}_1). \tag{4.3}$$

The proof that $\overline{Y}Score_{(C,V)}(c) \leq 2 \cdot \kappa(\mathcal{S}_1)$ will follow from Lemma 4.1.

Lemma 4.1. *For any λ with $3 \leq \lambda \leq \|\mathcal{S}_1\|$, let $V_\lambda \subseteq V$ be an arbitrary sublist such that*

(a) *V_λ contains exactly λ votes from group (2) or group (3) and*
(b) *c is a weak Condorcet winner in (C, V_λ).*

Then V_λ contains exactly λ votes from group (1), which represent pairwise disjoint sets from $\mathcal{S}_1$, and V_λ contains no votes from groups (4), (5), or (6).

Proof of Lemma 4.1. For fixed λ with $3 \leq \lambda \leq \|\mathcal{S}_1\|$, let V_λ satisfy (a) and (b). Each candidate $x_i \in B_1$ is preferred to c by the at least $\lambda - 1$ group (3) votes in V_λ. By (b), for each $x_i \in B_1$, at least $\lambda - 1 \geq 2$ of the votes in V_λ prefer c to x_i. By construction, these votes must be from group (1) or group (2). Since group (2) contains only one vote, V_λ must contain at least one vote from group (1), say v_E. Since v_E represents a nonempty set $E \in \mathcal{S}_1$, some candidate $x_j \in E$ is preferred to c in v_E. Note that x_j is also preferred to c by the at least $\lambda - 1$ group (3) votes in V_λ. To be a weak Condorcet winner in (C, V_λ), c must tie or defeat x_j and so needs at least $\lambda - 1 + 1 = \lambda$ votes in V_λ from groups (1) and (2) distinct from v_E. Again, since group (2) contains only one vote, V_λ must contain at least $\lambda - 1$ group (1) votes distinct from v_E. In total, V_λ thus contains at least λ votes from group (1). Again by (b), c must tie or defeat a in V_λ. Note that a is preferred to c in groups (1), (4), (5), and (6), and c is preferred to a in groups (2) and (3). By (a), V_λ has only λ votes from groups (2) and (3), so it can have no more than λ votes from groups (1), (4), (5), and (6). It follows that V_λ contains *exactly* λ votes from group (1) and no votes from groups (4), (5), or (6).

It remains to prove that the λ votes from group (1) in V_λ represent pairwise disjoint sets from $\mathcal{S}_1$. For a contradiction, suppose there is a candidate x_j who is preferred to c in more than one of the group (1) votes in V_λ. Then x_j is preferred to c in at least two votes from group (1) and in at least $\lambda - 1$ votes from group (3). On the other hand, c is preferred to x_j in at most $\lambda - 2$ votes from group (1) and in at most one vote from group (2). Summing up, c has no more than $\lambda - 2 + 1 = \lambda - 1$ votes in V_λ, whereas x_j has at least $2 + \lambda - 1 = \lambda + 1$ votes in V_λ, which contradicts (b). Thus, no x_i, $1 \leq i \leq m$, can occur in more than one of the sets corresponding to the λ group (1) votes in V_λ, which means that these votes represent pairwise disjoint sets from $\mathcal{S}_1$. ❑   Lemma 4.1

Continuing the proof of Theorem 4.5, let $k = \overline{YScore}_{(C,V)}(c)$ and let $\widehat{V} \subseteq V$ be a sublist of size k such that c is a weak Condorcet winner in $(C,\widehat{V})$. Suppose that $\widehat{V}$ contains $\lambda \leq \|\mathcal{S}_1\|$ votes from group (2) or group (3). Since c is a weak Condorcet winner of $(C,\widehat{V})$, c must in particular tie or defeat a, so $\lambda \geq \lceil k/2 \rceil$. Since $k \geq 2 \cdot \kappa(\mathcal{S}_1)$ by (4.3), our assumption that $\kappa(\mathcal{S}_1) > 2$ implies that $\lambda \geq 3$. By Lemma 4.1, $\widehat{V}$ contains exactly λ votes from group (1), which represent pairwise disjoint sets from $\mathcal{S}_1$, and $\widehat{V}$ contains no voters of the form (4), (5), or (6). Thus, $k = \|\widehat{V}\| = 2 \cdot \lambda$ is even and $k/2 = \lambda \leq \kappa(\mathcal{S}_1)$, which proves Equation (4.1). Equation (4.2) can be proven analogously. It follows that $\kappa(\mathcal{S}_1) \geq \kappa(\mathcal{S}_2)$ if and only if $\overline{YScore}_{(C,V)}(c) \geq \overline{YScore}_{(C,V)}(d)$, which completes the proof of Theorem 4.5. ❑

The above reduction from MSPC to Young-RANKING is now modified to yield a reduction from MSPC to Young-WINNER by replacing each candidate $e \in C - \{c,d\}$ by $\|V\|$ clones, $e^1, e^2, \ldots, e^{\|V\|}$, and by replacing each occurrence of e in the ith voter of V by:

$$e^{i \bmod \|V\|} > e^{i+1 \bmod \|V\|} > e^{i+2 \bmod \|V\|} > \cdots > e^{i+\|V\|-1 \bmod \|V\|}.$$

Let $(\tilde{C}, \tilde{V})$ be the resulting election. This modification leaves the Young scores of both distinguished candidates unchanged:

$$\overline{YScore}_{(\tilde{C},\tilde{V})}(c) = \overline{YScore}_{(C,V)}(c) \quad \text{and} \quad \overline{YScore}_{(\tilde{C},\tilde{V})}(d) = \overline{YScore}_{(C,V)}(d).$$

However, it decreases the Young scores of all other candidates so that they are smaller than those of either c or d in $(\tilde{C}, \tilde{V})$. Hence, there is no candidate $e \in \tilde{C}$ such that

$$\overline{YScore}_{(\tilde{C},\tilde{V})}(e) \geq \overline{YScore}_{(\tilde{C},\tilde{V})}(c) \quad \text{or} \quad \overline{YScore}_{(\tilde{C},\tilde{V})}(e) \geq \overline{YScore}_{(\tilde{C},\tilde{V})}(d).$$

It follows that the condition $\kappa(\mathcal{S}_1) \geq \kappa(\mathcal{S}_2)$ is equivalent to c being a weak Young winner of $(\tilde{C}, \tilde{V})$. That is, analogously to the $P_{\|}^{NP}$-completeness proof for StrongYoung-WINNER [825], we obtain the following theorem.

Theorem 4.6. *Young-WINNER is $P_{\|}^{NP}$-complete.*

As mentioned in Section 4.2.7.1, the winners in certain "homogeneous" variants (recall the definition of homogeneity from page 280) of the Dodgson and (Strong)Young voting systems can be determined efficiently (see the papers by Young [968] and Rothe, Spakowski, and Vogel [825]). For further details on Dodgson and Young elections, we refer to the book chapter by Caragiannis, Hemaspaandra, and Hemaspaandra [251].

4.3.2 Possible and Necessary Winners

In this section, we will define the notions of possible and necessary winner that have first been studied by Konczak and Lang [618]. Assuming that we have only incomplete information about the voters' preferences, a possible winner intuitively is a candidate that wins the election in *some* complete extension of the given incomplete preference profile, whereas a necessary winner is a candidate that wins the election in *all* its complete extensions.

After motivating these notions, we define the related decision problems and discuss their complexity for various voting systems. Finally, we will consider some variations of the possible winner problem.

4.3.2.1 Motivation

Closely related to the winner problem considered above is the problem of possible winners. So far, with the exception of approval and fallback voting, we always assumed that the voters' ballots are *complete* preference lists over the candidates. However, there are situations in which this is not the case.

Anna, Belle, and Chris want to spend their vacation together. They have the choice between four different ways of transportation for their journey: They can make their trip either by ship, or by car, or by plane, or by train.

"I'd love to fly," says Anna, "and my second choice would be the ship. The car is only next because you always get into traffic jams. And my least preferred choice is the train, as it is always late."

Belle does not have such definite preferences. "I think car and train in any case are better than either ship or plane," she says, "since the ship might sink and the plane might crash. But I am unable to decide between car and train, and if it is neither of them, I do not care whether we take the ship or the plane."

"My definite favorite is the trip by ship," utters Chris. "And if this is not possible, a trip by plane or by car would both be OK; I don't care which. But I really don't want to go by train. If the air conditioning breaks down again, I totally freak out!"

Figure 4.14 shows their preferences. In such a situation, the votes can also be *partial orders* on the set C of candidates. Mathematically, this is represented by a *partial preference relation* $\succ$, which is transitive and asymmetric but, unlike the preference relation $>$, not necessarily total. In such a *partial preference list* of a voter that relies on a partial order $\succ$, there may be pairs of candidates none of which is preferred to the other by this voter. But even in such a situation, one wants to determine whether a distinguished candidate

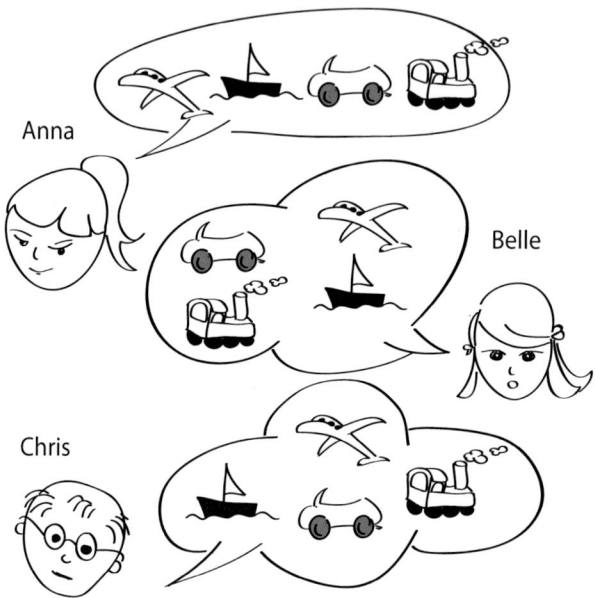

Fig. 4.14 Trip preferences of Anna, Belle, and Chris

still has a chance to win the election, i.e., whether he is a *possible winner*. Since most of the voting systems studied here require complete preference orders for winner determination, this question boils down to asking whether it is possible to extend the given partial preferences of the voters over the candidates to complete preference lists such that the distinguished candidate wins.

A linear preference list v over C is a *total extension* of a partial preference list w over C if for all $c, d \in C$ with $c \succ d$ in w, it holds that also in v candidate c is preferred to candidate d. In other words, such a total extension specifies a preference among each two candidates, in a way that is consistent with the original partial preference list. We say a profile $V = (v_1, \ldots, v_n)$ of linear preference lists over C is a *total extension* of a profile $W = (w_1, \ldots, w_n)$ of partial preference lists over C if v_i totally extends w_i for each i, $1 \leq i \leq n$.

The three friends agree to use Borda voting to make their decision. Since Anna really wants to fly, she wonders whether there is a chance at all for the trip by plane to win. She briefly ponders on this and then makes a suggestion.

"Belle," she says, "as you don't care if we take the plane or the ship, you might place the plane in position three and the ship in position four. Why not? And, other than that, you might also rank the car on top, ahead of the train in the second spot."

4.3 Complexity of Voting Problems

"You're right, why not?" agrees Belle, because she really doesn't care.

"That's true!" adds Chris. "If we want to use Borda, then every alternative must have exactly one position. Therefore, plane is my second and car my third choice."

"Perfect!" says Anna happily. "Then let's figure up the points to find out the Borda scores."

Due to this total extension of their partial preferences to linear preferences, the plane trip wins with six points under Borda. That means that with respect to the original partial preferences of the three voters, the plane trip is a possible winner. The trip by train, however, is not a possible winner, since it has a maximum Borda score of three points over all total extensions of these partial preferences, whereas the trip by plane has a Borda score of at least four points.

4.3.2.2 Possible and Necessary Winners

In an election with partial preferences, a *possible winner* is a candidate that has the possibility to win in some total extension of the partial preferences. This notion and the definition of the corresponding decision problem are due to Konczak and Lang [618], for any given voting system $\mathcal{E}$:

$\mathcal{E}$-POSSIBLE-WINNER
Given: An election (C,V), where the votes in V are represented as partial orders over C, and a distinguished candidate $c \in C$.
Question: Is c a possible $\mathcal{E}$ winner of (C,V), i.e., is it possible to fully extend every partial vote in V such that c is an $\mathcal{E}$ winner of the resulting election?

For each voting system whose winners can be determined in polynomial time, this problem is in NP, since in polynomial time one can nondeterministically guess total extensions of the given partial preference lists and then deterministically check whether the distinguished candidate wins the resulting election. This problem generalizes the manipulation problem, which we will study in Section 4.3.3 starting on page 306.

Depending on the voting system, the problem can be easy to solve or hard, i.e., POSSIBLE-WINNER is in P for some voting systems but is NP-complete for some others. Table 4.12 on page 302 provides an overview of known complexity results for this problem. As an example, we show that it is easy to decide whether a given candidate is a possible Condorcet winner. Note that Konczak and Lang [618] prove their result, stated here as Theorem 4.7, in a slightly more general model by allowing indifferences in partial preference orders, while we always consider only strict preferences in this chapter; the proof is essentially identical.

Theorem 4.7 (Konczak and Lang [618]). POSSIBLE-WINNER *is in* P *for Condorcet voting.*

Proof. For a profile T of linear orders over the set of candidates C and for any two candidates $x, y \in C$, let $D_T(x,y)$ denote the number of votes in T that prefer x to y minus the number of votes in T that prefer y to x. For a profile R of partial orders over C and for any two candidates $x, y \in C$, let $D_R^{max}(x,y)$ denote the maximal value of $D_T(x,y)$, taken over all total extensions T of the partial votes in R.

The proof will have two parts. We will first show that for a profile of partial orders $R = (R_1, \ldots, R_n)$ and any two candidates $x, y \in C$, we have

$$D_R^{max}(x,y) = \|\{i \mid \text{not}(y >_{R_i} x)\}\| - \|\{i \mid y >_{R_i} x\}\|. \tag{4.4}$$

This means that $D_R^{max}(x,y)$ equals the number of votes from R not preferring y to x (either because x is preferred to y or because they are incomparable in such a vote) minus the number of votes that prefer y to x. In the second part of the proof, we will show that

$$x \in C \text{ is a possible Condorcet winner for } R \iff \forall y \neq x,\ D_R^{max}(x,y) > 0.$$

Together with the first part, this shows that whether a candidate is a possible Condorcet winner can be decided in polynomial time.

We start by proving (4.4). If $y >_{R_i} x$ then it obviously holds that $y >_{T_i} x$ for every total extension T_i of R_i. Hence

$$\|\{i \mid y >_{T_i} x\}\| \geq \|\{i \mid y >_{R_i} x\}\|. \tag{4.5}$$

Furthermore, since T_i is an extension of R_i, $x >_{T_i} y$ implies $\text{not}(y >_{R_i} x)$. Thus,

$$\|\{i \mid x >_{T_i} y\}\| \leq \|\{i \mid \text{not}(y >_{R_i} x)\}\|. \tag{4.6}$$

Combining (4.5) and (4.6) yields

$$\|\{i \mid x >_{T_i} y\}\| - \|\{i \mid y >_{T_i} x\}\| \leq \|\{i \mid \text{not}(y >_{R_i} x)\}\| - \|\{i \mid y >_{R_i} x\}\|. \tag{4.7}$$

Since this holds for all total extensions T_i of R_i, we get

$$D_R^{max}(x,y) \leq \|\{i \mid \text{not}(y >_{R_i} x)\}\| - \|\{i \mid y >_{R_i} x\}\|.$$

For the converse inequality, consider a profile $R^x = (R_1^x, \ldots, R_n^x)$ of linear orders that are the best possible total extensions of the profile $R = (R_1, \ldots, R_n)$ with respect to candidate x, i.e., whenever the relation between x and some $z \in C$, $z \neq x$, is undetermined in R it holds that z will be placed behind x in R^x. Then,

$$\begin{aligned}
D_{R^x}(x,y) &= \|\{i \mid x >_{R_i^x} y\}\| - \|\{i \mid y >_{R_i^x} x\}\| \\
&= \|\{i \mid \text{not}(y >_{R_i^x} x)\}\| - \|\{i \mid y >_{R_i^x} x\}\| \\
&= \|\{i \mid \text{not}(y >_{R_i} x)\}\| - \|\{i \mid y >_{R_i} x\}\|.
\end{aligned}$$

Hence we have that $D_R^{max}(x,y) \geq \|\{i \mid \text{not}(y >_{R_i} x)\}\| - \|\{i \mid y >_{R_i} x\}\|$. This completes the proof of (4.4).

The second part of the proof is to show that a candidate x is a possible Condorcet winner for R if and only if $D_R^{max}(x,y) > 0$ for each $y \neq x$.

For the direction from right to left, assume that $D_R^{max}(x,y) > 0$ for each $y \neq x$. For a contradiction, suppose that x is not a possible Condorcet winner. Then, for all total extensions $T = (T_1, \ldots, T_n)$ of $R = (R_1, \ldots, R_n)$, there is a candidate y such that $\|\{i \mid x >_{T_i} y\}\| \leq \|\{i \mid y >_{T_i} x\}\|$. This also holds for the best possible extension R^x, so $D_{R^x}(x,y) \leq 0$, and hence $D_R^{max}(x,y) \leq 0$. But this is a contradiction to the assumption that $D_R^{max}(x,y) > 0$ for each $y \neq x$.

For the direction from left to right, assume that x is a possible Condorcet winner for R. Then there exists a total extension $T = (T_1, \ldots, T_n)$ of R such that for each candidate $y \neq x$, it holds that $\|\{i \mid x >_{T_i} y\}\| > \|\{i \mid y >_{T_i} x\}\|$. Hence we have $D_T(x,y) > 0$, and thus $D_R^{max}(x,y) > 0$. ❑

For the family of pure scoring protocols,[21] Betzler and Dorn [137] (and Baumeister and Rothe [122], solving the last missing case in Theorem 4.8) establish a *dichotomy result* for the POSSIBLE-WINNER problem, i.e., it distinguishes the easy from the hard cases:

Theorem 4.8 (Betzler and Dorn [137], Baumeister and Rothe [122]).
POSSIBLE-WINNER is in P *for plurality and veto (and the trivial rule with an all-zero scoring vector), and is* NP*-complete for all other pure scoring rules.*

If one is interested in the possible winners of an election, it also makes sense to ask whether a candidate is a *necessary winner* of an election with partial preference lists. Such a candidate has to win in *every* total extension of the given partial preference lists. The corresponding decision problem is denoted by $\mathcal{E}$-NECESSARY-WINNER, for each voting system $\mathcal{E}$.

The complexity of POSSIBLE-WINNER and NECESSARY-WINNER has been studied for a number of voting systems; see Table 4.12 for an overview of these results. The P results for Condorcet in Table 4.12 was shown by Konczak and Lang [618] and those for plurality and veto are due to Betzler and Dorn [137]. The NP- and coNP-completeness results for STV follow from the work of Bartholdi and Orlin [99]. All other NP-completeness results for POSSIBLE-WINNER and all other coNP-completeness results for NECESSARY-WINNER in this table are due to Xia and Conitzer [957].

---

[21] A scoring protocol (defined in Section 4.1.1 starting on page 235) is said to be *pure* if for each $m \geq 2$, the scoring vector for m candidates results from the scoring vector for $m-1$ candidates by inserting an additional score value α_i at any position i so that the condition $\alpha_1 \geq \alpha_2 \geq \cdots \geq \alpha_m$ is still satisfied.

Table 4.12 Complexity of POSSIBLE-WINNER and NECESSARY-WINNER

	POSSIBLE-WINNER	NECESSARY-WINNER
Condorcet	P	P
Plurality	P	P
Veto	P	P
Borda	NP-complete	P
Simpson	NP-complete	P
Bucklin (see Footnote 22)	NP-complete	P
Cup protocol	NP-complete	coNP-complete
Copeland	NP-complete	coNP-complete
STV	NP-complete	coNP-complete
Ranked pairs	NP-complete	coNP-complete

Note that unlike POSSIBLE-WINNER, the NECESSARY-WINNER problem can be solved in polynomial time for all scoring protocols [957]. This suggests that POSSIBLE-WINNER, in general, may be a harder problem than NECESSARY-WINNER, unless P = NP. Indeed, this is known to hold not only for scoring protocols such as Borda, but also for other voting systems such as Simpson and Bucklin voting.[22] POSSIBLE-WINNER in approval voting has been studied by Barrot et al. [97].

While the possible winner problem was originally defined for unweighted elections only, Walsh [927] was the first to investigate the weighted variant of POSSIBLE-WINNER and the restriction of this problem to single-peaked preferences (see Chapter 5), and Lang et al. [640] study it in the weighted case for Schwartz elections and the cup protocol. Pini et al. [781] investigate possible and necessary winners for STV and cup in the weighted and unweighted case and for a bounded and unbounded number of candidates. In particular, they show that finding an upper approximation bound of possible winners and a lower approximation bound of necessary winners are intractable tasks. They also establish sufficient conditions for tractability to hold: If a voting system satisfies certain properties (specifically, independence of irrelevant alternatives, here defined in Section 4.2.3 starting on page 272, and monotonicity, here defined in Section 4.2.6.3 starting on page 277), then it is easy to find the possible and necessary winners. Finally, they have shown how to use possible and necessary winners in order to decide when to terminate preference elicitation processes. For further details on voting with incomplete information, we refer to the book chapter by Boutilier and Rosenschein [174].

---

[22] The results of Xia and Conitzer [957] for "Bucklin" voting in Table 4.12 in fact refer to a simplified variant of this system: In their system, every candidate with the smallest Bucklin score is a "Bucklin" winner. By contrast, recall from Section 4.1.4.4 starting on page 256 that a Bucklin winner is a candidate that scores the most points among those with the smallest Bucklin score.

4.3.2.3 Variants of the Possible Winner Problem

In the original POSSIBLE-WINNER problem described above, there is no restriction on the structure of the ballots. We now describe some variants of this problem that either require certain restrictions of the partial votes or take into account some other type of uncertainty.

Possible Winner with New Alternatives. One such variant is POSSIBLE-WINNER-NEW-ALTERNATIVES. Here, some new alternatives may be added to an election, the votes are partial in the sense that they do not rank these new alternatives, and the question is whether these votes can be totally extended in a way that makes a given alternative a winner. This problem has first been studied by Chevaleyre et al. [292, 293], so as to model scenarios such as the following. Suppose a committee has to make a joint decision on how to fill a vacancy. Assume further that some committee members, after having reported their preferences, go on vacation where they cannot be reached by email or phone. However, several job applications arrive late due to a technical problem. In this situation it is quite natural to ask whether a distinguished candidate from the initial applicants can get the job by suitably inserting the additional applicants into the partial votes.

Table 4.13 Complexity of POSSIBLE-WINNER-NEW-ALTERNATIVES

	POSSIBLE-WINNER-NEW-ALTERNATIVES
Plurality	P
Veto	P
Borda	P
Simpson	NP-complete
Copeland^0	NP-complete
Bucklin	NP-complete

The complexity of POSSIBLE-WINNER-NEW-ALTERNATIVES has been studied for various voting rules by Chevaleyre et al. [292, 293] and Xia, Lang, and Monnot [962]. Some of their results are summarized in Table 4.13 (where the P results for plurality and veto follow immediately from those for the more general POSSIBLE-WINNER problem, see Table 4.12). Interestingly, while POSSIBLE-WINNER is NP-complete for Borda (again, see Table 4.12), Chevaleyre et al. [292] could show that POSSIBLE-WINNER-NEW-ALTERNATIVES for Borda is in P for adding any number of new alternatives. On the other hand, they show that this problem is NP-complete for the pure scoring protocol with vector $(3, 2, 1, 0, \ldots, 0)$ when only one new alternative is added.[23] Baumeister, Roos, and Rothe [120] slightly generalize the latter result to a class of pure scoring protocols. They also study the weighted

---
[23] Note that hardness of POSSIBLE-WINNER-NEW-ALTERNATIVES for adding some number of new alternatives does not in general imply hardness for adding any larger number

version of POSSIBLE-WINNER-NEW-ALTERNATIVES and show, in particular, that this problem is NP-complete even for a scoring rule as simple as plurality when only one candidate is added.

Possible Winner with Truncated Ballots. Another restriction on the form of the ballots are top- and/or bottom-truncated ballots. Here, the voters again report partial instead of linear votes but the partial votes all have a common structure. If the set of alternatives is too large, some lazy voters may be willing to specify only a ranking of their top and/or bottom candidates, as they care only about their most and/or least preferred candidates. Indeed, there are adaptions of the Borda rule to top-truncated ballots that are actually used for political elections; for example, the Irish Green Party uses the so-called *modified Borda count* to choose its leader (see [386]): Each voter ranks only k of the m candidates, a candidate ranked in position i gets $m-i$ points, and each of the unranked candidates gets $m-k-1$ points. In the corresponding three decision problems POSSIBLE-WINNER-TOP/BOTTOM/DOUBLY-TRUNCATED-BALLOTS (or, shorter, PW-TOP/BOTTOM/DOUBLY-TB), introduced by Baumeister et al. [117], we are given votes that either are top-truncated or bottom-truncated or both top- and bottom-truncated. The question is whether the given top/bottom/doubly-truncated ballots can be extended to linear orders such that a distinguished candidate wins the election. Baumeister et al. [117] show that these problems are closely related to some manipulation (see Section 4.3.3 starting on page 306) and bribery problems (see Section 4.3.5 starting on page 358), and that they are solvable in polynomial time for k-approval.

Building on their work, Menon and Larson [702] provide a fundamental study of the complexity of strategic behavior in elections with top-truncated ballots. In particular, they study constructive and destructive weighted manipulation (see Section 4.3.3 starting on page 306) for top-truncated ballots, establishing general results for all scoring rules, for certain elimination variants of all scoring rules, for plurality with run-off, for the family of Copeland$^\alpha$ rules, and for maximin. They also provide results for manipulative actions in electorates with top-truncated ballots that are single-peaked (see Chapter 5 for more background and results on manipulative actions in single-peaked societies).

Possible Winner with Uncertain Weights. In the possible winner problem and all its variants we have seen so far, uncertainty is associated with the votes being incomplete. The next variant is different: We now have complete votes (i.e., linear orders), which are weighted, and there is uncertainty about their weights. In the POSSIBLE-WINNER-UNCERTAIN-WEIGHTS problem, introduced by Baumeister et al. [121], the question is whether the weights of the voters can be set such that a distinguished candidate wins the election.

---

of new alternatives. Indeed, it may well be the case that hardness turns into easiness by adding sufficiently many new alternatives.

4.3 Complexity of Voting Problems

Table 4.14 Overview of complexity results for POSSIBLE-WINNER-UNCERTAIN-WEIGHTS with nonnegative integer weights (indicated by $\mathbb{N}$), with and without given upper bounds on the total weight (indicated by BW either occurring in the problem name or not) and with and without given ranges to choose weights from (indicated by RW either occurring in the problem name or not). NP-c. stands for "NP-complete" and k-AV stands for "k-approval."

	k-AV, $k \le 3$	k-AV, $k \ge 4$	k-veto, $k \le 2$	k-veto, $k \ge 3$	Plurality/Veto with run-off	Borda	Bucklin, Fallback	Copeland, Ranked pairs
$\mathbb{N}$	P	P	P	P	P	NP-c.	NP-c.	NP-c.
BW-RW-$\mathbb{N}$	P	NP-c.	P	NP-c.	P	NP-c.	NP-c.	NP-c.
BW-$\mathbb{N}$	P	NP-c.	P	NP-c.	P	NP-c.	NP-c.	NP-c.
RW-$\mathbb{N}$	P	P	P	P	P	NP-c.	NP-c.	NP-c.

Consider, for example, a university ranking based on different criteria (e.g., equipment, third-party funds, graduation rates, etc.). This can be seen as an election where the universities are the candidates and the criteria provide the votes, as each single criterion induces a linear order of the universities. Assume that the voting rule is fixed. It is then natural to ask whether it is possible to make a distinguished university win the aggregated ranking based on all criteria by carefully choosing their weights. Baumeister et al. [121] consider both nonnegative integer and nonnegative rational weights (and also several variants with regions from which to choose the weights or with bounds on the sum of the weights).

Formally, for a given voting system $\mathcal{E}$ and for $\mathbb{F} \in \{\mathbb{Q}^+, \mathbb{N}\}$, they define the general *possible winner with uncertain weights problem* as follows:

$\mathcal{E}$-POSSIBLE-WINNER-UNCERTAIN-WEIGHTS-$\mathbb{F}$
Given: A list of candidates C, a list of unit-weight votes V_1 over C, a list of votes V_0 over C with unspecified weights, and a distinguished candidate $c \in C$.
Question: Are there weights $w_i \in \mathbb{F}$, $1 \le i \le \|V_0\|$, for the votes in V_0 such that c is a winner of the weighted $\mathcal{E}$ election $(C, V_1 \cup V_0)$?

Note that for inputs with $V_0 = \emptyset$, one obtains the ordinary unweighted (i.e., unit-weight) winner problem for $\mathcal{E}$ that we dealt with in Section 4.3.1 starting on page 289. Note further that allowing weight zero for votes in V_0 in some sense corresponds to *control by deleting voters* [102, 541] in weighted elections, a problem that will be considered in Section 4.3.4.3 starting on page 333. In this general framework, Baumeister et al. [121] studied POSSIBLE-WINNER-UNCERTAIN-WEIGHTS for a variety of voting rules. The questions they left open—regarding, in particular, two variants of this problem for 3-approval and four variants for plurality with run-off—could later be solved by Neveling, Rothe, and Weishaupt [741], specifically via efficient integer linear programs that are technically rather involved and inspired by the work of Fitzsimmons and Hemaspaandra on high-multiplicity election problems [455]. The results of Baumeister et al. [121] and Neveling, Rothe, and Weishaupt [741] have then be merged to appear in a joint article [119]. Their results for nonnegative integer weights are presented in Table 4.14 while they establish P results for nonnegative *rational* weights in all combinations of problem variant and voting rule occurring in this table.

Possible Winner under Uncertain Voting System. Yet another variant is the POSSIBLE-WINNER-UNCERTAIN-VOTING-SYSTEM problem. Here, we again have complete votes and the source of uncertainty now lies in the voting rule used to aggregate the ballots. The motivation to study this problem is that uncertainty about the voting system may give the voters an incentive to vote truthfully, since reporting an insincere preference might result in a worse outcome for them. Consider, for example, an election with three candidates, A, B, and C, five honest voters and two strategic voters: There are three sincere votes $C > A > B$, two sincere votes $B > A > C$, and the true preference of the two strategic voters is $A > B > C$. If the strategic voters knew for sure that the election will be held under plurality, they would prefer reporting the insincere preference $B > A > C$ instead of their true one, because then candidate B instead of C would win the election. But if the election were held under the Borda rule, the strategic voters' favorite candidate, A, would win the election, whereas B would be the winner with the modified votes $B > A > C$ by the strategic voters. Hence uncertainty about the voting system may give a strong incentive to reveal the true preference. The corresponding possible winner problem has been introduced by Baumeister, Roos, and Rothe [120]. They show that this problem is NP-complete for a certain class of scoring rules, yet is solvable in polynomial time for Copeland$^\alpha$ elections and preference-based approval voting.

Obviously, some of the problems presented above are closely related to each other and hence complexity results may carry over from one problem to another. Figure 4.15 shows the relations between some of the above-mentioned possible winner problems and a variant of manipulation and of bribery, namely, COALITIONAL-MANIPULATION (see Section 4.3.3.2 starting on page 309) and SWAP-BRIBERY (see Section 4.3.5.5 starting on page 363). An arrow from X to Y means that there is a polynomial-time many-one reduction from X to Y.

4.3.3 Manipulation

As mentioned at the beginning of Section 4.3, the Gibbard–Satterthwaite theorem (see Theorem 4.2 on page 275) motivates to study the complexity of the manipulation problem in elections. Even though—according to this theorem—every reasonable voting system in principle is manipulable, computational hardness of computing a successful manipulative action (or merely of deciding whether this is possible for a given preference profile) may provide some protection against strategic voting.

4.3 Complexity of Voting Problems

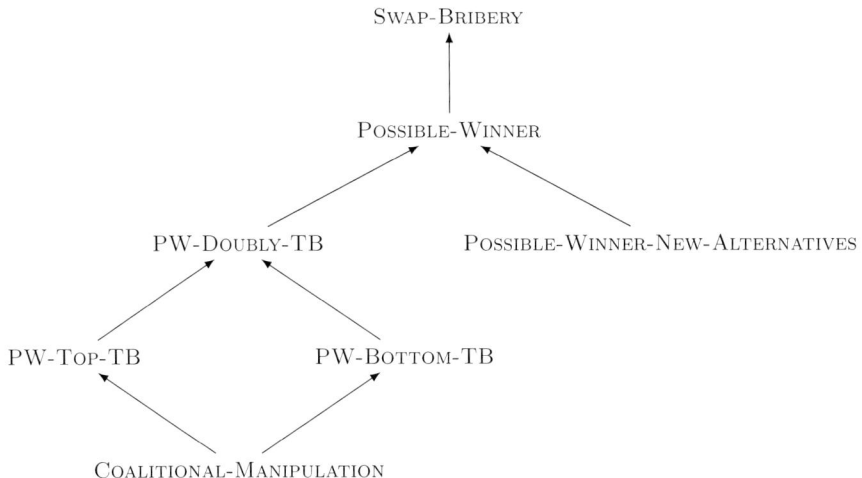

Fig. 4.15 Relations between some variants of the possible winner, manipulation, and bribery problem

As is common in the literature on manipulation, we assume that the manipulators (who seek to find strategic votes) have complete information about the other voters' ballots. This is a pretty restrictive assumption, as that is only rarely the case in reality (indeed, perhaps for very small elections only—in large-scale elections, one can merely rely on approximations via demoscopic poll ratings). However, if one can show that the manipulation problem is hard to solve even in the complete-information model, it obviously cannot become easier with less information. That is, assuming complete information gives conservative estimates regarding the computational complexity of manipulation problems.

As mentioned in Section 4.3.1.1 on page 290, the nonunique-winner model is more elegant and attractive than the unique-winner model, as it avoids blurring issues of winnership with issues of tie-breaking. Therefore, we will focus on the nonunique-winner model in the following.

There are various ways to formalize manipulation as decision problems,[24] and we start with the most restrictive variant.

---

[24] A notable exception to the common approach of studying the complexity of manipulation in terms of decision problems is due to Hemaspaandra, Hemaspaandra, and Menton [536], who instead consider the search problem of actually finding a successful manipulation action. Their main result is that there are (artificial) voting systems for which it is easy to decide whether or not manipulation is possible, and yet no successful manipulation action can be found in polynomial time, unless integer factoring were also easy (which is considered highly unlikely [819]).

In a follow-up work, Carleton et al. [257] explore the relationships between the search versions of collapsing pairs of electoral control types, a topic we will consider later on in Section 4.3.4.5 starting on page 336.

4.3.3.1 Single Manipulators

For any voting system $\mathcal{E}$, Bartholdi, Tovey, and Trick [100] define this problem as follows.

$\mathcal{E}$-Manipulation
Given: A set C of candidates and a list V of (sincere) voters that are given as linear preference lists over C, and a distinguished candidate $c \in C$.
Question: Is there a linear preference list s so that c is an $\mathcal{E}$ winner of $(C, V \cup \{s\})$?

In general, neither the number of candidates nor the number of voters is limited in this problem. We have defined it in the nonunique-winner model here; just as the Winner problem, Manipulation can be defined analogously in the unique-winner case—the complexity of both problem variants usually is the same. It is desirable to find voting systems for which winner determination is easy, but manipulation is hard. If it is possible to show for a system that Manipulation is NP-hard, but Winner is in P, then, assuming $P \neq NP$, the decision of whether or not a given election can be manipulated is strictly harder than to determine its winners.

Bartholdi, Tovey, and Trick [100] show that the Manipulation problem for second-order Copeland voting is NP-complete. This voting system works as the Copeland$^\alpha$ system, except that in case of a tie in a pairwise comparison the decision is made differently: Instead of both candidates receiving α points, only one candidate gets one point, namely whoever in his head-to-head contests defeated other candidates with a greater sum of Copeland scores.[25] This system (or, in fact, a slight modification thereof) is used, for example, in the tournaments of the Fédération Internationale des Échecs and the United States Chess Federation [100].

On the other hand, Bartholdi, Tovey, and Trick [100] show that Manipulation is easily solvable for many voting systems with easy winner determination. Considering the plurality rule, it is obvious that the only reasonable strategic vote has the distinguished candidate—who is to be made a winner—on top. Since all other candidates get no points at all in plurality, their order is irrelevant in this vote. It can be easily tested whether or not this action is successful. This idea can be extended to a much bigger class of voting systems, namely to all systems that assign points to the candidates according to a *scoring function* such that (a) the candidates with the maximum score are the winners and (b) the scoring function is monotonic. This includes all scoring protocols, Simpson voting, and the Copeland system, for example. A simple greedy algorithm solves Manipulation for these systems.

Since Copeland voting is efficiently manipulable, the NP-hardness proof of the manipulation problem for second-order Copeland mainly relies on the

---

[25] Of course, according to this tie-breaking rule it is again possible that ties occur, so an additional tie-breaking rule would have to be applied to determine unique winners in pairwise comparisons.

4.3 Complexity of Voting Problems

above-mentioned tie-breaking rule of this voting system. In contrast, ties do not play a fundamental role in the reduction by Bartholdi and Orlin [99], by which they show that STV-MANIPULATION is also NP-complete.[26]

4.3.3.2 Coalitions of Manipulators

Conitzer, Sandholm, and Lang [310] generalize the MANIPULATION problem to the COALITIONAL-MANIPULATION problem. Here, a group of manipulative voters seek to jointly select their votes such that the distinguished candidate wins the election. In general, by such a coordination of strategies the manipulators can achieve more than a single strategic voter. The above-defined problem MANIPULATION is the special case of COALITIONAL-MANIPULATION where this group consists of only a single voter. In particular, it is interesting to know what size of the manipulative coalition is required to turn COALITIONAL-MANIPULATION from an easy into a hard problem.

The complexity of COALITIONAL-MANIPULATION is still unknown for a number of voting systems; it is not even known for all scoring protocols [953, 963, 960]. Only in 2011, Betzler, Niedermeier, and Woeginger [140] and, independently, Davies et al. [334] have resolved a long-standing open problem by showing that COALITIONAL-MANIPULATION for Borda voting is NP-complete, even if there are only two manipulators. This result, along with other interesting results on using the Borda rule in collective decision making (including—in addition to computational social choice—coalition formation in hedonic games as described in Section 3.6 starting on page 198, and social welfare maximization in fair division of indivisible goods for agents with ordinal preferences, as to be described in Sections 9.6 and 9.7 starting on pages 635 and 649, respectively) are surveyed by Rothe [820] (see also [740]).

Since Borda is efficiently manipulable by *one* strategic voter, the exact threshold for which this problem becomes hard is known in this case. The first such result goes back to Faliszewski, Hemaspaandra, and Schnoor [432, 433], who show that COALITIONAL-MANIPULATION for Copeland$^\alpha$ elections is NP-complete if α is a rational in $[0,1]$ with $\alpha \neq 1/2$ and there are at least two manipulators. The complexity of this problem for $\alpha = 1/2$ is still unknown. Narodytska, Walsh, and Xia [726] study coalitional manipulation for Nanson and Baldwin voting and show that these two problems are NP-complete, even if there is only one manipulator.

---

[26] As noted by Rothe and Schend [824], the reduction in [99] in fact has a minor flaw. Looking at the proof of [99, Thm. 1], note that the distinguished candidate's direct opponent might gain too many points in the last two voter groups (see [99, pp. 9–10]); moving the "garbage candidates" in these votes to the third position fixes this flaw.

4.3.3.3 Coalitional Manipulation for Weighted Voters

Conitzer, Sandholm, and Lang [310] also consider another generalization of the manipulation problem: *weighted voters*. "Liberty, equality, fraternity" or "one man, one vote" ... are not always the principles on which an election is based: The votes of some voters may have more weight than those of others. Just think of voting inside your family (e.g., about the destination of your next vacation); only rarely are the children granted the same voting weight as their parents. Another example is the general assembly of a stock corporation where in principal every shareholder is entitled to vote, but their votes will certainly not have the same weight but are weighted by the number of shares each voter holds (see also the "weighted voting games" in Section 3.1.2.2 starting on page 146).

In a weighted election in which the candidates receive points from the voters, every vote is assigned a nonnegative integer as its weight. The points that a voter gives to a candidate are multiplied by the corresponding weight. The unweighted case (which we have always considered so far) is the special case of a weighted election in which all voters have unit weight.

The preference profile $P^{(k)} = (C, V^{(k)})$, which played a role in the property of homogeneity (see Section 4.2.7.1 on page 280), essentially means that every voter in the original election (C,V) has weight k in $P^{(k)}$, except that we were speaking of "multiplying" the votes there instead of weighing them. Unlike in $P^{(k)}$, though, it is now also allowed that distinct voters may have different weights. Note, however, that a voter with, say, weight k is something other than k voters with weight one each. The latter are more flexible: If $k = 3$, for example, then these three voters may have three different preference lists. By contrast, one voter with weight three can have only one preference list. This distinction is particularly important in the case of strategic voting, as it may make sense for the manipulators to use different strategic votes to achieve their common goal.

4.3.3.4 Constructive versus Destructive Manipulation

All variants of the manipulation problem considered so far aim to make a distinguished candidate a winner of an election. That is why one refers to this case as *constructive manipulation*. Conitzer, Sandholm, and Lang [310] also consider the analogous case of *destructive manipulation* in which the strategic voters seek to prevent a distinguished candidate from winning the election. Here, the manipulators do not care about who wins, all they care about is to prevent the given candidate's victory. In the example used to show that Borda is not strategy-proof (see Example 4.1 on page 274), a destructive manipulation was described: Since the manipulator could not make her most favorite candidate D (the top candidate of her sincere vote $D > C > B > A$)

win the election, she at least tried to prevent her most despised candidate A from winning and, therefore, casted the strategic vote $B > D > C > A$.

As noted by Hemaspaandra, Hemaspaandra, and Rothe [541],[27] for voting systems so strongly voiced as to always have at least one winner whenever there is at least one candidate (e.g., plurality voting is voiced in this sense, whereas Condorcet voting is not), any kind of *destructive* manipulation problem in the *unique-winner* model can be reduced in polynomial time to the corresponding *constructive* manipulation problem in the *nonunique-winner* model. Namely, to solve the former problem for some given candidate, it is enough to test whether a constructive manipulation (in the nonunique-winner model) is possible for any of the other candidates.

Such a reduction from a problem X to a problem Y is called a *(polynomial-time) disjunctive truth-table reduction* and is written $X \leq^{P}_{\text{d-tt}} Y$. A bit more formally, given an input x, we first compute several query strings $y_1, y_2, \ldots, y_k$ in polynomial time (in particular, this means that k and the size of each query is polynomially bounded in the size of x), and the answer to the question "$x \in X$?" is "yes" if and only if there is some i, $1 \leq i \leq k$, such that the answer to "$y_i \in Y$?" is "yes."

On the one hand, the $\leq^{P}_{\text{d-tt}}$-reducibility is a special case of the so-called *(polynomial-time) truth-table reducibility* in which arbitrary boolean expressions (instead of only disjunctions) are allowed to evaluate the answer tuple that the oracle returns on a list of query strings.[28] On the other hand, the $\leq^{P}_{\text{d-tt}}$-reducibility generalizes the $\leq^{P}_{\text{m}}$-reducibility that was introduced in Definition 1.2 on page 29, since $\leq^{P}_{\text{m}}$ is nothing other than $\leq^{P}_{\text{d-tt}}$ with only one query. Properties analogous to those of the $\leq^{P}_{\text{m}}$-reducibility stated in Lemma 1.1 on page 30 can be shown to hold for the $\leq^{P}_{\text{d-tt}}$-reducibility as well (see the original paper by Ladner, Lynch, and Selman [635] or, e.g., the textbook [819]). In particular, upper bounds are inherited downward by $\leq^{P}_{\text{d-tt}}$ (e.g., if $X \leq^{P}_{\text{d-tt}} Y$ and Y is in P, then X is in P). The above $\leq^{P}_{\text{d-tt}}$-reduction from a destructive, unique-winner manipulation problem to the corresponding constructive, nonunique-winner manipulation problem, both for the same strongly voiced voting system, thus implies that whenever this constructive manipulation problem can be solved in polynomial time, there is a polynomial-time algorithm solving this destructive manipulation problem.

4.3.3.5 Overview of the Complexity of Manipulation Problems

Summing up, the problems

---

[27] They actually observe this for control problems (see Section 4.3.4), but their insight applies to manipulation (this section) and bribery problems (Section 4.3.5) as well.

[28] Note that the complexity class $P^{NP}_{\|}$ (defined on page 291) is nothing other than the closure of NP under truth-table reductions, i.e., the class of problems that can be truth-table reduced to some set in NP.

- CONSTRUCTIVE-COALITIONAL-WEIGHTED-MANIPULATION (CCWM),
- DESTRUCTIVE-COALITIONAL-WEIGHTED-MANIPULATION (DCWM)

describe the most general manipulation scenarios considered here. We denote by CCM and DCM the unweighted special cases of CCWM and DCWM, respectively, and by CM and DM the special cases of CCM and DCM, respectively, in which there is only one manipulator instead of a coalition of manipulators. (We omit defining the weighted problems with a single manipulator because they haven't been considered separately in the literature.)

By Lemma 1.1 on page 30, upper complexity bounds are inherited downward with respect to $\leq_m^P$ (i.e., if $A \leq_m^P B$ and B in P, then A is in P as well), and lower complexity bounds are inherited upward with respect to $\leq_m^P$ (i.e., if $A \leq_m^P B$ and A is NP-hard, then B is NP-hard as well). For example, the above-mentioned NP-hardness of STV-CM proven by Bartholdi and Orlin [99] immediately implies that also STV-CCM and STV-CCWM are NP-hard. Conversely, since the most general manipulation problem, CCWM, is in P for plurality, its special cases are in P as well for this voting system.

Plurality voting and the trivial scoring protocol with vector $\boldsymbol{\alpha} = (0,\ldots,0)$, however, are the only scoring protocols for which CCWM can be solved in polynomial time, as Hemaspaandra and Hemaspaandra [535] have shown the following dichotomy result for scoring protocols (see also Theorem 5.4 on page 389 in the next chapter): The problem $\boldsymbol{\alpha}$-CCWM is NP-complete for all scoring protocols with vector $\boldsymbol{\alpha} = (\alpha_1, \alpha_2, \ldots, \alpha_m)$ satisfying that

$$\|\{\alpha_i \,|\, 2 \leq i \leq m\}\| \geq 2, \tag{4.8}$$

and otherwise it is in P. Condition (4.8) is also called *diversity of dislike*, since it describes all scoring protocols for which the "disliked candidates" (i.e., the candidates after the most preferred candidate) each receive one of at least two distinct scores (i.e., their scores are diverse). Since plurality and the trivial scoring protocol with an all-zero scoring vector are the only scoring rules not satisfying condition (4.8),[29] the problem CCWM can be solved in polynomial time only for these two scoring protocols (unless P = NP).

This dichotomy has also been observed by Conitzer, Sandholm, and Lang [310], for scoring protocols with three candidates. Furthermore, they study a variety of voting systems regarding their computational complexity for weighted, coalitional manipulation, i.e., for the problems CCWM and DCWM. They not only determine the complexity of these problems for an unlimited number of candidates, but they also identify the smallest number of candidates for which each of these problems becomes hard.

Table 4.15 summarizes the known results on the complexity of CCWM and DCWM for a number of voting systems. The results for Bucklin and fallback are due to Faliszewski et al. [439], the results on the constructive

---

[29] In particular, note that each scoring protocol with vector $\boldsymbol{\alpha} = (k, 0, \ldots, 0)$, $k > 1$, is equivalent to that with vector $\boldsymbol{\alpha} = (1, 0, \ldots, 0)$, i.e., to plurality.

4.3 Complexity of Voting Problems

Table 4.15 Complexity of manipulation in various voting systems. The entry marked by * means "NP-complete" in the nonunique-winner model and "P" in the unique-winner model. The entry marked by ? refers to an open question.

	CCWM			DCWM	
	\multicolumn{5}{c}{Number of candidates}
	≤ 2	3	≥ 4	≤ 2	≥ 3
Plurality	P	P	P	P	P
Cup protocol	P	P	P	P	P
Fallback	P	P	P	P	P
Simpson	P	P	NP-complete	P	P
Nanson	P	P	NP-complete	P	?
Copeland	P	NP-complete / P*	NP-complete	P	P
Bucklin	P	NP-complete	NP-complete	P	P
Veto	P	NP-complete	NP-complete	P	P
Borda	P	NP-complete	NP-complete	P	P
STV	P	NP-complete	NP-complete	P	NP-complete

case for Nanson are due to Narodytska, Walsh, and Xia [726] (who left the destructive case open, indicated by a question mark in the table; it is easy to see, though, that Nanson-DCWM is in P for two or fewer candidates), the results for Copeland in the nonunique-winner model are due to Faliszewski, Hemaspaandra, and Schnoor [432] and all other results in the table are due to Conitzer, Sandholm, and Lang [310] (who focus on the unique-winner model).

The polynomial-time algorithms for plurality, fallback, and the cup protocol in Table 4.15 work for any number of candidates. As one can see in the table, the destructive manipulation problem DCWM is actually never harder then the constructive manipulation problem CCWM. On the other hand, CCWM sometimes is really harder than DCWM (assuming $P \neq NP$). For example, CCWM is NP-complete for Bucklin, veto, and Borda with three candidates, but DCWM is in P for the same systems with three candidates. Gaspers et al. [493] show that Schulze-CCWM is in P when the number of candidates is a fixed constant, yet the complexity of this problem with an unbounded number of candidates remains open.

Our discussion of the complexity of these problems in general refers to the nonunique-winner model, though most results hold in the unique-winner model as well, and the proofs in both models usually differ only by minor tweaks. There are exceptions, though; for example, some voting systems (STV, Baldwin, etc.) are resolute by definition, i.e., they can have only one winner. Further, the discussion of nonunique versus unique winners is a bit subtle in the case of Copeland$^\alpha$-CCWM, and we refer to the original papers by Faliszewski, Hemaspaandra, and Schnoor [432, 434] for the details of this discussion. In particular, looking at the result marked by * in Table 4.15 (for Copeland-CCWM with three candidates), NP-completeness holds in the nonunique-winner model only [432], whereas this problem is in P in the unique-winner model ([310]; see also Theorem 4.9). Further, while the table gives the results only for Copeland-CCWM (i.e., for Copeland$^{1/2}$-CCWM), let us also mention that NP-completeness of Copeland$^\alpha$-CCWM holds for $\alpha < 1$ provided that there are at least four candidates; for $\alpha = 1$, NP-completeness is known only for an unbounded number of candidates. More precisely, for $\alpha = 1$ the problem is in P for up to four candidates [434], and its complexity is still unknown for other fixed numbers of candidates.

As an example for how to prove complexity results for manipulation problems, we now turn to Copeland voting in more detail.

4.3.3.6 Manipulating Copeland Elections with Three Candidates

According to Table 4.15, CCWM for Copeland elections with three candidates is solvable in polynomial time,[30] provided one adopts—as Conitzer, Sandholm, and Lang [310] do—the unique-winner model. Note that this is one of the rare cases where the complexity differs depending on whether one adopts the unique-winner or the nonunique-winner model: As we will see in Theorem 4.9 below, for elections with three candidates and for each rational number α with $0 < \alpha < 1$, Copeland$^\alpha$-CCWM is in P if the unique-winner model is employed; however, this problem is NP-complete in the nonunique-winner model by a result of Faliszewski, Hemaspaandra, and Schnoor [432].

Theorem 4.9 (Conitzer, Sandholm, and Lang [310]). *For elections with at most three candidates, Copeland-CCWM (in the unique-winner model) is in* P.

Proof. The key insight of this proof is that if there exists any successful way at all for the manipulators to cast their votes so as to achieve their goal, then they can do so by casting an identical, easy-to-determine vote. It is enough to consider the case of exactly three candidates. We are given an instance of (the unique-winner variant of) the problem Copeland-CCWM:

- a set $C = \{a, b, c\}$ of candidates, where c is the distinguished candidate the manipulators want to make a unique winner,
- the sincere voters' preference lists, collected in V, and a list of their weights, and
- a list of the weights of the manipulators, but not their preference lists (i.e., the strategic votes in S will only be set when the manipulation action is conducted).

Let $w : V \cup S \to \mathbb{N}$ be the weight function, and let $K = \sum_{s \in S} w(s)$ be the total weight of all manipulators.

For a sublist $U \subseteq V \cup S$ and any two candidates $x, y \in C$, we define

---

[30] Recall that Copeland means Copeland$^{1/2}$, that is, both candidates get half a point if there is a tie in a pairwise comparison. By the way, this is exactly the system where the proof method of Faliszewski, Hemaspaandra, and Schnoor [432, 433] fails to establish NP-completeness for manipulation in the unweighted case.

4.3 Complexity of Voting Problems

$$N_U(x,y) = \sum_{u \in U \,:\, x >_u y} w(u) \quad \text{and}$$

$$D_U(x,y) = N_U(x,y) - N_U(y,x).$$

$N_U(x,y)$ is the sum of the weights of those voters in U that prefer x to y. The difference $D_U(x,y)$ is positive if the total weight of x's supporters is greater than that of y's supporters, $D_U(x,y)$ is negative if y's supporters outweigh x's supporters (since $D_U(x,y) = -D_U(y,x)$), and $D_U(x,y) = 0$ if both groups balance each other out. Now consider the following four cases.

Case 1: $K > D_V(a,c)$ and $K > D_V(b,c)$. In this case, if all manipulators in S cast the vote $c > a > b$, c is the one and only winner of $(C, V \cup S)$.

Case 2: $K > D_V(a,c)$ and $K = D_V(b,c)$. It may be assumed, without loss of generality, that all manipulators put their favorite candidate c on top of their votes. However, who of them votes $c > a > b$ and who votes $c > b > a$? Since c is on top of every vote from S, we have from the case assumption that

$$D_{V \cup S}(c,a) = K - D_V(a,c) > 0, \tag{4.9}$$

$$D_{V \cup S}(c,b) = K - D_V(b,c) = 0. \tag{4.10}$$

Due to (4.9), c gets one point from the pairwise comparison with a in $(C, V \cup S)$, and a gets no points from this comparison. Due to (4.10), the pairwise comparison between b and c in $(C, V \cup S)$ ends up with a tie, so both get half a point. Without the last pairwise comparison, a versus b, c already has one and a half points in $(C, V \cup S)$, b half a point, and a has no point at all. In order to make c a unique winner in $(C, V \cup S)$, b must not get a whole point from the comparison with a, i.e., it must hold that $D_{V \cup S}(a,b) \geq 0$. This, however, is true exactly if $K \geq D_V(b,a)$. Also in this case, all manipulators cast the vote $c > a > b$, seeking to ensure that $D_{V \cup S}(a,b) \geq 0$. If this is not enough to make c a unique winner (namely, because $K < D_V(b,a)$), then there exists no successful manipulation in this case.

Case 3: $K = D_V(a,c)$ and $K > D_V(b,c)$. This case can be handled analogously to Case 2, with the roles of a and b reversed.

Case 4: $K < D_V(a,c)$ or $K < D_V(b,c)$ or $(K \leq D_V(a,c)$ and $K \leq D_V(b,c))$. In this case, the Copeland score of c in $(C, V \cup S)$ cannot be greater than 1, regardless of how the manipulators vote. Thus, they are doomed to fail: c cannot be made a unique winner.

In all four cases, we have that either

- no manipulation at all is possible (namely, if we have $K < D_V(b,a)$ in Case 2 or $K < D_V(a,b)$ in Case 3, or if we are in Case 4), or
- all manipulators can vote identically to achieve their goal (namely, $c > a > b$ in Cases 1 and 2, or $c > b > a$ in Case 3).

Thus, the polynomial-time algorithm only has to check which of the four cases applies and—according to the case at hand—to test whether the optimal strategy of the manipulators is successful. In Case 3, for example, it has to be checked whether c is the unique winner in $(C, V \cup S)$ if all manipulators cast the vote $c > b > a$. The input will be accepted exactly if this test is successful. ❏

4.3.3.7 Manipulating Copeland Elections with Four Candidates

Now we show that by adding only one more candidate Copeland-CCWM becomes NP-complete. To make use of the third statement of Lemma 1.1 from Section 1.5, we need an NP-complete problem that is $\leq_m^P$-reducible to Copeland-CCWM. For most of the weighted manipulation problems from Table 4.15, and in particular for Copeland-CCWM, the NP-complete problem PARTITION is especially well-suited. This problem has been defined in Section 3.5.2.1 starting on page 181 as follows:

PARTITION

Given: A nonempty sequence $(k_1, k_2, \ldots, k_n)$ of positive integers such that $\sum_{i=1}^{n} k_i$ is even.
Question: Is there a subset $A \subseteq \{1, 2, \ldots, n\}$ with $\sum_{i \in A} k_i = \sum_{i \in \{1,2,\ldots,n\} \setminus A} k_i$?

In the following result and its proof, we again adopt the unique-winner model. Do you see which details would have to be changed in the proof so as to establish the same result in the nonunique-winner model? (Hint: [432]).

Theorem 4.10 (Conitzer, Sandholm, and Lang [310]). *For elections with at least four candidates, Copeland-CCWM (in the unique-winner model) is* NP*-complete.*

Proof. Membership of the problem in NP is obvious. To show its NP-hardness, we will describe a $\leq_m^P$-reduction from PARTITION to Copeland-CCWM. Given a PARTITION instance $(k_1, k_2, \ldots, k_n)$ with $\sum_{i=1}^{n} k_i = 2K$, we construct the following instance of (the unique-winner variant of) the problem Copeland-CCWM:

- a set $C = \{a, b, c, d\}$ of candidates, where d is the distinguished candidate the manipulators want to make a unique winner,
- a list V of four sincere voters whose preferences and weights are shown in Figure 4.16(a), and
- the weights of the n strategic voters in S, where the ith manipulator has weight k_i, so the manipulators' total weight is $2K$.

Figure 4.16(b) shows the Copeland scores resulting from the pairwise comparisons in (C, V): the weight of the voters in favor of the row candidate minus

4.3 Complexity of Voting Problems 317

Weight	Preference list
$2K+2$	$d > a > b > c$
$2K+2$	$c > d > b > a$
$K+1$	$a > b > c > d$
$K+1$	$b > a > c > d$

(a) Sincere voters in V

	a	b	c	d	$C^{1/2}Score(\cdot)$
a	–	0	$2K+2$	$-2K-2$	1.5
b	–	–	$2K+2$	$-2K-2$	1.5
c	–	–	–	$2K+2$	1
d	–	–	–	–	2

(b) Pairwise comparisons and scores in (C,V)

Fig. 4.16 Reduction PARTITION $\leq_m^P$ Copeland-CCWM in the proof of Theorem 4.10

the weight of the voters in favor of the column candidate. As one can easily see, restricted to the sincere voters (i.e., to the election (C,V)), all but one of the pairwise comparisons are already decided, since the total weight of the manipulators, $2K$, is too low to flip the result of these pairwise comparisons.

The only as yet undecided comparison is the one between a and b. If a or b wins this comparison in $(C,V \cup S)$, then this candidate has the same Copeland score as d. However, the manipulators want to make d a *unique* winner, so they want to prevent that some of a and b outweighs the other. Therefore, it is possible for them to make their favorite candidate d a unique winner of $(C,V \cup S)$ if and only if the pairwise comparison between a and b ends up in a tie. They are tied already in the election (C,V) without the manipulators (this is the 0 entry in Figure 4.16(b)). This tie is preserved in $(C,V \cup S)$ exactly if

$$N_S(a,b) = \sum_{s \in S \,:\, a >_s b} w(s) = \sum_{s \in S \,:\, b >_s a} w(s) = N_S(b,a),$$

which in turn is equivalent to the equality $\sum_{i \in A} k_i = \sum_{i \in \{1,2,\ldots,n\} \setminus A} k_i$ for some subset $A \subseteq \{1,2,\ldots,n\}$, where $i \in A$ if and only if the ith manipulator prefers a to b. But that just says that $(k_1, k_2, \ldots, k_n)$ is a yes-instance of PARTITION. Summing up, this shows that the manipulators can make d a unique winner of $(C,V \cup S)$ if and only if $(k_1, k_2, \ldots, k_n)$ is in PARTITION. It follows that PARTITION $\leq_m^P$ Copeland-CCWM, which proves that Copeland-CCWM is NP-hard. ❑

4.3.3.8 Manipulation and Possible Winners

In Section 4.3.1 starting on page 289, the possible winner problem has been presented. This problem generalizes the constructive coalitional manipulation problem (CCM), since CCM can be seen as the following special case of POSSIBLE-WINNER (recall also Figure 4.15 on page 307): The votes of the nonmanipulative voters are complete linear preference lists and the votes of the manipulative voters are "extremely partial," namely empty, preference lists. Asking whether these empty votes of the manipulators can be set (i.e., whether they can be extended to complete votes) such that their desired candidate wins is nothing other than asking whether this candidate is a possible winner in this setting. This describes a (trivial) reduction from CCM to POSSIBLE-WINNER from which it follows, by Lemma 1.1 on page 30, that if CCM is NP-hard for a voting system, then POSSIBLE-WINNER is also NP-hard for this voting system. On the other hand, if POSSIBLE-WINNER is in P for a voting system, then CCM is also in P for this voting system.

As for the manipulation problems, one can also study weighted variants of POSSIBLE WINNER (see the references given in the last paragraph of Section 4.3.2.2 on page 302) that similarly generalize the constructive coalitional weighted manipulation problem (CCWM).

4.3.3.9 How Can Manipulators Cope with NP-Hardness?

The vast number of papers on the computational aspects of manipulation in voting has yielded NP-completeness or P results so far, as we have seen in this section (see also the survey by Faliszewski and Procaccia [438] and the book chapter by Conitzer and Walsh [312]). However, a manipulator who is faced with an NP-hard problem does not have to despair. NP-hardness, after all, only expresses the complexity of a problem in the *worst case*. A reduction often shows NP-hardness of a manipulation problem by constructing very special instances that cannot be solved in a reasonable time. However, this does not exclude that the problem may be efficiently solvable for many other instances—possibly for the overwhelming majority of instances occurring in practice. That is why efforts to analyze the hardness of manipulation for *typical* instances have been intensified, seeking to circumvent NP-hardness of such problems by heuristic algorithms that often—even though certainly not always—provide a correct solution.

These attempts, however, should not be mistaken for an analysis of these problems in the model of *average-case complexity* (which goes back to Levin [654], see also the survey by Goldreich [506] and the book chapters by Wang [933, 934]), even though they are sometimes—by mistake and misleadingly—termed like that in the literature. There are not many problems whose worst-case NP-hardness could be shown to transfer to hardness on the average, and among these there are no manipulation problems.

On the other hand, not a single NP-hard manipulation problem is currently known to be solvable in *"average-case polynomial time"* (AvgP). One reason for the difficulty to show that NP-hard manipulation problems are in AvgP is that the problem instances depend on the underlying probability distribution. And this raises a related question: What is a "typical" distribution of preference profiles in elections? Even for the uniform distribution (which is relatively easily manageable but in fact reflects real-world elections only rather insufficiently), not many results are known.

4.3 Complexity of Voting Problems

An interesting approach is due to Procaccia and Rosenschein [790], who study so-called "junta distributions" that put much weight on certain problem instances, namely as much weight as needed to make NP-complete manipulation problems for scoring protocols solvable in "heuristic polynomial time" (for a technical definition of this notion, see their paper [790]). Following up this line of research, Erdélyi et al. [409] show that in fact plenty of NP-hard problems can be solved in (this type of) heuristic polynomial time with high probability with respect to (a certain generalization of) junta distributions, even such problems as SAT (defined in Section 1.5.2 on page 24) that are not considered to be easily solvable under "natural" distributions. That means that solvability by heuristic polynomial-time algorithms for junta distributions does not seem to provide strong arguments for the efficient solvability of typical instances of NP-hard problems. Nevertheless, this quite interesting approach of Procaccia and Rosenschein [790] should be pursued, for adequately restricted variants of juntas.

In general, it is known from early work in theoretical computer science (to wit, from a conjecture by Schöning [841] that has been formally proven by Ogiwara and Watanabe [758]) that, assuming $P \neq NP$, no polynomial-time heuristic whatsoever can correctly solve any NP-hard problem on all but a *sparse* set of instances (i.e., with only a polynomial number of exceptions allowed at each input size). For a more detailed discussion, we refer the reader to the survey of Hemaspaandra and Williams [552].

Another idea for how to cope with NP-hardness is due to Homan and Hemaspaandra [555] (see also the closely related work of McCabe-Dansted, Pritchard, and Slinko [695]), who introduce the notion of *"frequently self-knowingly correct algorithm"* and show that, under reasonable assumptions, such an algorithm determines the Dodgson winners of an election with high probability. Though this actually is not a manipulation but a winner problem, it is an NP-hard (and, as mentioned in Section 4.3.1, even a $P_{\parallel}^{NP}$-complete) problem as well, and their result concerns the frequency of algorithmic correctness with regard to election instances chosen uniformly at random. Erdélyi et al. [408] related this to the theory of average-case complexity by showing that *every* problem from the complexity class AvgP with respect to the uniform distribution has such a frequently self-knowingly correct algorithm. The converse, however, is provably false [408].

Conitzer and Sandholm [308] analyze the hardness of manipulation problems beyond worst-case complexity measures in the following sense. Roughly speaking, they show that no voting rule satisfying two properties (namely, (a) *"weak monotonicity"* defined in a certain technical sense related to the manipulative votes, and (b) the property that the manipulators can make one of exactly two candidates win by strategic voting) can be hard to solve for "typical" instances. They justify and study these properties both theoretically and experimentally, for both monotonic voting systems (such as plurality, Borda, veto, Copeland, and maximin) and nonmonotonic voting systems (such as STV and plurality with run-off).

Friedgut et al. [473, 474] establish a *"quantitative version"* of the Gibbard–Satterthwaite theorem (see Theorem 4.2 on page 275). They show the following result for neutral voting rules that are "far" (in a certain technical sense[31]) from being a dictatorship: If the voters' preferences in a given election with three candidates are uniformly distributed, a single manipulator can successfully manipulate it with nonnegligible probability (i.e., with a probability that is at least inverse-polynomially small in the number of voters), by changing his preference into a randomly chosen one. Can this result be generalized to more than three candidates?

Pursuing this line of research, Xia and Conitzer [954] and Dobzinski and Procaccia [348] both establish versions of this result that do apply to four or more candidates, but require further properties to be satisfied or are otherwise limited. For example, the result of Dobzinski and Procaccia [348] requires the voting system to be Pareto-consistent and applies to only two voters. Can their result be generalized to more than two voters?

Solving one of the main open problems raised by Friedgut, Kalai, and Nisan [473], Isaksson, Kindler, and Mossel [568, 569] eventually succeeded in proving a general quantitative version of the Gibbard–Satterthwaite theorem for neutral voting systems that applies to more than three candidates and to more than two voters. Later on, Mossel and Rácz [714] generalized this result even further by proving that the assumption of neutrality in fact is not needed.

Continuing and generalizing the work started by Procaccia and Rosenschein [790] and others (see, e.g., the work of Slinko [875, 876, 877]), Xia and Conitzer [953] analyze the probability of a manipulative coalition being successful, depending on the number of voters and manipulators. They obtain results for a new class of voting systems, so-called *generalized scoring rules*, including many commonly used voting systems such as (regular) scoring rules, Copeland, STV, and ranked pairs.

Approximation algorithms are another approach to cope with the NP-hardness of manipulation problems. Although they do not always find optimal solutions of a problem, such algorithms can approximate it within a certain factor. Zuckerman, Procaccia, and Rosenschein [974] define an optimization variant of the decision problem CCM, which they call UNWEIGHTED-COALITIONAL-OPTIMIZATION (UCO). The objective is to determine the smallest number of manipulators for which the victory of a designated candidate can be guaranteed. They designed a greedy algorithm (called REVERSE) that is applicable to a number of voting systems so as to approximate UCO within a certain factor. For example, their algorithm approximates UCO within an additive constant of 1 for Borda (one additional manipulator is enough) and within a factor of 2 for Simpson (a.k.a. maximin) voting (twice the optimal number of manipulators suffices in the worst case). On the one hand, Zuckerman, Lev, and Rosenschein [973] later on improve the approx-

---

[31] A voting system $\mathcal{E}$ is ϵ-*far from a dictatorship* if the fraction of voter profiles (chosen uniformly at random) for which $\mathcal{E}$ differs from dictatorial systems is at least ϵ.

imation factor for Simpson-UCO to $5/3$; on the other hand, they show that this factor cannot be better than $3/2$ for any efficient approximation algorithm (unless $P = NP$). Xia, Conitzer, and Procaccia [960] reduce the manipulation problem to some scheduling problem so as to approximate both the weighted and unweighted manipulation problem for scoring protocols within an additive term of $m - 2$, where m is the number of candidates. That is, their algorithm is guaranteed to find a successful manipulation (whenever some exists for the given election) with at most $m - 2$ additional manipulators.

Another line of research deals with the experimental simulation and evaluation of heuristics and approximation algorithms to solve NP-hard manipulation problems. These studies have been initiated by Walsh [928, 929], whose empirical results show that NP-complete manipulation problems for veto and STV can be solved fast for many problem instances. Davies et al. [334] design two more approximation algorithms, LARGEST FIT and AVERAGE FIT, to find optimal manipulations for the Borda system and compare them both theoretically and experimentally with the above-mentioned algorithm REVERSE of Zuckerman, Procaccia, and Rosenschein [974].

The literature on how to cope with NP-complete manipulation problems is growing fast and provides many fascinating, technically deep results. Rothe and Schend [824] have surveyed and discussed such challenges to worst-case complexity results for manipulation, including—in addition to the results mentioned above—fixed-parameter tractability and domain restrictions such as restricting elections to so-called *"single-peaked"* electorates. The latter topic will be treated in more detail in Chapter 5 of this book.

4.3.3.10 Online Manipulation in Sequential Elections

So far we have always assumed that the votes in an election are cast *simultaneously* and that the manipulators know all the nonmanipulative votes. However, there are many situations where elections are held *sequentially* (that is, the votes come in one after the other—for comparison, recall the sequential two-player games from Section 2.3.1 where the players move in sequence), and in such a setting it is natural to assume that each manipulator, when it is her turn to move, does know the past votes, yet not the future ones.

Anna, Belle, Chris, David, and Edgar—as the members of the library committee of their school—are responsible for deciding on which books to acquire next. They have agreed on voting by the scoring protocol 2-approval (that gives one point each to the two top spots and no points to the others; recall this system from Section 4.1.1.3 on page 237) on the following four alternatives:

- J. R. R. Tolkien's *"Lord of the Rings"* trilogy plus *"The Hobbit,"*

- Cecily von Ziegesar's 13 *"Gossip Girl"* novels,
- J. K. Rowling's seven *"Harry Potter"* volumes, and
- Douglas Adams's five volumes of the (deliberately misnamed) *"Hitchhiker's Guide To The Galaxy Trilogy"* plus Eoin Colfer's *"And Another Thing ... "*

Anna and Belle want to make sure that *"Gossip Girl"* wins (and they don't care which other book series may win, too—the school has enough money to afford several winners; each purchase must be justified by sufficient support from the library committee, though; more precisely, only 2-approval winners will be purchased).

Unfortunately, the decision is due by the end of the year and the committee members are on New Year holidays already. That is why they have to vote by email. They agree that each email message containing a vote will be sent to all of them, so every voter—when it is his or her turn to vote—will know all the previous votes, but not the future ones. The order in which they will vote is fixed by their holiday schedules constraining the times when they will have internet access: Chris will vote first, then Anna, next David, then Belle, and Edgar is last. Before they vote, Anna calls Belle at home and they agree on their "ideal" preference order:

Gossip Girl > *Harry Potter* > *Hitchhiker's Guide* > *Lord of the Rings*.

Chris sends the first message. The two top choices are underlined.

From: Chris　　　　　　　　　　　Date: Tue, December 31, 2013
　To: Anna, Belle, David, Edgar　　Time: 4:23 pm
Subject: My vote on the books is ...

... <u>*Lord of the Rings*</u> > <u>*Harry Potter*</u> > *Hitchhiker's Guide* > *Gossip Girl*.
– Chris

Next, it is Anna's turn, and since Chris was approving of *Lord of the Rings* and *Harry Potter*, she must approve of the other two alternatives, to make sure that her favorite, *Gossip Girl*, draws level with Chris's two top choices and that these don't gain any further points.

From: Anna　　　　　　　　　　　Date: Tue, December 31, 2013
　To: Belle, Chris, David, Edgar　　Time: 4:31 pm
Subject: Re: My vote on the books is ...

... <u>*Gossip Girl*</u> > <u>*Hitchhiker's Guide*</u> > *Lord of the Rings* > *Harry Potter*.
– Anna

David is next. Just as Chris, he doesn't like *Gossip Girl* at all. But he loves the *Hitchhiker's Guide*.

4.3 Complexity of Voting Problems

From: David
To: Anna, Belle, Chris, Edgar
Subject: Re: My vote on the books is ...
Date: Tue, December 31, 2013
Time: 4:33 pm

... *Hitchhiker's Guide* > *Lord of the Rings* > *Harry Potter* > *Gossip Girl*.
– David

Now Belle has to cast her vote, and she looks at the previous votes first. Again, to make sure that her beloved *Gossip Girl* draws level with David's two top choices and that these don't gain any further points from her vote, she is forced to approve of the other two alternatives.

From: Belle
To: Anna, Chris, David, Edgar
Subject: Re: My vote on the books is ...
Date: Tue, December 31, 2013
Time: 4:46 pm

... *Gossip Girl* > *Harry Potter* > *Lord of the Rings* > *Hitchhiker's Guide*.
– Belle

Finally, it is Edgar's turn, a notorious *Harry Potter* hard-core fan.

From: Edgar
To: Anna, Belle, Chris, David
Subject: Re: My vote on the books is ...
Date: Tue, December 31, 2013
Time: 4:51 pm

... *Harry Potter* > *Hitchhiker's Guide* > *Lord of the Rings* > *Gossip Girl*.
– Edgar

Having counted the points, Anna sends the last email of the day.

From: Anna
To: Belle
Subject: Re: My vote on the books is ...
Date: Tue, December 31, 2013
Time: 4:52 pm

Drat!!! :(((
– Anna

Hemaspaandra, Hemaspaandra, and Rothe [543] formalize this setting, which they call an *online manipulation setting*, as a quintuple (C, u, V, σ, d) with the following components:

- C is a set of candidates (e.g., the four alternatives in the above example),
- u is a distinguished voter (e.g., Anna above); u will always be a member of the "manipulative coalition" (Anna and Belle in our example),
- V is a so-called *election snapshot for* C *and* u, which consists of all voters in the order they cast their votes, along with the vote of each voter before u, and for each voter after u there is some bit indicating whether or not that voter belongs to u's coalition; in the above example, the election snapshot for C and Anna would look like this:

1. Chris: $\underline{\text{Lord of the Rings}} > \underline{\text{Harry Potter}} > \underline{\text{Hitchhiker's Guide}} > \underline{\text{Gossip Girl}}$
2. Anna: **How do I vote at my *"magnifying-glass moment"*?**
3. David: 0
4. Belle: 1
5. Edgar: 0

where 0 means "doesn't belong to Anna's coalition" and 1 means "does belong to Anna's coalition." And two votes later, the election snapshot for C and Belle will look like this:

1. Chris: $\underline{\text{Lord of the Rings}} > \underline{\text{Harry Potter}} > \underline{\text{Hitchhiker's Guide}} > \underline{\text{Gossip Girl}}$
2. Anna: $\underline{\text{Gossip Girl}} > \underline{\text{Hitchhiker's Guide}} > \underline{\text{Lord of the Rings}} > \underline{\text{Harry Potter}}$
3. David: $\underline{\text{Hitchhiker's Guide}} > \underline{\text{Lord of the Rings}} > \underline{\text{Harry Potter}} > \underline{\text{Gossip Girl}}$
4. Belle: **How do I vote at my *"magnifying-glass moment"*?**
5. Edgar: 0

- σ is the preference order of u's coalition (*Gossip Girl* > *Harry Potter* > *Hitchhiker's Guide* > *Lord of the Rings* in our example), and
- $d \in C$ is a distinguished candidate (*Gossip Girl* in our example).

The key notion in this setting—and for the designated manipulator u in her election snapshot—is what Hemaspaandra, Hemaspaandra, and Rothe [543] call a *magnifying-glass moment*, the "moment of decision" for u who is aware of the previously cast votes (and which voters are still to come and in which order), but does not know the intentions, not to mention the votes, of those future voters not in her coalition. At this very moment, u must decide what the "best" vote is for her to cast (see the boldfaced lines in the election snapshots above). She can rely on support by the future voters belonging to her coalition: When it is their turn to vote (at their own magnifying-glass moments), they will use the knowledge they will then have so as to reach their common goal. And the common goal of u's coalition is to make d (or anyone even more preferred than d in their preference order σ) win the election. That is, the purpose of σ and d is just to have an order of candidates such that "d or someone better" according to this order is among the winners. Given a voting system $\mathcal{E}$, this can be formalized by the following decision problem.

ONLINE-$\mathcal{E}$-CONSTRUCTIVE-COALITIONAL-MANIPULATION (ONLINE-$\mathcal{E}$-CCM)

Given: An online manipulation setting (C, u, V, σ, d) as described above.

Question: Does there exist some vote for u to cast (where u can rely on support from the future manipulators in her coalition) such that whatever votes will be cast by the nonmanipulators after u, there exists some candidate $c \in C$ whom the manipulators like at least as much as d in their ranking σ and such that c is an $\mathcal{E}$ winner of the election?

This is the unweighted variant of the online constructive coalitional manipulation problem; the weighted variant, where each voter has a nonnegative integer weight, is defined analogously and is denoted by ONLINE-$\mathcal{E}$-CCWM. Let ONLINE-$\mathcal{E}$-CCM[k] and ONLINE-$\mathcal{E}$-CCWM[k], respectively, denote the restriction of ONLINE-$\mathcal{E}$-CCM and ONLINE-$\mathcal{E}$-CCWM, respectively, where the number of manipulators to come after u is bounded by k.

4.3 Complexity of Voting Problems

The destructive variants of these four problems are defined symmetrically. For example, in ONLINE-$\mathcal{E}$-DCM, given an online manipulation setting (C, u, V, σ, d), we ask whether there is some vote that u (relying on support from the future manipulators in u's coalition) can cast such that whatever votes will be cast by the nonmanipulators from u onward, the $\mathcal{E}$ winner set neither includes d nor any even more despised candidate. The other three destructive variants are defined analogously and denoted by ONLINE-$\mathcal{E}$-DCWM, ONLINE-$\mathcal{E}$-DCM[k], and ONLINE-$\mathcal{E}$-DCWM[k].

Each online manipulation problem generalizes the corresponding (standard) manipulation problem defined earlier in this section. For example, CCM is the special case of ONLINE-CCM with only one manipulator, who is the last one to cast a vote, and the designated candidate d is the manipulative coalition's most preferred candidate in σ (just as *Gossip Girl* is the most preferred alternative of Anna and Belle in our example). More to the point, the way ONLINE-CCM generalizes CCM (and the same applies to the other pairs of associated online and non-online manipulation problems) is that we have a sequence of alternating existential and universal quantifiers in the online problem: The question the current manipulator u must answer is whether she can cast a vote such that for each vote(s) cast by the next nonmanipulator(s), there will exist some vote(s) the subsequent manipulator(s) can cast such that ... the manipulators win (i.e., the winning set contains some candidate they like as much as d or more).

Does this sound familiar? As noted by Hemaspaandra, Hemaspaandra, and Rothe [543], their model is connected to (noncooperative) game theory: The online manipulation setting can be seen as a two-player combinatorial game, which (as we know from Section 2.3.1) is a special type of perfect-information sequential game. The two players are the manipulative coalition on the one hand and the nonmanipulative voters on the other hand,[32] and in an online manipulation problem we are asking whether there is a winning strategy for the manipulators. We have seen in Section 2.3.1.4 on page 91 that the problem of whether there is a winning strategy for the first player in the game of geography is PSPACE-complete.

And, indeed, the same can be shown for online manipulation with suitably constructed election systems. In fact, if we consider the restricted online manipulation problems where the number of manipulators to come after the current manipulator is bounded by k, we obtain completeness results for the even levels of the polynomial hierarchy (recall its definition from page 35 in Section 1.5.3.2).

Theorem 4.11 (Hemaspaandra, Hemaspaandra, and Rothe [543]).

1. *For each voting system $\mathcal{E}$ whose (weighted) winner problem is polynomial-time solvable, ONLINE-$\mathcal{E}$-CCWM (and thus also ONLINE-$\mathcal{E}$-CCM) is in* PSPACE.

---

[32] Calling some voters "manipulators" and the others "nonmanipulators" is a somewhat arbitrary choice. Both players can as well be seen as equal competitors in this game.

2. There is a voting system $\mathcal{E}$ whose (weighted) winner problem is solvable in polynomial time such that ONLINE-$\mathcal{E}$-CCM (and thus also ONLINE-$\mathcal{E}$-CCWM) is PSPACE-complete.
3. For each $k \geq 1$ and for each voting system $\mathcal{E}$ whose (weighted) winner problem is polynomial-time solvable, ONLINE-$\mathcal{E}$-CCWM[k] (and thus also ONLINE-$\mathcal{E}$-CCM[k]) is in Σ_{2k}^{p}.
4. Fix any $k \geq 1$. There is a voting system $\mathcal{E}$ whose (weighted) winner problem is solvable in polynomial time such that ONLINE-$\mathcal{E}$-CCM[k] (and thus also ONLINE-$\mathcal{E}$-CCWM[k]) is Σ_{2k}^{p}-complete.

The above approach to online manipulation in sequential elections is inspired by so-called "online algorithms" (see, e.g., [172]): Manipulation decisions have to be made based on only the partial information (about previously cast votes and the order of future voters) available to the current manipulator at her magnifying-glass moment, who pursues a "maxi-min strategy." What is meant by that is perhaps best explained by a literal quotation [543, p. 698]:

"[...] the idea that one may want to "maxi-min" things—*one may want to take the action that maximizes the goodness of the set of outcomes that one can expect regardless of what happens down the line from one time-wise*. For example, if the current manipulator's preferences are Alice > Ted > Carol > Bob and if she can cast a (perhaps insincere) vote that ensures that Alice or Ted will be a winner no matter what later voters do, and there is no vote she can cast that ensures that Alice will always be a winner, this maxi-min approach would say that that vote is a 'best' vote to cast."

A different approach to sequential elections with competing voters is due to Xia and Conitzer [955] (see also the related work of Desmedt and Elkind [339]), who investigate so-called *Stackelberg voting games* (a variant restricted to two players has been called *roll-call voting games* by Sloth [878]). In this scenario, the voters again cast their votes in sequence, but now the votes are *common knowledge* (recall this notion from Section 2.4.2). In particular, Xia and Conitzer [955] analyze Stackelberg voting games by computing a subgame-perfect Nash equilibrium using backward induction (recall these notions from Section 2.3.2.3 on page 101) and prove, under their model assumptions, that manipulation is in P for a bounded number of manipulators. This contrasts with Theorem 4.11 and reiterates that these are different models.

To briefly discuss some other related work, we mention that of Tennenholtz [911] on "dynamic voting" (or "transitive voting"). He also studies elections held sequentially, but the focus is more on axiomatic properties and specific voting systems than on manipulative coalitions in online manipulation settings as defined above. Parkes and Procaccia [770] investigate sequential decision-making under dynamically evolving preferences by means of Markov decision processes whose states are the preference profiles and whose (deterministic, stationary) policies correspond to social choice functions. Also remotely related to online manipulation is the work by Konczak and Lang [618] on possible and necessary winners that was introduced in

4.3 Complexity of Voting Problems

Section 4.3.2. One difference here is that possible or necessary winners involve only one (existential or universal) quantifier, whereas the number of quantifiers can grow with the input size in online manipulation and depends on the number of future manipulators for a given online manipulation setting (see Theorem 4.11). Finally, we mention the work of Xia, Conitzer, and Lang [956, 958, 959, 961] on multi-issue elections; a difference here is that even though multiple issues may be voted on in sequence, each single issue typically is voted on simultaneously and the votes are common knowledge.

Now turning back to online manipulation in sequential elections, we have seen rather general statements in Theorem 4.11, whose PSPACE- and Σ_{2k}^{p}-completeness results are proven via reductions from QBF (the quantified boolean formula problem; recall its definition from Section 2.3.1.4 on page 91) and its restriction with the number of alternating existential and universal quantifiers bounded by $2k$. These reductions are based on constructing suitable voting systems that, in a subtle way, encode quantified boolean formulas. Therefore, these are rather artificial voting systems. This raises the question of what is known for online manipulation in sequential elections for *natural* voting systems such as those defined in Section 4.1.

The following results are due to Hemaspaandra, Hemaspaandra, and Rothe [543]. First, they show that both constructive and destructive online manipulation for plurality can be solved efficiently in the weighted (and thus also in the unweighted) case, provided we are in the nonunique-winner model: ONLINE-plurality-CCWM, ONLINE-plurality-CCM, ONLINE-plurality-DCWM, and ONLINE-plurality-DCM each are in P. In the unique-winner model,[33] however, the complexity results sharply differ for these problems: We have NP-hardness (respectively, coNP-hardness) in the destructive (respectively, constructive) weighted case of online manipulation for plurality, even when restricted to just two candidates. This is one of the rare cases where it really matters whether we are in the nonunique-winner or in the unique-winner model.

Second, the dichotomy result for the (standard) manipulation problem CCWM that is due to Hemaspaandra and Hemaspaandra [535] (see condition (4.8) on page 312 and also Theorem 5.4 on page 389 in the next chapter) directly transfers to the online variant in the nonunique-winner model:[34] For all scoring protocols with vector $\boldsymbol{\alpha} = (\alpha_1, \alpha_2, \ldots, \alpha_m)$ satisfying that $\alpha_2 = \alpha_m$, ONLINE-$\boldsymbol{\alpha}$-CCWM is in P, and is NP-hard otherwise.

---

[33] In the unique-winner model, given an online manipulation setting (C, u, V, σ, d), we ask whether the current manipulator u (relying on support from the future manipulators) can guarantee that some candidate $c \in C$ who is liked by the manipulators at least as much as the distinguished candidate d is *the one and only* $\mathcal{E}$ winner of the election.

[34] In fact, Hemaspaandra and Hemaspaandra [535] prove their dichotomy result for $\boldsymbol{\alpha}$-CCWM in both the unique- and the nonunique-winner model. Hemaspaandra, Hemaspaandra, and Rothe [543], however, state their derived dichotomy result for ONLINE-$\boldsymbol{\alpha}$-CCWM in the nonunique-winner model only, as that is their default model.

Third, for each $k \geq 1$, ONLINE-k-approval-CCM and ONLINE-k-veto-CCM are both in P. An interesting point to note here is that the proof differs for $k = 1$ and $k > 1$. For 1-approval (i.e., plurality) and 1-veto (i.e., veto), the current manipulator u can already force success against the strongest actions of the remaining nonmanipulators (if that is possible at all) by setting her own and the future manipulators' votes, independently of the votes cast by the nonmanipulators after u.[35] By contrast, the actions of the manipulators coming after u can depend on the actions of intervening nonmanipulators for k-approval and k-veto if $k > 1$. Indeed, we have already observed such a case in the example given in the gray box at the beginning of this section: The library committee used 2-approval as their voting system in this example, and Anna's action depended on Chris's vote that she had to counter and Belle's action depended on David's vote that she had to counter as well.

Fourth, ONLINE-veto-CCWM is complete for P^{NP}, even when restricted to just two candidates. This appears to be the first P^{NP}-completeness result in computational social choice, at least with respect to voting problems. P^{NP} is the class of problems solvable by a P oracle machine with access to an NP oracle; it is similar to the class $\mathrm{P}^{\mathrm{NP}}_{\|}$ defined in Section 4.3.1.3 on page 291, except that the oracle access in a P^{NP} computation is not restricted to be parallel: All queries may be asked sequentially and may depend on the answers to previously asked queries. Moreover, when restricted to exactly three candidates, ONLINE-veto-CCWM is complete for $\mathrm{P}^{\mathrm{NP}[1]}$, the restriction of P^{NP} where only one oracle query is allowed.

4.3.4 Control

Manipulation is one possibility to influence the outcome of an election. As we have seen, this consists of strategic voters directly interfering with the election by casting insincere votes. Another possibility to influence the outcome of an election is called *electoral control*. In contrast to manipulation, in control it is not the voters themselves who interfere, but an *election chair* (or, shorter, the *chair*), who is able to change the structure of an election. The chair may have the power to delete candidates (alternatives) from a given election, or to add new candidates to it (cf. *independence of clones* in Section 4.2.5.2 on page 276 and the *possible winner with new alternatives* problem mentioned in Section 4.3.2.3 starting on page 303). The chair may also have the power to add new voters to an election, or to delete voters (cf. the *participation criterion* from Section 4.2.7.2 starting on page 282). Other structural changes the chair may make to influence the outcome of an election will be considered later. First, let us see how Mr. Slugger, the owner of the miniature golf facility, tries to control an election of Anna, Belle, and Chris.

---

[35] Do you see how?

4.3 Complexity of Voting Problems

Some days later, Anna, Belle, and Chris meet again and discuss what to do together today. Anna again suggests bicycling, playing miniature golf, and swimming, in this order (see Figure 4.1 on page 233), because her preferences haven't changed.

The owner of the miniature golf facility, Mr. Slugger, happens to come by and asks, "By which rule are you going to vote, kids?"

"Today by a quite simple one," Chris replies, "the plurality rule. And I still have the same preferences as Anna: Bicycling on top, else miniature golf, and if that's not possible either, swimming as a last resort."

"I actually prefer miniature golf the most," says Belle, "and I consider bicycling the worst. I still have sore muscles from our last trip. But since it is two of you who prefer bicycling the most, I'll go and get my bicycle."

"Just a moment, please!" Mr. Slugger hastily interjects, "not so fast! I see David and Edgar over there. Why don't you ask them if they want to join in?" Mr. Slugger knows how much the two like to play miniature golf because he sees them almost every day on his ground.

"Good idea!" exclaims Belle and happily beckons her little brother Edgar and his friend David to come over. As Mr. Slugger expected, David and Edgar vote for miniature golf, and miniature golf wins by three to two points ahead of bicycling. So the five friends go together to the miniature golf course and Mr. Slugger, smiling highly pleased, follows them.

By adding two more voters to the election, Mr. Slugger was able to make his favorite alternative win and so has achieved his goal. As with manipulation, it would be desirable to avoid the possibility of influencing the outcome of an election by structural changes of such kind. Are there voting systems that are immune to such control attacks? Or at least voting systems that make it hard for the chair to successfully exert control?

Bartholdi, Tovey, and Trick [102] introduce various types of electoral control and investigate the corresponding control problems for plurality and Condorcet elections. Each control type either changes the candidate set or the list of voters participating in the election by one of the following actions:

- adding or deleting either candidates or voters, and
- partitioning either the candidate set or the list of votes.

These common control types and the corresponding decision problems,[36] some of which come in several variants, will now be introduced in detail. We will define all control problems in the nonunique-winner model (see, e.g., [429]); alternatively, they can be defined in the unique-winner model as well (see, e.g., [541]); indeed, most papers on control focus on the latter.

As with manipulation, we will further distinguish between constructive and destructive control. While Bartholdi, Tovey, and Trick [102] considered only *constructive* control where the chair seeks to make a given candidate win, Hemaspaandra, Hemaspaandra, and Rothe [541] introduce *destructive* control types where the chair's goal is to prevent the victory of a given candidate. We will define these control types only for the constructive cases, as the corresponding destructive cases can be defined analogously. We start with the common types of candidate control.

4.3.4.1 Control by Adding or Deleting Candidates

An easy example for how to influence the outcome of an election by adding candidates has already been given when the property of independence of clones was presented (Tideman's example in Section 4.2.5.2). By introducing new candidates—"spoiler" candidates, so to speak—into the election, the chair can try to take votes away from the current champion, which then are conveyed to another candidate, or perhaps to several other candidates, who might benefit from that and might now win instead of the original winner.

A prominent political example of this control action is the US presidential election of 2000. The three frontrunners in this election were George W. Bush (Republicans), Al Gore (Democrats), and Ralph Nader (Green Party). Decisive for the election outcome was the result in Florida, and that was incredibly tight:

1. Bush received the most votes with $2,912,790$,
2. Gore was the runner-up with $2,912,253$ votes, and
3. Nader trailed with only $97,488$ votes in third position.

Had Nader not run for president, many of his voters most likely would have voted rather for Gore than for Bush. (Since Nader didn't have a realistic chance to win, anyway, it hadn't even been unreasonable for him not to run.)

---

[36] As with manipulation, Hemaspaandra, Hemaspaandra, and Menton [536] also consider the search problems of actually finding successful control actions and study the question of whether search reduces to decision. In contrast with manipulation, they show that for many control types (except for some types of "control by partition" to be defined in Sections 4.3.4.2 and 4.3.4.4 starting on pages 332 and 335, respectively), the search problem indeed *does* reduce to the decision problem. Thus, the bizarre behavior that it is easy to decide if control is possible, yet hard to actually find a successful control action does not occur in these cases. As mentioned earlier, Carleton et al. [257] investigate the relationships between the search versions of collapsing pairs of electoral control types (see Section 4.3.4.5 starting on page 336).

4.3 Complexity of Voting Problems

As Gore needed no more than 537 votes to draw level with Bush in Florida, Nader's decision to run for president very likely ensured Bush's second term of office as the President of the United States of America.

For a given voting system $\mathcal{E}$, define the constructive variant of the decision problem for control by adding candidates as follows.

$\mathcal{E}$-CONSTRUCTIVE-CONTROL-BY-ADDING-CANDIDATES ($\mathcal{E}$-CCAC)

Given: A set C of candidates and an additional set D of spoiler candidates, $C \cap D = \emptyset$, a list V of voters with votes over $C \cup D$, a distinguished candidate $c \in C$, and a positive integer $k \leq \|D\|$.

Question: Is there a subset $D' \subseteq D$ with $\|D'\| \leq k$ such that c is an $\mathcal{E}$ winner of election $(C \cup D', V)$?

If $D' \neq D$, the missing candidates in $D \smallsetminus D'$ are removed from the preference lists (or approval vectors) in V (which contains votes over $C \cup D$) in the election $(C \cup D', V)$. The same comment applies to the corresponding cases in all other control types.

In this form, the above control type, CCAC, has been introduced by Faliszewski et al. [427]. In the original definition of the problem due to Bartholdi, Tovey, and Trick [102], the parameter k limiting the size of D' is not given, and instead it is asked for *any* subset D' of D, no matter how large, whether adding it to the election results in a victory of the distinguished candidate. This original problem is denoted by $\mathcal{E}$-CONSTRUCTIVE-CONTROL-BY-ADDING-AN-UNLIMITED-NUMBER-OF-CANDIDATES ($\mathcal{E}$-CCAUC). The reason why Faliszewski et al. [427] introduce CCAC in addition to CCAUC is that also in other control problems (namely, control by deleting candidates and control by adding or by deleting voters), such a parameter k is given. Somewhat surprisingly, there are natural voting systems (namely, Copeland^0 and Copeland^1, i.e., Llull) for which the problems CCAC and CCAUC have different complexity (unless P = NP), see Table 4.17 on page 342 and the work of Faliszewski et al. [427, 429].

The motivation behind control by deleting candidates is analogous to that by adding candidate, since depending on which goal the chair has in mind, it can be equally advantageous for him to add some candidate d to an election (C, V) or to delete d from $(C \cup \{d\}, V)$. In the above example of the US presidential election of 2000, this means: Had it been possible to talk Nader into forgoing his presidential candidacy, Gore most likely would have won the election and would have become the 44th President of the United States of America.

Formally, this problem is defined as follows for a given voting system $\mathcal{E}$.

$\mathcal{E}$-CONSTRUCTIVE-CONTROL-BY-DELETING-CANDIDATES ($\mathcal{E}$-CCDC)

Given: An election (C, V), a distinguished candidate $c \in C$, and a positive integer $k \leq \|C\|$.

Question: Is there a subset $C' \subseteq C$ with $\|C \smallsetminus C'\| \leq k$ such that c is an $\mathcal{E}$ winner of election (C', V)?

When defining the destructive variant of this problem, note that it is not allowed to delete the distinguished candidate, for otherwise the chair's task in exerting this type of control would be trivial.

4.3.4.2 Control by Partition of Candidates

Two cases of this control scenario are distinguished: Partitioning the set of candidates with and without run-off. In both cases, a given election (C,V) is controlled by partitioning the candidate set C into C_1 and C_2 (i.e., $C = C_1 \cup C_2$ and $C_1 \cap C_2 = \emptyset$) and then performing two stages:

- In the first case (the scenario without run-off),
  1. the winners of the first-stage election (C_1, V) are determined and
  2. then run against all candidates of C_2 in the second (and final) stage to select the overall winners.

- In the second case (the scenario with run-off), however,
  1. the winners of the two first-stage elections (C_1, V) and (C_2, V) are determined in parallel and
  2. then run against each other in the second (and final) stage to select the overall winners.

Both scenarios occur in real life, in particular in elections but also in other contexts. Partitioning the candidate set without run-off corresponds to a preelection or qualification phase. For example, in big sports events such as the FIFA World Cup, the host countries often are qualified automatically for the final round of the competition and do not have to go through the qualifying round (in addition, the respective cup holders were qualified automatically for the final rounds of the FIFA World Cups from 1938 through 2002). Erdélyi et al. [405] give as a similar example the Eurovision Song Contest in which the active member countries of the European Broadcasting Union (EBU) are entitled to compete and whose winner is determined by voting. Most of the members have to qualify for the final round first, while the five largest funders of the EBU (namely, Germany, France, United Kingdom, Spain, and Italy) are qualified automatically for the final round.

By contrast, distinguishing the group stage from the knockout stage in a FIFA World Cup final round, may be seen as (an even more general form of) partitioning the candidate set with run-off. Two winners are determined in each of the single groups during the group stage, who then run against each other in the matches of the subsequent knockout stage to determine the overall winner—the new FIFA world champion. Of course, world champions are not determined by voting; however, the structure of a FIFA World Cup tournament nicely shows the structure of, and differences between, the above two types of control by partitioning candidates.

4.3 Complexity of Voting Problems

Since control by partition of candidates is defined through a two-stage election, it must be specified yet how to proceed in the case of a tie among two or more candidates in one of the first-stage pre-elections. Hemaspaandra, Hemaspaandra, and Rothe [541] propose two different approaches:

- *Ties eliminate* (TE) means that only a unique winner of a first-stage pre-election moves forward to the second and final stage. If there are several winners of a first-stage pre-election, none of them takes part in the final stage; they all are eliminated.
- *Ties promote* (TP), however, means that all winners of a first-stage pre-election move forward to the final stage.

These are *tie-handling rules*, not tie-breaking rules, because their purpose is not to break ties but to specify the conditions of participation in the final-stage election. In some sense, TE more naturally fits the unique-winner model, whereas TP is a better match for the nonunique-winner model.

Formally, we define the problems of control by partition of candidates for a given voting system $\mathcal{E}$ as follows. We start with the problem of partitioning candidates without run-off:

$\mathcal{E}$-CONSTRUCTIVE-CONTROL-BY-PARTITION-OF-CANDIDATES

Given: An election (C,V) and a distinguished candidate $c \in C$.
Question: Does there exist a partition of C into C_1 and C_2 such that c is an $\mathcal{E}$ winner of the two-stage election (with respect to the votes in V) in which the winners of the pre-election (C_1,V) who survive the tie-handling rule (TE and TP, respectively) run against the candidates from C_2?

The problem with run-off is defined as follows:

$\mathcal{E}$-CONSTRUCTIVE-CONTROL-BY-RUN-OFF-PARTITION-OF-CANDIDATES

Given: An election (C,V) and a distinguished candidate $c \in C$.
Question: Does there exist a partition of C into C_1 and C_2 such that c is an $\mathcal{E}$ winner of the two-stage election (with respect to the votes in V) in which the winners of the pre-elections (C_1,V) and (C_2,V) who survive the tie-handling rule (TE and TP, respectively) run against each other?

Distinguishing these two scenarios according to the TE and TP rule gives four problems in the constructive case for a given voting system $\mathcal{E}$, which we abbreviate by $\mathcal{E}$-CCPC-TE, $\mathcal{E}$-CCPC-TP, $\mathcal{E}$-CCRPC-TE, and $\mathcal{E}$-CCRPC-TP.

4.3.4.3 Control by Adding or Deleting Voters

Just as for candidates, we can define control by adding or deleting voters. The example mentioned at the beginning (see page 329) shows that adding voters can affect the outcome of a plurality election. In political elections, there are plenty of such examples, such as franchising or disfranchising groups of potential voters in part or completely, or seeking to otherwise encourage them to vote or else to keep them off the ballot boxes.

As a quite extreme example, women have the right to vote in Germany only since 1919—and there are even more shocking examples: In Liechtenstein, women are entitled to vote only since 1984, in the Swiss canton Appenzell Innerrhoden only since 1990, in Kazakhstan only since 1994, in Kuwait only since 2005, and Bhutan changed its previous "family voting system" to an individual voting system (thus granting women the full right to vote) only in 2008. Even today, Vatican City (which, as a theocracy, is a bit different, anyway) and Saudi Arabia deny women the right to vote completely. Another real-world scenario of control by deleting voters would be to spread rumors before election day that criminals would be arrested were they to occur at the polling place.

Conversely, a real-world scenario of control by adding voters is that the voting age in municipal elections has been decreased to 16 years in some federal states of Germany. In the USA, there are so-called *get-out-the-votes* drives, bringing in particular elderly people to the polling place at election day, for them to put their cross on the ballot (in the "right" place, as such drives are organized by parties that hope for the support of the elderly). Similarly, election campaigns typically place newspaper and TV ads tailor-made for particular target groups, for them to go to the polls or to stay away from them.

Given a voting system $\mathcal{E}$, we define the constructive variant of the decision problem for control by adding voters as follows.

$\mathcal{E}$-CONSTRUCTIVE-CONTROL-BY-ADDING-VOTERS ($\mathcal{E}$-CCAV)

Given: A set C of candidates, a list V of already registered voters with votes over C and a list U of (as yet) unregistered voters with votes over C, a distinguished candidate $c \in C$, and a positive integer $k \leq \|U\|$.

Question: Does there exist a list U' of no more than k voters from U such that c is an $\mathcal{E}$ winner of $(C, V \cup U')$?

Depending on what target a party is pursuing, it may make sense either to add a group of voters or to delete them, each with either constructive or destructive intent. Besides franchising, control by adding voters also refers to measures aiming to prompt the voters to cast their votes, including making it as easy as possible for them to do so, for example by voting by mail or online voting.

The problem of control by deleting voters is defined as follows:

$\mathcal{E}$-CONSTRUCTIVE-CONTROL-BY-DELETING-VOTERS ($\mathcal{E}$-CCDV)

Given: An election (C,V), a distinguished candidate $c \in C$, and a positive integer $k \leq \|V\|$.

Question: Does there exist a sublist V' of V with $\|V \smallsetminus V'\| \leq k$ such that c is an $\mathcal{E}$ winner of (C, V')?

4.3.4.4 Control by Partition of Voters

This control type models a practice that can sometimes be observed in the context of political elections and that is known as *"gerrymandering"*: the artful redistriction of election districts with the goal to give an advantage to one's favorite candidate by maximizing the effect of the votes of her supporters and minimizing the effect of her opponents' votes. Specifically, one seeks to reach this goal by two strategies (or their combination) known as *"packing and cracking."* Packing means the attempt to redistrict in a way that concentrates as many opponents of one's favorite candidate as possible in just a single election district so as to minimize their influence in the other election districts. That forces the opposing voters to waste many of their votes in this one district only, which then are missing in the other districts. *Cracking*, however, refers to the strategy that seeks to redistrict in such a way as to spreading the opposing voters over as many districts as possible, in an attempt to prevent them from having a decisive effect anywhere, for example, by missing to reach a majority of votes in all—or in as many as possible—election districts. This strategy, too, benefits from the opponents' wasted votes and thus is also known as the *"wasted votes effect."*

In the USA, gerrymandering is common practice to influence the outcome of elections. It has a long history there, dating back even to the time before the 1789 election of the First US Congress: In 1788, when anti-federalist allies were in control of the Virginia House of Delegates, they reshaped the boundaries of Virginia's 5th congressional district in an (unsuccessful) attempt to keep James Madison out of the US House of Representatives [634]. Although gerrymandering has been observed in many other countries as well, it is perhaps a bit less widely used there. For example, it has not been much of an issue in Australia because nonpartisan bodies such as the Australian Electoral Commission are responsible for drawing the boundaries of election districts there. It is also noteworthy that specific national election laws may boost the effect of gerrymandering, or else provide some protection against it. For example, Bundestag seats are allocated on a proportional basis in Germany, which means that large parties that easily overcome the 5% hurdle[37] are not affected much by gerrymandering attempts. However, a party may win extra seats (so-called *"Überhangmandate"*—excess mandates) by winning more election districts in any of the 16 federal states of Germany than its proportional share of the votes in this very state. In this regard, gerrymandering attempts may well become an issue. In the Bundestag election of 2009,

---

[37] In German elections, a political party must win at least 5% of the votes in order to enter the Bundestag. On a related note, while Bredereck et al. [228] study strategic campaign management in *apportionment elections* (whose goal it is to assign parliament seats to parties according to their vote counts), Laußmann, Rothe, and Seeger [645] consider *apportionment with thresholds* so as to also take thresholds such as the 3.25% hurdle in the Israel Knesset election election, the 4% hurdle in the Bulgarian parliament election, or the 5% hurdle in the German Bundestag election into account.

for example, the CDU/CSU congressional party of Angela Merkel gained the extraordinary number of 24 extra seats, whereas none of the other parties gained any extra seats, which led Germany's Federal Constitutional Court in Karlsruhe to declare the existing election laws invalid and to require the Bundestag to pass a new law that caps the number of such extra seats to no more than 15.

How can one model the scenario of gerrymandering? Bartholdi, Tovey, and Trick [102] do so by partitioning the voters. As gerrymandering is a rather involved issue, their model simplifies it in crucial ways. On the one hand, they restrict their model to partition voters into only two election districts; on the other hand, they abstract from spatial and geographical circumstances, which in real life of course do play a role for gerrymandering (consider, for example, exploiting the differences in the electoral behavior of voters in rural and in urban election districts).

Control by partition of candidates came in two variants, with and without run-off, modeled by two problems, CCRPC and CCPC. When the *voters* are partitioned, however, this distinction does not make any sense; one scenario is enough, namely that with run-off. If one were to try to define control by partition of voters *without run-off*, then—analogously with CCPC—the winners of the one pre-election would have to run against *all candidates* in the final stage. However, if all candidates take part in the final-stage election anyway, this pre-election is entirely wasted. As we thus are looking at one scenario only here, we omit "BY-RUN-OFF" in the problem name, even though the problem below is defined analogously with CCRPC.

Again, the election is held in two stages, and a tie-handling rule, TE or TP, is applied in both first-stage pre-elections to decide which of its winners, if any, will move forward to the second and final stage, the run-off. Formally, the constructive variant of the problem is defined as follows.

$\mathcal{E}$-CONSTRUCTIVE-CONTROL-BY-PARTITION-OF-VOTERS

Given: An election (C, V) and a distinguished candidate $c \in C$.

Question: Does there exist a partition of V into V_1 and V_2 such that c is an $\mathcal{E}$ winner of the two-stage election (with respect to the votes in V) in which the winners of the pre-elections (C, V_1) and (C, V_2) who survive the tie-handling rule (TE and TP, respectively) run against each other?

Depending on which tie-handling rule is applied, we obtain the two control problems $\mathcal{E}$-CCPV-TE and $\mathcal{E}$-CCPV-TP.

4.3.4.5 Overview of the Common Control Problems

We have defined our control problems in the nonunique-winner model, but they can also be defined in the unique-winner model, of course. In fact, most of the papers on electoral control use the latter model (probably due to the

4.3 Complexity of Voting Problems

fact that it was used in the first paper on control by Bartholdi, Tovey, and Trick [102]); some use both models (e.g., that by Faliszewski et al. [429]).

As already mentioned, in addition to the eleven previously defined constructive control problems (including the problem CCAUC and distinguishing between the two tie-handling rules, TE and TP, in the partitioning cases) we can define the corresponding *destructive* control problems. That would give a total of 22 common control problems. To summarize and for clarity, we have introduced above and will use throughout this section the following uniform notational scheme to abbreviate these control problems:

- The first two letters (CC and DC, respectively) stand for CONSTRUCTIVE-CONTROL and DESTRUCTIVE-CONTROL, respectively,
- the following one or two letters refer to the control type: A for ADDING, AU for ADDING-AN-UNLIMITED-NUMBER-OF, D for DELETING, P for PARTITION-OF, and RP for RUN-OFF-PARTITION-OF in the candidate control case,
- the next letter (C and V, respectively) indicates whether candidates or voters are controlled, and
- in the partitioning cases (with and without run-off), TE and TP, respectively, refers to the tie-handling rule used in these control types' pre-election(s).

Further, each such control problem depends on the voting system that precedes the problem name (as in, say, Bucklin-CCDV) and is used in each sub-election of the corresponding control type (e.g., in each of the two first-stage pre-elections and in the final-stage run-off in the case of, say, DCRPC-TP).

We mentioned above that there are 22 control problems in total, and indeed some papers have considered each of these 22 control problems for various voting rules—eight for voter control and the remaining 14 for candidate control. However, Hemaspaandra, Hemaspaandra, and Menton [536] have observed that, regardless of which voting system is used, some of the destructive partition problems in fact are just one and the same control problem. Which of these problems coincide depends on whether the unique-winner or the nonunique-winner model is applied. More concretely, they have shown that DCRPC-TE equals DCPC-TE in the unique-winner model, and that DCRPC-TE equals DCPC-TE and DCRPC-TP equals DCPC-TP in the nonunique-winner model—these are *pairs of collapsing electoral control types*. In a follow-up work, Carleton et al. [258] have shown that beyond these collapses of control cases, no other cases of control problems collapse in general (i.e., for all voting rules), even though there may occur additional collapses for some voting rules, and they specifically identify exactly which such additional collapses there are for plurality, veto, and approval voting. Table 4.16 gives an overview of all control problems considered so far.

As is common in the literature on control, we assume that the chair has complete information about the voters' preferences. If it is hard to decide for the chair whether control can be exerted successfully in the complete-

Table 4.16 Overview of the common control problems for voting system $\mathcal{E}$

Control by	Constructive	Destructive
Adding candidates	$\mathcal{E}$-CCAC	$\mathcal{E}$-DCAC
Adding an unlimited number of candidates	$\mathcal{E}$-CCAUC	$\mathcal{E}$-DCAUC
Deleting candidates	$\mathcal{E}$-CCDC	$\mathcal{E}$-DCDC
Partition of candidates	$\mathcal{E}$-CCPC-TE $\mathcal{E}$-CCPC-TP	$\mathcal{E}$-DCPC-TE[38] $\mathcal{E}$-DCPC-TP[38]
Partition of candidates with run-off	$\mathcal{E}$-CCRPC-TE $\mathcal{E}$-CCRPC-TP	$\mathcal{E}$-DCRPC-TE[38] $\mathcal{E}$-DCRPC-TP[38]
Adding voters	$\mathcal{E}$-CCAV	$\mathcal{E}$-DCAV
Deleting voters	$\mathcal{E}$-CCDV	$\mathcal{E}$-DCDV
Partition of voters	$\mathcal{E}$-CCPV-TE $\mathcal{E}$-CCPV-TP	$\mathcal{E}$-DCPV-TE $\mathcal{E}$-DCPV-TP

information model, it cannot be easy in the case of incomplete information. On the other hand, a control problem that can be solved efficiently in the complete-information model might become hard if the chair has only incomplete information. Our assumption of complete information thus leads to conservative estimates regarding the complexity of control problems.

4.3.4.6 Immunity, Susceptibility, Vulnerability, and Resistance

The complexity-theoretic investigation of manipulation problems is motivated by the Gibbard–Satterthwaite theorem (see Theorem 4.2 on page 275), which says that in principle every reasonable voting system is manipulable. No such impossibility theorem is known for the various types of electoral control defined above. Hence, there might be cases where for some voting system the chair's goal—turning a nonwinner into a winner in the constructive case and turning a winner into a nonwinner in the destructive case—can never be reached by a certain control action. Such a voting system is said to be *immune to this type of control*. For example, it is never possible for the chair to prevent a Condorcet winner's victory by deleting candidates; after all, a Condorcet winner defeats everyone else in their head-to-head contests, and will still defeat all remaining candidates after deleting some of them. That is, Condorcet voting is immune to destructive control by deleting candidates [102].

If a voting system is not immune to a control type, it is said to be *susceptible to it*. Only for a voting system that is susceptible to some control

---

[38] As mentioned earlier, Hemaspaandra, Hemaspaandra, and Menton [536] observe that DCRPC-TE = DCPC-TE in both the unique-winner and the nonunique-winner model, and that DCRPC-TP = DCPC-TP in the nonunique-winner model (see also [258]).

type does it make sense to consider the complexity of the corresponding control problem. Suppose that a given voting system is susceptible to a certain control type. If it turns out that the corresponding control problem can be solved in polynomial time, this voting system is said to be *vulnerable to this control type*. If the control problem is NP-hard, however, the voting system is said to be *resistant to this control type*.

As with manipulation, influencing the outcome of elections by electoral control is not desirable. Therefore, immunity is the most wanted property of voting systems in this regard. Unfortunately, most natural voting systems are not immune to many of the control types defined here, and many of them are even susceptible to all these control types. In these cases, resistance (i.e., NP-hardness of the respective control problem) provides protection to some extent: Assuming $P \neq NP$, the attacker will not be able to compute a successful control action in polynomial time, or even to decide in polynomial time if such an action is possible at all. However, one should also recall the discussion of *worst-case* hardness of NP-complete manipulation problems (see Section 4.3.3.9 on page 318) and how an attacker—here the chair—may try to cope with the problem's NP-hardness for "typical" problem instances (see Section 4.3.4.10 on page 354 for this discussion regarding control).

For each of our control types, every voting system is either immune or susceptible to it. To show immunity, we have to prove that the chair's goal in the respective control scenario can never be reached by a control action of this type. Susceptibility, however, is simply shown by a counterexample disproving immunity, i.e., one constructs a suitable election for which the chair's goal can be reached by exerting this type of control. Fortunately, to show the immunities and susceptibilities of a given voting system, it is not necessary to provide such a proof or counterexample for each single type of control. For example, it is obvious that a voting system is susceptible to constructive control by adding candidates if and only if it is susceptible to destructive control by deleting candidates. As there exist such links between the immunity/susceptibility results of certain control types, it is not needed to discuss every single control type separately.

Hemaspaandra, Hemaspaandra, and Rothe [541] establish the connections among control types that are shown in Figure 4.17. In addition to the control types, this figure mentions two other properties of voting systems, *"voiced"* and *"Unique-WARP"* (note that their paper [541] is deeply embedded in the unique-winner model), which are relevant for the immunity or susceptibility of voting systems in certain control scenarios and will be defined in the next two paragraphs. The control types CCAUC and DCAUC are not displayed in this figure, since they behave just as CCAC and DCAC, shown by analogous arguments. An arrow between two control types, say CCPV-TE $\rightarrow$ CCDC, indicates an implication: If a voting system is susceptible to CCPV-TE, then it is also susceptible to CCDC. A double arrow, such as CCAC $\leftrightarrow$ DCDC, indicates an equivalence between the corresponding control types'

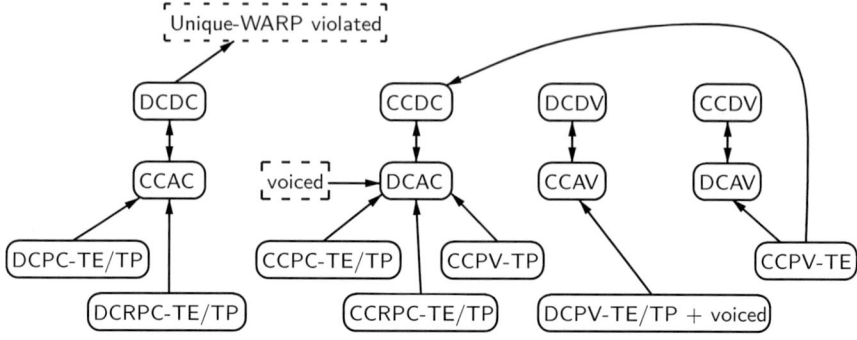

Fig. 4.17 Links between susceptibility results for various control types [541]

susceptibilities. Not all connections in Figure 4.17 are as obvious as this equivalence, but one can figure out the others relatively easily.

Hemaspaandra, Hemaspaandra, and Rothe [541] call a voting system *voiced* if in every election with only one candidate, this candidate wins. Obviously, all natural voting systems considered here satisfy this property. For a voiced voting system, c is the unique winner of election $(\{c\}, V)$. Similarly, d is the unique winner of election $(\{d\}, V)$. However, there can be at most one *unique* winner in $(\{c, d\}, V)$. Hence, every voiced voting system is susceptible to destructive control by adding candidates (DCAC). Moreover, voting systems that are both voiced and susceptible to destructive control by partition of voters in either the TE or TP model (DCPV-TE or DCPV-TP) are also susceptible to constructive control by adding voters (CCAV).

The *Weak Axiom of Revealed Preference (WARP)* says that a winner w of an election (C, V) must also win in each subelection (C', V) with $C' \subseteq C$ and $w \in C'$ (recall from Footnote 7 on page 251 that the votes in election (C', V) are those from V restricted to C'). Whenever we deal with control problems defined in the unique-winner model, we consider the *unique* variant of this axiom, Unique-WARP. A voting system satisfies *Unique-WARP* if a *unique* winner of an election always remains a unique winner in every subelection containing him. Bartholdi, Tovey, and Trick [102] (see also [541]) noticed that every voting system with this property is immune to CCAC (or, equivalently, to DCDC). In other words, the property Unique-WARP is violated whenever one can exert constructive control by adding candidates (CCAC)—equivalently, destructive control by deleting candidates (DCDC).

Now that we have discussed the general links between the susceptibility results for various control types given in Figure 4.17, one may wonder which specific voting systems are susceptible—and which are immune—to which control type. Table 4.17 in Section 4.3.4.7 will give an overview of control complexity results for various voting systems, and we will now discuss the relatively few immunities occurring in this table.

4.3 Complexity of Voting Problems

Condorcet, approval, and range voting each satisfy Unique-WARP and therefore are immune to CCAC, CCAUC, and DCDC. This immediately implies immunity of these three systems to the three cases of destructive control by partition of candidates (recall from Footnote 38 on page 338 that DCPC-TE and DCRPC-TE coincide in the unique-winner model). To see this, just reverse the arrows on the left-hand side of Figure 4.17 to state the contrapositives for the negated properties (i.e., for replacing immunity by susceptibility claims).

In addition, approval voting is immune to *constructive* control by partition of candidates (with or without run-off) whenever the TP tie-handling rule is used (i.e., for CCPC-TP and CCRPC-TP), both in the nonunique-winner and the unique-winner model. That is simply due to the fact that a candidate is a (unique) approval winner if and only if her approval score is (strictly) the largest, and because of the TP rule it is impossible for the chair to turn another candidate into a (unique) approval winner according to CCPC-TP or CCRPC-TP.

The immunities mentioned above are the only ones among the control types and voting systems in Table 4.17. In all other entries of the table, the respective voting system is susceptible to the considered control type. Two examples will be given later in more detail (see the proofs of Theorems 4.12 and 4.13 in Sections 4.3.4.8 and 4.3.4.9).

4.3.4.7 Control Complexity Results for Some Voting Systems

Table 4.17 gives an overview of the control complexity results for a number of voting systems, where the following abbreviations will be used:

- I for *immunity*,
- S for *susceptibility*,
- R for *resistance* (i.e., this voting system is susceptible to this control type and the corresponding control problem is NP-hard), and
- V for *vulnerability* (i.e., this voting system is susceptible to this control type and the corresponding control problem is in P).

The abbreviations of control problems can be found in Table 4.16 on page 338.

For each voting system, there are two columns in Table 4.17, one for the constructive and one for the destructive case. The corresponding symbol, C or D, is a prefix of the control type in the leftmost column of the table, which gives, e.g., the problems CCAUC and DCAUC in row CAUC.

We now describe the results of Table 4.17 in (a more or less) chronological order.

The first results on electoral control are due to Bartholdi, Tovey, and Trick [102], who introduce the crucial control scenarios and investigate them

Table 4.17 Overview of control complexity results for some voting systems

	Plurality	Condorcet	Approval	Bucklin	Fallback	SP-AV	Copeland$^\alpha$ $\alpha=0$	$0<\alpha<1$	$\alpha=1$ (Llull)	Range voting	NRV	Schulze
	C D	C D	C D	C D	C D	C D	C D	C D	C D	C D	C D	C D
CAUC	R R	R R	I V	I V	R R	R R	R R	V V	R V	V V	I V	R R	R S
CAC	R R	R R	I V	I V	R R	R R	R R	R V	R V	R V	I V	R R	R S
CDC	R R	R R	V I	V I	R R	R R	R R	R V	R V	R V	V I	R R	R V
CPC-TE	R R	V I	V I	R R	R R	R R	R V	R V	R V	V I	R R	R V
CPC-TP	R R	V I	I I	R R	R R	R R	R V	R V	R V	I I	R R	R V
CRPC-TE	R R	V I	V I	R R	R R	R R	R V	R V	R V	V I	R R	R V
CRPC-TP	R R	V I	I I	R R	R R	R R	R V	R V	R V	I I	R R	R V
CAV	V V	R V	R V	R V	R V	R V	R R	R R	R R	R V	R V	R V
CDV	V V	R V	R V	R V	R V	R V	R R	R R	R R	R V	R V	R V
CPV-TE	V V	R V	R V	R R	R R	R V	R R	R R	R R	R V	R R	R R
CPV-TP	R R	R V	R V	R S	R R	R R	R R	R R	R R	R V	R R	R R
References	[102] [541]	[102] [541]	[541]	[413] [406]	[414] [413] [406]	[412]	[429]	[429]	[429]	[703]	[703]	[771] [705] [690]

for plurality and Condorcet voting.[39] The tie-handling rules TE and TP of the partitioning cases have not been considered yet in their work. These rules have been introduced later on by Hemaspaandra, Hemaspaandra, and Rothe [541]. Note that it does not make sense to distinguish between TE and TP for Condorcet voting, since there are only unique Condorcet winners, if any. That is why there is only one entry in each partitioning case for both TE and TP in the Condorcet column of Table 4.17. Note further that the remark made in Footnote 38 on page 338 applies to the four cases of destructive control by partition of candidates, which drop down to three cases in the unique-winner model and to two cases in the nonunique-winner model.

Hemaspaandra, Hemaspaandra, and Rothe [539, 541] introduce the destructive control scenarios and investigate them for plurality and Condorcet

---

[39] The proofs of Bartholdi, Tovey, and Trick [102] showing that Condorcet voting is resistant to constructive control by deleting and by partition of voters (CCDV and CCPV) implicitly assume that voters can be indifferent with respect to several candidates (which actually is not allowed in their model), and therefore fail to work when the voters' preferences are linear orders. However, Faliszewski et al. [427, 429] show how to obtain these results also for the case of linear voter preferences.

4.3 Complexity of Voting Problems

voting. In addition, both in the constructive and the destructive case, they obtain the results for approval voting that are listed in Table 4.17.

Next, Hemaspaandra, Hemaspaandra, and Rothe [542] raised the question of whether an "impossibility theorem" (cf. the Gibbard–Satterthwaite theorem, see Theorem 4.2 on page 275) of the following type would be possible:

> For no voting system whose winners can be determined in polynomial time, each of the election problems from Table 4.16 is NP-hard.

The restriction to voting systems with efficient winner determination procedures is needed because the above-mentioned "impossibility theorem" should not hold simply due to including trivial cases. Namely, the (unique) winner problem can be trivially reduced to (the unique-winner version of) the four control problems CCAC, CCDC, CCAV, and CCDV. Therefore, each of these control problems is computationally at least as hard as the corresponding winner problem. For the voting systems of Dodgson, Young, and Kemeny, whose winner problems are $P_{\parallel}^{NP}$-complete both in the nonunique-winner and the unique-winner case,[40] this means that four control problems are trivially NP-hard. However, Hemaspaandra, Hemaspaandra, and Rothe [542] prove that such an "impossibility theorem" itself is impossible: There exists a voting system whose winners can be determined in polynomial time, yet whose control problems from Table 4.16 are each NP-hard. This is shown via a hybridization method (cf. Section 4.1.5) that combines several given voting systems to a new one such that,

- first, the new voting system inherits efficient winner determination from the constituent given systems and,
- second, the new system is resistant to a control type, provided that at least one of the constituent given systems is resistant to this type of control.

By hybridization of, e.g., plurality and Copeland, one thus obtains a *completely control-resistant* system, since at least one of these two voting systems has an "R" entry for each control type in Table 4.16. The disadvantage of this method, however, is that this hybrid system, though completely control-resistant, is a rather artificial voting system.

Turning again to specific natural voting systems, Faliszewski et al. [429] (see also the precursors [427, 428]) investigate the family of Copeland$^\alpha$ systems in terms of control. The results stated for Copeland$^\alpha$ in Table 4.17 show that the complexity of control problems can differ sharply depending on the value of α: For constructive control by adding an unlimited number of

---

[40] For $P_{\parallel}^{NP}$-completeness of these three winner problems in the nonunique-winner model, see the original papers by Hemaspaandra, Hemaspaandra, and Rothe [537], Rothe, Spakowski, and Vogel [825], and Hemaspaandra, Spakowski, and Vogel [550] (and see also Section 4.3.1). For these problems' $P_{\parallel}^{NP}$-completeness in the unique-winner model, see the paper by Hemaspaandra, Hemaspaandra, and Rothe [542, Theorem 2.3] and its technical report version [540] for the proof details.

candidates (CCAUC), Copeland$^\alpha$ is vulnerable for $\alpha \in \{0,1\}$, yet resistant for rational numbers α with $0 < \alpha < 1$. Interestingly, if $\alpha \in \{0,1\}$, though Copeland$^\alpha$ is vulnerable in the CCAUC case, it is resistant to the limited variant of this control type (CCAC). Note that Copeland^0 and Copeland^1 (Llull) are the *only* systems for which the complexity of CCAUC and CCAC is known to differ. Moreover, the first *natural* voting system with efficient winner determination that has been proven to be resistant to all types of *constructive* control is Copeland$^{1/2}$, the common Copeland system. Though, Copeland is still vulnerable to all cases of destructive candidate control. On the other hand, plurality is the first *natural* voting system with efficient winner determination that has been shown to be fully resistant to all types of *candidate* control, though it is vulnerable to six out of eight types of voter control.

This motivated Erdélyi, Nowak, and Rothe [412] to look for a natural voting system with an efficiently solvable winner problem that is vulnerable to even fewer control types than Copeland and plurality. They found the hybrid system SP-AV, which has been proposed by Brams and Sanver [201] and is defined on page 261; for this voting system they could show that it is vulnerable to only three (destructive voter) control types and is resistant to all others. There is a catch here too, though: In their original system, Brams and Sanver [201] require the voters to cast only (sincere and) *admissible* approval strategies, i.e., no voter is allowed to approve of either all candidates or no candidates at all. However, if in the course of a control action some candidates are deleted (be it in CCDC or in DCDC, be it in control by partition of candidates or voters where in the first-stage pre-elections certain candidates can be eliminated), then it is possible that in the correspondingly diminished preference profile a previously admissible vote becomes inadmissible.

Erdélyi, Nowak, and Rothe [412] avoid this problem by introducing a rule by which in such a case the approval line (recall Figure 4.11(a) on page 262) is moved by one position to the left or to the right in order to restore the votes' admissibility via changing them as little as possible and as much as needed. The effect of this rule is that this modified variant of SP-AV is a hybrid system combining plurality with approval voting, and indeed, SP-AV has exactly the resistances occurring in plurality and approval combined (see Table 4.17). And yet, this rule is somewhat problematic because it modifies the voters' ballots *after they have been cast*. Baumeister et al. [115] discuss in detail this and other problems with SP-AV in the modified variant due to Erdélyi, Nowak, and Rothe [412].

Later on, Erdélyi and Rothe [414] and Erdélyi, Piras, and Rothe [413] (see also the later work by Erdélyi et al. [406] that unifies and extends these earlier papers) succeeded in finding a natural voting system with easy winner determination, fallback voting, that has only two vulnerabilities among the control scenarios of Table 4.17, namely for destructive control by adding and by deleting voters (DCAV and DCDV) and is otherwise resistant. Among natural voting systems with efficient winner determination, this means that

4.3 Complexity of Voting Problems

fallback voting has the broadest control resistance currently known to hold. There is only one other such system known to have an equally broad control resistance, namely normalized range voting (NRV; see page 250 for its definition), as shown by Menton [703], who also study the control complexity of range voting itself (which turns out to behave exactly like approval voting in this regard, see Table 4.17). For Bucklin, which is a special case of fallback voting, as few as two vulnerabilities (again for DCAV and DCDV) and many resistances for the other control scenarios are known [406]. However, one case is still open: see the "S" entry for Bucklin-DCPV-TP in Table 4.17. Since Bucklin is a special case of fallback, it follows from Lemma 1.1 on page 30 that each resistance of a control problem for Bucklin voting implies the resistance of the corresponding control problem for fallback voting. In this sense, the resistance results for Bucklin slightly strengthen those for fallback. Conversely, Bucklin immediately inherits the P upper bounds for DCAV and DCDV from fallback.

As mentioned on page 311, Hemaspaandra, Hemaspaandra, and Rothe [541] noted that any type of *destructive* manipulation problem in the *unique-winner* model can be reduced in polynomial time to the corresponding *constructive* manipulation problem in the *nonunique-winner* model. And as pointed out in Footnote 27 on that page, they in fact made this observation for electoral control problems (though, of course, it applies to manipulation and bribery problems as well, see Sections 4.3.3 and 4.3.5). Looking at Table 4.17, we see that, indeed, it is never the case that there is a "V" entry for any of the considered constructive control problems, but an "R" entry for the corresponding destructive control problem. Granted, it may happen (such as for Condorcet and approval voting) that a "V" in some constructive case faces an "I" in the corresponding destructive case. However, this does not contradict what we said above, since when a voting system is immune to some control type, this simply means that this control action is always wasted: This control problem has only no-instances, and so—being represented by the empty set—is trivial. But a trivial problem, $\emptyset$, clearly is in P, just as all control problems with a "V."

Last but not least, Parkes and Xia [771] were the first to study strategic behavior (including manipulation, control, and bribery) for Schulze voting and ranked pairs in terms of their complexity (recall from Table 4.10 on page 286 in Section 4.2 that Schulze voting satisfies a large number of natural, desirable axiomatic properties [843], and the axiomatic properties of ranked pairs—specifically designed to satisfy independence of clones [914]—are similarly outstanding). Menton and Singh [705] continued to study the control complexity of Schulze voting; see Table 4.17 on page 342 for the currently known results on control in Schulze voting. In particular, they claimed Schulze voting to be vulnerable to destructive control by deleting candidates in the first version of their 2012 technical report [704], but later stated this problem as open [705] (and omitted it from their most recent arXiv version, v4, dated May 24, 2013). This P result for Schulze-DCDC, however, was recently re-established by Maushagen et al. [690] in the nonunique-winner model (leaving the complexity status of this problem open for the unique-winner model) by an efficient algorithm for the control problem that is based on a problem involving path-preserving s-t vertex cuts.

In addition, Maushagen et al. [690] re-establish another result in Table 4.17 by fixing a flaw in a reduction of Menton and Singh [705] by which they show that Schulze voting is resistant to constructive control by deleting candidates (in the nonunique-winner model, and their reduction can be modified to also work for the unique-winner model). Note that Schulze shares with Copeland, Bucklin, fallback, SP-AV, and NRV the property of being fully resistant to constructive control, but the only cases of destructive control Schulze is currently known to be resistant to concern the two cases of partition of voters. Otherwise, we have vulnerabilities and, in addition to the open case regarding the complexity of Bucklin-DCPV-TP in Table 4.17, two open cases for Schulze's system: It is known to be susceptible to destructive control by both variants of adding candidates (DCAUC and DCAC), but the complexity of these two control problems is unknown at the time of this writing.

In addition to the results summarized in Table 4.17, other popular voting systems have also been studied in terms of their control complexity for certain selected control types, such as adding or deleting candidates or voters. For example, Faliszewski, Hemaspaandra, and Hemaspaandra [424] obtained the first such results for maximin, and the complexity of selected control cases for Borda voting was studied by several authors [826, 377, 669, 668, 549, 282].

Since the first edition of this book, much progress has been made regarding the control complexity of veto and maximin by Maushagen and Rothe [691, 692, 693, 694], and regarding that of Borda voting by Neveling and Rothe [738, 737, 740] (see also the survey [820]). As to the latter, for instance, Table 4.18 gives an overview of all previous and new results on, respectively, constructive and destructive control in Borda elections, again with R standing for "resistance" and V for "vulnerability" and our standard notation for control types.

Table 4.18 Complexity of control in Borda elections (unique-winner model): Results are due to Russel (marked by §) [826], Elkind, Faliszewski, and Slinko ($) [377], Chen et al. (✻) [282], Hemaspaandra and Schnoor (♡) [549], Loreggia et al. (£) [669], and Neveling and Rothe (♣) [740].

	CAUC	CAC	CDC	CPC-TE	CPC-TP	CRPC-TE	CRPC-TP	CAV	CDV	CPV-TE	CPV-TP
Constructive	R♣	R$	R✻	R♣	R♣	R♣	R♣	R§	R♡	R♣	R♣
Destructive	V♣	V£	V£	V♣	R♣	V♣	R♣	V§	V§	V§	R♣

4.3 Complexity of Voting Problems

We have already argued why some of the voting systems are immune to certain control types in Table 4.17. Instead of proving the hundreds of other results from that table, we will now give one example each for how to prove vulnerability and resistance. By the way, the authors of the papers cited in this table usually do not prove each of their complexity results separately, but instead make use of relations between various control types and design general constructions establishing several complexity results at one fell swoop.

4.3.4.8 Condorcet Voting Is Vulnerable to Destructive Control by Partition of Voters (DCPV)

As an example for how to show vulnerability, we provide the proof of the following result for a destructive control type (the constructive counterpart of which is defined in Section 4.3.4.4 starting on page 335) and Condorcet voting (defined in Section 4.1.2.1 on page 239). Recall that Condorcet winners, whenever they exist, are always unique. In particular, this means that the distinction between the unique-winner and the nonunique-winner model is unneeded, and so are the tie-handling rules TE and TP for this voting system, as a tie can never occur.

Theorem 4.12 (Hemaspaandra, Hemaspaandra, and Rothe [541]).
Condorcet voting is vulnerable to destructive control by partition of voters.

Proof. First, we give an example proving that Condorcet voting is susceptible to this control type. Let's again have a look at the election from Table 4.2 on page 239 and the corresponding majority graph in Figure 4.3 on page 240, showing that alternative I ("ice cream") wins each pairwise comparison and thus is the Condorcet winner of this election. Now, partitioning the voters into two groups, Anna and Belle being in one and Chris, David, and Edgar being in the other group, there is no Condorcet winner in the first pre-election with voters Anna and Belle, since no alternative can beat all other alternatives in their head-to-head contest. In the second pre-election, however, where Chris, David, and Edgar vote, B ("barbecue") is the Condorcet winner. Since the first pre-election did not have a winner, only B takes part in the final round (where the votes are restricted to only B) and thus also wins the entire election. This means that alternative I could be prevented from winning by partition of voters. That is, Condorcet voting is susceptible to destructive control by partition of voters.

We now show that the problem Condorcet-DCPV is in P, i.e., one can decide in polynomial time whether it is possible to prevent a victory of a given candidate $c \in C$ in a given Condorcet election (C,V) by suitably partitioning the voters. Of course, we will not go through all possible partitions of V, checking one after the other, because their number is exponential in $\|V\|$. Instead, our polynomial-time algorithm for Condorcet-DCPV proceeds by executing the following three steps (or it stops as soon as an output is returned, without doing any further steps):

1. **Check the trivial cases.**
   a. **If C = {c}:** If only candidate c takes part in the election, c must win. Output: ✗ *control is impossible*.
   b. **Otherwise, if c already is not a Condorcet winner:** The trivial partition is successful in this case. Output: ✓ *control is possible with partition* $(V, \emptyset)$.
   c. **Otherwise, if $\|C\| = 2$:** Since c is the Condorcet winner (because the algorithm did not stop in Case 1b), a majority of voters prefers c to her rival. Thus, c wins at least one pre-election and consequently also the entire election. Output: ✗ *control is impossible*.

2. **Loop (if none of the trivial cases applies).**
   For each two distinct candidates $a, b \in C$ with $a \neq c \neq b$, check whether it is possible to partition V into (V_1, V_2) so that
   a. in pre-election (C, V_1), there is either a tie in the pairwise comparison of a and c, or a is preferred to c by a majority of voters, and
   b. in pre-election (C, V_2), there is either a tie in the pairwise comparison of b and c, or b is preferred to c by a majority of voters.

   Consider only the relations among a, b, and c in the votes and ignore the remaining candidates. We use the following notation:

   $$W_c = \text{the number of votes with } c > a > b \text{ or } c > b > a,$$
   $$L_c = \text{the number of votes with } a > b > c \text{ or } b > a > c,$$
   $$S_a = \text{the number of votes with } a > c > b, \text{ and}$$
   $$S_b = \text{the number of votes with } b > c > a.$$

   Now distinguish the following two cases.

   Case 1: $W_c - L_c > S_a + S_b$. These candidates a and b are hopeless (i.e., ✗ for this pair), and the next pair of candidates is to be checked.
   Case 2: $W_c - L_c \leq S_a + S_b$. Partition V into (V_1, V_2) as follows:
   a. V_1 contains
      i. all S_a voters with votes $a > c > b$ and
      ii. $\min(W_c, S_a)$ of the voters who place c above a and b.
      Since a is then preferred to c in S_a votes and c is preferred to a in $\min(W_c, S_a) \leq S_a$ votes, c cannot be a Condorcet winner in this pre-election.
   b. $V_2 = V \smallsetminus V_1$ contains the remaining votes. Then b is preferred to c in $S_b + L_c$ votes and c is preferred to b in $W_c - \min(W_c, S_a)$ votes. Hence, for c to not defeat b in pre-election (C, V_2), we must have $S_b + L_c \geq W_c - \min(W_c, S_a)$, or, equivalently:

$$S_b + \min(W_c, S_a) \geq W_c - L_c. \qquad (4.11)$$

  i. If $S_a \leq W_c$, we have $\min(W_c, S_a) = S_a$ and inequality (4.11) is satisfied due to the case assumption $W_c - L_c \leq S_a + S_b$.
  ii. On the other hand, if $S_a > W_c$, inequality (4.11) immediately follows from the obvious fact that $S_b + L_c \geq 0$.

  Therefore, c cannot win this second pre-election either.

  Since c does not win any pre-election, she does not take part in the final round and thus cannot win the entire election. That is, we have found a successful partition of V. Output: ✓ control is possible with (V_1, V_2).

3. **Termination.** If a successful partition of V (i.e., a pair of candidates, a and b, making such a partition of V possible that dethrones c as described above) cannot be found in any execution of the loop, the algorithm stops with the output: ✗ control is impossible.

Since each step can be done in polynomial time (in particular, the number of loop executions is quadratic in the number of candidates), this algorithm runs in polynomial time. Its correctness follows easily from the comments made above. ❏

Note that the above algorithm not only decides whether c's victory can be prevented (by outputting ✓ if this is the case, and ✗ otherwise), but in the ✓ case also explicitly determines a successful control action (a suitable partition of V) that prevents c from winning. This property is called *certifiable vulnerability* by Hemaspaandra, Hemaspaandra, and Rothe [541], and it applies to most "V" entries in Table 4.17 (recall also Footnote 36 on page 330).

4.3.4.9 Bucklin Voting Is Resistant to Constructive Control by Partition of Voters (CCPV-TE and CCPV-TP)

As an example for how to show resistance, we now prove the following result for a constructive control type (again, see Section 4.3.4.4 starting on page 335) and Bucklin voting (see page 256). Unlike Condorcet winners, Bucklin winners are not always unique, though there always exists at least one. It thus makes sense to distinguish between the unique-winner and the nonunique-winner model, and between the TE and TP tie-handling rules in the case of control by partition. Theorem 4.13 is shown for both tie-handling rules and in the unique-winner model. Here is a question for the advanced reader: Does the same proof work for the nonunique-winner model as well? If you think so, why? If you don't think so, what would you change to make it work in that model, too?

Theorem 4.13 (Erdélyi, Piras, and Rothe [413]; see also [406]).
Bucklin voting is resistant to constructive control by partition of voters, both in the TE and in the TP tie-handling model.

Proof. We first will show that Bucklin voting is susceptible to these two control types.[41] Consider the Bucklin election with four candidates, A, B, C, and D, and six voters whose preferences are given in Figure 4.9 on page 257. As shown in this figure, C and D are the two level 2 Bucklin winners of this election. Now partition the six voters into two groups, as shown in Figure 4.18. C is the Bucklin winner of the first pre-election (Figure 4.18(a)) and A is the Bucklin winner of the second pre-election (Figure 4.18(b)), so A and C run against each other in the final round (Figure 4.18(c)) with all six votes restricted to the candidate set $\{A,C\}$. Note that all three subelections have been decided on the first level already.

$A > D > C > B$		$A > C$
$C > D > B > A$		$C > A$
$C > D > B > A$		$C > A$
	$B > D > A > C$	$A > C$
	$A > C > D > B$	$A > C$
	$A > C > B > D$	$A > C$
(a) First pre-election	(b) Second pre-election	(c) Final round

Fig. 4.18 Bucklin voting is susceptible to constructive control by partition of voters

Since both pre-elections have a unique winner, TE and TP don't differ for this election: A and C move forward under both rules. Since A (who didn't win originally in the election from Figure 4.9) is the Bucklin winner of this run-off, it follows that Bucklin voting is susceptible to constructive control by partition of voters in both tie-handling models.

It remains to show NP-hardness of the problems Bucklin-CCPV-TE and Bucklin-CCPV-TP. We do so by providing a reduction from the following well-known NP-complete problem (see, e.g., [483]):

Exact-Cover-by-Three-Sets (X3C)
Given: A set $B = \{b_1, \ldots, b_{3m}\}$, $m \geq 1$, and a family $\mathcal{S} = \{S_1, \ldots, S_n\}$ of subsets $S_i \subseteq B$ with $\|S_i\| = 3$, $1 \leq i \leq n$.
Question: Does there exist an *exact cover* $\mathcal{S}' \subseteq \mathcal{S}$ of B (i.e., does each element of B occur in exactly one subset $S_i \in \mathcal{S}'$)?

To provide a reduction from X3C to our control problems, we construct from a given X3C instance $(B, \mathcal{S})$ an election (C, V) and specify a distin-

---
[41] In fact, susceptibility implicitly follows from the NP-hardness reductions to be shown later on for these two problems. However, for concreteness and to be explicit, we provide the direct, easy susceptibility proofs.

4.3 Complexity of Voting Problems

guished candidate $w \in C$ such that

$$(B, \mathcal{S}) \in \text{X3C} \iff (C, V, w) \in \text{Bucklin-CCPV-TE/TP}, \quad (4.12)$$

where—recall that we are in the unique-winner model—for (C, V, w) to be a yes-instance of Bucklin-CCPV-TE/TP, w must be a *unique* winner of the election resulting from a suitable partition of voters.

As mentioned right before this theorem, unlike in Condorcet voting, the pre-elections in a control-by-partition-of-voters scenario may have several Bucklin winners. However, our reduction will ensure that this will not happen. Therefore, we will show resistance for TE and TP simultaneously.

Let $(B, \mathcal{S})$ be a given X3C instance with $B = \{b_1, \ldots, b_{3m}\}$, $m \geq 1$, and let $\mathcal{S} = \{S_1, \ldots, S_n\}$ be a given family of three-element sets $S_i \subseteq B$, $1 \leq i \leq n$. For each element $b_j \in B$, let $\ell_j = \|\{S_i \in \mathcal{S} \mid b_j \in S_i\}\|$ denote the number of sets from $\mathcal{S}$ in which b_j occurs. Define the election (C, V) with candidate set

$$C = B \cup \{c, w, x\} \cup D \cup E \cup F \cup G,$$

where w is the distinguished candidate that the chair wants to make a unique Bucklin winner by partition of voters. D, E, F, and G are merely sets of "padding candidates," needed to make sure that c, w, and x cannot gain Bucklin points too early, but only on later levels. These padding candidates will not be given explicitly here; for the reduction to be polynomial-time computable, it will be enough to make sure that there are at most polynomially many padding candidates. We will need the following numbers of padding candidates: $\|D\| = 3nm$, $\|E\| = (3m-1)(m+1)$, $\|F\| = (3m+1)(m-1)$, and $\|G\| = n(3m-3)$. Thus, $\|C\| = 3m(2n+2m+1) - 3n + 1$ is polynomial in the input size.

Define n possibly overlapping subsets of the given set B in our X3C instance (in the first item below) and partition the sets of padding candidates into disjoint blocks (in the other items below):

- $B_i = \{b_j \in B \mid i \leq n - \ell_j\}$ for each i, $1 \leq i \leq n$. For example, if any $b_j \in B$ happens to occur in all $S_i \in \mathcal{S}$, then it is contained in no B_i at all, since we have $n - \ell_j = 0$ in this case. However, if b_j happens to occur in no $S_i \in \mathcal{S}$, then it must be in each B_i, since we have $n - \ell_j = n$ in this case. In general, every $b_j \in B$ occurs in exactly $n - \ell_j$ of the subsets B_i.
- $D = \bigcup_{i=1}^{n} D_i$ with $\|D_i\| = 3m - \|B_i\|$ for each i, $1 \leq i \leq n$. The purpose of these sets is to ensure that $\|B_i \cup D_i\| = 3m$ for each i, $1 \leq i \leq n$.
- $E = \bigcup_{j=1}^{m+1} E_j$ with $\|E_j\| = 3m - 1$ for each j, $1 \leq j \leq m+1$.
- $F = \bigcup_{k=1}^{m-1} F_k$ with $\|F_k\| = 3m + 1$ for each k, $1 \leq k \leq m - 1$.
- $G = \bigcup_{i=1}^{n} G_i$ with $\|G_i\| = 3m - 3$ for each i, $1 \leq i \leq n$.

V consists of a total of $2n + 2m$ votes that are given in Table 4.19, subdivided into four groups. The first two groups consist of n votes each, where the ith vote in the first group depends on S_i and G_i and the ith vote in the

Table 4.19 Reduction X3C $\leq_m^P$ Bucklin-CCPV in the proof of Theorem 4.13

Group	For each ...	Preference list
1	$i \in \{1,\ldots,n\}$	c S_i G_i $(G \setminus G_i)$ F D E $(B \setminus S_i)$ w x
2	$i \in \{1,\ldots,n\}$	B_i D_i w G E $(D \setminus D_i)$ F $(B \setminus B_i)$ c x
3	$j \in \{1,\ldots,m+1\}$	x c E_j F $(E \setminus E_j)$ G D B w
4	$k \in \{1,\ldots,m-1\}$	F_k c $(F \setminus F_k)$ G D E B w x

second group depends on B_i and D_i. In the third group with $m+1$ votes, the jth vote depends on E_j, and in the fourth group with $m-1$ votes, the kth vote depends on F_k. The candidates in the voters' preference lists in Table 4.19 are ordered from left to right, where the leftmost candidate is the most preferred one and, to save space, we omit stating the preference relation $>$ explicitly. If a set of candidates (e.g., S_i or G_i or $(G \setminus G_i)$, etc.) occurs in a vote, this means that the candidates of such a set are ordered according to a fixed ordering of candidates (which ordering exactly doesn't matter for the construction).

Notice that the padding candidates from any of the sets D_i, E_j, F_k, and G_i score at most one Bucklin point each up to level $3m + 1$. Moreover, it follows from the definition of the sets B_i and from the construction of the votes in V that every candidate $b_j \in B$ occurring in the votes of the first or second voter group scores exactly n Bucklin points up to level $3m$. In the thus constructed election (C, V), c is the unique Bucklin winner, since c scores $maj(V) = n + m + 1$ points up to the second level (with the votes from the first and third voter group), which no other candidate achieves that early.

Now, we prove equivalence (4.12), first from left to right. Let $\mathcal{S}' \subseteq \mathcal{S}$ be an exact cover for B. Partition V into (V_1, V_2) as follows:

1. V_1 consists of

   a. the m votes from the first group corresponding to the exact cover:

   $$c \quad S_i \quad G_i \quad (G - G_i) \quad F \quad D \quad E \quad (B - S_i) \quad w \quad x$$

   for each $S_i \in \mathcal{S}'$, $1 \leq i \leq m$, and
   b. all $m+1$ votes from the third group:

   $$x \quad c \quad E_j \quad F \quad (E - E_j) \quad G \quad D \quad B \quad w.$$

2. $V_2 = V \setminus V_1$ consists of

   a. the remaining $n - m$ votes from the first group,
   b. all n votes from the second group, and
   c. all $m - 1$ votes from the fourth group.

Since V_1 contains a total of $2m + 1$ votes, a candidate needs at least $maj(V_1) = m + 1$ points to win. Candidate x is the only candidate to score

4.3 Complexity of Voting Problems

$m+1$ points already on the first level from the votes of the third group and thus is the unique Bucklin winner of pre-election (C, V_1).

In the second pre-election, (C, V_2), there are $2n-1$ votes, so $maj(V_2) = n$ points are required for a candidate to achieve an absolute majority. Candidate w is the first one to reach this threshold on level $3m+1$ with the votes from the second group, since $\|B_i \cup D_i\| = 3m$ for each i, $1 \le i \le n$. Up to this level, x scores no points at all and c scores only m points from the votes of the first group in (C, V_2). By definition of the sets B_i, a candidate $b_j \in B$ can score no more than n points from the votes of the first or second voter group up to level $3m+1$. However, since the votes from the first group in V_1 correspond to an exact cover for B, each candidate from B can score at most $n-1$ points up to level $3m+1$ in (C, V_2). Thus, w is the unique Bucklin winner of the second pre-election.

It follows that w and x face each other in the final round. However, since x is preferred to w only by the $m+1$ voters of the third voter group, w is the unique winner of the final run-off. Thus, w has been turned into a unique Bucklin winner by this partition of the votes in V.

Conversely, we show the implication from right to left in equivalence (4.12). Suppose that V can be partitioned into V_1 and V_2 such that w is the unique Bucklin winner of the resulting two-stage election. We have to show that there exists an exact cover for the set B in the X3C instance $(B, \mathcal{S})$. For w to become the only Bucklin winner, w must prevail in at least one pre-election, say in (C, V_1). Moreover, c must not win in any pre-election, as otherwise c would win the run-off. For w to win in (C, V_1), V_1 must contain votes from the second voter group, since only in these votes w will score earlier than on the last two levels. Thus, w can win in (C, V_1) only on level $3m+1$, with the votes of the second voter group. However, since c scores points by the votes from the first and third group already on the first two levels, c's victory in (C, V_2) can be prevented only by x or any of the candidates in B. On the other hand, V_1 must contain votes from the second group in order to make sure that w proceeds to the run-off—this decisively weakens the candidates from B in (C, V_2). Hence, only x can prevent c's victory in (C, V_2), namely already on the first level with the votes from the third voter group, since c is preferred to x in all other groups.

Candidate w scores his points from the second voter group in the first pre-election, (C, V_1), on level $3m+1$. But all other candidates that receive more than one point up to this level score these points on levels other than $3m+1$. Therefore, no other candidate can tie with w in (C, V_1), and w is the unique Bucklin winner of this pre-election. Since both pre-elections have unique winners, it doesn't matter whether we use the TE or TP tie-handling rule.

It remains to show that $\mathcal{S}$ has an exact cover for B. Since w must win in (C, V_1) by the votes from the second voter group, not all the n votes of the first group can belong to V_1, since otherwise it would be impossible for w, who can get at most n votes up to level $3m+1$, to be a unique winner in

(C, V_1)—either w misses the absolute majority or c wins already on the first level. Moreover, at most m votes of the first voter group can belong to V_2, since otherwise it would be impossible for x to win on the first level against c in (C, V_2). Now, to ensure that no candidate from B scores the same number n of points as w in (C, V_1) on an earlier level than w (i.e., on a level $i < 3m+1$), *exactly* m votes of the first voter group must be contained in V_2 such that each candidate from B occurs in exactly one of these votes. These m votes, however, correspond to an exact cover for B. This concludes the proof. ❑

4.3.4.10 How Can a Chair Cope with Resistance to Control?

In Section 4.3.3.9 starting on page 318, we have discussed how manipulators can cope with NP-hard manipulation problems. In principle, the arguments of this discussion also apply to NP-hard control problems, so a chair facing a resistance to some control type does not need to despair. Though, there are not as many specific results on that for electoral control as there are for manipulation. And some approaches, such as the approximation algorithms of Zuckerman, Procaccia, and Rosenschein [974, 973], are suitable for coping with hard manipulation problems, but less so for coping with hard control problems. Experimental investigations of NP-hard control problems (similarly to those due to Walsh [928, 929] and Davies et al. [334] for manipulation) have been done by Rothe and Schend [823] for Bucklin, fallback, and plurality voting; see also the journal version [407] and their survey [824], which in addition to experimental findings elaborates on results with respect to fixed-parameter tractability and domain restrictions such as single-peakedness for control problems. Regarding the latter, see Section 5.2 in the next chapter for more details.

4.3.4.11 Outlook and Related Work

Future research on electoral control might, for example, be concerned with new models of control scenarios. Below we mention some previous work on related models of control.

First, while most work on control considers unweighted elections, Faliszewski, Hemaspaandra, and Hemaspaandra [426] study the case of *weighted electoral control*. They explore the complexity of the weighted variants of CCAV and DCAV for scoring protocols (including Borda, k-approval, and k-veto), Llull and other Condorcet-consistent voting rules. They also provide approximation algorithms for k-approval and k-veto.

Next, a model where *candidates* (and not voters, as in manipulation) act strategically and may choose to either enter or exit an election, which has the flavor of control by either adding or deleting candidates, has been consid-

4.3 Complexity of Voting Problems

ered by Dutta, Jackson, and Le Breton [366]. Roughly speaking, their results indicate that it is impossible to avoid strategic candidacy, i.e., for every nondictatorial voting system satisfying *unanimity*,[42] there must exist a candidate who is able to influence the outcome of the election by entering or exiting it. They also study game-theoretic properties in this voting framework, exploring Nash equilibria (recall this notion from Chapter 2), and then focus on a procedure known as *voting by successive elimination*.

Another interesting approach has been proposed by Puppe and Tasnádi [794], who unlike our partition-of-voters problems model gerrymandering in consideration of the given geographic conditions and show that a correspondingly defined problem for "packing and cracking" (see page 335 for an explanation of these notions) is NP-complete.

Also very interesting is an extended control model, called *multimode control*, which has been proposed and investigated by Faliszewski, Hemaspaandra, and Hemaspaandra [424]: While we have assumed so far that the chair's attack is restricted to be *one* control scenario only, it is very plausible that he exerts various attacks simultaneously; for example, he might exert control by deleting candidates and partition of voters at the same time. Later on, multimode control has been studied, for instance, by Loreggia et al. [666, 667, 668, 669] for plurality, Copeland, and iterative voting, by Erdélyi et al. [411] for a variety of voting rules (see Table 4.20 on page 357), and by Maushagen et al. [690] for Schulze voting and ranked pairs.

Fitzsimmons, Hemaspaandra, and Hemaspaandra [457] take this idea a level higher by combining electoral control with manipulation scenarios. They study both cooperative and competitive scenarios, the former scenario modeling cases where the chair is allied with the manipulators, acting in concert with them, and the latter scenario modeling cases where the manipulators seek to counter the chair's action. The obtained complexity results (showing, specifically, completeness in the lower levels of the polynomial hierarchy) in particular indicate that the order of play matters in the competitive case, and they also show that in natural voting systems, such as plurality and approval voting, the complexity can drop down to as low a class as P, while for Borda voting competition between the chair and the manipulators will raise the complexity, unless NP equals coNP (which is considered highly unlikely [819]). Relatedly, Dorn and Krüger [354] study voter deterrence by deletion of candidates, which in some sense combines control by deleting candidates and control by deleting voters, since in their setting the removal of certain candidates from the election might cause some voters to abstain from voting.

Loreggia et al. [666, 667, 668, 669] study electoral control by *replacing candidates or voters*. This control scenario can be viewed as a combination of control by adding and deleting them, with the additional constraint that

---

[42] Unanimity requires that if the same candidate is most preferred by all voters, then this candidate wins. Note the similarity to the property of Pareto consistency from Section 4.2.2.2 starting on page 268.

the same number of either candidates or voters must be added as have been deleted. The currently most recent work on this topic is due to Erdélyi et al. [411]. They describe the subtle key difference between replacement control and multimode control combining the standard control types of adding and deleting candidates or voters, which can often be handled separately. This enables resistances of voting rules to certain types of standard control to transfer trivially to multimode control, which in general fails to hold for the corresponding types of replacement control. In particular, Erdélyi et al. [411] fill many gaps in the literature and provide an almost[43] complete picture on a variety of voting rules, as displayed in Table 4.20.

Turning now to another topic, recall that we have considered online manipulation in sequential elections in Section 4.3.3.10 starting on page 321. A similar approach to model *online control in sequential elections* is due to Hemaspaandra, Hemaspaandra, and Rothe, both for online voter control [545] and for online candidate control [544].

In *online voter control* [545], the voters enter the election in sequence, one after the other, and the chair does not have complete knowledge of the voters' preferences, but knows only the previously cast votes and not the future ones at a given point in time. Nonetheless, the chair must instantaneously make the decision on which control action to take. That is, the chair has a use-it-or-lose-it ability to exert control, each time a new vote is coming in. Hemaspaandra, Hemaspaandra, and Rothe [545] prove that in this setting, online control problems can be much harder to solve than ordinary (non-online) control problems: There are (artificial) voting systems, whose winners can be determined efficiently, yet for which online control by deleting, adding, or partitioning voters is PSPACE-complete, even for only two candidates. For certain restricted cases, they obtain completeness for NP or coNP, and for plurality voting even P results for online control by deleting or by adding voters, and coNP-hardness for online control by partitioning voters.

In *online candidate control* [544], the candidates come in sequentially, one after another, and the voters gradually extend their preferences according to which candidates have been presented so far. Again, there is a "moment of decision" for the chair (who only knows the partial votes ranking the candidates up to this moment), for him to take action either instantaneously or never. For example, when a new candidate appears, the chair—with his limited knowledge about the voters' preferences—must decide right on the spot whether or not to delete this candidate, and if he doesn't do so at that very moment, this candidate cannot be removed later on. In this setting, Hemaspaandra, Hemaspaandra, and Rothe [544] again show PSPACE-completeness of online candidate control problems for appropriately designed voting systems.

---

[43] The complexity of CCRV for 2-approval—marked by $\mathbf{V}^{\heartsuit}$ in Table 4.20—was left open by Erdélyi et al. [411] but could recently be solved by Fitzsimmons and Hemaspaandra [456]: They showed vulnerability via a very clever weighted matching algorithm.

4.3 Complexity of Voting Problems

Table 4.20 Overview of results on the complexity of control by adding, deleting, and replacing either candidates or voters in various voting rules: **Results by Erdélyi et al. [411] and Fitzsimmons and Hemaspaandra [456] (the latter marked by ♡) are in boldface.** All previous results in gray are due to Bartholdi, Tovey, and Trick [102], Faliszewski et al. [424, 429, 541], Lin [656], Loreggia et al. [667, 669], and Menton [703]. Again, we use R for "resistance," V for "vulnerability," I for "immunity," and our standard notation for control types; in addition, CCRC/DCRC/CCRV/DCRV stand for constructive/destructive control by replacing candidates/voters.

(a) Constructive control

	CCAV	CCDV	CCRV	CCAC	CCDC	CCRC
Copeland$^\alpha$	R	R	R	R	R	**R**
Maximin	R	R	**R**	R	V	**R**
Plurality	V	V	V	R	R	R
2-Approval	V	V	**V**$^\heartsuit$	R	R	R
3-Approval	V	R	R	R	R	R
k-Approval, $k \geq 4$	R	R	R	R	R	R
Veto	V	V	V	R	R	R
2-Veto	V	V	**V**	R	R	**R**
k-Veto, $k \geq 3$	R	R	**R**	R	R	**R**
Plurality with run-off	**V**	**V**	**V**	R	R	**R**
Veto with run-off	**V**	**V**	**V**	R	R	**R**
Condorcet voting	R	R	**R**	I	V	**V**
Fallback voting	R	R	**R**	R	R	**R**
Range voting	R	R	**R**	I	V	**V**
Normalized range voting	R	R	**R**	R	R	**R**

(b) Destructive control

	DCAV	DCDV	DCRV	DCAC	DCDC	DCRC
Copeland$^\alpha$	R	R	**R**	V	V	**V**
Maximin	R	R	**R**	V	V	**V**
Plurality	V	V	V	R	R	R
2-Approval	V	V	V	R	R	R
3-Approval	V	V	V	R	R	R
k-Approval, $k \geq 4$	V	V	V	R	R	R
Veto	V	V	V	R	R	R
2-Veto	V	V	V	R	R	**R**
k-Veto, $k \geq 3$	V	V	V	R	R	**R**
Plurality with run-off	**V**	**V**	**V**	R	R	**R**
Veto with run-off	**V**	**V**	**V**	R	R	**R**
Condorcet voting	V	V	**V**	V	I	**V**
Fallback voting	V	V	**V**	R	R	**R**
Range voting	V	V	**V**	V	I	**V**
Normalized range voting	V	V	**V**	R	R	**R**

4.3.5 Bribery

Besides manipulation (Section 4.3.3) and electoral control (Section 4.3.4), bribery is a third way of influencing the outcome of an election. The standard bribery problem for elections has been introduced by Faliszewski, Hemaspaandra, and Hemaspaandra [421, 422], who were the first to study the complexity of bribery for various voting systems, followed by, e.g., Faliszewski et al. [427, 418, 428, 429] and Elkind, Faliszewski, and Slinko [376]; see also the book chapter by Faliszewski and Rothe [440] for an overview.

4.3.5.1 The Simplest Form of Bribery

Bribery shares with manipulation the feature that preference lists are being changed, and with electoral control the feature that there is an external actor, here the briber, conducting these changes. There are different possibilities as to what kind of changes the briber is allowed to conduct in the various bribery scenarios. We start with the constructive case where, in the simplest variant of bribery, the briber's goal is to make a distinguished candidate a winner of a given election by changing at most k votes. For any election system $\mathcal{E}$, this problem is formally defined as follows.

$\mathcal{E}$-BRIBERY

Given: An election (C,V), a distinguished candidate $c \in C$, and a nonnegative integer $k \leq \|V\|$.

Question: Is it possible to make c an $\mathcal{E}$ winner of the election that results from changing no more than k votes in V?

If k is sufficiently large (e.g., if $k = \|V\|$), the briber will always be successful, at least for each reasonable voting system that satisfies citizens' sovereignty, see Section 4.2.4.2 on page 274. "Immunity" as in control scenarios (see Section 4.3.4.6 on page 338) does not apply here. However, computational barriers—such as NP-hardness of the BRIBERY problem for certain voting systems—might still be useful to protect elections against bribery attacks. The simplest variant of bribery, as defined above, is closely related to (coalitional, unweighted) manipulation, as in both cases votes are being modified during the attack. A difference is that it is not known right from the start which votes to change: The briber has to thoughtfully pick the "right" votes to change so as to achieve his goal of making his favorite candidate a winner. In that sense, bribery is also somewhat akin to control where the chair, for example, has to pick the right votes to add or to delete. A difference between bribery and control, on the other hand, is that the votes can be modified in a bribery, yet not in a control scenario.

4.3 Complexity of Voting Problems 359

4.3.5.2 Priced and Weighted Bribery

In a natural extension of the original bribery problem, it is assumed that a certain budget is available to the briber and each voter is willing to change her vote only in exchange for a certain price. The scene in the gray box on page 329, used there—at the beginning of Section 4.3.4—to introduce electoral control, could have taken another twist if Mr. Slugger would have intervened as a briber rather than a chair.

Some days later, Anna, Belle, and Chris meet again and discuss what to do together today. Anna again suggests bicycling, playing miniature golf, and swimming, in this order (see Figure 4.1 on page 233), because her preferences haven't changed.
 The owner of the miniature golf facility, Mr. Slugger, happens to come by and asks, "By which rule are you going to vote, kids?"
 "Today by a quite simple one," Chris replies, "the plurality rule. And I still have the same preferences as Anna: Bicycling on top, else miniature golf, and if that's not possible either, swimming as a last resort."
 "I actually prefer miniature golf the most," says Belle, "and I consider bicycling the worst. I still have sore muscles from our last trip. But since it is two of you who prefer bicycling the most, I'll go and get my bicycle."
 "Just a moment, please!" Mr. Slugger hastily interjects, "not so fast!" He looks around, hoping to see David and Edgar somewhere perhaps, in order to introduce them as additional voters (who, as he knows for sure, prefer miniature golf) into the election. Unfortunately, however, they don't show up this time. Therefore ...
 "Chris, come over here, please!" Mr. Slugger takes him aside. "If you vote for miniature golf," he confidentially whispers to him, "I will give you a brand-new golf club! Whatcha think? How cool is that?"
 Chris thinks. "Two!" he eventually demands. "I want to have two clubs. Otherwise, I won't change my vote. And I want to have a golf ball in addition!"
 "Are you out of your mind, that's way too expensive!" Mr. Slugger responds, disappointed. "That would totally exceed my budget!"
 Chris turns around with a shrug and goes back to the girls to start with them for their bicycle trip, while Mr. Slugger is speculating whether he may have had more luck with making his offer to Anna.

Mr. Slugger was not able to reach his goal (that the children elect playing miniature golf on his ground) via bribery because the price demanded by Chris exceeded Mr. Slugger's budget. In the *priced bribery problem*, every voter specifies not only a preference list but also a price for which he would be willing to change his true preferences, and the question is whether it is possible for the briber to make a desired candidate win without exceeding the given budget. However, it is well possible that a voter might be willing to change her vote in one way for a certain price, but refuses to change it in another way for the same price. For example, it might be that in an election with ten candidates, a voter would agree to swap the candidates in the fifth and sixth positions of her vote for 10 bucks, but would not agree to put her most despised candidate on top for the same price. For such a severe change in her preferences, she perhaps would demand 100 bucks, and even 150 bucks to actually *swap* the candidates in the first and last positions of her vote.

Therefore, it would only be reasonable to assume that the price function of a voter depends on the vote the briber has in mind for this voter when bribing her. However, there are $m!$ possible linear rankings for m candidates. Thus, if we were to represent price functions by specifying a price for each possible vote, we would get into a hell of a mess algorithmically. That is why one usually focuses on price functions that can be represented succinctly.[44] The most common families of price functions are:

- The family of *discrete price functions* consists of 0-1-valued functions: Changing a vote costs 1, and leaving it unchanged costs 0.
- The family of *$discrete price functions* consists of functions with two possible values: Changing a vote costs c, where c is a positive integer (and may be a different integer for each voter), and leaving it unchanged costs 0.

The unpriced problem BRIBERY defined in Section 4.3.5.1 is the special case of the priced problem where all voters have unit price (i.e., have a discrete price function) and the briber has a given budget of k (and so can change up to k votes). Note that we have defined only *constructive* variants of bribery so far, and will focus on them in the overview in Section 4.3.5.3; the destructive variants of bribery will be briefly discussed in Section 4.3.5.4 on page 362.

Faliszewski, Hemaspaandra, and Hemaspaandra [421] also introduce a variant of bribery, called *negative bribery*, where the briber is not allowed to make the distinguished candidate c (whose victory he desires) the top choice in a bribed voter's preference (unless c was this voter's top candidate already before the bribe). The intuition behind that is that the briber intents to be inconspicuous and unsuspicious with his attack: He doesn't want to get caught supporting his favorite candidate directly. For approval voting, Brandt et al. [210] modify this notion to mean that the approval vector of a bribed voter may approve of c after the bribe only if this voter's vector before the bribe approved of c, and they define *strongnegative bribery* to mean that a bribed voter is not allowed to approve of c *at all* after the bribe. We denote the corresponding problems by NEGATIVE-BRIBERY and STRONGNEGATIVE-BRIBERY; they will be considered later in Section 5.4 starting on page 393.

---

[44] Similar representation issues have been discussed, e.g., for simple games in Section 3.4 starting on page 171, and will be discussed later on, e.g., in Section 9.3 starting on page 613 with respect to preference elicitation for the allocation of indivisible goods.

4.3 Complexity of Voting Problems

Table 4.21 Complexity of bribery problems for some voting systems

	Bribery				Reference
	$	⚖	⚖ $		
Scoring rules (m fixed)					
$\alpha_1 = \cdots = \alpha_m$	P	P	P	P	[422]
$\alpha_1 > \alpha_2 \neq \alpha_m$	P	P	P	NP-comp.	
$\alpha_2 \neq \alpha_m$	P	P	NP-comp.	NP-comp.	
Borda	NP-comp.	NP-comp.	NP-comp.	NP-comp.	[238]
Approval	NP-comp.	NP-comp.	NP-comp.	NP-comp.	[422]
Range voting	NP-comp.	NP-comp.	NP-comp.	NP-comp.	[from above]
Copeland$^\alpha$, $0 \leq \alpha \leq 1$	NP-comp.	NP-comp.	NP-comp.	NP-comp.	[429]
Simpson	NP-comp.	NP-comp.	NP-comp.	NP-comp.	[424]
Ranked pairs	NP-comp.	NP-comp.	NP-comp.	NP-comp.	[952]
STV	NP-comp.	NP-comp.	NP-comp.	NP-comp.	
Schulze	NP-comp.	NP-comp.	NP-comp.	NP-comp.	[771]
Bucklin	NP-comp.	NP-comp.	NP-comp.	NP-comp.	[439]
Fallback	NP-comp.	NP-comp.	NP-comp.	NP-comp.	
Cup	NP-comp.	NP-comp.	NP-comp.	NP-comp.	[800]

4.3.5.3 Overview of Complexity Results for Constructive Bribery

Table 4.21 gives an overview of complexity results for the bribery problem and its variants for scoring protocols, approval, Copeland$^\alpha$, and a number of other voting systems. Here, $ refers to the variant in which voters have price tags, and ⚖ to the case of weighted voters. In addition, the weighted *and* priced variant of bribery (indicated by "⚖ $" in the table) has also been studied by Faliszewski et al. [422, 429]. By Lemma 1.1 on page 30, NP-hardness is inherited upward from the more special to the more general variant, and, conversely, upper bounds like membership in P are inherited downward from the more general variant to its special cases.

Note that the three "scoring rules" rows in Table 4.21 hold for a fixed number m of candidates and provide dichotomy results for bribery problems. For example, the "⚖ $" column of the table in these three rows gives a *dichotomy result for priced, weighted bribery*: For a fixed number m of candidates, this problem is in P for the trivial scoring vector $\boldsymbol{\alpha} = (\alpha_1, \ldots, \alpha_m)$ with $\alpha_1 = \cdots = \alpha_m$, but is NP-complete for all other scoring vectors. By contrast, the result for Borda in Table 4.21 holds for any (unbounded) number of candidates and so is not subsumed by the three "scoring rules" rows above.

There are also further such results. For example, Faliszewski, Hemaspaandra, and Hemaspaandra [422] show that, for an unbounded number of candidates, BRIBERY (even if it is either weighted or priced) is in P, yet is NP-complete if it is both weighted and priced. Furthermore, Lin [657] shows that BRIBERY for 2-approval is in P, but for k-approval it is NP-complete whenever $k \geq 3$, and that for k-veto it is in P if $k \leq 3$ but NP-complete if $k \geq 4$.

Note that for each voting system, the unweighted manipulation problem reduces to the (unweighted) priced bribery problem, and that the weighted manipulation problem reduces to the weighted, priced bribery problem [422].

4.3.5.4 Destructive Bribery and the Margin of Victory

As with manipulation and control, also bribery has its destructive variants. For example, NP-completeness is known to hold for the four destructive cases of bribery (with and without weights and with and without prices) in Copeland$^\alpha$ elections, $0 \leq \alpha \leq 1$, due to Faliszewski et al. [429], in Schulze elections due to Parkes and Xia [771], and in cup elections due to Reisch, Rothe, and Schend [800]. For Bucklin and fallback elections, Faliszewski et al. [439] have shown that destructive, weighted, priced bribery is NP-complete as well, but the three other destructive cases each are in P.

Interestingly, destructive bribery is closely related to the *margin of victory*, a critical measure for how robust a voting system is in terms of changing election outcomes that are due to errors in the ballots or fraud in using electronic voting machines. More formally, given a voting system $\mathcal{E}$ and an election (C, V), the *margin of victory* is defined to be the smallest integer k such that the winner set can be changed by modifying k of the votes in V, while all the other votes remain unchanged. Clearly, the higher the margin of victory is, the more robust is the election.

Among the applications of this notion are *risk-limiting post-election audits* by which the trust in the correctness of the outcomes of political election can be restored, namely, by means of so-called "verifiable paper records" [756]: If too many mismatches are found in these records, an extremely costly recount of all votes is in order. By contrast, *risk-limiting* post-election audit methods (studied by, e.g., Stark [887, 888] and Sarwate, Checkoway, and Shacham [834]) do not require to recount *all* the votes, but do limit the risk that the result might still be wrong. The margin of victory is a critical parameter in this approach. Much of the previous work on this notion has employed statistical methods and has focused on scoring rules (mainly, plurality voting [885, 886]), approval and range voting, and STV [834, 259, 676].

Xia [952] initiated the study of the margin of victory in terms of its computational complexity for various voting systems,[45] including all scoring protocols (and also the *generalized scoring rules* mentioned in Section 4.3.3.9 starting on page 318), plurality with run-off, Copeland, maximin, STV, Bucklin, and ranked pairs. He obtains both NP-completeness and approximation

---

[45] For other robustness notions in elections that have also been studied from a computational point of view, we refer to the work of Procaccia, Rosenschein, and Kaminka [791] and Shiryaev, Yu, and Elkind [862].

4.3 Complexity of Voting Problems

results for these voting systems. Reisch, Rothe, and Schend [800] expand this study by introducing the *exact variant* of the margin of victory problem (asking whether the margin of victory is exactly equal to some given value k), and they show that this problem is DP-complete (recall this complexity class from Section 1.5.3.3 on page 37) for Schulze, cup, and Copeland elections.

Relatedly, Baumeister and Hogrebe [118] and Boehmer et al. [159] study the robustness of elections via considering swap and shift bribery for counting problems, a topic we will turn to now.

4.3.5.5 Swap Bribery and Other Variants of Bribery

In the BRIBERY problem variants considered so far, it doesn't matter how a vote is changed by the briber, who has "bought" the entire preference list of this voter and may now do with it as he likes. As mentioned earlier, however, voters may have different prices for different target votes of the briber (so a small change might cost less than a strongly invasive one), which cannot be expressed by discrete or $discrete price functions. For example, swapping two adjacent candidates in the middle of a vote may be less expensive to buy for the briber than swapping the top and the bottom candidate. This is the motivation behind the *swap-bribery price functions* introduced by Elkind, Faliszewski, and Slinko [376], which are defined as follows: For each two distinct candidates $c, d \in C$, there is a cost that a voter demands for swapping c and d in her vote (from $c > d$ to $d > c$, or vice versa), and the cost of changing this (original) vote into the vote desired by the briber is the sum of the single costs of swaps needed to transform the original vote into the briber's target vote. Note that only swaps of adjacent candidates are allowed; if the briber wishes to swap two more distant candidates in some vote, he must do so via a sequence of swaps, each between two then adjacent candidates.[46] For example, when there are m candidates, it takes $m-1$ swaps to move the bottom candidate to the top of a preference list. Note that swap-bribery price functions can still be represented compactly. This model of swap bribery gives rise to defining the following problem, for any given voting system $\mathcal{E}$.

$\mathcal{E}$-SWAP-BRIBERY

Given: An election (C, V), a swap-bribery price function for each voter in V, a distinguished candidate $c \in C$, and a nonnegative integer budget B.

Question: Is it possible to turn c into an $\mathcal{E}$ winner by a sequence of allowed swaps of candidates in the votes such that the sum of the prices for all these swaps does not exceed the budget B?

Again, weighted and destructive variants of the above-defined swap bribery problem can be defined in a straightforward way.

---

[46] This is somewhat reminiscent to how the Dodgson voting system works based on swapping adjacent candidates until, eventually, a Condorcet winner emerges; recall the definition of this voting system from Section 4.1.2.3 starting on page 242.

As observed by Elkind, Faliszewski, and Slinko [376], SWAP-BRIBERY generalizes MANIPULATION and even POSSIBLE-WINNER. We have already seen in Section 4.3.3.8 on page 317 that MANIPULATION is a special case of POSSIBLE-WINNER. This problem, in turn, is the special case of SWAP-BRIBERY in which the already linearly ordered pairs of candidates are so costly that swapping them would exceed the briber's budget, but a swap between any two candidates that are not yet linearly ordered (in the given partially ordered preference profile) is available for free. This means that, by Lemma 1.1 on page 30, for each voting system, NP-hardness of MANIPULATION is inherited by POSSIBLE-WINNER, and NP-hardness of POSSIBLE-WINNER, in turn, is inherited by SWAP-BRIBERY. Conversely, if SWAP-BRIBERY is in P for some voting system, then so are MANIPULATION and POSSIBLE-WINNER for this system. However, due to SWAP-BRIBERY being such a general problem, there are not many voting systems for which this problem is in P; this is the case only for systems as simple as plurality [376].

On the other hand, there are lots of hardness results for SWAP-BRIBERY. For example, Elkind, Faliszewski, and Slinko [376] have shown that it is NP-complete for Borda, Copeland, maximin, and k-approval whenever $k \geq 3$, and its NP-completeness for 2-approval is due to Betzler and Dorn [137] (note that 1-approval is just the above-mentioned P result for plurality). For Bucklin and fallback elections, Faliszewski et al. [439] have shown NP-completeness in each of the constructive/destructive, weighted/unweighted, and unique-winner/nonunique-winner variants of swap bribery. For Copeland and cup elections, NP-completeness of constructive swap bribery, even in the unweighted case, is a direct consequence of the corresponding result for POSSIBLE-WINNER due to Xia and Conitzer [957]. The corresponding NP-completeness results of the destructive cases of swap bribery in these two systems follow again from the work of Xia and Conitzer [957], along with a result due to Shiryaev, Yu, and Elkind [862], who prove that coNP-hardness of the *necessary winner problem* implies NP-hardness of destructive, unweighted swap bribery. Furthermore, note that (respectively, constructive and destructive) weighted swap bribery is equivalent to (respectively, constructive and destructive) weighted, priced bribery in elections with two candidates; hence, the two weighted variants of swap bribery for Schulze voting are NP-complete by the results of Parkes and Xia [771]. The two unweighted cases for Schulze are currently still open.

Because SWAP-BRIBERY has so many hardness results, Elkind, Faliszewski, and Slinko [376] also introduce a restricted variant, called SHIFT-BRIBERY, where only swaps that involve the distinguished candidate are allowed. Although even in this special variant they show NP-completeness for Borda, Copeland, and maximin, they were able to provide an efficient approximation algorithm for Borda. This latter result was later extended to all scoring protocols by Elkind and Faliszewski [373], who also obtained somewhat weaker approximations for Copeland and maximin voting. Faliszewski, Manurangsi, and Sornat [437] even provided a corresponding polynomial-time approximation scheme. Schlotter, Faliszewski, and Elkind [839] prove that SHIFT-BRIBERY is in P for Bucklin and fallback voting (see also [439]).[47] In this paper, they also study another variant of swap bribery, called *support bribery*, which can be seen as a refinement of multimode control by adding and deleting voters (which is due to Faliszewski, Hemaspaandra, and Hemaspaandra [424] and was mentioned in Section 4.3.4.11 starting on page 354) for systems combining approval voting with other systems, such as SP-AV and fallback voting.

Brederect et al. [230] investigate *combinatorial* shift bribery. Moreover, Faliszewski et al. [439] study *extension bribery* for fallback voting, a bribery variant introduced by Baumeister et al. [117] in the context of "truncated ballots." Without defining it formally here, let us mention that this variant is based on the idea that the briber (or campaign manager) seeks to bribe (or convince) certain voters to include the distinguished candidate at the end of the ranking of approved candidates, a rather noninvasive bribery (or campaign) action.

Kaczmarczyk and Faliszewski [582] introduced and studied *destructive shift bribery*. Maushagen et al. [689] proved NP-completeness of SHIFT-BRIBERY and its destructive variant for a number of iterative voting rules, including STV, the voting systems due to Nanson [724] and Baldwin [82] as well as plurality with run-off and veto with run-off. Brederect et al. [225] were the first to consider shift bribery from the perspective of parametrized complexity theory. Knop, Koutecký, and Mnich [614] solved a related long-standing open problem regarding the parameterized complexity of bribery (and, in particular, of shift bribery), and we refer to the survey by Brederect et al. [224] for a more comprehensive discussion of this problem. Finally, Brederect et al. [231] explore the complexity of shift bribery in committee elections, a topic we will cover in great detail in Chapter 6.

Faliszewski [418] investigated yet another variant of bribery, called NONUNIFORM-BRIBERY, for various voting systems.

Last but not least, Faliszewski et al. [429] introduce the problem MICRO-BRIBERY and study it for Copeland$^\alpha$ elections with "irrational" voters. The ballot of an *irrational voter* is not a linear preference list, but it can contain cycles, thus violating transitivity of the underlying preference order. Irrational preferences often occur in real life when a voter's preferences are based on multiple criteria. For example, Anna likes bicycling (B) more than miniature golf because working out is a big deal for her. Also, she prefers miniature golf (M) to swimming (S), since technical skills in a sport are also important to her. So far, her preferences are completely rational and transitive (see Figure 4.1 on page 233). However, for the second reason—that she appreciates technical skills in a sport—she actually prefers swimming to bicycling. Overall, she thus has the irrational (or, cyclic) preference: $B > M > S > B$. Such

---

[47] Note that some papers use the term *"campaign management"* to refer to bribery in order to emphasize the positive aspects of this type of influence on elections.

an irrational vote can be represented by a preference table in which each voter compares any two alternatives and sets her preferences accordingly.

MICROBRIBERY for irrational voters is the problem of whether one can make a desired candidate win by changing no more than k entries in the given preference tables of the voters. As with BRIBERY, also MICROBRIBERY can be associated with a price function and/or with weighted voters.

To show that Copeland$^\alpha$-MICROBRIBERY for $\alpha \in \{0,1\}$ is in P, Faliszewski et al. [427, 429] used flow networks (that are very common in computer science; recall also the notion of *network flow game* from Section 3.4.2.2 starting on page 176) and applied the min-cost flow theorem. Flow networks have later also been applied, for example, by Betzler and Dorn [137] to show that certain variants of POSSIBLE-WINNER can be solved in polynomial time.

Remotely related to bribery are the problems OPTIMAL-LOBBYING introduced by Christian at al. [294], and PROBABILISTIC-LOBBYING introduced by Binkele-Raible et al. [149]. Also recall that we have already seen some results on bribery in path disruption games in Section 3.4.2.3 on page 177.

Recall that we discussed online manipulation in Section 4.3.3.10 starting on page 321 and online control in Section 4.3.4.11 starting on page 354. Similarly, Hemaspaandra, Hemaspaandra, and Rothe [546] studied *online bribery in sequential elections*. Again, they show that even for (artificially constructed) voting rules whose winners can be easily determined, the complexity of online bribery in sequential elections may be jump up to completeness for high levels of the polynomial hierarchy or even PSPACE. On the other hand, they also show that such a dramatic complexity increase does not occur for some natural, central voting rules, including plurality and other scoring protocols as well as approval voting, and they pinpoint the complexity of the corresponding online bribery problems.

To conclude this chapter that has focused on voting systems and their properties and on the complexity of the associated problems, let us mention that preference aggregation has been studied also from other angles, not always restricted to voting scenarios. For example, motivated by issues of preference elicitation, Pini et al. [781] consider incomplete preferences and incomparability in preference aggregation; Procaccia, Rosenschein, and Kaminka [791] study the robustness of preference aggregation in noisy environments; Xia, Conitzer, and Lang [956, 958, 959, 961] investigate preference aggregation in multi-issue domains; and in order to calibrate the aggregated scores of biased reviewers of scientific papers in peer-reviewing, Roos, Rothe, and Scheuermann [817] apply quadratic programming and Kuhlisch et al. [624] propose a statistical approach to this task.

Finally, Brandes, Laußmann, and Rothe [207] employ voting in network science: Exploiting the parallels between social choice theory and network science, they propose a novel approach to define network centrality indices based on voting rules, including plurality, Borda, and Copeland voting. In their model, nodes in a network vote on the other nodes in the network based on their shortest-path distance, which yields a preference profile on

4.3 Complexity of Voting Problems

which a voting rule can be applied to determine the winners as the "most important" nodes in the network. For example, it is shown that, in this sense, plurality voting corresponds to degree centrality. However, the centrality rankings induced by other voting rules, such as Borda or Copeland, differ from the rankings induced by classical centrality indices from network science. A similar approach has been taken by Skibski [869] who shows that closeness centrality is induced by the Condorcet principle.

In the next three chapters, we will continue to study manipulation, control, and bribery: with respect to single-peaked elections in Chapter 5, with respect to multiwinner voting in Chapter 6, and with respect to judgment aggregation in Chapter 7.

Chapter 5
The Complexity of Manipulative Actions in Single-Peaked Societies

Edith Hemaspaandra · Lane A. Hemaspaandra · Jörg Rothe

Anna and Belle, who now have become so aware of their role as guides in this book that they can even refer to the book's content, meet on what will be a very exciting day for them. Let's listen in on their conversation.

"I have two pieces of wonderful news," says Belle.

"Tell me, tell me," says Anna.

"First, as shown in the previous chapter, many types of manipulative attacks on elections are computationally intractable. Like *wow*!"

"I'm glad that you're excited by that, but it isn't doing much for me. What is the second piece of good news?"

"Today is the annual charity Pumpkin Pie Taste-Off! You remember it well, I'm sure. Tables and tables of pumpkin pies are set out, and are compared based on their taste, competing for the coveted honor of being chosen as a best-tasting pumpkin pie."

"Like *wow*! That is my favorite event of the entire fall season. I love pumpkin pie, at least when it tastes just right. But it is hard to get it just right. As everyone knows, the key is getting the sweetness level to be exactly right."

"Absolutely. Everyone I've ever met agrees that the way to judge pumpkin pies is by their level of sweetness. We seem to be in perfect agreement, as we usually are."

Anna and Belle rush down to the Pumpkin Pie Taste-Off, a yearly charity event of their town (see Figure 5.1). They find that there are 26 amateur and professional bakers competing in this year's contest. Each baker has brought a large, fresh-baked pumpkin pie, so that people can taste the 26 pies and then cast their votes—by a strict, linear ordering of the 26, of course! By a remarkable coincidence, it turns out that this year the 26 bakers have the last names Adams, Brown, Chavez, Dylan, ..., Young, Zimmerman. By an even more amazing coincidence, it turns out that Adams baked the least sweet pie,

370 5 The Complexity of Manipulative Actions in Single-Peaked Societies

Fig. 5.1 Anna, Belle, Chris, and Edgar at the annual charity Pumpkin Pie Taste-Off

Brown baked the second least sweet pie, and so on through the alphabet up to Zimmerman, who baked the sweetest of all the pies.

It sounds as if this contest might be a run-away for Zimmerman, but let us listen in some more. By now, Anna, Belle, all their friends and family, and many others have tasted all the pies and voted.

Anna, Belle, Chris, and Edgar all say, simultaneously, "Well, that was an easy decision. My vote was based on the most important thing about pumpkin pies, their level of sweetness."

"That is great," says Anna. "Clearly we all gave our top spot for Zimmerman, whose pies are the sweetest; yummy! So my vote was Zimmerman > Young > ⋯ > Dylan > Chavez > Brown > Adams, of course—the only reasonable vote."

"Now hold on a minute," says Belle. "When I said that what matters about pumpkin pies is their level of sweetness, I obviously meant that the sweetness level had to be not too sweet and not too tart. Among these 26 entrants, King's pie is the one that best matches, to my taste, that point of perfect balance. And to my taste, if one has to miss that point of perfect balance, it is better to miss on the side of being overly sweet, although not by too much. That is why, after tasting all the pies, my own vote put King first, Larsen second, Martinez third, Norton fourth, and Juarez fifth."

"Taste must run in families," says Edgar, Belle's sister. "Like Belle, King's pie to me is the ideal one. But big sis is wrong about which ones come right after that. To my taste, if one has to miss the perfect balance point, it is better to miss on the side of being a bit less sweet, although not by too much. That is why the top spots on my ballot went to, in this order, King, Juarez, Iverson, Heck, Larsen, and Gilchrist."

"You've all got it wrong," says Edgar's friend Chris. "It is Thibodeaux's pie that gets the sweetness just right. I gave it the top spot in my vote."

"Well, this is a fine mess," comments Belle. "Even though we all are rating pie based on their sweetness, we each have differing views on what the ideal level of sweetness is. And we also have different views as to how our liking for pies drops off as they diverge from our ideal sweetness point in one direction or the other, although for each of us, clearly between two pies that are sweeter than our ideal sweetness point, we'll prefer the one that is the less sweet of the two, and similarly, between two pies that are less sweet than our ideal sweetness point, we'll prefer the one that is the sweeter of the two."

Everyone answers, "Certainly, there is no doubt about that."

After short pause, in which they all think about this, Belle, who is very smart, exclaims, "Oh no! My first piece of wonderful news was that many types of manipulative actions on elections are computationally intractable. But our discussion of pies now has me very worried as to whether that is truly so. My worry is this: The intractability proofs were based on constructions that allowed arbitrary collections of voters. But we've just seen here that some electorates may cast only certain patterns of votes. For example, does anyone here think that someone might cast a vote that put Zimmerman first and Adams second?"

Everyone replies in unison, "Impossible! Unthinkable! No one with the ability to taste food could possibly cast such a bizarre vote."

"So," continues Belle, "perhaps the complexity of manipulative actions on elections that have restrictions on what votes can be cast—or on what collections of votes can be cast—might be far lower than the complexity is regarding the case where there are no restrictions on what votes can be cast? Perhaps those intractability results that so raised my spirits may turn to dust in this case, leaving the elections open to perfect polynomial-time attack algorithms? These delicious pieces of pie are costing me quite a bit of my peace of mind!"

Belle and her friends (whose preferences are graphically displayed as Figure 5.2) have touched upon a tremendously important point. As Belle realized, it is possible that if the collections of votes that can be cast are restricted, the complexity of manipulative attack problems may change. In this chapter, we will explore this for the case of the most famous and important restriction on electorate behavior. This restriction is known as *single-peaked* electorates, and it essentially is just the type of situation that Belle and her friends have innocently stumbled upon as they discussed the nature of their preferences regarding pumpkin pies.

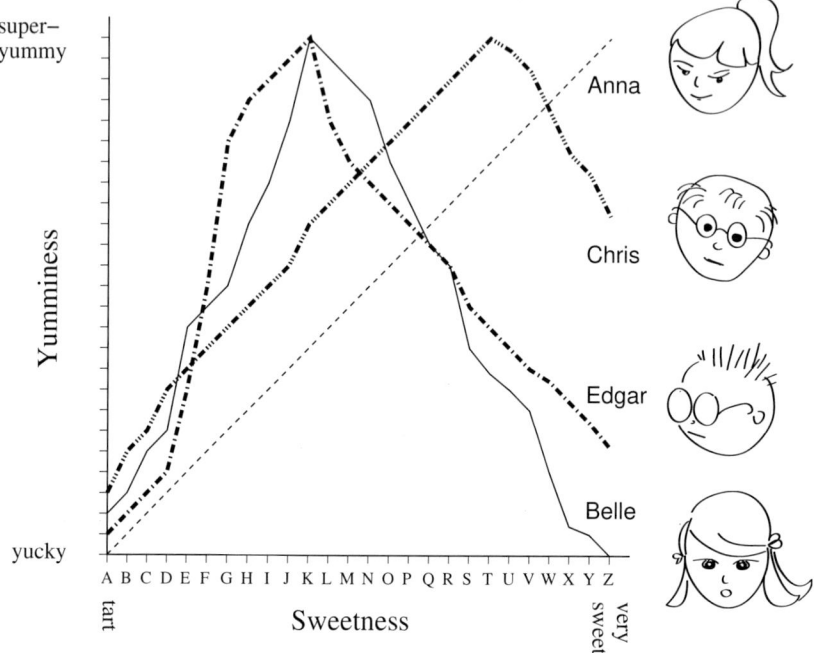

Fig. 5.2 Preferences regarding sweetness of pumpkin pie

We will see that single-peakedness often does lower the complexity of attacks on elections, just as Belle feared it would. However, we'll see that sometimes single-peakedness does not change the complexity of attacks on elections. And we will even see that, although this might seem so obviously "impossible" that Belle above did not even imagine that it could happen, there are cases where looking at the special case of single-peakedness increases the complexity of attacks on elections.

Our study of single-peaked elections will be structured as follows. Section 5.1 will more formally define single-peaked electorates, will discuss and further motivate them, and will mention how the study of single-peakedness is integrated into the key manipulative-action problems that were introduced in the previous chapter. Sections 5.2, 5.3, and 5.4 will cover some examples of control, manipulation, and bribery in the context of single-peaked electorates. Finally, Section 5.5 will hear from Helena, who has very surprising preferences regarding pumpkin pie. This will lead us to more generally consider what happens in electorates that are *nearly single-peaked*. That is, they may contain a few "maverick" voters who vote in ways potentially having nothing to do with the single-peakedness of the setting, e.g., voters who judge pumpkin pies based on the crust or the color of the filling. We will see that in some cases, the presence of even one such maverick can make the complexity of manipulative action problems jump back up to intractability.

5.1 Single-Peaked Electorates

We will now more formally define single-peaked electorates (for both voting by preferences and voting by approval vectors—recall these notions from Section 4.1), will discuss single-peakedness and further motivate it, and will explain how single-peakedness can be integrated into the key manipulative-action problems that were presented in Section 4.3 to model control, manipulation, and bribery scenarios.

Black [155, 156] introduced the notion of single-peaked preferences in order to model societies that are heavily focused on a single issue, such as the level of sweetness in the Pumpkin Pie Taste-Off described above. Clearly, in the political world there often is a dominant issue on which the electorate is heavily focused and on which voter preferences are naturally single-peaked, be it level of taxation, breadth of the social welfare network, or degree of participation in an overseas military action. Even when there is no one salient issue, political parties as well as politicians themselves can often be linearly ordered according to their position on a left-right spectrum, where left-wing (right-wing) parties/politicians take a more liberal (conservative) position. Thus it is not at all surprising that single-peakedness is one of the key concepts of political science, and is central in the study of elections. Gailmard, Patty, and Penn, who studied Arrow's impossibility theorem (see Theorem 4.1 on page 273) in the context of single-peaked electorates, described single-peakedness as "*the canonical setting for models of political institutions*" [477].

We now formally define this notion both for electorates whose preferences are linear rankings and for electorates using approval vectors.

Definition 5.1 (single-peaked preferences). Let C be a set of candidates.

1. A list V of votes over C, each vote in V being a linear order $>_i$, is said to be *single-peaked* if there exists a linear order L over C (which we will refer to as the *societal axis*) such that for each triple of candidates, a, b, and c, if $a L b L c$ or $c L b L a$, then for each i it holds that $a >_i b$ implies $b >_i c$.

2. A list V of approval vectors over C is said to be *single-peaked* if there exists a linear order L over C such that for each triple of candidates, a, b, and c, if $a L b L c$ then whenever a vote in V approves of a and c, it must also approve of b.

Anna wants to know, "What does this mean, actually?"

"It means," explains Edgar, "that whenever you take any three candidates who are ordered consistently with the societal axis (like, for example, Adams, King, Larsen or Larsen, King, Adams in Figure 5.2), then in each individual vote, whenever the middle candidate of the three is ranked below one candidate, it must be ranked above the other candidate."

"I still don't get it!"

"Rule of thumb: *'Never rank the middle candidate last!'* For example, among Adams, King, and Larsen in their societal order of Figure 5.2, if one of us were to put Larsen first, Adams second, and King last, then we wouldn't be single-peaked with respect to this societal axis. This is because if you prefer, say, Larsen to King, just as you and Chris do in Figure 5.2, Definition 5.1 *requires* you to prefer King to Adams. On the other hand, it is absolutely fine to prefer King to *both* Adams and Larsen, as Belle and I do; that doesn't contradict Definition 5.1. And remember, that applies to *all* triples of candidates, not just to Adams, King, and Larsen, and it also applies to *each* of us voters."

"Another way to put it is," Belle adds, "that for each of us, with respect to the societal axis, our preference-based utilities rise to a peak and then fall, or just rise, or just fall. That is why it is called *single-peaked*. For example, Anna, your preferences in Figure 5.2 'just rise.'"

"If we aren't single-peaked with respect to some given societal axis (like the alphabetical order of Figure 5.2), does this mean we cannot be single-peaked at all with our preferences?"

"No," Belle replies, "there might be another societal axis for which our preferences indeed are single-peaked. All that matters is that there *exists* at least one such axis. Actually, I wonder how difficult it is to find out whether a given list of votes, as linear rankings, in fact are single-peaked. After all, there are $m!$ ways to order m candidates on a societal axis, and that is a huge number of possible axes to check!"

"That's easy!" Edgar claims. "Give me your list of votes and I'll tell you whether they are single-peaked in no time at all."

Edgar is right that this is an easy problem in the sense that it can be solved efficiently (though not "in no time at all," as he claims, but rather in *polynomial time*—recall the foundations of complexity theory outlined in Section 1.5). Indeed, Bartholdi and Trick [98] show that, given a list of linear rankings over the candidates, it can be decided in polynomial time whether they are single-peaked, and that when they are, one can also find one societal axis—in fact, even (in implicit form) *all* societal axes—witnessing the single-peakedness. They show this by transforming this problem in polynomial time into the problem of determining whether a matrix has the so-called "consecutive ones property." The result then follows from the work from Fulk-

5.1 Single-Peaked Electorates

erson and Gross [476, Sections 5 and 6] and Booth and Lueker [168, Theorem 6]. Doignon and Falmagne [350] (see also the work of Escoffier, Lang, and Öztürk [415], Fitzsimmons and Lackner [458], and Elkind, Lackner, and Peters [382]) give a direct (and faster) polynomial-time algorithm for this problem.

"But what about a given list of approval vectors?" Anna then asks. "What does single-peakedness mean in that case?"

"As a rule of thumb, this simply means: *'Never leave a gap in your approvals!'* Of course, this again refers to a societal axis that works for the complete list of approval vectors. When a voter goes along the societal axis, say from left to right, and approves of a first candidate, the voter may then keep approving of further candidates, but as soon as the voter next disapproves of anyone, the voter can't go back to approving. That is, there is just a single peak consisting of a contiguous (possibly empty) interval of approved candidates. Pretty simple!"

"One could also say," Belle adds, "that with respect to the societal axis, we each rise to a peak where we approve and then fall back to disapproval, or we always approve, or never approve. But all this talk makes me wonder how difficult it is to find out whether a given list of approval vectors in fact is single-peaked."

"That's easy, too!" Edgar exclaims. "Give me your list of approval vectors, and I'll tell you whether they are single-peaked in no time at all."

Again, Edgar is right. As pointed out by Faliszewski et al. [431, Section 2], the work of Fulkerson and Gross [476, Sections 5 and 6] and Booth and Lueker [168, Theorem 6] shows that, given a list of approval vectors, it can be decided in polynomial time (in fact, in a certain natural sense even in linear time) whether they are single-peaked, and if so, one can also find one societal axis—in fact, even (in implicit form) *all* societal axes—witnessing the single-peakedness. In effect, testing whether a given list of approval vectors is single-peaked is the same as testing whether a matrix whose columns are those approval vectors has the consecutive ones property.

In the following sections we will study problems modeling control, manipulation, and bribery scenarios—recall these notions from the previous chapter, in particular from Sections 4.3.4, 4.3.3, and 4.3.5—when restricted to single-peaked electorates. In each of these restricted problem variants, it is important to note that a societal axis L witnessing the single-peakedness of the given electorate is part of the problem instance. (So inputs that don't contain a valid such axis are not "Yes" instances of the given problem.) Also, it is important to note that the electorates both *before* and *after* the manipulative action must be single-peaked with respect to that same—i.e., the given—societal axis L. For example, in the single-peaked restrictions of control problems such as constructive control by adding voters (CCAV, as defined on page 334 in Section 4.3.4), we require that the entire list of votes, including even votes of any unregistered voters, be single-peaked with respect to L. That strong requirement itself immediately ensures single-peakedness both before and after this control action. We will denote the single-peaked restriction of CCAV by SP-CCAV, etc. For manipulation problems (where votes are being specified) and bribery problems (where votes are being outright changed), we similarly require that both the initial vote set and the final vote set be single-peaked with respect to the axis L that was provided as part of the problem's input.

5.2 Control of Single-Peaked Electorates

Let us recall Belle's insightful comment.

> "Perhaps the complexity of manipulative actions on elections that have restrictions on what votes can be cast—or on what collections of votes can be cast—might be far lower than the complexity is regarding the case where there are no restrictions on what votes can be cast? Perhaps those intractability results that so raised my spirits may turn to dust in this case, leaving the elections open to perfect polynomial-time attack algorithms? These delicious pieces of pie are costing me quite a bit of my peace of mind!"—Belle

Let us start right in, with a theorem showing precisely this, for an important voting system, and what is probably the most important type of control. To make clear why the following theorem really is showing a case where single-peakedness reduces complexity (unless P = NP), it is important to keep in mind that constructive control by adding voters for approval voting is NP-complete (see Table 4.17 on page 342 in Section 4.3.4).

Theorem 5.1 (Faliszewski et al. [431]). *For the single-peaked case, approval voting is vulnerable to constructive control by adding voters.*

The above result holds in both the nonunique-winner model and the unique-winner model, and holds both in the standard model of input, in which each vote appears on a distinct ballot, and in the so-called "succinct" input model, in which the input is a list of the distinct preference orders cast by at least one voter and each such preference order is paired with a nonnegative integer coded in binary that indicates how many voters cast that preference as their vote.

We won't include a full, formal proof of this theorem. Rather, using an extended example we'll convey the idea of the proof. In particular, in the example we'll show how the polynomial-time algorithm for this problem works.

5.2 Control of Single-Peaked Electorates

In our example, we'll use the nonunique-winner model and the succinct input model (although for better readability, in Figures 5.3 through 5.9 we'll write numbers in base 10 rather than in binary).

While Belle relaxes to restore her peace of mind, our guides in the following extended example will be our emerging experts on single-peakedness, Anna and Edgar.

"I think I do understand Theorem 5.1," says Anna, "even though I'm not sure *why* it actually holds. Do you have any idea, Edgar? Would you please clarify this all for me?"

"Sure," Edgar replies, "it's easy. We can establish the theorem by giving a polynomial-time algorithm solving the constructive-control-by-adding-voters problem for approval voting in the single-peaked case. Now, what's the input to our algorithm?"

"I know, I know!" Anna exclaims. "According to the problem definition of CCAV on page 334 in Section 4.3.4, we are given some votes (which are approval vectors) over the candidates, including our preferred candidate p. And we have some additional votes (again approval vectors). We know that all votes are single-peaked with respect to the given societal axis—also part of the input. And our goal will be to check whether we can make p a winner by adding less than or equal to a certain number of the additional votes; that number, the so-called addition limit, itself is also given as part of the problem."

Edgar says approvingly, "Exactly! We want to solve the problem SP-CCAV for approval voting. Let's assume that we are in the succinct input model, so the input doesn't explicitly list all the votes but only has, written down in binary, how often each vote occurs that has been cast at least once. This compact representation can only make it harder for our algorithm to run in time polynomial in the input size, so if our algorithm runs in polynomial time in the succinct case, it surely also does so in the case of the standard input model."

"But how does the algorithm work?"

Edgar replies, "I'll give you an example. Let us look at Figure 5.3!"

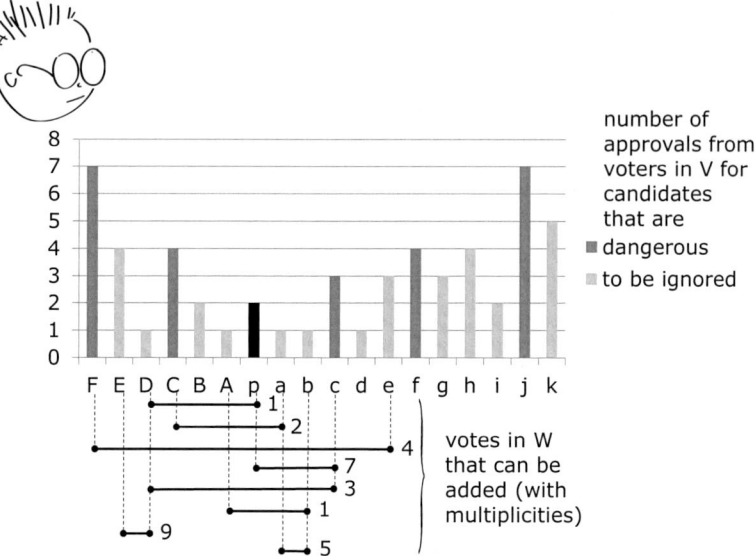

Fig. 5.3 Proving Theorem 5.1 by example: input to SP-CCAV

"What you see here," Edgar explains, "are 18 candidates ordered from left to right along our societal axis, F, E, D, C, B, and A to the left of p, then p, and then $a, b, \ldots, k$ to the right of p. The diagram shows the number of approvals each candidate has from the already registered voters (those in V); for example, j has seven approvals but p has only two. And as mentioned above, keep in mind that the input must also give us the limit on how many votes we may add."

"Why are some of the candidates called *'dangerous,'* while others are *'to be ignored'*?" interrupts Anna.

"Wait a minute, and I'll answer that later. First, do you know why the votes from W—those that may be added (and each coming with a number saying how often it occurs)—are all intervals? For example, note the two votes approving of C, B, A, p, and a?"

"Of course, I know that!" says Anna, brimming with indignation. "It is because they are single-peaked approval vectors. So they cannot have gaps between their approvals!" She pauses to ponder for a second. "But, which types of vote should we add to the election, especially if two votes are *incomparable*? For example, if you look at the two votes approving of C through a versus the seven votes approving of p through c in Figure 5.4: Both vote types approve of p, and that is great. But they also approve of different other candidates. The former but not the latter helps p relative to b and c; the latter but not the former helps p relative to C, B, and A. If we have lots of such choices to make, I

5.2 Control of Single-Peaked Electorates   379

foresee a combinatorially explosive number of collections of votes that
we need to consider as possible choices for the set of votes we should
add, and so I would not be surprised if this whole thing ended up
being NP-complete. Or do you have some clever way of avoiding that
combinatorial explosion, by wisely deciding which should be added?"

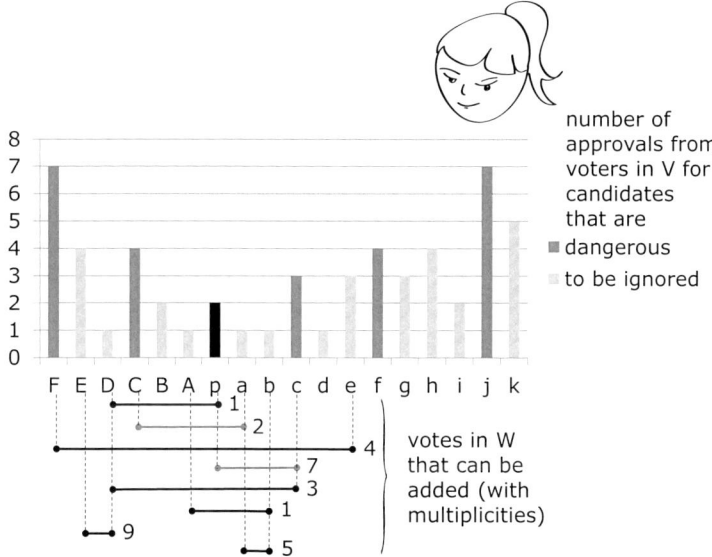

Fig. 5.4 Proving Theorem 5.1 by example: two incomparable votes

"You have identified the heart of the matter! But have no fear.
We'll handle this—and avoid any combinatorial explosion—by a 'smart
greedy' algorithm letting us make such choices in a decisive way that as-
sures us that if either of the choices can lead to success, then the choice
we make will lead to success," says Edgar. "I'll explain that algorithm
later."

"You always postpone answering my questions!" Anna is not amused.
"A minute ago you similarly avoided explaining to me why some of the
candidates are called 'dangerous,' namely F, C, c, f, and j, while all
others are 'to be ignored.' So please tell me now why you have labeled
them in these ways!"

"OK. First, each added vote of course will be an interval including p;
it would be insane to add votes whose interval does not include p. So
we drop all other votes. Figure 5.5 shows the result of doing so in our

example: The nine votes approving of only E and D and the five votes approving of only a and b have been dropped. All remaining votes include p.

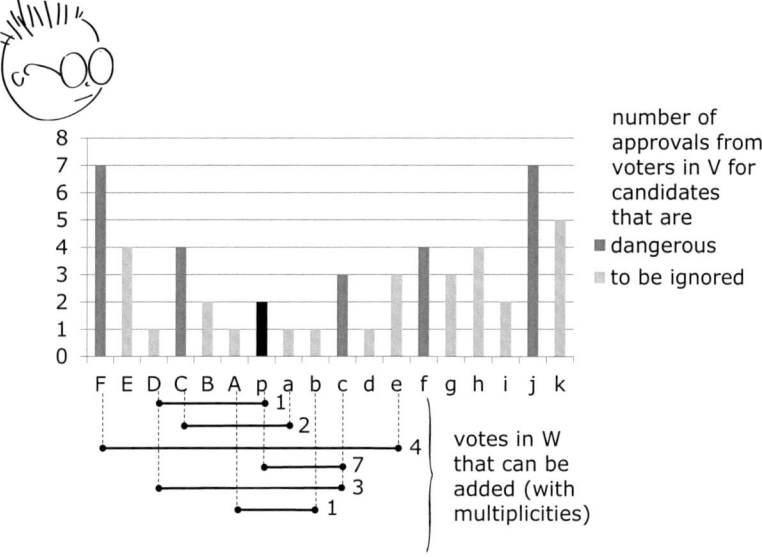

Fig. 5.5 Proving Theorem 5.1 by example: dropping all votes not approving of p

"That doesn't tell me why, for instance, c is a 'dangerous' candidate," says Anna.

"Well, if adding votes from (what remains of) W causes p to draw level with c in terms of approvals, then—since all remaining votes include p—p must at least draw level with a and b. That is since, due to our interval property, every vote that approves of p and c also approves of a and b, yet p as you can see starts this part of our algorithm with at least as many approvals as a and b."

"Agreed."

"Thus c is a dangerous rival for p, and a and b can safely be ignored. Likewise, f is dangerous for p but d and e can safely be ignored. And similarly, j is dangerous for p but g, h, and i can safely be ignored."

"Hey, why do you do that step by step? Just say j is dangerous for p, and ignore a, b, c, d, e, f, g, h, and i! Figure 5.6 shows what I mean, both for the candidates to the left and to the right of p. And while I'm complaining, let me mention that although a and b started with fewer

5.2 Control of Single-Peaked Electorates

approvals than p, e for example starts with more, and so the reasoning you applied above as to why a and b can be ignored seems to me to not apply to e."

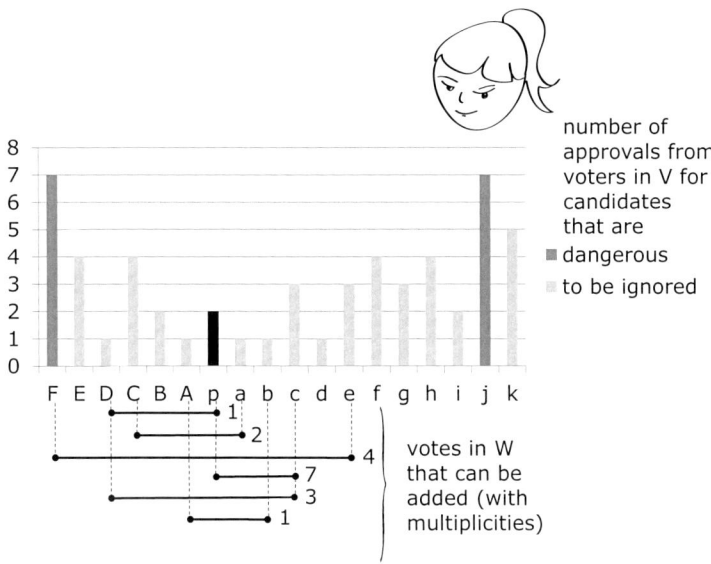

Fig. 5.6 Proving Theorem 5.1 by example: Anna suggests to not go step by step

"The two points you just made are deeply intertwined, as both are connected to the importance of this algorithm going step by step," Edgar says as, startled, he jumps over to Figure 5.7. "Let us consider your suggestion that we say that j is the only dangerous candidate to the right of p. Look what happens if we add, say, five of the seven votes approving of p through c. Then the number of these votes in W is reduced to two, and each of p, a, b, and c get five more approvals. No doubt, p has now drawn level with j, both having seven approvals now, but c was riding the wave and was boosted to even eight approvals! So if we don't go step by step, p might well draw level with j but still is not a winner."

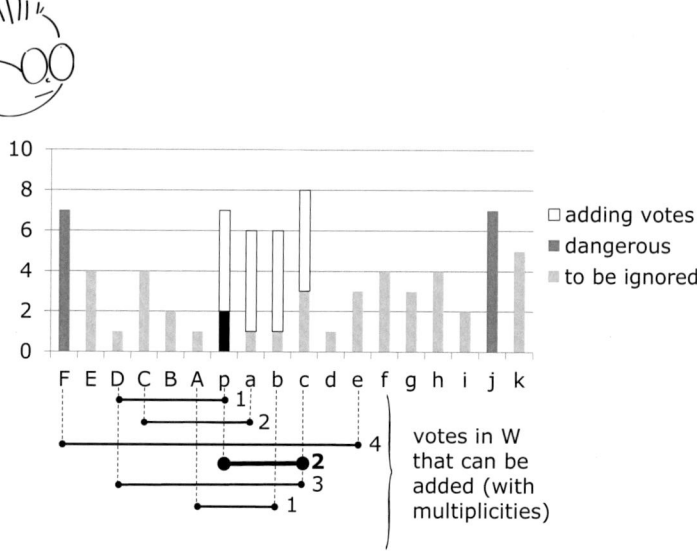

Fig. 5.7 Proving Theorem 5.1 by example: What happens if we don't go step by step?

"I see," Anna concedes. "That means the first dangerous candidate to the right of p is the leftmost candidate to the right of p that is approved by more voters from V than p, namely c in the figure, and the second dangerous candidate to the right of p is the leftmost candidate to the right of c that is approved by more voters from V than c, namely f, and so on. And we can define this analogously for the candidates to the left of p. So, as indicated in Figure 5.8, we indeed get the dangerous candidates C and F to the left and c, f, and j to the right of p, and we ignore all other candidates. And knowing this, I understand the right answer to my second point above—the one where I pointed out that e for example starts with more approvals than p. I now see that that is true but, crucially, *after* we have made p tie or beat c in approvals—by adding only votes that approve of p but don't approve of c (and thus also don't approve of e!)—at *that* point p will surely tie or beat all candidates to c's right that started with no more approvals than c. So f indeed is the next dangerous candidate; e is not a worry at all!"

5.2 Control of Single-Peaked Electorates

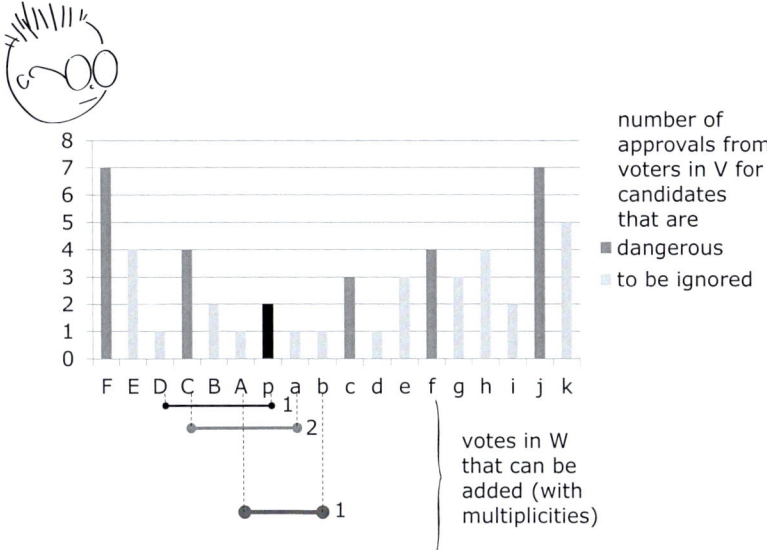

Fig. 5.8 Proving Theorem 5.1 by example: Which votes can help in "smart greedy"?

"Exactly! Of course, what we have just been discussing is the notion of *dangerous candidate for p* in the nonunique-winner model. But what change do you think we'd want to make if we are in the unique-winner model?"

"Then I'd say, when looking for the first dangerous candidate on for example the right side of p, we take whichever candidate is the first to have *at least as many approvals* as p, all else being the same."

"That's right," Edgar agrees. "If we want to make p a *unique* winner, even having the same number of approvals as p already makes a candidate a dangerous rival for p; we must ensure that p strictly beats this candidate. That is, B too would be dangerous for p in the example of our figure."

Anna nods her agreement, and then suggests, "However, let's stay in the nonunique-winner model, which seems to be more natural. Now tell me, how does your 'smart greedy' algorithm work? How does it find the right votes from W to add?

"In the 'smart greedy' algorithm, we need to eat through all dangerous rivals to the right of p, starting with the leftmost, c. To become a winner, p in particular must draw level with c. However, only votes (i.e., intervals) in W whose right endpoints fall into $[p, c)$ can help."

"I see. These are exactly the votes from W that still are shown in Figure 5.8."

"Let X be the set of these votes," Edgar continues. "Now, the key insight of the algorithm is that we will be choosing votes from X *starting with those having the rightmost left endpoint*. In our example of Figure 5.9, we start by adding the voter—shown by a fat line—approving of A through b. As shown in Figure 5.9, this is already enough for p to draw level with c, so p's first dangerous rival has been taken care of. And the key point is that this has been achieved by an easy (in the sense of polynomial time) yet perfectly safe strategy!"

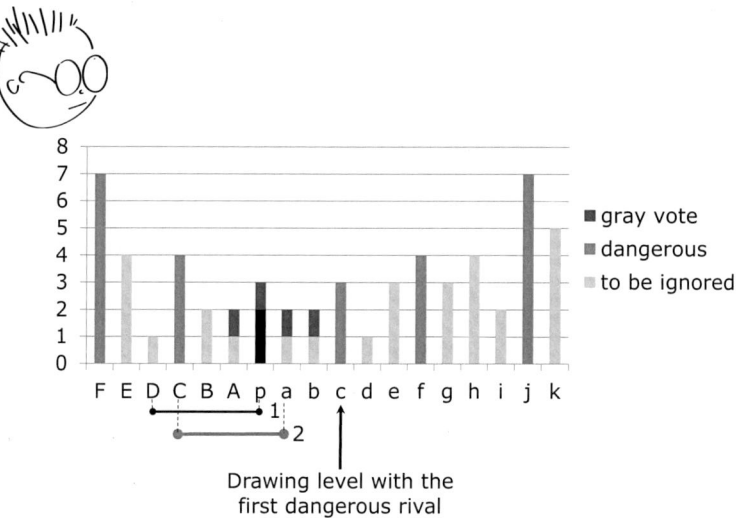

Drawing level with the first dangerous rival

Fig. 5.9 Proving Theorem 5.1 by example: drawing level with the first dangerous rival

"Why?" asks Anna.

"Because if there is *any* way at all to choose votes to add such that p draws level with his first dangerous rival (and eventually can become a winner), our strategy of starting with those votes from X that have the rightmost left endpoint will succeed, too. (And if there is no such way, then of course this strategy cannot succeed either.) And among all such choices, note that crucially our choice is at least as good regarding how it leaves us relative to other dangerous candidates—most crucially those to the left of p—since it is approving of as few of them as possible. Briefly put, the extra candidates on the right that the 'fat' vote helped do us no harm, and the extra candidates on the left that the nonfat votes would have helped might do us harm. (By the way, before moving

5.2 Control of Single-Peaked Electorates

on, I should mention that if we had two or more copies of the fat vote, and p needed at that point two or more approvals relative to c, then we would in a single step have added as many copies of the fat vote as needed, or if that is more than the number of copies that we had available, then would have added all its copies—unless either of those took us beyond our addition limit, in which case we'd have to admit defeat. The reason I mention this is to make clear that we really are handling the succinct case—and so we can't add votes one at a time, but rather add them drawing on the right 'multiplicity' of the given vote to support our progress.)"

"Now that p's first dangerous rival has been taken care of," says Anna, "what do we do next?"

"We iterate. That is, updating the approvals of all candidates and the votes that may be added as in Figure 5.9, we apply the same procedure to handle the next dangerous candidate, here f, as long as our allowed number of votes to be added hasn't been used up. If we run out of dangerous candidates on the right-hand side of p, we reverse the societal order, and we finish off the remaining dangerous candidates (which have been mirrored from the left to the right of p to make the same procedure applicable to them) in exactly the same way until we either succeed in making p a winner and so can output 'yes,' or reach the addition limit without having achieved our goal. In the latter case, we know for sure that no strategy whatsoever could possibly make p win by adding the allowed number of votes from W, so we can safely output 'no,' i.e., that success is not possible."

That concludes our extended example sketching the polynomial-time algorithm for constructive control by adding voters under approval voting when one is dealing with a single-peaked electorate.

But Anna can be a tough person to convince of anything.

Anna: Thank you for that example. I do believe your polynomial-time claim for constructive control by adding voters in the single-peaked case. But I worry: Maybe that is the only control type where single-peakedness helps, and maybe approval is the only voting system showing this behavior.

Edgar: Have no fear, they are not alone!

The following result gives some examples of what Edgar is referring to—other control cases that are NP-complete in the general case but have polynomial-time algorithms for the case of single-peaked electorates. Recall the definitions of these control problems from Section 4.3.4 starting on page 328.

Theorem 5.2 (Faliszewski et al. [431]).

1. *For the single-peaked case, approval voting is vulnerable to constructive control by deleting voters.*
2. *For the single-peaked case, plurality voting is vulnerable to constructive and destructive control by adding candidates, by adding an unlimited number of candidates, and by deleting candidates.*

Having seen this theorem, which we mention holds in both the nonunique-winner model and the unique-winner model, Anna has been convinced—but perhaps a bit *too* convinced, as the following shows.

> Anna: Wow. Those additional cases make it clear to me that restricting our focus to single-peaked electorates lowers the complexity. I'll bet that this approach will undercut all existing NP-hardness result regarding all election problems.
>
> Edgar: Not so fast! In fact, for the devilishly complex system STV (which is defined on page 252 in Section 4.1.4), Walsh [927] has noted that even when restricted to single-peaked electorates, the possible winner problem remains NP-complete and the necessary winner problem remains coNP-complete (see Section 4.3.2 for the problem definitions); sometimes, hard things stay hard even under single-peaked preferences.

5.3 Manipulation of Single-Peaked Electorates

Our guides Anna and Belle (the latter of whom has through resting recovered her peace of mind) are chatting again, and the chat takes a shocking turn.

> Anna: In our discussion of control, we saw that restricting our focus to single-peaked electorates sometimes lowers the complexity. And Edgar mentioned to me an example, regarding possible and necessary winners, where restricting our focus to single-peaked electorates fails to lower the complexity. Clearly, that covers all the possibilities, since restricting ourselves to single-peaked electorates obviously cannot ever raise the complexity.
>
> Belle: I disagree. I claim that restricting ourselves to single-peaked electorates *can* raise the complexity!
>
> Anna: That's clearly not possible. Anyone who knows the basics of complexity knows that if a problem is easy, any easily identified restricted case of it is also easy. In this case, since every single-peaked electorate is an electorate, it follows that if the problem has a polynomial-

5.3 Manipulation of Single-Peaked Electorates

time algorithm for all electorates, then it has a polynomial-time algorithm for single-peaked electorates.

Belle: No, my dear friend. I understand what you're thinking, and your error is a quite subtle one. The error is hidden in your words "the problem" above. You are assuming that "the problem" is the same in both cases. If that were true, your claim about subcases inheriting polynomial-time algorithms would be fine. But the problems in question do *not* differ merely on whether the electorate must be single-peaked.

Anna: I don't see any other way in which they differ.

Belle: That is where the "subtle" comes in. Recall that in defining our problems in the single-peaked context, we required that the electorates be single-peaked (with respect to the given societal axis) even *after* the manipulation.

Anna: Yes, that is natural, but what does it have to do with some difference in the problems.

Belle: The difference is that for the single-peaked case of manipulation, we are asking whether (the input is single-peaked with respect to the societal axis L and) there is some set of votes by the manipulators *under which the election is still single-peaked with respect to the axis L* and p is a winner of the election. In contrast, the general-case is merely asking whether there is some set of votes by the manipulators such that p is a winner of the election.

Anna: Then the single-peaked case gives fewer options to the manipulators as to what votes they can cast, and so the problem is a subcase, and so as I said before it can only be simpler.

Belle: No. It is a different problem. The "subcases only reduce complexity" argument line only refers to restrictions of the problem domain. If the actions inside the problem can differ, even if they are more restrictive, that is a whole different issue. It is possible that for the less restrictive set of actions a manipulation problem is computationally easy, even though it would be computationally hard for a more restrictive set of actions, such as being limited to manipulations that leave the electorate single-peaked.

Anna: Huh?

Belle: Let me try to give you a bit of intuition as to how this might happen. Let us consider constructive size-3-coalitional unweighted manipulation, and our model will be that votes are approval vectors. However, our voting system won't be approval voting. In fact, suppose our voting system has the property that when the electorate isn't single-peaked, then it is easy to manipulate successfully. As an extreme example, consider a voting system that when the electorate isn't single-peaked makes all candidates be winners, and thus makes the preferred candidate p be a winner; and if the electorate is single-peaked this system chooses some winner in some different and very complex-to-manipulate way. So a coalition of three or more manipulators can achieve success—even if we for a moment jump out of the model where the societal axis is fixed and given—simply by having each of the three ballots (which among themselves form a "Condorcet-cycle"-like pattern) $a > b > c$, $b > c > a$, and $c > a > b$ (where a, b, and c represent the names of the three lexicographically smallest candidates) be cast by at least one manipulator. And if the number of candidates is less than three, we in our election system just have everyone always win. Note that there is no axis that makes any vote collection with the just mentioned three votes be single-peaked. Clearly, manipulation for this problem is in polynomial time for the general case. But the single-peaked case can't use this approach, since it isn't allowed to manipulate votes in such a way as to violate single-peakedness; and in fact, the single-peaked case has no such easy, obvious path to successful manipulation. Indeed, one can specify a voting system of this sort in such a way that the manipulation problem for the single-peaked case is NP-complete.

Anna: I certainly don't see all the details, since you didn't specify them, but I do see the general flavor. Your example counterintuitively makes single-peakedness's more limited set of legal electorates a complexity-increasing disadvantage, rather than a complexity-lowering advantage. And I see that that isn't paradoxical, because the single-peaked case not only limits the set of legal inputs, but also limits the set of legal manipulative actions the coalition can take. So we aren't merely a special case of a problem; we're a slightly different problem, since the single-peakedness in some sense penetrates the problem to its very core.

Belle: Well put; I see a real future for you as a complexity theorist when we grow up.

Anna: Heaven forfend! Anyway, I've always thought that your little brother Edgar was the most likely of any of us to live *that* life.

Belle: Heaven forfend! He's already insufferable enough as it is, and I hear that complexity theorists are beyond insufferable.

Anna: I've heard that too.

The theorem that Belle was outlining is the following result. Its detailed proof, which we won't give here, is a bit twisty, especially regarding achieving the NP-completeness part. But Belle's description of the proof strategy is in fact spot-on.

Theorem 5.3 (Faliszewski et al. [431]). *There exists a voting system $\mathcal{E}$, whose votes are approval vectors, for which constructive size-3-coalition unweighted manipulation is in polynomial time for the general case but is NP-complete in the single-peaked case.*

5.3 Manipulation of Single-Peaked Electorates

Anna: But now that I'm at least dabbling at thinking at things complexity-theoretically, I'm wondering whether that strange complexity-raising behavior can ever happen for systems that I'm familiar with. In particular, if you can show me any scoring protocol under which we get this complexity-raising behavior, I'll give you my next ten slices of pumpkin pie.

Belle: I love pie but, alas and alack, I cannot show you an example. But neither can anyone else, since no such example can exist!

Let us see what Belle—who despite her protestations seems well on her way to becoming a complexity theorist—is referring to. Recall from Chapter 4 that for scoring protocols (defined in Section 4.1.1) there is a dichotomy theorem (stated in Section 4.3.3) for the constructive coalitional weighted manipulation problem. In particular, we mentioned there what in effect is the following result.

Theorem 5.4 (Hemaspaandra and Hemaspaandra [535]). *For each m and each scoring protocol $\boldsymbol{\alpha} = (\alpha_1, \ldots, \alpha_m)$, the constructive coalitional weighted manipulation problem is NP-complete if $\alpha_2 > \alpha_m$, and is in P otherwise.*

For scoring protocols, there also is a dichotomy theorem for the constructive coalitional weighted manipulation problem in the single-peaked case.

Theorem 5.5 (Brandt et al. [210]). *Consider an m-candidate scoring protocol $\boldsymbol{\alpha} = (\alpha_1, \alpha_2, \ldots, \alpha_m)$.*

1. *If $m \geq 2$ and $\alpha_2 > \alpha_{\lfloor (m-1)/2 \rfloor + 2}$ and there exist integers $i > 1$ and $j > 1$ such that $i + j \leq m + 1$ and $(\alpha_1 - \alpha_i)(\alpha_1 - \alpha_j) > (\alpha_i - \alpha_{i+1})(\alpha_j - \alpha_{j+1})$, then the constructive coalitional weighted manipulation problem for the single-peaked case is NP-complete.*
2. *If $m \geq 2$ and $\alpha_2 = \alpha_{\lfloor (m-1)/2 \rfloor + 2}$ and $\alpha_1 > \alpha_2 > \alpha_m$ and ($\alpha_2 > \alpha_{m-1}$ or $\alpha_1 - \alpha_m > 2(\alpha_2 - \alpha_m)$), then the constructive coalitional weighted manipulation problem for the single-peaked case is NP-complete.*
3. *In all other cases, the constructive coalitional weighted manipulation problem for the single-peaked case is in P.*

The above result, Theorem 5.5, truly is a dichotomy theorem; it proves that every scoring protocol is either NP-complete or in P. In fact, this dichotomy theorem even has a very easy-to-check condition that tells for a given scoring protocol which case holds for it, thus making this theorem very easy to apply to actual, natural protocols. For example, from the theorem we can immediately see that for each m it holds that for the single-peaked case of the constructive coalitional weighted manipulation problem, m-candidate plurality and m-candidate veto are in P; and so is m-candidate Borda for $m < 4$. The theorem also makes clear that m-candidate Borda is NP-complete for

390   5 The Complexity of Manipulative Actions in Single-Peaked Societies

each $m \geq 4$. Given this theorem's ease of application, one probably cannot fairly complain about how very much more involved the theorem statement is than the analogous and also easy-to-apply theorem for the general case, Theorem 5.4. If anything, one should—while thanking the universe for the fact that the characterization is easy to apply—blame the universe for making the single-peaked case have such a complex-looking characterization. What isn't complex to observe is what Belle was commenting about: Every P case from the general-case dichotomy of Theorem 5.4 clearly remains a P case in the single-peaked dichotomy of Theorem 5.5. So, just as Belle claimed, in the setting of scoring protocols and constructive coalitional weighted manipulation, restricting attention to the single-peaked case never raises complexity. As to the NP-complete cases from the general-case dichotomy of Theorem 5.4, in the single-peaked dichotomy of Theorem 5.5 some of those cases remain NP-complete and some fall to P.

> Belle: I've now shown you why no example of the sort you requested can exist.
> Anna: True, but since you didn't provide the requested example, I'm still keeping those ten future slices of yummy pie.
> Belle: Grrrrr!

Theorem 5.5 hides, underneath its complexity, a quite broad range of cases. And some of those at first are surprising. For example, the behavior that people typically expect for voting systems, viewed at each fixed number of candidates, is that as one increases the number of candidates, the problem either stays the same in complexity or increases in complexity. But let us apply Theorem 5.5 to the case of 3-veto. Recall that the scoring vector for 3-veto is given by $(1,\ldots,1,0,0,0)$; for example, for $m = 5$ candidates this is $(1,1,0,0,0)$ and for $m = 6$ candidates it is $(1,1,1,0,0,0)$. We get the following strange behavior, which was first noticed and proven by Faliszewski et al. [431].

Theorem 5.6 (Faliszewski et al. [431]). *For the single-peaked case, the constructive coalitional weighted manipulation problem for m-candidate 3-veto is in* P *for* $m \in \{3,4,6,7,8,\ldots\}$, *and is* NP-*complete for* $m = 5$.

So this is a case where, between $m = 5$ and $m = 6$, the complexity *drops*. Let us get a sense of how this kind of unexpected behavior can arise. 3-veto for $m = 3$ is in effect triviality, and so everyone always wins. 3-veto for $m = 4$ is just plurality (1-approval), and the single-peakedness is irrelevant since every 1-approval vote is trivially single-peaked. So these two cases are certainly in P.

The $m = 5$ case is shown NP-hard by a standard type of reduction, which we will now describe, again in the nonunique-winner model. (Membership of the problem in NP is obvious, so we in fact have NP-completeness once NP-hardness is shown.) To prove NP-hardness, we will give a reduction from

5.3 Manipulation of Single-Peaked Electorates

the NP-complete problem PARTITION, which has been defined on page 182 in Section 3.5.2.1 as the set of all nonempty sequences $(k_1,\ldots,k_n)$ of positive integers summing up to an even number, $2K = \sum_{i=1}^{n} k_i$, that can be partitioned into two subsequences, each summing up to the same value K. Suppose, we are given an input $(k_1,\ldots,k_n)$ of PARTITION, and for concreteness let us say we have $n=3$ and we will consider two particular cases for illustration:

1. $(k_1,k_2,k_3) = (1,2,3)$, so $K=3$. Since $1+2=3$, this is a yes-instance of PARTITION.
2. $(k_1,k_2,k_3) = (1,2,5)$, so $K=4$. Note that this is a no-instance of PARTITION.

(Of course, just handling how to reduce from these very special instances of PARTITION does not establish NP-hardness, since these two cases are quite trivial. However, although we will use these cases as illustrations, we in fact will be quietly giving the general case of this reduction.)

We construct an instance of the constructive coalitional weighted manipulation problem from this PARTITION instance as follows. Our candidates are Adams, Brown, Chavez, Dylan, and the preferred candidate Pearl that the manipulators wish to make a winner. We also fix the following societal axis L:

$$\text{Chavez } L \text{ Adams } L \text{ Pearl } L \text{ Brown } L \text{ Dylan.}$$

There are two nonmanipulators with weight K each, Anna and Belle, and Anna votes

$$\text{Chavez} >_{\text{Anna}} \text{Adams} >_{\text{Anna}} \text{Pearl} >_{\text{Anna}} \text{Brown} >_{\text{Anna}} \text{Dylan},$$

while Belle votes

$$\text{Dylan} >_{\text{Belle}} \text{Brown} >_{\text{Belle}} \text{Pearl} >_{\text{Belle}} \text{Adams} >_{\text{Belle}} \text{Chavez}.$$

As is sometimes the case for best friends, Anna and Belle have completely opposite preferences.[1] Obviously, both votes are single-peaked with respect to L, with Anna's preference-based utility "just falling" and Belle's preference-based utility "just rising." In addition, there are n manipulators where the ith manipulator has weight k_i; in our example with $n=3$ manipulators, Chris has weight k_1, David weight k_2, and Edgar weight k_3. The manipulative boys want to see Pearl win.[2] We will now show that they can reach their goal by suitably setting their preferences if and only if $(k_1,\ldots,k_n)$ is a yes-instance of PARTITION. In particular, they can ensure Pearl's victory in the first case $((k_1,k_2,k_3) = (1,2,3))$, but not in the second $((k_1,k_2,k_3) = (1,2,5))$.

---

[1] They are not voting here on who bakes the best pumpkin pie, but rather, let us say, on which of the bakers is the most beautiful person—again, tastes differ.

[2] Because Pearl promised them pumpkin pie for life if Pearl wins. But that bribe notwithstanding, the computational problem here still is about manipulation; bribery will be handled in the next section.

Suppose there is a successful partition of $(k_1,\ldots,k_n)$, i.e., suppose we have a set $A \subseteq \{1,\ldots,n\}$ such that $\sum_{i \in A} k_i = \sum_{i \in \{1,\ldots,n\} \setminus A} k_i = K$ (e.g., $1+2=3$ as in our first case). Then all manipulators whose weight is in the set $\{k_i \,|\, i \in A\}$ (so Chris and David in our first case) can set their preferences to be

$$\text{Pearl} >_i \text{Adams} >_i \text{Brown} >_i \text{Chavez} >_i \text{Dylan},$$

while all other manipulators (namely, Edgar in our first case) can choose the preference order

$$\text{Pearl} >_i \text{Brown} >_i \text{Adams} >_i \text{Chavez} >_i \text{Dylan}.$$

Note that these manipulative votes are single-peaked with respect to L. Recall that for $m=5$ candidates our scoring vector is $(1,1,0,0,0)$. Thus, in the election with both the manipulative and the nonmanipulative votes, Adams, Brown, and Pearl each score $2K=6$ points, while Chavez and Dylan get only $K=3$ points each, so Pearl is a winner.

For the converse, suppose now that there is no partition of $(k_1,\ldots,k_n)$ (such as in our second case). Can the manipulators, who are obliged to cast single-peaked votes with respect to L, still make Pearl a winner? Seeking a contradiction, let us assume that the answer is yes. Note that in each such single-peaked vote, whenever Pearl scores a point, Adams or Brown does so also. Thus, among the manipulative votes, Pearl's score is bounded above by the sum of the scores of Adams and Brown. Note that Pearl doesn't get any points from the nonmanipulators, but Adams and Brown receive K points each from them (in our example, Adams gets K points from Anna and Brown gets K points from Belle). Since we assumed that Pearl wins the election whose voter set includes both the manipulators and the nonmanipulators, it follows that among the manipulative votes alone, Pearl must score at least K points more than Adams does and at least K points more than Brown does. It follows that from the manipulative votes Pearl's score is $2K$, and Adams and Brown score K points each. However, this implies that the weights of the manipulators ranking Adams in their top two positions sum to K (and the same applies to the manipulators ranking Brown in their top two positions), that is, there is a partition of $(k_1,\ldots,k_n)$, a contradiction. This concludes our informal proof that our reduction for the $m=5$ case is correct. Thus the constructive coalitional weighted manipulation problem is NP-hard.

Anna: Wait a minute! What if we are in the unique-winner model? Does this reduction apply to that case, too?

Belle: Almost. Just change the weights of the two nonmanipulators from K to $K-1$.

Turning now to a discussion of the $m>5$ cases in Theorem 5.6, we get the at-first-surprising drop in complexity. However, we claim that that drop has a quite clear source. Let's consider the $m=7$ case, to see how it can possibly be simpler than the $m=5$ case. Note that for $m=7$ and 3-veto voting, each vote cast is a 4-approval vote. So whichever candidate is the middle (i.e., fourth) one among the 7 candidates, along the societal axis L, certainly must be approved of by every voter, since the votes are single-peaked, and thus each vote's approved-of candidates must be contiguous within L. So that middle candidate is always a winner. And each other candidate, a, is a winner exactly if every voter approves of a. Since we can certainly make all the manipulators approve of a, the only issue we need to look at to efficiently decide whether a can be made a winner is whether all nonmanipulative voters approve of a. We have just given a polynomial-time algorithm for the constructive coalitional weighted manipulation problem for 7-candidate 3-veto, in the single-peaked case. And the algorithm makes clear the type of effect at issue here—an effect that clearly will hold for all $m > 7$ also. Namely, we have that one or more candidates are forced to be approved of by every voter, and so the only real issue in this problem is whether a given candidate is approved of by all nonmanipulative voters. One can also argue that the $m=6$ case is in P, though one has to be a bit more careful.

5.4 Bribery of Single-Peaked Electorates

In our study of control, we saw that some problems that are NP-complete in the general case fall all the way down to polynomial time for the single-peaked case. Where this behavior can be found depends on the exact setting: what the manipulative action is and what the voting system is.

For bribery, we have similarly mixed behavior. Under the single-peaked restriction, some NP-complete bribery cases fall from being NP-complete to being in P, and some do not. Whether such a complexity drop occurs is sensitive to both the type of bribery and the voting system.

Let us give some examples of these behaviors. For approval voting, the following theorem gives three cases where types of bribery that are known (some by Faliszewski, Hemaspaandra, and Hemaspaandra [422] and some by Brandt et al. [210]) to be NP-complete in the general case fall into polynomial time for the single-peaked case. Before looking at the theorem, recall the various notions of bribery defined in Section 4.3.5 starting on page 358; in particular, the basic problem variant BRIBERY, which can be weighted (indicated by ⚖ in the problem name) and/or priced (indicated by $), and the notions of negative and strongnegative bribery.

Theorem 5.7 (Brandt et al. [210]).

1. For the single-peaked case, approval-BRIBERY is in P.
2. For the single-peaked case, approval-NEGATIVE-BRIBERY is in P.
3. For the single-peaked case, approval-STRONGNEGATIVE-BRIBERY is in P.

Of course, that is just one voting system. Can we cast a wider net as to understanding when bribery problems fall in polynomial time for the single-peaked case? The answer is a "yes," although as we'll soon see, it is a somewhat qualified "yes."

Recall the notion of weak Condorcet winner from page 239 in Section 4.1.2. In particular, recall that a voting system is said to be *weakCondorcet-consistent* if whenever there are candidates that tie-or-beat all other candidates in head-on-head pairwise contests, the winner set is exactly the set of candidates having that property. It turns out that for the single-peaked case we can in one fell swoop capture the bribery complexity, under all eight standard types of bribery, of all weakCondorcet-consistent voting systems. Five of the bribery types are simple and three are complex.

Theorem 5.8 (Brandt et al. [210]). *Let $\mathcal{E}$ be any voting system that is weakCondorcet-consistent, or even that merely is always weakCondorcet-consistent on single-peaked electorates.*

1. *For the single-peaked case, $\mathcal{E}$-⚖-\$BRIBERY, $\mathcal{E}$-NEGATIVE-⚖-BRIBERY, and $\mathcal{E}$-NEGATIVE-⚖-\$BRIBERY are each NP-complete.*
2. *For the single-peaked case, $\mathcal{E}$-BRIBERY, $\mathcal{E}$-\$BRIBERY, $\mathcal{E}$-⚖-BRIBERY, $\mathcal{E}$-NEGATIVE-BRIBERY, and $\mathcal{E}$-NEGATIVE-\$BRIBERY are each in polynomial time.*

Many important systems are weakCondorcet-consistent. For example, the voting systems—each defined in Section 4.1—Llull, maximin, Young, weak-Dodgson (by weak we mean Dodgson altered so that the goal is to by adjacent-exchanges make a candidate tie-or-beat each other candidate in head-on-head contests), and weakBlack (with weak analogously interpreted) are weakCondorcet-consistent [454, 210]. And as noted by Brandt et al. [210], the voting systems of Kemeny, Schwartz, Llull, and two variants of Nanson are weakCondorcet-consistent when restricted to single-peaked electorates. So Theorem 5.8 applies to all the just-mentioned systems, and classifies all its types of bribery.

Are any of the P results obtained in that way examples of complexity being reduced due to single-peakedness? Absolutely. For example, for Llull elections, bribery, \$BRIBERY, ⚖-BRIBERY, and ⚖-\$BRIBERY, are each NP-complete [429], but by Theorem 5.8 each of these cases is in P for the single-peaked setting. For Kemeny the drop is even more dramatic. Each of the eight standard types of bribery is $P_{\parallel}^{NP}$-hard for Kemeny elections [210]. Yet by Theorem 5.8, in the single-peaked case three of those $P_{\parallel}^{NP}$-hardness bounds change to NP-completeness results and five change to P results.

On the other hand, let us come back to our earlier comment about the "yes" regarding the wider net being a qualified "yes." What we meant by that is that part of what is underpinning Theorem 5.8 is the fact that in single-peaked electorates (with the voters voting by linear orders), there always is a weak Condorcet winner, namely, a candidate who ties-or-beats each other candidate. And that means that all the voting systems we are discussing here (and more generally, all weakCondorcet-consistent voting systems) become the same as each other for the case of single-peaked electorates, namely, the winner set in that case for each is exactly the collection of all weak Condorcet winners. On one hand, that might be viewed as disappointing, since the systems are all becoming the same system, for the case that Theorem 5.8 is speaking of. On the other hand, the more interesting and important points to focus on are how very varied those systems are in the general case and how hard their bribery problems often are in the general case, and yet despite that how single-peakedness removes so many of those hardness results for those systems.

5.5 Do Nearly Single-Peaked Electorates Restore Intractability?

Let us return to the Pumpkin Pie Taste-Off. Although everyone seemed to be in agreement that level of sweetness was what anyone who had the ability to taste food uses to judge pumpkin pies, a surprising twist is about to occur.

"Did I just hear you say that no one with the ability to taste food could possibly put Zimmerman first and Adams second, given that Zimmerman makes the sweetest pumpkin pie, Adams makes the least sweet, and the twenty-four other contestants' pies fall between them in sweetness," says a quiet voice from off to the side, belonging to Helena. "That in fact is exactly what I would have cast as my vote had I been here in time to vote."

"Impossible! Unthinkable! Have you no sense of taste?"

"I do," replies Helena, "but I have celiac disease, and so energetically avoid eating the protein known as gluten. And only the pies of Zimmerman and Adams are gluten-free, thanks to the ingredients in their crusts being made respective of rice flour and amaranth flour. When I judge a pie, the crust's ingredient set is the issue that I use."

Quick-witted Belle, who is still aware that she is helping guide us through this book, immediately says, "You may just have made my day, you wonderful maverick! As I mentioned earlier, I was overjoyed that complexity might provide a shield against attacks on elections. Then when pumpkin pie led us to discuss single-peaked elections, I became worried that in that natural setting the protections might evaporate. The past few sections of this book in large part showed my fears to be well-grounded. We've seen that single-peakedness often does sidestep existing complexity-theoretic protections against attacks on elections. But you have opened my eyes to a new hope. Those sidestepping results were based on the assumption that the electorates in question are single-peaked. However, it seems to me that pretty much no electorate will be perfectly single-peaked. There will always be at least a few mavericks. At this Taste-Off, although we all thought it obvious that every person judges pumpkin pie based on the sweetness, we found that you judge based on the crust. And in large political elections where almost everyone is voting based on the candidates' positions on some important spectrum—perhaps liberal versus conservative—there surely will be a few people who see things differently. Perhaps some people are libertarians and so care about an aspect not even captured by that axis, and perhaps others are influenced by issues such as a candidate's religion or a candidate's charisma. It seems to me at least possible that for cases where there are some mavericks amidst a largely single-peaked society, some of the complexity-theoretic protections against manipulative attacks may still hold. At the very least this is worth looking into ... although only after I reward myself with another slice of King's pumpkin pie!"

Belle makes an excellent point, and in this section we'll see examples of the type of behavior that she is imagining. Indeed, we'll even see that in some cases, the presence of *a single maverick* can jump a problem's complexity from P back up to NP-completeness!

On the other hand, we'll also explore cases where manipulative-action problems remain polynomial-time solvable even if the electorate has a few mavericks. Each such result is, of course, stronger than the analogous polynomial-time claim for the single-peaked case. Such results are typically proven by showing how we can efficiently handle the chaos added by mavericks.

Anna: Where did the term "maverick" come from?
Belle: Samuel Maverick, a colorful Texan who lived from 1803 to 1870, refused to brand his cattle. You can see one of his unbranded cows in Figure 5.10. Unbranded cattle came to be called "mavericks," and the term "maverick" also came to be applied to anyone who is individualistic and unorthodox.

Faliszewski, Hemaspaandra, and Hemaspaandra [425] defined many notions of nearness to single-peakedness, and studied the complexity-theoretic behavior of control and manipulation for electorates of the given nearness types. In this section, for clarity, we will limit ourselves to a sampling of results about two of the more attractive nearness notions that they studied. (Additional nearness measures and discussion of how hard it is to evaluate nearness to single-peakedness of an electorate can be found in [227, 410, 425], and for the complexity of consistency testing and axis-production for the

5.5 Do Nearly Single-Peaked Electorates Restore Intractability?

Fig. 5.10 A ruminating cow (left) and a maverick (right)

(pure) single-peaked case see [98, 167, 350, 415, 476] and the discussion in Section 2.2 of [431] and Section 4 of [425]. Briefly put, although for the single-peaked case this is easy, for "nearly single-peaked" cases such issues can become hard.)

5.5.1 K-Maverick-Single-Peakedness

The first of the two nearness-to-single-peakedness notions that we will study is the notion of a *k-maverick-SP* (for *"k-maverick-single-peaked"*) electorate. As always, our problems will come with a societal axis, L, as part of the input. And a collection of votes is said to be a *k-maverick-SP* electorate if all but at most k of those votes are consistent with the societal axis. When studying manipulative actions problems for the k-maverick-SP case, we require both the input and the after-the-manipulative-action state to be k-maverick-SP electorates. So one cannot in a manipulation problem make so many manipulators be maverick-like that the total number is greater than k. For control-by-adding-voters problems, the entire collection of input votes—both the registered voters and the unregistered ones, viewed together as one big collection—must be a k-maverick electorate.

The motivation for looking at nearly single-peaked electorates is quite compelling. Often electorates are very heavily focused on some issue, such as the sweetness of pumpkin pie in a pie contest, or the degree to which candidates for political office want to redistribute wealth. However, as we saw in the case of Helena, it is perhaps too much to hope that every single person in any reasonably large society will have preferences that mesh with that axis. Helena, due to a medical condition, cared not about sweetness but about the ingredients in the crust. In political elections, even if it seems there is a single clear, salient axis/issue, it is possible that at least a few voters may refuse to vote for the candidate whose position on the axis's issue best matches the voter's, perhaps because the voter has biases, such as refusing to vote for any candidate of a certain religion. Or perhaps the voter perceives differently than others the positions of the candidates on the society's most salient issue. Or perhaps the voter simply doesn't see as the most important issue the same issue that almost everyone in the society sees as their vote-controlling issue. The truth is, although single-peakedness is a very natural notion, probably the right claim to make about it is that in many real-world settings electorates are quite close to being single-peaked. Perfect and pure single-peakedness is too much to hope for in the chaos, confusion, and noise of the real world.

5.5.2 Swoon-Single-Peakedness

The second model of nearness to single-peakedness that we'll study is the notion of a *swoon-SP* (for *"swoon-single-peaked"*) electorate. In that we require that, for each voter, if one removes the top choice of that voter from that voter's ballot and also from the societal axis L, then the resulting ballot is consistent with the resulting axis. This models the case where each voter is perfectly single-peaked along the societal axis, except the voter's top choice may be determined not due to the axis but because the voter has some perhaps emotional, irrational reason to "swoon" for that person. Of course, the model merely allows voters to swoon—it does not force all voters to swoon.

For example, perhaps almost everybody agrees that taxes are the most important issue facing the country. Yet the swoon-SP model would allow some (or all) voters to cast votes where all but the voter's top candidates were ordered in a single-peaked-like fashion, except the top spot in the voter's vote went, for example, to Scarlett Johansson or Arnold Schwarzenegger (Figure 5.11) for some unfathomable swoon-related reason.

Let us now stop swooning and get back to the most salient current issue: learning about the complexity of manipulative actions in nearly single-peaked electorates.

Earlier in this chapter, a remarkably involved dichotomy condition was given, as Theorem 5.5, telling which scoring protocols had their constructive coalitional weighted manipulation complexity in P and for which that problem was NP-complete. For the special case of 3-candidate elections, that theorem simplifies to the following result, which was obtained earlier than Theorem 5.5.

5.5 Do Nearly Single-Peaked Electorates Restore Intractability?

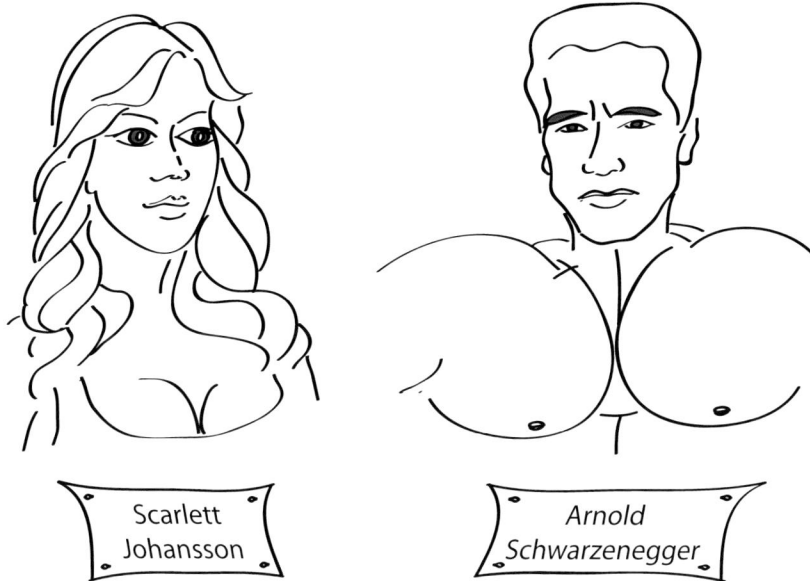

Fig. 5.11 Beneficiaries of swooning

Theorem 5.9 (Faliszewski et al. [431]). *For each scoring protocol $\alpha = (\alpha_1, \alpha_2, \alpha_3)$, the constructive coalitional weighted manipulation problem for the single-peaked case is* NP*-complete if $\alpha_1 - \alpha_3 > 2(\alpha_2 - \alpha_3) > 0$, and is in* P *otherwise.*

In contrast, for the case of 1-maverick-SP societies, we have the following dichotomy theorem.

Theorem 5.10 (Faliszewski, Hemaspaandra, & Hemaspaandra [425]). *For each scoring protocol $\alpha = (\alpha_1, \alpha_2, \alpha_3)$, the constructive coalitional weighted manipulation problem for 1-maverick-SP societies is* NP*-complete if $\alpha_2 > \alpha_3$, and is in* P *otherwise.*

Note that this characterization is quite different than the single-peaked case's characterization. For example, veto elections were NP-complete in the general case, dropped to P in the single-peaked case, but are NP-complete in the 1-maverick-SP case. In fact, since the characterization condition of Theorem 5.10 is identical to the characterization condition of Theorem 5.4, *every* 3-candidate scoring-protocol case that dropped from NP-completeness to P due to single-peakedness is restored to NP-completeness for the case of 1-maverick-SP societies. Even a *single* maverick can result in a tremendous difference in this problem's complexity!

This type of behavior is not limited to the case of just three candidates, or just to the case of 1-maverick-SP. For example, keeping our focus still on veto elections, the following was shown by Faliszewski, Hemaspaandra, and Hemaspaandra [425].

Theorem 5.11 (Faliszewski, Hemaspaandra, & Hemaspaandra [425]). *Let $m \geq 0$. For all $k \geq 0$, the constructive coalitional weighted manipulation problem for $(m+3)$-candidate veto elections for k-maverick-SP societies is NP-complete if $k > m$, and is in P otherwise.*

For example, for 10-maverick-SP societies, the constructive coalitional weighted manipulation problem for m-candidate veto is in P for $m \in \{0, 1, 2, 13, 14, 15, \ldots\}$ and is NP-complete for $m \in \{3, 4, 5, 6, 7, 8, 9, 10, 11, 12\}$. In contrast, veto is in P for any number of candidates for single-peaked societies, and is NP-complete for any number of candidates in the general case. (This behavior can be seen at smaller candidate cardinalities too. For 3-candidate elections in 1-maverick-SP societies the constructive coalitional weighted manipulation problem is NP-complete, but for 4-candidate elections in 1-maverick-SP societies the constructive coalitional weighted manipulation problem is in P.)

Again, we are seeing a type of behavior that is highly unexpected, namely, we are seeing the complexity *drop* as the number of candidates increases from 12 to 13. However, fortified with what we learned from Theorem 5.6, we no longer view this type of behavior as inherently precluded or impossible, and so we don't need to be *too* surprised by it.

That is a good thing, because the swoon-SP case of veto shows the same type of behavior. We have NP-completeness for the 4-candidate case but polynomial-time algorithms for the 5-candidate case.

Theorem 5.12 (Faliszewski, Hemaspaandra, & Hemaspaandra [425]). *For the swoon-SP case, the constructive coalitional weighted manipulation problem for m-candidate veto is NP-complete for $m \in \{3, 4\}$, and is in P for $m \geq 5$.*

(For $m < 3$, every set of votes is swoon-SP with respect to any societal axis L, so those cases are in fact the general case in disguise.)

So far in this section we have been looking only at manipulation. Does control also show interesting, varied behavior for the nearly single-peaked case? The answer is yes. For brevity, we will give a sense of what results hold not by stating a number of theorems, but rather by giving Table 5.1, which is a restriction to the cases we are interested in of a table from Faliszewski, Hemaspaandra, and Hemaspaandra [425].

Although we won't prove the results of this table, let us briefly mention a key idea behind how one shows that a constant number of mavericks can be tolerated (and for some cases one can even extend this to a logarithmic number of mavericks, see [425]). That idea is from Faliszewski, Hemaspaandra, and Hemaspaandra [425], and they call it *"demaverickification."* Basically, this process takes every maverick voter and shatters him or her into a collection of votes, each of which votes for precisely one of the candidates that that

5.5 Do Nearly Single-Peaked Electorates Restore Intractability?

Table 5.1 Control complexity results comparison table (adapted from [425]) between two types of nearly single-peaked electorates: the maverick-free single-peaked case and the general case. The "t-approval" column holds for each $t \geq 2$ unless otherwise noted; $t = 1$ is the "plurality" column. N/A means "not applicable" and NPC means "NP-complete."

Control problem	Complexity results for			References
	plurality	t-approval	approval	
CCAC and CCDC				
general case	NPC	NPC	P	[102, 377, 657, 541]
single-peaked	P	P	P	[431, 425]
k-maverick-SP	P for each fixed k	P for each fixed k	P	[425]
swoon-SP	NPC	NPC	N/A	[425]
CCAV				
general case	P	P for $t < 4$ and NPC for $t \geq 4$	NPC	[102, 657, 541]
single-peaked	P	P	P	[102, 431]
k-maverick-SP	P	P for each fixed k	P for each fixed k	[102, 425]
swoon-SP	P	P	N/A	[102, 425]
CCDV				
general case	P	P for $t < 3$, and NPC for $t \geq 3$	NPC	[102, 657, 541]
single-peaked	P	P	P	[102, 431]
k-maverick-SP	P	P for each fixed k	P for each fixed k	[102, 425]
swoon-SP	P	P for $t < 3$ and "2-approximable" for $t \geq 3$	N/A	[102, 657, 425]

voter approved of (and note that these new votes are, crucially, not maverick votes). Of course, there may be a number of mavericks in the pool of additional votes, and we must decide which ones, if any, to add. When one wraps that up together with the demaverickification process, one ends up with a so-called polynomial-time disjunctive truth-table reduction (as was defined on page 311) from the k-maverick-SP case to the single-peaked case. That is, we can turn one k-maverick-SP instance into a large—but not too large—number of maverick-free single-peaked instances, such that the original problem has the answer "yes" if and only if at least one of our new instances has the answer "yes." This in some sense lets the k-maverick-SP case ride on the coattails of the (maverick-free) single-peaked case, although with quite a bit of work being done to make this possible. However, the bottom line is that this is enough to cast many problems into P for the k-maverick-SP case.

"This has been fun," says Belle to her friends, "but I think we now deserve an extra treat. Let's all go and have a yummy meal of 'pumpkin-pie surprise,' which of course is pumpkin pie surprisingly topped with *more* pumpkin pie!" Belle's friends' eyes open wide with delight, and—now with a better understanding of the complexity of manipulative actions in single-peaked elections—they all happily race off to eat.

Chapter 6
Multiwinner Voting

Dorothea Baumeister · Piotr Faliszewski · Jörg Rothe✉ · Piotr Skowron

6.1 Introduction

The headmaster of the school that Anna, Belle, and Chris attend decided that students should have more say in how the school is ran. As the first step, the students would decide what new books the school's library should buy. This was quite a devious plan on the part of the headmaster since both reading statistics and student involvement with school affairs were dropping, and, with this approach, he hoped to improve both. By some strange coincidence, completely unrelated to being late one day, Anna, Belle, and Chris were asked to form the selection team that would decide which books to buy.

"I think we should simply buy all the comic books that we can find!" says Chris, who realizes that such an opportunity may never show up again.

"No way! Not everyone enjoys comics and we have responsibility to all the pupils! We cannot just buy whatever we like . . . ," Anna opposes.

"Exactly! But we already know how to run elections, right?" Belle chimes in. "Except now we are not choosing just one winner, but a whole set of them . . . You know, we can simply get the list of available books and ask the students to rank them from the most to the least desirable one."

"That does sound like a lot of ranking!" Chris responds, not yet ready to give up on the comic book idea.

"But it will be so much fun! We will have multiwinner elections!" both girls exclaim, and Chris can do nothing, even though he doesn't fully approve of the idea.

6.2 Preferences as Rankings

The goal of a multiwinner election is to select a subset of candidates that, in some sense, reflect the preferences of the voters. To represent these preferences, we use the same formalism as in the single-winner case (handled in Chapters 4 and 5), that is, an election is a pair (C,V), where C is a set of candidates and V is a list of votes. For now, we focus on ordinal elections, where each vote is a preference order ranking the candidates from the most to the least desirable one. Later we will also consider approval preferences, where each voter indicates which candidates he or she appreciates enough to include among the winners, or cardinal profiles, where each voter can specify numerical values quantifying their level of appreciation toward different candidates.

"Actually, why don't we just let the students rank sets of books that they would like to buy?" Anne wonders. "Then we could use one of the single-winner election methods, that we already know, to pick the winning set!"

"Oh, no, no!" Chris opposes vehemently. "The headmaster said we can only choose three books from the catalog and this means we would have ... a lot of sets of books to rank!" Chris frowns and starts counting. "The catalog has 80 books ... and we can choose three ... so this is ..."

"So this is $\binom{80}{3} = 82,160$ sets and Chris does have a point," Belle rescues Chris. "Maybe the students won't be able to express their preferences perfectly by only ranking the books instead of their packages, but at least they will finish ranking before we finish school ..."

"I see," Anna agrees, "but we still have to narrow down the set of possibilities. You know, the catalog has eight genres, with ten books for each, so let's just take ... one of the books for each genre! The students will vote which three of those to get, OK?"

We refer to sets of candidates as *committees*. While it may be somewhat strange to speak of a winning committee of books, this terminology often simplifies the discussion and is commonly adopted. Typically, we are interested in finding winning committees of a given, fixed size. A *multiwinner voting system* (or, a *multiwinner voting rule*) is a function that given an election (C,V) and the size k of the desired committee, outputs a family of size-k committees that tie as winners. As in the single-winner case, practical election systems need to provide some tie-breaking rule as, eventually, we are interested in a single size-k committee, but many theoretical considerations are simpler and cleaner if we disregard this issue.

6.2.1 k-Borda and Individual Excellence

After a long discussion, Anna, Belle, and Chris managed to choose eight books for the students to rank. However, since they still did not decide which voting system to use, they code-named the books with letters, so that they would select the best voting system, and not the one that selects the books they like themselves (of course, the students who were voting knew the coding). All in all, the candidate set is $C = \{a,b,c,d,e,f,g,h\}$ and the collected preferences of the students are as follows:

	7	6	5	4	3	2	1	0
8 students report:	$e > \mathbf{c} > \mathbf{d} > \mathbf{b} > a > g > f > h$,							
6 students report:	$h > \mathbf{d} > \mathbf{b} > \mathbf{c} > g > f > a > e$,							
5 students report:	$a > \mathbf{b} > e > \mathbf{c} > \mathbf{d} > f > g > h$,							
5 students report:	$f > a > \mathbf{b} > e > \mathbf{c} > \mathbf{d} > g > h$,							
4 students report:	$g > a > \mathbf{d} > \mathbf{c} > \mathbf{b} > e > f > h$,							
1 student reports:	$h > e > \mathbf{b} > \mathbf{c} > a > \mathbf{d} > f > g$.							

One of the easiest ways to perform a multiwinner election is to simply use a single-winner voting rule that either assigns scores to the candidates, or that produces their ranking (such systems are known as social welfare functions). Then we simply choose the top k candidates. For example, the *k-Borda multiwinner rule* operates exactly in this fashion: One first computes the Borda scores of all the candidates and then selects k of those with the highest scores. Recall from Section 4.1.1.4 on page 237 that the Borda score of a candidate in an m-candidate election is the sum of the points that he or she receives from each of the voters, where each voter gives $m-1$ points to his or her top-ranked candidate, $m-2$ points to the second one, and so on. In the book election, we have the Borda scores shown in the table below:

	a	b	c	d	e	f	g	h
Borda score	122	**134**	**127**	**123**	115	70	72	49

Thus the *winning Borda committee of size three* is $W = \{b,c,d\}$. In this case, there is just one winning Borda committee, but ties may happen. For example, consider a smaller election, with candidate set $C = \{a,b,c,d\}$, committee size $k=2$, and the following three votes:

$$a > b > d > c,$$
$$b > a > d > c,$$
$$d > c > a > b.$$

In this election, the Borda score of a is 6, both b and d have Borda score 5, and c has Borda score 2. Thus the family of winning Borda committees consists of $\{a,b\}$ and $\{a,d\}$. As we can see, under k-Borda a tie can regard only those candidates that have the same Borda score as the lowest-scoring committee member(s).

> Anna, Belle, and Chris consider the results they get with k-Borda. "I don't know ...," wonders Belle, "it seems that everyone thinks that these books are OK, but no one is really excited."
> "That's true, and it seems that everyone ranks b, c, and d close to each other. Maybe these are the three novels we should choose then?" Anna points out that the catalog contains genres such as historical novels, sci-fi novels, and adventure novels, in addition to textbooks, biographies, maps, encyclopedias, and do-it-yourself books, and then she adds, "It would make sense that similar books would be ranked similarly, right?"
> "Well, yes ... but each book should be judged on its own merit, shouldn't it? It is a sort of a contest, I guess?" responds Chris, who thinks that novels are by far a better choice than textbooks, biographies, and all the other genres (but graphical novels would be even better yet!).

Indeed, as Anna, Belle, and Chris observe, multiwinner rules derived from social welfare functions tend to treat all the candidates individually. After all, social welfare functions are often used for selecting a single winner, so they do not pay attention to interactions among the candidates. Further, in single-winner elections we—quite clearly—want to select the best, *the most excellent*, candidate. Thus, multiwinner rules based on social welfare functions are often referred to as choosing a committee of *individually excellent* candidates, or as striving for the goal of *individual excellence*. Using such multiwinner rules is natural, for instance, if the ultimate goal is to select a single candidate and we are choosing a committee to narrow down the set of candidates. For example, we may use such rules when short-listing candidates for a job, or if we are selecting finalists of a competition.

While we have derived the idea of individual excellence from the properties of a particular type of multiwinner rules, we can also take a different route. Indeed, if our goal were to short-list job candidates (based on their evaluation on a number of criteria) or to select the finalists of a competition (based on how a number of judges evaluated the participants) then what properties would we want our multiwinner rule to have? Below we provide three very basic ones:

- If there were two candidates who performed similarly, then either both of them should be included in the committee, or both of them should be excluded (modulo the fact that the committee has a fixed size). Indeed, if each candidate is evaluated individually, then similar performances should lead to similar outcomes. Of course, it is not clear what "similar performance" means and each multiwinner voting system that aims to select committees of individually excellent candidates has to answer

this question on its own. The k-Borda rule equates a candidate's performance with his or her Borda score (which is equivalent to his or her average position in the votes).
- If a candidate's performance improves (e.g., if he or she gets ranked higher by some voter, but nothing else changes) and he or she would have been included in the winning committee prior to this improvement, then he or she should still be included after the improvement. Indeed, this is a basic *monotonicity* condition (cf. Section 4.2.6.3 starting on page 277): If someone is good enough to be interviewed for a job, or is good enough to be included in a competition's final, then this person should still be good enough after improving in some way (while all others stay as they were).
- If we extend the committee size then everyone included in a smaller committee should still be included in a larger one. Indeed, it would be rather strange if someone were invited to a job interview, but a week later got a letter saying that since the company can interview 11 people rather than 10, he or she is no longer welcome.

This last feature—formally referred to as *committee enlargement monotonicity*—limits multiwinner rules focused on individual excellence to those based on social welfare functions. To see why this is the case, consider a multiwinner rule f that is committee enlargement monotonic and some election $E = (C, V)$ with $m = \|C\|$ candidates. We will use the notation $[z] = \{1, \ldots, z\}$ for each positive integer z. For the sake of simplicity, let us assume that f always provides a single committee. Consider a sequence of committees:

$$f(E, 1), f(E, 2), f(E, 3), \ldots, f(E, m).$$

For each $i \in [m-1]$, committee enlargement monotonicity implies that $f(E, i) \setminus f(E, i-1)$ contains just one candidate (we assume that $f(E, 0) = \emptyset$). Let us call this candidate c_i. Then we can order the candidates as

$$c_1 > c_2 > c_3 > \cdots > c_m,$$

thus deriving a social welfare function from a committee enlargement monotonic multiwinner rule. On the other hand, we have already argued how a social welfare function generates a multiwinner rule (for more details on this relation, see the work of Elkind et al. [375]).

6.2.2 The Chamberlin–Courant Rule and Diversity

While k-Borda is a very natural extension of the single-winner Borda rule to the multiwinner world, it is not the only one. In this section we explore another possibility, introduced by Chamberlin and Courant [270]. Interestingly,

Anna and Belle come up with the exactly same idea. For brevity, we will refer to the *Chamberlin–Courant (multiwinner) rule* as *CC*.

Even though Chris is quite satisfied with using the k-Borda rule, Anna and Belle keep looking for a different approach. "You know," says Anna, "I just think that the b, c, d books are simply not very diverse. But what can we do?"

"Oh, I think I have an idea!" Belle suddenly exclaims. "Maybe we should select the books so that everyone finds at least one great book, instead of having many average ones?"

"That's a wonderful idea! After all, many pupils rent only one book during the whole semester, so the one they take should be as good as possible, and all the others don't matter! We just need to describe this rule formally!" Anna responds, and the two girls immediately set to work (with a brief moment of bewilderedness on Anna's side during which she realizes that she actually was not used to say things such as "describe formally" until rather recently).

Let us introduce some additional notation. Given a preference order v and a candidate c, we write $\text{pos}_v(c)$ to denote the position of c in v. That is, if v ranks c on top then $\text{pos}_v(c) = 1$, if v ranks c as the second best candidate then $\text{pos}_v(c) = 2$, and so on. If i is a position of a candidate in an m-candidate preference order, then we let

$$\beta_m(i) = m - i$$

be the Borda score that a candidate ranked in this position would obtain.

The CC rule is based on the idea of representation. Given a preference order v and some committee W, we say that the member of W ranked highest is v's representative in W. This convention is more natural if one thinks of v not as a preference order, but as a voter with that preference order: The voter wants to be represented by the highest-ranked member of the committee. In our case, the voter's representative would be his or her favorite book among those selected, that is, the one that the voter is most likely to read.

Now, if v is a preference order and W is a committee, then we define the *CC-Score* that v assigns to W as the Borda score of v's representative. Formally, we denote this as follows (recall that m is the number of candidates in the election):

$$CC\text{-}Score_v(W) = \max_{c \in W} \big(\beta_m(\text{pos}_v(c))\big). \tag{6.1}$$

Then the *CC-Score* of a committee W in election (C,V) is the sum of its *CC-Scores* over all the voters:

6.2 Preferences as Rankings

$$CC\text{-}Score_{(C,V)}(W) = \sum_{v \in V} CC\text{-}Score_v(W).$$

The *winning CC committees* are those with the highest *CC-Score*. For example, in the book election, the winning CC committee is $\{a,e,h\}$. Indeed, all but 9 voters rank some member of this committee in the top position, and these remaining 9 voters rank some member of the committee on the second position. Thus its $CC\text{-}Score$ is $7 \cdot (8+6+5+1) + 6 \cdot (5+4) = 194$. By contrast, the k-Borda winning committee $\{b,c,d\}$ has $CC\text{-}Score$ equal to 164.

Unfortunately, finding a winning committee under CC is not as easy as under k-Borda. Indeed, in Section 6.5 starting on page 447 we will see that deciding whether there is a committee with at least a given $CC\text{-}Score$ is NP-complete. On the positive side, there are quite a number of different ways of circumventing this hardness result and in practice it is possible to compute (or at least to approximate) CC winning committees in a reasonable time even for fairly large elections.

"Chris, look at the committee we found! It is $\{a,e,h\}$ according to our rule!" Belle is very excited to announce the results of her work with Anna.

"Well, ... you're right, every student seems to be covered ... ," Chris reluctantly admits, "at least in the sense that every student really likes at least one book ... "

As Chris observed, the CC rule tries to "cover" the interests of each voter, assuming that each voter is only interested in his or her representative. This assumption is natural, for instance, if we are seeking a set of items of which each voter will choose only one to "use." In the book example, the girls argue that each student rents only one book during a semester, so the CC rule is appropriate (we assume here that if a student rents a book, then he or she returns it quickly enough for many other students to rent it as well; alternatively, we could imagine that there are multiple copies of each book available, for example as digital e-books).

Another situation where the CC rule would be appropriate is selecting representatives to discuss some issue. For example, a housing cooperative might want to discuss building new facilities and may want to hear its members' opinions regarding which ones are the most useful and least troublesome. While families may prefer building a playground, some people may rather insist to have a skating park, or a fitness area, or a parking lot. Since it would be infeasible to hear everybody's individual opinion, the cooperative may run a CC election to choose a committee whose members would present the pros and cons of all the possible projects. The committee does not need to represent the interests of the cooperative's members proportionally, but rather has to represent as many diverse opinions as possible. This way it can provide an assessment of each possible project from many—hopefully diverse—points of view (and the final decision on what facility to choose could be made via a different vote, after all the community members familiarized themselves with the assessment).

The connection between the CC rule and diversity is strong but not direct. As we argued above, the rule tries to "cover" the preferences of each voter, by selecting such a committee that each voter ranks some member as highly as possible. To see why this may lead to a diverse committee, let us consider candidate x, who is ranked highly by a large majority of voters, candidate y, who is ranked highly by the same voters as x, but typically comes just a bit lower than x, and candidate z, who is ranked highly by the remaining voters (the majority that supports x and y ranks z at the very bottom). There are also some other candidates and the goal is to choose a committee of size two. The k-Borda rule would, likely, select x and y, because they both perform very well. The CC rule would, however, choose x and z. Indeed, x is selected because he or she is very popular and doing so covers the interests of many voters; z would be selected because he or she covers the interests of a whole new, as yet not satisfied, group (and y would not be selected, because he or she would improve a situation of only a few voters, for whom x already is a good representative). Thus, while the CC rule is not directly defined to find diverse committees, it typically does so.

As in the case of the individual excellence goal, we also provide some basic properties that we expect from the rules designed to cover the interests of all the voters (and since covering the interests of the voters implies finding a diverse committee, these properties are also key from the point of view of diversity-oriented rules):

- If it is possible to find a committee such that each voter ranks a member of this committee in the top position, then such a committee should be selected. Indeed, if everyone's preferences can be covered perfectly, then one should do so.
- If W is a winning committee and some voter moves his or her highest-ranked member of W even higher, then W should still be winning. Indeed, if someone is "even more covered" while everyone else is "as covered as before" then there is no reason why the winning committee should change.

The above two features are known as *narrow-top consistency* and *top-member monotonicity* and, as we will see in the next section, are typical for rules related to CC.

6.2.3 Committee Scoring Rules

The k-Borda and CC rules are two extreme ways in which the classical, single-winner Borda rule can be extended to the multiwinner setting, but there are

6.2 Preferences as Rankings

also many other approaches. Indeed, Chris decided to argue that the CC rule is perhaps too extreme.

Chris ponders the CC rule proposed by Anna and Belle. It all made sense, but he still hopes that, perhaps, there is a way to ensure that at least one of the b, c, d books would be selected. Then an idea occurs to him. "You know," he says, "I would expect that among the students who vote, all would rent at least one book, but some might rent more."

"Actually, that's true ... I think my average is at least three books per semester," Anna replies, while Bella smirkingly adds, "with two from me, and one from Chris, this means two books on average ... but some other students rent only one, so there is only about a 50% chance that a student rents a second book."

"Right, we should take this into account!" Chris nearly shouts, highly pleased to see hope that at least one of his favorite options might get selected after all.

To capture the spectrum of rules between k-Borda and CC, we define the class of *committee scoring rules* which, roughly speaking, correspond to the class of *positional scoring rules* from the single-winner setting (see Section 4.1.1 starting on page 235). As in the single-winner case, these rules assign scores based on positions, but instead of looking at positions of individual candidates, they consider positions of whole committees.

Let v be a preference order over some candidate set $C = \{c_1, \ldots, c_m\}$ and let $W = \{w_1, \ldots, w_k\} \subseteq C$ be some size-k committee. By $\text{pos}_v(W)$ we mean the k-tuple $(i_1, \ldots, i_k)$ obtained by sorting the elements of the set

$$\{\text{pos}_v(w_i) \mid i \in [k]\}$$

in increasing order (note that, for each $i \neq j$, we have $\text{pos}_v(w_i) \neq \text{pos}_v(w_j)$; so, indeed, speaking of an increasing order is well-defined). For example, if we take the preference order

$$v\colon e > c > d > b > a > g > f > h$$

and let $W = \{d, e, f\}$ be a size-3 committee, then $\text{pos}_v(W) = (1, 3, 7)$. Given two positive integers, m and k, $k \leq m$, we write $[m]_k$ to denote the set of all increasing length-k sequences of numbers from $[m]$, and we interpret it as the set of all possible positions of size-k committees in votes over m candidates. Indeed, we will often refer to these tuples as *committee positions*.

A *committee scoring function* $\gamma = (\gamma_{m,k})_{k \leq m}$ is a family of functions that map committee positions to scores. Typically, the number of candidates, m, and the committee size, k, are fixed before collecting ballots, thus we are interested in one specific function $\gamma_{m,k}$. We require that if one committee position is "clearly" at least as good as another one, then its assigned score has to be at least as high as that of the "worse" committee. Formally, if γ is a family of committee scoring functions and we have two committee positions $I = (i_1, \ldots, i_k)$ and $J = (j_1, \ldots, j_k)$ such that $i_1 \leq j_1$, $i_2 \leq j_2$, ..., $i_k \leq j_k$ (denoted by $I \geq J$), then we require that $\gamma_{m,k}(I) \geq \gamma_{m,k}(J)$.[1] With a committee scoring function γ in hand, we can define the score of a committee in a natural way. Consider an election (C, V) with m candidates and some size-k committee W. We say that its γ-score in this election is

$$\gamma\text{-}Score_{(C,V)}(W) = \sum_{v \in V} \gamma_{m,k}(\text{pos}_v(W)).$$

Committee scoring function γ defines a *committee scoring rule* which for a given election and desired committee size outputs exactly those committees that achieve the highest γ-score in this election. The next theorem says that, up to affine transformations, there is only one way to define each committee scoring rule.

Theorem 6.1 (Faliszewski et al. [442]). *Let f and g be two committee scoring rules, defined by committee scoring functions γ^f and γ^g. If $f = g$ then for each m and k, $k \leq m$, there are two numbers, $a_{m,k}$ and $b_{m,k}$, such that for each committee position $I \in [m]_k$, we have that $\gamma^f_{m,k}(I) = a_{m,k} \cdot \gamma^g_{m,k}(I) + b_{m,k}$.*

In the following sections, we discuss several subclasses of committee scoring rules. Throughout this discussion, we let m be the number of candidates and k be the committee size. For each $t \in [m]$, by α_t we mean a function such that $\alpha_t(i) = 1$ for $i \leq t$, and $\alpha_t(i) = 0$ for $i > t$. We refer to α_t as the *t-approval scoring function* (note that it is not a committee scoring function because it is focused on a single position; cf. Section 4.1.1.3 on page 237 for single-winner t-approval voting). Further, recall that $\beta_m(i) = m - i$ is the *Borda scoring function*.

6.2.3.1 Weakly Separable Rules

To express the k-Borda rule as a committee scoring rule, it suffices to take committee scoring function

$$\gamma^{k\text{-Borda}}_{m,k}(i_1, \ldots, i_k) = \beta_m(i_1) + \beta_m(i_2) + \cdots + \beta_m(i_k). \quad (6.2)$$

---

[1] The notation $I \geq J$ may seem confusing given that we have $i_1 \leq j_1, \ldots, i_k \leq j_k$ (i.e., the inequalities are in the opposite direction). Our notation mimics the fact that if some voter v prefers some candidate c to some other candidate d, then we write this as $v: c > d$. In other words, this means that position I is "(weakly) preferred" to position J. Another reason for using this notation is that even though it has some degree of inconsistency, it is already established in the literature.

6.2 Preferences as Rankings

Indeed, given an election (C,V) with m candidates, the score of a committee $W = \{w_1,\ldots,w_k\} \subseteq C$, according to this function, is

$$\sum_{v \in V} \gamma_{m,k}^{k\text{-Borda}}(\text{pos}_v(W)) = \sum_{v \in V} \sum_{w_t \in W} \beta_m(\text{pos}_v(w_t)) = \sum_{w_t \in W} \sum_{v \in V} \beta_m(\text{pos}_v(w_t)).$$

That is, we consider each candidate independently and sum up his or her Borda score. As a consequence, the highest-scoring committee consists of the k candidates with the highest individual Borda scores, which is exactly the definition of the k-Borda rule. An analogous procedure applies to every committee scoring rule whose scoring function is the sum of single-winner scoring functions. Formally, if a committee scoring rule is defined via a scoring function of the form

$$\gamma_{m,k}(i_1,\ldots,i_k) = \delta_{m,k}(i_1) + \delta_{m,k}(i_2) + \cdots + \delta_{m,k}(i_k),$$

where $(\delta_{m,k})_{k \leq m}$ is a family of single-winner scoring functions, then we say that this rules is *weakly separable*. This indicates that we can compute the δ-score separately for each candidate and take those with the highest ones. If for each m it holds that $\delta_{m,1} = \delta_{m,2} = \cdots$, i.e., if the single-winner scoring functions do not depend on k, then we say that the rule is *separable*. Hence, k-Borda is not only weakly separable but even separable.

There are a number of other (weakly) separable rules, of which we discuss the two most famous ones, the *single nontransferable vote* rule (the SNTV rule) and the *bloc* rule. Under the former, each voter names his or her favorite candidate and the k candidates mentioned most frequently are selected (up to ties) as a *winning SNTV committee*. Formally, this rule is defined via the following committee scoring function:

$$\gamma_{m,k}^{\text{SNTV}}(i_1,\ldots,i_k) = \alpha_1(i_1) + \alpha_1(i_2) + \cdots + \alpha_1(i_k).$$

Due to its similarity to the single-winner plurality rule (recall Section 4.1.1.1 on page 236), SNTV is often used in practice, even though for many applications one can find more suitable rules. In the books example, committees $\{a,e,h\}$ and $\{e,f,h\}$ both tie as winners under SNTV. Interestingly, we can also express SNTV in a different way, namely as

$$\gamma_{m,k}^{\text{SNTV}}(i_1,\ldots,i_k) = \alpha_1(i_1).$$

This formulation is equivalent to the original one because $\alpha_1(i_t) = 0$ for each $t \geq 2$. Indeed, since we have that $1 \leq i_1 < i_2 < \cdots < i_k$, it is impossible for such i_t to be equal to 1 and, hence, $\alpha_1(i_t)$ must be 0. Committee scoring rules whose functions depend only on i_1 (in addition to the number of candidates and the committee size) are called *representation-focused*, and we will discuss them in the next section. SNTV is both separable and representation-focused (and, in fact, it is the only committee scoring rule satisfying these two properties simultaneously). Note that this does not contradict Theorem 6.1: Both ways of writing the scoring function for SNTV define the same mathematical object.

The *bloc* rule is quite similar to SNTV, but now each voter provides a list of k candidates that form this voter's favorite committee.[2] Those k candidates that appear in the largest number of favorite committees form (up to ties) a *winning bloc committee*. In the language of committee scoring rules, we express this rule via the following committee scoring function:

$$\gamma_{m,k}^{\mathrm{bloc}}(i_1,\ldots,i_k) = \alpha_k(i_1) + \alpha_k(i_2) + \cdots + \alpha_k(i_k). \quad (6.3)$$

Due to its simplicity and intuitive appeal, bloc is very often used in practice (e.g., in many local elections). Indeed, it feels very natural to let each voter name all the prospective committee members (and not just their favorite ones, as in SNTV) and let the most popular ones win. However, it can also behave in an unexpected way. For example, it is not committee enlargement monotonic, so if some candidate belongs to a bloc committee of size k, there is no guarantee that he or she would also belong to a bloc committee of size $k+1$ (we show an example a bit later). In the books example, bloc selects committees $\{b,d,e\}$ and $\{a,b,d\}$. It is an example of a rule that is weakly separable but not separable. Indeed, it uses k-approval functions, which, by definition, depend on k.

Let us discuss committee enlargement monotonicity in more detail. Consider an election with candidate set $C = \{a,b,c,d\}$ and the following twelve votes:

$a > b > c > d,$	$a > c > b > d,$	$d > b > c > a,$	$b > c > a > d,$
$a > b > c > d,$	$a > c > b > d,$	$d > b > c > a,$	$b > c > a > d,$
$a > b > c > d,$	$a > c > b > d,$	$d > c > b > a,$	$c > b > a > d.$

Under SNTV, the winning committees of sizes 1, 2, 3, and 4 are, respectively,

$$\{a\}, \{a,d\}, \{a,b,d\}, \text{ and } \{a,b,c,d\}.$$

For k-Borda, these committees are

$$\{b\}, \{a,b\}, \{a,b,c\}, \text{ and } \{a,b,c,d\}.$$

While these committees are different in both cases—which is expected given that the two rules are quite different—they illustrate that both rules satisfy committee enlargement monotonicity: Each winning committee extends each smaller one. However, in case of bloc the sequence is

---

[2] In our formalism, we still reason as if each voter provided a ranking of the candidates. However, given the nature of the rule and its scoring function, we see that it suffices for each voter to provide his or her top k candidates, in an arbitrary order.

$\{a\}, \{b,c\}, \{a,b,c\}$, and $\{a,b,c,d\}$.

We see that even though $\{a\}$ is the unique winning bloc committee of size one, the only winning bloc committee of size two does not include a. Thus bloc is not committee enlargement monotonic. The reason why this happens is that the scoring functions used by bloc change with the committee size and candidates b and c get different scores for committees of size one and two. This is exactly the difference between separable and weakly separable rules.

It turns out that committee enlargement monotonicty is exactly what distinguishes separable rules from all the other committee scoring rules. To express this fact, we first need to define this notion formally. The definition is somewhat more involved than one might expect at first, due to the fact that multiwinner rules can provide sets of tied winning committees as output.

Definition 6.1 (Elkind et al. [375]). We say that a multiwinner voting rule f is *committee enlargement monotonic* if for every election (C,V), the following two conditions hold for every $k \in [\|C\|-1]$:

1. If committee W belongs to $f(E,k)$ then there is some candidate $c \in C \setminus W$ such that $W \cup \{c\}$ belongs to $f(E,k+1)$.
2. If committee W belongs to $f(E,k+1)$ then there is some candidate $c \in W$ such that $W \setminus \{c\}$ belongs to $f(E,k)$.

The first condition ensures that if there is some winning committee W of size k, then there is also a winning committee of size $k+1$ that is obtained by extending W. The second condition requires that every winning committee of size $k+1$ can be obtained by extending some smaller winning committee. To see that the second condition is useful, let us consider a scenario where $\{a\}$ is the unique winning committee of size one, while $\{a,b\}$ and $\{c,d\}$ are two tied winning committees of size two. These committees satisfy the first condition because $\{a,b\}$ is obtained by extending $\{a\}$, but not the second one because $\{c,d\}$ is not an extension of any size-1 winning committee.

With this definition in hand, we are ready to characterize separable committee scoring rules as exactly those that are committee enlargement monotonic (it is not hard to convince oneself that all separable committee scoring rules are, indeed, committee enlargement monotonic; the reverse direction takes much more work, though).

Theorem 6.2 (Faliszewski et al. [442]). *A committee scoring rule is committee enlargement monotonic if and only if it is separable.*

As the reader may recall, committee enlargement monotonicity is the third of the conditions that we have put forward in Section 6.2.1 starting on page 405 in the context of individual excellence. Theorem 6.2 states that among committee scoring rules, only the separable ones fulfill this desideratum. The second requirement listed there was that if a candidate belongs to a winning committee, but then improves his or her position, then this candidate should still belong to the winning committee. Formally, we express this property as follows.

Definition 6.2 (Elkind et al. [375]). We say that a multiwinner voting rule f is *candidate monotonic* if for every election (C,V), every committee size $k \in [\|C\|]$, every winning committee $W \in f((C,V),k)$, and every member c of W, the following holds: If in one of the votes in V we swap c with the preceding candidate (leaving everything else the same), then c still belongs to some winning committee under f.

It turns out that this property is satisfied by all committee scoring rules.

Theorem 6.3. *Every committee scoring rule is candidate monotonic.*

Proof. Consider a committee scoring rule f, defined via committee scoring function γ. Further, let (C,V) be some election, let $W \in f((C,V),k)$ be one of its winning committees of size k, and let c be a member of W. Let v be some vote in V, where c is not ranked first, and let v' be an identical vote, except that c is pushed one position higher (note that if c were ranked first in all the votes in V then speaking of candidate monotonicity would be immaterial). Finally, form an election (C,V') that is identical to (C,V), except that v is replaced with v'.

We claim that c belongs to some f-winning committee of (C,V'). To see why this is the case, let W be some size-k committee that does not include c. We observe that $\gamma\text{-}Score_{(C,V')}(W) \leq \gamma\text{-}Score_{(C,V)}(W)$. This is so because for each vote in V', the positions of the members of W are either the same as or worse than in the corresponding vote in V: The only change is that a member of W might have been overtaken by c. By analogous reasoning, for every committee W' that includes c, we have that $\gamma\text{-}Score_{(C,V')}(W') \geq \gamma\text{-}Score_{(C,V)}(W)$. Hence, since W had at least as high a score as every other size-k committee in (C,V), in (C,V') it still has a higher score than every committee not including c. Thus either W or some other committee including c is winning. □

Note that while Definition 6.2 ensures that the candidate whose position improves remains a winner, it makes no promises about the other members of a winning committee. For example, consider the following SNTV election:

$$a > b > c > d, \quad a > c > d > b, \quad b > a > d > c, \quad c > a > b > d.$$

It has two size-2 winning committees, $\{a,b\}$ and $\{a,c\}$. However, if in the third vote we push a one position forward, then only $\{a,c\}$ remains winning. Committee $\{a,b\}$ no longer wins because this swap takes a point away from b and gives it to a. If we forbid such operations then we obtain the notion of *noncrossing monotonicity*.

6.2 Preferences as Rankings

Definition 6.3 (Elkind et al. [375]). We say that a multiwinner voting rule f is *noncrossing monotonic* if for every election (C,V), every committee size $k \in [\|C\|]$, and every committee $W \in f((C,V),k)$ the following holds: If in one of the votes in V we swap a member of W with the preceding candidate, who does not belong to W (leaving everything else the same), then W remains a winning committee under f.

It turns out that among committee scoring rules, exactly the weakly separable ones have the noncrossing monotonicity property.

Theorem 6.4 (Faliszewski et al. [442]). *A committee scoring rule is noncrossing monotonic if and only if it is weakly separable.*

Proof. Let f be some weakly separable committee scoring rule defined via a family of single-winner scoring functions $\{\delta_{m,k}\}_{k \leq m}$. We will show that f is noncrossing monotonic. Consider some election (C,V) over m candidates and some winning size-k committee W. Further, let w be some member of W and let v be a vote in the election, where w is ranked in some position $i > 1$, and where candidate c, who is ranked directly ahead of w, is not a member of W. (If there is no such vote then this means that members of W are ranked on top k positions in each vote and the scenario is irrelevant.) If we push w by one position forward in v then the score of each resulting winning size-k committee W' changes as follows:

1. If W' contains w but not c (as in the case of W) then the score of W' increases by $\delta_{m,k}(i-1) - \delta_{m,k}(i) \geq 0$. (This value is nonnegative by definition of a committee scoring function).
2. If W' contains both w and c then its score remains the same because w and c swap and, therefore, $\text{pos}_v(W')$ does not change.
3. If W' contains c but not w, then its score decreases by $\delta_{m,k}(i-1) - \delta_{m,k}(i)$ which, as we have already seen, is nonnegative.

Since W belongs to the first category and its score was the highest prior shifting w forward, its score is also highest after the shift. Hence, f is noncrossing monotonic.

For the other direction, let f be some noncrossing monotonic committee scoring rule, defined by committee scoring function γ. We will show that f is, in fact, weakly separable. Fix some set C of m candidates and let (C,V) be an election where every possible vote appears exactly once (hence, we have that $\|V\| = m!$). Fix committee size k. By the symmetry of (C,V), every size-k committee is winning in this election under f.

Fix some vote v in V and a number $p \in [k] \setminus \{1\}$. Further, let I and J be two distinct committee positions such that both I and J include position p but not $p-1$. Next, let $W(I)$ and $W(J)$ be the committees that consist of the candidates that in vote v appear, respectively, on position I and J. By symmetry, both $W(I)$ and $W(J)$ are winning.

Let c be the candidate that appears on position p in vote v. By the choice of I and J, we know that c belongs to both $W(I)$ and $W(J)$, but the candidate right before c does not. Hence, by noncrossing monotonicity, if we shift c one position forward, then $W(I)$ and $W(J)$ remain winning. In particular, their score increases by the same value (or stays unchanged). However, since the choice of I and J is arbitrary, this value depends only on p and we denote it as $\alpha_{m,k}(p-1)$.[3] Intuitively, $\alpha_{m,k}(p-1)$ is the amount by which a score of the committee increases if we shift its member from position p to $p-1$, without passing other members of this committee.

Next, we define a single-winner scoring function $\delta_{m,k}$ so that for each $i \in [m-1]$ we have

$$\delta_{m,k}(i) - \delta_{m,k}(i+1) = \alpha_{m,k}(i),$$

and

$$\delta_{m,k}(m-k+1) + \delta_{m,k}(m-k+2) + \cdots + \delta_{m,k}(m)$$
$$= \gamma_{m,k}(m-k+1, m-k+2, \ldots, m).$$

It is easy to verify that, indeed, $\delta_{m,k}$ is defined in a unique way. Further, since all values of $\alpha_{m,k}$ are nonnegative, $\delta_{m,k}$ is a proper (nonincreasing) single-winner scoring function. We will show that for each committee position $R = (r_1, \ldots, r_k)$, it holds that

$$\gamma_{m,k}(r_1, \ldots, r_k) = \delta_{m,k}(r_1) + \cdots + \delta_{m,k}(r_k)$$

and, hence, f is weakly separable.

Indeed, fix committee position $R = (r_1, \ldots, r_k)$. Let $R' = (r_1, r'_2, r'_3, \ldots, r'_k)$ be a committee position where $r'_2 = m-k+2$, $r'_3 = m-k+3$, ..., $r'_k = m$. By definition of $\delta_{m,k}$ we know that

$$\delta_{m,k}(m-k+1) + \delta_{m,k}(r'_2) + \cdots + \delta_{m,k}(r'_k) = \gamma_{m,k}(m-k+1, r'_2, \ldots, r'_k).$$

But we also know that

$$\delta_{m,k}(m-k) - \delta_{m,k}(m-k+1)$$
$$= \alpha_{m,k}(m-k)$$
$$= \gamma_{m,k}(m-k, r'_2, \ldots, r'_k) - \gamma_{m,k}(m-k+1, r'_2, \ldots, r'_k).$$

This means that

$$\delta_{m,k}(m-k) + \delta_{m,k}(r'_2) + \cdots + \delta_{m,k}(r'_k) = \gamma_{m,k}(m-k, r'_2, \ldots, r'_k).$$

By repeating this observation to decrease the first position sufficiently many times, we arrive at the following equality:

---

[3] Naturally, we could have used p as the argument instead of $p-1$, but the current notation is more convenient.

6.2 Preferences as Rankings

$$\delta_{m,k}(r_1) + \delta_{m,k}(r_2') + \cdots + \delta_{m,k}(r_k') = \gamma_{m,k}(r_1, r_2', \ldots, r_k').$$

We can then do the same to decrease, one by one, the remaining positions, arriving at

$$\delta_{m,k}(r_1) + \delta_{m,k}(r_2) + \cdots + \delta_{m,k}(r_k) = \gamma_{m,k}(r_1, r_2, \ldots, r_k).$$

This concludes the proof. ❑

Together, Theorems 6.2 and 6.4 provide a somewhat surprising corollary: If a committee scoring rule is committee enlargement monotonic then it also is noncrossing monotonic. This is so because every separable rule is also weakly separable. Altogether, we conclude that among committee scoring rules, separable ones seem best suited for selecting committees of individually excellent candidates.

"I don't get it, how come such different rules as k-Borda and SNTV can *both* be good for individual excellence?" Chris complains.

"Well ... maybe such axiomatic criteria are just one aspect of choosing a good rule?" Anna tries to help. "Or situations where we look for individually excellent candidates vary among each other?"

"That indeed must be the case," Belle adds. "Maybe if we were looking for top Formula 1 drivers over the course of the last 50 years, we would simply count how many world championships each of them had won, but if we were looking for the best five drivers of a given season, then we would use k-Borda?"

"I don't even see votes here ...," Chriss frowns, but Belle quickly explains: "Oh, it's easy! In the first setting, each season is a vote, where the season's champion is the candidate ranked first, which is all we need for SNTV. In the second one, each race is a vote, ranking the candidates in the order in which they finish."

6.2.3.2 Representation-Focused Rules and Beyond

The idea of representation-focued rules is that each voter only cares about this member of the committee that he or she ranks highest. In other words, these are variants of the CC rule and, by definition, use committee scoring functions of the form:

$$\gamma_{m,k}(i_1, \ldots, i_k) = \delta_{m,k}(i_1). \tag{6.4}$$

To obtain the classical CC rule, we take $\delta_{m,k} = \beta_m$, and to obtain SNTV we take $\delta_{m,k} = \alpha_1$.

Representation-focused rules have two main properties. First, if there is a committee such that every voter ranks some member of this committee in the top position, then this committee must be winning. This property is known as *narrow-top consistency*. The second characteristic feature is *top-member monotonicity*, which is a form of noncrossing monotonicity, restricted to each voter's top-ranked committee member. Together, narrow-top consistency and top-member monotonicity exactly characterize the class of representation-focused rules among the committee scoring rules (see the work of Faliszewski et al. [442] for formal definitions and arguments). Since in Section 6.2.2 starting on page 407 we argued that these properties are important for identifying diverse committees, if one is looking for a diverse committee scoring rule, then a representation-focused one would be a good choice.

There are also many committee scoring rule that go beyond the classes of (weakly) separable or representation-focused rules. For example, let us consider the *harmonic Borda (HB)* rule defined via the following committee scoring function:

$$\gamma_{m,k}^{HB}(i_1,\ldots,i_k) = \beta_m(i_1) + \tfrac{1}{2}\beta_m(i_2) + \tfrac{1}{3}\beta_m(i_3) + \cdots + \tfrac{1}{k}\beta_m(i_k).$$

Intuitively, HB implements a certain compromise between k-Borda and CC. On the one hand, it puts focus on the higher-ranked committee members, but on the other hand, it uses all of them to derive the score. The HB rule has not been used much in practice, but we show it here as an example. The coefficients 1, $1/2$, $1/3$, ... are inspired by the PAV rule, that we will discuss in Section 6.3.2 starting on page 430. In the books example, HB chooses the winning committee $\{a,b,c\}$.

"See! Let's just use the harmonic Borda rule! It has the best name!" Chris is getting very excited when he sees that this rule would select the committee $\{a,b,c\}$, which includes the books b and c that he suspects to be novels. "I mean, certainly everyone wants to live in harmony, right?"

"Well ... but it does feel a bit ad hoc ...," Anna worries, "we do want to pick a rule that works well, but it should follow some sensible principles and not just return the result we want to have in some specific election ... "

The fact that harmonic Borda implements a compromise between k-Borda and the Chamberlin–Courant rule is visualized in Figure 6.1. The first column illustrates two distributions of points in a two-dimensional Euclidean space. The upper one consists of four Gaussian distributions concentrated around four different points, and the lower one is simply a uniform distribution on a square. We interpret the points as voters and/or candidates—their positions in the 2D-Euclidean space reflect their opinions in the left-right political spectrum. For example the x-coordinate of a candidate or a voter can tell whether he or she is more left-wing or right-wing with respect to the economic issues,

6.2 Preferences as Rankings

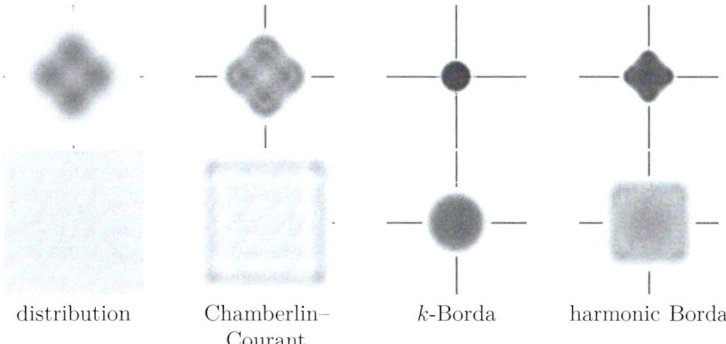

Fig. 6.1 Visualization of the outcomes of the selected committee scoring rules rules for Euclidean preferences [374]

and the y-coordinate might be interpreted as his or her position with respect to the social freedoms. The preferences of the voters over the candidates are induced by the distances between the corresponding points: A voter prefers closer candidates, as those have more similar ideological views.

Thus, given a distribution of points, we can sample elections: First, we sample points corresponding to the voters and the candidates, and then we construct the voters' preferences based on the distances between the respective points. For a sampled election, we can use a multiwinner rule to select a winning committee, and highlight the points that correspond to the elected candidates. Next, we can sample another election and put points that correspond to the elected candidates on the same diagram. After repeating the sampling a large number of times, we obtain a histogram, where the intensity of a color corresponds to the probability that a candidate in a given position will be chosen as a member of a winning committee. These histograms, obtained for the three aforementioned rules, are depicted in the consecutive columns in Figure 6.1. Indeed, we can see that the distribution of winning candidates for the Chamberlin–Courant rule resembles the initial distribution of the voters, and is most diverse. The committee selected by k-Borda consists of similar "central" candidates, the individually "most excellent" candidates, and the outcomes returned by harmonic Borda are in-between: The elected committee is more representative in comparison to the one returned by k-Borda, but does not only contain candidates with the most extreme views.

Many other families of committee scoring rules, with different properties and applications, have also been considered in the literature, but their analysis goes beyond the scope of this book chapter.

6.2.4 STV and Proportionality for Solid Coalitions

As we have seen, within the class of committee scoring rules there are voting methods lying in-between CC and k-Borda. One can suspect that such rules exhibit a behavior somewhat similar to that of these two rules, i.e., that they follow—to some extent—the principles of diversity and individual excellence. In some contexts, however, one would be interested in a behavior that cannot be described using the intuition of diversity and individual excellence.

> "We found a number of interesting rules that are suitable for many types of elections, but I want something else," Belle tries to explain why she feels they should continue their research. "In harmonic Borda there is too little focus on the top positions in the voters' preference orders. It is too much about finding consensus candidates rather than about finding those which are liked most by some groups of voters."

Indeed, we can support Belle's claim with the following example.

Example 6.1. Consider the following profile with four voters and eight candidates:

$$v_1: e > \mathbf{a} > \mathbf{b} > \mathbf{d} > \mathbf{c} > f > h > g,$$
$$v_2: f > \mathbf{b} > \mathbf{a} > \mathbf{c} > \mathbf{d} > g > e > h,$$
$$v_3: g > \mathbf{c} > \mathbf{d} > \mathbf{b} > \mathbf{a} > h > f > e,$$
$$v_4: h > \mathbf{d} > \mathbf{c} > \mathbf{a} > \mathbf{b} > e > g > f.$$

If the size of the committee equals $k = 4$, harmonic Borda would pick $\{a,b,c,d\}$. In comparison, representation-focused rules would select $\{e,f,g,h\}$ (indeed, for CC and SNTV this would be the unique winning committee; a rule defined by committee scoring function $\gamma_{m,k}(i_1,\ldots,i_k) = \alpha_2(i_1)$ would have more winning committees).

> "I like CC and representation-focused rules, but they can treat large and small groups of voters the same way," Belle continues. "I think that larger groups of voters with similar preferences should be able to decide about more committee members!"

In fact, Belle has informally expressed that she is interested in a proportional rule that puts emphasis on the top positions in the voters' preference rankings. This intuition can be formally captured as follows.

Definition 6.4 (proportionality for solid coalitions). Given an election $E = (C,V)$ with $n = \|V\|$ voters, we say a committee W of size k satisfies

6.2 Preferences as Rankings

proportionality for solid coalitions if for each $\ell \in [k]$, each group of voters $S \subseteq V$ with $\|S\| \geq \ell \cdot \frac{n}{k}$, and each subset of candidates $T \subseteq C$ such that each voter from S prefers each candidate from T to each candidate from $C \smallsetminus T$, it holds that $\|W \cap T\| \geq \min(\ell, \|T\|)$.

Let us consider this definition intuitively. If there are n voters and we are choosing a committee of size k, then a candidate supported by n/k voters should be able to enter the committee. Put differently, if there is a group of at least n/k like-minded voters, then these voters should be able to control one position in the committee. Indeed, this is their proportional share of the power.[4] More generally, if a group of $\ell \cdot \frac{n}{k}$ can agree on up to ℓ candidates—that is, if there is a set T of up to ℓ candidates that these voters rank ahead of all the other ones—then these candidates deserve to be selected. For instance, in Example 6.1 proportionality for solid coalitions requires committee $\{e, f, g, h\}$ to be selected.

While at first it is not clear that satisfying proportionality for solid coalitions is even possible, there is multiwinner rule that is used for electing the parliaments in Australia, Ireland, and Malta (and for local elections in many countries, including United States, Canada, New Zealand, and the United Kingdom) that does have this property. This rule is called *single transferable vote* (STV).[5] Note that we already defined a variant of STV for the single-winner setting in Section 4.1.4.2 starting on page 252.

Definition 6.5 (STV). Given an election (C, V) with $n = \|V\|$ voters, the *single transferable vote (STV)* rule operates as follows. Initially, each voter v is given a weight of 1: $w(v) = 1$. For each candidate c, we count the total weight of the voters who put this candidate in the top position of their ranking; let us denote this value as $w(c)$. Until k candidates are selected, we repeat:

1. If there exists a candidate c for whom $w(c) \geq n/k$, then we select this candidate and evenly decrease the total weight of the voters supporting c by n/k. Formally, for each voter v who ranks c first, we update $w(v)$ to the value $w(v) \cdot \frac{w(c) - n/k}{w(c)}$. We additionally remove c from the election.
2. If there exists no candidate c with $w(c) \geq n/k$, we remove a candidate c with the lowest value $w(c)$ (possibly breaking ties in some way).

The k selected candidates form a *winning STV committee*.

For example, consider the following preference profile and assume $k = 2$:

$$a > b > c > d, \qquad a > b > c > d, \qquad a > c > b > d, \qquad d > c > a > b.$$

---

[4] See Definitions 8.1 and 8.2 on pages 514 and 514 for the notion of proportionality in the context of fair division of divisible goods (cake-cutting) and see Section 9.5.4 starting on page 629 for the same notion in the context of fair division of indivisible goods.

[5] Note that there are many variants of STV that differ in numerous details and we only provide one of them.

In the first round we have $w(a) = 3, w(d) = 1$, and $w(b) = w(c) = 0$. Candidate a is elected, and the preferences in the second round are:

$$b > c > d, \qquad b > c > d, \qquad c > b > d, \qquad d > c > b.$$

The first three voters now have a weight of $1/3$ each, and the last voter has a weight of 1. Thus, $w(b) = 2/3$, $w(c) = 1/3$, and $w(d) = 1$. Since no candidate tips the scales at a total weight of at least 2, the candidate with the lowest weight, c, is eliminated. The new election is:

$$b > d, \qquad b > d, \qquad b > d, \qquad d > b.$$

In the last round, b and d have the same total weight of 1, so either of them can be eliminated, depending on tie-breaking. As a result, there are two winning committees: $\{a,b\}$ and $\{a,d\}$.

Theorem 6.5. *STV satisfies proportionality for solid coalitions.*

Proof. Fix an election $E = (C,V)$ with $n = \|V\|$ voters, a committee size k, and a group of voters $S \subseteq V$ with $\|S\| \geq \ell \cdot \frac{n}{k}$. Let $T \subseteq C$ be such that each voter $v \in S$ prefers each candidate from T to each candidate from $C \setminus T$.

We consider two cases. First, let us assume that some members of T remain in the election after STV completes (i.e., some members of T are neither selected as winning committee members nor eliminated). Yet, when STV finishes then the total weight of the voters is zero: Initially the total weight of the voters is n and each time a candidate is selected to be in the winning committee, the total weight decreases by n/k. Since some member of T remains in the election, the weight of the voters from S could only have been decreased by including members of T in the winning committee. This means that the number of candidates from T that were selected is at least $\frac{\|S\|}{n/k} \geq \ell$.

For the second case, let us assume that all members of T were either included in the winning committee or deleted. If all of them were included in the winning committee, then proportionality for solid coalitions is not violated. Hence, let us assume that at least one member of T were removed and consider the last iteration when this happens. Let y be the number of candidates from T that were at this point included in the winning committee, and let x be the number of those that will be included in it during the following iterations (this is the last iteration where a member of T is removed, so all the other remaining members of T must be included in the winning committee). If $x + y \geq \ell$, then proportionality for solid coalitions is not violated; so, for the sake of contradiction, let us assume that $x + y < \ell$. Since, prior to the current iteration, y members of T were included in the winning committee, the weight of the voters in S is:

$$\|S\| - y \cdot \frac{n}{k} \geq (\ell - y) \cdot \frac{n}{k}.$$

6.2 Preferences as Rankings

We have assumed that $x+y < \ell$, which implies that $\ell - y > x$ and, as ℓ and y are integers, $\ell - y \geq x+1$. By the pigeonhole principle, this means that in the current iteration one of the remaining $x+1$ members of T is ranked first by members of S of total weight at least n/k, which contradicts the fact that a member of T is removed in this iteration. Thus it could not have been the case that $x+y < \ell$ and the proof is complete. ❑

STV has also other appealing properties, such as, for example, resilience to cloning candidates. Informally, this means that if in an election there would appear a candidate c that is very similar to an already winning candidate c_w, then after such an appearance either c or c_w would still be winning. This property is often considered important since it removes incentives for the candidates to strategically withdraw from elections in order to improve chances of similar candidates. In the books example, it might mean that Chris would not be tempted to withdraw one of the sci-fi novels to increase the chances of another one. The property of independence of clones, which we now define, has already been briefly mentioned for the single-winner setting in Section 4.2.5.2 on page 276.[6]

Definition 6.6 (independence of clones). Consider two elections, $E = (C, V)$ and $E' = (C \cup \{c\}, V)$, such that for some $a \in C$ each voter ranks c and a next to each other (either a comes right before c or vice versa). A rule satisfies *independence of clones* if for each such pair of elections, if a is winning in E then c or a must be winning in E'.

Let us note that the above definition is somewhat simplified and it only deals with the case where candidates are cloned once. More generally, given an election $E = (C, V)$ and a subset $T \subseteq C$ of candidates, we say that T is a *clone set* if all voters rank the members of T consecutively in any, not necessarily the same, order (as an extreme case, the set of all candidates always is a clone set). For the sake of simplicity, our variant of independence of clones does not go beyond two candidates.

Theorem 6.6. *STV satisfies independence of clones.*

Proof. Fix some committee size k. For the sake of contradiction, consider elections $E = (C, V)$ and $E' = (C \cup \{c'\}, V)$, such that c' is a clone of some candidate $c \in C$ and STV selects c in E, but neither c nor c' is selected in E'. Note that both elections include the same $n = \|V\|$ voters, but in E' they additionally include c' next to c in their preference orders.

Let r be the first round where either c is selected in E or one of c and c' is removed in E'. Until round r, STV made the same decisions both in E

---

[6] Neveling and Rothe [739] transfer the model of cloning due to Elkind, Faliszewski, and Slinko [377] (see also [378]) from single-winner to multiwinner elections. They study the problems of *possible* and *necessary cloning* (which are inspired by the notions of possible and necessary winner—recall Section 4.3.3.8 starting on page 317) in the so-called *zero-cost*, the *unit-cost*, and the *general-cost model* in terms of their computational complexity.

and in E'. Indeed, every candidate $d \in C \smallsetminus \{c\}$ is ranked first in E by the same voters as in E', and round r is the first one where STV acts on either of c and c'. If in round r, one of c and c' is removed in E', then E' in round $r+1$ will look exactly the same as E in round r (up to the fact that if c is removed, then c' takes its role). Thus, STV will continue in the same way for both elections, and will choose the same sets of candidates. However, since we assumed that STV selects c in E, but neither c nor c' is selected in E', this is impossible.

Hence, we know that in round r candidate c is selected in E. This means that at the beginning of round r there are voters in E' with total weight at least n/k who rank either c or c' first. At some point one of these two candidates will be removed (indeed, those voters cannot decrease their weight due to adding a different candidate than c or c' to the winning committee). Without loss of generality, assume that c is removed first. Then all voters who ranked c or c' first will now rank c as their most preferred candidate and the total weight of these voters will be at least n/k. Thus c' will be elected in one of the following rounds. This gives the desired contradiction and completes the proof. ❏

On the other hand, STV has also some drawbacks. For example, it fails candidate monotonicity (recall Definition 6.2), even for $k = 1$.

There exists a rule called *expanding approval voting*, which satisfies proportionality for solid coalitions and committee enlargement monotonicity, yet is not independent of clones. For more discussion on this rule, we refer the reader to the paper by Aziz and Lee [58].

6.3 Preferences as Approvals

While Chris and Anna are enthusiastically discussing the concept of proportionality, Belle's face is still expressing some concerns: "But why did we pick exactly these eight books to choose from? Shouldn't we present the pupils with more options and allow them to choose from, say, twenty books?" she asks.

"I was thinking the same!" Anna promptly replies. "But can we expect the pupils to rank as many as 20 books? I've heard that ranking more than just a few candidates is a hard task for any human, and it might take a lot of time and effort [707]. You know, voters tend to be lazy! If you don't believe me, ask Menon and Larson [702] or Baumeister et al. [117]. They consider voting ballots with truncated rankings." Anna sees Chris nodding, and she is frightened—and highly pleased—

6.3 Preferences as Approvals

how easy it is to convince others when you use proper references to the literature.

"You are right!" Chris joins the discussion. "Who said we *have* to use rankings? Why can't we just ask the pupils to simply tick the books that they like instead of arranging them in a ranking?"

"A marvelous idea!" Anna and Belle shout at once. "This way the pupils will also be able to separate the books they like from those which they do not want to read. We will kill two birds with one stone!"[7]

The friends ask the students to cast their ballots again, but now in the form of approval ballots, and they observe the following results:

8 students approve: $\{b,c,d,e\}$ 6 students approve: $\{b,d,h\}$
3 students approve: $\{a,b,e\}$ 2 students approve: $\{a,b,c,e\}$
2 students approve: $\{a,b,c,d,e,f\}$ 2 students approve: $\{a,f\}$
1 student approves: $\{f\}$ 4 students approve: $\{a,b,c,d,g\}$
1 student approves: $\{h\}$

Recall the method of approval voting in the single-winner setting, due to Brams and Fishburn [191] from Section 4.1.3 starting on page 249 where the voters do not rank the candidates but instead either approve or disapprove of them. This voting method can also be used in the multiwinning setting for electing a committee, and we will see several variants of it.

Figure 6.2 shows an example of an approval ballot. An approval preference profile is a function $A\colon V \to 2^C$ that for each voter $v \in V$ specifies a subset of candidates, $A(v)$, that v approves of.

Fig. 6.2 An approval ballot with three approvals for the eight books

Belle notices that the ideas of individual excellence and diversity naturally carry over to the case of approval elections. For example, if we were focused on individual excellence, we should pick those k candidates who are approved by most voters. This simple rule is commonly known as *multiwinner approval voting*. Belle counts the number of approvals each candidate received and obtains the following results:

---

[7] Disclaimer: No stone was hurt while writing this book.

	a	b	c	d	e	f	g	h
# approvals	13	**25**	16	**20**	15	5	4	7

She rightfully claims the approach based on individual excellence would suggest picking $\{b,c,d\}$ in the above election, and this would be indeed the result returned by multiwinner approval voting as its winning committee of size three for this election. Yet, if we picked such a committee, there would be four students who are not happy at all with the outcome of the election, since none of them approves of any of the selected books.

6.3.1 Approval Chamberlin–Courant and Diversity

On the contrary, if we picked $\{b,f,h\}$ then each student could find a book that she likes in the elected committee. This would be the unique winning committee returned by the *approval Chamberlin–Courant (approval CC)* rule. Formally, given an election (C,V), this rule selects the committees W that maximize the following score:

$$ACC\text{-}Score_{(C,V)}(W) = \sum_{v \in V} \min(1, \|A(v) \cap W\|).$$

In other words, in the approval setting CC considers a voter to be satisfied if he or she approves of at least one committee member (who could act as the voter's representative), and the rule seeks a committee that maximizes the number of satisfied voters.

Belle is very excited by the fact that she managed to adapt the Chamberlin–Courant rule to the approval model. She further makes the following point.

> "Hey, look! The solution returned by approval Chamberlin–Courant is even fair!" she exclaims. "Assuming we have n students, $\lceil n/k \rceil$ of them should have the right to decide about one book out of k. I can prove that by using approval Chamberlin–Courant we can guarantee that each group of voters is justifiably represented!"

Belle's intuition of fairness can be stated as a formal axiom, called *justified representation*.

Definition 6.7 (justified representation [50]). Given an election $E = (C,V)$ with $n = \|V\|$ voters and a committee size k, we say that a group of voters $S \subseteq V$ is *cohesive* if (1) it is large enough, i.e., $\|S\| \geq n/k$, and (2) it agrees on at least one candidate, i.e., $\|\bigcap_{v \in S} A(v)\| \geq 1$.

6.3 Preferences as Approvals

We say that a committee W, $\|W\| = k$, satisfies *justified representation (JR)* if in each cohesive group of voters $S \subseteq V$ there exists a voter $v \in S$ who approves of some candidates from W, i.e., $A(v) \cap W \neq \emptyset$.

A voting rule f satisfies *justified representation (JR)* if for each election $E = (C, V)$ and size k, each winning committee $W \in f(E, k)$ satisfies justified representation.

At first, the above definition may seem quite strange. Indeed, why should it be so desirable that, possibly, just one voter in each cohesive group approves a member of the selected committee? The reasoning here follows the idea of Belle: If there were $\lceil n/k \rceil$ voters who all approved of some candidate c, but none of these candidates approved of a member of the selected committee, then these voters would have a right to complain. After all, they would form a $1/k$-fraction of the society but they would not have the power to control a $1/k$-fraction of the committee, even though they would know what to do with this power (namely, they would put c into the committee). By requiring that at least one member of each cohesive group approves of a member of the committee, we ensure that such disenfranchised groups do not appear or, more precisely, are too small to be eligible to unconditionally demand to control a place in the committee.[8]

Proposition 6.1. *Approval CC satisfies justified representation.*

Proof. Consider an election $E = (C, V)$, a committee size k, and assume toward a contradiction that a committee W returned by the approval Chamberlin–Courant rule does not satisfy justified representation. Let S be a group of voters, such that $\|S\| \geq n/k$, $\|\bigcap_{v \in S} A(v)\| \geq 1$, and no voter in S approves of any candidates in W.

Consider those voters $v \in V$ who approve of at least one candidate in W. Let us assign each such voter v one candidate c from $W \cap A(v)$. We will call c the *representative* of v, and, conversely, we will say that c represents v in W. Since no voter in the group S has a representative, we infer that at least one candidate $c_r \in W$ represents at most $\frac{n - n/k}{k} < n/k$ voters. Thus, if we remove c_r from W, then $ACC\text{-}Score(W)$ will decrease by less than n/k (the voters who are represented will still contribute the value of one to the score of W).

On the other hand, let c_a be some candidate approved by all voters in S (i.e., $c_a \in \bigcap_{v \in S} A(v)$). If we add c_a to the committee W, then $ACC\text{-}Score(W)$ will increase by at least n/k (each voter from S will obtain a representative). Consequently,

---

[8] By an "unconditional demand" we mean that the group would ask for a place in the committee irrespective of what other groups exist in the electorate. One could argue that if there are k groups of $\lceil n/k \rceil - 1$ voters each, each approving of a distinct candidate, then these k candidates should form a winning committee. While this may be reasonable, it is not captured by the axiom of justified representation.

$$ACC\text{-}Score((W \smallsetminus \{c_r\}) \cup \{c_a\}) > ACC\text{-}Score(W) - \frac{n}{k} + \frac{n}{k}$$
$$= ACC\text{-}Score(W).$$

This shows that W could not have been selected by the approval Chamberlin–Courant rule. We obtained a contradiction, which completes the proof. ☐

Justified representation ensures that a voting rule cannot ignore minorities of voters with common interests. Yet, this axiom is rather weak and should be viewed as a necessary rather than sufficient condition for a rule that aims at ensuring proportional representation. This can be observed by examining the following simple example, which is suggested by Anna.

Example 6.2. Consider the following election with eight candidates, $C = \{a, b, \ldots, h\}$. Assume the voters' preferences are as follows:

> 67 voters approve of: $\{b, c, d, e\}$,   30 voters approve of: $\{f, g\}$,
> 2 voters approve of: $\{b, f, g\}$,   1 voter approves of: $\{h\}$.

Further, assume that our goal is to pick a committee of size $k = 3$. In this example, approval CC would select a committee that consists of one candidate from $\{b, c, d, e\}$, one candidate from $\{f, g\}$, and h.

Anna rightfully points out that the outcomes returned by approval CC in Example 6.2 are not exactly proportional. She notices that the axiom of JR should be satisfied not only by rules that aim at achieving proportional representation, but also by those where the goal is to select committees representing a diverse set of predominant opinions. Indeed, in Example 6.2 approval CC picks h in order to ensure each voter is represented, instead of selecting a second representative for the group of 67 voters.

"Look at Example 6.2!" Anna whispers. "The first group of voters forms 2/3 of the society, yet they decide only about a single candidate. This is not proportional!"

6.3.2 Extended Justified Representation and Proportional Approval Voting

Motivated by Example 6.2, Anna suggests a related stronger property. Given that the axiom she suggested implies JR, she decides to call it *extended justified representation (EJR)*.

6.3 Preferences as Approvals

Definition 6.8 (extended justified representation [50]). Consider an election $E = (C, V)$ and a committee size k. We say that a group of voters $S \subseteq V$ is ℓ-*cohesive*, $\ell \in [k]$, if (1) $\|S\| \geq \ell \cdot n/k$, and (2) $\|\bigcap_{v \in S} A(v)\| \geq \ell$.

A size-k committee W satisfies *extended justified representation (EJR)* if in each ℓ-cohesive group of voters S there exists a voter $v \in S$ who approves of at least ℓ members of the committee, i.e., $\|A(v) \cap W\| \geq \ell$.

A voting rule f satisfies *extended justified representation (EJR)* if for all elections E and all committee sizes k, it returns only winning committees that satisfy EJR.

Anna's intuitive explanation of Definition 6.8 is the following. Since there are n voters and we want to select k candidates, each group of n/k voters should have the right to decide about one committee member. Consequently, a group of $\ell \cdot \frac{n}{k}$ voters should be able to decide about ℓ candidates in the elected committee. However, such a group of voters might have incompatible preferences, in which case it is not possible for them to influence the decision (for example, this is the case when each of these voters approves of a different candidate). This is why we can only require that groups of voters with *cohesive* preferences are able to influence the decision. (Note that we can also adapt the explanation provided for JR under Definition 6.7 to apply to EJR.)

Chris looks at Definition 6.8 and raises some concerns. "Why do we require only a single voter in group S to have ℓ representatives?" he complains. "I would rather like each voter from S to have at least ℓ approved candidates in the elected committee." Anna nods her head and replies: "So do I, but this is too much to demand."

Anna supports her claim by presenting the following example.

Example 6.3. Consider $m = 3$ candidates, a, b, and c, and $n = 6$ voters with the following preferences:

$$A(v_1) = \{a, c\}, \quad A(v_3) = \{a, b\}, \quad A(v_5) = \{b, c\},$$
$$A(v_2) = \{a\}, \quad A(v_4) = \{b\}, \quad A(v_6) = \{c\}.$$

If our goal is to select $k = 2$ candidates, then the groups $\{v_1, v_2, v_3\}$, $\{v_3, v_4, v_5\}$, and $\{v_5, v_6, v_1\}$ are all cohesive. If we require that each voter from a cohesive group should be represented, then we would need to select candidates a, b, and c. This is impossible, however, because we can select only two of them.

"Okay!" Chris agrees with Anna. "But I would like to ensure at least that the average number of representatives of the members of a cohesive group is large enough ..."

> Anna smiles, since she already knew that EJR implies this kind of behavior. "Be my guest!" she says mildly.

Indeed, it is important to note that Definition 6.8 applies to *each* ℓ-cohesive group of voters. For example, consider a committee W, and an ℓ-cohesive group of voters S. EJR implies that in S there is a voter v with $\|A(v) \cap W\| \geq \ell$. Yet, the same axiom applies to the group $S \setminus \{v\}$ and it ensures that there must exist a voter $v' \in S \setminus \{v\}$ such that $\|A(v) \cap W\| \geq \ell - 1$. We can further apply EJR to $S \setminus \{v, v'\}$ and so on, until we assess the average number of representatives of the voters from S. More specifically, we first observe the following (see the explanations below):

$$\frac{1}{\|S\|} \cdot \sum_{v \in S} \|A(v) \cap W\| > \frac{1}{\|S\|} \sum_{i=1}^{\ell-1} \frac{n}{k} \cdot i. \tag{6.5}$$

On the left-hand side, we have the average number of committee members approved by the voters from S. On the right-hand side, we have the bound that follows from the reasoning presented above. In more detail, since S is an ℓ-cohesive group, it includes at least one voter who approves of ℓ members of W. Let us call this voter v_1. After we remove v_1 from consideration, we have an $(\ell-1)$-cohesive group of size at least $\ell \cdot \frac{n}{k} - 1$. This group contains at least one voter, call him or her v_2, who approves of at least $\ell - 1$ committee members. After removing v_2 from consideration, we obtain an $(\ell-1)$-cohesive group of size at least $\ell \cdot \frac{n}{k} - 2$. We repeat this process n/k times, obtaining a sequence $v_1, \ldots, v_{n/k}$ of voters from S who all approve of at least $\ell - 1$ members of W (v_1 certainly approves of at least one more member of W, but we disregard this). Together, these voters have at least $(\ell - 1) \cdot n/k$ approvals for members of W. Next, we see that $S \setminus \{v_1, \ldots, v_{n/k}\}$ is an $(\ell-1)$-cohesive group of voters and, by repeating the just-described process for it, we get a group of new n/k voters from S who jointly have $(\ell - 2) \cdot \frac{n}{k}$ approvals for members of W. Altogether, we can continue this process to find $(\ell - 1)$ groups of n/k voters, whose total numbers of approvals for members of W are $(\ell - 1) \cdot \frac{n}{k}, (\ell - 2) \cdot \frac{n}{k}, \ldots, 1 \cdot \frac{n}{k}$. By summing these values and averaging over $\|S\|$ voters, we get the right-hand side of inequality (6.5).

Next, we see that the right-hand side of inequality (6.5) can be further bounded by

$$\frac{1}{\|S\|} \sum_{i=1}^{\ell-1} \frac{n}{k} \cdot i \geq \frac{k}{n\ell} \sum_{i=1}^{\ell-1} \frac{n}{k} \cdot i = \frac{1}{\ell} \cdot \frac{(\ell-1)\ell}{2} = \frac{\ell-1}{2}.$$

Consequently, if a committee W satisfies EJR, then on the average the voters in each ℓ-cohesive group approve of at least $\frac{\ell-1}{2}$ candidates from W. Then we say that the committee W has the *proportionality degree* of $\frac{\ell-1}{2}$ (see [871]).

6.3 Preferences as Approvals

"I'm still wondering about Example 6.3 ...," Belle joins the discussion. "Can we perhaps construct a similar example, showing that there exists no rule satisfying EJR?"

"No, we can't, fortunately!" Anna is still smiling.

Anna suggests the following voting rule.

Definition 6.9 (proportional approval voting, PAV). Given an election $E = (C, V)$ and a committee size k, *proportional approval voting (PAV)* selects committees W that maximize the following score:

$$\text{PAV-Score}(W) = \sum_{v \in V} \text{H}(\|A(v) \cap W\|), \quad \text{where} \quad \text{H}(r) = \sum_{i=1}^{r} \frac{1}{i}.$$

Example 6.4. If we apply PAV to the election from Example 6.2 then we find that each winning committee of size three consists of candidate b, one of the candidates from $\{c, d, e\}$, and one candidate from $\{f, g\}$. For example, the *PAV-Score* of committee $\{b, c, f\}$ is 133.5. The first 67 voters who have approval set $\{b, c, d, e\}$ and the two voters with approval set $\{b, f, g\}$ contribute $1 + \frac{1}{2}$ points each, the 30 voters with approval set $\{f, g\}$ contribute one point each, and the single voter with approval set $\{h\}$ gives zero points.

At first, using the harmonic numbers, $\text{H}(\cdot)$, in Definition 6.9 might seem counter-intuitive, yet this exact formula makes PAV particularly interesting in terms of proportionality. In particular, Aziz et al. [53] have shown that PAV has the optimal proportionality degree of $\ell - 1$ (that is, PAV has a proportionality degree of $\ell - 1$, and for each $\epsilon > 0$, there exists no rule that has a proportionality degree of $\ell - 1 + \epsilon$).

Theorem 6.7 (Aziz et al. [53]). *Let W be a committee returned by PAV for an election $E = (C, V)$ and committee size k. Let S be an ℓ-cohesive group of voters. Then, on the average, the voters from S approve of more than $\ell - 1$ members of W.*

Proof. For the sake of contradiction, assume that $\frac{1}{\|S\|} \cdot \sum_{v \in S} \|A(v) \cap W\| \leq \ell - 1$. Note that there exists a candidate c_a who is approved by all voters from S but not elected, i.e., $c_a \notin W$. Indeed, since S is an ℓ-cohesive group, if such a candidate would not exist then each voter from S would approve of at least ℓ members of W, which would contradict our assumptions. Let $W' = W \cup \{c_a\}$ and let $\lambda(c_a)$ denote the value by which the *PAV-Score* of W would increase if c_a were added:

$$\lambda(c_a) = \text{PAV-Score}(W') - \text{PAV-Score}(W)$$
$$\geq \sum_{v \in S} \Big(\text{H}(\|A(v) \cap W\| + 1) - \text{H}(\|A(v) \cap W\|) \Big) = \sum_{v \in S} \frac{1}{\|A(v) \cap W\| + 1}.$$

Next, we will use the inequality between the arithmetic and harmonic means. This inequality says that if $x_1, \ldots, x_s$ are positive real numbers, then

$$\frac{x_1 + x_2 + \cdots + x_s}{s} \geq \frac{s}{\frac{1}{x_1} + \frac{1}{x_2} + \cdots + \frac{1}{x_s}}.$$

Multiplying both sides by s, we obtain the following inequality:

$$x_1 + x_2 + \cdots + x_s \geq \frac{s^2}{\frac{1}{x_1} + \frac{1}{x_2} + \cdots + \frac{1}{x_s}}.$$

By applying it to $\lambda(c_a)$, we get

$$\lambda(c_a) \geq \sum_{v \in S} \frac{1}{\|A(v) \cap W\| + 1} \geq \frac{\|S\|^2}{\sum_{v \in S}(\|A(v) \cap W\| + 1)}$$

$$\geq \frac{\|S\|^2}{(\ell-1)\|S\| + \|S\|} = \frac{\|S\|}{\ell} \geq \frac{n\ell}{k\ell} = \frac{n}{k}.$$

Next, for each committee member $c \in W'$, by $\Delta(c)$ we denote the value by which the score of committee W' would decrease if we removed c from W':

$$\Delta(c) = \textit{PAV-Score}(W') - \textit{PAV-Score}(W' \smallsetminus \{c\}).$$

Now, observe that

$$\sum_{c \in W'} \Delta(c) = \sum_{c \in W'} \sum_{v:\, c \in A(v)} \Big(H(\|A(v) \cap W'\|) - H(\|A(v) \cap W'\| - 1) \Big)$$

$$= \sum_{c \in W'} \sum_{v:\, c \in A(v)} \frac{1}{\|A(v) \cap W'\|}$$

$$= \sum_{v:\, A(v) \cap W' \neq \emptyset} \sum_{c \in A(v) \cap W'} \frac{1}{\|A(v) \cap W'\|}$$

$$= \sum_{v:\, A(v) \cap W' \neq \emptyset} 1 \leq \|V\| = n.$$

Hence, there exists a candidate $c_r \in W'$ such that $\Delta(c_r) \leq \frac{n}{k+1}$. As a result, we have that

$$\textit{PAV-Score}\Big((W \cup \{c_a\}) \smallsetminus \{c_r\}\Big) \geq \textit{PAV-Score}(W) + \frac{n}{k} - \frac{n}{k+1}$$

$$> \textit{PAV-Score}(W).$$

This means that W could not be selected by PAV, which is a contradiction. $\square$

6.3 Preferences as Approvals

As a consequence of Theorem 6.7, we also get that PAV satisfies EJR. Indeed, if on average each voter from a given ℓ-cohesive group approves of more than $\ell - 1$ members of the committee, then, by the pigeonhole principle, there must be at least one voter in this group who approves of at least ℓ committee members. This suffices for EJR to be satisfied. Note that Theorem 6.7 gives a stronger proportionality degree of $\ell - 1$ than that implied by EJR (i.e., a proportionality degree of $\frac{\ell-1}{2}$).

The three friends are really impressed by the fact that PAV satisfies EJR, a strong notion of proportionality. They have a closer look at the definition of PAV to make sure that they did not miss any hidden defects of the rule. It is Chris who notices the first potential problem: "Yeah, but how can we compute the outcome of the rule if the numbers of voters and candidates are large? The only thing that comes to my mind is to examine each k-element subset of the candidates and compute its score. Yet, there are $\binom{m}{k}$ such subsets! This can be computationally difficult!"

Chris is right—it is NP-hard to compute the winning committees for PAV [872, 54]. However, for reasonably small elections, it can be solved effectively by integer linear programming (ILP). Indeed, modern ILP solvers can deal with PAV elections with hundreds, or even thousands, of candidates and voters. For larger elections, it can be approached by applying approximation algorithms, similarly to those that we will discuss in Section 6.5.3 starting on page 456 (see also [872, 361]).

6.3.3 The Phragmén Sequential Rule

While searching for a rule that is both proportional and computable in polynomial time, Chris comes up with the following idea.

"Let's give the voters some virtual tokens and allow them to 'buy' the candidates! The outcome should be fair since each voter would be initially given the same amount of tokens!"

This idea leads to particularly interesting voting rules.

Definition 6.10 (Phragmén sequential rule). The *Phragmén sequential rule* can be described as the following process. We assume that the voters earn virtual money with a constant speed of one dollar per time unit, and that buying a candidate costs n dollars. In each time moment (we assume that time is continuous), when there is a group of voters who approve of an as yet not elected candidate c and who have n dollars in total, the voters buy the candidates greedily, that is, the budget of each voter approving of c is reset to zero, and the candidate c is added to the winning committee. The rule stops when k candidates are selected.

Example 6.5. Consider the election from Example 6.2. Recall that we have $n = 100$ voters, of whom 67 have approval ballot $\{b,c,d,e\}$, two have approval ballot $\{b,f,g\}$, 30 have approval ballot $\{f,g\}$, and one has approval ballot $\{h\}$. At time $t_1 = \frac{100}{69}$, the first 69 voters collect exactly $\$100$. Thus, the rule selects candidate b, and these voters lose their money. At time $t_2 = t_1 + \frac{100}{67}$, each of the first 67 voters has $\frac{100}{67}$ dollars. They are able to buy an additional candidate, say c (the rule can also select d or e, depending on the tie-breaking rule). At this time, there are two voters with $\frac{100}{67}$ dollars left, and the remaining voters have $\frac{100}{67} + \frac{100}{69}$ dollars each, since none of them paid for any candidates so far. Finally, at time t_3 such that $30t_3 + 2(t_3 - t_1) = 100$, the 32 voters who approve of f and g have enough money to buy their representative, say f. The rule stops and returns the winning committee $\{b,c,f\}$.

Note that Phragmén produces the same output on this election as PAV. This is hardly surprising, as the election has a very simply structure, where it is clear what a proportional outcome should be. We will soon see that these two rules can also produce quite different outcomes.

The Phragmén sequential rule does not satisfy EJR, yet it is still a rule with good proportionality properties. For example, it satisfies the axiom of *proportional justified representation* (*PJR*) [829], a weaker variant of EJR, and the average number of representatives that the voters from an ℓ-cohesive group can have in a committee returned by the Phragmén sequential rule is at least $\frac{\ell-1}{2}$ (i.e., its proportionality degree is only two times lower than that of PAV [871]).

It can further be argued that the Phragmén sequential rule and PAV interpret proportionality in a different manner, which is explained in the following example.

Example 6.6. Assume there are $n = 6$ voters, $v_1, \ldots, v_6$, and $m = 15$ candidates, $c_1, \ldots, c_{15}$; the size of the committee to be elected is $k = 12$. The voters' ballots are illustrated in the diagram below: Each candidate is represented as a box, and a voter approves of the candidates whose corresponding boxes are above the voter. For example, voter v_1 approves of $\{c_1, c_2, c_3, c_4\}$ and voter v_4 approves of $\{c_7, c_8, c_9\}$. The candidates selected by the Phragmén sequential rule and PAV are shaded on the diagrams below.

Here, we can see that the two rules interpret proportionality quite differently. The approval sets of the first three voters are disjoint from the approval sets of the rest of the society. These voters form half of the society, so intuitively they shall be allowed to decide about half of the committee members. This is reflected in the outcome returned by the Phragmén sequential rule.

6.3 Preferences as Approvals

On the other hand, PAV realizes that if we removed candidate c_4 from the outcome returned by the Phragmén sequential rule, and added c_9 instead, then the inequality in the voters' satisfaction would decrease. PAV is concerned with the distribution of the voters' satisfaction. Thus, it selects an outcome where each voter is happy with three committee members, even though in such a case the first half of the society decides only about a quarter of the committee members.

c_4	c_5	c_6					c_4	c_5	c_6			
c_3			c_9	c_{12}	c_{15}		c_3			c_9	c_{12}	c_{15}
c_2			c_8	c_{11}	c_{14}		c_2			c_8	c_{11}	c_{14}
c_1			c_7	c_{10}	c_{13}		c_1			c_7	c_{10}	c_{13}
v_1	v_2	v_3	v_4	v_5	v_6		v_1	v_2	v_3	v_4	v_5	v_6

(a) Phragmén sequential rule   (b) PAV

Further, in comparison to PAV, the Phragmén sequential rule satisfies the axiom of committee enlargement monotonicity (recall Definition 6.1). Together with its good proportionality degree, and the fact that it can be computed in polynomial time, it makes it a good alternative for PAV.

The reader may wonder what happens if the voters will not earn money over time, but rather if we give them their money upfront. This idea leads to an interesting rule, called the *method of equal shares*. This rule satisfies EJR, is computable in polynomial time, and extends to more general types of voters' preferences. We will discuss it in more detail in Section 6.4.

6.3.4 Different Interpretations of Approvals

Belle thinks about all the possibilities for approval voting they have discussed so far, and eventually comes up with a new idea.

"All our ideas rely on finding a set of books that in some sense are liked by the students. But as a consequence, we somehow ignore that we have also a set of books each student did *not* approve of and consequently *dis*likes."

Anna agrees: "You are right. Among the voters there are also the students from the reading club. They read more than three books per semester, but they have special preferences. I know that they will be very disappointed when we choose new books for the library that they don't like. So perhaps, instead of maximizing the voters' satisfaction, we could try to minimize their dissatisfaction."

As presented so far, the focus of approval voting is often on the approved candidates. An opposite interpretation has been proposed by Brams, Kilgour, and Sanver [199] under the name *minimax approval voting*. The basic idea of this voting procedure is to find a committee where the maximal dissatisfaction of the voters is minimized. The voters still report a regular approval preference profile which, given some fixed order over the candidates, will then be interpreted as a 0-1 vector for each voter, where a 1 stands for approval and a 0 for disapproval of the respective candidate, just as the approval vectors in Section 4.1.3 starting on page 249.

A committee can again be interpreted as a 0-1 vector, where exactly those candidates receiving a 1 are elected. Then the Hamming distance will be used to determine a winning committee. For a given committee and a given vote, both represented as 0-1 vectors, the *Hamming distance* between these two is the number of candidates on which they disagree, i.e., it counts +1 for each disapproved candidate who is in the committee and for each approved candidate who is not in the committee. Finally, the winning committees are those where the maximum Hamming distances to all votes are minimal. In its original form, the size of a winning committee in minimax elections is not fixed in advance but is determined through the method.

Example 6.7. Consider a committee election with three candidates, $C = \{a,b,c\}$, and three voters having the following approval ballots:

$$A(v_1) = \{a,c\}, \quad A(v_2) = \{b,c\}, \quad A(v_3) = \{a\}.$$

As approval vectors, this can be written in the following form:

	a	b	c
v_1:	1	0	1
v_2:	0	1	1
v_3:	1	0	0

In multiwinner approval voting, the committee $\{a,c\}$ is among the committees that win this election, since both candidate a and candidate c receive two approvals, whereas b has only one. To determine a winning committee under minimax approval voting, we calculate the Hamming distance for each possible committee to the votes.

6.3 Preferences as Approvals

Committee	Hamming distance to			Max
	v_1	v_2	v_3	
000	2	2	1	2
001	1	1	2	2
010	3	1	2	3
011	2	0	3	3
100	1	3	0	3
101	0	2	1	2
110	2	2	1	2
111	1	1	2	2

Among the eight possible committees, five committees have a maximum Hamming distance of 2 to the three approval ballots, which is the minimal value; hence, they are the winning committees under minimax approval voting.

While this example shows a simple variant of the minimax procedure, other variants using so-called proximity weights have been proposed by Brams, Kilgour, and Sanver [199]. The intuition behind the minimax procedure is to elect a committee such that no voter is really far away from the elected consensus.[9] Unfortunately, it happens quite often that there is a large number of tied winning committees, as shown in the above example. This is one of the reasons why minimax approval voting rules are not often used in practice. Another drawback of the minimax procedure is that the associated winner problem is NP-complete and can only be efficiently approximated [648, 647]. For a deeper analysis of the parameterized complexity of the related problems and further approximation results, we refer to the work of Cygan et al. [321].

Anna, Belle, and Chris agree that the minimax procedure has too many drawbacks.

"I have yet another idea of how to interpret an approval vote differently," says Anna and explains: "When considering approval committee elections, a voter that approves of plenty of candidates has more influence on the election outcome. So, instead of giving every approved candidate

---

[9] The minimax approach to minimize voter dissatisfaction has also been applied to the generalized preference types by Baumeister et al. [108], who propose to combine approval ballots with linear rankings of the candidates: A *generalized preference ballot* takes the form of several disjoint groups of candidates, which may also be empty, and these groups are ranked by the voter. Yet another modification of approval votes are the *bounded approval ballots* due to Baumeister et al. [107] by which they seek to balance expressiveness and simplicity for the ballots used in multiwinner approval elections: Voters can express incompatibilities, dependencies, and/or substitution effects among the alternatives while their votes are still almost as easy as regular approval ballots to cast.

the value 1, we might give each voter a value of 1 in total that will then be equally distributed among all candidates approved of by this voter."

Chris thinks about this suggestion and then approves of it: "That might be a good idea, let's try it."

This idea was initially proposed by Brams, Kilgour, and Sanver [199] under the name *satisfaction approval voting (SAV)*. The intention is to find a more diverse set of winners by maximizing voters' satisfaction, which is somewhat similar to PAV. However, in SAV, the satisfaction of a voter with a candidate in a committee is $1/p$, where p is the number of candidates this voter approves of. The main difference between SAV and PAV is that in the former, a voter's satisfaction with electing a single candidate may be computed independently of other elected candidates. Hence, the total voter satisfaction in SAV is an additive function that can be easily computed.

Considering again the multiwinner election with books as candidates on approval ballots that Anna, Belle, and Chris have their fellow students hold at the beginning of Section 6.3 starting on page 426, the total voter satisfaction with electing each of the candidates can be calculated as follows:

Ballot	Total voter satisfaction with electing candidate							
	a	b	c	d	e	f	g	h
$8 \times \{b,c,d,e\}$	0	$8 \cdot 1/4$	$8 \cdot 1/4$	$8 \cdot 1/4$	$8 \cdot 1/4$	0	0	0
$6 \times \{b,d,h\}$	0	$6 \cdot 1/6$	0	$6 \cdot 1/6$	0	0	0	$6 \cdot 1/6$
$3 \times \{a,b,e\}$	$3 \cdot 1/3$	$3 \cdot 1/3$	0	0	$3 \cdot 1/3$	0	0	0
$2 \times \{a,b,c,e\}$	$2 \cdot 1/4$	$2 \cdot 1/4$	$2 \cdot 1/4$	0	$2 \cdot 1/4$	0	0	0
$2 \times \{a,b,c,d,e,f\}$	$2 \cdot 1/6$	$2 \cdot 1/6$	$2 \cdot 1/6$	$2 \cdot 1/6$	$2 \cdot 1/6$	$2 \cdot 1/6$	0	0
$2 \times \{a,f\}$	$2 \cdot 1/2$	0	0	0	0	$2 \cdot 1/2$	0	0
$1 \times \{f\}$	0	0	0	0	0	$1 \cdot 1/1$	0	0
$4 \times \{a,b,c,d,g\}$	$4 \cdot 1/5$	$4 \cdot 1/5$	$4 \cdot 1/5$	$4 \cdot 1/5$	0	0	$4 \cdot 1/5$	0
$1 \times \{h\}$	0	0	0	0	0	0	0	$1 \cdot 1/1$
Sum	$\frac{109}{30}$	$\frac{169}{30}$	$\frac{109}{30}$	$\frac{124}{30}$	$\frac{115}{30}$	$\frac{70}{30}$	$\frac{24}{30}$	$\frac{60}{30}$

An *SAV winning committee* consists of those candidates with the highest total voter satisfaction. Here, Anna, Belle, and Chris would choose to buy books b, d, and e, in contrast to the approval winning committee consisting of books b, c, and d. In this example, the diversity of the SAV winning committee can be seen by looking at the three voters with approval set $\{a,b,e\}$ and the four voters with approval set $\{a,b,c,d,g\}$. The choice of candidate e instead of c is reasonable with respect to diversity, since when e is elected, two of their approved candidates are already in the SAV winning committee $\{b,d,e\}$ for the second group of voters, while candidate e would be only the second approved candidate in it for the first voter group.

We stress that SAV still has many features of a multiwinner voting rule focused on individual excellence. Most importantly, this is so because each candidate is (largely) evaluated independently from the other ones.

Finally, we mention the *attribute multiwinner approval elections* introduced and studied by Baumeister and Boes [106]. The goal in such elections is to select winning committees of candidates such that each candidate satisfies certain attributes from various categories. Every voter specifies which attributes from which category are desirable for a candidate, and each candidate might satisfy only some of these attributes. This approach is somewhat akin in spirit to the *hedonic expertise games* (recall the notion of hedonic games from Section 3.6 starting on page 198) introduced by Caskurlu, Kizilkaya, and Ozen [260] to model a variety of situations where agents with complementary qualities and skills aim to form coalitions. Baumeister and Boes [106] investigate the *distortion* in attribute multiwinner approval elections, i.e., the ratio of the voters' satisfaction with the worst possible committee and their satisfaction with an actually winning committee, given their attribute approval ballots.

6.4 Different Costs of Candidates and Score Voting

Anna, Belle, and Chris finally feel that they have found two appealing proportional committee voting rules—PAV and the Phragmén sequential rule—and that the choice between these two boils down mainly to considerations about their computational complexity, that is, the size of an election that they would like to solve. At the same time, they realize that their model of committee elections is a bit too simplistic to capture the problem of buying the books.

"Guys, books may have different costs," Anna expresses her concerns. "We are not required to buy a fixed number of copies. We can spend our budget on buying fewer expensive books, or more cheaper ones."

Chris makes a further remark: "Indeed, since the books have different costs, it is harder to interpret approval ballots. If a student approves of a book, does this mean he or she thinks this is a good book independently of its cost, or does this mean he or she thinks the book is worth the money it costs?"

"It seems we need a different model of ballots which would allow the voters to express different levels of appreciation towards different candidates," Belle concludes.

We assume that each candidate $c \in C$ has a *cost*, denoted by $\text{cost}(c)$, and that the goal is to select a subset of candidates whose total cost does not exceed a predefined budget B. Thus, a committee W is *feasible* if

$$\sum_{c \in W} \text{cost}(c) \leq B.$$

Additionally, we assume the cardinal model of utilities, where the voters assign nonnegative numbers to the candidates: By $u_v(c)$ we denote the *score* (also called the *utility*) that a voter $v \in V$ assigns to a candidate $c \in C$—the higher the value, the more the voter likes the candidate. This model is most general; for example, approval ballots can be encoded in this model, using two different interpretations:

1. We can define the utility of a voter as the total amount of funds spent on the candidates that the voter approves of. This corresponds to setting $u_v(c) = \text{cost}(c)$ if voter v approves of candidate c, and $u_v(c) = 0$ otherwise.
2. The utility of a voter v can be alternatively defined as the *number* of approved candidates that have been selected. Then we should set $u_v(c) = 1$ for candidates c that are approved of by v, and $u_v(c) = 0$ for the other candidates.

We extend the notation to sets, writing

$$\text{cost}(T) = \sum_{c \in T} \text{cost}(c) \quad \text{and} \quad u_v(T) = \sum_{c \in T} u_v(c),$$

for all $v \in V$ and $T \subseteq C$.

In the remainder of this section we explain how to adapt the axioms of proportionality and design a proportional rule in this generic framework.

The axiom of EJR can be formulated in the cardinal model, yet this requires specifying what it means for a group of voters to have cohesive preferences. We start from the following observation: For a single candidate $c \in C$ and a subset S of voters, the members of S agree that the value of c equals at least $\min_{v \in S} u_v(c)$. Similarly, for a subset of candidates $T \subseteq C$, the members of S agree that these candidates have the combined value of at least $\sum_{c \in T} \min_{v \in S} u_v(c)$. Suppose that the voters in S are given a proportional share of the budget, that is, $B \cdot \frac{\|S\|}{n}$. We say that S is β-*cohesive* if (1) its voters can afford some subset of candidates T, i.e., $\text{cost}(T) \leq B \cdot \frac{\|S\|}{n}$, and (2) they view the combined value of these candidates as at least β, i.e., $\sum_{c \in T} \min_{v \in S} u_v(c) \geq \beta$. EJR requires that for every β-cohesive subset of voters S there is a voter $v \in S$ who obtains at least utility β from the outcome of the election.

For example, assume there is a group of 20% voters and two candidates, a and b, whose total cost is lower than 20% of the budget. Assume all voters from S agree that the utility of a is at least 7 ($u_v(a) \geq 7$ for each $v \in S$) and that the utility of b is at least 11. Then EJR requires the rule to elect a subset W such that $u_v(W) \geq 18$ for at least one voter $v \in S$.

6.4 Different Costs of Candidates and Score Voting

Definition 6.11 (EJR for additive utilities [777]). Given an election (C, V) and a budget B, we say that a group of voters $S \subseteq V$ is β-cohesive if there exists a subset of candidates $T \subseteq C$ (called proposal) such that

1. $\mathrm{cost}(T) \leq B \cdot \frac{\|S\|}{n}$ and
2. $\sum_{c \in T} \min_{v \in S} u_v(c) \geq \beta$.

We say that a committee W satisfies *extended justified representation (EJR) for additive utilities* if for each β and each β-cohesive group of voters S, there exists a voter $v \in S$ with $u_v(W) \geq \beta$.

A voting rule satisfies *extended justified representation (EJR) for additive utilities* if for each election, the returned winning committees satisfy EJR.

In the cardinal model, EJR is very demanding. For example, even if we have a single voter, finding a committee that satisfies EJR for additive utilities is equivalent to the knapsack problem, which is NP-hard. This observation excludes the existence of a polynomial-time computable rule that satisfies EJR for additive utilities. Yet, we can relax this axiom a bit and provide a corresponding voting rule that is computationally efficient. In the definition below, we use the notation from Definition 6.11, and assume that T is the set witnessing that the group is β-cohesive.

Definition 6.12 (EJR-up-to-one [777]). Given an election (C, V), a committee W satisfies *EJR-up-to-one*, if for each β-cohesive group of voters S with proposal T, there is a voter $v \in S$ and a candidate $c \in T$ with $u_v(W \cup \{c\}) \geq \beta$.

In other words, EJR-up-to-one says that for each cohesive group of voters the condition of EJR for additive utilities would be satisfied if we gave this group at most one additional candidate. In the subsequent part, we define a polynomial-time computable voting rule that satisfies EJR-up-to-one.

Definition 6.13 (method of equal shares [777]). The *method of equal shares* starts with an empty set of candidates $W = \emptyset$ and adds elements to W sequentially. Each voter is initially given an equal part of the budget, that is, B/n. If a candidate c is selected, then the voters pay for c its cost; let $p_v(c)$ denote the amount that voter v pays for c. We extend this notation to sets: $p_v(W) = \sum_{c \in W} p_v(c)$. At each time, we say that a candidate c is q-*affordable* if the voters can buy c so that each voter pays at most q per unit of satisfaction. In other words, candidate c is q-*affordable* if the following equation holds (by b_v we mean the amount of money held by voter v in this particular moment, i.e., $b_v = \frac{B}{n} - p_v(W)$):

$$\sum_{v \in V} \min(b_v, u_v(c) \cdot q) = \mathrm{cost}(c). \quad (6.6)$$

In each round, if there exists no q-affordable candidate for any q, then the rule stops, and returns W. Otherwise, the rule picks the candidate c that is q-affordable for the lowest value of q, asks each voter v to pay $p_v(c) = \min(b_v, u_v(c) \cdot q)$, and adds c to the committee by updating W to $W \cup \{c\}$.

Let us explain this rule by the following example.

Example 6.8 (method of equal shares). Assume there are 200 voters, and the budget is $B = 2,000$. There are eight candidates, $a, b, \ldots, h$, and the cardinal utilities of the voters are presented in the figure below. For example, the first 65 voters assign utility 30 to candidate b, utility 10 to candidates e and g, and utility zero to the remaining candidates.

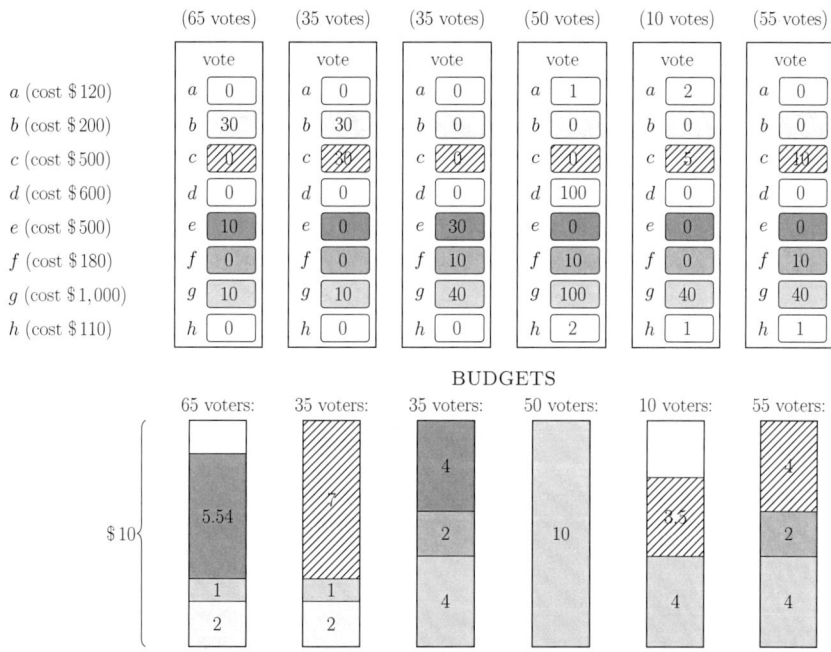

The budget is split evenly among the voters, hence each of them gets 10 units of budget. First, the rule picks candidate b, which is $\frac{1}{15}$-affordable: The voters from the first two groups pay two dollars for it and get a utility of 30. It is easy to check that the other candidates' affordability is worse, i.e., they are q-affordable for a value of q higher than $1/15$. For example, candidate g received the total number of 10,000 utility points from all the voters, and costs 1,000 dollars. We could split g's cost so that each voter would pay $1/10$ per unit of satisfaction: The voters from the first and the second group would pay one dollar for this candidate, those from the fourth group would pay ten dollars, and the remaining voters would pay four dollars each. Thus, candidate g is $\frac{1}{10}$-affordable, so g is less affordable than b since $1/10 > 1/15$.

In fact, candidate g will be selected in the second round. In the third round, candidate f is selected. Note that the fourth group of voters ran out of money already, so the cost of f is shared only by the voters from the third and the sixth group. Those voters pay $1/5$ per unit of satisfaction. In the next two rounds, e and c are selected. After that, there are no q-affordable candidates for any q, and the rule returns $\{b,c,e,f,g\}$.

Observe that the budget constraint would allow to also add a or h, but the groups of voters who approve of either of these two candidates do not have enough money to afford buying them. This shows that the outcome of the method of equal shares can be nonexhaustive, and sometimes it makes sense to complement it using some other rule.

Theorem 6.8 (Peters et al. [777]). *The method of equal shares is computable in polynomial time and satisfies EJR-up-to-one. For approval committee elections, it satisfies EJR.*

Proof. We prove the EJR-up-to-one part only, leaving the EJR part as an exercise for the reader.

Fix an election instance and let W denote the outcome returned by equal shares. Consider a β-cohesive group S of voters and let $T \subseteq C$ be a subset of candidates such that (1) $\text{cost}(T) \leq B \cdot \frac{\|S\|}{n}$, and (2) $\sum_{c \in T} \min_{v \in S} u_v(c) \geq \beta$. We must show that there exists a candidate $c \in T$ and a voter $v \in S$ for whom $u_v(W \cup \{c\}) \geq \beta$.

Imagine that the voters from S have no budget restrictions with respect to the candidates from T, that is, they can exceed their initial endowment of B/n, but only when buying candidates from T. Let us analyze how the method of equal shares would work in this case. Consider the first round r where some voter $v \in S$ spent in total B/n or more. Assume that this happened when some candidate a was bought. There are two cases:

1. If $a \notin T$, then v in round r must have spent in total exactly B/n, and until the end of round r, the rule worked exactly the same as the standard variant of the method of equal shares (i.e., the variant where the voters from S cannot exceed their endowments).
2. The other case is when $a \in T$; here, v could spend strictly more than B/n. In this case, the rule works the same way as its standard variant at least until the end of round $(r-1)$.

Let us define a function f such that for each number x, $x \geq 0$, $f(x)$ is the price that voter v paid for the first x units of utility that he or she received. For the sake of this function's definition, we consider the candidates to be divisible: For example, if v paid y dollars for the first candidate and got the utility of x, we assume that $f(\gamma x) = \gamma y$ for each $\gamma \in [0,1]$. If $a \in T$, then f is convex. This follows from the fact that the method of equal shares always picks the candidate that minimizes the price per utility q, and that by our assumption the voter is always able to pay $q \cdot u_v(a)$. If $a \notin T$, then the function is convex except for the segment that corresponds to buying a (note that this is because the last purchase of a voter can be more efficient; he or she can pay less than he or she should if he or she does not have enough funds). This function is depicted in Figure 6.3 (the lower line in the plot).

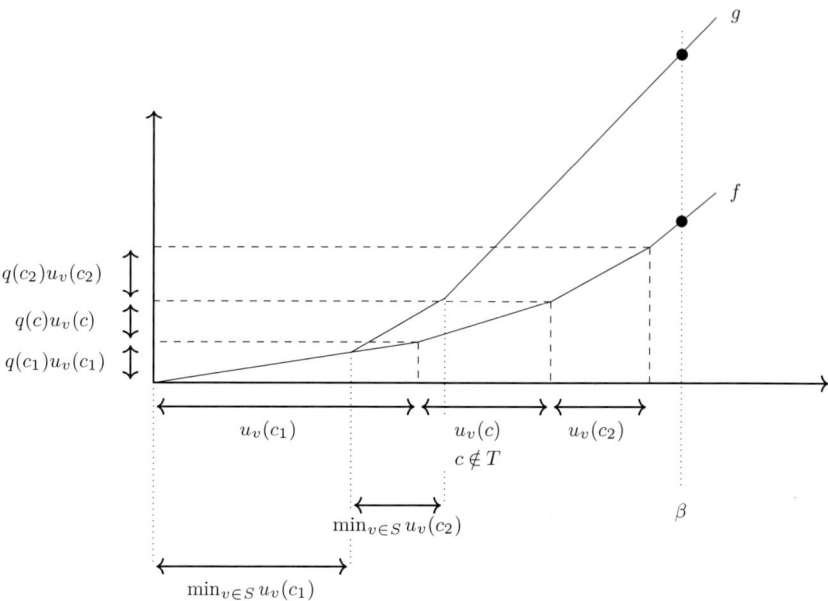

Fig. 6.3 A plot of a function describing the cost of utility used in the proof of Theorem 6.8, assuming that c_1 and c_2 belong to T

Since we assumed that the voters from S can exceed their initial budgets, each candidate $c \in T$ that is supported by at least one voter from S will be chosen at some point. Let us write $q(c)$ for the minimal value q such that c was q-affordable at the point when it was selected. Let us now upper-bound the value of the function f by taking the value of another function g which is built from parts of "shorter" segments of f (the upper line in the plot). The segments for all candidates from outside of T are removed from g, and for each $c \in T$ we only consider utility up to value $\min_{v \in S} u_v(c)$ that all members of S are guaranteed to derive from c. Thus, we estimate $f(\beta)$ as follows:

$$f(\beta) \leq f\left(\sum_{c \in T} \min_{v \in S} u_v(c)\right) \leq \underbrace{\sum_{c \in T} \min_{v \in S} u_v(c) q(c)}_{\text{this is the value of the } g \text{ function}}.$$

Since for each candidate c, it holds that $\sum_{v \in S} u_v(c) q(c) \leq \mathrm{cost}(c)$ (cf. Equation (6.6)); equivalently, we have

$$q(c) \leq \frac{\mathrm{cost}(c)}{\sum_{v \in S} u_v(c)}.$$

6.5 Complexity of Computing Winning Committees

By plugging this inequality into the just-derived upper bound on $f(\beta)$, we get that

$$f(\beta) \leq \sum_{c \in T} \min_{v \in S} u_v(c) \cdot \frac{\text{cost}(c)}{\sum_{v \in S} u_v(c)} \leq \frac{\sum_{c \in T} \text{cost}(c)}{\|S\|} \leq \frac{B}{n}.$$

This means that after spending B/n units of budget (that is, after buying at most one additional candidate from T) voter v will have at least β units of utility. This completes the proof. ❏

One potential disadvantage of the method of equal shares is that it can sometimes select a subset of candidates with total cost considerably lower than the available budget. There are a few possible remedies to this:

1. We can gradually increase the initial endowment of the voters starting with the value of B/n (each time increasing it by one unit of currency). After each such increase, we run the method of equal shares from scratch. We increase the endowment until the moment when the next increase would make us exceed the available budget B. This approach guarantees that the returned set of candidates is priceable.
2. If all utilities are strictly positive then the method of equal shares is exhaustive (that is, we cannot add another candidate to the returned set without exceeding the budget). Thus, for all $v \in V$ and $c \in C$ with $u_v(c) = 0$, we can override the utility with $u_v(c) = \varepsilon$, for some fixed small value of ε; this way we would obtain an exhaustive rule.
3. We can use the remaining budget to pursue other aims or to select candidates with maximum utilitarian social welfare, thereby obtaining an outcome with high utilitarian welfare that still satisfies EJR.

To summarize, using the method of equal shares we can efficiently find proportional committees even for the setting where the candidates have costs and we are limited by a budget. This setting is particularly useful, e.g., in the context of *participatory budgeting*: The idea is that a city allocates a part of its budget for citizen projects, the citizens first submit the projects that they would like to see funded (each project, viewed in our nomenclature as a candidate, may have a different cost), and then they vote on which ones to actually fund. While, in practice, many cities use a simple greedy rule to choose the projects (i.e., they iteratively keep choosing the projects that received the most approvals and are still within the remaining budget), the method of equal shares would give more proportional, and fairer, outcomes.

6.5 Complexity of Computing Winning Committees

In the preceding sections, we have seen quite a number of different multiwinner voting rules, ranging from very simple ones, such as SNTV, k-Borda, bloc, or multiwinner approval voting, through more involved ones, such as STV, Phragmén, or Chamberlin–Courant, to fairly complicated ones, such as harmonic Borda, the method of equal shares, or PAV. Yet, irrespective of their apparent complexity, many of these rules can be computed in polynomial time. Indeed, even for the method of equal shares, a rule with rather strong axiomatic properties, we can easily compute a winning committee. Still, for many other rules this task is intractable.

In this section, we will mostly focus on PAV—although we will also mention other rules occasionally—and we will show that while it is NP-hard to compute PAV winning committees, there are many ways of going around this issue.

"Guys, what's the point of trying to compute PAV winning committees if we have these equal shares, which are better anyway?" Chris expresses his never-ending desire to acquire new knowledge.

"Well ... how do you know that equal shares are better?" Anna replies and smiles.

"What you mean, how?" Chris looks suprised. "Didn't we show all these proportionality properties?"

"But, you know, ... if the candidates have no costs, and we want to simply select k candidates, then PAV has equally good properties, if not better ones! Besides, these are only theoretical guarantees. We should at least look at several election instances and compare the outcomes returned by the two rules. And for this ... we do need to know how to compute such outcomes!" All of a sudden, Anna feels that her argument is even more compelling than she initially thought.

"Besides, we might learn techniques that go beyond PAV," Belle adds quietly.

"OK, fine with me!" Chris answers with his usual grace.

6.5.1 The Problem of Winner Determination

Recall from Section 4.3.1 starting on page 289 that in the case of single-winner voting rules, the problem of computing an election winner (or the election winners) can be expressed in quite a straightforward way. For example, given an election E, we might ask for the set of tied winners under a particular rule or, if we want to be formally correct and specify a decision problem, we may ask if a given candidate p is among the winners. Both approaches are computationally equivalent: If we can solve the former then clearly we can solve the latter, and if we can solve the latter then, by asking about all the candidates one by one, we can also solve the former.

6.5 Complexity of Computing Winning Committees

In the case of multiwinner voting, however, the situation is more involved because there are exponentially many committees that might be winning (with respect to the number of candidates). Hence, even for the simplest of all rules, computing *all* winning committees may be infeasible, simply because there are too many of them.

Example 6.9. Consider an election with candidate set $C = \{a, b, c_1, \ldots, c_{100}\}$ and two voters, u and v, with the following preference orders:

$$u\colon a > b > c_1 > c_2 > \cdots > c_{99} > c_{100},$$
$$v\colon a > b > c_{100} > c_{99} > \cdots > c_2 > c_1.$$

We are interested in k-Borda winning committees of size $k = 50$. We see that candidates a and b have higher Borda scores than each of the candidates $c_1, \ldots, c_{100}$, who all have equal score. Hence, there are $\binom{100}{48} \approx 10^{29}$ winning committees. Naturally, this is far too many to output. However, it is easy to observe that each of these committees consists of candidates a and b, and of 48 arbitrary candidates among $c_1, \ldots, c_{100}$.

Indeed, for simple rules such as SNTV, bloc, k-Borda, or multiwinner approval voting, we know that each winning committee consists of some highest-scoring candidates and an arbitrary subset of those candidates with the next highest score, so that the committee has the required size k.

The approach from the above example works for rules that associate scores with candidates and choose committees that consist of the highest-scoring ones. For rules that associate whole committees with scores (and choose those with the highest one), a natural way of capturing the complexity of winner determination is by using the following problem: Given an election E, committee size k, and score value s, is there a committee whose score is at least s? If this problem is NP-hard for a given voting rule—and it is possible to compute the score of a given committee in polynomial time—then it cannot be easy to compute any winning committee under this rule.[10]

Based on the above discussion, we can show that winner determination for the PAV rule is intractable.

Theorem 6.9 (Aziz et al. [54], Skowron, Faliszewski, and Lang [872]).
The problem of deciding whether for a given approval election $E = (C, V)$, there is a size-k committee with PAV-Score at least s is NP-complete.

---

[10] Indeed, if doing so were easy (i.e., doable in polynomial time), we could compute some winning committee and compare its score with s, thus solving an NP-hard problem in polynomial time. On the other hand, if our problem can be solved in polynomial time then we can compute the score of a winning committee in polynomial time (assuming that it can be written using polynomially many bits) but, in principle, we may not be able to compute a committee that achieves this score. Fortunately, natural voting rules do not seem to exhibit such strange behavior.

Proof. First, we see that the problem belongs to NP: Indeed, it suffices to guess a size-k committee and verify that its score is at least s.

Next, we will show that the problem is NP-hard by providing a reduction from the well-known NP-complete problem INDEPENDENT SET [483]. In this problem, we are given an undirected graph G with vertex set $V(G) = \{v_1, \ldots, v_n\}$ and edge set $E(G) = \{e_1, \ldots, e_m\}$, and an integer k. We ask whether there is a set of k vertices no two of which are connected by an edge (i.e., we ask whether there is an *independent set* of size k).

Given such an instance of INDEPENDENT SET, for each vertex v_i we write $d(v_i)$ to denote its degree (i.e., the number of edges that touch it) and we write $\delta = \max_{v_i \in V(G)} d(v_i)$ to denote the highest degree of the vertices in G. We form an election (C,V) with candidate set $C = V(G)$ and the following voters:

1. For each edge $e = \{v_i, v_j\} \in E(G)$, there is a voter who approves of exactly the candidates v_i and v_j.
2. For each vertex $v_i \in V(G)$, there are $\delta - d(v_i)$ voters who only approve of candidate v_i.

We claim that this election contains a size-k committee with *PAV-Score* at least $\delta \cdot k$ if and only if G has an independent set of size k.

First, let us assume that G has an independent set of size k and let $S = \{v_{i_1}, \ldots, v_{i_k}\}$ be one such set. If we consider S as a set of candidates in C, then we can easily verify that its score is $\delta \cdot k$. Indeed, each member of S is approved of by exactly δ voters and no two voters approve of the same two members of S (otherwise, S would not be independent).

For the other direction, let us assume that there is a committee S that contains candidates $v_{i_1}, \ldots, v_{i_k}$, such that *PAV-Score*$(S) \geq \delta \cdot k$. We claim that S corresponds to an independent set of G. Indeed, each candidate in C is approved of by exactly δ voters, so the score of S is at most $\delta \cdot k$, because each candidate receives at most one point from each voter that approves of him or her. If there were a voter that approved of two members of S (i.e., if in G there were an edge connecting the corresponding vertices) then this voter would contribute score $1 + \frac{1}{2}$ to the *PAV-Score* of S (recall that each voter approves of at most two candidates). We could say that such a voter would give one point to one of the approved candidates, but only half a point to the other one. In consequence, *PAV-Score*(S) would be less than $\delta \cdot k$.

As the reduction can obviously be computed in polynomial time, this completes the proof. ❏

The above proof works equally well for quite a large family of rules that are defined similarly to PAV but with different score functions. Specifically, consider an election (C,V), committee size k, and a sequence $w = (w_1, w_2, w_3, \ldots)$ of numbers. For a size-k commitee $W \subseteq C$, we define the *w-Score* of W to be:

6.5 Complexity of Computing Winning Committees

$$w\text{-}Score(W) = \sum_{v \in V} \sum_{i=1}^{\|A(v)\|} w_i.$$

A rule that always selects a size-k committee with the highest w-$Score$ is called a w-*Thiele rule*. In particular, multiwinner approval voting is a w_{mav}-Thiele rule for the sequence $w_{\text{mav}} = (1, 1, 1, \ldots)$, approval Chamberlin–Courant is a w_{aCC}-Thiele rule for the sequence $w_{\text{aCC}} = (1, 0, 0, \ldots)$, and PAV is a w_{PAV}-Thiele rule for the sequence $w_{\text{PAV}} = (1, 1/2, 1/3, \ldots)$. With a bit of care, the reader can verify that Theorem 6.9 works for all w-Thiele rules such that $w_1 > w_2 \geq w_3 \geq w_4 \geq \cdots$ (and, in fact, by adapting it a bit, it also works for w-Thiele rules where $w_1 \geq w_2 \geq w_3 \geq \cdots$ such that there is an i for which $w_i > w_{i+1}$).

In particular, the problem of deciding whether there is a committee of a given size that achieves at least a given score according to the approval CC rule is NP-complete. For the case of approval CC, there is also an in-depth analysis of other forms of winner determination problems. For example, deciding whether a particular committee is winning is coNP-complete, and deciding whether a particular candidate belongs to some winning committee is even $P_\|^{\text{NP}}$-complete (recall the definition of the complexity class $P_\|^{\text{NP}}$—known as "parallel access to NP"—from Section 1.5.3.2 starting on page 35).

Theorem 6.10. *The following statements about the approval Chamberlin–Courant rule hold:*

1. *The problem of deciding whether in a given election with a given committee size there is a committee with at least a given score is NP-complete [792, 141]*
2. *The problem of deciding whether a given committee of a given size is winning in a given election is coNP-complete [881].*
3. *The problem of deciding whether there is a committee of a given size in a given election that contains a given candidate is $P_\|^{\text{NP}}$-complete [229, 881].*

A further issue related to winner determination in multiwinner voting regards tie-breaking. While in the single-winner setting there are multiple natural strategies to employ (e.g., use lexicographic tie-breaking based on candidate names, select one candidate uniformly at random from the set of tied winners, select a random voter whose preference order breaks the tie, etc.), the issues is more involved in the multiwinner setting. In this chapter, we omit this problem and, instead, we focus on ways of computing at least one winning committee. For readers interested in a deeper understanding, we point out that the issue of detecting ties in (approval-based) multiwinner elections has been studied by Janeczko and Faliszewski [577].

$$\text{maximize} \quad \sum_{i=1}^{n} \sum_{\ell=1}^{k} \frac{1}{\ell} \cdot z_{i,\ell} \qquad (6.7)$$

$$\text{subject to:} \quad \sum_{c \in C} x_c = k \qquad (6.8)$$

$$z_i = \sum_{c \in A(i)} x_c \qquad \text{for } i \in [n] \qquad (6.9)$$

$$\sum_{\ell=1}^{k} z_{i,\ell} = z_i \qquad \text{for } i \in [n] \qquad (6.10)$$

$$x_c \in \{0,1\} \qquad \text{for } c \in C \qquad (6.11)$$

$$z_{i,\ell} \in \{0,1\} \qquad \text{for } i \in [n], \ell \in [k] \qquad (6.12)$$

Fig. 6.4 An integer linear program (ILP) for computing PAV on election $E = (C, V)$, where $C = \{c_1, \ldots, c_m\}$, $V = (v_1, \ldots, v_n)$, and where the committee size is k

6.5.2 Exact Agorithms

For moderately small elections, using efficient exponential-time winner determination algorithms can still be possible in practice. For example, one can express PAV as an integer linear program (ILP) and apply one of the classic ILP solvers, such as CPLEX or Gurobi. This approach is usually quite efficient and can work with hundreds of candidates and voters.[11]

Figure 6.4 gives an ILP formulation of PAV. In this program, we have two types of variables. For each candidate c, we have a binary variable x_c which indicates whether candidate c is part of the winning committee. Constraint (6.8) ensures that we select exactly k candidates. Next, for each voter v_i, we have a variable z_i that counts the number of selected candidates that the voter approves of. This is explicitly enforced by Constraint (6.9). Additionally, for each integer $\ell \in [k]$, we have a variable $z_{i,\ell}$ which is set to one only if v_i approves of at least ℓ members of the winning committee. This is ensured by Constraint (6.10), and to understand this, we need to look at the goal function. Let us fix some $i \in [n]$. In the goal function, we compute a weighted sum of variables $z_{i,1}, z_{i,2}, \ldots, z_{i,k}$, where the first variable has weight 1, the next one has weight $1/2$, the third one has weight $1/3$, and so on, until weight $1/k$. Further, by Constraint (6.10), we know that $z_{i,1} + z_{i,2} + \cdots + z_{i,k}$ is equal to z_i (which is an integer). The only way to maximize the value of our weighted sum under this condition is to set $z_{i,1} = \cdots = z_{i,z_i} = 1$ and $z_{i,z_i+1} = \cdots = z_{i,k} = 0$.

---

[11] If one is willing to spend a bit more time, dealing with even thousands of voters might be possible. The biggest downside of this approach is the necessity of using highly optimized commercial software. Indeed, at the time of writing this chapter, free ILP solvers were often too slow even for relatively small election instances.

6.5 Complexity of Computing Winning Committees

This way the variables $z_{i,\ell}$ obtain exactly the values as described in our interpretation, and the formula (6.7) corresponds exactly to the PAV score of the winning committee.

The above approach works very well in practice for elections that are not too large. However, if we know that the election has very few candidates or very few votes, we can use tools from the realm of parameterized complexity theory (for more background, we refer to the textbooks by Downey and Fellows [355], Niedermeier [753], Flum and Grohe [461], and Cygan et al. [320]). Specifically, we say that a (parameterized) problem is *fixed-parameter tractable* (is FPT) for a given parameter (such as the number of candidates or the number of voters) if there is an algorithm that on input of size n with the parameter value t, has running time of the form:

$$f(t) n^{\mathcal{O}(1)},$$

where f is some arbitrary (possibly superpolynomial) computable function. In other words, an FPT algorithm is required to run in polynomial time with respect to the input size, *except* that it is allowed to have arbitrary dependence on the parameter value. For example, an algorithm that runs in time that is exponential with respect to the number of candidates, but is polynomial with respect to the rest of the input, is said to be FPT with respect to the parameterization by the number of candidates.

As an immediate example, it suffices to consider a winner determination algorithm for PAV that tries all committees of size k and chooses one with the highest score. If there are m candidates and n voters, then the running time of such an algorithm is

$$\mathcal{O}\left(\binom{m}{k} nm\right) = \mathcal{O}(m^k nm).$$

Indeed, there are $\binom{m}{k}$ committees to try and computing the score of each of them takes $\mathcal{O}(nm)$ time. Since the committee size cannot be larger than the number of candidates, we can express this running time as $\mathcal{O}(m^m nm)$. This means that the problem of winner determination for PAV is FPT with respect to the parameterization by the number of candidates.

Theorem 6.11. *For approval CC, PAV, and every w-Thiele rule,[12] there is an FPT algorithm (with respect to the parameterization by the number of candidates) that outputs their winning committees.*

Naturally, the above algorithm is just a glorified form of a brute-force search, but some FPT algorithms are both more intricate and more useful. To see this, let us consider PAV and the parameterization by the number of voters. The algorithm uses two main observations:

---

[12] We assume here that the sequence w is such that we can compute committee scores in FPT time with respect to the number of candidates. This assumption is satisfied by all classical rules; indeed, the committee scores can typically be computed even in polynomial time.

$$\text{maximize} \quad \sum_{i=1}^{n}\sum_{\ell=1}^{k} \frac{1}{\ell} \cdot z_{i,\ell} \qquad (6.13)$$

$$\text{subject to:} \quad \sum_{U \subseteq [n]} x_U = k \qquad (6.14)$$

$$\sum_{U \subseteq [n]:\ i \in U} x_U = z_i \quad \text{for } i \in [n] \qquad (6.15)$$

$$\sum_{\ell=1}^{k} z_{i,\ell} = z_i \quad \text{for } i \in [n] \qquad (6.16)$$

$$0 \le x_U \le \|\mathcal{T}(U)\| \quad \text{for } U \subseteq [n] \qquad (6.17)$$

$$0 \le z_{i,\ell} \le 1 \quad \text{for } i \in [n],\ \ell \in [k] \qquad (6.18)$$

Fig. 6.5 A mixed integer linear program (MILP) used by the FPT algorithm computing PAV, parameterized by the number of voters

- First, we note that if an approval election contains n voters then each candidate belongs to one of 2^n groups, depending which voters approve of him or her. Candidates approved of by the same voters are "interchangeable."
- Second, we use the fact that there is an FPT algorithm for solving mixed ILPs (or, MILPs, for short), parameterized by the number of integer variables, invented by Lenstra [651]. A *mixed ILP* is an ILP where some variables are required to be integers but some others are free to assume rational values. (We should, however, mention that Lenstra's algorithm is not practical and actual MILP solvers use other approaches.)

Theorem 6.12 (Faliszewski et al. [441] and, more explicitly, Yang and Wang [966]). *There is an FPT algorithm (with respect to the parameterization by the number of voters) that, given an election E and committee size k, outputs one of the PAV winning committees of size k for E.*

Proof. Let $E = (C, V)$ be our input election, where $C = \{c_1, \ldots, c_m\}$ is a set of candidates and $V = (v_1, \ldots, v_n)$ is a collection of votes. We are also given the committee size k. For each subset U of $[n]$, by $\mathcal{T}(U)$ we mean the set of those candidates that are approved of exactly by the voters in the set $\{v_i \mid i \in U\}$.

Our algorithm proceeds by forming and solving the MILP presented in Figure 6.5. Note that this program has exponentially many variables and constraints, but this exponential function depends only on the number of voters. Hence, using Lenstra's algorithm [651], it can be solved in FPT time (with respect to the parameterization by the number of voters).

6.5 Complexity of Computing Winning Committees

Throughout the rest of the proof we explain how this MILP works and why it is correct. Let us first describe its variables. For each set $U \subseteq [n]$, we have an integer variable x_U which specifies how many candidates from $\mathcal{T}(U)$ are in the winning committee, and for each $i \in [n]$, we have an integer variable z_i that specifies how many candidates approved of by voter v_i are in the winning committee. To compute the PAV score of the winning committee, we use a third set of variables: For each $i \in [n]$ and $\ell \in [k]$, we have a *rational* variable $z_{i,\ell}$ that takes values between 0 and 1 and whose intended meaning is as follows: It has value 1 if v_i approves of at least ℓ members of the winning committee, and otherwise it has value 0 (formally, it can also take values between 0 and 1, but the idea of our MILP is to ensure that this does not happen for optimal solutions).

Next we analyze the MILP's constraints. Constraint (6.14) requires that the number of selected candidates—as described by variables x_U—is exactly k. (Note that in Constraint (6.17) we require that for each set U, we use at most as many candidates from $\mathcal{T}(U)$ as are available in the election. Consequently, the variables x_U describe some size-k committee; or, in fact, a whole family of size-k committees, since for each U, we are free to choose an arbitrary size-x_U subset of $\mathcal{T}(U)$ to be included in the committee.) Then Constraint (6.15) links the variables z_i with the x_U variables. More precisely, we express the number of selected candidates that a given voter v_i approves of by summing over all sets $U \subseteq [n]$ that include i.

Finally, in Constraint (6.16) we link the variables $z_{i,\ell}$ with the z_i variables the same way we did in the ILP from Figure (6.4). This way the goal function expresses the PAV score of the committee described by the variables x_U.

We conclude by noting that we can read off one of the winning committees from the values of the x_U variables (note here that we speak of committees in plural not only because the x_U variables possibly describe a whole set of committees, but also because there could be substantially different committees that also achieve the same PAV score and are winning). ❑

"This MILP is so much different from the one from the beginning of the section, I wonder if it is somewhat better too!" Anna is very excited by seeing two similar but quite different ways of solving the same problem.

"Oh no, I guess it is not," Belle worries. "I mean, look, it has exponentially many variables with respect to the number of voters ..."

Anna counters this by pointing out: "But if there were just ten voters ... or maybe even three, like us, and we had to proportionally distribute the contents of a box of sweets, then it might be faster!"

Chris adds: "Yes, that's right, but it does feel like a theoretical toy only ... Also, I wonder why they don't come up with an FPT algorithm parameterized by the committee size!"

Unfortunately, there is a good answer for Chris's question: The reason why we do not present an FPT algorithm parameterized by the committee size is that, most likely, such an algorithm does not exist. Indeed, there is a theory of parameterized hardness, analogous to the theory of NP-hardness [483], based on the so-called W-hierarchy that consists of the classes W[1], W[2], and so on. Specifically, the problem of deciding whether there is a PAV committee of a given size and with at least a given score is W[1]-hard with respect to the parameterization by the committee size (if one is familiar with parameterized complexity, this is already established by the proof of Theorem 6.9, as INDEPENDENT SET is W[1]-complete when parameterized by k). While we will not venture on the definition of the class W[1], the practical consequence is that an FPT algorithm for the problem is widely believed to not exist. We again point readers interested in more details of parameterized complexity to the textbooks by Downey and Fellows [355], Niedermeier [753], Flum and Grohe [461], and Cygan et al. [320].

6.5.3 Approximation Algorithms

"Hmm ... okay, computing winning PAV committees is hard. But how about using a polynomial-time algorithm that would be similar to how PAV works? Can we find such an algorithm?" Anna asks, and immediately starts explaining her idea of an algorithm.

There exists an intuitive and natural approximation algorithm for PAV. We start with an empty committee $W = \emptyset$ and in consecutive steps we add candidates to W. In each step, we add a candidate $c \notin W$ that increases the PAV score the most, that is, one that maximizes $PAV\text{-}Score(W \cup \{c\})$. This algorithm can be viewed as an independent stand-alone rule, and is often called *sequential PAV* (occasionally in the literature, it is also called *greedy PAV*).

The proof of the next result, Theorem 6.13, follows from the fact that the $PAV\text{-}Score$ function is submodular and monotonic. Let us recall that a (scoring) function f is *submodular* if for all committees $W', W \subseteq C$, we have

$$f(W' \cup W) + f(W' \cap W) \leq f(W') + f(W).$$

Analogously to Equation (3.2) on page 145 in Section 3.1.2.1 (which shows the same for supermodularity), one can show that this condition is equivalent to

$$f(W \cup \{c\}) - f(W) \leq f(W' \cup \{c\}) - f(W')$$

6.5 Complexity of Computing Winning Committees

for all committees W and W', $W' \subseteq W$, and $c \in C \setminus W$. Indeed, one can easily verify that adding a candidate to a larger committee increases the *PAV-Score* less than adding the same candidate to a smaller committee. Monotonicity means that $f(W') \leq f(W)$ for $W' \subseteq W$.

Theorem 6.13. *Sequential PAV is a 0.63-approximation algorithm for PAV.*

Proof. Let W_{OPT} be some committee with the maximal PAV score, and let W_i be the set of candidates selected in the first i steps of sequential PAV. In particular, $\|W_i\| = i$, and W_k is the committee returned by sequential PAV.

We will first show by induction that

$$PAV\text{-}Score(W_i) \geq \left(1 - \left(1 - \frac{1}{k}\right)^i\right) PAV\text{-}Score(W_{\text{OPT}}).$$

Let us rename the candidates so that $W_{\text{OPT}} \setminus W_i = \{c_1, c_2, \ldots, c_{k-i}\}$.[13] We obtain the following (see below for explanations):

$$PAV\text{-}Score(W_{\text{OPT}}) - PAV\text{-}Score(W_i)$$
$$= \sum_{j=1}^{k-i} \left(PAV\text{-}Score(W_i \cup \{c_1, \ldots c_j\}) - PAV\text{-}Score(W_i \cup \{c_1, \ldots c_{j-1}\})\right)$$
$$\leq \sum_{j=1}^{k-i} \left(PAV\text{-}Score(W_i \cup \{c_j\}) - PAV\text{-}Score(W_i)\right)$$
$$\leq (k-i)\left(PAV\text{-}Score(W_{i+1}) - PAV\text{-}Score(W_i)\right).$$

The first equality is simply a telescoping sum, the next inequality follows by submodularity, and the last inequality follows from the fact that sequential PAV selects the candidate that increases the *PAV-Score* of the committee the most. Clearly, a weaker version of this inequality also holds:

$$PAV\text{-}Score(W_{\text{OPT}}) - PAV\text{-}Score(W_i)$$
$$\leq k\left(PAV\text{-}Score(W_{i+1}) - PAV\text{-}Score(W_i)\right).$$

Consequently, using the inductive assumption, we have that:

---

[13] For the sake of simplicity, we here assume that W_{OPT} and W_k are disjoint, but of course, if they shared some members, this would not hurt the analysis.

$$PAV\text{-}Score(W_{i+1}) \geq \frac{PAV\text{-}Score(W_{\text{OPT}})}{k} + PAV\text{-}Score(W_i)\left(1 - \frac{1}{k}\right)$$

$$\geq \frac{PAV\text{-}Score(W_{\text{OPT}})}{k} + \left(1 - \left(1 - \frac{1}{k}\right)^i\right) PAV\text{-}Score(W_{\text{OPT}}) \left(1 - \frac{1}{k}\right)$$

$$= PAV\text{-}Score(W_{\text{OPT}}) \left(\frac{1}{k} + 1 - \frac{1}{k} - \left(1 - \frac{1}{k}\right)^{i+1}\right)$$

$$= PAV\text{-}Score(W_{\text{OPT}}) \left(1 - \left(1 - \frac{1}{k}\right)^{i+1}\right).$$

This proves the inductive step, and thus the inductive hypothesis. In particular, we get that:

$$PAV\text{-}Score(W_k) \geq \left(1 - \left(1 - \frac{1}{k}\right)^k\right) PAV\text{-}Score(W_{\text{OPT}})$$

$$\geq \left(1 - \frac{1}{e}\right) PAV\text{-}Score(W_{\text{OPT}}) \geq 0.63 \cdot PAV\text{-}Score(W_{\text{OPT}}).$$

This completes the proof. ☐

Belle and Chris are listening to Anna's explanations with a growing interest. Suddenly, Belle exclaims with fascination: "The proof does not use anything specific to PAV! Analogously, we can define sequential CC and show that it is a 0.63-approximation algorithm for the Chamberlin–Courant rule!"

Indeed, a similar algorithm can be defined for other Thiele methods. Theorem 6.13 is, in fact, a special case of a more general result on approximating submodular set functions [733].

More complex approximation algorithms for PAV have been proposed by Byrka, Skowron, and Sornat [247] and Dudycz et al. [361]. In particular, an algorithm based on rounding a fractional solution returned by a linear program for PAV gives approximation guarantee of 0.7965 and, unless P = NP, this is the best possible approximation ratio that is attainable in polynomial time [361].

A natural question that arises is whether approximation algorithms, such as sequential PAV, share the good properties of the rules that they aim to approximate. In case of sequential PAV, it has a relatively good proportionality degree [871] and satisfies committee monotonicity. On the other hand, it fails the axiom of justified representation and, thus, might mistreat small cohesive groups of voters. For this reason, the Phragmén sequential rule and the method of equal shares might seem slightly better alternatives to be used for large elections. For the time being, there are no works on axiomatic properties of more involved approximation algorithms, even though such an analysis would be of great value.

6.5.4 Other Approaches

There is also a number of other approaches to computing the outcomes of multiwinner voting rules. For example, instead of using approximation algorithms with provable guarantees, we might use other heuristics, such as local search, simulated annealing, or approaches based on clustering. For example, Faliszewski et al. [436] show that both simulated annealing and variants of the sequential rules from the preceding section perform very well for a number of committee scoring rules. Interestingly, they have shown a variant of sequential CC (for the ordinal, Borda-based variant) that was later proven by Munagala, Shen, and Wang [719] to achieve optimal approximation ratio for this rule. An interesting feature of this algorithm is that it views committee selection as a cooperative game (in the sense of Chapter 3 starting on page 139) where candidates are players, and greedy selection is based on their Banzhaf power indices (see Section 3.3.2 starting on page 168).

A different approach, already discussed in Chapter 4 starting on page 233 for the case of single-winner rules, is to consider structured domains of preferences. For example, if we assume that all the voters have single-peaked preferences (recall Chapter 5 starting on page 369) then we can compute a winning committee according to the Chamberlin–Courant rule in polynomial time. For the ordinal setting, this was shown by Betzler, Slinko, and Uhlmann [141] and for the approval setting, it follows, e.g., from the work of Peters and Lackner [776]. Indeed, Peters and Lackner provide a very general approach that applies also to PAV and to harmonic Borda. Briefly put, their idea is to express the winner determination problem as solving an ILP instance. They argue that even though solving ILPs in general is an NP-hard task, if the preferences of the voters are single-peaked then it is possible to express the ILP in such a way that it is totally unimodular, which implies the ability to solve it in polynomial time.

All in all, even though many multiwinner voting rules are NP-hard to compute, in many settings we can deal with this issue effectively and high computational complexity does not preclude the use of these rules.

6.6 Further Topics Related to Committee Elections

We conclude this chapter by giving a brief outlook on important other topics related to multiwinner voting that we decided not to cover in more depth.

Note that we are now moving back to also allow the ordinal setting—that is, we no longer focus on the approval setting alone.

6.6.1 Robustness

We have seen that the actual choice of a committee election rule may have an immense impact on the winning committees. Therefore, it is very important to identify the aim of the election before choosing a rule. For some of these rules, the winning committees may be determined easily; for others, the corresponding decision problems are NP-hard. Another very important question besides winner determination, however, is how *robust* the outcome of a rule in fact is.

> Belle starts rethinking her own preferences and she realizes that maybe she made a mistake in her ranking: "I just read a review of one of the books we have proposed. My initial ranking wasn't good; instead, I should have swapped this book forward by one position."
>
> Anna replies: "This is such a small change, it shouldn't have an effect on the outcome."
>
> Chris isn't convinced, though: "Are you sure? There may be situations where even a small change in the preferences may lead to different outcomes. And perhaps this might also depend on which voting rule was used?"

The investigation of robustness in committee elections, initiated by Bredereck et al. [229], studies exactly questions like this: What is the effect on the winning committees if two adjacent candidates are swapped? In particular, a committee election has a *robustness level of* ℓ if the swap of two adjacent candidates may lead to replacing at most ℓ candidates in the winning committee and there is an actual example of an election where ℓ candidates indeed are replaced. The following formal definition also takes ties into account.

Definition 6.14 (Bredereck et al. [229]). Define the *robustness level of a committee election rule* as the smallest number ℓ such that for each pair of elections, E and E', where the only difference between E and E' is a single swap of two adjacent candidates in one vote, and for each winning committee W for election E, there exists a winning committee W' for election E' such that $\|W \cap W'\| \geq k - \ell$.

A very interesting observation is that for some of the multiwinner voting rules considered here, the robustness level is one, which means that at most one candidate in the winning committee can be changed through a single swap of adjacent candidates, whereas the robustness level of some other rules

6.6 Further Topics Related to Committee Elections

equals the full size of the winning committee: Every single member of it may have to be replaced. We first show that the robustness level is one for a large group of committee scoring rules.

Proposition 6.2 (Brederick et al. [229]). *Let f be a given committee scoring rule that assigns points to candidates and returns a committee consisting of k candidates with the highest score. If a swap of two adjacent candidates in one vote only influences the scores of at most two candidates, then f has a robustness level of 1.*

Proof. Consider a size-k committee election $E = (C,V)$ with $C = \{c_1,\ldots,c_m\}$ and a winning committee W under committee scoring rule f. Let $s(c)$ denote the score of candidate $c \in C$ in election E using rule f. Rename the candidates such that it holds that

$$s(c_1) \geq s(c_2) \geq \cdots \geq s(c_m),$$

where the first k candidates (possibly using some tie-breaking rule) form the winning committee W. Let $E' = (C,V')$ be the election that is obtained from E by making one single swap of two adjacent candidates in one vote of V. This swap can affect only the scores of at most two candidates, possibly increasing one candidate's score, say that of c_i, and possibly decreasing the other candidate's score, say that of c_j. We distinguish the following two cases, each with two subcases:

Case 1: If $i \leq k$, the score of a candidate in W may have increased.

  Case 1(a): If $j > k$, the score of a candidate not in W may have decreased, so W is still a winning committee for E'.
  Case 1(b): If $j \leq k$, the score of a candidate in W may have decreased. Then
  1. either W nonetheless is still a winning committee for E',
  2. or there is a winning committee $W' = \{c_1,\ldots,c_{j-1},c_{j+1},\ldots,c_{k+1}\}$ for E', resulting from W by replacing c_j with c_{k+1}.

Case 2: If $i > k$, the score of a candidate not in W may have increased.

  Case 2(a): If $j > k$, the score of a candidate not in W may have decreased. Then
  1. either W nonetheless is still a winning committee for E',
  2. or there is a winning committee $W' = \{c_1,\ldots,c_{k-1},c_i\}$ for E', resulting from W by replacing c_k with c_i.
  Case 2(b): If $j \leq k$, the score of a candidate in W may have decreased. Then
  1. either W nonetheless is still a winning committee for E',
  2. or there is a winning committee $W' = \{c_1,\ldots,c_{k-1},c_i\}$ for E', resulting from W by replacing c_k with c_i,

3. or there is a winning committee $W'' = \{c_1, \ldots, c_{j-1}, c_{j+1}, \ldots, c_{k+1}\}$ for E', resulting from W by replacing c_j with c_{k+1},
4. or there is a winning committee $W''' = (W \setminus \{c_j\}) \cup \{c_i\}$ for E', resulting from W by replacing c_j with c_i.

Since at most one candidate in the winning committee has changed in all cases, the robustness level of f is equal to one. ❑

The above proposition applies to some of the rules presented here.

Corollary 6.1 (Bredereck et al. [229]). *The robustness level for SNTV, bloc, and k-Borda is one.*

In contrast, there are other rules that are not at all robust, since a single swap may lead to a *completely* different committee.

Proposition 6.3 (Bredereck et al. [229]). *The robustness level for STV and CC equals k, the committee size.*

In addition to the robustness level, also the *robustness radius of a committee election rule* can be considered. This refers to the smallest number of swaps of adjacent candidates needed so as to change a winning committee. Interestingly, the corresponding decision problem is easy for SNTV, bloc, and k-Borda, while it is hard for STV and CC. The corresponding robustness problems for approval ballots have been studied by Faliszewski, Gawron, and Kusek [420, 494].

6.6.2 Manipulation, Control, and Bribery

Finally, we give some literature pointers to three topics on strategic behavior in multiwinner voting—manipulation, control, and bribery—that have been studied in depth for single-winner voting in Sections 4.3.3, 4.3.4, and 4.3.5 starting on page 306 and, in particular, for single-peaked societies in Sections 5.2, 5.3, and 5.4 starting on page 376; and that will be studied for judgment aggregation in Sections 7.3.3, 7.3.4, and 7.3.5 starting on page 489. Note further that, starting on page 189 in Chapter 3 on cooperative game theory, control by deleting or adding players in weighted voting games has also been discussed.

Meir et al. [700] were the first to study strategic behavior in committee elections by establishing complexity results on manipulation and control for various multiwinner voting rules, including SNTV, bloc, and approval voting. They assume the manipulator or election chair to have a utility function over the candidates and to aim at achieving a set of winners with total utility above some threshold. For all considered utility functions, they establish P

6.6 Further Topics Related to Committee Elections

results for control by either adding or deleting voters in SNTV, yet NP-completeness in bloc and approval elections; for control by either adding or deleting candidates, they establish NP-completeness in SNTV, yet P results in bloc and approval elections;[14] and they obtain P results on manipulation for each of these three multiwinner voting rules.

Bredereck, Kaczmarczyk, and Niedermeier [236] focus on the goal of shortlisting the candidates using bloc voting with group evaluation functions (e.g., egalitarian versus utilitarian) and tie-breaking mechanisms that model either pessimistic or optimistic manipulators. In particular, they show that, regardless of the tie-breaking rule used, strategic voting can be computationally intractable in an egalitarian setting. Further, they establish a pretty comprehensive picture of computational complexity results in this scenario.

Mehrizi and D'Angelo [698] consider a scenario where voters can be influenced in a social network so as to change their mind on either supporting or opposing a political party. They study this problem for various graph structures of the social network and for various diffusion models, with the goal of maximizing or minimizing the number of winners in a target party, and they establish results showing that the problem can be hard to approximate. On the other hand, they provide a dynamic programming approach to design efficient algorithms for certain restricted cases, e.g., when the voting system is a variation of straight-party voting or when voters form a tree in the social network.

Botan [173] studies strategyproofness versus manipulability based on party-list profiles. Distinguishing three types of manipulation—namely, "subset-manipulation," "superset-manipulation," and "disjoint-set-manipulation"—she focuses on the class of irresolute Thiele rules. For these three types of manipulation, she establishes strategyproofness of Thiele rules on party-list profiles for various well-known preference extensions—functions that extend agents' preferences over committees to preferences over sets of committees, such as those due to Gärdenfors [481], Kelly [601], and Fishburn [453]. She further shows that Thiele methods are not manipulable in any way on party-list profiles for the "optimistic" preference extension, i.e., Thiele rules are fully strategyproof on these profiles for optimistic agents.

Turning now to bribery, note that bribery problems are also particularly helpful in the multiwinner setting because they allow one to evaluate the individual performance of candidates. Indeed, many rules either only assign scores to whole committees (as in the cases of, e.g., CC, HB, or PAV) or do not assign scores at all (as in the cases of, e.g., STV, Phragmén, or the method of equal shares). Then the amount of bribery we need to perform to ensure a given candidate's victory is a measure of how close this candidate was to winning. Such an information is useful, as many candidates would like to know how they did. Alternatively, the amount of bribery needed to

---

[14] Some of these results on control are attributed to the work of Bartholdi, Tovey, and Trick [102] and Hemaspaandra, Hemaspaandra, and Rothe [541], respectively, as they hold even in single-winner elections (recall Section 4.3.4 starting on page 328).

preclude a candidate from winning is a measure of its margin of victory. This interpretation of bribery also applies to the single-winner setting, but in the multiwinner setting it is particularly important (indeed, among single-winner rules it is common to have scores of individual candidates that can be used to evaluate their performance, even if bribery-based measures might be more appealing).

Bredereck et al. [232] studied shift bribery (which was mentioned for single-winner voting in Section 4.3.5.5 starting on page 363) for multiwinner voting rules in terms of their (classical and parameterized) computational complexity. Recall that in the shift bribery problem, we are given an election, a preferred candidate p, and a budget, and we ask whether—without exceeding the budget—we can make p win the election by shifting p upwards in some of the votes, where depending on the vote the extent of shifting exacts a price. Focusing on the SNTV, bloc, k-Borda, and Chamberlin–Courant rules (and also on approximate variants of the Chamberlin–Courant rule, since the original rule is NP-hard to compute), they show that the shift bribery problem tends to be harder for multiwinner than for single-winner voting rules.

Faliszewski, Skowron, and Talmon [443] consider various approval-based multiwinner voting rules, including SAV, PAV, approval CC, and others, and study bribery for them where the bribery actions are limited to adding approvals to, deleting them from, or moving them within votes. They obtain results ranging from polynomial-time algorithms to NP-hardness with constant-factor approximations, and even to outright inapproximability. In addition, they study the parameterized complexity of these problems, parameterized by the numbers of voters or candidates.

Kusek et al. [632] study the complexity of bribery for *structured* multiwinner approval elections, i.e., elections are assumed to either have the *candidate interval* property (which corresponds to *single-peakedness*; recall Chapter 5 starting on page 369) or the *voter interval* property (which corresponds to the related property of *single-crossingness* [709, 812]) before and after the bribery action. Given such an election, the question is whether a given candidate can join a winning committee by adding approvals to, deleting them from, or swapping them within some of the votes, where each such bribery action exacts a price and a given budget must not be exceeded. While manipulative attacks usually are easier in structured elections than in general elections (again, see Chapter 5 for examples), also examples showing the opposite behavior are given here.

Finally, Yang [965] studies the *destructive* counterparts of some of the results mentioned earlier that focused on *constructive* bribery where the briber aims at making a preferred candidate win. In destructive bribery, however, the goal of the briber is to prevent a despised candidate from winning. Yang [965] focuses on five approval-based multiwinner voting rules (including SAV, PAV, approval CC, and two more) and on five specific bribery actions (including adding approvals to and deleting them from votes) and he obtains (classical

6.6 Further Topics Related to Committee Elections

and parameterized) complexity results for each of the resulting 25 combinations of voting rule and bribery action.

Chapter 7
Judgment Aggregation

Dorothea Baumeister · Gábor Erdélyi · Ronald de Haan · Jörg Rothe

In Chapters 4 – 6, we were concerned with making collective decisions by voting, i.e., with methods for how to aggregate the voters' individual preferences so as to determine the winning alternative(s) as a collective consensus. In the present chapter, we turn to the closely related topic of judgment aggregation, i.e., to methods for how to aggregate the individual judgments of a number of judges so as to determine the joint judgment(s) as a collective consensus.

However, there is a crucial difference between the preferences that voters cast as their ballots in an election and the judgments that are reported by the judges in a judgment aggregation procedure. A preference list reflects the *personal, subjective preferences* of a voter. For example, Anna likes playing chess more than playing poker (see Figure 1.2 on page 6). This is a personal inclination that Belle can (or even should) accept, even if she herself dissents from Anna's preference and likes poker better than chess. This is not the case for judgments (even though they are personal and subjective as well), since if, say, Anna and Belle report different—or even contrary—judgments, they both may think that the other one is wrong and that, at least in certain cases, one can even objectively decide who of the two is right and who is wrong. For example, if Anna opines that chess is a board game, while Belle holds the view that chess is a deadly dull card game, it is clear who is right and who is wrong. That is, we can assign truth values to the propositions or opinions reported by the judges. Note, however, that this is only one possible view at judgment aggregation; there are other perspectives on this as well.

While in preference aggregation various candidates are up for election and a voting system is used to choose the winner(s) among them, in judgment aggregation a number of propositions are under consideration, each of which can be evaluated to be either true or false, and they are often logically connected. This gives rise to an issue that doesn't matter at all in voting: The aggregated judgment is required to be *logically consistent*.

For a situation where a judgment—in the literal sense of the word—is to be pronounced, think of a court proceedings. Suppose there are three judges who have to adjudge a defendant either guilty or innocent. There is complete agreement among them that, by law, the defendant is guilty if he has committed a crime. They may have different opinions, though, as to whether the defendant indeed committed this offense and whether this offense indeed is a crime.

Several shots have been heard at night on the premises of Mr. Slugger's miniature golf facility. No sign of an attacker has been found, nor of the weapon involved in a possible attack, and there are neither dead bodies nor eyewitnesses! The police was merely able to seize some of the fired 9 mm caliber projectiles. By an unfortunate series of events, whose detailed description would take us too far afield, Chris has attracted the attention of the investigators and is now standing trial.

"Please raise from your seats!" David, the court usher, loudly exclaims. "The Members of the Jury are entering the courtroom!"

Anna, Belle, and Edgar step in, take their seats, and everyone else, including Chris, sits back down. After some hours of tough proceedings, judges Anna, Belle, and Edgar have formed their individual judgments about the following central questions:

1. "Did the defendant commit the offense?"
2. "Is the offense a crime?"
3. "Is the defendant guilty?"

"Without any doubt, one has to answer the first two questions in the affirmative," Anna starts. "Because Chris is having a *motive*. It is no secret that he isn't fond of miniature golf. Therefore, I believe he has committed this offense. And, of course, this offense is a crime: To go on the rampage using firearms is not a peccadillo in this country! I conclude from these assessments that the defendant is guilty."

"I beg to differ," disagrees Belle. "I certainly do agree that Chris is having a motive to damnify Mr. Slugger's property and therefore is highly suspicious to have committed this offense. However, I strongly doubt that this offense—or shall we call it, this *incident*?—can indeed be seen to fulfill the requirements of a crime according to law. After all, as far as we know, all that has happened is that another hole has been shot into the miniature golf facility. The truth of the matter is that this actually makes the golf course only the more attractive, as it is now possible to putt more easily. I conclude that the defendant is innocent and doesn't belong behind bars."

"I absolutely do agree with this conclusion," Edgar now catches the speaker's eye. "My reasons to come to this conclusion, though, are different. First, I do not agree that firing at a miniature golf facility may be trivialized. That is a criminal damage and must be prosecuted! Second, I do not believe that it was Chris who committed the crime. The

7 Judgment Aggregation

'motive' that he is not fond of miniature golf doesn't convince me on two accounts:

(a) After all, he prefers miniature golf to swimming (even though he likes bicycling even more), as you can see by looking at, *piece of evidence number one*: Figure 4.1 on page 233, and
(b) Anna, too, prefers riding the bicycle to playing miniature golf (again, have a look at the piece of evidence number one)—according to this, she would be just as suspicious as Chris!

Therefore, I argue the case for *not guilty*!

And even if this does not belong into *this* trial where Chris has been accused and stands in court: Has anyone actually checked if Mr. Slugger perhaps has an unusually high insurance coverage for criminal damages done to his miniature golf facility?"

Fig. 7.1 Three individual judgments

Every judge gives a logically consistent assessment of each of the three central questions; their individual judgments are depicted in Figure 7.1. A common decision is now to be reached by majority. Two of the three judges think that the defendant did commit the offense, which results in a "yes" by majority. The same applies to the question of whether the offense indeed is a crime: Two "yes" and only one "no" give a "yes" again. The conclusion, however, reveals a paradoxical situation that is shown in Figure 7.2 (and also in Table 7.1): Since two of the three judges think that the defendant is not guilty, the last question is answered "no." That is, from *"Yes, the defendant committed the offense!"* and *"Yes, the offense is a crime!"* they conclude, *"No, the defendant is not guilty!"* In other words, the outcome of the majority decision is no longer logically consistent, even though each individual judgment was logically consistent in the first place.

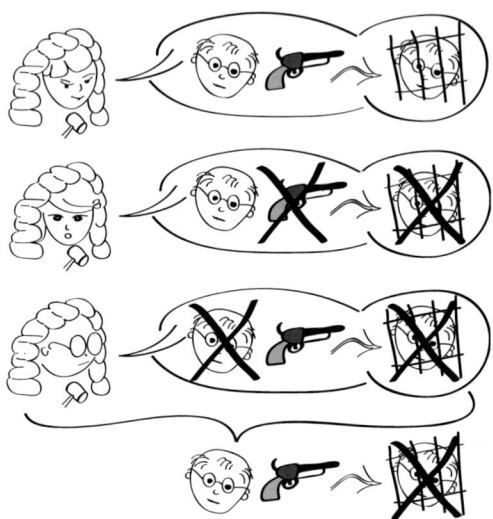

Fig. 7.2 Doctrinal paradox

Kornhauser and Sager [620] were the first to describe this situation, which they called the *"doctrinal paradox."* Later on, Pettit [778] introduced a more general variant as the *"discursive dilemma,"* which often goes under the name of *"doctrinal paradox"* as well (see the work of Mongin [711] for a detailed discussion of how they actually differ). This paradox is akin to the Condorcet paradox (see Section 1.2 and, in particular, Figure 1.3 on page 8), which says that preference aggregation by the majority rule based on pairwise comparisons may result in an intransitive joint preference order, even though all individual preferences are transitive.

The goal of judgment aggregation is to find a collective judgment of a number of propositional formulas that are either true or false. Just as preference aggregation by voting, processes of judgment aggregation play a central role in collective decision-making within democratic societies. For example, these

Table 7.1 Doctrinal paradox

Judge	Did Chris commit the offense?	Is the offense a crime?	Is Chris guilty?
Anna	yes	yes	yes
Belle	yes	no	no
Edgar	no	yes	no
Majority	yes	yes	no

processes can be used to merge dissenting opinions of individual experts on important questions from various areas. That is why judgment aggregation has been investigated in many disciplines, as diverse as law, economics, and the political sciences, and more recently also in computer science and mathematics, in particular in relation to logic and computational social choice.

The foundations of judgment aggregation will be explained in Section 7.1. A number of important judgment aggregation procedures and their properties, as well as properties of the agenda (i.e., the set of propositional formulas that are to be judged), will be introduced in Section 7.2. As in voting (recall Section 4.2 starting on page 264), we will again encounter some impossibility results, stating that certain properties cannot be satisfied simultaneously, and we will also see some "possibility" results that provide characterizations of certain properties being simultaneously satisfied. Finally, we turn to the algorithmic and complexity-theoretic aspects of judgment aggregation in Section 7.3, again focusing on winner determination, manipulation, control, and bribery as in voting (recall Section 4.3 starting on page 287) but also on an issue called "safety of the agenda" that is specific to judgment aggregation.

7.1 Foundations

List and Pettit [663] introduced the formal foundations of judgment aggregation. Here, we restrict ourselves to judgment aggregation for propositional logic and use the standard connectives (i.e., boolean operations) described in Section 1.5.2.1 starting on page 24. It is also possible, though, to define judgment aggregation for more general logics (see, e.g., the work of Dietrich [343]).

We will begin in Section 7.1.1 by introducing the framework of judgment aggregation as originally defined, where the issues that are to be judged are modeled as propositional logic formulas. This is also the framework that we will use to discuss judgment aggregation in the remainder of this chapter. Then, in Section 7.1.2, we will briefly discuss a different, more recently introduced framework of judgment aggregation, where the issues are modeled as propositional variables, and there is an additional propositional logic formula that expresses the logical relation between these issues. Moreover, we will explain how these two different models relate to each other.

7.1.1 Judgment Aggregation

The set of *judges* (sometimes called *referees*) is denoted by $N = \{1, 2, \ldots, n\}$, i.e., each judge is assigned a number between 1 and n (occasionally, as in the example above, they will have names instead of numbers). Further, let P be a set of atomic propositions (i.e., boolean variables) and let $\mathcal{L}_P$ be the set of all boolean formulas in propositional logic that can be formed from the atomic propositions in P by the common logical connectives (again, see Section 1.5.2.1 starting on page 24). In more detail, $\mathcal{L}_P$ contains all atomic propositions from P, the boolean constants 1 and 0 (representing *true* and *false*), and for all $\varphi, \psi \in \mathcal{L}_P$ the expressions $\neg\varphi$, $(\varphi \wedge \psi)$, $(\varphi \vee \psi)$, $(\varphi \Longrightarrow \psi)$, and $(\varphi \Longleftrightarrow \psi)$ are contained in $\mathcal{L}_P$ as well. Recall that $\neg$ refers to *negation*, $\wedge$ to *conjunction* (the boolean connective *and*), $\vee$ to *disjunction* (the boolean connective *or*), $\Longrightarrow$ means *implication*, and $\Longleftrightarrow$ *equivalence*. We also assume that $\mathcal{L}_P$ is *closed under negation*, i.e., $\neg\varphi$ is in $\mathcal{L}_P$ for each $\varphi \in \mathcal{L}_P$. To avoid double negation, we let $\overline{\alpha}$ denote the *complement of a boolean formula* α, where $\overline{\alpha} = \neg\alpha$ if α is not negated; otherwise, $\overline{\alpha} = \beta$ for $\alpha = \neg\beta$.

The set of propositional formulas to be judged is said to be the *agenda* and is denoted by Φ, and we require that $\Phi \subseteq \mathcal{L}_P$. The agenda is assumed to be finite and nonempty, to not contain doubly negated formulas, and to be *closed under complement*, i.e., for each $\varphi \in \Phi$, we have $\overline{\varphi} \in \Phi$. Further, it is ruled out that the agenda contains *tautologies* (formulas that evaluate to *true* for each truth assignment, such as $(p \vee \neg p)$) or *contradictions* (formulas that evaluate to *false* for each truth assignment, such as $(p \wedge \neg p)$). By the way, the formula $(p \vee \neg p)$ just mentioned represents the *principle of the excluded third* and the formula $(p \wedge \neg p)$ represents the *principle of excluded contradiction*.

Every subset of the agenda is said to be a *judgment set*. In particular, each judge's *individual judgment set* consists of those propositional formulas in the agenda he approves of because he personally considers them to be true. Of course, no one else except he himself can tell whether his reported individual judgment set gives his truthful opinion; later on we will also consider manipulative judges who may report untruthful judgment sets.

By applying a *judgment aggregation procedure* (or *JA procedure*, for short) the individual judgment sets are aggregated so as to obtain a *collective judgment set*, which contains those elements of the agenda the group of judges collectively approves of according to this procedure.

It is common to require the individual judgment sets to be complete, complement-free, and consistent.

Definition 7.1. Let Φ be an agenda. A judgment set $J \subseteq \Phi$ is said to be

1. *complete* if for each $\varphi \in \Phi$, J contains φ or $\overline{\varphi}$,
2. *consistent* if there exists a truth assignment that makes all formulas in J true, and
3. *complement-free* if for each $\varphi \in \Phi$, J contains at most one of φ and $\overline{\varphi}$.

For each formula in Φ, a complete and complement-free judgment set contains either the formula itself or its complement. A complete and consistent judgment set is said to be *rational*; obviously, every rational judgment set is complement-free. Let $\mathcal{J}(\Phi)$ denote the set of all rational subsets of Φ. As with preference profiles in voting (see Section 4.1 starting on page 234), we write the n given individual judgment sets as a profile $\mathbf{J} = (J_1, \ldots, J_n)$, where $J_i \in \mathcal{J}(\Phi)$ represents the ith judgment set.

7.1 Foundations

A formal description of the example illustrating the doctrinal paradox (see Figure 7.2 and Table 7.1 on page 470) looks as follows: $N = \{1,2,3\}$ is the set of the three judges and $\Phi = \{p, \neg p, q, \neg q, (p \wedge q), \neg(p \wedge q)\}$ is the agenda whose formulas have the following meaning:

$$p \mathrel{\hat{=}} \text{``The defendant committed the offense.''}$$
$$q \mathrel{\hat{=}} \text{``The offense is a crime.''}$$
$$(p \wedge q) \mathrel{\hat{=}} \text{``The defendant committed a crime.''}$$

Note that $(p \wedge q)$ means that the defendant is guilty. The meaning of the negated propositions, $\neg p$, $\neg q$, and $\neg(p \wedge q)$, is clear. The profile of the three individual judgment sets is $\mathbf{J} = (J_1, J_2, J_3)$ with

$$J_1 = \{p,\ q,\ (p\wedge q)\}, \quad J_2 = \{p,\ \neg q,\ \neg(p\wedge q)\}, \quad \text{and} \quad J_3 = \{\neg p,\ q,\ \neg(p\wedge q)\}.$$

These three individual judgment sets are each rational.

To complete the formal description, we need a JA procedure

$$\mathcal{F}: \mathcal{J}(\Phi)^n \to 2^\Phi,$$

represented by a function mapping every profile of n individual judgment sets to a subset of the agenda Φ, the collective judgment set. (Recall that 2^Φ denotes the *power set of* Φ, i.e., the set of all subsets of Φ.)

In our example of the doctrinal paradox, the majority rule is used (the corresponding function $\mathcal{F}$ will be formally defined in Section 7.2.1.1 on page 478), which maps the above individual judgment sets to the collective judgment set $\mathcal{F}(J_1, J_2, J_3) = \{p,\ q,\ \neg(p \wedge q)\}$. Just as the given individual judgment sets, this collective judgment set is complete and complement-free. However, unlike those, it is not consistent, as there is no truth assignment that satisfies all its formulas at the same time. It is this loss of consistency caused by the majority rule what constitutes the doctrinal paradox.

A JA procedure $\mathcal{F}$ is said to be, respectively, *complete, complement-free,* and *consistent* if the collective judgment set $\mathcal{F}(\mathbf{J})$ possesses the corresponding property for each profile $\mathbf{J} \in \mathcal{J}(\Phi)^n$. If a JA procedure is complete and consistent, it is also said to be *rational*. We will restrict ourselves to JA procedures having a so-called *universal domain*, i.e., their domain is all of $\mathcal{J}(\Phi)^n$, the set of all profiles of n consistent and complete judgment sets with respect to the given agenda Φ.

As we have seen, the doctrinal paradox occurs already in very simple examples. How can this problem be avoided? So far, we have considered the majority rule only, and we have seen that it can produce inconsistent collective judgment sets from consistent individual judgment sets. Is this true only for the majority rule? Does there perhaps exist a different judgment aggregation procedure that is rational? And what influence does the agenda have on collective judgment sets being possibly inconsistent? In Sections 7.2

and 7.3, we will look at these and other questions and will try to find answers. Before doing so, we will make a short digression and discuss some other formal frameworks that have been used to model the setting of judgment aggregation.

7.1.2 Other Frameworks for Judgment Aggregation

In addition to the framework of judgment aggregation that we just introduced, several other frameworks have been considered in the literature. In the remainder of this section, we will give a brief overview of several of these frameworks, and how they relate to each other. After having done so, we will return to the framework introduced in Section 7.1.1 and use that for the remainder of the chapter.

The framework of judgment aggregation that we saw in Section 7.1.1 is based on the idea that each issue that is to be judged is modeled as a propositional logic formula, and that the logical relations between issues arise as the logical relations between the different formulas representing the issues. The frameworks that we will discuss now are based on the following, slightly different idea: The issues are represented by propositional variables, and the logical relations between the issues are captured using a propositional logic formula called the *integrity constraint*.

This approach, introduced by Grandi [515] and Endriss and Grandi [516], is also known under the name *binary aggregation with integrity constraints*.

7.1.2.1 Binary Aggregation with Integrity Constraints

In the framework of binary aggregation with integrity constraints, the agenda Φ consists of a set of atomic propositions (i.e., boolean variables) and their negations. In addition, there is an *integrity constraint* $\Gamma \in \mathcal{L}_P$ in the form of a propositional logic formula over the atomic propositions in Φ. The notions of *completeness* and *complement-freeness* of judgment sets $J \subseteq \Phi$ are defined exactly as in the formula-based framework. The notion of consistency is slightly different: A judgment set $J \subseteq \Phi$ is *consistent* if the set $J \cup \{\Gamma\}$ is logically consistent, i.e., if there exists a truth assignment that makes all formulas in $J \cup \{\Gamma\}$ true. *Rational* judgment sets are—as in the formula-based framework—judgment sets that are complete and consistent. Let $\mathcal{J}(\Phi,\Gamma)$ denote the set of all rational subsets of Φ, where Γ denotes the integrity constraint. A profile is a sequence $\mathbf{J} = (J_n,\ldots,J_n)$, where $J_i \in \mathcal{J}(\Phi,\Gamma)$ represents the ith judgment set, and JA procedures $\mathcal{F} : \mathcal{J}(\Phi,\Gamma)^n \to 2^{\Phi}$ map every profile of n individual judgment sets to a collective judgment set.

We can formalize the example illustrating the doctrinal paradox (see Figure 7.2 and Table 7.1 on page 470) as follows: $N = \{1,2,3\}$ is the set of the

7.1 Foundations

three judges and $\Phi = \{p, \neg p, q, \neg q, r, \neg r\}$ is the agenda whose formulas have the following meaning:

$p \,\hat{=}\,$ "The defendant committed the offense."
$q \,\hat{=}\,$ "The offense is a crime."
$r \,\hat{=}\,$ "The defendant committed a crime."

The following integrity constraint captures the logical relations between the issues in the agenda:
$$\Gamma = r \iff (p \land q).$$
The profile of the three individual judgment sets is $\mathbf{J} = (J_1, J_2, J_3)$ with
$$J_1 = \{p, q, r\}, \quad J_2 = \{p, \neg q, \neg r\}, \quad \text{and} \quad J_3 = \{\neg p, q, \neg r\}.$$
These three individual judgment sets are each consistent (since they are logically consistent with Γ) and complete, and thus rational.

The majority rule used in our example of the doctrinal paradox maps the above individual judgment sets to the collective judgment set $\mathcal{F}(J_1, J_2, J_3) = \{p, q, \neg r\}$. This collective judgment set is not consistent, as there is no truth assignment that satisfies all its formulas and the integrity constraint Γ at the same time.

7.1.2.2 Variants of the Frameworks

The framework of binary aggregation with integrity constraints differs from the original framework of judgment aggregation in two ways: (i) the issues in the agenda are required to be atomic propositions (or their negations), and (ii) there is an additional integrity constraint impacting which judgment sets are consistent. For both of these aspects, one could consider several variants, leading to various variations of the framework. Some of these variants are more interesting than others. For example, by restricting the issues in the agenda to atomic propositions or their negations, but without adding an integrity constraint, there are no logical relations between the issues. In this restricted variant, there are no inconsistent outcomes, which means that the doctrinal paradox never occurs, but also that virtually all interesting cases cannot be modeled in this variant.

Some other natural variants of the two frameworks have also been studied. To illustrate the range of choices that can be made when modeling the setting of judgment aggregation, we mention a few examples.

- To the original framework of judgment aggregation, where the issues in the agenda are arbitrary propositional formulas in $\mathcal{L}_P$, one can add an additional integrity constraint $\Gamma \in \mathcal{L}_P$ that needs to be satisfied as well

for a judgment set to be consistent. This variant has been considered by de Haan and Slavkovik [524], for example.
- In the framework of binary aggregation with integrity constraints, one can allow the integrity constraint to contain propositional atoms that do not appear in the agenda Φ.
- In the framework of binary aggregation with integrity constraints, one can also consider different integrity constraints on the opinions reported by the individual judges (the constraint on these is typically called a *rationality constraint*) and the collective outcome that is produced by the judgment aggregation rule (the constraint for this is called a *feasibility constraint*). This makes sense, for instance, in settings where there are constraints (e.g., budgetary constraints) that are to be taken into account only for the aggregated, collective opinion. Endriss [389] introduced this variant of the framework.

We have seen the (original) framework of judgment aggregation and the framework of binary aggregation with integrity constraints, and several of its variants. These different frameworks can all be used to model similar settings (for example, all of them can capture the doctrinal paradox of Figure 7.2 and Table 7.1 on page 470). Naturally, then, some questions arise. Some of the framework variants are restricted cases of others. Is any of these frameworks strictly more expressive than the others, or are they all interchangeable? Can we translate from one framework to the other? What complexity-theoretic impact does the choice of framework have?

7.1.2.3 Relation Between the Different Frameworks

The available variety of (variants of) frameworks for modeling the setting of judgment aggregation—that we discussed in Section 7.1.2.2—has sparked research into clarifying the relation between these variants (this started with the work of Endriss et al. [391]). In this section, we will explain how the different variants relate to each other.

One main way in which these frameworks are similar is that they can all express every possible agenda. Take any collection of issues, and take any set of combinations of these issues that you want to allow to occur as consistent combinations. Then you can model this setting in all of the frameworks in a way that the logical structure of the formulas in the agenda and/or the logical structure of the integrity constraint yields exactly these combinations as consistent judgment sets. Put differently, all frameworks are *fully expressive*.

There is a caveat to this general similarity: In the framework with separate rationality (for the individual opinions) and feasibility constraints (for the collective opinion), situations can be modeled that cannot be modeled in the other frameworks. This is the case simply because in all other frameworks, the set of allowed combinations of issues are always the same for the individuals and the collective: There is only one agenda (and possibly one

7.1 Foundations

integrity constraint) that specifies what is consistent (at both the individual and the collective level).

The full expressiveness of all of the frameworks means that one can translate from any of them to any other of them. But some of these translations can be done more efficiently than others. For example, there exists a polynomial-time algorithm that takes an agenda and an integrity constraint in the setting of binary aggregation with integrity constraints and produces an equivalent agenda in the original judgment aggregation framework.[1] On the other hand, there exists no polynomial-time algorithm that can do the reverse translation (unless the polynomial hierarchy collapses to the second level; recall Section 1.5.3.2 starting on page 35).

That translations in some directions are easier than in others is related to differences in *compactness* between the different frameworks. All of the frameworks are fully expressive, which means that one can express any situation in any of them. However, in certain frameworks some situations can be expressed in a more succinct (or more compact) way than in others. For example, there are agendas in the original framework of judgment aggregation that can only be expressed equivalently in the framework of binary aggregation with integrity constraints using exponentially long integrity constraints.

Very roughly, one can divide (the variants of) the frameworks that we mentioned into two groups, based on their compactness and their ease of translating back and forth.[2] In the less compact group—let's call it Group I—lies (the standard variant of) the framework of binary aggregation with integrity constraints, where the integrity constraint may only contain propositional variables that appear in the agenda. In the more compact group—let's call it Group II—lie all of the other frameworks that we mentioned in Section 7.1.2.2. All the frameworks in Group II are similarly compact, and one can translate easily from one to the other. One can also translate easily from any framework in Group II to the single framework in Group I, and translating from the framework in Group I to any of the frameworks in Group II cannot be done in polynomial-time (unless the polynomial hierarchy collapses to its second level).

Interestingly, these differences in compactness and efficiency of translating back and forth, seem to have no effect on the complexity-theoretic aspects of computing outcomes for different judgment aggregation rules (see, e.g., the work of Endriss et al. [396]). We will discuss the complexity-theoretic aspects of computing outcomes in Section 7.3.

From a purely pragmatic point of view, there are also important differences between the frameworks. For example, in some settings, the agenda and allowed combinations of issues are more naturally expressed using the original framework of judgment aggregation. A typical example of such a setting is that where issues have "local" relations to other issues—such as in the

---

[1] To be exact, this algorithm also needs to receive a single rational judgment set.
[2] Here we are considering only the frameworks where the allowed opinions at the individual and collective levels are the same.

example of the doctrinal paradox where the third issue is readily expressed as the conjunction $(p \wedge q)$ of the first two (p and q). In other settings, the framework of binary aggregation with integrity constraints allows an easier way of modeling the scenario. A typical example of such a setting is that where there are "global" constraints on the combinations of issues—such as in the case where each issue is associated with a cost, and there is a total maximum budget.

Another pragmatic difference is that in the framework of binary aggregation with integrity constraints—where there is only a single logical formula—it seems often easier to devise and describe restrictions that enable more efficient algorithms, for example for computing outcomes (see, e.g., the work of de Haan [523]).

Having now discussed various different frameworks for judgment aggregation, for the sake of simplicity and readability we will now go back to using only the original framework of judgment aggregation (as defined in Section 7.1.1) for the remainder of the chapter.

7.2 Judgment Aggregation Procedures and Their Properties

In this section, we will first introduce a number of specific judgment aggregation procedures and will then turn to their properties, some impossibility results, and some characterizations.

7.2.1 Some Specific Judgment Aggregation Procedures

We now describe some specific judgment aggregation procedures, such as quota rules, premise-based and distance-based procedures, and—most basically—the majority rule. The selection of specific judgment aggregation procedures that we are going to present only gives a limited taste of the different procedures that have been considered and studied. After this tasting, we will discuss some further ingredients that can be (and have been) used to define additional judgment aggregation procedures (in Section 7.2.1.5). The first dish on the menu, that we will turn to now, is the majority rule.

7.2.1.1 Majority Rule

An intuitive description of the *majority rule* has already been given in Section 7.1.1. Formally, it is defined as follows: If there are n judges, an element of the agenda is contained in the collective judgment set if and only if strictly

7.2 Judgment Aggregation Procedures and Their Properties

more than half (i.e., at least $\lceil (n+1)/2 \rceil$) of the judges agree with this element,[3] i.e., declare this formula to be *true*. For an odd number of judges, this JA procedure is complete and complement-free, yet—as we have seen in the example of the doctrinal paradox—it is not consistent.

7.2.1.2 Quota Rules

Dietrich and List [345] introduced the *quota rules* as a natural extension of the majority rule. To each formula φ in the agenda we assign a *quota*—a rational number q_φ between 1 and n that indicates how many of the n judges have to agree with it for φ to be contained in the collective judgment set. If the same quota q is assigned to all elements of the agenda, this JA procedure is called a *uniform quota rule*. The uniform quota rule with quota $q = \lceil (n+1)/2 \rceil$ is nothing other than the majority rule. For a quota rule to be complete, it must satisfy $q_\varphi + q_{\overline{\varphi}} \leq n+1$ for each $\varphi \in \Phi$, and for consistency the inequality

$$\sum_{\varphi \in \Phi} q_\varphi > n(\|Z\| - 1)$$

is required to hold for each minimally inconsistent subset $Z \subseteq \Phi$. A subset of Φ is said to be *minimally inconsistent* if it is inconsistent, but each of its strict subsets is consistent.

7.2.1.3 Premise-Based Judgment Aggregation Procedures

To avoid the doctrinal paradox, so-called *premise-based* and *conclusion-based* methods have been proposed. In such JA procedures, the agenda is subdivided into the premises and the conclusions, and the judges give their judgment of either only the premises or only the conclusions. We focus on the premise-based JA procedures.[4] Which of the conclusions are contained in the collective judgment set can then be logically derived from which of the premises are contained in it, thus preserving consistency.

To define a premise-based JA procedure, subdivide the agenda into two disjoint subsets, $\Phi = \Phi_p \uplus \Phi_c$, where Φ_p is the *set of premises* and Φ_c is the *set of conclusions* and both Φ_p and Φ_c are *closed under complement* (i.e., if $\varphi \in \Phi_p$ then $\overline{\varphi} \in \Phi_p$, and if $\varphi \in \Phi_c$ then $\overline{\varphi} \in \Phi_c$). Given a JA procedure $\mathcal{F}$, the associated *premise-based JA procedure* is a function $\mathcal{P}_\mathcal{F} : \mathcal{J}(\Phi)^n \to 2^\Phi$ defined by

$$\mathcal{P}_\mathcal{F}(\mathbf{J}) = \mathcal{F}(\mathbf{J}_{|\Phi_p}) \cup \{\varphi \in \Phi_c \mid \mathcal{F}(\mathbf{J}_{|\Phi_p}) \text{ satisfies } \varphi\},$$

---

[3] For a real number x, $\lceil x \rceil$ denotes the smallest integer not dropping below x.
[4] In the literature, premise-based methods are often combined with the majority principle. Of course, they can be combined with any other JA procedure as well.

where $\mathcal{F}(\mathbf{J}_{|\Phi_p})$ is the outcome of the JA procedure $\mathcal{F}$ applied to $\mathbf{J}_{|\Phi_p}$, the profile $\mathbf{J}$ of judgment sets restricted to the premises of the agenda. If the premises are logically independent, the collective judgment set is guaranteed to be consistent. If, furthermore, the premises form a logical basis for the agenda (i.e., the premises additionally determine the truth value for every conclusion), the outcome will also be complete. By contrast, for *conclusion-based JA procedures*, which can be defined analogously to the premise-based JA procedures, it is not always possible to derive a decision for all premises. Therefore, such a JA procedure can have incomplete collective judgment sets. Since this is undesirable, we won't consider conclusion-based JA procedures any further.

Table 7.2 shows how to avoid the doctrinal paradox from our previous example given in Figure 7.2 and Table 7.1 by applying a premise-based JA procedure, i.e., by applying the majority rule to the premises only and then logically deriving the outcome for the conclusion.

Table 7.2 Avoiding the doctrinal paradox by a premise-based JA procedure

	Premises		Conclusion
Judge	p	q	$p \wedge q$
Anna	1	1	1
Belle	1	0	0
Edgar	0	1	0
Majority	1	1 $\Rightarrow$	1

For the majority rule, one obtains a rational premise-based JA procedure if, first, the agenda is *closed under propositional variables* (i.e., if every variable occurring in any formula of Φ is itself contained in Φ) and, second, the set of premises coincides with the set of literals of the agenda. In addition, the number of judges must be odd. This procedure will simply be called the *premise-based procedure* (*PBP*, for short).

It is also possible to define a rational premise-based JA procedure for the majority rule with an even number of judges. To achieve this, one needs a tie-breaking rule that specifies for each element of the agenda which of φ and $\overline{\varphi}$ is preferred in case of a tie.

A generalization of *PBP* to an arbitrary (even or odd) number of judges and different quotas is the class of *uniform premise-based quota rules*. Here, in addition to subdividing the agenda $\Phi = \Phi_p \uplus \Phi_c$, also the set Φ_p of premises is subdivided into two disjoint subsets, $\Phi_p = \Phi_1 \uplus \Phi_2$, such that exactly one member of each pair $\{\varphi, \overline{\varphi}\} \subseteq \Phi_p$ is contained in Φ_1 and the other one in Φ_2. Assign a quota $q \in \mathbb{Q}$, $0 \leq q < 1$, to all literals in Φ_1; we thus have the quota $q' = 1 - q$ for every literal in Φ_2. For $\mathbf{J} = (J_1, J_2, \ldots, J_n)$, the function

$UPQR_q(\mathbf{J}) = \Delta \cup \{\psi \in \Phi_c \mid \Delta \text{ satisfies } \psi\}$, where

$$\Delta = \{\varphi \in \Phi_1 \mid \|\{i \mid \varphi \in J_i\}\| > nq\} \cup \{\varphi \in \Phi_2 \mid \|\{i \mid \varphi \in J_i\}\| \geq nq'\},$$

then defines the *uniform premise-based quota rule for quota q*. To guarantee rational outcomes, we again require Φ to be closed under propositional variables and to consist of all literals.

The definition of Δ implies that a literal $\varphi \in \Phi_1$ (respectively, $\varphi \in \Phi_2$) has to be in at least $\lfloor nq+1 \rfloor$ (respectively, $\lceil nq' \rceil$) individual judgment sets of profile $\mathbf{J}$ for φ to be in the collective judgment set of $UPQR_q(\mathbf{J})$. Note that $UPQR_q$ is complete: Since $\lfloor nq+1 \rfloor + \lceil nq' \rceil = n+1$, it is ensured that for every $\varphi \in \Phi$, either $\varphi \in UPQR_q(\mathbf{J})$ or $\overline{\varphi} \in UPQR_q(\mathbf{J})$.

Note further that since the quota $q=1$ would require that some $\varphi \in \Phi_1$ occurs in at least $n+1$ individual judgment sets to be accepted according to $UPQR_q$, this quota is not allowed. By contrast, the quota $q=0$ merely requires that $\varphi \in \Phi_1$ is in at least one individual judgment set to be collectively accepted; therefore, $q=0$ is allowed. The premise-based procedure is nothing other than the uniform premise-based quota rule with quota $1/2$, restricted to an odd number of judges.

7.2.1.4 Distance-Based Judgment Aggregation Procedures

Pigozzi [780] (see also the work of Miller and Osherson [708]) introduced a JA procedure that has some similarities with Kemeny's voting system, which has been presented on page 246 in Section 4.1.2.6. Given an agenda Φ and a profile $\mathbf{J} = (J_1, \ldots, J_n)$ in $\mathcal{J}(\Phi)^n$, the *distance-based JA procedure* $\mathcal{D}_H : \mathcal{J}(\Phi)^n \to \mathcal{J}(\Phi)$ selects the following collective judgment sets:

$$\mathcal{D}_H(\mathbf{J}) = \underset{J \in \mathcal{J}(\Phi)}{\operatorname{argmin}} \sum_{i=1}^{n} H(J, J_i), \tag{7.1}$$

where $H(J, J_i)$ is the Hamming distance between the two judgment sets J and J_i. The *Hamming distance* between two 0-1 vectors of equal length is defined as the number of positions in which they differ. A judgment set J can be seen as the 0-1 vector having the entry 1 for those elements of the agenda contained in J, and the entry 0 for those elements of the agenda not contained in J. That is, the *Hamming distance between two complete judgment sets* over the same agenda Φ is the number of *positive* formulas in Φ for which the two judgment sets disagree.

According to (7.1), the distance-based JA procedure selects all those judgment sets from $\mathcal{J}(\Phi)$ for which the sum of the Hamming distances to all individual judgment sets is minimal. That means that the number of disagreements between the judges is minimized. The distance-based JA procedure is rational, since its outcomes are required to be from $\mathcal{J}(\Phi)$, the set of all complete, consistent judgment sets over Φ. Note that, in general, it returns not only a single collective judgment set (because several judgment sets can minimize the number of disagreements with the individual judgment sets); therefore, if a unique result is required, a tie-breaking rule has to be applied.

The JA procedure $\mathcal{D}_H$ is the prototypical example of a distance-based procedure, but there are more distance-based procedures. Roughly, there are two ways in which distance-based procedures vary: (i) the choice of distance measure, and (ii) how the distances to the individual judgment sets in the profile are combined. Besides the Hamming distance, other (pseudo-)distance measures have been studied (see, e.g., the work of Lang et al. [639] for an overview).

The other way in which one can vary distance-based procedures is the way in which to combine several distances into one. For the procedure $\mathcal{D}_H$, the individual distances are summed, as described in (7.1). This has the effect that the JA procedure takes a utilitarian point of view: The overall sum of distances is minimized. Alternatively, one can adopt an egalitarian point of view, for example by taking the maximum of the individual distances instead of the sum. This yields the procedure $\mathcal{D}_{\max,H} : \mathcal{J}(\Phi)^n \to \mathcal{J}(\Phi)$, which selects the following collective judgment sets:

$$\mathcal{D}_{\max,H}(\mathbf{J}) = \operatorname*{argmin}_{J \in \mathcal{J}(\Phi)} \max_{i=1}^{n} H(J, J_i). \qquad (7.2)$$

7.2.1.5 Other Judgment Aggregation Procedures

In addition to the JA procedures that we discussed in Sections 7.2.1.1–7.2.1.4, many more concrete procedures have been studied in the literature. These procedures are based on a variety of conceptual motivations and have been studied for their normative properties. (For example, in Section 7.2.1.4 we already briefly mentioned the difference between utilitarianism and egalitarianism that is visible in the difference between the procedures $\mathcal{D}_H$ and $\mathcal{D}_{\max,H}$.)

To illustrate the range of principles that have been called upon to motivate different judgment aggregation procedures, we sketch a few of them:

- The distance-based procedures described in Section 7.2.1.4 ensure that a consistent collective outcome is produced by selecting consistent judgment sets that minimize some (combined) distance to the entire profile. Instead, one could select consistent judgment sets that minimize some distance to the (possibly inconsistent) outcome of the majority rule (as described in Section 7.2.1.1).
- Instead of measuring (and minimizing) the distance from the collective outcome to the profile (or to the majority outcome), one could also consider the distance (under various notions of distance) from the profile to another (hypothetical) profile for which the majority procedure gives a consistent outcome. Two notable examples of instantiations of this idea are the following, again inspired by the corresponding voting systems:

7.2 Judgment Aggregation Procedures and Their Properties

Young voting (see Section 4.1.2.5 starting on page 245) and Dodgson voting (see Section 4.1.2.3 starting on page 242). The *Young JA procedure* selects the judgment sets that can be obtained as consistent outputs of the majority procedure after deleting a minimum number of judges from the profile. The *Dodgson JA procedure* selects the judgment sets that can be obtained as consistent outputs of the majority procedure after making a minimum number of modifications to the profile (where a modification consists of changing some judge's reported opinion on some issue in a way that leaves this judge's judgment set consistent).
- Another way to obtain a collective opinion that is consistent is to go over the issues in the agenda in some order, one-by-one, and for each issue selecting the majority outcome unless this creates an inconsistency. This is known as the *ranked agenda rule*, and there are various variants of this (e.g., using some fixed order over the issues, or taking the union of all outcomes for each order over the issues in the agenda).

Good starting points to further explore the wide variety of concrete judgment aggregation procedures that have been studied, are the work of Lang et al. [639] and that of Endriss et al. [396].

7.2.2 Properties, Impossibility Results, and Characterizations

Similarly to the properties of voting systems described in Section 4.2 starting on page 264, we now present a number of desirable properties of JA procedures, which can be used to evaluate and compare them. So far, we know the properties of completeness, complement-freeness, and consistency for judgment sets and JA procedures (as explained in Section 7.1.1). There are many further natural properties that one might want a JA procedure to satisfy. We will define and discuss several of them below. (For the interested reader, a good starting point for learning about further such properties is the overview by Lang et al. [639].)

7.2.2.1 Unanimity

If all judges have the same judgment of an element of the agenda, the collective judgment of this element should coincide with it. Formally, a JA procedure $\mathcal{F}$ is *unanimous* if $\varphi \in \mathcal{F}(\mathbf{J})$ for each profile $\mathbf{J} = (J_1, \ldots, J_n)$ with $\varphi \in J_i$, $1 \leq i \leq n$. This property corresponds to unanimity in voting (see Footnote 42 on page 355) and is similar to a weak form of Pareto consistency for voting systems (recall this notion from page 268).

7.2.2.2 Anonymity

Anonymity says that the collective judgment should not depend on which of the judges reports which judgment set. Formally, a JA procedure $\mathcal{F}$ is said to be *anonymous* if for each profile $\mathbf{J} = (J_1, \ldots, J_n)$ and for each permutation $\sigma : N \to N$, we have $\mathcal{F}(J_1, \ldots, J_n) = \mathcal{F}(J_{\sigma(1)}, \ldots, J_{\sigma(n)})$.

7.2.2.3 Neutrality

If every judge thinks the same about any two formulas (i.e., considers them either both to be true or both to be false), then these two formulas are either both contained in the collective judgment set of a neutral JA procedure, or they both are not contained in it. Formally, a JA procedure $\mathcal{F}$ is said to be *neutral* if for all $\varphi, \psi \in \Phi$ and for each profile $\mathbf{J} = (J_1, \ldots, J_n)$ in $\mathcal{J}(\Phi)^n$ with $\varphi \in J_i \iff \psi \in J_i$ for all i, $1 \leq i \leq n$, we have $\varphi \in \mathcal{F}(\mathbf{J}) \iff \psi \in \mathcal{F}(\mathbf{J})$.

7.2.2.4 Independence

The criterion of independence says that if two distinct profiles with the same number of judges coincide in their individual judgments of a formula, then also the collective judgments of this formula that result from these profiles should coincide. Formally, a JA procedure $\mathcal{F}$ is said to be *independent* if for each formula $\varphi \in \Phi$ and for any two profiles $\mathbf{J} = (J_1, \ldots, J_n)$ and $\mathbf{J}' = (J'_1, \ldots, J'_n)$ in $\mathcal{J}(\Phi)^n$ with $\varphi \in J_i \iff \varphi \in J'_i$ for all i, $1 \leq i \leq n$, we have that $\varphi \in \mathcal{F}(\mathbf{J}) \iff \varphi \in \mathcal{F}(\mathbf{J}')$.

7.2.2.5 Systematicity

The criterion of *systematicity* is satisfied by exactly those JA procedures that are neutral and independent. Formally, a JA procedure $\mathcal{F}$ is *systematic* if for all formulas $\varphi, \psi \in \Phi$ and for all profiles $\mathbf{J} = (J_1, \ldots, J_n)$ and $\mathbf{J}' = (J'_1, \ldots, J'_n)$ in $\mathcal{J}(\Phi)^n$ with $\varphi \in J_i \iff \psi \in J'_i$ for all i, $1 \leq i \leq n$, we have that $\varphi \in \mathcal{F}(\mathbf{J}) \iff \psi \in \mathcal{F}(\mathbf{J}')$.

7.2.2.6 Nondictatorship

The collective judgment set of a nondictatorial JA procedure must not solely depend on a single judge's individual judgment set (provided that there are at least two judges). Formally, a JA procedure $\mathcal{F}$ is *nondictatorial* if whenever $\|N\| \geq 2$, there is no $i \in N$ such that $\mathcal{F}(\mathbf{J}) = J_i$ for each profile $\mathbf{J} \in \mathcal{J}(\Phi)^n$.

7.2.2.7 Monotonicity

This property says that if

1. a judge changes her judgment of a formula from *false* to *true*,
2. the other judges stick to their judgment of this formula, and
3. this formula was contained in the collective judgment set before the judge has changed her mind,

then this formula is also contained in the collective judgment set after the change. In other words, a formula should not be judged worse collectively due to getting more individual support. Formally, a JA procedure $\mathcal{F}$ is said to be *monotonic* if for all $\varphi \in \Phi$ and for any two profiles

$$\mathbf{J} = (J_1, \ldots, J_{i-1}, J_i, J_{i+1}, \ldots, J_n) \quad \text{and} \quad \mathbf{J}' = (J_1, \ldots, J_{i-1}, J'_i, J_{i+1}, \ldots, J_n)$$

with $\varphi \notin J_i$ and $\varphi \in J'_i$, we have that $\varphi \in \mathcal{F}(\mathbf{J})$ implies $\varphi \in \mathcal{F}(\mathbf{J}')$.

The corresponding property has been presented on page 277 for voting systems, and it will be presented for preferences in the context of allocation procedures in Section 9.3.1.1 on page 615.

7.2.2.8 Impossibility Results and Characterizations

There are a number of connections between these properties; some are straightforward, others are the subject of hard theorems. As an example of a straightforward connection between properties, by definition, a JA procedure is systematic if and only if it is neutral and independent. It is also easy to see that anonymity of a JA procedure always ensures that it is not a dictatorship.

Intuitively, all these properties—including completeness, consistency, and complement-freeness—are desirable, and we therefore are interested in finding JA procedures satisfying them. Similarly to Arrow's impossibility theorem [29] in voting (see Theorem 4.1 on page 273), however, List and Pettit [663] have established a first impossibility theorem for JA procedures, which shows that certain desirable properties cannot be satisfied simultaneously, and one thus is forced to make compromises.

Theorem 7.1 (List and Pettit [663]). *If the agenda contains at least two distinct atomic propositions (say, p and q) and also their conjunction $(p \wedge q)$, disjunction $(p \vee q)$, or implication $(p \Longrightarrow q)$, then there is no JA procedure with a universal domain that is systematic, anonymous, and rational.*

This impossibility theorem shows that the doctrinal paradox is not to blame on just the majority principle. It would not be possible to preserve consistency (which is contained in the requirement of rationality) by any other JA procedure, at least not without losing any of the other properties mentioned in Theorem 7.1. Simply choosing another JA procedure is not enough to avoid the paradox.

There are also a number of other impossibility theorems for JA procedures. For example, Pauly and van Hees [772] strengthened the original result by replacing the requirement of anonymity in Theorem 7.1 by the weaker requirement of nondictatorship.

Inconsistency of the collective judgment set according to the doctrinal paradox cannot occur for every agenda, though, for example not if the agenda contains only premises. A characterization of agendas for which the majority rule is consistent is given by Nehring and Puppe [732] through the median property. An agenda Φ satisfies the *median property* if every inconsistent subset of Φ itself has an inconsistent subset of size less than or equal to two.

Theorem 7.2 (Nehring and Puppe [732]). *If there are at least three judges, the majority rule is consistent for a given agenda Φ if and only if Φ satisfies the median property.*

Furthermore, they also show that for a given agenda there is a judgment aggregation procedure that satisfies certain criteria if and only if the agenda satisfies the median property.

Theorem 7.3 (Nehring and Puppe [732]). *If there are at least three judges, there is a rational judgment aggregation procedure that satisfies the axioms of neutrality, independence, monotonicity, and nondictatorship for a given agenda Φ if and only if Φ satisfies the median property.*

Various further properties of the agenda with respect to consistent collective judgment sets have been studied by List and Puppe [664], Nehring and Puppe [731], Dietrich and List [344], and Dokow and Holzman [351].

7.3 Complexity of Judgment Aggregation Problems

As we did for voting problems in Section 4.3 starting on page 287 (and also in Chapter 5 where we focused on voting problems in single-peaked domains), we now turn to judgment aggregation problems (*JA problems*, for short) and study them in terms of their computational complexity. This research field is much younger for judgment aggregation than it is for voting; while the latter has been an extremely active area since more than 25 years now, starting with the early work of Bartholdi, Tovey, and Trick [100, 101, 102] (see also the work of Bartholdi and Orlin [99]) and yielding plenty of outstanding results (see, e.g., Section 4.3 and Chapter 5 of this book and also Chapters 3–10 in the *"Handbook of Computational Social Choice"* [209, 452, 251, 312, 440, 383, 641, 174] for an overview), the complexity-theoretic study of JA problems has been initiated by Endriss, Grandi, and Porello [392, 393, 394].

Before having a closer look at the complexity of JA problems, some technical remarks are in order. In contrast to the assumptions made in Section 7.1

7.3 Complexity of Judgment Aggregation Problems

starting on page 471 where we formally described the foundations of judgment aggregation, we from now on *will allow* that the agenda may contain tautologies and contradictions because we are now studying the complexity of JA problems. The reason is that identifying a tautology or a contradiction are difficult problems themselves, from a computational point of view. For example, the problem TAUTOLOGY ("Is a given formula a tautology?") is known to be coNP-complete (see, e.g., the books by Garey and Johnson [483], Papadimitriou [764], and Rothe [819]). However, if this requirement cannot be tested in polynomial time (unless P would equal coNP, which is equivalent to P = NP and, hence, is widely considered to be very unlikely), then any results on the complexity of JA problems would not be meaningful. Therefore, it is reasonable to now drop the assumption that the agenda must contain neither tautologies nor contradictions.

In the remainder of this section, we will study JA problems related to winner determination, safety of the agenda, manipulation, bribery, and control.

7.3.1 Winner Determination in Judgment Aggregation

As for the WINNER problem in elections, it is desirable for a JA procedure $\mathcal{F}$ that its "winner(s)" (namely, the collective judgment set(s) output by $\mathcal{F}$) can be determined in polynomial time. The corresponding decision problem asks whether or not a specific formula from the given agenda for a given profile of individual judgment sets is in the collective judgment set output by $\mathcal{F}$ (where we assume that there is only one such set):

$\mathcal{F}$-JA-WINNER

Given: An agenda Φ, a profile $\mathbf{J} \in \mathcal{J}(\Phi)^n$, and a formula $\varphi \in \Phi$.
Question: Is φ contained in the collective judgment set $\mathcal{F}(\mathbf{J})$?

If $\mathcal{F}$-JA-WINNER is solvable in polynomial time for a JA procedure $\mathcal{F}$, then the collective judgment set can also be determined in polynomial time by checking for every $\varphi \in \Phi$ whether or not it is contained in the collective judgment set.

Endriss, Grandi, and Porello [394] observe that JA-WINNER is solvable in polynomial time for the premise-based procedure: For each premise in the agenda, it can be checked in polynomial time whether enough judges have approved of them, and the decision for the conclusions can also be derived in polynomial time from the decision for the premises. For premise-based quota rules, too, JA-WINNER is easily seen to be solvable in polynomial time.

Since the distance-based procedure (without a tie-breaking rule) does not always yield a single collective judgment set, we here do not ask whether a specific formula $\varphi \in \Phi$ is contained in the collective judgment set, but rather we ask whether there is a collective judgment set J produced by the distance-based procedure that contains a given subset of formulas. Endriss, Grandi, and Porello [394] show by a reduction from a problem related to the problem Kemeny-WINNER (see page 291) that the thus modified decision problem for the distance-based procedure $\mathcal{D}_H$ is $P_{\parallel}^{NP}$-complete. As for the JA-WINNER problem, it is possible to create a collective judgment set according to $\mathcal{D}_H$, by solving the thus modified JA-WINNER problem for the distance-based procedure $\mathcal{D}_H$ polynomially many times. But since this problem is $P_{\parallel}^{NP}$-complete, this does not yield a polynomial-time algorithm for creating collective judgment sets according to $\mathcal{D}_H$. Moreover, producing a collective judgment set according to $\mathcal{D}_H$ by iterating the $P_{\parallel}^{NP}$ algorithm polynomially many times does not correspond to a polynomial-time algorithm with parallel access to an NP oracle. In other words, from the results for the modified JA-WINNER decision problem for the distance-based procedure, we cannot directly conclude that the search problem of producing a collective judgment set has similar complexity (it might be harder). To address this question, Endriss and de Haan [395] studied the complexity of this search problem (using a variant of the computational complexity toolbox that is suited for search problems). They show that the search problem of producing collective outcomes for the distance-based procedure $\mathcal{D}_H$ is complete for a suitable search variant of the class $P_{\parallel}^{NP}$.

The computational complexity of the winner determination problem for a list of judgment aggregation procedures has been established (see, e.g., the overview by Endriss et al. [396]). Interestingly, for most judgment aggregation procedures that have been studied in the literature—including, for example, the procedure $\mathcal{D}_{\max,H}$—the winner determination problem is $P_{\parallel}^{NP}$-complete. Interestingly also, for judgment aggregation procedures that have been seriously considered in the literature, the choice of which judgment aggregation framework to use (as discussed in Section 7.1.2) has no effect on the complexity of computing outcomes—even though in theory there exist judgment aggregation procedures for which computing outcomes is easier in one framework than in another (as shown by Endriss et al. [391]).

7.3.2 Safety of the Agenda

We have already seen that a collective judgment set may be inconsistent, even though all its underlying individual judgment sets are consistent. The impossibility result of List and Pettit [663] (see Theorem 7.1 on page 485) shows that, even if the agenda satisfies only some very mild requirements, inconsistent collective judgment sets may occur for any JA procedure with a universal domain satisfying quite fundamental properties such as systematicity and anonymity. In particular, this applies to the majority rule. However, as a positive result, we have also seen that for a given agenda the majority

7.3 Complexity of Judgment Aggregation Problems

rule is consistent if and only if the agenda satisfies the median property (see Theorem 7.2 on page 486).

Endriss, Grandi, and Porello [394] call an agenda *safe* for judgment aggregation procedures that satisfy some given criteria if they will always be consistent when the individual judgment sets are restricted to this agenda. In particular, Endriss, Grandi, and Porello [394] define the decision problem MEDIAN-PROPERTY, which asks whether a given agenda satisfies the median property (see page 486) and show that this is a highly intractable problem: It is complete for the second level of the polynomial hierarchy (recall this hierarchy from Section 1.5.3.2 starting on page 35).

Theorem 7.4 (Endriss, Grandi, and Porello [394]). MEDIAN-PROPERTY *is* Π_2^p*-complete.*

Other properties of the agenda—each a variant of the median property—lead to possibility results in the style of Theorem 7.2, too. For the so-called "simple median property," the "simplified median property," the "syntactic simplified median property," and the "k-median property," Endriss, Grandi, and Porello [394] also show Π_2^p-completeness of the corresponding decision problems, and we refer to their work for further details. Endriss, de Haan, and Szeider [397] study agenda safety in terms of parameterized complexity.

7.3.3 Manipulation in Judgment Aggregation

Besides the winner problem, the manipulation problem has also been studied for JA procedures. If one considers manipulation in JA procedures to be undesirable, NP-hardness of the corresponding problem is to be viewed as a worthwhile property: If it is computationally hard for a manipulative judge to determine his insincere individual judgment set (or even to decide whether such a judgment set exists), this would offer some kind of protection against strategic judging.

In elections the goal of a manipulation is to make a designated candidate a winner of the election. How can such a scenario be modeled for judgment aggregation where there is no "winner"? Strategic judging has first been studied by List [661] and Dietrich and List [346]. The first complexity-theoretic work on manipulation in judgment aggregation is due to Endriss, Grandi, and Porello [394]. In the definition of the manipulation problem, they assume that the sincere individual judgment set of a manipulative judge is also his *desired* collective judgment set, the desired outcome of the JA procedure. They suggest that, from the manipulative judge's point of view, the collective judgment set should be as similar to his individual judgment set as possible. According to that, a manipulation would be successful if, by reporting an insincere judgment set, the manipulator achieves that the collective judgment set is "more similar" to his sincere judgment set than if he had reported his sincere judgment set itself. Endriss, Grandi, and Porello [394] measure the similarity between two (complete) judgment sets by their Hamming distance, that is, by the number of positive formulas in which these two judgment sets differ (recall this definition from Section 7.2.1.4 starting on page 481).

The Hamming distance is not the only way of measuring better or worse outcomes in JA procedures. Of course, knowing the manipulator's preferences over all possible judgment sets would allow for the best comparison between two judgment sets. Unfortunately, such a complete ranking of all possible judgment sets is usually not given, as this information is exponentially large in the number of formulas in the agenda. This is an issue similar to the one we were facing when we studied cooperative games and their succinct representations in Section 3.4 starting on page 171 and to the issue we will encounter later on in Section 9.3 starting on page 613 when we will be concerned with the compact representation of preferences in the context of fair division of indivisible goods. For preferences in judgment aggregation, Dietrich and List [346] circumvent this problem by deriving preference assumptions from a single given judgment set J. To this end, they define so-called unrestricted, top-respecting, and closeness-respecting J-induced preferences, as explained informally below, and for completeness we will also define Hamming-distance-respecting J-induced preferences.

But first let us extend the notion of Hamming distance to *incomplete* judgment sets: Given two possibly incomplete judgment sets J and J' over Φ, their *Hamming distance* (denoted by $HD(J, J')$) is defined as the number of disagreements on the positive formulas of the agenda that occur in both judgment sets. Now consider the following cases:

- In the case of *unrestricted J-induced preferences*, we do not have any information on the link between the manipulator's sincere judgment set J and his preferences over all possible judgment sets. Informally stated, we cannot compare any two judgment sets; basically, every preference is possible, even a preference where the manipulator's sincere judgment set J is not his most preferred judgment set.
- The notion of *top-respecting J-induced preferences* captures the case where the manipulator prefers his own sincere judgment set J over all other judgment sets; but otherwise nothing is known.
- In *closeness-respecting J-induced preferences*, given the manipulator's sincere judgment set J, he prefers judgment set X to judgment set Y if the set of propositions coinciding in Y and J is a proper subset of the propositions coinciding in X and J (i.e., if $Y \cap J \subset X \cap J$).
- Finally, in *Hamming-distance-respecting J-induced preferences*, given the manipulator's sincere judgment set J, he prefers judgment set X to judgment set Y if the Hamming distance between X and J is smaller than the Hamming distance between Y and J (i.e., if $HD(X, J) < HD(Y, J)$).

For an illustration of these notions of preference assumptions, consider the following example.

7.3 Complexity of Judgment Aggregation Problems

Example 7.1. For the variables a, b, and c, let the agenda contain the formulas a, b, c, $a \wedge b$, $a \vee c$, $b \vee c$, and their negations, where the premises are a, b, c, and their negations and where the remaining formulas are the conclusions. The individual judgment sets of the three judges are J_i for judge i, where the premises in J_1 are a, b, and $\neg c$, those in J_2 are $\neg a$, b, and $\neg c$, those in J_3 are $\neg a$, $\neg b$, and c, and they each contain the accordingly derived conclusions. Indicating by a "0" that the negation of a formula is in the judgment set and by a "1" that the formula itself is contained in the judgment set, these individual judgment sets are represented as follows:

Judge	Premises				Conclusions		
	a	b	c		$a \wedge b$	$a \vee c$	$b \vee c$
1	1	1	0		1	1	1
2	0	1	0		0	0	1
3	0	0	1		0	1	1
PBP	0	1	0	$\Rightarrow$	0	0	1

The outcome of the premise-based procedure is also given above:

$$J_0 = \{\neg a,\ b,\ \neg c,\ \neg(a \wedge b),\ \neg(a \vee c),\ b \vee c\}.$$

Now assume that the third judge is not reporting his honest judgment set J_3, consisting of $\neg a$, $\neg b$, c, and the corresponding conclusions, but instead is trying to manipulate and reports the untruthful individual judgment set J_3^*, consisting of a, $\neg b$, c, and the corresponding conclusions. Then the manipulated collective outcome by PBP is

$$J_0^* = \{a,\ b,\ \neg c,\ a \wedge b,\ a \vee c,\ b \vee c\}.$$

Now we try to compare the original outcome J_0 with the manipulated outcome J_0^* according to the types of J_3-induced preferences informally described above, seeking to find out if we can conclude whether the manipulator has achieved a "better" result by his action:

- If the manipulator has unrestricted J_3-induced preferences, we do not know whether or not he prefers this new outcome, J_0^*, to the original one, J_0.
- If he has closeness-respecting J_3-induced preferences, we again do not know whether he prefers the new outcome J_0^* to the original one, J_0, since the agreements of J_0 and J_3 on $\neg a$ and $\neg(a \wedge b)$ are

no longer given for J_0^* and J_3. However, if he is interested only in the two formulas $a \vee c$ and $b \vee c$ (i.e., only in an *incomplete* collective judgment set), then we know that he does prefer the new outcome to the original one, since the intersection $\{a \vee c, b \vee c\} \cap J_0 = \{b \vee c\}$ is a proper subset of the intersection $\{a \vee c, b \vee c\} \cap J_0^* = \{a \vee c, b \vee c\}$.

- The same holds for top-respecting J_3-induced preferences: If the manipulator is interested in the whole collective judgment set, we do not know which outcome is better for him, but restricted to the two formulas $a \vee c$ and $b \vee c$, the new outcome equals his initial individual judgment set and thus is preferred to all other outcomes.

- If the manipulator has Hamming-distance-respecting J_3-induced preferences, we know that the new outcome is not an improvement to the old outcome and thus is not preferred by the manipulator, since the Hamming distance between the collective outcome and J_3 equals 3 before and after the manipulation.

Adopting the notation used by Baumeister et al. [113], we will now give the formal definitions of the above notions of induced weak preferences over judgment sets.

Definition 7.2 (J-induced preferences of certain types). Let Φ be an agenda and let $\mathcal{J}(\Phi)$ be the set of all rational judgment sets over Φ.

1. Let U denote the set of all weak orders over $\mathcal{J}(\Phi)$.[5] For a weak order $\succeq$ over $\mathcal{J}(\Phi)$ and for all $J, J' \in \mathcal{J}(\Phi)$, define $J \succ J'$ by $J \succeq J'$ and $J' \not\succeq J$, and define $J \sim J'$ by $J \succeq J'$ and $J' \succeq J$. We say that J *is weakly preferred to* J' if $J \succeq J'$; J *is preferred to* J' if $J \succ J'$; and there is *indifference between* J *and* J' if $J \sim J'$.
2. Letting J be a fixed (possibly incomplete) individual judgment set, define the *set of*
    a. *unrestricted J-induced weak orders* $\succeq$ in U, denoted by U_J, as the set of weak orders $\succeq$ in U satisfying that for all $X, Y \in \mathcal{J}(\Phi)$, we have $X \sim Y$ whenever $X \cap J = Y \cap J$;
    b. *top-respecting J-induced weak preferences*, $TR_J \subseteq U_J$, as the set of weak orders $\succeq$ in U_J satisfying that $\succeq \in TR_J$ if and only if for all $X \in \mathcal{J}(\Phi)$ with $X \cap J \neq J$, we have $J \succ X$;
    c. *closeness-respecting J-induced weak preferences*, $CR_J \subseteq U_J$, as the set of weak orders $\succeq$ in U_J satisfying that $\succeq \in CR_J$ if and only if for all $X, Y \in \mathcal{J}(\Phi)$ with $Y \cap J \subseteq X \cap J$, we have $X \succeq Y$; and

---

[5] A *weak order* (commonly denoted by $\succeq$) is a transitive and total—thus, in particular, a reflexive—relation (recall these notions from page 199 in Section 3.6).

7.3 Complexity of Judgment Aggregation Problems

d. *Hamming-distance-respecting J-induced weak preferences*, $HD_J \subseteq U_J$, as the set of weak orders $\succeq$ in U_J satisfying that $\succeq \, \in HD_J$ if and only if for all $X, Y \in \mathcal{J}(\Phi)$, we have $X \succeq Y$ if and only if $HD(X, J) \leq HD(Y, J)$.

In order to cope with the distinction between known and unknown preferences relations, Baumeister et al. [113] (see also [112]) introduced the following notions that are inspired by the notions of possible and necessary winners in voting (recall them from Section 4.3.2 starting on page 297).

Definition 7.3 (possible and necessary strategy-proofness). Let J be a given (possibly incomplete) judgment set, let X and Y be judgment sets for the same agenda, and let $T_J \in \{U_J, TR_J, CR_J\}$ be a type of J-induced weak preferences.

1. A judge *possibly weakly prefers* X *to* Y *for J-induced weak preferences of type* T_J if there is some $\succeq \, \in T_J$ with $X \succeq Y$.
2. A judge *necessarily weakly prefers* X *to* Y *for J-induced weak preferences of type* T_J if $X \succeq Y$ for all $\succeq \, \in T_J$.
3. Analogously, replacing $\succeq$ by $\succ$, define the notions of *possible* and *necessary J-induced preferences of type* T_J.
4. A JA procedure $\mathcal{F}$ is *possibly* (respectively, *necessarily*) *strategy-proof for J-induced weak preferences of type* T_J if for all profiles $(J_1, J_2, \ldots, J_n)$, for each i, $1 \leq i \leq n$, and for each $J_i^* \in \mathcal{J}(\Phi)$, judge i possibly (respectively, necessarily) weakly prefers the outcome $\mathcal{F}(J_1, J_2, \ldots, J_n)$ to the outcome $\mathcal{F}(J_1, J_2, \ldots, J_{i-1}, J_i^*, J_{i+1}, \ldots, J_n)$ for J_i-induced weak preferences of type T_{J_i}.

The above notion of necessary strategy-proofness has been simply called "strategy-proofness" by Dietrich and List [346]. Note further that, for each judgment set J desired by the manipulator, the Hamming-distance-respecting J-induced weak preferences are a complete relation. Therefore, we say that $\mathcal{F}$ is *strategy-proof for Hamming-distance-respecting J-induced weak preferences* if each individual judge weakly prefers the actual outcome to any outcome obtained by reporting another individual judgment set, both measured in terms of their Hamming distance to J, omitting the distinction into "possible" and "necessary" strategy-proofness of this type.

Now we are ready to give our formal definition of manipulation for a JA procedure $\mathcal{F}$ and a given preference type $T \in \{U, TR, CR\}$ induced by a fixed judgment set.

$\mathcal{F}$-T-JA-Possible-Manipulation

Given: An agenda Φ, a profile $\mathbf{J} = (J_1, \ldots, J_n) \in \mathcal{J}(\Phi)^n$, and a consistent (possibly incomplete) judgment set $J \subseteq J_n$ desired by the manipulator.

Question: Is there a judgment set $J^* \in \mathcal{J}(\Phi)$ such that

$$\mathcal{F}(J_1, J_2, \ldots, J_{n-1}, J^*) \succ \mathcal{F}(J_1, J_2, \ldots, J_{n-1}, J_n)$$

for some $\succeq \, \in T_J$?

In the above problem definition, J is the manipulative judge's desired judgment set (which does not need to coincide with J_n completely, to allow for manipulators interested in only an incomplete desired judgment set) and J^* is his reported insincere judgment set by which he seeks to alter the outcome of the JA procedure so that the new outcome is preferred to the original outcome for J-induced preferences of type T_J. Note further that $\succeq$ induces $\succ$ (recall the first item of Definition 7.2 on page 492).

We define the problem $\mathcal{F}$-T-JA-NECESSARY-MANIPULATION analogously, with the question replaced by asking whether there exists a judgment set $J^* \in \mathcal{J}(\Phi)$ such that

$$\mathcal{F}(J_1, J_2, \ldots, J_{n-1}, J^*) \succ \mathcal{F}(J_1, J_2, \ldots, J_{n-1}, J_n)$$

for *all* $\succeq \in T_J$. As mentioned above, for Hamming-distance-respecting induced preferences, the relation between any two judgment sets is always known. Therefore, following the notation of Baumeister et al. [113, 112], we will denote this problem by $\mathcal{F}$-HD-JA-MANIPULATION, omitting "POSSIBLE" and "NECESSARY" in the problem name.

Endriss, Grandi, and Porello [394] show that HD-JA-MANIPULATION is NP-complete for the premise-based procedure. Baumeister, Erdélyi, and Rothe [116] extend this result by showing that HD-JA-MANIPULATION is NP-complete for every uniform quota rule with a rational quota. Furthermore, they study the parameterized complexity of this problem: It remains hard even when the number of judges or the number of changes in the manipulative judge's premises are bounded by a fixed constant.

De Haan [522] showed that the complexity of HD-JA-MANIPULATION for the distance-based procedure (with any fixed tie-breaking rule) is complete for Σ_2^p, the second level of the polynomial hierarchy.

Recall from Section 1.5.3.2 starting on page 35 that every problem in $\Sigma_2^p = \text{NP}^{\text{NP}}$ is accepted by a nondeterministic polynomial-time oracle Turing machine that has access to an NP oracle (e.g., SAT). Unlike the class $\text{P}_{\|}^{\text{NP}}$ (see page 291), the oracle does not need to be accessed in parallel: Which queries will be asked next may depend on the answers for the preceding queries. However, on every path of the computation tree of the NP oracle machine, the number of queries as well as their sizes must be at most polynomial in the size of the input, since they all must be queried in polynomial time. In the whole computation tree, however, there may be an exponential number of queries, each polynomially length-bounded. For an input to be accepted it is required that there is at least one accepting path in the computation tree of the NP oracle machine (accessing its NP oracle) on the given input. Clearly, $\text{P}_{\|}^{\text{NP}}$ is contained in Σ_2^p, and thus NP is contained in Σ_2^p as well. It is widely believed that the classes NP and Σ_2^p differ (and that $\text{P}_{\|}^{\text{NP}}$ differs from them both), but an actual proof of this presumed complexity class separation is still elusive, just as a proof for settling the famous "P = NP?" question.

7.3 Complexity of Judgment Aggregation Problems

The attempts to model manipulation in JA procedures so far have been targeted at better outcomes only. However, it is possible that a manipulator would like to not just improve the outcome but to *exactly* get the outcome he desires. Introduced by Baumeister et al. [113, 112], $\mathcal{F}$-EXACT-JA-MANIPULATION captures this problem, where for the same input as in the manipulation problems above the question now is whether there is some judgment set $J^* \in \mathcal{J}(\Phi)$ such that

$$J \subseteq \mathcal{F}(J_1, \ldots, J_{n-1}, J^*).$$

Furthermore, they provided the following generic relations between the various manipulation problems for uniform premise-based quota rules.

Theorem 7.5 (Baumeister et al. [113, 112]). *For each uniform premise-based quota rule with rational quota q, $0 \leq q < 1$, it holds that*

1. EXACT-JA-MANIPULATION $\leq_m^P$ T-JA-POSSIBLE-MANIPULATION for each type $T \in \{U, TR, CR\}$ of induced weak preferences,
2. EXACT-JA-MANIPULATION $\leq_m^P$ T-JA-NECESSARY-MANIPULATION for each type $T \in \{TR, CR\}$ of induced weak preferences, and
3. EXACT-JA-MANIPULATION $\leq_m^P$ HD-JA-MANIPULATION.

Table 7.3 gives an overview of the complexity of manipulation problems in judgment aggregation with uniform premise-based quota rules. Almost all of the results of Table 7.3 are due to Baumeister et al. [113] (see also [112]), who attribute the NP-completeness result on possible manipulation for complete desired sets with respect to closeness-respecting preferences to Selker's thesis [850] and the NP-completeness result for HD-JA-MANIPULATION in the special case of the premise-based procedure to Endriss, Grandi, and Porello [394].

7.3.4 Bribery in Judgment Aggregation

Inspired by bribery in voting systems (see Section 4.3.5 starting on page 358), Baumeister, Erdélyi, and Rothe [116] have introduced various bribery problems for JA procedures and have studied their complexity. A basic variant of these problems is defined for a JA procedure $\mathcal{F}$ as follows.

$\mathcal{F}$-JA-BRIBERY	
Given:	An agenda Φ, a profile $\mathbf{J} \in \mathcal{J}(\Phi)^n$, a not necessarily complete, consistent set $J \subseteq \hat{J} \in \mathcal{J}(\Phi)$, and a positive integer k.
Question:	Is it possible to change up to k individual judgment sets in $\mathbf{J}$ such that for the resulting new profile $\mathbf{J}'$ it holds that $HD(\mathcal{F}(\mathbf{J}'), J) < HD(\mathcal{F}(\mathbf{J}), J)$?

Table 7.3 Complexity of JA manipulation with uniform premise-based quota rules. DS means "desired set," SP means "strategy-proof," and NPC means "NP-complete."

	U	TR	CR	HD	Exact
JA-Possible-Manipulation for incomplete DS	NPC	NPC	NPC	NPC	NPC
JA-Necessary-Manipulation for incomplete DS	possibly SP	NPC	NPC		
JA-Possible-Manipulation for complete DS	in P	in P	NPC	NPC	SP
JA-Necessary-Manipulation for complete DS	possibly SP	possibly SP	possibly SP		

In the above problem definition, a briber—whose desired judgment set is J—tries to influence at most k judges by "buying" their individual judgment sets, and he is successful if he can change the profile of individual judgment sets so that the resulting collective judgment set is more similar to his desired set J than the collective judgment set resulting from the original individual judgment sets. Similarity between two judgment sets is defined in terms of the Hamming distance—again, the standard definition of the Hamming distance can be extended to *incomplete* judgment sets (see page 490). Among other results, Baumeister et al. [113, 116] show that JA-Bribery is NP-complete for the premise-based procedure, and they also obtain parameterized complexity results for this problem: It remains hard even if the number n of judges or the number k of bribed judges is bounded by a fixed constant.

To give a flavor of NP-hardness proofs for bribery problems related to JA procedures, we will now provide a detailed proof of (a special case of) this theorem.

Theorem 7.6 (Baumeister et al. [113, 116]). *PBP-JA-Bribery is NP-complete, even when the total number of judges ($n \geq 3$ odd) or the number of judges that can be bribed is a fixed constant.*

Proof. Membership in NP is easy to see: Simply guess a bribery and check whether it is successful; both can be done in polynomial time. In order to prove NP-hardness, we will describe a reduction from the NP-complete problem k-Dominating-Set, which is formally defined as follows:

7.3 Complexity of Judgment Aggregation Problems

---
k-DOMINATING-SET

Given: An undirected graph $G = (V, E)$ and a positive integer $k \leq \|V\|$.

Question: Is there a dominating set of size at most k in G (i.e., a subset $V' \subseteq V$ with $\|V'\| \leq k$ such that every vertex $d \in V \setminus V'$ is adjacent to at least one vertex in V')?

We will only give the proof for the special case where the briber is allowed to bribe exactly one judge; the general case can easily be obtained by slight modifications (see the original paper [113] for the details).

As an instance of k-DOMINATING-SET, let a graph $G = (V, E)$ and a positive constant k be given. Let us fix the following notation: The vertices of G will be denoted by $v_1, v_2, \ldots, v_n$ and each vertex v_i along with its neighbors will also be written as $v_i^1, v_i^2, \ldots, v_i^{j_i}$ for some j_i. We will now construct an instance of PBP-JA-BRIBERY from (G, k). Let the agenda Φ contain

- the variables $v_1, v_2, \ldots, v_n, y$, and their negations,
- for each i, $1 \leq i \leq n$, the formula $\varphi_i = v_i^1 \vee v_i^2 \vee \cdots \vee v_i^{j_i} \vee y$ and its negation and $n-1$ syntactic variations of each of these formulas and their negations (which can be obtained, for example, by an additional conjunction with the constant 1), and
- the formula $\rho = v_1 \vee v_2 \vee \cdots \vee v_n$, its negation, and $n^2 - k - 2$ syntactic variations of this formula and their negations.

In this construction, we have three judges, $N = \{1, 2, 3\}$, a profile $\mathbf{J} = (J_1, J_2, J_3)$ with the individual judgment sets J_1, J_2, and J_3, the briber's desired set J, and the collective judgment set $PBP(\mathbf{J})$ shown in Table 7.4. Note that the Hamming distance between J and $PBP(\mathbf{J})$ is $n^2 + 1$.

Table 7.4 Construction for the proof of Theorem 7.6

Judgment set	v_1 v_2 $\cdots$ v_n	y	φ_1 φ_2 $\cdots$ φ_n	ρ
J_1	1 1 $\cdots$ 1	0	1 1 $\cdots$ 1	1
J_2	0 0 $\cdots$ 0	0	0 0 $\cdots$ 0	0
J_3	0 0 $\cdots$ 0	0	0 0 $\cdots$ 0	0
PBP	0 0 $\cdots$ 0	0 $\Rightarrow$	0 0 $\cdots$ 0	0
J	0 0 $\cdots$ 0	1	1 1 $\cdots$ 1	0

We claim that there is a successful bribery action by bribing one judge if and only if there is a dominating set of size at most k in G.

From right to left: Suppose that there is a dominating set V' of size k in G.[6] Without loss of generality, assume that the briber changes judgment set J_3 by flipping the zeros to ones in all variables $v_i \in V'$ in the premises. Now,

---
[6] Note that if the size of the dominating set V' is less than k, we can add any $k - \|V'\|$ vertices to create a dominating set of size exactly k.

the collective outcome contains the variables $v_i \in V'$. Furthermore, since V' is a dominating set, each formula φ_i, $1 \leq i \leq n$, in the conclusions evaluates to 1, and so do their syntactic variations; thus all these formulas belong to the collective outcome after the bribery. Finally, the formula ρ and its syntactic variations are in the collective outcome, too. Summing up, after the bribery the briber's desired judgment set differs from the collective judgment set in the variables $v_i \in V'$, in y, and in the $n^2 - k - 1$ formulas ρ and its syntactic variations. This implies that the Hamming distance after the bribery is $k + 1 + (n^2 - k - 1) = n^2$, which is one less than before the bribery, so the bribery action has been successful.

From left to right: Assume that the briber can successfully bribe one judge. In order to get a variable into the collective judgment set, it needs to be in at least two judgment sets. Changing any variable in judgment set J_1 will not change anything in the outcome. Thus the only possibility for a successful bribery is to bribe either the second or the third judge, i.e., to change either of the judgment sets J_2 and J_3. Since J_2 and J_3 are identical, let us assume the briber changes J_3. Since none of the judges gave a 1 to variable y, changing it in the bribed judgment set will not change the outcome; hence, the changes must be among the variables $v_1, \ldots, v_n$; assume that the briber needs to flip h variables of them from zero to one for reaching his goal. Let ℓ be the number of formulas φ_i that will also have changed from zero to one by this bribery action. This implies that the new collective judgment set $PBP(\mathbf{J}')$, where $\mathbf{J}'$ is the profile resulting from the bribery, and the briber's desired set J differ in the h changed entries of v_i, in the variable y, in overall $n^2 - \ell n$ formulas φ_i, $1 \leq i \leq n$, and their syntactic variations (namely, $n - \ell$ formulas for each φ_i), and in the $n^2 - k - 1$ formulas ρ and its syntactic variations. Thus the Hamming distance between J and $PBP(\mathbf{J}')$ is the sum

$$h + 1 + (n^2 - \ell n) + (n^2 - k - 1).$$

Now, since the bribery is assumed to be successful, this Hamming distance must be less than $n^2 + 1$, which is the Hamming distance between J and the collective judgment set before the bribery, $PBP(\mathbf{J})$. It follows that

$$n^2 \geq h + 1 + (n^2 - \ell n) + (n^2 - k - 1).$$

After reorganizing this inequality, we get

$$k - n(n - \ell) \geq h.$$

If $\ell < n$, it follows that

$$k - n \geq k - n(n - \ell) \geq h.$$

This, however, is a contradiction, since $k \leq n$ and $h > 0$. Hence, $\ell = n$ and thus $k \geq h$. Therefore, the h vertices corresponding to the variables v_i form a dominating set of size at most k. This concludes the proof for the case where the briber can bribe one judge only; as mentioned earlier, it is easy to modify this construction and the argument so as to prove the general case. ❑

Furthermore, Baumeister et al. [113, 116] have studied additional bribery scenarios and problems with respect to their classical and parameterized complexity:

- $\mathcal{F}$-JA-MICROBRIBERY models a situation in which a briber seeks to make the collective judgment set of a JA procedure $\mathcal{F}$ more similar to his own desired set by buying up to k individual judgments in the individual judgment sets of the judges. This corresponds to the MICROBRIBERY problem in voting mentioned on page 366.
- In $\mathcal{F}$-EXACT-JA-BRIBERY the briber does not only try to maximize the similarity of the collective judgment set to his desired set, but he tries to *exactly* obtain his desired set as a subset of the collective judgment set of the JA procedure $\mathcal{F}$ by buying up to k individual judgment sets.

Baumeister et al. [113, 116] studied the computational complexity of bribery in judgment aggregation. For the problems defined above, they provide NP-completeness and parameterized complexity results for the premise-based procedure, and polynomial-time algorithms for certain restrictions of the combined $\mathcal{F}$-EXACT-JA-MICROBRIBERY problem. Their results on the premise-based procedures are summarized in Table 7.5, where the symbol ✗ indicates that the corresponding problem/parameter pair is not compatible, NPC means "NP-complete," and W[2] is a "parameterized" complexity class capturing fixed-parameter intractability (see the book by Downey and Fellows [355] for the formal definition). The question mark in the table stands for the only open problem: the complexity of exact JA bribery with the premise-based procedure for a fixed number of judges.

Table 7.5 Complexity of JA bribery with the premise-based procedure

Problem	General problem	# of judges	# of bribes	# of microbribes
JA-BRIBERY	NPC	NPC	NPC	✗
EXACT-JA-BRIBERY	NPC	?	W[2]-hard	✗
JA-MICROBRIBERY	NPC	NPC	✗	NPC
EXACT-JA-MICROBRIBERY	NPC	NPC	✗	NPC

Several variants of the bribery problem for the distance-based procedure $\mathcal{D}_H$ (with any fixed tie-breaking rule) have been shown to be Σ_2^p-complete by de Haan [522].

7.3.5 Control in Judgment Aggregation

Just as in voting (see Section 4.3.4 starting on page 328), control in JA procedures is a way of influencing their outcome by making structural changes to the judgment process. However, not all control actions in elections make sense in judgment aggregation and, conversely, there are control actions that are natural in the setting of judgment aggregation but less so in elections. Similarly to manipulation and bribery for JA procedures where the manipulator or briber have their own desired sets, in control the *JA chair* (or, shorter, the *chair*) has her own desired set, too.

Control in judgment aggregation has been introduced by Baumeister et al. [114] (see also [111, 112]) who have studied the corresponding decision problems in terms of their computational complexity—as in manipulation and bribery, computational hardness may provide some kind of protection against control in JA procedures. In particular, they consider control by adding judges (and the corresponding decision problem JA-CONTROL-BY-ADDING-JUDGES) and control by deleting judges (JA-CONTROL-BY-DELETING-JUDGES), the judgment aggregation analogues of control by adding voters and control by deleting voters in elections. Furthermore, Baumeister et al. [114] introduce the new, JA-specific actions of control by replacing judges and control by bundling judges,[7] giving real-world examples of these types of control from the regulations on implementing powers in the European Commission or the Council of the European Union. We will describe the latter two control actions here in detail, starting with the definition of control by replacing judges.

$\mathcal{F}$-JA-CONTROL-BY-REPLACING-JUDGES

Given: An agenda Φ, complete profiles $\mathbf{J} \in \mathcal{J}(\Phi)^n$ and $\mathbf{K} \in \mathcal{J}(\Phi)^m$, a positive integer k, and a consistent, not necessarily complete set $J \subseteq \Phi$.
Question: Are there subprofiles $\mathbf{J}' \subseteq \mathbf{J}$ and $\mathbf{K}' \subseteq \mathbf{K}$ with $\|\mathbf{J}'\| = \|\mathbf{K}'\| \leq k$ such that

$$HD(J, \mathcal{F}((\mathbf{J} \smallsetminus \mathbf{J}') \cup \mathbf{K}')) < HD(J, \mathcal{F}(\mathbf{J}))?$$

In this setting, the chair is trying to achieve an outcome as close as possible to her desired set J by replacing no more than k judges by the same number of new judges from a set $\{n+1, n+2, \ldots, n+m\}$ of possible judges specified by their given individual judgment sets in $\mathbf{K}$.

---

[7] Remotely related bundling problems in judgment aggregation have been studied by Alon et al. [16]. However, their setting is different from ours. They consider judgment aggregation over independent variables, and only the variables are bundled in their bundling attacks. It is assumed that then all judges decide over all bundles by deciding uniformly for all variables contained in the same bundle. Furthermore, the goal in their model is to always accept all positive variables, that is, a complete desired judgment set. This setting in fact covers a restriction of judgment aggregation known as optimal lobbying (recall the definition from Section 4.3.5.5 starting on page 363; see also the papers by Christian et al. [294], Binkele-Raible et al. [149], and Bredereck et al. [226]).

7.3 Complexity of Judgment Aggregation Problems   501

  Before giving the formal definition of our second JA-specific control problem, namely control by bundling judges, let us ask Chris, David, Edgar, Felix, George, and Helena for their help in illustrating this model.

Biathlon is a popular and demanding winter sports. Each athlete needs to have cross-country skiing skills, and after each round they each have to shoot with a rifle on five targets. The Biathlon World Cup is taking place in the sports park just outside the city, and Edgar, Felix, George, and Helena are the official referees on this weekend. Before the competition can start, it is their job to decide whether the conditions are good enough for the competition to take place. To this end, they have to check the quality of the slopes and the shooting range, and they have to check the weather conditions, which are crucial both for skiing and for shooting.
  The event organizers, Chris and David, are really afraid of cancelling the competition, as that would be a financial disaster for them.
  "I am pretty sure that when all four referees will aggregate their judgments, the event will be cancelled because the weather actually is not good enough for biathlon today. *And we will definitely lose all our investments!*" Chris cries out in utter despair.
  "Don't worry, Chris!" David calms him down. "I know a way to make sure that we won't lose any money! Felix, George, and Helena have all been biathlon athletes a few years ago and they indeed are experts on judging the critical conditions, but—lucky for us!—Edgar doesn't have the slightest concept of biathlon! So let's make him our 'weather expert,' and let us make sure that the others will decide on the technical issues only, the slopes and the shooting range, which in fact are fine."
  No sooner said than done! According to the event organizers' decision, only Edgar's judgment regarding the weather will count, and the other three judges' evaluation will be relevant only for the other two issues.
  "Fortunately, the fog disappeared in the last hours, it is a bit windy though, but I'm pretty sure that's nothing the athletes wouldn't be able to handle," Edgar explains to the others. And, quietly, he adds to himself, "The slope and the shooting range look good to me as well, even though I've been told that I'm actually not here to judge them!"
  Unlike Edgar, who never was really good at sports, Felix, George, and Helena have been doing biathlon themselves. With this background they now are discussing the technical issues. Felix is the first to speak up.
  "The organizers have done a very good job: The slopes are perfectly prepared and, in accordance with the guidelines, the shooting range is fulfilling all the necessary conditions!" he says enthusiastically. "But still," Felix continues, suddenly frowning, "I'm afraid it may be too windy, the weather conditions in fact are irregular, and I personally

Table 7.6 Example for control by bundling judges

Referee	Weather	Slope	Shooting range		All conditions fulfilled: Race!
Edgar	1	1	1		1
Felix	0	1	1		0
George	0	1	0		0
Helena	0	1	1		0
$UPQR_{1/2}$	1	1	1	$\Rightarrow$	1

would not start the race. Though, we've been told that's none of our business..."

"I agree with you regarding the slopes and the weather," adds George, "but I am a bit skeptical about the shooting range. I think the wall behind it is not high enough!"

"I don't think the height of the wall will give cause for serious concern," analyzes Helena. "You see, there is a river directly behind the shooting range and at this time of the year it is very icy, so there are no boats on it. Even in the unlikely case that an athlete might aim too high, the other side of the river is far out of range. I agree with both of you on the great quality of the slopes and on the irregular weather conditions." Turning to Edgar, she adds, "Edgar, you've heard our doubts regarding the weather, so let me ask you: Are you really sure about your judgment of the weather?"

"Yes, I am!" Edgar sticks with his earlier opinion. "And I'm sick of being lectured by people who haven't been asked to judge the weather conditions. That's *my* job!"

The referees' individual judgments are given in Table 7.6, where 1 and 0 represent "yes" and "no" answers, respectively. Furthermore, we mark the individual judgment for a single variable in the premises with 1 or 0 if it is the judgment of a referee not belonging to the group who decides over this variable. Assume that the referees decide each issue by the uniform premise-based quota rule with quota $q = 1/2$.[8] If we would count each referee's judgments of all issues, the competition would not take place (much to Chris's and David's regret), since a strict majority of the four referees think the weather is not good enough for the race to take place. The reader can easily verify, however, that after bundling the judges, all three conditions have a strict majority of approvals among those referees that are responsible for them; thus the competition will take place. This is an example of a successful control action

---

[8] Note that we don't use the premise-based procedure here, as it is defined only for an odd number of judges, but we consider cases with both an odd and an even number of judges in this example.

7.3 Complexity of Judgment Aggregation Problems

by bundling judges. Formally, the problem is defined as follows for the uniform premise-based quota rule with quota q, $UPQR_q$, where we will use the notation

$$\Delta = \bigcup_{1 \leq i \leq k} UPQR_q(\mathbf{J}|_{\Phi_p^i, N_i}),$$

with $UPQR_q(\mathbf{J}|_{\Phi_p^i, N_i})$ being the collective judgment set obtained by restricting the premises of the agenda, Φ_p, to its Φ_p^i part in a partition according to the problem definition below and by restricting the set of judges to $N_i \subseteq N$.

$UPQR_q$-JA-CONTROL-BY-BUNDLING-JUDGES

Given: An agenda $\Phi = \Phi_p \uplus \Phi_c$, subdivided into the premises and the conclusions, where the set Φ_p of premises is partitioned into k subsets $\Phi_p^1, \ldots, \Phi_p^k$, a complete profile $\mathbf{J} \in \mathcal{J}(\Phi)^n$, and a consistent, not necessarily complete set $J \subseteq \Phi$.

Question: Is there a partition $\{N_1, \ldots, N_k\}$ of the set N of n judges such that

$$HD(J, \Delta \cup \{\varphi \in \Phi_c \mid \Delta \text{ satisfies } \varphi\}) < HD(J, \mathcal{F}(\mathbf{J}))?$$

As for bribery in JA procedures, one can define the exact variants of the previous control problems. Again, this can be done by asking whether it is possible to achieve a Hamming distance of zero by a given control action.

Baumeister et al. [114] show that for the uniform premise-based quota rules the exact variants of the above four control problems polynomial-time many-one reduce to their Hamming-distance-respecting variants. For example, $UPQR_q$-EXACT-JA-CONTROL-BY-ADDING-JUDGES $\leq_m^P$ $UPQR_q$-JA-CONTROL-BY-ADDING-JUDGES, and the analogous statements apply to the cases of control by deleting judges, replacing judges, and bundling judges.

Similarly to control in elections, we say that a JA procedure $\mathcal{F}$ is *immune to a control type* if it is never possible for the chair to successfully exert this type of control. On the other hand, $\mathcal{F}$ is said to be *susceptible to a type of control* if it is not immune to this control type. If $\mathcal{F}$ is susceptible to some control type and the corresponding decision problem is in P, we say that $\mathcal{F}$ is *vulnerable to this type of control*, whereas $\mathcal{F}$ is said to be *resistant to a type of control* if it is susceptible to this control type and the corresponding decision problem is NP-hard.

Baumeister et al. [114] show that the premise-based procedure is resistant both to exact control and to control by all four control types mentioned above (i.e., to control by adding, by deleting, by replacing, and by bundling judges), and they also obtain some results on uniform premise-based quota rules.

For the distance-based procedure $\mathcal{D}_H$, the complexity has been studied for variants of the control problem where—instead of adding or deleting judges—issues can be added to the agenda or removed from the agenda. In this variant of the control problem, there is a larger agenda (a "superagenda," if you will), and all judges have an opinion on this larger agenda (in the form of a judgment set over this superagenda). The judgment aggregation scenario initially involves an agenda that is a subset of the superagenda, and the judgment set of each judge is restricted to this agenda. The control problem then involves deleting issues from the agenda and/or adding issues (and their negation) from the superagenda to the agenda, in such a way that applying the judgment aggregation procedure to the resulting agenda yields a more preferable outcome than when applied to the original agenda. De Haan [522] showed that for the distance-based procedure $\mathcal{D}_H$ (with any fixed tie-breaking rule) several variants of this control problem are Σ_2^p-complete.

7.4 Concluding Remarks

Judgment aggregation is a rapidly evolving field and the related literature is growing fast. Although it is not possible to cover all of it in this short chapter, we refer the reader to some further papers.

Alon et al. [15] have studied the following problems that are remotely related to judgment aggregation. There is a set of proposals and a group of voters want to make a collective decision as to which ballot of proposals should be accepted. Every voter has a favorite ballot containing those proposals she accepts, while she rejects the other ones. A voter *accepts a ballot* if more than half of its proposals are contained in her favorite ballot. The problem is to decide whether there exists a ballot Q that (a) satisfies a certain ballot called the *agenda* (which is unrevealed to the voters and is not to be confused with what "agenda" means in judgment aggregation) in the sense that Q has a given number of proposals in common with the agenda, and (b) Q is accepted either by all voters or by a strict majority of voters. This gives rise to two problems, UNANIMOUSLY-ACCEPTED-BALLOT and MAJORITYWISE-ACCEPTED-BALLOT, which are studied in terms of their (parameterized) complexity (with respect to a number of natural parameters such as the number of proposals, the number of voters, the size of the agenda, the size of the solution ballot, and the maximum size of favorite ballots).

For further background on judgment aggregation, we refer to the surveys by List [662] and List and Puppe [664] and to the book chapters by Endriss [387] and Baumeister, Rothe, and Selker [123].

Part III
Fair Division

Chapter 8
Cake-Cutting: Fair Division of Divisible Goods

Claudia Lindner · Jörg Rothe

8.1 How to Have a Great Party with only a Single Cake

Everyone knows that whenever there is a party, there is also a variety of tastes. Each to his own, you may say—but what if you are the host? Wishing to keep everyone happy and to avoid arguments amongst the guests, a good host would offer a wide choice of food. But what if there is only a single wedding cake? Well, even a single cake may serve well to account for all the different tastes. There may be several layers of delicious sponge, tasty fresh strawberries, and even loads of cream and chocolate splits. It is easy to see that this cake would offer something for everyone—the fruit lover as well as the chocolate fanatic. The crucial bit, though, is that the host will have to divide the cake such that each guest receives a piece she is satisfied with. Even though this may sound easy at a first glance, it will soon become apparent that this is not necessarily the case. It can actually be quite a challenge to cut a single cake in a way such that everyone is happy with the piece received and does not envy anyone else for their portion.

Cutting a cake (or, *cake-cutting*) is a metaphor for dividing an infinitely divisible resource (or, good) amongst several players (or, agents). As everyone knows, people's preferences do not only differ with respect to taste but to all kind of things. Hence, the problem of fair division of a heterogeneous divisible resource also applies to areas such as economics & law (e.g., the division of a property in a divorce or inheritance case), science & technology (e.g., the division of computing time among several users sharing a single computer, or the division of bandwidth when sharing a network), and even politics.

A rather historical example is the division of Germany among the allied powers—the USA, Great Britain, France, and Russia—as a result of the Second World War. Here, the aim was not only to divide the former German Empire (from 1945 to 1949 known as "Germany as a whole") into four occupation zones but to also divide the capital Berlin. Even though this might have been just a coincidence in terms of time, it is interesting to note that the first scientific results in the area of cake-cutting were published at around the same time, at the end of the 1940s, starting with the pioneering work *"The Problem of Fair Division"* by Steinhaus [889].

As just mentioned, cake-cutting describes the problem of fairly dividing a divisible resource. The decisive question, though, is what "fair" might mean in this context. What does it take for a division to be fair? Fairness in this regard is not related to the sheer size of a player's share but rather to the player's preferences. The crucial point is how much a player *values* a certain share, irrespective of its size. What aspects need to be taken into account when evaluating the quality of a division? After providing a more formal description of the cake-cutting problem in Section 8.2, we will try to find answers to all these questions, and beyond, in Section 8.3. Once we have got more of an idea of the different valuation criteria that may play a role in fairly dividing a cake, we will look at the cake-cutting problem from a practical point of view. That is, we will ask: *How* can we fairly divide a cake? Therefore, in Section 8.4 we will present and discuss a number of cake-cutting protocols some of which are fairly sophisticated.

8.2 Basics

Let us begin with the definition of some fundamental terms and with outlining some key ideas. To represent an infinitely divisible resource, we will use cake $X = [0,1]$, defined by the unit interval $[0,1]$ of real numbers (see Figure 8.1). At a first glance, it may seem that there is no need to restrict X to the unit interval $[0,1]$. However, this definition is justified because even in $[0,1]$ real numbers are uncountable. Using bigger intervals such as $[0,117]$ or even higher-dimensional intervals like $[0,1]^2$ does not add anything to the description of the problem. Obviously, X does not necessarily need to be a cake. In fact, interval X represents *any* infinitely divisible resource that is to be divided among n players.

Given n players, $p_1, p_2, \ldots, p_n$, the aim is to perform cuts to divide cake X into n *portions* (or, *shares*), $X_1, X_2, \ldots, X_n$, where player p_j receives portion $X_j \subseteq X$ and it holds that $X = \bigcup_{j=1}^{n} X_j$ and $X_i \cap X_j = \emptyset$ for $1 \leq i < j \leq n$. Such a partition of cake X into n portions, which are assigned to the n players, is called a *division of X*. Note that a portion is not necessarily a single, contiguous piece of cake, but it can also be a collection of disjoint, possibly noncontiguous pieces of X (cf. Definition 8.10 on page 544). In other words, to obtain a division of X, more than $n-1$ cuts may be made and hence more than n pieces may be cut.

For $1 \leq i \leq n$, player p_i's preferences over the pieces of X are measured using an individual valuation function v_i that maps every piece of the cake to some real number in $[0,1]$. Note that every player knows only the value of (arbitrary) pieces of X corresponding to her own valuation function. Which

8.2 Basics

subsets of the cake are allowed to be "pieces of cake"? Without going into detail here, we mention that Kern et al. [609, 610] survey the commonly used definitions in the cake-cutting literature and highlight their strengths and weaknesses. Looking through the lens of measure theory at cake-cutting, they also make some recommendations on what definitions would be most reasonable, depending on what type of cake-cutting protocol one is considering.

It is worth pointing out that distinct players may value one and the same piece of cake differently, i.e., their individual valuation functions will in general be distinct. Moreover, subintervals of $[0,1]$ having equal size can be valued differently by the same player. Anna, for example, loves strawberries but hates cream. Whenever her grandmother serves strawberry cream cake at a birthday party, she is always hunting for a piece with loads of strawberries and as little cream as possible. For Anna, the strawberry cream cake is a *heterogeneous* resource. It is not all about its size; Anna may value a larger piece even less than a smaller one depending on the topping.

We will not go into detail of the division of a *homogeneous* cake, as this is only an extremely simplified version of the problem.[1] In this case, due to the resource being homogeneous, a single player values all pieces of equal size the same. To fairly divide the cake, a simple and straightforward procedure would be to assign a piece of size $1/n$ to each of the n players. This trivially solves the problem, since every player feels that every other player received a piece of exactly the same value as she herself.

Each valuation function v_i is assumed to satisfy the following properties:

1. *Normalization:* $v_i(X) = 1$ and $v_i(\emptyset) = 0$.
2. *Positivity:*[2] For all pieces $A \subseteq X$, $A \neq \emptyset$, we have

$$v_i(A) > 0.$$

3. *Additivity:* For all pieces $A, B \subseteq X$, $A \cap B = \emptyset$, we have

$$v_i(A \cup B) = v_i(A) + v_i(B).$$

---

[1] Note, however, that Feige and Tennenholtz [446] studied the problem of dividing a *homogeneous* divisible resource among n players, where every player's private, nonnegative valuation function only depends on the amount of the resource that she receives. They define the *fair share* of a player to be the average value a player could receive if all the players had the very same valuation function as this player. A division is said to be *fair* if every player receives a fair share. Further, they present a randomized allocation procedure in which every player that follows her dominant strategy is guaranteed to maximize her expected utility (i.e., is guaranteed to get an expected utility of at least $n/(2n-1)$ of her fair share, where n is the number of players). This is optimal in the sense that there is a collection of valuation functions for which no allocation procedure whatsoever can guarantee a larger fraction of the fair share. In certain special cases, however, this allocation procedure offers a larger fraction than the players' fair shares.

[2] The literature is a bit ambiguous regarding the property of positivity, which is assumed, for example, by Brams and Taylor [204]. Some authors (e.g., Robertson and Webb [814]) instead require *nonnegativity:* $v_i(A) \geq 0$ for all pieces $A \subseteq X$ with $A \neq \emptyset$.

4. *Divisibility:* For all pieces $A \subseteq X$ and all real numbers α, $0 \leq \alpha \leq 1$, there exists a piece $B \subseteq A$, such that

$$v_i(B) = \alpha \cdot v_i(A).$$

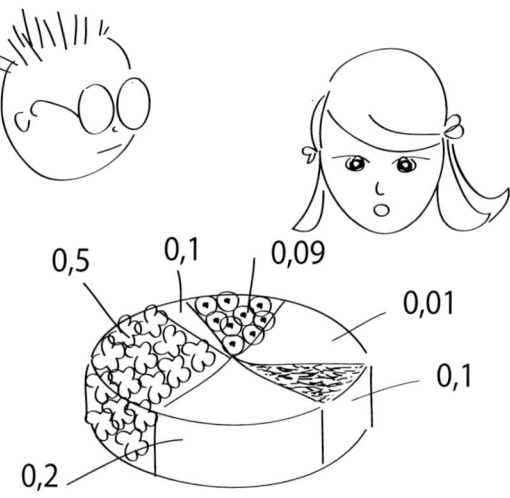

Fig. 8.1 A (normalized) cake as a metaphor of a heterogeneous resource

Sometimes, the valuation functions are required to satisfy additional requirements. We will point these out where applicable.

> "Thanks for putting down these properties." says Edgar, once his sister Belle has finished writing. After looking at them again, he adds, "I don't have a clue, though, what they might mean ..."
>
> "OK. Let me try to explain. Normalization means that you fit your valuation for the whole cake into some range, here $[0, 1]$. For example, if you would assign the whole cake the value 10 then after normalization you would value the whole cake only 1. The same applies to every single piece. If before normalization you value a piece 2 then after normalization you would value the same piece only 0.2. Fitting every valuation function into a standardized range allows to easily compare divisions of different cakes. In addition, you should not assign any nonzero value to a piece of size zero, which is also called the *empty* piece.
>
> Edgar nods his head: "I see. And what about the other three properties?"

8.2 Basics

Belle is a bit in a hurry because she is going to meet with Anna, so she is keeping her explanations briefer: "Let's see. Positivity means that you value every proper—or *nonempty*—piece more than nothing. That is, no matter how tiny a proper piece is, you will assign a value greater than zero to it. The only piece that is valued zero is the empty piece. Additivity basically describes that the value you would assign to the union of two pieces equals the sum of your valuation of both the single pieces. Now, divisibility means ... wait, I got it: Let's assume you have a piece that you would want to share."

"Why would I want to do *that*?" baffled Edgar interrupts her.

"Fine! Let's assume that *I* have a piece I would want to share with *you*. Divisibility says that I can always cut my piece in a way such that I value the share to be given to you whatever I would want it to be. By the way, divisibility and additivity imply that you can cut a piece as often as you want, and you will never lose any value. Got it, Edgar? Bye ... "

Belle turns around and off she goes. Edgar is yelling after her: "What about all the crumbs? Doesn't that mean that I do lose value?" But Belle is already gone.

In Section 8.3, we will discuss various fairness and other criteria that are to be taken into account when evaluating a division. First of all, what about the *existence* of a solution: *Is there always a division that does fulfill a certain criterion (e.g., fairness—which we will specify below)?* Many of the questions about the existence of a solution have been answered by economists and mathematicians. In this book, however, where we focus on the *computational* aspects of problems, we also want to answer the following question: *How can we achieve such a division (e.g., a "fair" division)?* The answer to this question requires algorithmic solutions, which will be given by so-called cake-cutting protocols.

A *cake-cutting protocol* describes an interactive procedure recommended to be followed to divide cake X among n players. The protocol itself does not hold any information about the players' valuation functions. However, the protocol may ask some player to provide his valuation of a specific piece of cake. Based on the player's answer, the protocol may then recommend how to continue (e.g., for this player to either accept the piece, or to trim it if he valued it above a certain threshold). Robertson and Webb [814] formalize this interaction between the protocol and the players by distinguishing two kinds of requests:

1. *evaluation requests* of the form $\texttt{eval}_i(S)$ for $S \subseteq X = [0,1]$ require player p_i to return the value $v_i(S)$, and
2. *cut requests* of the form $\texttt{cut}_i(S, \alpha)$ for $S \subseteq X = [0,1]$ and $0 \leq \alpha \leq 1$ require player p_i either to return the value $x \in S = [s_1, s_2]$ such that $v_i([s_1, x]) = \alpha$

(that is, the protocol asks p_i to cut—or make a marking—at point x to produce (or mark) a subpiece of value α), or to announce that this is impossible because no such x exists in S.

We will present a variety of cake-cutting protocols in Section 8.4.

Every cake-cutting protocol is characterized by a set of rules and a set of strategies. The *rules* determine the course of action by giving general instructions to the players on what to do next. These instructions are independent of the players' valuation functions and their execution is always verifiable, for example: "Cut the cake into two pieces!" As described in the chapter on noncooperative game theory (see Section 2.1), a *strategy* is a complete and precise policy defining an action for every contingency (i.e., for every situation a player may encounter in the course of a game), and every player has a set of such (rule-consistent) strategies to choose from. (Note that cake-cutting can be viewed as a game as well.) However, referring to cake-cutting protocols, a strategy does not provide actions for the entirety of contingencies but only focuses on actions required to achieve a *fair* and/or *efficient* solution. For example, a cake-cutting protocol may guarantee every player following its recommended strategies to end up with a proportional share, independently of this player's valuation function and independently of the actions undertaken by the other players.

It's Edgar's birthday and his grandmother made him one of her delicious strawberry cream cakes. By the time the grandmother arrives, though, Edgar had already snacked on a bunch of sweets, which is why he is happy to share his birthday cake with his sister Belle. Their grandmother suggests the following rules to cut the cake: "Edgar, please cut the cake into two pieces to start with, and Belle, you will then choose any of the two pieces."

Edgar places the knife to start cutting, but then suddenly he stops and shouts: "That's not fair! Just because I agreed to share my cake with Belle, that does not mean that I would not want to have at least half of it for my own. Why is Belle the one to choose a piece?"

"Edgar, don't you worry. It *is* fair!" the grandmother tries to calm Edgar down. "The two of you just need to follow *this* strategy: Edgar, you will have to cut the cake into two pieces such that both pieces are of equal value to you. And Belle, you will then choose the piece that you value the most."

"I see ... What a great idea, grandma!" says Edgar and starts cutting.

This example demonstrates the difference between rules and strategies when using cake-cutting protocols. This shows that it is not sufficient to provide rules when aiming to guarantee fairness. As we will see in Section 8.4, this example describes the well-known "Cut & Choose" protocol for two players—one player cuts and the other one chooses.

When stating concrete cake-cutting protocols, as is common in this area, we will in general not distinguish explicitly between rules and strategies, but instead will give the protocol in its entirety as in this example. It will always be fairly easy to tell the rules apart from the strategies.

8.3 Valuation Criteria

There is a variety of cake-cutting protocols described in the literature (see, e.g., the textbooks [204, 814] or Section 8.4 for an overview), and it is probably fairly easy for you to design one yourself.

In the following, we will present and discuss aspects of fairness (Section 8.3.1), efficiency (Section 8.3.2), manipulation (Section 8.3.3), and runtime (Section 8.3.4) of cake-cutting protocols. As always, nothing less than perfect is good enough. Hence, it is desirable to aim for cake-cutting protocols that fulfill as many valuation criteria as possible.

8.3.1 Fairness

The one valuation criterion of cake-cutting protocols everyone talks about is fairness. Though everyone has probably a good idea of what "fairness" might mean, there is a big subjective factor going into this term. There are probably quite a few different interpretations of fairness among people, and this is not only the case when dealing with cake-cutting protocols but also in other areas of life. For example, the way a returned verdict is perceived by the convicted offender is likely to be very different from the victim's perception. Newly industrialized countries, such as China, India, or Brazil, are likely to have a very different view on global CO_2-emission regulations than already highly developed countries, such as the USA or Germany. Thinking of the historic Wembley Goal during the England vs Germany game in the 1966 FIFA World Cup,[3] the whistling of a soccer referee may kick off a long discussion about the interpretation of the rules of the game.

Aiming to guarantee a fair division of a cake, the first idea that might come to mind is to give a piece of equal size to every player. However, this will provide a fair division only for a homogeneous cake where—as mentioned above—every player values equally sized pieces the very same. When cutting a heterogeneous cake, though, the players' individual preferences need to be taken into account. The latter are reflected in the players' valuation functions,

---

[3] In the authors' personal opinion, the ball was definitely *not* behind the line!

which may assign different values to pieces of equal size and which almost certainly differ among the players. Therefore, the players' valuation functions play an important role when aiming at a fair division of a cake.

Nonetheless, the approach outlined above does indicate a first criterion for the degree of fairness of a division: proportionality. In the following, we define and explain proportionality along with two related fairness criteria.

Definition 8.1 (proportionality, super-proportionality, exactness).
Let $v_1, v_2, \ldots, v_n$ be the valuation functions of the n players. For each i, $1 \leq i \leq n$, the share X_i of player p_i is said to be

1. *proportional* (or, *simply fair*) if $v_i(X_i) \geq 1/n$;
2. *super-proportional* (or, *strongly fair*) if $v_i(X_i) > 1/n$;
3. *exact* if $v_i(X_i) = 1/n$.

A division of cake $X = \bigcup_{i=1}^{n} X_i$ is said to be, respectively, *proportional* (or, *simply fair*), *super-proportional* (or, *strongly fair*), and *exact* if every player receives, respectively, a proportional, a super-proportional, and an exact share.

So far, we have only talked about what it means for a *share* or a *division* to be proportional, super-proportional, or exact. We now define what it means for a *cake-cutting protocol* to guarantee any of these properties.

Definition 8.2. A cake-cutting protocol is said to be, respectively, *proportional*, *super-proportional*,[4] and *exact* if every player who follows the strategies of the protocol is guaranteed to receive, respectively, a proportional, a super-proportional, and an exact share of the cake—regardless of the players' valuation functions and regardless of whether or not the other players follow their recommended strategies.

We will now discuss these three properties in a bit more detail.

8.3.1.1 Proportionality

Proportionality is perhaps the most obvious criterion for evaluating the fairness of a division. All players ought to consider their portion to be a proportional share of the cake. That is, when there are two players, both of them ought to feel to have received at least half of the cake; with three players, everyone should value their share to be worth at least one third; and so on.

---

[4] Regarding "super-proportional protocols," we will relax the requirement stated in this definition, since super-proportionality cannot be *guaranteed*: For players having the same valuation function, a super-proportional division is impossible to achieve. In Section 8.4.3, we will specify conditions under which certain protocols do provide super-proportional divisions, and—slightly abusing notation—we will refer to them as super-proportional protocols, even though super-proportionality is not guaranteed for all valuation functions.

8.3.1.2 Super-Proportionality

Although super-proportionality is quite similar to proportionality, it is a more stringent criterion: Every super-proportional division is also proportional, but a proportional division does not need to be super-proportional.

Compared to designing proportional cake-cutting protocols, designing a cake-cutting protocol that (under certain conditions) achieves a super-proportional division is somewhat more challenging. Note that a super-proportional division can only be obtained if the players' valuation functions are not identical—therefore, super-proportionality cannot be guaranteed (recall Footnote 4). Though there is no need for the valuation functions to be "completely different," they do have to differ for at least the tiniest bit of cake. If this is the case, every player could receive even more than a proportional share (see Section 8.4.3). However, this is merely a rather general statement on what is required for a super-proportional division to exist. When designing an algorithmic solution (i.e., some super-proportional cake-cutting protocol), it might become necessary to specify in what way the players' valuation functions need to differ in order for this specific protocol to provide a super-proportional division.

8.3.1.3 Exactness

Exactness, too, is a more stringent criterion than proportionality: Every exact division is also proportional, but a proportional division does not need to be exact. For example, a super-proportional division is proportional but cannot be exact. However, for proportional divisions exactness and super-proportionality are not complementary (i.e., it does not hold that a proportional division is exact if and only if it is not super-proportional). Consider, for example, a proportional division that is not super-proportional. This implies that *at least one* of the players receives a portion that she values to be an exact share. An exact division, in contrast, would require *every* player to value their portion to be an exact share. Hence, there are proportional divisions that are neither super-proportional nor exact.

Exactness is a fairness criterion that has only been scarcely investigated. Here, we only include it for the sake of completeness. Its under-investigation may be due to the fact that Robertson and Webb [813] have shown that no "finite" cake-cutting protocol can guarantee an exact division. (In Section 8.3.4, we will explain the difference between finite and continuous cake-cutting protocols, see Definition 8.7.)

As described above, proportionality, super-proportionality, and exactness do provide a certain degree of fairness. The question, though, is whether these criteria give the full picture, or whether there are even stronger criteria. The criteria given in Definition 8.1 require the players to evaluate their own portion only; players do not care about other players' portions. In the following, we will see what happens when players start to worry not only about their own share but also about the other players' shares.

8.3.1.4 Looking Beyond One's Own Plate

Fig. 8.2 Belle and Edgar share a cake without envying each other

Edgar and Belle just finished cutting grandmother's delicious strawberry cream cake in a fair way—i.e., proportionally (see Figure 8.2). Suddenly, the doorbell is ringing. It is Belle's friend Anna. Edgar is completely unimpressed by Anna visiting and is about to bite off his piece of cake when his grandmother suddenly shouts: "Edgar, Anna is here! Don't you want to share your birthday cake with her, too?"

Edgar seems slightly confused and puts his piece of cake back—before he gets the chance to bite something off. After throwing an angry look at his grandmother, he puts on a big wide grin and says: "I would love to. Please, grandmother, tell us how to cut the cake such that every single one of us is happy with their portion." After a moment of silence he adds: "As there are now three of us, I'd like to be guaranteed a third of *my* birthday cake."

After briefly thinking about it, the grandmother says: "Well, I have an idea of how we can cut the cake such that each of you receives at least a third of the cake. Edgar, please ..."

8.3 Valuation Criteria

Suddenly, Edgar interrupts her saying: "What do you exactly mean by *at least* a third? As everyone should know, the cake only consists of exactly three thirds. Well, at least that is what they tell us in school."

"I am glad that you are paying attention in school, Edgar," says the grandmother, "and yes, you are right—from your point of view, the cake does consist of exactly three thirds. But, a piece that you would value exactly a third might be worth a lot less—or perhaps more—to Anna. Therefore, it could be that each of you, based on your own valuations, receives a piece that is worth more—or less—than a third."

Edgar nods and pretends to have understood what his grandmother was saying but tacitly he thinks to get back to that later. He just doesn't want to make a fool of himself in front of the girls.

"Anyway," the grandmother continues, "I suggest that each of you, Edgar and Belle, cuts their piece into three pieces of equal value. So you should value each of the new pieces to be worth exactly a third of your current piece. And once that is done, Anna chooses her best piece from both of your plates." To avoid any further arguments, the grandmother quickly adds: "Your pieces are far too big for you anyway."

No sooner said than done! Edgar and Belle cut their pieces into three pieces each, and Anna chooses the piece she likes the most from each of them.

Suddenly though, Anna bursts out: "Well, I am only a guest in your house, but isn't that a bit unfair? Why will I have to choose one piece from Belle and one from Edgar? Can't I choose two from Belle's plate instead?"

Once Anna broke the silence, Edgar doesn't seem very happy either. Acting like a prima donna, he looks at Anna's piece and says: "Why would *you* complain? You already got Belle's best piece, which is a lot better than any of mine!"

"I am quite happy with what I have got and I will be having my portion now," says Belle placidly, "before anyone else is popping up ... Bon appetit!"

The scenario just described (see also Figure 8.3) refers to the Lone Chooser protocol, which will be discussed in more detail in Section 8.4. We will see that using this protocol each of the players is guaranteed to receive a proportional share. However, this does not necessarily mean that all players will be happy with their portion. In our example, the players are envying each other! Anna envies Belle, and Edgar envies Anna; only Belle does not envy anyone. Envy-freeness is another important fairness criterion, which we will introduce and discuss along with three further criteria.

Fig. 8.3 Anna, Belle, and Edgar share a cake, and envy raises its ugly head

Definition 8.3 (envy-freeness, super-envy-freeness, equitability).
Let $v_1, v_2, \ldots, v_n$ be the valuation functions of the n players. For each i, $1 \leq i \leq n$, the share X_i of player p_i is said to be

1. *envy-free* if $v_i(X_i) \geq v_i(X_j)$ for each j, $1 \leq j \leq n$;
2. *super-envy-free* if $v_i(X_j) < 1/n$ for each j, $1 \leq j \leq n$ and $i \neq j$;
3. *equitable* if $v_i(X_i) = v_j(X_j)$ for each j, $1 \leq j \leq n$.

A division of cake $X = \bigcup_{i=1}^{n} X_i$ is said to be, respectively, *envy-free*, *super-envy-free*, and *equitable* if every player receives, respectively, an envy-free, a super-envy-free, and an equitable share.

Again, we extend the properties stated in Definition 8.3 from shares and divisions to cake-cutting protocols.

Definition 8.4. A cake-cutting protocol is said to be, respectively, *envy-free*, *super-envy-free*, and *equitable* if every player who follows the strategies of the protocol is guaranteed to receive, respectively, an envy-free, a super-envy-free, and an equitable share of the cake—regardless of the valuation functions of the players and regardless of whether or not the other players follow their recommended strategies.

We will now discuss these three properties in a bit more detail.

8.3 Valuation Criteria

8.3.1.5 Envy-Freeness

An envy-free division is a division where none of the players envies any of the other players. In Section 8.4.5, we will introduce cake-cutting protocols that guarantee an envy-free division.

It is rather obvious that envy-freeness implies proportionality—though the converse implication does not hold in general. To show that every envy-free division is also proportional, we use a proof by contrapositive: If a division is *not* proportional, there must be at least one player p_i for which it holds that $v_i(X_i) < 1/n$. Thus, there must also exist a portion X_j with $v_i(X_j) > 1/n$. The latter is due to the properties of additivity and normalization:

$$\sum_{k=1}^{n} v_i(X_k) = v_i \left(\bigcup_{k=1}^{n} X_k \right) = v_i(X) = 1.$$

Hence, player p_i envies player p_j for his share.

In the case of $n = 2$ players, the proof is even simpler. When there are no more than two players—and only in this case—a division is envy-free if and only if it is proportional (see Proposition 8.1 on page 530).

As we will see in Section 8.4.5, it is quite a challenge to find a way to obtain an envy-free division for more than three players. And even for only three players it is not that simple and does require some clever ideas to be able to guarantee an envy-free division in every case (see the Selfridge–Conway protocol in Section 8.4.5).

8.3.1.6 Super-Envy-Freeness

This criterion was introduced by Barbanel [87] and is a stricter criterion than envy-freeness. A super-envy-free division gives every player the feeling to have obtained more than every other player.[5] It is easy to see that super-envy-freeness implies envy-freeness and hence proportionality. However, the converse does not hold: Not every envy-free division is super-envy-free.

As with exactness, there are only very few results on this criterion yet. This might not only be because it is very difficult to develop a cake-cutting protocol that guarantees a super-envy-free division, but also because this criterion has been introduced only fairly recently. In terms of how challenging this criterion really is, Barbanel [87] shows that a super-envy-free division exists if and only if the valuation functions of all players are linearly independent. Webb [936] has developed the very first super-envy-free cake-cutting protocol.

---

[5] Due to the properties of additivity and normalization, it holds that if a player p_i considers the portions of all other players to be *sub-proportional* (i.e., $v_i(X_j) < 1/n$ for $j \neq i$, cf. Definition 8.3), then the same player must consider her own portion to be *super-proportional* (i.e., $v_i(X_i) > 1/n$). Therefore, for a super-envy-free division it holds that $v_i(X_i) > v_i(X_j)$ for all i and j, where $1 \leq i, j \leq n$ and $i \neq j$.

8.3.1.7 Equitability

Equitability means that all players are equally happy with their portion of the cake. For example, if two players apply the well-known Cut & Choose protocol (in which one player cuts and the other one chooses), the cutter will get exactly one half of the cake's value from her point of view, while it might happen that the chooser receives more than that from his perspective. Therefore, Cut & Choose is not equitable. Brams, Jones, and Klamler [194], however, have proposed a cake-cutting protocol for two players that, even though it fails to guarantee equitability in an absolute sense, does satisfy it in a relative sense, i.e., in terms of *proportional equitability*: After guaranteeing each of the two players a share of exactly one half of the cake in their valuation, it in addition guarantees them both the same proportion of the remaining cake, again in each player's own valuation.

In an equitable division, it is possible that everyone receives a very large value of the cake and it is also possible that everyone receives a very small value of the cake, from every player's own perspective. All that matters for a division to be equitable is that all players feel the same degree of happiness, i.e., they all value their own portion the same. For example, a division among three players each of which gets 5/6 of the cake is equitable, and so is a division among three players where everyone gets only 1/6 of the cake. In particular, this means that equitability and proportionality are incomparable: Not every equitable division is proportional (as shown by the previous example), and not every proportional division is equitable (as seen for the division obtained by Cut & Choose above). Moreover, Brams, Jones, and Klamler [194] show that equitability and envy-freeness are incompatible for more than two players, and they provide an equitable cake-cutting protocol for n players that guarantees all players the maximal equal share they all can achieve.

8.3.1.8 Overview of Valuation Criteria

Note that for all the valuation criteria described and discussed above, it is necessary for the cake to be divided such that there are no leftovers—only in this case will we call it a *division*: $X = \bigcup_{i=1}^{n} X_i$. If we would allow leftovers (a piece R that is not assigned to any of the players, i.e., $X \supset \bigcup_{i=1}^{n} X_i$ allows a nonempty leftover $R = X - \left(\bigcup_{i=1}^{n} X_i\right)$), then an envy-free but wasteful "division" would not even need to be proportional. Note, however, that some authors have also considered cake-cutting with leftovers. For example, Arzi, Aumann, and Dombb [33] show that discarding some part of the cake and fairly dividing only the remaining cake may be socially preferable to every possible fair division of the entire cake.

Figure 8.4 illustrates the relationships between all as yet defined valuation criteria. Here, an arrow ($A \to B$) indicates an implication: If a division fulfills criterion A then it also fulfills criterion B.

8.3 Valuation Criteria

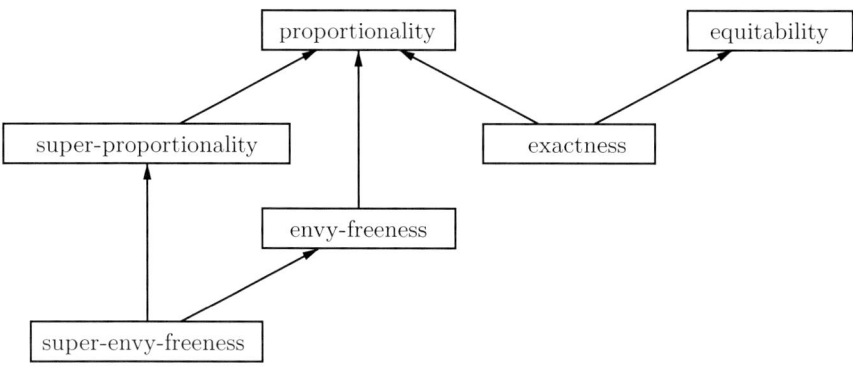

Fig. 8.4 Implications between the valuation criteria for divisions

Figure 8.4 does not show all relationships. For example, it does not show that there are proportional divisions that are neither super-proportional nor envy-free. Therefore, we also illustrate the set-theoretic inclusions among the sets of divisions satisfying certain properties (or valuation criteria) in Figure 8.5 as a Venn diagram. The big area in Figure 8.5 to the left represents the set consisting of all proportional divisions, and the smaller areas contained in it are proper subsets of this set. For example, the small area in the lower left-hand side of Figure 8.5 represents the set of all super-envy-free divisions. On the one hand, this set is a proper subset of all super-proportional divisions and, on the other hand, it is also a proper subset of all envy-free divisions. Moreover, it can be seen in this figure that, for example, the set of super-envy-free divisions and the set of exact divisions are disjoint.

In Figure 8.5, a subset relationship $A \subseteq B$ is displayed by area A being completely included in area B (and if these areas do not coincide, it is a strict inclusion: $A \subset B$). This way it is easy to see that, for example,

- all exact divisions are both proportional and equitable, but there are proportional divisions that are not exact (and not even equitable) and there are equitable divisions that are not exact (and not even proportional),
- all super-envy-free divisions are both envy-free and super-proportional, but there are envy-free divisions that are not super-envy-free (and not even super-proportional) and there are super-proportional divisions that are not super-envy-free (and not even envy-free),
- there are envy-free divisions that are neither super-envy-free nor equitable, and
- that super-proportionality excludes exactness and vice versa.

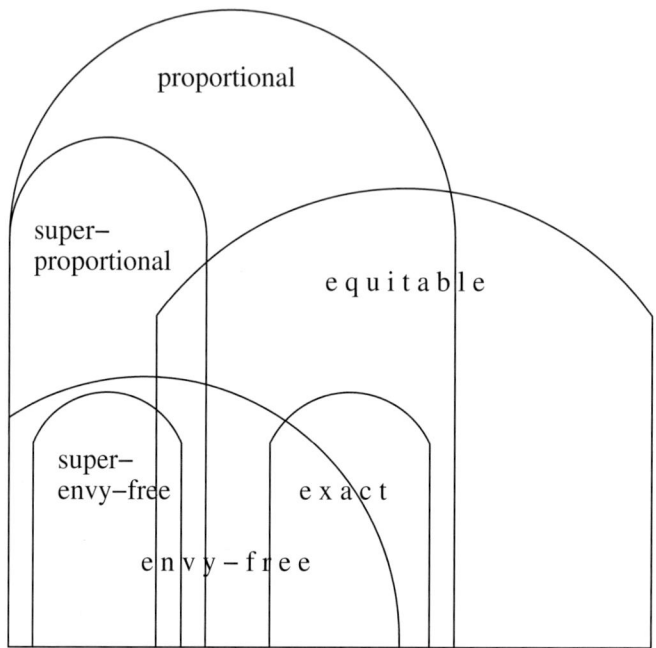

Fig. 8.5 Inclusions between the valuation criteria for divisions as a Venn diagram

8.3.2 Efficiency

Another valuation criterion for cake-cutting protocols is efficiency, which when applied to cake-cutting protocols means to find a "best possible" division. Best possible in the sense that for the same cake and the given valuation functions there is no other division that would make at least one player better off without making any of the other players worse off. We say a player is *better off* if she considers the new portion to be more valuable according to her valuation function. This notion of efficiency is also called *Pareto efficiency* (or, *Pareto optimality*) and it is based on the corresponding game-theoretic notion (recall Definition 2.3 from Section 2.1), applied to the cake-cutting setting.

Definition 8.5 (Pareto optimality or Pareto efficiency).

1. Let $v_1, v_2, \ldots, v_n$ be the valuation functions of the n players. A division of cake $X = \bigcup_{i=1}^{n} X_i$ (where X_i is the portion of player p_i) is said to be *Pareto-optimal* (or, *Pareto-efficient*) if there is no other division $X = \bigcup_{i=1}^{n} Y_i$ (where Y_i is the portion of player p_i) such that

    a. $v_i(Y_i) \geq v_i(X_i)$ for all players p_i and
    b. there is at least one player p_j with $v_j(Y_j) > v_j(X_j)$.

2. A cake-cutting protocol is said to be *Pareto-optimal* (or, *Pareto-efficient*) if every division of cake X obtained using this protocol (irrespective of the players' valuation functions) is Pareto-optimal.

It is quite obvious that (Pareto) efficiency of a cake-cutting protocol does not make any statement on fairness. A Pareto-optimal division does neither need to be proportional nor envy-free. For example, take the case that when dividing a cake among only two players one of the players gets the whole cake, whereas the other player ends up empty-handed. This division is clearly Pareto-optimal because if the player who received the whole cake would share it (i.e., give a nonempty piece to the other player), she would be worse off. In this scenario, it is not possible to make one of the players better off without making the other player worse off. However, this division by far is not fair.

This example may give the impression that you will have to decide between having *either* a fair *or* an efficient division of the cake. But is this really the case, or are there maybe ways to achieve both? The good news is that there might be hope: Weller [939] proved the existence of divisions that are both efficient and envy-free (and thus proportional). However, no concrete cake-cutting protocol is currently known to guarantee such a division for any number of players.

8.3.3 Manipulability

It's summer time. The sun is shining and Belle is shopping with her friend Anna. As shopping can be quite stressful, they decide to take a break and relax in a nearby ice cream parlor. The decision on what to get is an easy one for the two friends. They go for a Belgian waffle with ice cream and hot cherry sauce, topped with whipped cream. As both of them are quite conscious about watching their waistline, they order only one waffle to share. Fortunately, it doesn't take long for the delicious waffle plate to be served.

"Looking at these waffles makes my mouth water," Belle bursts out.

"Go ahead and cut it into two portions," Anna replies generously. — Although, it isn't really generosity that happens here. Anna just knows that the one who is in the position to choose any of the two portions is slightly better off, as in most cases she will value her portion to be more than just a half. At least, that's Anna's plan but she did overlook one thing ...

Just when Belle starts cutting speedily, anger rises in Anna and her face flushes: "That's not fair! Belle, you know that I really don't like cherries, they are worth almost nothing to me. You did that on purpose!"

"Well, go for the portion without cherries then," says Belle, confident of victory.

"But this would mean that you not only get your beloved cherries but also just a tiny bit less of the ice cream than me," replies Anna accusatorily. "If you had divided it fairly, you would have cut it so that the portion with the cherries has only a little ice cream and whipped cream and the other portion has almost all of them."

"*You* were the one who suggested for *me* to divide it into two portions," Belle puts it straight. Thereupon, Anna folds her arms and turns grumpy and moody: *"THAT'S NOT FAIR!"*

"Now you call *this* unfair!" replies Belle snappishly. "You only wanted me to cut the waffle so that you can choose the portion that appears to be the biggest and best to you!"

Suddenly, they both look at each other and start laughing. They then order a second waffle, and this time Anna is the one to cut it.

It is easy to see that the Cut & Choose protocol is an envy-free cake-cutting protocol: None of the two players would like to swap portions with the other one. If this is really the case, though, why is it that in the example given above Anna is not happy at all? Sure enough, being the chooser she can make sure to get a portion she values to be at least one half of the waffle. In the example given above, however, Belle is taking advantage of the fact that she knows that Anna does not like cherries. Belle does not divide the waffle in the ratio of 1:1 according to *her* valuation (as the envy-free Cut & Choose protocol would require), but instead she divides it so that Anna (provided that she sticks to the protocol strategies and indeed chooses the portion that is worth the most to her) is forced to choose the portion that Belle values significantly less than 50%. Doing this, Belle secures herself a portion that she values significantly more than 50% of the whole waffle.

Despite the fact that Belle was cheating, both Belle and Anna receive a portion they would not want to swap. Hence, the division is still envy-free. However, Anna is peeved at Belle because she played her off to secure herself significantly more than one half of the waffle (according to Belle's own valuation), whereas Anna ends up with only a tiny bit more than one half.

Even when a player is cheating, the Cut & Choose protocol guarantees an envy-free (and thus proportional and in that sense fair) share for all players playing truthfully. However, Belle's attempt to defraud could have gone badly wrong, in the sense that she herself could have ended up with less than half of the waffle. For example, Anna's preferences could have changed such that she now loves cherries. In that case, Belle would have misjudged Anna's preferences and Belle herself would receive the portion that she values less than 50%. Being greedy and aiming for more than a fair share, Belle was risking to end up with even less.

8.3 Valuation Criteria

The following definition captures what it means for a cake-cutting protocol to be immune against such manipulation attempts.

Definition 8.6 (immunity to manipulation).
1. Given a division of cake $X = \bigcup_{i=1}^{n} X_i$, where X_i is player p_i's portion, $v_i(X_i)$ is the *payoff of player p_i*.
2. A player is said to be *risk-averse* if she seeks to maximize her payoff in the *worst case* (i.e., with respect to all possible valuation functions or actions of all players).
3. A cake-cutting protocol is said to be *immune to manipulation for risk-averse players* if all players maximize their payoff in the worst case only by playing truthfully.

A player's payoff is given by the value she assigns to her portion after a cake-cutting protocol has been executed. Which portion she gets (and thus her payoff) depends on her and all other players' actions during the execution of the protocol. The players' *actions* result from their valuation functions and, of course, they depend on the rules and the strategies of the protocol. Which actions a player performs can usually be observed by the others (unless the protocol requests the player to do something in private). While the rules have to be followed by all players, the strategies can—but don't have to—be followed; and when some action of a player is observed,[6] one can never be sure if he really has followed the proposed strategy and indeed has cut two halves that are worth the same in his valuation.

A risk-averse player is very conservative and always expects the worst to happen. His goal therefore is to maximize his payoff even in the worst case. He doesn't take chances. He just wants to be on the safe side, rather letting some potential gain slip through his fingers than to jeopardize his best worst-case payoff. His motto is: *"A bird in the hand is worth two in the bush!"*

The point of Definition 8.6 is that only playing truthfully (i.e., playing according to the strategies proposed by the protocol) can give some player his maximum payoff in the worst case. Whenever a player tries to cheat by departing from the protocol's strategies, he is no longer guaranteed his maximum payoff, i.e., in the worst case he will get less than that. That does not exclude that he might be lucky in other cases than the worst case (i.e., for certain other, fortunate valuations/actions of the other players) in which he might increase his payoff by cheating. A bit more formally,

- let $A_i = \sigma_i(v_1, \ldots, v_n)$ denote the sequence of actions of player p_i, $1 \leq i \leq n$, that result from the sequence of strategies σ_i that the procotocol proposes for p_i when applied to the valuation functions $v_1, \ldots, v_n$, and
- let $A'_i = \sigma'_i(v_1, \ldots, v_n)$ denote the sequence of actions of player p_i, $1 \leq i \leq n$, that result from another sequence of strategies σ'_i when applied to the valuation functions $v_1, \ldots, v_n$.

---

[6] Such as Edgar cutting the cake into two halves in the Cut & Choose protocol mentioned in the example near the end of Section 8.2; see also Figure 8.8 on page 529.

Immunity to manipulation for risk-averse players in a cake-cutting protocol means:
$$(\forall i)\,(\forall \sigma'_i)\,(\exists v_{-i})\,[A'_i \neq A_i \implies v_i(X'_i) < v_i(X_i)], \tag{8.1}$$
where

- v_{-i} denotes the valuation functions of all other players, i.e., v_{-i} consists of $v_1, \ldots, v_{i-1}, v_{i+1}, \ldots, v_n$,
- X'_i is the portion p_i receives when playing the actions A'_i, and
- X_i is the portion p_i receives when playing the (truthful) actions A_i.

In other words, a protocol is immune to manipulation for risk-averse players if for all i and for all sequences of strategies σ'_i, there exists some worst case (caused by the other players' valuation functions v_{-i}) in which p_i is punished (because of $v_i(X'_i) < v_i(X_i)$) for playing the actions $A'_i = \sigma'_i(v_{-i}, v_i)$, provided they differ from the truthful actions $A_i = \sigma_i(v_{-i}, v_i)$ proposed by the protocol.[7]

Negating (8.1), we obtain the notion of *manipulability for risk-averse players*, which is formally expressed as follows:
$$(\exists i)\,(\exists \sigma'_i)\,(\forall v_{-i})\,[A'_i \neq A_i \wedge v_i(X'_i) \geq v_i(X_i)]. \tag{8.3}$$

That is, a protocol is manipulable by risk-averse players if for some i and for some sequence of strategies σ'_i, whatever valuation functions v_{-i} the other players have, p_i cannot be punished (because of $v_i(X'_i) \geq v_i(X_i)$) for playing actions $A'_i = \sigma'_i(v_{-i}, v_i)$ different from the truthful actions $A_i = \sigma_i(v_{-i}, v_i)$ proposed by the protocol.

More concretely, applying Definition 8.6 to proportional cake-cutting protocols only (such as those to be presented in Section 8.4.2), immunity to manipulation for risk-averse players means that by playing not truthfully a cheater jeopardizes receiving a proportional share of the cake.[8] This provides an incentive for the players to play truthfully.

The above notion of (non)manipulability is specific to cake-cutting protocols where the focus is on worst-case guarantees. In particular, this notion differs from *"strategy-proofness"* as defined in the context of voting (see page 274 in Section 4.2 and also Section 4.3.3). We note in passing that one could also

---

[7] Deleting the word "only" in item 3 of Definition 8.6 would mean to replace (8.1) by
$$(\forall i)\,(\forall \sigma'_i)\,(\exists v_{-i})\,[A'_i \neq A_i \implies v_i(X'_i) \leq v_i(X_i)], \tag{8.2}$$
i.e., all (risk-averse) players would then maximize their payoff in the worst case by playing truthfully, but playing untruthfully could also give them maximum payoff. Immunity to manipulation in the sense of Definition 8.6 is the stronger of the two notions. The difference between them somewhat resembles the difference between the unique-winner and the nonunique-winner model in voting (see page 290 in Chapter 4).

[8] On the other hand, it is clear that in a proportional cake-cutting protocol a cheater cannot harm any of the other players (provided they do play truthfully) with respect to proportionality.

consider *"strategy-proofness"* from a game-theoretic point of view, in the sense that playing truthfully is a "dominant strategy" (see Definition 2.2 in Section 2.1.1) for the players in the "cake-cutting game."

8.3.4 Runtime

The number of decisions made is a well-established measurement to estimate the runtime of a cake-cutting protocol. But what do we mean by a *decision*? Here, we do not only count the actual cuts of the cake as "decisions" but also the valuations and markings that may be made in the course of the protocol without actually cutting the cake. The aim is to provide a runtime measurement that can reliably be used across all cake-cutting protocols. When taking cake-cutting protocols into practice, runtime efficiency is an important factor. It is not sufficient to just know *whether* a protocol will provide a division with certain properties, but it is also important to know *when* it will provide this division (i.e., how many decisions will be required to obtain this division) in the worst case.

Generally speaking, one distinguishes *finite* from *continuous* cake-cutting protocols. While a finite protocol always provides a division after only a finite number of decisions have been made, a continuous protocol requires infinitely (in fact, even uncountably) many decisions to be made by the players. Formally, decisions are viewed as being either cut or evaluation requests (see Section 8.2).

Definition 8.7 (finite vs continuous cake-cutting protocol).

1. A cake-cutting protocol is said to be *finite* if it always (i.e., independently of the players' valuation functions) terminates with a solution after only a finite number of decisions have been made.
2. A cake-cutting protocol is said to be *continuous* (or, a *moving-knife protocol*) if a solution can only be achieved by the players continuously making decisions.

Among finite protocols one can further distinguish between *finite bounded* and *finite unbounded* ones.

Definition 8.8 (finite bounded vs unbounded cake-cutting protocol).
A finite cake-cutting protocol is said to be *finite bounded* if a finite bound on the number of decisions (which may depend on the number of players) required to solve the problem is known in advance, irrespective of the players' valuation functions; otherwise, it is said to be *finite unbounded*.

Continuous cake-cutting protocols are also known as *moving-knife protocols*. Why? The idea is that there is a knife being held above a rectangular cake. The knife gradually and continuously moves across the cake in a horizontal axis and in parallel to the left and right edges of the cake; this is visualized in Figure 8.6. Unless indicated otherwise, the starting position of the knife is always the left edge of the cake. In order to cut the cake at a certain position, a player calls "Stop!" when the knife is above this position. To determine the right moment to call "Stop!," every player needs to *continuously*—at any point in time—make decisions, and hence an (uncountably) infinite number of decisions are being made.

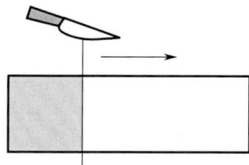

Fig. 8.6 Schematic representation of the knife movement in a moving-knife protocol

8.4 Cake-Cutting Protocols

In the following, we will present a number of known cake-cutting protocols. This will not only introduce existing cake-cutting protocols but will also show that for some valuation criteria it is very challenging to design even a single cake-cutting protocol. Table 8.11 on page 603 gives an overview of all presented cake-cutting protocols and their properties. For more details and further cake-cutting protocols, the reader is also referred to the textbooks by Brams and Taylor [204] and by Robertson and Webb [814].

Given: Cake $X = [0, 1]$ and players p_1, p_2, and p_3 with valuation functions v_1, v_2, and v_3.

Step 1: p_1 cuts X into three pieces of equal value, S_1, S_2, and S_3, i.e.,

$$v_1(S_1) = v_1(S_2) = v_1(S_3) = \frac{1}{3}.$$

Step 2: p_3 chooses one of the three pieces that is most valuable to her.
Step 3: p_2 chooses one of the remaining two pieces.
Step 4: p_1 receives the remaining piece.

Fig. 8.7 A nonproportional cake-cutting protocol for three players

8.4 Cake-Cutting Protocols

What does a cake-cutting protocol look like? Figure 8.7 shows a very short and simple example of a cake-cutting protocol for three players. Being so simple also implies that it does not fulfill many evaluation criteria. It is quite obvious that this protocol is not envy-free because player p_2 may envy player p_3 for her portion. Looking closer, it becomes apparent that this protocol is not even proportional: Though players p_1 and p_3 each receive proportional shares (p_1 because he cuts the cake into three pieces of equal value, and p_3 because she is the first one to choose), player p_2 is not guaranteed to receive a proportional share, for example not in the case when $v_2(S_1) = 1/2$ and $v_2(S_2) = v_2(S_3) = 1/4$ and player p_3 happens to choose S_1 in Step 2. In this case, player p_2 has the choice between two pieces both of which are worth less than one third of the cake to her.

8.4.1 Two Envy-Free Protocols for Two Players

Let us start with the simplest case when there are only two players. As mentioned in Sections 8.2, two players could simply apply the Cut & Choose protocol.

8.4.1.1 The Cut & Choose Protocol

The name of the protocol gives it all away: "One player cuts the cake into two pieces, and the other player chooses one of them." This procedure is probably known to everyone who grew up with a sibling and did not want to miss out on a yummy piece of cake. A more formal description is given in Figure 8.8.

Given: Cake $X = [0, 1]$ and players p_1 and p_2 with valuation functions v_1 and v_2.
Step 1: p_1 executes one cut to divide the cake into two pieces, S_1 and S_2, that he considers equal halves: $v_1(S_1) = v_1(S_2) = 1/2$. It holds that $X = S_1 \cup S_2$.
Step 2: p_2 chooses her best piece.
Step 3: p_1 receives the remaining piece.

Fig. 8.8 Cut & Choose protocol for two players

This protocol, too, looks rather short and simple. However, this one is fair: It is proportional and even envy-free. The latter follows immediately from Proposition 8.1 below, which states that in the case of $n = 2$ players proportionality implies envy-freeness. This is because for both players, if they consider their own piece to be worth at least one half, then they must value the other piece at most one half. Therefore, none of the two players receiving a proportional share will envy the other player. As mentioned in Section 8.3, this argument fails to work for $n \geq 3$ players.

Proposition 8.1. *Every proportional cake-cutting protocol for two players is envy-free.*

As mentioned in the introductory example of Section 8.3.3, when applying the Cut & Choose protocol the cutting player always receives *exactly* one half of the cake (according to his valuation function), whereas the other player always receives *at least* one half of the cake (according to her valuation function) and possibly more than one half, i.e., this protocol is not equitable. This means that Cut & Choose gives an advantage to the choosing player. However, this potential advantage of the chooser is a rather random event. It is not a *guaranteed* advantage, as the chooser herself has no influence on whether this situation is going to occur or not. On the other hand, we have seen that the cutter does have an advantage in the case that he knows the valuation function of the chooser. This means that even though the Cut & Choose protocol is proportional and even envy-free (and thus appears to be *very* fair), it does not treat the players equally. This raises the question of whether the Cut & Choose protocol really is as fair as it seems to be.

This question is less a criticism of the Cut & Choose protocol itself than it is of fairness criteria in general. Can we call a cake-cutting protocol fair just because it always outputs a fair division? What about the *way* this division was achieved? Should we not only call a cake-cutting protocol fair if both the way of achieving the division as well the division itself are fair? The reader is referred to the paper by Holcombe [554] for a detailed discussion of this issue. On the basis of this rather general approach to fairness, Nicolo and Yu [752] present an iterative version of the Cut & Choose protocol which aims at a fair way of dividing as well as at a fair division. And regarding other variants of Cut & Choose and other notions of fairness, as mentioned in Section 8.3.1.7 on page 520 already, Brams, Jones, and Klamler [194] have proposed a cake-cutting protocol for two players that—even though it is not equitable—is *proportionally* equitable.

The literature on fair division generally focuses on the fairness criteria presented in Section 8.3.1 and this is why, in the following, we will set the focus on the same. When dealing with theoretical models, though, keep in mind that a model is just a model. It is always possible to question its applicability and it is worth discussing its limitations, rather than to trust the model blindly.

8.4.1.2 Box Representations of Valuation Functions

In this chapter, we will make use of the box representation to specify the valuation function of a player. Figure 8.9 visualizes this representation: On

8.4 Cake-Cutting Protocols

the left, you see Belle's valuation function and, on the right, Edgar's valuation function.

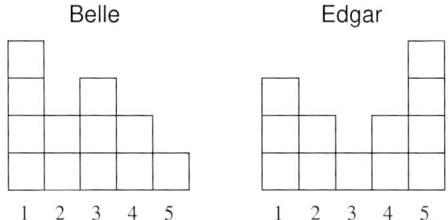

Fig. 8.9 Box representations of two valuation functions

In this example, the cake consists of a total of 12 boxes. The number of boxes per column specifies the valuation of this part of the cake, i.e., every column represents a distinct portion of the cake. Here, Belle values the right edge of the cake significantly less than Edgar does, and Edgar shows only little interest in the middle part of the cake. More specifically, Figure 8.9 shows that the left fifth of the cake is worth $4/12 = 1/3$ to Belle and $3/12 = 1/4$ to Edgar.

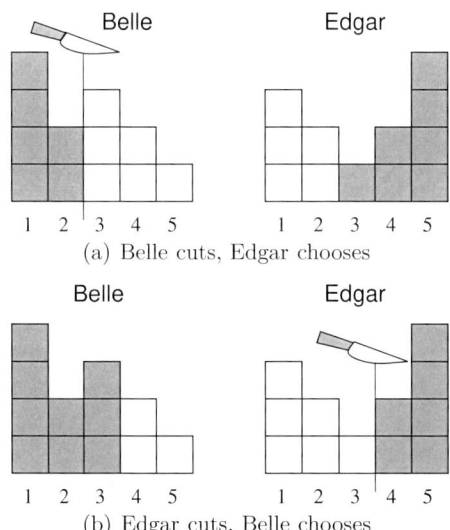

Fig. 8.10 Cut & Choose protocol: Two examples

Figure 8.10 demonstrates the step-by-step procedure of the Cut & Choose protocol using the valuation functions from Figure 8.9. In Figure 8.10(a) Belle cuts and Edgar chooses, whereas in Figure 8.10(b) Edgar cuts and Belle chooses. Dark gray shaded boxes indicate the portion assigned to the given player. It is easy to see that using the valuation functions in this example, there is a difference in outcome depending on who is cutting and who is choosing. If Belle cuts, she herself receives 6 boxes (i.e., exactly half of the cake) and Edgar receives 7 boxes (i.e., 7/12 of the cake, more than half of it). In contrast, if Edgar cuts, he himself receives 6 boxes (i.e., exactly half of the cake) and Belle even receives 9 boxes (i.e., 9/12 = 3/4 of the cake). In addition to this dependency on the order in which the players take their turn, it is easily observed that the Cut & Choose protocol is not Pareto-optimal. Figure 8.11 demonstrates an alternative division of the cake which compared to the division shown in Figure 8.10(a) assigns a more valuable portion to Belle without making Edgar worse off.

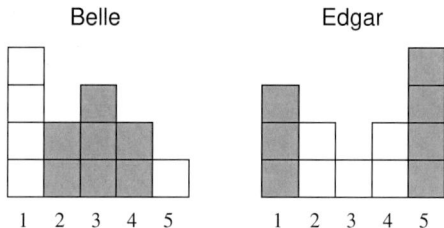

Fig. 8.11 Alternative cake division: The Cut & Choose protocol is not Pareto-optimal

It is important to realize that the box representation is a very coarse and simplified representation of the players' valuation functions: There are valuation functions where the box representation fails to adequately describe them. However, the box representation is a very handy tool to visualize examples and to demonstrate step-by-step the procedure of a protocol for a given set of valuation functions. Markings and cuts (the latter of which will be labeled by a knife) can only be done in a vertical manner. In this chapter, most examples using the box representation are designed such that cuts are always made where two columns join (one exception is Figure 8.30). This is for the sake of clarity only; in general, it is allowed to vertically cut a box.

8.4.1.3 Austin's Moving-Knife Protocol

There is a moving-knife protocol that is directly analogous to the (finite bounded) Cut & Choose protocol. However, we will give this one a miss. Instead, Figure 8.12 shows an extended version of this Cut & Choose moving-knife analogue. It was introduced by Austin [36] and is a protocol for two players that involves *two* knives. It is an extended version in the sense that it guarantees every player a portion of *exactly* one half of the cake (according to their valuation functions). Hence, this version is not only envy-free but

8.4 Cake-Cutting Protocols

even exact and also equitable. Though, it is easy to see that this cake-cutting protocol, too, is not Pareto-optimal.

With respect to moving-knife protocols, every player's valuation function needs to be continuous. Intuitively, this means that there are no "gaps."

Definition 8.9 (continuity). The valuation function v_i of a player p_i is said to be *continuous* if we have the following: Given real numbers α and β where $v_i([0,a]) = \alpha$ and $v_i([0,b]) = \beta$ for a and b with $0 < a < b \leq 1$, for every real number $\gamma \in [\alpha, \beta]$ there exists some $c \in [a, b]$ with $v_i([0, c]) = \gamma$.

Given: Cake $X = [0,1]$ and players p_1 and p_2 with continuous valuation functions v_1 and v_2.

Step 1: A knife is being held above the cake and is moved gradually and continuously from the left to the right until any of the players, say p_1, calls: "Stop!" When this happens, the knife position divides the cake into two pieces, S_1 and S_2 with $X = S_1 \cup S_2$, such that $v_1(S_1) = v_1(S_2) = 1/2$. We will call this player the *step-1-caller*. If both players call "Stop!" at the same time, a random tie-breaker is applied (for the sake of simplicity, we assume the outcome to be in favor of player p_1).

Step 2: The step-1-caller, p_1, takes a second knife and holds it above the left edge of the cake. Now, p_1 moves both knives in parallel gradually and continuously from the left to the right of the cake such that p_1 values the part of the cake that is in-between the two knives to always be exactly half of the cake. We will refer to this (over time t constantly changing) piece in-between the knives as S_1^t.

Step 3: p_2 is calling "Stop!" as soon as $v_2(S_1^{t_0}) = 1/2$ (we will refer to this point in time as t_0) and both knives will cut the cake immediately at their positions. Now, p_2 either chooses piece $S_1^{t_0}$ or the two adjacent pieces which give $X - S_1^{t_0}$, and p_1 receives the remaining portion.

Fig. 8.12 Austin's moving-knife protocol for two players

Theorem 8.1. *Austin's moving-knife protocol for two players (shown in Figure 8.12) is*

1. *exact,*
2. *not immune to manipulation for risk-averse players (in the sense of Definition 8.6),*
3. *though it does satisfy the weaker notion (8.2) stated in Footnote 7 on page 526.*

Proof. To prove exactness, look at Figure 8.12. The question is whether it is guaranteed that there always is a point in time t_0 where $v_2(S_1^{t_0}) = 1/2$ and hence p_2 calls: "Stop!" (As in Figure 8.12, we will assume that p_1 is the step-1-caller.) At the time when in Step 2 player p_1 starts to move both knives to the right (i.e., at $t = 0$), it obviously holds that $v_2(S_1^0) \leq 1/2$, as otherwise p_2 would have called "Stop!" in Step 1 before p_1 did so. In the case that $v_2(S_1^0) = 1/2$ at $t = 0$,[9] player p_2 calls "Stop!" right at the beginning of Step 2 before p_1 has a chance to start moving the knives. In this case, we have $S_1^0 = S_1$ and thus $v_1(S_1^0) = v_2(S_1^0) = 1/2$.

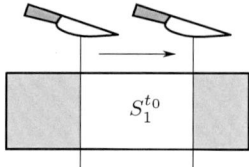

Fig. 8.13 Austin's moving-knife protocol: p_2 calls "Stop!" at time t_0

Otherwise, we have $v_2(S_1^0) < 1/2$ at time $t = 0$. Let us assume that the right knife reaches the right edge of the cake at time $t = 1$. In that case, S_1^1, the piece between both knives at that point in time, is the complement of S_1^0, i.e., $S_1^1 = X - S_1^0$. Hence, it holds that $v_2(S_1^1) > 1/2$. Because we have $v_2(S_1^0) < 1/2 < v_2(S_1^1)$ (and because we assume the valuation functions to be continuous), there must exist a time t_0, $0 < t_0 < 1$, when $v_2(S_1^{t_0}) = 1/2$.

We leave the formal proof of the second and third item—that Austin's moving-knife protocol is not immune to manipulation for risk-averse players (in the sense of Definition 8.6, see also (8.1) and (8.3) on page 526), though it does satisfy the weaker notion (8.2) stated in Footnote 7 on page 526—as an exercise for the reader.

Hint for the second item (nonimmunity to manipulation): Suppose that player p_2 is cheating by deviating from the protocol strategy as follows: Regardless of the value to the left of the knife in Step 1, p_2 will not call "Stop!" (and so will not be the step-1-caller). Assume that p_1 plays sincerely (otherwise, if no one calls "Stop!" in Step 1, the protocol will be executed anew). Show that no matter which valuation function p_1 has, p_2 will obtain at least an exact share by this cheating strategy. (Note that, by contrast, in the worst case the step-1-caller would jeopardize receiving a proportional share by deviating from the protocol strategy.) ❑

The description of the protocol in Figure 8.12 could be extended by adding that (in case p_2 does not call "Stop!" in Step 3) at the time when the right knife reaches the right edge of the cake, the left knife must be in the very same position as the right knife was at the beginning of Step 2 (i.e., before the knives started moving). However, this is just an additional note and is not required as long as the players follow the protocol's strategies.

Here is another small exercise for the reader: What would happen if Step 3 of the protocol in Figure 8.12 would be modified such that the step-1-caller, here p_1, could choose which of the two portions $S_1^{t_0}$ or $X - S_1^{t_0}$ to receive?

---

[9] This would have been the case if in Step 1 both players would have called "Stop!" at the same time but p_1 won the bid due to the tie-breaking rule.

8.4.2 Proportional Protocols for n Players

How can the idea behind the Cut & Choose protocol be transferred to three or even more players? There are actually lots of options to do so, some of which we will present in the following. All protocols in this section represent a generalization of the Cut & Choose principle to more than two players—all of them are proportional, but none is envy-free.

8.4.2.1 The Last Diminisher Protocol

Figure 8.14 shows the Last Diminisher protocol, which has been developed by Banach and Knaster and was first presented in a paper by Steinhaus [889]. This protocol works as follows: One of the n players cuts a piece that she considers to be worth exactly $1/n$ in her measure. All other players (still) in the game then value this piece in turn according to their valuation functions. If any of these players considers the piece to be super-proportional (i.e., to be worth more than $1/n$), this player trims the piece to exactly $1/n$ according to his measure before passing it on to the next player. When the last player has evaluated this piece, it is given to the player who was the last trimming it, or to the player who cut it in the first place if no trimmings have been made. The player receiving the piece leaves the game and the trimmings are reassembled with the remainder of the cake. The same procedure is then applied to the $n-1$ remaining players and the reassembled remainder of the cake. This process is repeated until only two players remain who then apply the simple Cut & Choose protocol as described before.

Given: Cake $X = [0,1]$ and players $p_1, p_2, \ldots, p_n$, where p_i has valuation function v_i, $1 \leq i \leq n$. Set $N := n$.

Step 1: Player p_1 cuts piece S_1 such that $v_1(S_1) = 1/N$.

Step 2: The cut piece is passed successively to $p_2, p_3, \ldots, p_n$, each of whom trims the piece as appropriate. In more detail, let S_{i-1}, $2 \leq i \leq n$, be the piece player p_{i-1} passes on to p_i.

- If $v_i(S_{i-1}) > 1/N$, player p_i trims piece S_{i-1} and passes on the trimmed piece S_i where $v_i(S_i) = 1/N$.
- If $v_i(S_{i-1}) \leq 1/N$, player p_i passes on the untrimmed piece $S_i = S_{i-1}$.

The last player that either cut or trimmed this piece receives S_n and drops out.

Step 3: Reassemble the remainings of the cake to $X := X - S_n$, rename the remaining players to $p_1, p_2, \ldots, p_{n-1}$, and set $n := n - 1$.

Step 4: Repeat Steps 1 to 3 until $n = 2$. The remaining two players, p_1 and p_2, apply the Cut & Choose protocol shown in Figure 8.8.

Fig. 8.14 Last Diminisher protocol for n players

Let us consider a specific example. Belle and Edgar invited Chris and David for a pizza. The preferences of the four players will be given using the box representation and the players will take their turn in alphabetical order. The valuation functions of the players are represented using 20 boxes each. In the first round, Belle cuts a slice of the pizza that she values exactly $1/4$; in Figure 8.15, this is represented by the five light gray boxes in the first two columns of Belle's box representation. She then passes the slice on to Chris who values it to be $7/20 > 1/4$. Thus, Chris trims the slice to exactly $1/4$ according to his measure and passes on the trimmed slice (the five dark gray boxes in the first column of Chris's box representation) to David. Since David values the trimmed slice only $3/20 < 1/4$ according to his measure, he passes the slice on as it is (i.e., without trimming it). Edgar is the last one to evaluate the trimmed slice and considers it to be worth $2/20 < 1/4$. As Chris was the last one trimming the slice, he is the one receiving it and drops out. In the following figures, we will represent the portion a player is receiving by dark gray boxes.

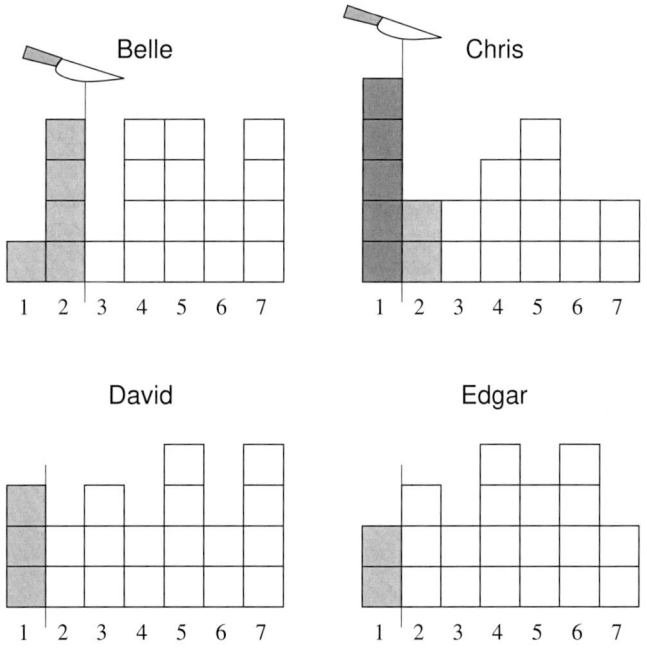

Fig. 8.15 Last Diminisher protocol for four players: 1st round

In the second round (see Figure 8.16), columns 2 to 7 are left to be shared. Here, the hatched boxes represent the part of the pizza that is no longer available to a particular player: The first column was assigned to Chris which is why this column is no longer available to Belle, David, or Edgar. Further,

8.4 Cake-Cutting Protocols

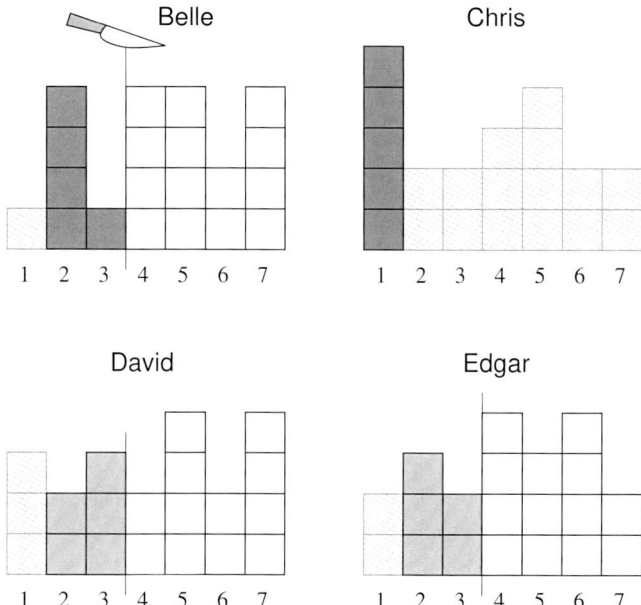

Fig. 8.16 Last Diminisher protocol for four players: 2nd round

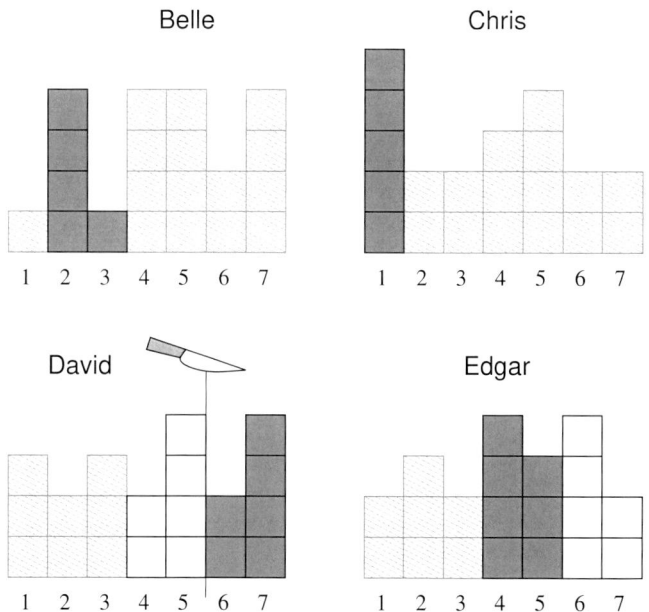

Fig. 8.17 Last Diminisher protocol for four players: 3rd and final round

as Chris has already received his portion and dropped out, columns 2 to 7 are not available to him. For any of the remaining players (Belle, David, and Edgar), columns 2 to 7 sum up to *at least* (in this example even more than) three quarters of the whole pizza. This is because each of them valued Chris' portion *at most* (in this example even less than) one quarter of the whole pizza. Again, Belle starts and cuts a slice of the pizza that she values exactly $1/4$; this corresponds to the five boxes in columns 2 and 3. She then passes the slice on to David who, too, values this slice to be $1/4$ of the whole pizza. David then passes the slice on to Edgar who also considers this slice to be $1/4$ of the whole pizza. Since in this round Belle was the last one cutting, she receives the slice represented by the dark gray shaded boxes in columns 2 and 3 and drops out.

In the final round, only David and Edgar are remaining. They share the remaining pizza (i.e., columns 4 to 7), using the Cut & Choose protocol (see Figure 8.17). Since none of them trimmed any of the previous slices, we know that both of them must value the remaining pizza to be worth *at least* (in this example even more than) half of the whole pizza. This time, David starts and cuts the remaining pizza into two halves of equal value. Then it is Edgar's turn to choose: He chooses columns 4 and 5 (seven boxes), as these are more valuable to him than columns 6 and 7 (six boxes). David receives columns 6 and 7. The whole pizza has been shared and each of the players received a proportional share (Belle and Chris) or even more (David and Edgar).

Theorem 8.2. *The Last Diminisher protocol is*

1. *proportional, but not envy-free,*
2. *finite bounded,*
3. *immune to manipulation for risk-averse players, and*
4. *not Pareto-optimal.*

Proof. 1. Proportionality of the Last Diminisher protocol follows from the fact that in each round (except for the final one) the player dropping out receives a proportional portion according to her measure, and that in the final round the remaining two players apply the proportional Cut & Choose protocol. Every player is guaranteed to receive a proportional share because none of the remaining players considers the piece that is going to get assigned to the player dropping out to be worth more than $1/n$ of the whole cake.

However, it is quite clear that the Last Diminisher protocol is not envy-free. None of the players already dropped out is involved in any way in the assignment of future portions. Players already dropped out cannot raise an objection if they consider any of the future pieces to be worth more than $1/n$. For instance, in our example Belle envies David, as she values his portion (columns 6 and 7 in Figure 8.17) to be worth $6/20 = 3/10$, whereas she considers her own portion (columns 2 and 3 in Figure 8.17) to be worth only $5/20 = 1/4$. In addition, Chris envies Edgar because he values Edgar's portion to be worth $7/20$, but considers his own portion to be worth only $5/20 = 1/4$. Can you find any more envy-relations in this example?

8.4 Cake-Cutting Protocols

2. To prove that the protocol is finite bounded, note that there are a total of $n-1$ rounds for n players: In each of the first $n-2$ rounds, Steps 1 through 3 are executed and one player drops out with her portion, and in the final round, the last two players divide the remainings of the cake by Cut & Choose. In round i, $1 \leq i \leq n-1$, each of the $n-i+1$ players still in the game makes exactly one decision.[10] In total, this gives

$$\sum_{i=1}^{n-1}(n-i+1) = n+(n-1)+\cdots+2 = \frac{n(n-1)}{2}+n-1 = \frac{(n+2)(n-1)}{2}.$$

3. To show that the Last Diminisher protocol is immune to manipulation for risk-averse players, recall Definition 8.6 on page 525 and its formal explanation from (8.1):

$$(\forall i)(\forall \sigma_i')(\exists v_{-i})[A_i' \neq A_i \implies v_i(X_i') < v_i(X_i)].$$

Consider an arbitrary player p_i. We need to show that in whatever way p_i may depart from the proposed strategies σ_i of the protocol using some σ_i' instead, there exists some worst case (caused by the other players' valuation functions v_{-i}) such that whenever p_i's actions $A_i' = \sigma_i'(v_{-i}, v_i)$ differ from the truthful actions $A_i = \sigma_i(v_{-i}, v_i)$, p_i is punished by receiving less than his best worst-case payoff: $v_i(X_i') < v_i(X_i)$, where X_i is the portion p_i receives when playing A_i and X_i' is p_i's portion when playing A_i'.

How can p_i possibly deviate from the strategy proposed by the protocol?

First, if p_i is the chooser when Cut & Choose is executed in Step 4, the only way for him to deviate from the protocol strategy is to choose a piece that is *at most* as valuable as the other for him. Clearly, this action would be punished in the worst case where the cutter in Cut & Choose cuts in a way that the piece chosen by p_i is less valuable than the other piece for p_i.

Otherwise, p_i gets one of the following cut requests:

- $\text{cut}_i(X, 1/N)$ in Step 1, where $i=1$ and X is the cake (possibly reassembled from the remainings of the previous round) at the start of this round, or
- $\text{cut}_i(S_{i-1}, 1/N)$ in Step 2, where p_i, $i > 1$, is to evaluate piece S_{i-1} and to trim it if it is super-proportional for him, and to pass it on otherwise, or
- $\text{cut}_i(X, 1/2)$ in Step 4, where $i=1$, p_i is the cutter in Cut & Choose, and X is the cake (possibly reassembled from the remainings of the previous round) at the start of this final round.

---

[10] Step 2 in round $i \leq n-2$ requires each player to make only one decision (even though one might be tempted to think that one evaluation request and, possibly, one additional cut request would be needed for each player), since one cut request is enough: $\text{cut}_j(S_{j-1}, 1/N)$.

Suppose that p_i deviates from the protocol strategy in response to this cut request. According to (8.1), we only need to show that he is punished if his resulting action differs from the truthful action (i.e., only if $A_i' \neq A_i$).[11]

Case 1: p_i cuts/trims the piece (call it X_i') to a smaller value in A_i' than in A_i, or he trims it in A_i' but passes it on in A_i. Setting the other players' valuation functions v_{-i} so that

- p_i is the last player to cut/trim this piece X_i' in Step 1 or 2, or
- the chooser in Cut & Choose of Step 4 chooses the other piece,

p_i receives X_i' and drops out with it, thus being punished with a payoff of $v_i(X_i') < v_i(X_i)$.

Case 2: p_i cuts/trims the piece (call it T) to a greater value in A_i' than in A_i, or he passes it on in A_i' but trims it in A_i.

Case 2.1: If this happens in Cut & Choose of Step 4 in round $N-1$, set the other players' valuation functions v_{-i} so that the chooser chooses this piece T, which is worth more than $1/2$ of the remaining cake of value $2/N$ for p_i. Hence, p_i receives a portion of value less than $(1/2)(2/N) = 1/N$ for himself.

Case 2.2: If this happens in Step 1 or Step 2 of round, say, $j < N-1$ (where N is the original number of players), set the other players' valuation functions v_{-i} so that

- at the start of this round, the (possibly reassembled, remaining) cake to be shared among the $N-j+1$ (remaining) players is of value $(N-j+1)/N$ for p_i, and
- p_i is not the last player to cut/trim this piece T in this round, but some other player drops out with a piece that p_i values to be worth more than $1/N$.

Thus, p_i does not receive T but remains in the game after round j, so the reassembled new cake to be shared among the $N-j$ remaining players in the following rounds is worth less than $(N-j)/N$ for p_i. Hence, we can set the valuation functions v_{-i} of the other players still in the game so that p_i eventually must receive a portion of value less than $1/N$.

In all cases, p_i is punished with a payoff of $v_i(X_i') < v_i(X_i)$.

---

[11] To understand the difference between deviating from the protocol strategy and the corresponding action, consider the following deviating strategy of a cheater p_i: "If p_i is the last player to evaluate S_{i-1} in Step 2 of some round and $v_i(S_{i-1}) = \alpha > 1/N$, then let p_i pretend to value S_{i-1} to be worth a tiny bit less than α (to be able to trim it) but still more than $1/N$ (to receive a super-proportional share)." Sure enough, if p_i is lucky, he might get a super-proportional share that way. However, the point is that in the *worst case* he will have $v_i(S_{i-1}) = 1/N$, and even though the above strategy differs from that of the protocol, his corresponding *action* will be identical to the truthful action in the worst case, i.e., (8.1) is true simply due to $A_i' = A_i$.

8.4 Cake-Cutting Protocols

4. It remains to show that the Last Diminisher protocol is not Pareto-optimal. This is obvious, for the reason alone that it includes the non-Pareto-optimal Cut & Choose protocol. ❑

The above proof to show immunity to manipulation for risk-averse players was a bit involved, even though not very difficult. Therefore, we refrain from formally showing this property again for the other protocols of this section, leaving this task to the reader.

8.4.2.2 The Dubins–Spanier Moving-Knife Protocol

On the basis of the finite bounded Last Diminisher protocol as presented above, Dubins and Spanier [360] presented a proportional moving-knife protocol. The Dubins–Spanier moving-knife protocol is shown in Figure 8.18. It is by and large a direct analogue of the Last Diminisher protocol, the only differences being that it is a continuous rather than a finite bounded protocol and that it deals differently with the situation when there are only two players remaining. The Dubins–Spanier moving-knife protocol does not use the moving-knife analogue of the Cut & Choose protocol in this situation. Instead, the first $n-1$ portions are all assigned in the same way and the last remaining player receives all the remaining cake. It is easy to see that the Last Diminisher protocol could be adapted in a similar way while still being proportional. Though, this would mean that solely the last remaining player enjoys the advantage of maybe getting more than a proportional share. In contrast, when applying the original Last Diminisher protocol, both the last two remaining players enjoy this advantage.

Theorem 8.3. *The Dubins–Spanier moving-knife protocol is:*

1. *proportional, but not envy-free,*
2. *continuous,*
3. *immune to manipulation for risk-averse players, and*
4. *not Pareto-optimal.*

We omit the proof of Theorem 8.3, which is similar to that of Theorem 8.2, but we make some comments on the Dubins–Spanier moving-knife protocol not being envy-free. Assume that three players, Anna, Belle, and Chris, share a cake using this protocol, and that Belle is the first one to call "Stop!" In this case, it is guaranteed that neither Anna nor Chris envy Belle (as otherwise some of them would have called "Stop!" before Belle did). However, it is possible that Belle envies either Anna or Chris—she cannot envy both of them at the same time. This would be the case if Anna and Chris divide the remaining cake (which Belle values 2/3) not in a way such that Belle considers both halves to be of equal value. In addition, envy can occur between Anna and Chris: The last remaining player is almost always going to be envied

---

Given: Cake $X = [0,1]$ and players $p_1, p_2, \ldots, p_n$ with continuous valuation functions.

Step 1:

- A knife is being held above the cake and is moved gradually and continuously from the left to the right until any of the players calls "Stop!" because she considers the piece of cake to the left of the knife to be worth $1/n$.
- In the case that several players call "Stop!" at the same time, a random tie-breaker is applied (for the sake of simplicity, we assume the outcome to be in favor of player p_i, where i is the smallest index amongst all players calling "Stop!").
- The knife cuts the cake at the current position.
- The player who called "Stop!" receives the left piece of the cake and drops out.

Step 2, 3, ..., $n-1$: Repeat Step 1 for all remaining players and the remaining cake.

Step n: The last remaining player receives the remaining cake.

Fig. 8.18 The Dubins–Spanier moving-knife protocol for n players

by the second-to-last player, the only exception being if the second-to-last player happens to receive a piece that she values exactly the same as she values the last remaining piece of cake. Being aware of this, there are options to decrease the envy created by this protocol: After the third-to-last player dropped out with his portion, the remaining two players call "Stop!" only if they consider the knife to be in a position that would cut the cake into halves of equal value. Following this strategy, the last two players will not envy each other (and they will not envy any of the other players either). Though, even in this case it is still possible for the last two players to be envied by any of the players dropped out before.

8.4.2.3 The Lone Chooser Protocol

The Lone Chooser protocol is a proportional and finite bounded cake-cutting protocol for any number of players. It was proposed by Fink [451] in 1964 and is presented in Figure 8.19. This cake-cutting protocol, too, generalizes the idea of the Cut & Choose approach—though in a different way than the Last Diminisher protocol does. The name of the protocol results from the fact that following the rules of the protocol, in every round, all but one player cut the cake. The non-cutting player then chooses wisely from all the pieces available—she is the *"lone chooser."* Unlike the Last Diminisher protocol, this protocol requires the number of players to *increase* with every round and it assigns all portions in the very last round; no player is dropping out before. Adding the players round-by-round as well as assigning all portions in the

8.4 Cake-Cutting Protocols

last round has the advantage that the protocol can cope with an unknown number of players.

Given: Cake $X = [0,1]$ and players $p_1, p_2, \ldots, p_n$, where p_i has valuation function v_i, $1 \leq i \leq n$.

Round 1: p_1 and p_2 apply the Cut & Choose protocol with p_1 being the one cutting the cake. Let S_1 be the piece player p_1 receives and S_2 be the piece p_2 chooses. It holds that $X = S_1 \cup S_2$, $v_1(S_1) = 1/2$, and $v_2(S_2) \geq 1/2$.

Round 2: p_3 shares S_1 with p_1 and S_2 with p_2 as follows:

- p_1 cuts S_1 into S_{11}, S_{12}, and S_{13} such that
$$v_1(S_{11}) = v_1(S_{12}) = v_1(S_{13}) = \frac{1}{6}.$$

- p_2 cuts S_2 into S_{21}, S_{22}, and S_{23} such that
$$v_2(S_{21}) = v_2(S_{22}) = v_2(S_{23}) \geq \frac{1}{6}.$$

- p_3 chooses a most valuable piece from $\{S_{11}, S_{12}, S_{13}\}$ and another most valuable piece from $\{S_{21}, S_{22}, S_{23}\}$.

$\vdots$

Round $n-1$: For each i, $1 \leq i \leq n-1$, player p_i so far has a share T_i with $v_i(T_i) \geq 1/(n-1)$ and cuts T_i into n pieces $T_{i1}, T_{i2}, \ldots, T_{in}$ such that for each j, $1 \leq j \leq n$,
$$v_i(T_{i1}) = v_i(T_{i2}) = \cdots = v_i(T_{in}) \geq 1/n(n-1).$$

Player p_n then chooses for every i, $1 \leq i \leq n-1$, one piece out of $T_{i1}, T_{i2}, \ldots, T_{in}$ that is most valuable according to her valuation function v_n.

Fig. 8.19 Lone Chooser protocol for n players

Theorem 8.4. *The Lone Chooser protocol is*

1. *proportional, but not envy-free,*
2. *finite bounded,*
3. *immune to manipulation for risk-averse players, and*
4. *not Pareto-optimal.*

Proof Sketch. We will only show that the Lone Chooser protocol is finite bounded, leaving the proof of the other properties to the reader.

For $n \geq 1$ players, there are $n-1$ rounds. In round i,

- each of the dividers $p_1, \ldots, p_i$ needs to make i decisions (namely, p_j, for $1 \leq j \leq i$, needs to respond to i cut requests for dividing his current share T_j into $i+1$ pieces of equal value, which is possible since p_j knows the value $v_j(T_j)$),

- if $i=1$, it is enough for the lone chooser p_2 to make just one evaluation, since this is Cut & Choose and she knows the value of the whole cake: $v_2(X) = 1$, and
- if $i \geq 2$, the lone chooser p_{i+1} of this round needs to make $i(i+1) - 1$ decisions (namely, to evaluate the $i(i+1)$ pieces of the dividers to choose a best piece from each divider's plate, where one evaluation can be saved, since she knows the value of the whole cake: $v_{i+1}(X) = 1$).

In total, we obtain

$$\left(\sum_{i=1}^{n-1} i^2\right) + 1 + \left(\sum_{i=2}^{n-1} (i(i+1) - 1)\right)$$
$$= \left(\sum_{i=1}^{n-1} 2i^2\right) + \left(\sum_{i=2}^{n-1} (i-1)\right) = \frac{(n-1)n(2n-1)}{3} + \frac{(n-2)(n-1)}{2}$$

decisions, which completes the proof of finite boundedness. ❏

For the Last Diminisher protocol it holds that the last two players (the ones that play Cut & Choose at the end) will never envy any of the $n-2$ other players. This is not true for the Lone Chooser protocol. In fact, the Lone Chooser protocol does not guarantee for any of the players that they will not envy any of the $n-1$ other players. Moreover, it is worth mentioning that the very first player, p_1, is slightly disadvantaged in the sense that he is only guaranteed to receive an exact share, whereas all other players (depending on their valuation functions) might receive even more than that, i.e., a superproportional share.

Note that when applying the Lone Chooser protocol, players are not guaranteed to receive a contiguous portion.[12] This is a difference, for example, between the Lone Chooser and the Cut & Choose or the Dubins–Spanier moving-knife protocol.[13]

Definition 8.10 (contiguous portion). Let a division of cake $X = \bigcup_{i=1}^{n} X_i$ be given with X_i being the portion assigned to player p_i. A portion X_i is called *contiguous* if there exist $x_1, x_2 \in \mathbb{R}$ with $0 \leq x_1 \leq x_2 \leq 1$ such that the following holds: $X_i = [x_1, x_2] = \{x \in \mathbb{R} \mid x_1 \leq x \leq x_2\}$.

---

[12] Here, we disregard the cuts that a contiguous portion might contain, i.e., a contiguous portion might consist of several adjacent pieces. This is valid without loss of generality because of the divisibility property introduced in Section 8.2: Cuts do not reduce the value of a piece, i.e., they do not create a "gap."

[13] In the Last Diminisher protocol, leftovers might be created in all but the last round whenever some player decides to trim a piece. When the trimmings are reassembled with the remainder of the cake for the next round, one may consider this "new" cake to be contiguous, but only if the protocol specifies that trimmings always have to be made from the same side.

8.4 Cake-Cutting Protocols

The Lone Chooser protocol demonstrates that a player's portion does not need to be contiguous but might be a collection of noncontiguous pieces. This is a rather important aspect, considering that Stromquist [895] has shown that there is no *finite* envy-free cake-cutting protocol for more than two players when only allowing contiguous portions. From this it also follows that there is no finite envy-free cake-cutting protocol for $n \geq 3$ players that requires no more than $n - 1$ cuts (see also Section 8.4.7).

Austin [36] points out that his moving-knife protocol as described in Section 8.4.1 can be combined with the Lone Chooser protocol such that the resulting division is not only guaranteed to be proportional but even exact. This combined version, too, is a moving-knife protocol.

8.4.2.4 The Lone Divider Protocol

Kuhn [625] presented the Lone Divider protocol, which is a proportional and finite bounded cake-cutting protocol. His protocol generalizes a method introduced by Steinhaus for three players to any number of players. The procedure is fairly straightforward in the case of three players. However, when increasing the number of players beyond three, the protocol relies on sophisticated mathematical theories.[14] Dawson [335] introduced a recursive implementation of the Lone Divider protocol.

We will not present the general Lone Divider protocol for any number of players but will focus on the selected cases of three and four players only. Let us start with the case of three players: Belle, Chris, and David. In contrast to the Last Diminisher protocol and the Lone Chooser protocol, the Lone Divider protocol only entitles a single player to cut the cake—he is the *"lone divider."* The players may toss the dice to decide on which player is going to be the lone divider—in our example, let us assume it is Belle. She cuts the cake into as many equally valued (according to her valuation function) pieces as there are players; in our case into three pieces, A, B, and C. The next step is to get all other players to evaluate these pieces according to their valuation functions. All players indicate which of the pieces they find acceptable, where a player considers a piece to be *acceptable* if and only if she considers it to be a proportional share. In our example, pieces valued at least one third are marked as acceptable. It is easy to see that all three pieces are acceptable for Belle, whereas for both the other players at least one piece must be acceptable. Here, we need to distinguish the following two cases.

Case 1: For Chris or David (let us say, for Chris), at least two pieces are acceptable. In this case, David chooses any of the pieces he considers to

---

[14] The original proof by Kuhn [625] is using the Frobenius–König theorem on the permanents of 0-1-matrices. Other sources that discuss Kuhn's protocol (e.g., the book by Robertson and Webb [814]), use graph-theoretic and combinatorial ideas such as Hall's marriage theorem on the existence of a perfect matching in a graph.

be acceptable. Then, as Chris initially considered at least two pieces to be acceptable, there must be at least one acceptable piece left for him. Chris chooses any of these. Belle receives the remaining piece, which to her is as valuable as any of the other two pieces.

Case 2: Both Chris and David consider only one of the pieces to be acceptable. Let us assume Chris considers A to be acceptable and David considers B to be acceptable. (Note that the same argument applies if both Chris and David would find the very same piece acceptable.) In this case, there must be one piece (or even two pieces in the case that both players consider the very same piece to be acceptable), say C, that both Chris and David consider to be unacceptable (i.e., both Chris and David value this piece less than one third of the whole cake). Hence, Belle receives C, as she considers all three pieces to be of equal value. This also means that both Chris and David consider the remainings of the cake, $A \cup B$, to be worth more than two thirds of the whole cake. Chris and David now share $A \cup B$ using the Cut & Choose protocol and, hence, are each guaranteed to receive more than $(1/2) \cdot (2/3) = 1/3$ of the whole cake.

This shows that in the case of three players, every player is guaranteed to receive a proportional share. However, the Lone Divider protocol is not envy-free. For example, in Case 1 Chris might value the two acceptable pieces differently. If David now chooses the piece that is most valuable to Chris, then Chris will receive a proportional share but will envy David for his piece. And in Case 2 it might happen that Belle envies one of the two other players. This would be the case if Belle considers the two pieces $A \cup B$ is cut into not to be of equal value according to her valuation function. David, however, is guaranteed to envy none of the other players, neither in Case 1 nor in Case 2.

As mentioned above, the procedure gets more complicated when there are more than three players and mathematical tools such as the Frobenius–König theorem or Hall's marriage theorem are required for proving the general case. However, for the cases of having either four or five players, there is a clever *direct* way to only assign acceptable pieces to all the players. This approach was pointed out by Custer [319], see also [204]. In the following, we will demonstrate it for the case of four players: Belle (the lone divider), Chris, David, and Edgar. Again, Belle cuts the cake, this time into four pieces of equal value according to her valuation function: A, B, C, and D. All other players then evaluate all four pieces and decide which pieces they consider to be acceptable, i.e., to be worth at least one quarter of the cake. Again, Belle finds all four pieces acceptable, and for all other players at least one piece must be acceptable. Here, we need to distinguish the following three cases.

Case 1: There is at least one piece (let us say A) that only Belle considers to be acceptable, i.e., all other players consider A to be unacceptable. If we now assign piece A to Belle, all remaining players will consider the remainings of the cake, $B \cup C \cup D$, to be worth more than three quarters of the whole cake. Hence, Chris, David, and Edgar can now

8.4 Cake-Cutting Protocols

apply the procedure for three players described above to fairly share the remainings of the cake. This guarantees each of them to receive more than $(1/3) \cdot (3/4) = 1/4$ of the whole cake.

Case 2: There is at least one piece (let us say A) that only Belle and one other player (let us say Chris) consider to be acceptable. If we now assign piece A to Chris, all remaining players will consider the remainings of the cake $B \cup C \cup D$ to be worth at least three quarters of the whole cake (for Belle exactly three quarters; for David and Edgar even more than three quarters, as they consider A to be unacceptable). Hence, Belle, David and Edgar can now apply the procedure for three players described above to fairly share the remainings of the cake. This guarantees each of them to receive at least $(1/3) \cdot (3/4) = 1/4$ of the whole cake.

Case 3: Each of the pieces A, B, C, and D is considered acceptable by Belle and at least two other players.

Table 8.1 Case 3 of Custer's procedure [319]

A	B	C	D
Belle	Belle	Belle	Belle
?	?	?	?
?	?	?	?

Table 8.1 shows all four pieces and for each piece the three players that consider this piece to be acceptable. (For the sake of convenience and without loss of generality, we ignore the possibility that a piece could also be acceptable for all four players.) At the beginning of the procedure, all we know is that Belle considers all four pieces to be acceptable. However, there must be at least one other player (let us say Chris) who considers at least three of the pieces to be acceptable, as otherwise only $2+2+2=6$ of the eight question marks in Table 8.1 could be replaced by the names of the players. But even if Chris considers all four pieces to be acceptable, either David or Edgar (let us say David) must consider at least two of the pieces to be acceptable, as otherwise only $4+1+1=6$ of the eight question marks in Table 8.1 could be replaced by the names of the players. We conclude that Belle considers all four pieces to be acceptable, Chris considers at least three pieces to be acceptable, David considers at least two pieces acceptable, and Edgar considers at least one piece acceptable. Hence, in this case, too, every player is guaranteed to receive a proportional share if the players are to choose a piece in reverse order: first Edgar, second David, then Chris, and finally Belle.

It is easy to see that for four players, too, the protocol is not envy-free. Here is a small exercise for the reader: In which of the cases and between which players could envy occur in this procedure for four players?

8.4.2.5 Cut Your Own Piece Protocol

The Cut Your Own Piece protocol, presented by Steinhaus [891], is very suitable to be used for dividing a property. So let us assume uncle Ian sadly passed away. Ian was a very rich guy and left behind a lovely lakeside property to be divided amongst his three heirs, Felix, George, and Helena. As with cakes, parts of the property having equal size can be valued differently by the same player, and also distinct players may value one and the same part of the property differently. For example, Felix would love to receive the forestry bit on the left-hand side of the property. George, however, is very enthusiastic about sailing and hence is very keen on receiving the middle bit with the private jetty. And Helena, who cares about neither the forest nor the boats, would instead enjoy having a nice herb and flower garden as well as a bit of beach, preferably the sheltered bay on the right-hand side of the property.

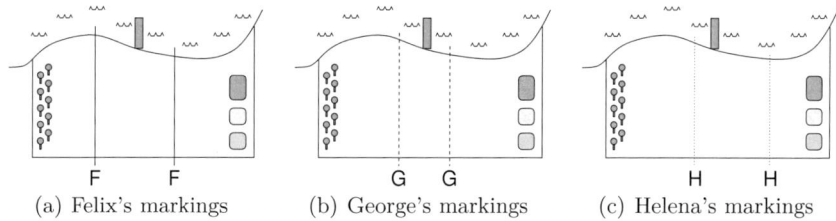

(a) Felix's markings   (b) George's markings   (c) Helena's markings

Fig. 8.20 Cut Your Own Piece protocol: individual markings

The Cut Your Own Piece protocol asks each of the three heirs to make two markings on a separate transparent sheet of paper—each of which shows a true-to-scale drawing of the property. Each marking needs to be in parallel to the left and right edge of the property and divides the property into three parts (see Figure 8.20). Markings are made such that the players consider each of their marked parts to be of the same value (according to their valuation functions). This implies that the more valuable a part of the property is to a player, the smaller the land area will be. For example, if George does not manage to receive his preferred middle part with the jetty on it (see Figure 8.20(b)) then, to make up for his loss, he would only accept a fairly large area to the left or right of this. None of the players knows or shall know the markings of the other players—which is why every single one of them is using a separate transparent sheet of paper.

The idea now is to assign to every player a part of the property that he or she marked, where for obvious reasons the three assigned parts need to be disjoint. Steinhaus [891] has shown that there always is a way to assign the marked parts of the property in a proportional manner. Being interested merely in the existence of such an allocation, he did not explicitly point out *how* to assign them. One method of doing so is given in Figure 8.23 (see

8.4 Cake-Cutting Protocols

also [659]): Assign the parts starting from the left edge of the property and keep moving to the right (where ties can be broken arbitrarily).

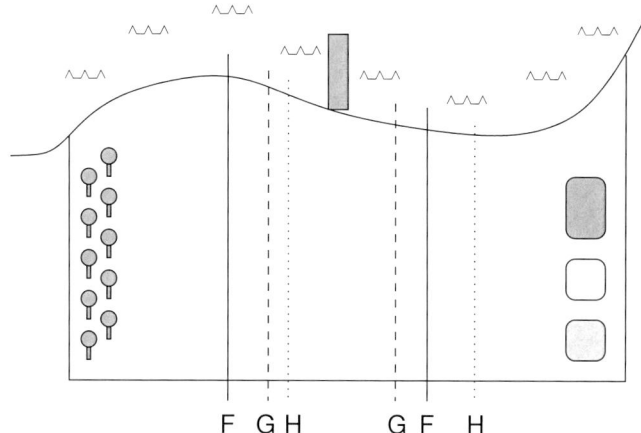

Fig. 8.21 Cut Your Own Piece protocol: all markings

Going back to uncle Ian's inheritance, Felix, George, and Helena place their three transparent sheets of paper on top of each other resulting in the picture shown in Figure 8.21. The leftmost part of the property is assigned to Felix, as he made the leftmost marking. (In the case that several players would have placed a mark at the very same position, a tie-breaker would have been applied.) That is, Felix receives a proportional share according to his measure and drops out (see Figure 8.22(a)).

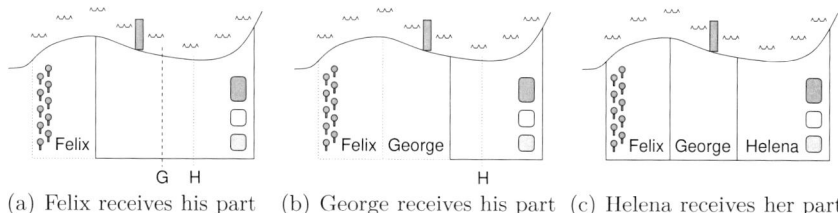

(a) Felix receives his part (b) George receives his part (c) Helena receives her part

Fig. 8.22 Cut Your Own Piece protocol: assignment of the shares

Now, all markings made by Felix as well as the leftmost markings of both the other players will be removed and the same procedure will be repeated for the remaining part of the property and the remaining players. Since George's and Helena's leftmost marks were to the right of Felix' leftmost mark, both George and Helena consider the remaining part of the property to be worth at least (in this example even more than) two thirds of the whole property.

The next player to get assigned a part of the property is George (see Figure 8.22(b)), as his originally rightmost mark (which now is his leftmost mark) is to the left of Helena's. Helena then receives the remaining part of the property (see Figure 8.22(c)). While Felix has got an exact share, George and Helena even receive a super-proportional share each. Figure 8.23 specifies the proportional Cut Your Own Piece protocol for any number of players.

Given: Cake $X = [0,1]$ and players $p_1, p_2, \ldots, p_n$.

Step 1: Every player makes $n-1$ markings to divide the cake into n pieces of value $1/n$ (according to their valuation function). All $n(n-1)$ markings are required to be in parallel to the left and right edge of the cake, and no player knows the markings of any other player.

Step 2: Identify the leftmost marking.

- The piece of cake between the left edge of the cake and the leftmost marking is assigned to any of the players that placed a marking at this position, and this player drops out.
- Remove all markings of the player just dropped out as well as the leftmost markings of all remaining players.

Step 3: Repeat Step 2 for all remaining players and the remaining cake until all markings have been removed.

Step 4: The last remaining player receives the remaining cake.

Fig. 8.23 Cut Your Own Piece protocol for n players

Theorem 8.5. *Suppose that at least two of the n players in the Cut Your Own Piece protocol have made distinct markings. Then the cake can be divided such that*

1. *all players receive a piece that they themselves have marked,*
2. *no two of these pieces overlap, and*
3. *a nonempty piece remains.*

Proof. We prove this result by induction on n. The induction base, $n = 2$, is easy to see: If two players disagree by making distinct markings, we can assign them disjoint pieces of the cake that they have marked themselves, and a nonempty piece remains between the two markings.

For the induction step, assume the claim to be true for n players. To show it for $n+1$ players, suppose that $p_1, p_2, \ldots, p_{n+1}$ have made their markings according to the Cut Your Own Piece protocol shown in Figure 8.23. Cut at the leftmost marking and assign the piece left of the cut to the player who made this marking, where ties can be broken arbitrarily. Without loss of generality, assume that p_1 gets this piece and drops out. Now, remove all markings of p_1 and also the leftmost markings of all other players p_i, $i \neq 1$. By induction hypothesis, all the players $p_2, p_3, \ldots, p_{n+1}$ receive a piece that they themselves have marked, and no two of these pieces overlap.

To prove that a nonempty piece of the cake remains, consider the following two cases.

Case 1: There are players p_i and p_j, $2 \leq i,j \leq n+1$ and $i \neq j$, whose remaining markings do not coincide. By induction hypothesis, a nonempty piece of the cake remains.

Case 2: For all players p_i and p_j, $2 \leq i,j \leq n+1$ and $i \neq j$, all remaining markings coincide. Consider the following two subcases.

  Case 2.1: The disagreement of our original assumption for $n+1$ players concerned the leftmost markings only. Thus, there exists some i, $2 \leq i \leq n+1$, such that p_i would get more than she originally has marked. Assign to p_i her originally marked piece instead. A nonempty piece of the cake remains. (All other pieces can be arbitrarily assigned to the other players, due to their markings being identical.)

  Case 2.2: The leftmost markings of the players coincided. That is, the disagreement of our original assumption for $n+1$ players concerned their other (not their leftmost) markings only. In that case, assign the first (the leftmost) piece not to p_1, but to, say, p_2. We then are again in Case 2.1.

This completes the proof. ❑

It is fairly easy to see that the Cut Your Own Piece protocol is proportional and immune to manipulation for risk-averse players. All players are guaranteed to receive a piece that they marked themselves. Thus, every truthful player is guaranteed to receive a proportional share. However, if a player does try to cheat, she is jeopardizing to receive a proportional share herself (but this does not have any impact on any other player's guaranteed proportional share). For example, a cheating player might place his marks such that some of his marked pieces is worth more than $1/n$, hoping to receive this particular piece. However, this also means that at least one of his marked pieces is worth less than $1/n$. Because the players cannot directly influence which piece they are going to get assigned (as this always also depends on the other players' markings, which are unknown), the cheating player may end up with even less than a proportional share.

8.4.2.6 Divide & Conquer Protocol

Even and Paz [416] presented a cake-cutting protocol that makes use of a principle known from the Roman times: *"Divide et impera!"* The Divide & Conquer protocol is based on the idea that sharing the whole cake X among n players can be reduced to the problem of sharing two disjoint parts partitioning the cake, A and B with $X = A \cup B$, among two disjoint subgroups of players with roughly $n/2$ players each. Here, it is important to notice that a fair division of parts A and B leads to a fair division of the whole cake X if and only if parts A and B as well as the two subsets of players were determined in a fair way.

When applying the Divide & Conquer protocol, it is essential to distinguish between the number of players n being even or odd. If the number of players is even then the group of players will be divided into two subgroups with an equal number of players. However, if the number of players is odd then the group of players will be divided into two groups of roughly the same size, namely one subgroup with an even number of players and another subgroup with an odd number of players. Following this rule, the cake is recursively divided into smaller and smaller disjoint parts, and so are the groups of players. The procedure stops when every subgroup consists of one player only, who considers the part of the cake to be shared in this "subgroup" to be a proportional share of the whole cake. Figure 8.24 specifies the details of the Divide & Conquer protocol for any number of players.

For $n = 2$ players, the Divide & Conquer protocol from Figure 8.24 is just the Cut & Choose protocol (see Figure 8.8 on page 529).

For $n = 3$ players (e.g., Belle, Chris, and David), the Divide & Conquer protocol works as follows. Belle and Chris each divide the cake in the ratio of $1 : 2$ according to their valuation functions using parallel markings (i.e., both of them make a marking where they consider the cake would be cut into two pieces, the left one being worth one third and the right one being worth two thirds). Then it is David's turn to evaluate the piece between the left edge of the cake and the leftmost marking; let us call this piece A. If David considers piece A to be worth at least one third of the whole cake, he accepts this piece and drops out. In this case, both Belle and Chris value the remaining part of the cake, $X - A$, to be worth at least two thirds. If Belle and Chris now share this part of the cake using the Cut & Choose protocol, each of them is guaranteed to receive at least one third of the whole cake. On the other hand, if David considers piece A to be worth less than one third, he shares piece $X - A$ (using the Cut & Choose protocol) with the player whose marking is the rightmost.[15] In this case, the player whose marking separates piece A from the remaining cake receives A and drops out. Again, every player is guaranteed to receive a proportional share, the player that receives piece A because she marked it herself, and the other two players because each of them considers piece $X - A$ to be worth at least two thirds and they apply the proportional Cut & Choose protocol (i.e., the case of $n = 2$ players in the Divide & Conquer protocol) to share it.

To give an example with an even number $n > 2$ of players, let us now assume we have $n = 4$ players (say Belle, Chris, David, and Edgar). It is easy to see that in this case the divide-and-conquer principle can also be applied

---

[15] In the case that both Belle and Chris made their markings at the very same position, a random tie-breaker will be used. This also applies to the general protocol given in Figure 8.24 in the case that more than one marking happen to be at the very same position but a decision on the "kth marking" needs to be made.

8.4 Cake-Cutting Protocols

Given: Cake $X = [0,1]$ and players $p_1, p_2, \ldots, p_n$, where p_i has valuation function v_i, $1 \leq i \leq n$.

Case 1: If $n = 1$ then player p_1 receives the whole cake X.

Case 2: If $n = 2k$ for some $k \geq 1$, then the following steps will be applied.

Step 2.1: $p_1, p_2, \ldots, p_{n-1}$ divide the cake using parallel markings in the ratio of $k : k$ according to their valuation functions. Let A be the piece of cake left of the kth marking. For example, if $n = 8 = 2 \cdot 4$, this may result in:

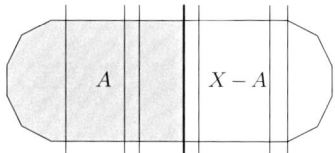

Step 2.2: If $v_n(A) \geq k/n = 1/2$, then p_n chooses piece A, and otherwise p_n chooses piece $X - A$.

Step 2.3: Recursively applying the Divide & Conquer protocol for k players,
- p_n shares the chosen piece with the $k-1$ players whose markings fall within this piece and are not the kth marking;
- the other k players share the other piece.

Case 3: If $n = 2k+1$ for some $k \geq 1$, then the following steps will be applied.

Step 3.1: $p_1, p_2, \ldots, p_{n-1}$ divide the cake using parallel markings in the ratio of $k : k+1$ according to their valuation functions. Let A be the piece of cake left of the kth marking. For example, if $n = 9 = 2 \cdot 4 + 1$, this may result in:

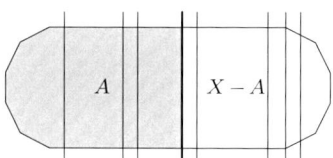

Step 3.2: If $v_n(A) \geq k/n = k/(2k+1)$, then p_n chooses piece A, and otherwise p_n chooses piece $X - A$.

Step 3.3:
- If p_n chose piece A, then he shares this piece using the Divide & Conquer protocol for k players with the $k-1$ players whose markings fall within piece A and are not the kth marking.
- If p_n chose piece $X - A$, then he shares this piece using the Divide & Conquer protocol for $k+1$ players with the k players whose markings fall within piece $X - A$ and are not the kth marking.

In both cases, the other $k+1$ (or k, respectively) players share the other piece using the Divide & Conquer protocol for $k+1$ (or k, respectively) players.

Fig. 8.24 Divide & Conquer protocol for n players

and that proportionality will be "passed on" from one recursion stage (or, round) to the next. To start with, Belle, Chris, and David each cut the cake in half, i.e., they divide the cake in the ratio of $1:1$ (according to their valuation functions) using parallel markings. This time, the middle marking is used to divide the cake into pieces A and $X - A$ (recall Footnote 15 on how to deal with markings at the very same position).

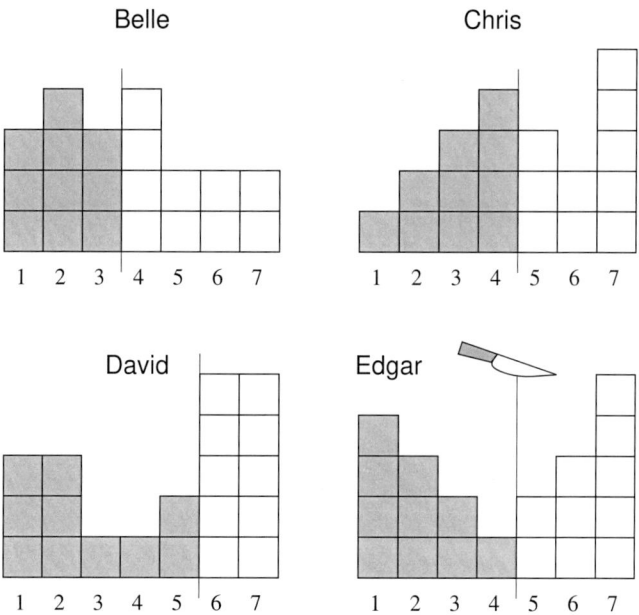

Fig. 8.25 Divide & Conquer protocol for four players: 1st round

Figure 8.25 shows a specific example. The preferences of the four players are given using the box representation. For every player, there are 20 boxes to represent the whole cake from their point of view. In this example, Belle makes a marking after the third column, Chris after the fourth column, and David after the fifth column—for each of them, the left half contains 10 boxes and is shaded in light gray. Edgar then cuts the cake using the middle marking (the one made by Chris) into pieces A, consisting of columns 1 to 4, and $X - A$, consisting of columns 5 to 7. Because Edgar considers piece A to be worth at least as much as half of the cake, he chooses A and shares this in the next round with Belle using the Cut & Choose protocol (see Figure 8.26(a)). Belle also considers piece A to be worth at least as much as (in this example even more than) half of the cake, as her marking falls within A. Simultaneously, the remaining part of the cake, piece $X - A$, is shared among Chris and David using the Cut & Choose protocol (see Figure 8.26(b)). Since both Chris's and David's markings fall within $X - A$, both of them consider this piece to be

8.4 Cake-Cutting Protocols

worth at least as much as half of the cake. Therefore, every player ends up with at least a quarter of the whole cake.

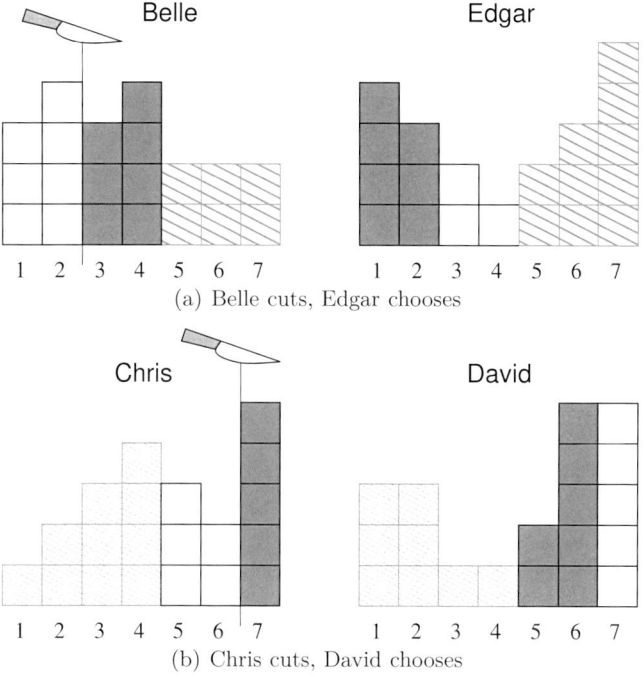

Fig. 8.26 Divide & Conquer protocol for four players: 2nd round

Theorem 8.6. *The Divide & Conquer protocol is*

1. *proportional, but not envy-free,*
2. *finite bounded,*
3. *immune to manipulation for risk-averse players, and*
4. *not Pareto-optimal.*

Proof Sketch. In terms of proportionality, the same reasoning as used above for the cases of three and four players can also be applied to any number of players $n \geq 3$. We can conclude that the Divide & Conquer protocol is a proportional cake-cutting protocol. However, it is not envy-free: In the example for four players given above, Chris envies Belle, as he values his own portion exactly one quarter of the cake (i.e., 5 out of 20 boxes), but he considers Belle's portion to be worth more than a quarter (i.e., 7 out of 20 boxes). Can you find any more envy-relations in this example?

We omit the proof of finite boundedness (but see (8.7) and Theorem 8.10 on page 587 for the minimal number of cut requests) and of immunity to manipulation for risk-averse players, leaving them to the reader. That the Divide & Conquer protocol is not Pareto-optimal again follows from the fact that it includes Cut & Choose, which is not Pareto-optimal. ❑

Given: Cake $X = [0, 1]$ and players $p_1, p_2, \ldots, p_n$, where p_i has valuation function v_i, $1 \leq i \leq n$.

Case 1: If $n = 1$ then player p_1 receives the whole cake X.

Case 2: If $n = 2k$ for some $k \geq 1$, then the following steps will be applied.

Step 2.1: $p_1, p_2, \ldots, p_n$ divide the cake using parallel markings in the ratio of $k : k$ according to their valuation functions. Let A be the piece from the left boundary of the cake to the middle between the kth and the $(k+1)$st marking. For example, if $n = 6 = 2 \cdot 3$, this may result in:

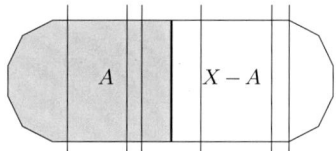

Step 2.2: Recursively applying the BJK Divide & Conquer protocol for k players,
- A is divided among the k (first) players whose markings fall within this piece;
- the other piece, $X - A$, is divided among the k remaining players.

Case 3: If $n = 2k+1$ for some $k \geq 1$, then the following steps will be applied.

Step 3.1: $p_1, p_2, \ldots, p_n$ divide the cake using parallel markings in the ratio of $k : k+1$ according to their valuation functions. Let A be the piece from the left boundary of the cake to the middle between the kth and the $(k+1)$st marking. For example, if $n = 7 = 2 \cdot 3 + 1$, this may result in:

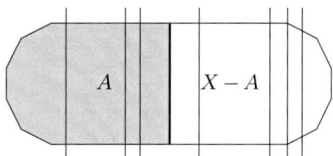

Step 3.2: Recursively applying the BJK Divide & Conquer protocol,
- A is divided among the k (first) players whose markings fall within this piece;
- the other piece, $X - A$, is divided among the $k+1$ remaining players.

Fig. 8.27 BJK Divide & Conquer protocol for n players due to Brams, Jones, and Klamler [195]: If the kth and the $(k+1)$st marking are distinct in the first round, then it provides a super-proportional division.

8.4 Cake-Cutting Protocols

Brams, Jones, and Klamler [195] presented an interesting variant of the Divide & Conquer protocol (see Figure 8.27) that minimizes the number of players the most envious player envies. The main difference in this improved version lies in the idea that after all markings have been made to divide the cake in the ratio of $k:k$ (if the number of players is even) or of $k:k+1$ (if the number of players is odd), player p_n does not cut the cake at the kth marking but between two adjacent markings. That is, the area between the two adjacent markings is used to reduce the number of envy-relations. Furthermore, Brams, Jones, and Klamler [195] consider the option to add another step at the end of the protocol, which would allow all players in an envy cycle to trade their portions in order to reduce the number of envy-relations.[16]

The BJK Divide & Conquer protocol from Figure 8.27 will be further discussed in Section 8.4.3 in terms of super-proportionality, and the number of envy-relations and guaranteed envy-relations of various cake-cutting protocols will be discussed in more detail in Section 8.4.8.

8.4.3 Super-Proportional Protocols for n Players

For the proportional protocols from Section 8.4.2, one could often observe that players even received a super-proportional share. For example, when applying the Divide & Conquer protocol, Belle, David, and Edgar received a share of $7/20$ each (see Figure 8.26), only Chris had to make do with one quarter of the cake. Is it possible to provide all players with a super-proportional share? What conditions need to be fulfilled to allow super-proportional divisions? An (informal) answer to this question about super-proportionality has already been given by Steinhaus [889, pp. 102–103]:

> "It may be stated incidentally that if there are two (or more) partners with different estimations, there exists a division giving to everybody more than his due part (Knaster); this fact disproves the common opinion that differences in estimations make fair division difficult."

This "Knaster property" hints at the condition that needs to be fulfilled to make a super-proportional division of the cake possible at all: disagreement among the players regarding their valuation of the cake. Obviously, a super-proportional division would not be possible if all players had identical valuation functions. Dubins and Spanier [360] expressed this insight as follows (see also Theorem 8.5 on page 550, which specifies this general insight for the Cut Your Own Piece protocol).

---

[16] For example, in the case that Belle envies Chris, Chris envies David, David envies Edgar, and Edgar envies Belle, the number of envy-relations can be reduced by every envied player passing on her portion to the player she is envied by.

Theorem 8.7 (Dubins and Spanier [360]). *If the players' valuation functions differ for at least one part of the cake, there always exists a super-proportional division.*

In essence, one can give a super-proportional variant of every proportional cake-cutting protocol, assuming that the players' valuation functions differ for at least one part of the cake. However, the adjustments needed to make this work are often specific to the protocol at hand. That is, it depends on the protocol we wish to make super-proportionally *in what way* the players' valuation functions must differ to ensure a super-proportional division as mentioned in Theorem 8.7; in particular, we may need to have disagreement on a *specific* part of the cake.

For example, the Divide & Conquer protocol from Figure 8.24 can be adjusted to provide a super-proportional division whenever the kth and the $(k+1)$st marking of the players are not identical in the first round (irrespective of the other players' markings in this round and of any markings in later rounds), in order to allow a cut lying strictly between these two specific marks. The BJK Divide & Conquer protocol by Brams, Jones, and Klamler [195] presented in Figure 8.27 is this modified variant of the protocol. It also differs from the original protocol in Figure 8.24 by requiring *all* n players (not only the first $n-1$ players) to provide a marking dividing the cake in the ratio of $k:k$ (for an even number of players) or in the ratio of $k:k+1$ (for an odd number of players). This difference between the two protocols is due to the fact that the last player, p_n, is not choosing which of the pieces to share with whom, but now simply participates as an equal among equals.

The disagreement of the two players making the kth and the $(k+1)$st marking can then be exploited by cutting the cake not along the kth marking (as in the original protocol from Figure 8.24), but strictly—e.g., exactly in the middle—between the kth and the $(k+1)$st marking. Thus, the resulting piece A (see Figure 8.27) is somewhat more valuable than the k players whose markings fall into A would actually need for a proportional division (i.e., A is more valuable than k/n for each of these players), due to the property of positivity. Similarly, the other part of the cake, $X - A$, is somewhat more valuable than the k (if n is even) or $k+1$ (if n is odd) remaining players would actually need for a proportional division (i.e., for each of these players, $X - A$ is more valuable than k/n if n is even, and is more valuable than $(k+1)/n$ if n is odd).

It is enough—and also required—for the disagreement of the two players making the kth and the $(k+1)$st marking to occur in the first round. Due to the recursive nature of the BJK Divide & Conquer protocol, this means that in the subsequent rounds a proportional protocol for k or $k+1$ players is applied to a piece of cake valued to be worth more than k/n or $(k+1)/n$ of the cake by each of these players. Thus, every player is guaranteed a super-proportional share, even if in the subsequent rounds no further disagreement would occur and the cuts would superpose the other markings of the players. Note further that if the condition that the kth and the $(k+1)$st marking are

8.4 Cake-Cutting Protocols

distinct is not satisfied in the first round of the protocol from Figure 8.27, then this protocol does still guarantee a proportional, even though not superproportional, division of the cake.

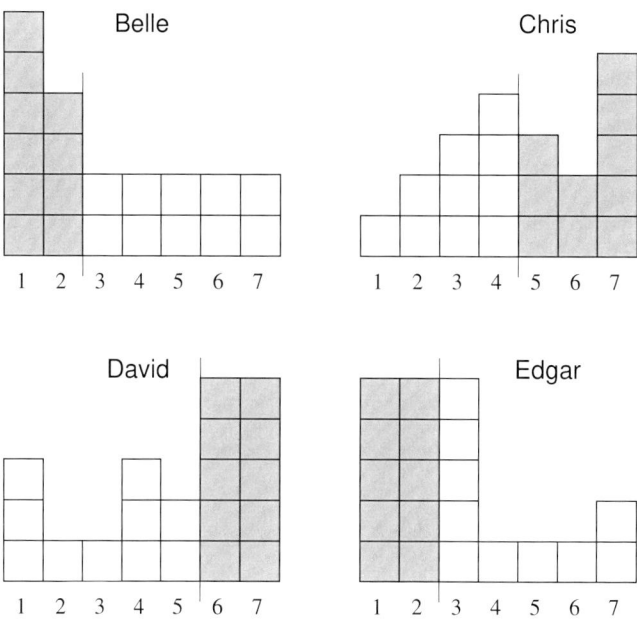

Fig. 8.28 BJK Divide & Conquer protocol for four players: marking in the 1st round

If we would apply the protocol from Figure 8.27 to the valuation functions of the four players given in Figure 8.25, the condition required for a superproportional division would not be satisfied: In this example, we have $k = 2$ and the second and third marking from the left (namely, those made by Chris and Edgar) are identical (namely, both are just between the fourth and the fifth column). Therefore, we slightly modify the valuations of Belle, David, and Edgar in this example and obtain the box representation of these four players' valuation functions that is shown in Figure 8.28. Now, the second marking from the left (right after column 2) is distinct from the third marking from the left (right after column 4). According to the BJK Divide & Conquer protocol from Figure 8.27, a cut in the middle between these two markings (right after column 3) divides the cake into the pieces A and $X - A$. Belle and Edgar, whose markings fall into A, share this piece with each other by applying the BJK Divide & Conquer protocol for two players (see Figure 8.29(a)). Simultaneously, Chris and David, whose markings fall into $X - A$, share this other piece with each other by applying the BJK Divide & Conquer protocol for two players as well (see Figure 8.29(b)).

Figure 8.30(a) shows the second round for Belle and Edgar: They both mark one half of piece A, which consists of columns 1–3, according to their

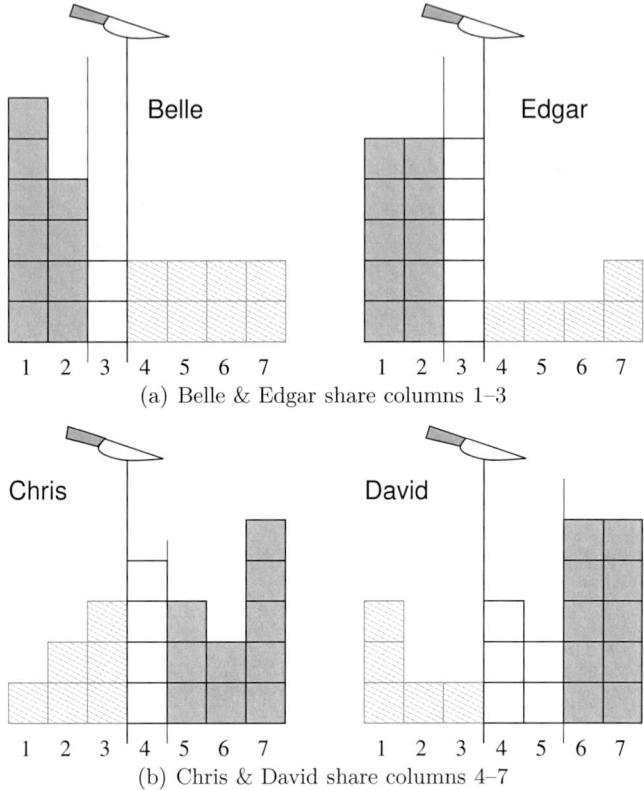

Fig. 8.29 BJK Divide & Conquer protocol for four players: cutting in the 1st round

valuation functions. Belle's mark is right after column 1 and Edgar's mark is exactly in the middle of column 2. Cutting in the middle between these two marks gives Belle the first column and the first fourth of the second column, while Edgar gets the rest of the second and all of the third column.

Figure 8.30(b) shows the same for Chris and David, who are sharing $X - A$, i.e., columns 4–7: Chris's mark is right after column 5 and David's mark is exactly in the middle of column 6. Now, cutting in the middle between these two marks gives Chris the fourth and the fifth column and the first fourth of the sixth column, while David gets the rest of the sixth and all of the seventh column.

Unlike in the division from Figure 8.26, each player here receives more than a proportional share: Belle receives $7/20 > 1/4$, Chris receives $7.5/20 > 1/4$, and David and Edgar each receive $8.75/20 > 1/4$ of the cake.

In a similar way, all other proportional protocols can be adjusted to yield a super-proportional division of the cake, provided that the players' valuation functions differ for a suitable part of the cake. For example, Woodall [947]

8.4 Cake-Cutting Protocols

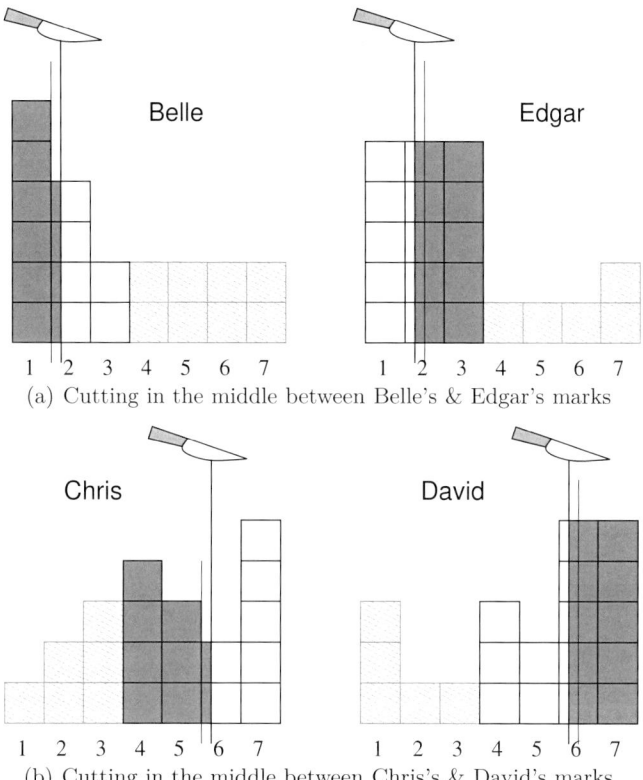

Fig. 8.30 BJK Divide & Conquer protocol for four players: 2nd round

shows that a modified variant of the Lone Chooser protocol achieves a superproportional division, provided that, first, there is a piece that two players value differently and, second, the values of all other players for this piece are known. Brams and Taylor [204] propose a simplification of the original method of Woodall [947], which again is based on the Lone Chooser protocol.

We leave it as an exercise for the reader to modify the other proportional protocols so that, under suitable conditions, a super-proportional division can be achieved. Usually, it is enough that merely two players disagree on some part of the cake, and this little difference in their valuations can then be given to all players in addition to their actual proportional portions.

8.4.4 A Royal Wedding: Dividing into Unequal Shares

So far we were concerned only with the case that every player is entitled to receive the same share of the cake. However, that is not always what we are out for. Under certain circumstances, one might reasonably consider it fair for some players to receive more and for other players to receive less than a proportional share. By contrast, the following story, which goes back to Aesop's fable, tells us that sometimes fairness is not in the focus at all when a cake is divided into unequal shares.

"Grandmother, please tell me a story!" Edgar is begging.
"What do you want it to be about?" grandmother wants to know.
"About sharing a cake?"
"No, about animals," desires Edgar, "about *wild* animals, like a lion!"
"Or about animals sharing a cake?" grandmother suggests.
"Since when do lions eat cake?"
"There you go, that's right, of course," grandmother admits, puts down her knitting needles and starts telling a story.
"A lion, a fox, and a donkey were hunting together. Asked by the lion, the donkey divides the prey into three equal shares. Foaming with rage, the lion rends the donkey, devours it, and asks the fox to divide the prey anew. The fox huddles everything together, except for a tiny morsel. Very pleased by this division, the lion asks: 'Who has taught you, my dearest friend, the art of dividing?' The fox replies: 'I've learned that from the donkey, by witnessing its fate.'"

We will not be concerned with the lion's share,[17] which the stronger one claims from the weaker one and gets it by browbeating, but we will be interested in a *fair*—even though not proportional—division into unequal shares. However, in this case, too, it must be ensured that the allocation of portions to the players is *guaranteed* to be according to their higher or lower requirements, irrespective of their valuation functions and of the other players' strategies that possibly deviate from the strategies recommended by the protocol.

There are many situations where it is justified for a player to assert a claim of more than a proportional share, or of less than that. For example, one might imagine a wedding with 30 guests where both the bride and the groom are entitled to a more valuable share of the wedding cake than any of

---

[17] This fable is also told by Brams and Taylor [204] on page vii. In a similar fable by Aesop, the lion again is hunting together with other animals, divides the prey himself into four equal shares, and assigns them with the following words: "The first quarter belongs to me because I am the king of all animals; the second quarter is mine because I am the judge; another share goes to me for my efforts in the hunt; and what the fourth quarter concerns, I'd like to see who among you dares to put a paw on it."

8.4 Cake-Cutting Protocols

Fig. 8.31 Unequal shares: the lion's share

the guests. In particular, it might be the case that each guest is to get only an ordinary proportional share, while the bride and the groom are entitled to a threefold value of a proportional share (each one according to their own valuation function, of course). That would mean for each guest to receive a portion of value $1/36$ of the cake, but the bride and the groom would both receive a portion they each value to be $3/36 = 1/12$ of the cake.

The situation can be much more complicated yet. At a royal wedding, for example, there might be many more differentiations into gradually distinct values of shares. The bride and the groom are again entitled to a better share than any of the guests, of course. But we now may have to differentiate amongst the guests as well, depending on their nobility rank—as an earl, or a duchess, or a count, etc.—and depending on whether or not they belong to the royal family. How should we proceed in such a situation? How can we ensure that all players are guaranteed to receive a (not necessarily proportional) share they are entitled to?

There is a straightforward method to adjust any proportional cake-cutting protocol so that it can be used to assign unequal shares, namely simply by introducing pseudo-players (or clones). This only works, though, if the shares the players are entitled to are rational numbers and so can be represented by fractions a/b for nonnegative integers a and b with $1 \leq a \leq b$.[18]

First, we reduce all (possibly unequal) shares to a common denominator, b, to make it easier to compare them. The ith player, p_i, is then entitled to

---

[18] Of course, it is also conceivable that a player is entitled to an irrational share of the cake, say $1/\pi$, where π is the well-known irrational number $3,14159\cdots$. This case is more difficult to deal with (see, e.g., the book by Robertson and Webb [814]).

receive a share of a_i/b. To this end, we introduce $a_i - 1$ additional clones of p_i, call them $p_i^1, \ldots, p_i^{a_i-1}$, whose valuation functions are identical to that of p_i. Players entitled to just a share of value $1/b$ are not accompanied by clones. For n original players, we have

$$\frac{a_1}{b} + \frac{a_2}{b} + \cdots + \frac{a_n}{b} = 1,$$

since the entire cake has a value of 1 due to our normalization requirement. This implies $a_1 + a_2 + \cdots + a_n = b$, which equals the total number of original players and their clones. Hence, a proportional share of players and clones is $1/b$. Now we apply the proportional cake-cutting protocol at hand to the cake and all players and their clones, which guarantees each player and each clone a share of $1/b$ of the cake. Finally, we assign to all original players their own share as well as that of their clones. Thus, every original player is guaranteed to receive the desired share of a_i/b. The clones, sadly enough, go away empty-handed and are destroyed after having served their purpose.

8.4.5 Envy-Free Protocols for Three and More Players

All proportional protocols given in Section 8.4.2 have an unpleasing disadvantage: They allow envy to occur among the players as soon as there are more than two of them. However, an envy-free division of the cake for any number of players always exists, as has been shown by Steinhaus [890]. That is merely an existence result, though. How can one actually achieve such a division? In the next section, we will exhibit such a protocol for three players, the famous Selfridge–Conway protocol. This protocol is based on a simple, yet eminently sophisticated, idea.

Despite intense efforts over decades, until 2016, no one succeeded in finding a *finite bounded* cake-cutting protocol that *guarantees envy-freeness* for more than three players. Even for four players, this used to be an open problem—perhaps *the* central open problem in the field of cake-cutting. In 2016, Aziz and Mackenzie finally achieved a breakthrough by providing the very first finite bounded, envy-free cake-cutting protocol, first for four players [61], then for any number of players [60]. Their protocol is fairly complicated and technically demanding (which can be seen from the fact that it runs in time $\mathcal{O}\left(n^{n^{n^{n^{n^n}}}}\right)$ for n players) and will not be presented here. For an accessible presentation of their protocol and a high-level description of their proof of correctness, we refer to their article [62].

8.4 Cake-Cutting Protocols

8.4.5.1 The Selfridge–Conway Protocol

The oldest known envy-free cake-cutting protocol for more than two players is the Selfridge–Conway protocol, a finite bounded protocol for three players. This protocol was proposed by John L. Selfridge and John H. Conway around 1960: Both of them were having the same idea independently, none of them made it public. It is probably due to the simplicity and elegance of this envy-free protocol that it still found its way to the public and has been widely spread ever since—first of all by Richard K. Guy; see also, for example, the papers by Gardner [482], Stromquist [894], Woodall [946], and Austin [36], and later on the books by Brams and Taylor [204] and Robertson and Webb [814]. The Selfridge–Conway protocol is given in Figure 8.32.

Given: Cake $X = [0,1]$ and players p_1, p_2, and p_3 with valuation functions v_1, v_2, and v_3.

Step 1: p_1 cuts the cake X into three pieces of equal value and p_2 sorts them as S_1, S_2, and S_3 according to their value in his measure:

$$v_1(S_1) = v_1(S_2) = v_1(S_3) = \frac{1}{3},$$
$$v_2(S_1) \geq v_2(S_2) \geq v_2(S_3).$$

Step 2: If $v_2(S_1) > v_2(S_2)$, p_2 trims S_1 such that he values the resulting piece $S_1' = S_1 - R$ to satisfy $v_2(S_1') = v_2(S_2)$ and obtains some remainder $R = S_1 - S_1' \neq \emptyset$. If $v_2(S_1) = v_2(S_2)$, set $S_1' = S_1$ and $R = \emptyset$.

Step 3: p_3, p_2, and p_1 (in this order) choose one most valuable piece each from $\{S_1', S_2, S_3\}$. If $R \neq \emptyset$ and p_3 did not take S_1', p_2 is obliged to choose S_1'.

Step 4 (if and only if R is nonempty): Either p_2 or p_3 has chosen S_1'. Call this player p_A and the other one p_B. Now, p_B cuts the remainder R into three pieces of equal value, R_1, R_2, and R_3, i.e., they satisfy

$$v_B(R_1) = v_B(R_2) = v_B(R_3) = (1/3) \cdot v_B(R),$$

among which the players p_A, p_1, and p_B (in this order) choose one most valuable piece each.

Fig. 8.32 Selfridge–Conway protocol for three players

Theorem 8.8 (Selfridge and Conway). *The Selfridge–Conway protocol from Figure 8.32 is finite bounded and envy-free, yet not Pareto-optimal.*

Proof. There are no more than nine evaluation requests and no more than five cut requests in the Selfridge–Conway protocol. The number of cuts can be seen immediately from the protocol in Figure 8.32. For the detailed analysis of the number of evaluations, notice that, after p_1's two cut requests in Step 1, p_2 makes two evaluation requests to determine $v_2(S_1)$ and $v_2(S_2)$ (and thus knows $v_2(S_3) = 1 - v_2(S_1) - v_2(S_2)$). In Step 2, if $v_2(S_1) > v_2(S_2)$, then p_2

makes one cut request to determine a subpiece S_1' of S_1 he values to be equal to $v_2(S_2)$ and he also knows $v_2(R) = v_2(S_1) - v_2(S_2)$. In Step 3, p_3 makes three evaluation requests in order to find out which of the pieces S_1', S_2, and S_3 is of highest value to her—here, only two evaluation requests do not suffice if $R \neq \emptyset$—and p_3 now also knows $v_3(R) = 1 - v_3(S_1') - v_3(S_2) - v_3(S_3)$. Players p_1 and p_2 do not have to make any evaluations in this step, since they still know their values for the remaining pieces from earlier steps. In Step 4 (which is done if and only if the remainings R are nonempty), p_B (which is either p_2 or p_3, who both know their own value of R) makes two cut requests in order to partition R into three pieces R_1, R_2, and R_3, each of value $(1/3) \cdot v_B(R)$. Finally, both p_A (which again is either p_2 or p_3, distinct from p_B, and so knows $v_A(R)$) and p_1 make two evaluation requests (p_A to choose a most valuable one among the pieces R_1, R_2, and R_3 and p_1 to choose a most valuable one among the two remaining pieces), and p_B takes the last remaining piece. Summing up, we have at most five cut and at most nine evaluation requests.

Thus, the Selfridge–Conway protocol is finite bounded. Now, why is it envy-free?

First, consider the part of the cake divided among the players in Step 3, $X - R$. We have the following:

1. Player p_1 is the last one to choose from $\{S_1', S_2, S_3\}$. If S_1 was trimmed (i.e., if $v_1(S_1') < v_1(S_1) = 1/3$), then he receives either one of the pieces S_2 or S_3, which he has cut himself and which both have the value $1/3$ to him. It is crucial here to note that p_2 has to choose the possibly trimmed piece S_1', provided it is still available after p_3 has made her choice in Step 3—this piece thus must go to either p_2 or p_3 if it was trimmed, and otherwise it has the same value of $1/3$ to p_1 as the other two pieces. Hence, p_1 envies none of the other players with respect to their portion of $X - R$.

2. For player p_2, who sorted the pieces in Step 2 according to their value to him and who possibly trimmed S_1, the pieces S_1' and S_2 have the same value and both are at least as valuable as S_3. Since only p_3 chooses before p_2, either S_1' or S_2 is still available to p_2 when it is his turn to choose. Hence, also p_2 envies none of the other players with respect to their portion of $X - R$.

3. Player p_3, finally, is the first one to choose one among the pieces S_1', S_2, and S_3 and hence does not envy any of the other players with respect to their portion of $X - R$ either.

Thus, there is no envy with respect to $X - R$, and in case R is empty, we are done with the proof of envy-freeness at this point. However, if R is nonempty, the cake is not completely allocated yet. If we would now apply Steps 1 through 3 to R instead of X, it might happen that another leftover remains. And if we would continue in this way, it might be that a nonempty leftover would be created again and again, and so it wouldn't be possible

8.4 Cake-Cutting Protocols

to guarantee an envy-free, complete division of the cake by a *finite* (not to mention by a *finite bounded*) protocol.

Step 4 in the protocol from Figure 8.32, however, is different. It does not simply repeat the first three steps. Namely, something important has changed by dividing $X - R$, which makes a crucial difference between R and X, at least from the point of view of one of the three players: p_1 now has an "*irrevocable advantage*" with respect to the remainder R, an advantage p_1 cannot lose again and that he did not have with respect to the whole cake X. This advantage is that R in p_1's valuation is completely contained in S_1, a piece that p_1 values to be one third of the cake. And p_1 already has received a piece of the cake, either S_2 or S_3, that he values to be one third of the cake. This implies that even if all of R would go to the player who received S_1' (i.e., to p_A), p_1 would not envy this player. This is the ingenious insight of Selfridge and Conway for how to guarantee an envy-free, complete division of the cake by a finite bounded protocol.

In more detail, we make the following considerations to argue why there will be no envy with respect to the whole cake X, even if there remains a nonempty leftover after the first three steps, so Step 4 needs to be executed:

1. After assigning the remainder pieces R_1, R_2, and R_3, p_1 does not envy any of the other two players with respect to X. He doesn't envy p_A, since he doesn't envy p_A with respect to $X - R$ and—as we have just seen—he has an irrevocable advantage with respect to R. He doesn't envy p_B either, since he can choose before p_B in Step 4. As he doesn't envy p_B with respect to R, nor with respect to $X - R$, it follows from additivity of the valuation functions that p_1 doesn't envy p_B with respect to X either.
2. p_A envies neither p_1 nor p_B for their shares from R, since p_A may choose first in Step 4. Since p_A envies neither p_1 nor p_B with respect to $X - R$ either, it again follows from additivity of the valuation functions that p_A envies neither p_1 nor p_B with respect to X.
3. p_B, finally, envies neither p_A nor p_1 for their shares from R, since p_B has divided R into pieces of equal value. Since p_B envies neither p_A nor p_1 with respect to $X - R$, it again follows from additivity of the valuation functions that p_B envies neither p_A nor p_1 with respect to X.

The proof that the Selfridge–Conway protocol is not Pareto-optimal is not given here, but will be provided at the end of the following example. ❑

Figure 8.34 shows how the pieces S_1', S_2, and S_3 are assigned in Step 3:

1. Helena chooses first, namely her most valuable piece S_3, which has seven boxes, so $v_{\text{Helena}}(S_3) = 7/18$.
2. Since Helena didn't take it, George must take the piece S_1' now (which, recall, is equally valuable as S_2 to him).
3. Felix receives the remaining piece, S_2.

For illustration, consider the example in Figure 8.33. Felix, George, and Helena, whose valuation functions are given in the box representation (with 18 boxes each for the cake), apply the Selfridge–Conway protocol. Figure 8.33(a) shows the first step of the protocol: Felix divides the cake into three pieces of equal value and George sorts them as

- S_1 (columns 1 through 4 with $v_{\text{George}}(S_1) = 10/18 = 5/9$),
- S_2 (columns 5 and 6 with $v_{\text{George}}(S_2) = 4/18 = 2/9$), and
- S_3 (columns 7 and 8 with $v_{\text{George}}(S_3) = 4/18 = 2/9$)

according to their value to him. The different shades of gray indicate this partition of the cake. The second step of the protocol is shown in Figure 8.33(b): George trims S_1 to obtain S_1' (column 1) and the remainder R (columns 2 through 4) to ensure that S_1' and S_2 have the same value by his measure, i.e., both contain four boxes: $v_{\text{George}}(S_1') = v_{\text{George}}(S_2) = 2/9$. The remainder R with $v_{\text{George}}(R) = v_{\text{George}}(S_1) - v_{\text{George}}(S_1') = (5-2)/9 = 3/9$ is put aside for the moment.

The chosen pieces are shaded dark gray in Figure 8.34.

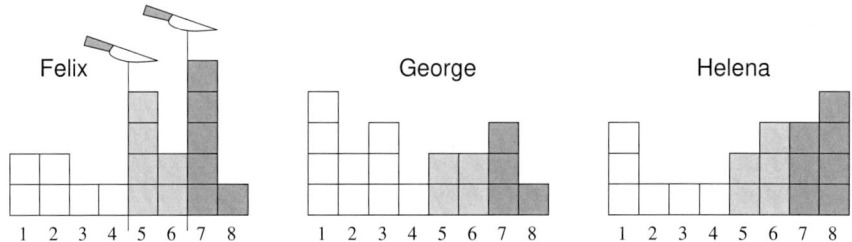
(a) Step 1: Felix divides the cake into three pieces of equal value to him, and George sorts them by their value to him.

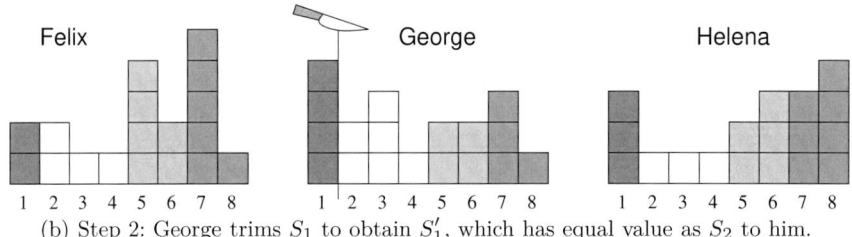
(b) Step 2: George trims S_1 to obtain S_1', which has equal value as S_2 to him.

Fig. 8.33 Selfridge–Conway protocol: Steps 1 and 2

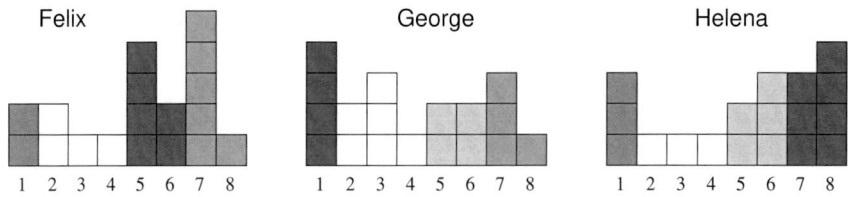

Fig. 8.34 Selfridge–Conway protocol: Step 3

8.4 Cake-Cutting Protocols

Now we still have to assign the remainder R, shown by white boxes in Figure 8.34. Those parts of the cake that are no longer available for the players are shaded light gray in Figure 8.35. According to Step 4 in the protocol, Helena divides R into three equally valuable pieces by her measure:

- R_1 (column 2 with $v_{\text{Helena}}(R_1) = 1/18$),
- R_2 (column 3 with $v_{\text{Helena}}(R_2) = 1/18$), and
- R_3 (column 4 with $v_{\text{Helena}}(R_3) = 1/18$).

Then, the players choose their part of the remainder in the order given in Step 4 of the protocol:

1. Since George received S_1', he chooses first, namely the piece R_2, which with three boxes is most valuable to him.
2. Next, Felix chooses the piece R_1, which he values higher (two boxes) than the other still remaining piece, R_3, consisting of only one box for him.
3. Finally, Helena receives the remaining piece, R_3.

The final allocation of portions to the players is shown in Figure 8.35 by dark gray boxes.

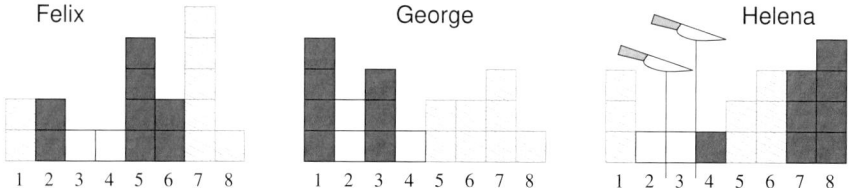

Fig. 8.35 Selfridge–Conway protocol: Step 4

Table 8.2 Envy-freeness in the Selfridge–Conway protocol

	... the share of		
	Felix	George	Helena
Felix values ...	**8/18**	3/18	7/18
George values ...	6/18	**7/18**	5/18
Helena values ...	6/18	4/18	**8/18**

Table 8.2 shows how each of the players values all players' portions in the above example (see Figure 8.35). For example, in Felix's valuation Helena's portion (columns 4, 7, and 8) has seven boxes, i.e., $v_{\text{Felix}}(S_3 \cup R_3) = 7/18$. Each player's own share of the cake is shown boldfaced in Table 8.2, and one can see that there is no envy among Felix, George, and Helena.

To see that the Selfridge–Conway protocol from Figure 8.32 is not Pareto-optimal, as claimed in Theorem 8.8, look again at the division in Figure 8.35

that is obtained by this protocol in our example. This division gives both Felix and Helena a share of 8/18 and George a share of 7/18 of the cake. Now, if we suppose that Felix and Helena swap columns 6 and 7, all else being equal, then neither George nor Helena would be worse off; they still would have a share of 7/18 and 8/18, respectively. However, Felix would be better off under this new allocation, as he would increase his share from 8/18 to 11/18 of the cake. This proves that the Selfridge–Conway protocol is not Pareto-optimal.

8.4.5.2 Stromquist's Moving-Knife Protocol

Another envy-free cake-cutting protocol for three players is Stromquist's moving-knife protocol [894] that is shown in Figure 8.36 (see also Figure 8.37). It is based on a different idea than the Selfridge–Conway protocol, but is very elegant as well. Of course, being a moving-knife protocol it is inferior to the Selfridge–Conway protocol, as it is not finite bounded (and not even finite). Moreover, whoever wishes to apply the Stromquist protocol must be armed to the teeth when dividing the cake: A sword and three knives are needed. Further envy-free moving-knife protocols for three players can be found, e.g., in the textbooks by Brams and Taylor [204] and Robertson and Webb [814].

Given: Cake $X = [0,1]$ and players p_1, p_2, and p_3 with valuation functions v_1, v_2, and v_3.

Step 1:

1. A sword is being held above the cake and is moved gradually and continuously from the left to the right, thus (hypothetically) dividing the cake into a left piece L and a right piece R, where $X = L \cup R$ and $L \cap R = \emptyset$ (see Figure 8.37(a)).
2. Each of the three players holds an own knife in parallel with the sword and moves it with the moving sword so that R is (hypothetically) divided in half according to the player's valuation.
3. The middle knife of the three—breaking ties arbitrarily if needed—(hypothetically) divides R into two pieces S and T, i.e., $R = S \cup T$ and $S \cap T = \emptyset$ (see Figure 8.37(b)).

Step 2:

1. As soon as one player thinks that L is at least as valuable as each of S and T, he calls: "Stop!" Again, ties are broken arbitrarily when they occur.
2. The sword and the middle knife cut at their positions.
3. The player who called "Stop!" receives L and drops out.
4. The player (among the two still in the game) whose knife is closest to the sword (where ties are broken arbitrarily if needed) receives S and drops out.
5. The last remaining player receives T.

Fig. 8.36 Stromquist's moving-knife protocol for three players

8.4 Cake-Cutting Protocols

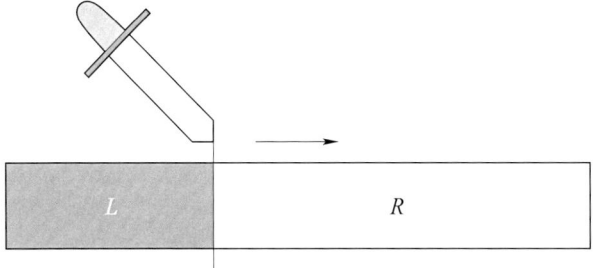

(a) A sword is moved from the left to the right above the cake, (hypothetically) dividing it into a left and a right piece.

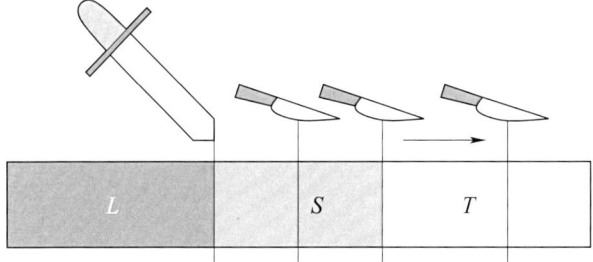

(b) The players move their knives in parallel with the moving sword, each (hypothetically) dividing R in half according to their valuation. The middle knife's position determines the pieces S and T partitioning R.

Fig. 8.37 Illustration of Stromquist's moving-knife protocol for three players

Why is Stromquist's moving-knife protocol from Figure 8.36 envy-free? Suppose that Felix, George, and Helena have applied it to divide a cake. As explained in Figure 8.36, a tie (that can occur if two or three players call "Stop!" simultaneously, or if more than one knife is in one and the same position when the first player calls "Stop!") can be broken arbitrarily. Without loss of generality, assume that ties are broken alphabetically: If they call "Stop!" simultaneously or hold their knives at the same positions, Felix wins against George and Helena, and George wins against Helena. Also, let us say that Felix is the first player to call "Stop!" in Step 2 (or has won the tie-breaking) and then drops out with L. Suppose further that George has received S (because his knife was closest to the sword when Felix called "Stop!"—i.e., when the pieces L, S, and T are cut—or, possibly, because he has won the tie-breaking against Helena) and Helena was assigned the piece T to the right of the middle knife that still was available. (If knives are at the same position, the "left," "middle," and "right" knife refers to the result of the tie-breaking.) That none of the players envies any of the other players can be seen as follows:

1. Felix envies neither George nor Helena, since he called "Stop!" because he thinks his share L is worth at least as much as George's share S and as Helena's share T.
2. Since George received S, he must have held either the left or the middle knife. If he held the left knife, his share S is at least as valuable as Helena's share T according to his measure. If he held the middle knife, S and T are equally valuable to George. In both cases, George does not envy Helena. George does not envy Felix either, since otherwise he would have called "Stop!" before him. As he didn't, L is at most as valuable for George as his own portion S.
3. Since Helena received T, she must have held either the middle or the right knife. If she held the middle knife, S and T are equally valuable for Helena. If she held the right knife, however, T is at least as valuable as S for her. In both cases, Helena does not envy George.
   Helena does not envy Felix either, for the same reason why George doesn't envy Felix: Otherwise, she would have called "Stop!" before Felix.

Thus, we have shown the following result.

Theorem 8.9 (Stromquist [894]). *Stromquist's moving-knife protocol from Figure 8.36 is envy-free.*

8.4.5.3 The Moving-Knife Protocol of Brams, Taylor, and Zwicker

As already mentioned at the beginning of this section, Aziz and Mackenzie [60, 61, 62] were the first to provide a finite bounded, envy-free cake-cutting protocol for any number of players. Prior to their breakthrough result, an obvious question was raised: Can the idea underlying the Selfridge–Conway protocol from Section 8.4.5.1 starting on page 565 perhaps be transferred to four players such that envy-freeness is preserved?

A solution to this problem is due to Brams, Taylor, and Zwicker [206] and is shown in Figure 8.38. The price to pay for four players being guaranteed envy-freeness in this protocol is the loss of finite boundedness—and even of finiteness. The protocol of Brams, Taylor, and Zwicker combines in a clever way the envy-free Selfridge–Conway protocol for three players (see Figure 8.32 on page 565) with Austin's exact moving-knife protocol for two players (see Figure 8.12 on page 533).

The analysis of envy-freeness in the Brams–Taylor–Zwicker protocol is done similarly to that in the Selfridge–Conway protocol (see Theorem 8.8)—simultaneously exploiting the exactness of Austin's protocol. Therefore, we omit this analysis here and leave the detailed argumentation to the reader.

No generalization of the moving-knife protocol of Brams, Taylor, and Zwicker from Figure 8.38 to an envy-free moving-knife protocol for more than four players is known to date. However, Brams and Taylor [203] (see also [204]) succeeded in designing a *finite* envy-free cake-cutting protocol for

> **Given:** Cake $X = [0,1]$ and players p_1, p_2, p_3, and p_4 with valuation functions v_1, v_2, v_3, and v_4.
> **Step 1:** Applying Austin's exact moving-knife protocol from Figure 8.12 three times, p_1 and p_2 cut four pieces that they both consider, according to their valuation functions, to have equal value each, and p_3 sorts them as S_1, S_2, S_3, and S_4 according to their value in her measure. Thus, we have:
> $$v_1(S_i) = v_2(S_i) = \frac{1}{4} \quad \text{for } 1 \leq i \leq 4,$$
> $$v_3(S_1) \geq v_3(S_2) \geq v_3(S_3) \geq v_3(S_4).$$
> **Step 2:** If $v_3(S_1) > v_3(S_2)$, p_3 trims S_1 such that she values the resulting piece $S_1' = S_1 - R$ to satisfy $v_3(S_1') = v_3(S_2)$ and obtains some remainder $R = S_1 - S_1' \neq \emptyset$.
> If $v_3(S_1) = v_3(S_2)$, set $S_1' = S_1$ and $R = \emptyset$.
> **Step 3:** p_4, p_3, p_2, and p_1 (in this order) choose one most valuable piece each from $\{S_1', S_2, S_3, S_4\}$. If $R \neq \emptyset$ and p_4 did not take S_1', p_3 is obliged to choose S_1'.
> **Step 4 (if and only if R is nonempty):** Either p_3 or p_4 has chosen S_1'. Call this player p_A and the other one p_B. Again applying Austin's exact moving-knife protocol from Figure 8.12 three times, p_2 and p_B divide the remainder R into four pieces that they both consider, according to their valuation functions, to have equal value each, R_1, R_2, R_3, and R_4 with
> $$v_2(R_1) = v_2(R_2) = v_2(R_3) = v_2(R_4) = (1/4) \cdot v_2(R),$$
> $$v_B(R_1) = v_B(R_2) = v_B(R_3) = v_B(R_4) = (1/4) \cdot v_B(R).$$
> Now, p_A, p_1, p_B, and p_2 (in this order) choose one most valuable piece each from $\{R_1, R_2, R_3, R_4\}$.

Fig. 8.38 The moving-knife protocol of Brams, Taylor, and Zwicker for four players

any number of players (even though it is not finite bounded). It is a sophisticated protocol that we only mention here without presenting it in detail.

8.4.6 Oversalted Cream Cake: Dirty-Work Protocols

Belle has been invited to her friend Anna's birthday party. To surprise her, she wants to bake a delicious cream cake with cherries and wild berries according to her grandmother's famous recipe. To make sure the birthday child doesn't miss out, she even bakes two cream cakes—a small one with a candle on top only for Anna and a big one for the guests to share.

Anna is entranced when she sees her small cream cake: "Doesn't that one look yummy?! Thanks so much, Belle, that is so nice of you!" She cannot resist to taste it right away. Suddenly, Anna makes a horri-

> fied grimace: "Belle, can it perhaps be that you have mistaken salt for sugar?"
>
> Belle turns pale. Quickly Anna adds: "But that's not *so* bad! It doesn't taste *that* bad ... only ... somehow ... *interesting*! Come on, guys, all of you, taste it yourself!"
>
> Anna puts the big cream cake on the table and grabs the kitchen knife to divide it fairly among her guests.

That was not what the guests were hoping for. Of course, they did have to share the entire oversalted cream cake nonetheless; after all, none of them was going to insult Belle. However, it is more than understandable that none of the guests would like to go for a super-proportional share. On the contrary, they all would be more than happy with getting a piece no bigger than courtesy requires them to take. The smaller, the better, as in "less is more." Though, the division should still be fair. Interpreting fairness as proportionality, the goal now is to divide the oversalted cream cake in a way that guarantees every player to receive *no more* than a proportional share.

Other situations where players would like to go for a share as small as possible in a division are known in many domains, for example when assigning household chores—or any other type of "dirty work" that is to be done jointly by a group. Therefore, Gardner [482] refers to this kind of allocation problem as the *dirty-work problem*; another term commonly used is *"chore division."* Accordingly, cake-cutting protocols for solving these problems are referred to as *dirty-work protocols*.

8.4.6.1 The Dirty-Work Last Diminisher Protocol

How does one proceed when faced with a dirty-work problem? How can one fairly divide a disliked good among the players? If every player is to be guaranteed to receive *at most* a proportional share, the answer to this question is fairly simple: Each of the proportional protocols presented in Section 8.4.2 can be transformed into a proportional dirty-work protocol, usually with only minor modifications. The required adjustments are obvious in most cases. For example, considering the Last Diminisher protocol (see Figure 8.14 on page 535) where in every round one piece of the cake is passed on from one player to the next to be evaluated by everyone, instead of the players *trimming* this piece to a proportional share of value $1/n$, they now *increase* it to a proportional share of value $1/n$ if possible.[19] At the end of each round, the

---

[19] If at some point there is not enough cake left to increase the current piece to value $1/n$ (or to cut a piece of value $1/n$ at the start of some round), the current player increases to whatever is left of the cake and passes this piece on.

8.4 Cake-Cutting Protocols

piece is given to the player who was the last one to increase it or who cut it in the first place.

In Figures 8.39, 8.40, and 8.41, an example is given for how to divide the oversalted cream cake among Anna's four guests—Belle, Chris, David, and Edgar—by the dirty-work Last Diminisher protocol. The cake is represented by a total of 20 boxes to show the valuation functions of the four players.

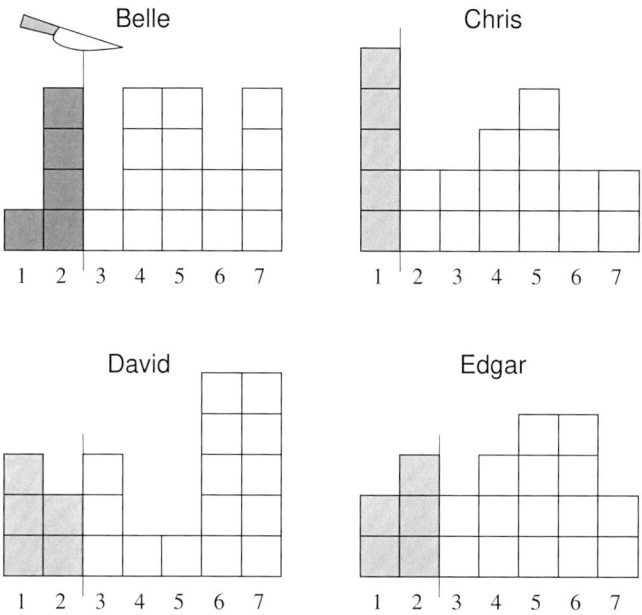

Fig. 8.39 Dirty-work Last Diminisher protocol for four players: 1st round

In the first round (see Figure 8.39), Belle cuts a piece she values to be worth one quarter of the cake (the five boxes of the first two columns). This piece is then passed on to Chris, David, and Edgar (in this order) for them to evaluate it and to increase it if possible. For all three players, Belle's piece is worth no less than one quarter of the cake according to their valuation functions (for Chris, it is worth even more than one quarter); the light gray boxes indicate one quarter of the cake for each of these players. Since none of them has increased it, it goes to Belle who drops out with her share.

In the second round (see Figure 8.40), the hatched boxes represent for each of the players which part of the cake is no longer available to them. This time, Chris starts and cuts one quarter of the cake according to his valuation function (the five boxes of the columns 3 and 4). This piece is now passed on to be evaluated by the two other players, first by David, who increases it by column 5 to reach his quarter of the cake. Finally, Edgar evaluates the increased piece and notices that his quarter of the cake (just as Chris's five boxes of columns 3 and 4) is contained in David's piece. Therefore, this piece goes to David, the last player who increased it.

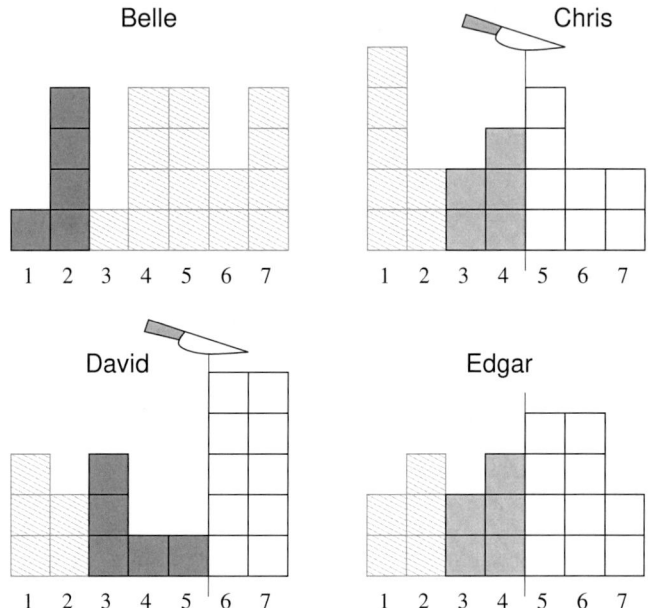

Fig. 8.40 Dirty-work Last Diminisher protocol for four players: 2nd round

The only players left are Chris and Edgar. Applying the dirty-work Cut & Choose protocol, they share the remaining cake (columns 6 and 7), where Chris is the cutter and Edgar is the chooser (see Figure 8.41). As one can see, every player has received *at most* a proportional share, Chris and Edgar even less than that.

We omit an explicit algorithmic presentation of the dirty-work Last Diminisher protocol, since its procedure is obvious from the above comments and the concrete application shown in Figures 8.39, 8.40, and 8.41.

8.4.6.2 The Dirty-Work Divide & Conquer Protocol

As an example for the formal, algorithmic presentation of a dirty-work protocol, we give the dirty-work Divide & Conquer protocol in Figure 8.42. (Identical markings are treated as described in Footnote 15 on page 552.)

The adaption of further (super-)proportional protocols from Sections 8.4.2 and 8.4.3 to the dirty-work problem are left to the reader as an exercise.

Much more extensive and challenging—if possible at all—are the dirty-work adaptions required for envy-free cake-cutting protocols. There are still many questions open, in particular for envy-free dirty-work protocols with

8.4 Cake-Cutting Protocols

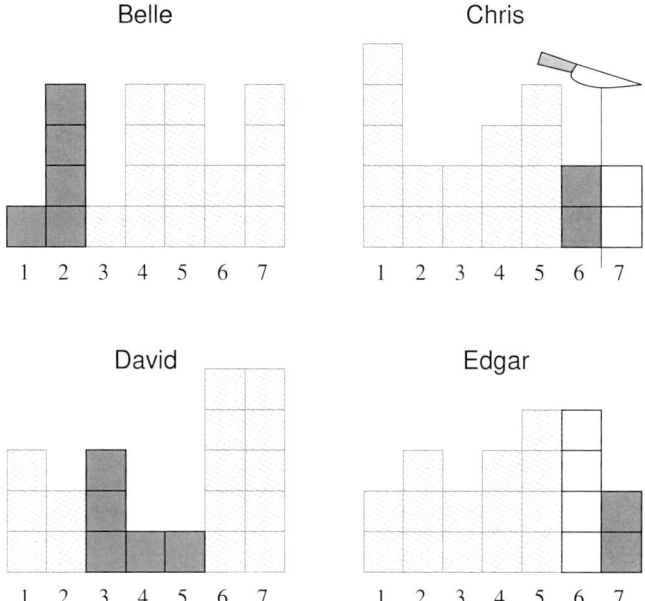

Fig. 8.41 Dirty-work Last Diminisher protocol for four players: 3rd round

respect to unequal shares (see Section 8.4.4) or for more than three players. An example for an envy-free protocol for three players that can be adapted to solve the dirty-work problem is Stromquist's moving-knife protocol from Figure 8.36 (see [814, Exercise 5.11]). Further information about the dirty-work problem can be found in the book by Robertson and Webb [814].

8.4.7 Avoiding Crumbs: Minimizing the Number of Cuts

How many cuts are required to proportionally divide a cake among n players? How many cuts are needed to guarantee envy-freeness? These questions are important both in a practical and in an esthetical sense. When one divides a real (not a metaphorical) cake, it is utterly clear that one doesn't want to hash it up by overly many cuts into tiny crumbs before putting them onto the plates. But also in the figurative sense—when dividing an arbitrary heterogeneous resource whose parts do not lose any value due to the divisibility axiom—it is more efficient and saves some unneeded effort if one can make do with as few cuts as possible. Moreover, this means finding a mathematically beautiful solution to a challenging problem.

Given: Cake $X = [0,1]$ and players $p_1, p_2, \ldots, p_n$, where p_i has valuation function v_i, $1 \leq i \leq n$.

Case 1: If $n = 1$ then player p_1 receives the whole cake X.

Case 2: If $n = 2k$ for some $k \geq 1$, then the following steps will be applied.

  Step 2.1: $p_1, p_2, \ldots, p_{n-1}$ divide the cake using parallel markings in the ratio of $k : k$ according to their valuation functions. Let A be the piece of cake left of the kth marking. For example, if $n = 8 = 2 \cdot 4$, this may result in:

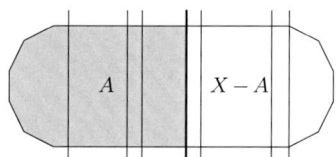

  Step 2.2: If $v_n(A) \leq k/n = 1/2$, then p_n chooses piece A, and otherwise p_n chooses piece $X - A$.

  Step 2.3: Recursively applying the dirty-work Divide & Conquer protocol for k players,
  - p_n shares the chosen piece with the $k - 1$ players whose markings fall within the other piece and are not the kth marking;
  - the other k players share the other piece.

Case 3: If $n = 2k + 1$ for some $k \geq 1$, then the following steps will be applied.

  Step 3.1: $p_1, p_2, \ldots, p_{n-1}$ divide the cake using parallel markings in the ratio of $k : k + 1$ according to their valuation functions. Let A be the piece of cake left of the $(k+1)$st marking. For example, if $n = 9 = 2 \cdot 4 + 1$, this may result in:

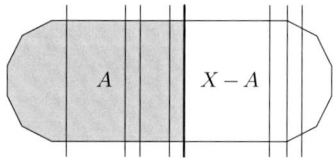

  Step 3.2: If $v_n(A) \leq k/n = k/(2k+1)$, then p_n chooses piece A, and otherwise p_n chooses piece $X - A$.

  Step 3.3:
  - If p_n chose piece A, then he shares this piece using the dirty-work Divide & Conquer protocol for k players with the $k - 1$ players whose markings fall within piece $X - A$ and are not the $(k+1)$st marking.
  - If p_n chose piece $X - A$, then he shares this piece using the dirty-work Divide & Conquer protocol for $k + 1$ players with the k players whose markings fall within piece A and are not the $(k+1)$st marking.

  In both cases, the other $k + 1$ (or k, respectively) players share the other piece using the dirty-work Divide & Conquer protocol for $k + 1$ (or k, respectively) players.

Fig. 8.42 Dirty-work Divide & Conquer protocol for n players

8.4 Cake-Cutting Protocols

But, what does count as a "cut"? In the protocols introduced so far, we distinguished between markings and actual cuts. However, if we would maintain this distinction and would count real cuts only and not markings, we would be able to minimize the required number of cuts quite easily: Instead of headily cutting pieces, the players would first make markings only and would make actual cuts just immediately prior to assigning the portions to the players. Using this way of counting, the Last Diminisher protocol (see Figure 8.14 on page 535), for example, would require no more than the optimal number of $n-1$ cuts for n players (only one cut per round), although there might be many markings in every round. Thus, if we just ignored these markings (which may be seen as cuts that are not executed), we would have found merely the trivial solution to a trivial problems. Certainly, this would not be the meaning Steinhaus [889] had in mind when he wrote:

> "Interesting mathematical problems arise if we are to determine the minimal number of cuts necessary for fair division."

Therefore, throughout this section we treat markings just like actual cuts—both will count when we determine the minimal number of cuts required for fair division.

The analysis of this minimal number of cuts of a cake-cutting protocol is closely related to its running time. However, a bounded number of cuts required for fair division does not guarantee that the protocol is finite bounded, and not even that it is finite. The reason is that when analyzing the minimal number of cuts, we consider only cut requests (including those made for markings) but we neglect evaluation requests (that may be needed, e.g., to compare pieces when making a decision). In particular, moving-knife protocols for n players usually require exactly $n-1$ cuts (see, e.g., the moving-knife protocol of Dubins and Spanier in Figure 8.18 on page 542), and that is optimal. However, we know that moving-knife protocols require *infinitely many* (even *uncountably many*) decisions in order to make these cuts at the right positions. That is why we will not include moving-knife protocols in the following analysis of the minimal number of cuts, but we will consider *finite* cake-cutting protocols only.

Obviously, the minimal number of cuts required for fair division depends both on the number of players and on the protocol used. Let us first define what exactly we mean by the minimal number of cuts (as a function of the number of players) required by some cake-cutting protocol.

Definition 8.11 (minimal number of cuts). Let Π be a cake-cutting protocol for n players. The *minimal number of cuts required by Π* is the number $k(n)$ satisfying that:

1. Π always terminates with at most $k(n)$ cuts (including markings), and
2. in the worst case (with respect to the valuation functions of the players), at least $k(n)$ cuts (including markings) are needed for Π to terminate.

That is, a cake-cutting protocol for n players requires *exactly* $k(n)$ cuts in the worst case, but may make do with fewer than that for suitably chosen valuation functions. For example, if the valuation functions of the three players in the Selfridge–Conway protocol (see Figure 8.32 on page 565) are chosen so that there remains no leftover after the second step (because the second player views his two most valuable pieces to be worth the same), this protocol needs no more than merely two cuts (made by the first player in the first step when he divides the cake into three pieces of equal value according to his valuation function). In the worst case, however, the second player trims his (unique) most valuable piece and obtains a nonempty remainder, which is to be divided into three pieces by two more cuts in the fourth step. Hence, the minimal number of cuts required by this protocol is five, not two.[20] That this number is not a function of the number of players is due merely to the fact that the Selfridge–Conway protocol is defined for a constant number of players, namely, for three players.

It is quite obvious that no proportional cake-cutting protocol for $n \geq 1$ players can require a minimal number of fewer than $n-1$ cuts: All players must receive nonempty portions that they value to be worth at least $1/n > 0$ each. Thus, $n-1$ is a lower bound on the minimal number of cuts required by proportional cake-cutting protocols. As we will see below, however, this is not the best lower bound for such protocols. If proportionality is to be guaranteed, this lower bound of $n-1$ does not coincide with the upper bound on the minimal number of cuts required by such protocols (given that only finite protocols are considered here), and it is possible to prove a lower bound greater than $n-1$.

8.4.7.1 The One-Cut-Suffices Principle

First, however, let us investigate the minimal number of cuts required by specific proportional cake-cutting protocols, and let us start with the Last Diminisher protocol (see Figure 8.14 on page 535) for the simple case of three players, Belle, Chris, and David. Suppose that Belle cuts piece S_1 from the cake, which she values to be worth one third but which is worth more than one third to Chris. Chris then trims S_1 and passes the smaller piece S_2, which he values to be worth exactly one third, on to David. Suppose further that S_2 is worth less than one third to David, so this piece goes to Chris who drops out with it. Now, Belle and David play Cut & Choose for the reassembled remainings of the cake, $R = X - S_2$, which consists of *two* pieces: $A = X - S_1$ and $B = S_1 - S_2$ (see Figure 8.43).

Do both pieces, A and B, have to be cut separately? No, one cut suffices! This insight is known as the *one-cut-suffices* principle, and is often used in the analysis of the minimal number of cuts required by cake-cutting protocols.

---

[20] That is why one sees a total of five knives in Figures 8.33, 8.34, and 8.35; there are no markings.

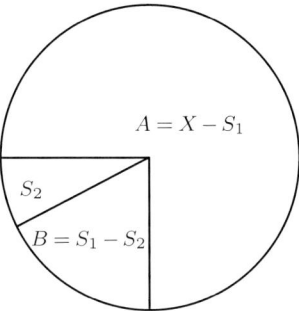

Fig. 8.43 One cut suffices for the division of A and B with Cut & Choose

More specifically, suppose that in our example Belle values piece B to be worth $1/6$ (and A to be worth two thirds, of course). Then Belle, the cutter in Cut & Choose, ought to receive a piece of value $(1/2) \cdot (2/3 + 1/6) = 5/12$ according to her valuation function. So she cuts the more valuable of the two pieces, A, into two parts, A_1 of value $5/12$ and A_2 of value $2/3 - 5/12 = 3/12$ according to her valuation function. By additivity, $A_2 \cup B$ has the same value for Belle as A_1, namely, $3/12 + 1/6 = 5/12$, and David now has the choice between A_1 and $A_2 \cup B$. Thus, the Last Diminisher protocol requires no more than a total of three cuts for these valuations by the three players. Note, however, that this is not a worst-case scenario. The worst case with respect to the minimal number of required cuts would have occurred when also David would have trimmed the piece in the first round. Thus, the minimal number of cuts required by the Last Diminisher protocol for three players is four.

The one-cut-suffices principle can be generalized to any number of pieces, $S_1, S_2, \ldots, S_m$. Suppose that S_i has the value s_i, $1 \leq i \leq m$, to Belle, and her task is to divide $S = \bigcup_{1 \leq i \leq m} S_i$ in the ratio of $x : y$, where $x + y = s_1 + s_2 + \cdots + s_m$. To this end, she first determines the subscript t for which it holds that

$$s_1 + s_2 + \cdots + s_t \leq x < s_1 + s_2 + \cdots + s_t + s_{t+1}$$

and cuts piece S_{t+1} into two parts according to her valuation function:

$$R_{t+1} \text{ of value } x - (s_1 + s_2 + \cdots + s_t) \text{ and}$$
$$T_{t+1} \text{ of value } s_{t+1} - x + (s_1 + s_2 + \cdots + s_t).$$

Obviously, Belle values $S_{t+1} = R_{t+1} \cup T_{t+1}$ to be worth

$$x - (s_1 + s_2 + \cdots + s_t) + s_{t+1} - x + (s_1 + s_2 + \cdots + s_t) = s_{t+1},$$

so everything is alright. Moreover, Belle values

- piece $A = S_1 \cup S_2 \cup \cdots \cup S_t \cup R_{t+1}$ to be worth

$$s_1 + s_2 + \cdots + s_t + x - (s_1 + s_2 + \cdots + s_t) = x$$

and
- piece $B = T_{t+1} \cup S_{t+2} \cup S_{t+3} \cup \cdots \cup S_m$ to be worth

$$s_{t+1} - x + (s_1 + s_2 + \cdots + s_t) + s_{t+2} + s_{t+3} + \cdots + s_m = s_1 + s_2 + \cdots + s_m - x = y.$$

Thus, as desired, she has divided S into A und B in the ratio of $x:y$ using just a single cut. We state this principle as a lemma, for later reference.

Lemma 8.1 (one-cut-suffices (OCS) principle). *Let $S_1, S_2, \ldots, S_m$ be m given pieces. A player who values S_i to be worth s_i, $1 \le i \le m$, can divide $S = \bigcup_{1 \le i \le m} S_i$ in the ratio of $x:y$ using just a single cut, where $x + y = s_1 + s_2 + \cdots + s_m$.*

8.4.7.2 How Many Cuts Are Required by the Last Diminisher Protocol?

Let us turn again to the minimal number of cuts required by the Last Diminisher protocol for any number of players. If there is only a single player, she gets the whole cake—no cut is needed. From the one-cut-suffices principle in Lemma 8.1 we know that one cut is enough (and required) for two players. We have also seen already that four cuts are enough (and required) for three players. In general, we need to take into account $n-1$ rounds for n players. In the ith round, $1 \le i \le n-2$, each of the $n-i+1$ participating players makes one cut in the worst case. The last round must be considered separately, as it differs from the previous rounds: The last two players simply play Cut & Choose. By the one-cut-suffices principle, again, one cut is enough (and required) in the last round. Summing up the cuts made in all rounds, we obtain for the Last Diminisher protocol a total of

$$\left(\sum_{i=1}^{n-2} n-i+1 \right) + 1 = (n + (n-1) + (n-2) + \cdots + 3) + 1 = \frac{n^2 + n - 4}{2}$$

cuts needed in the worst case, which we state in the following proposition.

Proposition 8.2. *The minimal number of cuts required by the Last Diminisher protocol is $(n^2+n-4)/2$.*

One might also modify the Last Diminisher protocol as follows.[21]

Modified Last Diminisher protocol:

---

[21] This modified Last Diminisher protocol is named *"trimming algorithm"* by Robertson and Webb [814].

- In each execution of the second step, when one piece is being passed from one player to the next and is possibly being trimmed each time, the last player does *not* trim it if it is super-proportional to him, and drops out with this larger portion instead.
- The first three steps of the protocol are repeated $n-1$ times instead of $n-2$ times according to the same scheme, i.e., the last two players do not apply the Cut & Choose protocol.

Thus, for n players, these minor modifications would save one cut in each of the first $n-2$ rounds, without spoiling proportionality. Hence,

$$\left(\sum_{i=1}^{n-1} n-i\right) = (n-1) + (n-2) + \cdots + 1 = \frac{n(n-1)}{2}$$

cuts would be required by this modified protocol in the worst case.

Proposition 8.3. *The minimal number of cuts required by the modified Last Diminisher protocol is* $n(n-1)/2$.

With $n = 50$ players, we have $1\,225$ cuts required in the worst case, and with $n = 100$ players, we have 4950 cuts. Certainly, one wouldn't want to divide a real cake like that. Are there any proportional protocols that can make do with fewer cuts in the worst case?

8.4.7.3 How Many Cuts Are Required by the Lone Chooser Protocol?

The Lone Chooser protocol (see Figure 8.19 on page 543) strikingly illustrates that the one-cut-suffices principle can drastically decrease the minimal number of required cuts. Without this principle, this number can be estimated as follows for this protocol. In the first round, players p_1 and p_2 apply the Cut & Choose protocol: one cut. In the $(n-1)$st round, $n \geq 3$, when player p_n joins in, each of the players $p_1, p_2, \ldots, p_{n-1}$ already has $(n-2)!$ pieces and divides each of these pieces by $n-1$ cuts into n subpieces, for p_n to choose one of them. In total, for n players we thus have

$$(n-2)!(n-1)n = n!$$

pieces, which (ignoring the one-cut-suffices principle) require $n!-1$ cuts, many more than the Last Diminisher protocol does. For example, Table 8.4 on page 587 shows that already for $n = 6$ players the Last Diminisher protocol requires no more than 19 (and the modified Last Diminisher protocol even only 15) cuts in the worst case, whereas the Lone Chooser protocol (in the analysis without the OCS principle) needs 719 cuts. That is a huge difference, which acceleratingly increases with a growing number of players, since the factorial function, $n!$, is known to grow exponentially in n, which is much faster than the growth of the quadratic functions $(n^2+n-4)/2$ and $(n^2-n)/2$.

Using the one-cut-suffices principle in the analysis of the minimal number of cuts required by the Lone Chooser protocol, however, gives a drastically decreased number. For two players, one cut is enough (and required) again. Now, if a third player joins in, the first two players divide their pieces into three equally valued parts using two cuts—in total, we have $1+2\cdot 2 = 5$ cuts, which still coincides with the analysis without the OCS principle. The first difference occurs with four players in the third round: The first three players have two pieces each, which they are to divide into four equally valued parts according to their valuation functions. Instead of $3\cdot 2 = 6$ cuts for each of these three players (summing up to 18, in total), however, three cuts per player are enough (and required), which sums up to nine new cuts. Up to the third round, the cuts total $1+4+9=14$, nine cuts fewer than in the analysis without the OCS principle. Analogously, in the fourth round when the fifth player joins in, we obtain $4\cdot 4 = 16$ new cuts, which totals $1+4+9+16=30$ cuts, instead of 119 without the OCS principle (see Table 8.4 on page 587). In general, using the one-cut-suffices principle in the analysis of the minimal number of required cuts, we obtain the formula stated in the following proposition.

Proposition 8.4. *The minimal number of cuts required by the Lone Chooser protocol is*
$$\sum_{i=1}^{n-1} i^2 = \frac{(n-1)n(2n-1)}{6}.$$

This function is in $\mathcal{O}(n^3)$, i.e., it by far grows slower than $n!-1$, even though it grows a bit faster than the quadratic functions stated above for the minimal number of cuts required by the (modified) Last Diminisher protocol.

Can one do better than this upper bound for the minimal number of cuts required by any proportional cake-cutting protocol?

8.4.7.4 How Many Cuts Are Required by the Divide & Conquer Protocol?

While in the (modified) Last Diminisher protocol only one player per round drops out of the game and the other players keep executing the protocol, the Divide & Conquer protocol (see Figure 8.24 on page 553) proceeds in a more clever way: It recursively divides each of the present groups of players (as well as the parts of the cake currently to be divided among them) into two new subgroups of players (as well as two new subparts of the cake to be divided) having approximately *the same size*, and lets them divide their parts of the cake in parallel. We will see that this approach greatly improves—in comparison with the (modified) Last Diminisher protocol—the minimal number of required cuts, and that this improvement is essentially optimal.

8.4 Cake-Cutting Protocols

First, let us consider the simple cases with up to four players.

Case 1: $n = 1$. If there is only one player, she rakes in the whole cake—no cut is needed.

Case 2: $n = 2$. Two players apply the Cut & Choose protocol, making one cut.

Case 3: $n = 3$. With three players, the clever idea of the divide-and-conquer principle comes into play for the first time. As described on page 552, this allows to recursively reduce this case to the simpler cases given above:

- Two of the three players divide the cake in the ratio of $1 : 2$ according to their valuation functions;
- the third player
  - either chooses the left one of the two pieces (if this is worth at least one third to him) and drops out with it, making no cut, while the first two players share the other piece (which is worth at least two thirds to them both) by Cut & Choose, making one cut;
  - or he shares the other piece (which then must be worth more than two thirds to him) with one of the other players (for whom it is worth at least two thirds as well) by Cut & Choose, making one cut, while the remaining player receives the left piece (which is worth one third to her), without making a cut.

Either way, $2 + 0 + 1 = 3$ cuts are enough and needed.[22]

Case 4: $n = 4$. With four players, we again reduce to the simpler cases above.

- Each of the first three players divides the cake into equal halves according to their valuation function. The group of four players is then divided into two groups with two players each as follows (see also Figure 8.24 on page 553):
  - Among the two pieces separated by the second marking, the fourth player chooses one that she values to be worth no less than the other, i.e., at least $1/2$. She plays together with the player whose marking falls within the chosen piece (and who, in case of identical markings, did not make the second marking, see also Footnote 15 on page 552).
  - The player who made the second marking plays together with the remaining player, sharing the other piece, which they both value to be worth at least $1/2$.

  Both groups of two players now apply the Cut & Choose protocol, making one cut each.

---

[22] Recall that the modified Last Diminisher protocol requires a minimal number of three cuts for three players as well. This is optimal: We will see later that there can be no finite cake-cutting protocol for three players that guarantees each of the players a proportional share with only two cuts.

In total, the Divide & Conquer protocol for four players needs the three markings (which we count as cuts) of the first three players, and one cut each in the two Cut & Choose applications, which sums up to $3+1+1=5$ cuts. This is one cut fewer than required by the modified Last Diminisher protocol for four players in the worst case.

Analogously, the case of five players can be reduced to the two previous cases, as this big group is divided into two smaller groups, one with two, the other with three players. This leads to $4+1+3=8$ required cuts.

This way, we keep reducing the given case to two suitable smaller cases. In general, let $D(n)$ denote the minimal number of cuts required by the Divide & Conquer protocol for n players. Table 8.3 shows the values of $D(n)$ for $1 \leq n \leq 10$ and in general. For n players, the statement "$n-1$ cuts reduce to the cases $\lfloor n/2 \rfloor$ & $\lceil n/2 \rceil$" in the middle column means that the first $n-1$ players make $n-1$ markings (that is, cuts) according to the protocol and the group of all n players is then divided into two groups of sizes $\lfloor n/2 \rfloor$ and $\lceil n/2 \rceil$ for which smaller minimal numbers of required cuts (namely, $D(\lfloor n/2 \rfloor)$ and $D(\lceil n/2 \rceil)$) are already known.

Table 8.3 Minimal number of cuts required by the Divide & Conquer protocol

n	Method	$D(n)$
1	no cut needed	0
2	Cut & Choose	1
3	2 cuts reduce to the cases 1 & 2	$2+0+1 = 3$
4	3 cuts reduce to the cases 2 & 2	$3+1+1 = 5$
5	4 cuts reduce to the cases 2 & 3	$4+1+3 = 8$
6	5 cuts reduce to the cases 3 & 3	$5+3+3 = 11$
7	6 cuts reduce to the cases 3 & 4	$6+3+5 = 14$
8	7 cuts reduce to the cases 4 & 4	$7+5+5 = 17$
9	8 cuts reduce to the cases 4 & 5	$8+5+8 = 21$
10	9 cuts reduce to the cases 5 & 5	$9+8+8 = 25$
$\vdots$	$\vdots$	$\vdots$
n	$n-1$ cuts reduce to the cases $\lfloor n/2 \rfloor$ & $\lceil n/2 \rceil$	$D(n) = n \lceil \log n \rceil - 2^{\lceil \log n \rceil} + 1$

Being a recursive algorithm, the Divide & Conquer protocol for n players gives rise to the following recurrences, where we separately consider the cases $n \in \{1,2,3\}$ and, for $n \geq 4$, an odd or even number of players:

$$D(1) = 0$$
$$D(2) = 1$$
$$D(3) = 3$$
$$D(2k) = 2k - 1 + 2D(k) \qquad (8.4)$$
$$D(2k+1) = 2k + D(k) + D(k+1) \qquad (8.5)$$

8.4 Cake-Cutting Protocols

for $k \geq 2$. Equations (8.4) and (8.5) can be combined into just one recurrence for $n \geq 4$:

$$D(n) = n - 1 + D(\lfloor n/2 \rfloor) + D(\lceil n/2 \rceil), \qquad (8.6)$$

where for a real number r, as is common, $\lfloor r \rfloor$ denotes the greatest integer not exceeding r, and $\lceil r \rceil$ the least integer not smaller than r.

Recurrence (8.6) can be solved by induction on n (see, e.g., the book by Robertson and Webb [814] for this proof), and we obtain:

$$D(n) = n \cdot \lceil \log n \rceil - 2^{\lceil \log n \rceil} + 1. \qquad (8.7)$$

This result is stated in the following theorem.

Theorem 8.10 (Even and Paz [416]). *The minimal number of cuts required by the Divide & Conquer protocol from Figure 8.24 is $D(n) \in \mathcal{O}(n \log n)$.*

8.4.7.5 Overview of the Minimal Numbers of Cuts Required by Some Proportional Cake-Cutting Protocols

Table 8.4 summarizes the minimal numbers of cuts required by those cake-cutting protocols for which we have determined these numbers so far. We leave it as an exercise for the reader to do this analysis for the other finite, proportional cake-cutting protocols from Section 8.4.2, namely, the Lone Divider protocol and the Cut Your Own Piece protocol.

Table 8.4 Minimal numbers of cuts required by some proportional protocols

Protocol	Number of players						
	2	3	4	5	6	$\cdots$	n
Divide & Conquer	1	3	5	8	11	$\cdots$	$n \lceil \log n \rceil - 2^{\lceil \log n \rceil} + 1$
Last Diminisher	1	4	8	13	19	$\cdots$	$(n^2 + n - 4)/2$
Last Diminisher (modified)	1	3	6	10	15	$\cdots$	$(n^2 - n)/2$
Lone Chooser (without OCS)	1	5	23	119	719	$\cdots$	$n! - 1$
Lone Chooser (with OCS)	1	5	14	30	55	$\cdots$	$(n-1)n(2n-1)/6$

Among the cake-cutting protocols in Table 8.4, the Divide & Conquer protocol comes off best. Thus, it so far provides the best upper bound on the minimal number of cuts required to guarantee proportionality. Is there a finite, proportional protocol that beats this upper bound? Or is there a matching lower bound that would prove that this upper bound is optimal?

8.4.7.6 Upper and Lower Bounds for the Minimal Number of Cuts

While upper bounds on the minimal number of cuts, such as those stated in Propositions 8.2–8.4, Theorem 8.10, and also in Table 8.4, are commonly presented more or less informally, the proof of a lower bound on that number requires a precise model that, in particular, specifies which operations are allowed. Robertson and Webb [814] formalize such a model. Recall from Section 8.2 that in their model two kinds of operations are allowed:

1. *evaluation requests* by which the protocol can gain information about how much a certain player values a certain piece of the cake, and
2. *cut requests* by which the protocol can suggest where a certain player ought to make a cut (or marking).

In this model, Woeginger and Sgall [945] proved a lower bound of $\Omega(n \log n)$ for the number of such operations in *any* finite, proportional cake-cutting protocol—under the condition, though, that the protocol assigns a *contiguous* portion to each of the n players (see Definition 8.10 on page 544). This is a relatively severe restriction. Without requiring this condition, Edmonds and Pruhs [370] were able to prove the same lower bound of $\Omega(n \log n)$ for *every* finite, proportional protocol—but by "approximating" both fairness (i.e., the value of proportional portions) and the positions of cuts in cut requests. Note that, although we considered only cut requests in estimating the upper bound stated in Theorem 8.10 for the Divide & Conquer protocol, ignoring evaluation requests, it is easy to see that this upper bound of $\mathcal{O}(n \log n)$ remains valid when counting both cut and evaluation requests in the analysis of this protocol for n players.

This immediately raises the question of how many cuts are required for envy-free protocols. At the beginning of this section (on page 580) we have argued why the Selfridge–Conway protocol for three players requires five cuts in the worst case. But how quickly does the number of cuts grow with the number of players in finite, envy-free protocols for n players? In Section 8.4.5 we mentioned that Brams and Taylor [203] proposed a finite, envy-free cake-cutting protocol for any number of players—which, unfortunately, is not finite bounded. The minimal number of cuts required by this protocol is not bounded, not even for four players. For finite, envy-free cake-cutting protocols with any number of players, no upper bound on the minimal number of required cuts is as yet known.[23] This is one of the most important open problems in the area of cake-cutting.

On the other hand, Procaccia [788] proved, again in the model of Robertson and Webb [814], a lower bound of $\Omega(n^2)$ for the number of cut and evaluation requests in finite, envy-free cake-cutting protocols. His result highlights

---

[23] Even though we actually disregard moving-knife protocols in this section, we mention that the Brams–Taylor–Zwicker protocol for four players (see Figure 8.38 on page 573) is known to require a bounded number of cuts only, but no envy-free moving-knife protocol with a bounded number of required cuts is known for more than four players.

the difference between proportionality and envy-freeness: An upper bound of $\mathcal{O}(n \log n)$ contrasting with a lower bound of $\Omega(n^2)$ indicates a *qualitative* discrepancy between these two concepts and perhaps explains—at least partially—the notorious difficulty inherent in the open problem mentioned above regarding finite, envy-free cake-cutting protocols for any number of players.

8.4.7.7 Minimal Number of Cuts Required for a Proportional Division

Let us come back to the minimal number of cuts required by proportional protocols. Upper bounds like $\mathcal{O}(n \log n)$ and lower bounds like $\Omega(n \log n)$ describe the asymptotic growth of the minimal number of cuts required by proportional protocols depending on the number n of players. Asymptotic estimates, however, are not precise and actually are sensible for large values of n only. In the following, we will therefore be concerned with the *exact* number of cuts required for proportionality to be guaranteed, and we explicitly focus on small values of n.

Let $P(n)$ denote the minimal number of cuts required for a finite cake-cutting protocol to guarantee each of the n players a proportional share. This value $P(n)$ is not specific to a concrete protocol, but it is defined over *all* finite, proportional protocols. As we have seen before, among such protocols the Divide & Conquer protocol appears to be the best one in terms of minimal number of required cuts. For all n, $1 \leq n \leq 16$, Table 8.5 compares the known values for the minimal number of cuts $D(n)$ required by the Divide & Conquer protocol with the best known upper bound on the number $P(n)$ defined above, see also [814]. Boldfaced entries indicate that this number is exact, i.e., coincides with the entries for the corresponding lower bound.

Table 8.5 Comparison of $D(n)$ and the best known upper bound on $P(n)$

Number n of players	1	2	3	4	5	6	7	8	9	10	11	12	13	14	15	16
$D(n)$ in Divide & Conquer	0	1	3	5	8	11	14	17	21	25	29	33	37	41	45	49
Upper bound on $P(n)$	**0**	**1**	**3**	**4**	**6**	**8**	13	15	18	21	24	27	33	36	40	44

Further information regarding the upper bounds on $P(n)$—and about their proofs—can be found in the book by Robertson and Webb [814]. For example, they modify the divide-and-conquer idea in a way that divides groups of players not into subgroups of approximately the same size, but into subgroups of different sizes, and they discuss in what sense this division can be superior to the one obtained by the Divide & Conquer protocol.

8.4.7.8 Two Cuts Are Not Enough for Three Proportional Shares

We will refrain from proving asymptotic lower bounds such as those mentioned above, which are due to Edmonds and Pruhs [370], Woeginger and Sgall [945], and Procaccia [788]. Instead, we will now briefly discuss the proof methods used to establish lower bounds in cake-cutting. Specifically, we will prove the assertion made already in Footnote 22 on page 585: Two cuts are provably too few to guarantee three players a proportional share of the cake.

But, how does one prove that a certain number of cuts (here two) are not enough to guarantee a certain property (here proportionality)? One would have to show that *every* finite[24] cake-cutting protocol for three players requires at least three cuts in the worst case in order to assign a proportional share to each player. However, it wouldn't make sense to indeed consider every single finite cake-cutting protocol for three players, one after the other, to prove this property. On the one hand, there are infinitely many of such protocols; on the other hand, one would also have to consider those protocols that haven't even been found yet. Instead, we will use other arguments to show that no finite protocol for three players can guarantee every player *more than a quarter—and thus, in particular, a third—of the cake with only two cuts.*

Why is it just *"more than a quarter"*? This is simply due to the fact that there exists a finite protocol that does guarantee each of the three players a quarter of the cake with only two cuts. This protocol is shown in Figure 8.44.

Proposition 8.5. *The Quarter protocol for three players from Figure 8.44 guarantees each of the three players a quarter of the cake with only two cuts.*

Proof. As one can see, two cuts are made in each case: One cut is made by p_1 in the first step, and another cut is added when executing the Cut & Choose protocol in each case of the second step.

To see that each of the three players indeed is guaranteed a quarter of the cake, suppose that Belle, Chris, and David apply the Quarter protocol for three players. In the first step, Belle divides the cake in the ratio of $1:2$ and thus creates the pieces S_1 and S_2, where S_1 is worth one third to her and S_2 is worth two thirds. Then, according to the case distinction in the second step (which obviously is exhaustive), we obtain the following:

1. In the first case, S_2 is worth at least one half of the cake to one of Chris and David (say, to Chris), and S_1 is worth at least one quarter of the cake to the other player (i.e., to David). S_1 goes to David, who thus has received at least one quarter. Belle and Chris now share S_2 using the Cut & Choose protocol. Hence, Belle (the cutter) receives a share of $(1/2) \cdot (2/3) = 1/3 > 1/4$, and Chris a share of at least $(1/2) \cdot (1/2) = 1/4$.

---

[24] The restriction to finite protocols is crucial here: For example, the moving-knife protocol of Dubins and Spanier (see Figure 8.18 on page 542) succeeds in providing a proportional division for three players using only two cuts.

8.4 Cake-Cutting Protocols

Given: Cake $X = [0, 1]$ and players p_1, p_2, and p_3 with valuation functions v_1, v_2, and v_3.

Step 1: p_1 divides the cake in the ratio of $1 : 2$ according to her valuation function, thus creating pieces S_1 and S_2 with:

$$v_1(S_1) = \frac{1}{3} \text{ and } v_1(S_2) = \frac{2}{3}.$$

Step 2: Consider the following three cases.

Case 1: S_2 is worth at least one half of the cake for one of p_2 and p_3 (say, for p_2), and S_1 is worth at least one quarter of the cake for the other player (i.e., for p_3). In this case ($v_2(S_2) \geq 1/2$ and $v_3(S_1) \geq 1/4$), S_1 goes to p_3, and p_1 and p_2 share S_2 using the Cut & Choose protocol. The other case ($v_3(S_2) \geq 1/2$ and $v_2(S_1) \geq 1/4$) is treated analogously.

Case 2: S_2 is worth at least one half of the cake for one of p_2 and p_3 (say, for p_2), and S_1 is worth less than one quarter of the cake for the other player (i.e., for p_3). In this case ($v_2(S_2) \geq 1/2$ and $v_3(S_1) < 1/4$), S_1 goes to p_1, and p_2 and p_3 share S_2 using the Cut & Choose protocol. The other case ($v_3(S_2) \geq 1/2$ and $v_2(S_1) < 1/4$) is treated analogously.

Case 3: S_2 is worth less than one half of the cake for both p_2 and p_3 (i.e., we have $v_2(S_2) < 1/2$ and $v_3(S_2) < 1/2$). In this case, S_2 goes to p_1, and p_2 and p_3 share S_1 using the Cut & Choose protocol.

Fig. 8.44 The Quarter protocol for three players

2. In the second case, S_2 is worth at least one half of the cake to one of Chris and David (say, to Chris), and S_1 is worth less than one quarter of the cake to the other player (i.e., to David). In this case, David cannot be satisfied with S_1. Instead, S_1 now goes to Belle, who values it to be worth $1/3 > 1/4$. Now, Chris and David share S_2 using the Cut & Choose protocol, so Chris (the cutter) receives a share of at least $(1/2) \cdot (1/2) = 1/4$, and David receives a share of more than $(1/2) \cdot (3/4) = 3/8 > 1/4$.
3. In the third case, both Chris and David consider S_2 to be worth less than one half of the cake. Thus, they wouldn't get their quarter when they would share it using the Cut & Choose protocol; hence, S_2 must go to Belle. But since S_2 is worth less than one half of the cake to Chris and David, they both consider S_1 to be worth *more than one half*. Sharing S_1 with the Cut & Choose protocol thus guarantees them a share of more than $(1/2) \cdot (1/2) = 1/4$ each.

The least valuable share that all three players are guaranteed to receive is one quarter. ❑

But why can no finite protocol guarantee each of the three players *more than* one quarter of the cake with only two cuts? The following proof answers this question.

Theorem 8.11. *No finite cake-cutting protocol for three players guarantees each of the players more than one quarter—and thus, in particular, one third—of the cake with only two cuts.*

Proof. Suppose there were a finite cake-cutting protocol for three players (say, Belle, Chris, and David) that guarantees each of the players more than one quarter of the cake with only two cuts. One of the players (say, Belle) must make the first cut, thus dividing the cake into two pieces, S_1 and S_2. Of course, it is beyond the control of the other two players, Chris and David, what values S_1 and S_2 will have for them. It might happen, for example, that S_2 is worth exactly one half of the cake to one of them. Since only one more cut can be made, the following case distinction shows that it is impossible to make this cut in a way such that all three players—for all possible valuations of the resulting pieces of cake—receive more than a quarter of the cake.

Case 1: Belle also makes the second cut, thus dividing S_1 or S_2 into two subpieces. No matter which one she has divided; since S_2 (thus S_1 as well) is worth exactly one half of the cake to each of Chris and David, it is possible that they both consider each of the two resulting subpieces to be worth one quarter. Hence, one of them, Chris or David, must receive one of these subpieces, and therefore merely a quarter of the cake.

Case 2: Chris or David (say, Chris) makes the second cut, thus dividing S_1 or S_2 into two subpieces. No matter which one he has divided; since S_2 (thus S_1 as well) is worth exactly one half of the cake to Chris, (at least) one of the two resulting subpieces can be worth no more than one quarter to him. Since Belle and David did not cut, it is possible that they both consider this new subpiece to be worth one quarter. And since one of the three players must receive this subpiece, not all of them are guaranteed to receive more than a quarter of the cake.

This completes the proof. ❑

Of course, it is crucial in this argumentation that not all of the players are *guaranteed* more than one quarter of the cake. For some fortunately chosen valuation functions, it is certainly possible that all players might receive more than their quarter. The point, however, is that this is not true in the worst case.

8.4.7.9 Four Cuts Can Guarantee Four Proportional Shares

As we have seen above, three cuts are necessary to guarantee each of three players to receive one third of the cake. That is why the entry "3" for the upper bound of $P(3)$ in Table 8.5 on page 589 is boldfaced: This upper bound for $P(3)$ is exact, that is, it coincides with the lower bound for $P(3)$. Also the upper bound for $P(4)$ is exact, and the corresponding entry in the table, "4,"

8.4 Cake-Cutting Protocols

is therefore boldfaced as well. That is, a proportional division for four players can be guaranteed with four, yet not with three cuts. Figure 8.45 shows a protocol that is due to Even and Paz [416] and that achieves a proportional division for four players by four cuts in the worst case.

As one can also see in Table 8.5, the Divide & Conquer protocol requires five cuts to guarantee a proportional division for four players. Even though this protocol—with respect to a growing number n of players—is asymptotically nearly optimal (namely, up to constant factors in the $\mathcal{O}$ notation), it still can be improved for specific numbers of players. The Even–Paz protocol from Figure 8.45 illustrates this for $n = 4$ players.

The idea of the Even–Paz protocol can be described as follows. Suppose that some player divides cake X into four pieces, Y_1, Y_2, Z_1, and Z_2, such that Y_2 is at least as valuable as Y_1 to him, and Z_1 is at least as valuable as Z_2 to him. This implies that $Y_2 \cup Z_1$ is at least as valuable as one half of the cake to this player. Hence, he would be willing to share $Y_2 \cup Z_1$ with anyone else using the Cut & Choose protocol, as that would guarantee him one quarter of the cake. Let us first introduce some notation.

Definition 8.12. *For any player p_i with valuation function v_i and for any two pieces of the cake, A and B, we say that:*

1. *p_i prefers A to B if $v_i(A) \geq v_i(B)$,*
2. *p_i strictly prefers A to B if $v_i(A) > v_i(B)$, and*
3. *A is acceptable to p_i if $v_i(A) \geq 1/n$, where n is the number of players.*

Theorem 8.12 (Even and Paz [416]). *The Quarter protocol for four players from Figure 8.45 guarantees each of the four players a quarter of the cake with only four cuts.*

Proof. Why does the Quarter protocol of Even and Paz [416] from Figure 8.45 achieve what is claimed in this theorem? Let us have a closer look at the single cases and their subcases in the second step of the protocol:

1. In Case 1, each of the four players receives an acceptable portion with only three cuts. This is because p_1 needs one cut to create Y and Z, and p_2 and p_3 as well as p_1 and p_4 need one more cut each when executing the Cut & Choose protocol. It is obvious that everyone receives an acceptable share.
2. In Case 2, we first count two cuts: p_1 first divides X into Y and Z, and then Y into Y_1 and Y_2. Now, we need further cuts according to the subcases considered in this case.

    a. In Case 2.1, a third cut is added when p_3 and p_4 divide Z using the Cut & Choose protocol. Due to $v_2(Y_i) \geq 1/4$, Y_i is acceptable to p_2, and due to $v_1(Y_1) = v_1(Y_2) = 1/4$, the other piece, Y_j, $j \neq i$, is acceptable to p_1. Also p_3 and p_4 receive acceptable portions (in fact, even more than that), since they both consider Z to be worth more than one half of the cake.

Given: Cake $X = [0,1]$ and players p_1, p_2, p_3, and p_4 with valuation functions v_1, v_2, v_3, and v_4.

Step 1: p_1 divides cake X into two pieces, Y and Z, of equal value according to his valuation function. We have: $X = Y \cup Z$, $Y \cap Z = \emptyset$, and $v_1(Y) = v_1(Z) = 1/2$.

Step 2: Consider the following cases.

  Case 1: Not all of p_2, p_3, and p_4 strictly prefer the same piece to the other. We thus may assume that $v_2(Y) \geq 1/2$, $v_3(Y) \geq 1/2$, and $v_4(Z) \geq 1/2$. Then, p_2 and p_3 share Y and p_1 and p_4 share Z, both pairs using the Cut & Choose protocol.

  Case 2: Each of p_2, p_3, and p_4 strictly prefer the same piece to the other. We thus may assume that $v_i(Z) > 1/2$ for $i \in \{2,3,4\}$. Then, p_1 divides Y into two pieces, Y_1 and Y_2, of equal value according to his valuation function. We have: $Y = Y_1 \cup Y_2$, $Y_1 \cap Y_2 = \emptyset$, and $v_1(Y_1) = v_1(Y_2) = 1/4$.

   Case 2.1: Y_1 or Y_2 is acceptable to one of p_2, p_3, and p_4 (say, to p_2). (This is possible, even if we have $v_i(Y) < 1/2$ for each $i \in \{2,3,4\}$.) Hence, $v_2(Y_i) \geq 1/4$ for $i = 1$ or $i = 2$. Then p_2 receives Y_i, p_1 receives Y_j, $j \neq i$, and p_3 and p_4 share Z using the Cut & Choose protocol.

   Case 2.2: Neither Y_1 nor Y_2 is acceptable to any of p_2, p_3, and p_4. However, since each of p_2, p_3, and p_4 prefers one of the pieces Y_1 and Y_2 to the other, two of these players (say, p_2 and p_3) must prefer the same (say, Y_2) to the other. Hence, we have $v_i(Y_2) \geq v_i(Y_1)$ for each $i \in \{2,3\}$.
   Then p_1 receives Y_1 and drops out, while p_2, p_3, and p_4 share $X - Y_1 = Y_2 \cup Z$ among each other. To this end, p_2 divides Z into two pieces, Z_1 and Z_2, of equal value according to her valuation function. We have: $Z = Z_1 \cup Z_2$, $Z_1 \cap Z_2 = \emptyset$, and

   $$v_2(Z_1) = v_2(Z_2) > \frac{1}{4}.$$

   Let us assume that p_3 prefers Z_1 to Z_2.

    Case 2.2.1: Z_2 is acceptable to p_4. Then Z_2 goes to p_4, and p_2 and p_3 share $Y_2 \cup Z_1$ using the Cut & Choose protocol.

    Case 2.2.2: Z_2 is not acceptable to p_4. Then p_4 must prefer Z_1 to Z_2. In this case, Z_2 goes to p_2, and p_3 and p_4 share $Y_2 \cup Z_1$ using the Cut & Choose protocol.

Fig. 8.45 The Quarter protocol of Even and Paz [416] for four players

b. In Case 2.2, p_1 first drops out with Y_1, which is acceptable to him. Then, p_2, p_3, and p_4 share $X - Y_1 = Y_2 \cup Z$ among each other. But wait! Didn't we show earlier that two cuts (and we are not allowed to make any more cuts at this point) are *not* enough to guarantee a proportional division among three players? Doesn't the division of the remaining cake among p_2, p_3, and p_4 thus require *three* cuts, which would give a total of *five* cuts?

No! The piece to be divided among the three, $X - Y_1 = Y_2 \cup Z$, is not just any piece of cake for the players p_2, p_3, and p_4, but we arrived in our Case 2.2 only because they have certain valuations of the pieces Y_2 and Z, and that is what we are going to exploit now. Namely, if p_2

has made the third cut, thus creating from Z two pieces, Z_1 and Z_2, that satisfy $v_2(Z_1) = v_2(Z_2) > 1/4$, we know at least in part what p_2, p_3, and p_4 think about Y_1, Y_2, Z_1, and Z_2.

Table 8.6 What do p_2, p_3, and p_4 think about Y_1, Y_2, Z_1, and Z_2 in Case 2.2?

	Y_1	Y_2	Z_1	Z_2	Key
p_2	I	♡I	♡A	♡A	A : acceptable (of value $\geq 1/4$)
p_3	I	♡I	♡A		I : inacceptable (of value $< 1/4$)
p_4	I	I			♡ : prefers Y_i to Y_j or Z_i to Z_j

Table 8.6 shows what we know about these three players' views right at the beginning of Case 2.2 (immediately before distinguishing Cases 2.2.1 and 2.2.2). For example, the "I" and the "♡" in p_2's row and Y_2's column indicates that p_2 considers Y_2 to be inacceptable, but nonetheless prefers this piece to Y_1.

That p_3 prefers Z_1 to Z_2 may be assumed, since one of $v_3(Z_1) \geq v_3(Z_2)$ and $v_3(Z_2) \geq v_3(Z_1)$ must hold, and in case $v_3(Z_1) \geq v_3(Z_2)$ does not hold, we can simply use the analogous arguments with Z_1 and Z_2 playing reverse roles. This is possible because the only other entries already existing in the columns of Z_1 and Z_2 (namely, the two "♡A" in p_2's row) are identical. Furthermore, we have $v_3(Z) > 1/2$ by the assumption of Case 2. Now, since p_3 prefers Z_1 to Z_2, this implies that Z_1 must be acceptable to p_3.

What might the still missing entries in Table 8.6 look like? To answer this question, we need to make a final case distinction.

i. The assumption of Case 2.2.1 that Z_2 is acceptable to p_4 adds another entry to Table 8.6 (see Table 8.7).

Table 8.7 What do p_2, p_3, and p_4 think about Y_1, Y_2, Z_1, and Z_2 in Case 2.2.1?

	Y_1	Y_2	Z_1	Z_2
p_2	I	♡I	♡A	♡A
p_3	I	♡I	♡A	
p_4	I	I		A

Since Z_2 is acceptable to p_4, she can drop out with it. Since p_2 and p_3 prefer Y_2 to Y_1 and Z_1 to Z_2, they both receive an acceptable portion by sharing $Y_2 \cup Z_1$ with each other using the Cut & Choose protocol.[25] Thus, all players have received a portion they

---

[25] This argument is the crucial point of the Even–Paz protocol that has been explained right above Definition 8.12.

consider to be acceptable. In total, they needed only four cuts for this division.

ii. The assumption of Case 2.2.2, that Z_2 is inacceptable to p_4 adds three other entries to Table 8.6 (see p_4's row in Table 8.8), the first one being the entry "I" in the column of Z_2, of course. Moreover, since we in particular have $v_4(Z) > 1/2$ by the assumption of Case 2, p_4 must

A. prefer Z_1 to the inacceptable piece Z_2, and
B. view Z_1 as being acceptable,

which explains the two entries "♡A" in p_4's row and Z_1's column.

Table 8.8 What do p_2, p_3, and p_4 think about Y_1, Y_2, Z_1, and Z_2 in Case 2.2.2?

	Y_1	Y_2	Z_1	Z_2
p_2	I	♡I	♡A	♡A
p_3	I	♡I	♡A	
p_4	I	I	♡A	I

Now, p_2 drops out with Z_2, which she thinks is acceptable. The leftover, $Y_2 \cup Z_1$, is shared this time by p_3 and p_4 using the Cut & Choose protocol. As above, the decisive idea of the Even–Paz protocol can be applied again, this time only to player p_3, who prefers Y_2 to Y_1 as well as Z_1 to Z_2, and therefore receives an acceptable portion. But also p_4 receives a portion she thinks is (even more than) acceptable, since the other two pieces, Y_1 and Z_2, are inacceptable for her in this case, which gives $v_4(Y_2 \cup Z_1) > 1/2$. Thus, in this subcase, too, all players have received a portion they each find acceptable. Again, only four cuts were needed.

This completes the proof. ❑

The statement of Theorem 8.12 is optimal because one can also show that no finite cake-cutting protocol can guarantee all four players more than one sixth—and thus, in particular, one quarter—of the cake with only three cuts.

8.4.7.10 How Many Cuts Guarantee How Many Players Which Share of the Cake?

To conclude, let us summarize—and generalize—the previously mentioned results on minimizing the number of cuts required for divisions with certain properties. We have seen that in finite cake-cutting protocols:

- **with one cut**, two players can be guaranteed to receive at least one half of the cake (see the Cut & Choose protocol in Figure 8.8 on page 529), and that this is optimal, as zero cuts obviously do not accomplish this;

8.4 Cake-Cutting Protocols

- **with two cuts**, each of three players can be guaranteed to receive one quarter of the cake (see the Quarter protocol for three players in Figure 8.44 on page 591 and Proposition 8.5), and that this is optimal, as two cuts cannot guarantee the three players to receive a more valuable share of the cake (see Theorem 8.11);
- **with three cuts**, each of three players can be guaranteed to receive one third of the cake (see, e.g., the modified Last Diminisher protocol mentioned on page 582 and Proposition 8.3, or the Divide & Conquer protocol from Figure 8.24 on page 553, which both are listed in Table 8.4 on page 587), and that this again is optimal, as two cuts in particular cannot guarantee three players one third of the cake (again, see Theorem 8.11);
- **with four cuts**, each of four players can be guaranteed to receive one quarter of the cake (see Theorem 8.12 and the Quarter protocol for four players that is due to Even and Paz [416] and is shown in Figure 8.45 on page 594), and that this is optimal by the remark after the proof of Theorem 8.12 that three cuts can guarantee each of four players one sixth, but no more than one sixth of the cake.

These special results raise a more general question: How many cuts guarantee how many players which share of the cake? Denoting by $M(n,k)$ the most valuable share of the cake that each of the n players can be guaranteed by k cuts in a finite cake-cutting protocol, the above results can be expressed as follows:

$$M(2,1) = 1/2,\ M(3,2) = 1/4,\ M(3,3) = 1/3,\ M(4,3) = 1/6,\ \text{and}\ M(4,4) = 1/4.$$

As one may have guessed, these values follow a pattern, and Robertson and Webb [814] presented this pattern as follows.

Theorem 8.13 (Robertson and Webb [814]).

1. $M(n, n-1) = 1/(2n-2)$ for $n \geq 2$.
2. $M(3,3) = 1/3$ and $M(n,n) = 1/(2n-4)$ for $n \geq 4$.
3. $M(n, n+1) \geq 1/(2n-5)$ for $n \geq 5$.

The results stated in Theorem 8.13 as well as a number of further results from the literature give the values summarized in Table 8.9 (again, see the book by Robertson and Webb [814]). Boldfaced entries in this table indicate that the given value of $M(n,k)$ is optimal. Some entries have been omitted in the table for the following simple reason: If n players can be guaranteed a share of $1/n$ using k cuts, then this is also possible with more than k cuts.

Table 8.9 How many cuts guarantee how many players which share of the cake?

Number of cuts	Number of players								
	2	3	4	5	6	7	8	⋯	n
$n-1$	1/2	1/4	1/6	1/8	1/10	1/12	1/14	⋯	$1/(2n-2)$
n	–	1/3	1/4	1/6	1/8	1/10	1/12	⋯	$1/(2n-4)$
$n+1$	–	–	–	1/5	1/7	1/9	1/11	⋯	$1/(2n-5)$
$n+2$	–	–	–	–	1/6	1/8	1/10	⋯	?

8.4.8 Degree of Guaranteed Envy-Freeness

No doubt, without envy our world would be a better place. Envy often causes conflicts, even violent ones. Therefore, one of the most important goals in the area of cake-cutting is to find procedures satisfying that when dividing some resource, all players are guaranteed to receive a share that they each consider to be worth at least as much as everyone else's share. For the sake of practical feasibility, it would be most desirable if such cake-cutting protocols were finite bounded. Do there exist envy-free, finite bounded cake-cutting protocols? For three players, the Selfridge–Conway protocol from Figure 8.32 on page 565 solves this problem. As noted earlier, extending this result to four and more players turned out to be an utmost stubborn, challenging problem that eventually—after decades of effort—could be solved by Aziz and Mackenzie [60, 61, 62].

If it is so difficult to reach the ideal of envy-freeness for finite bounded cake-cutting protocols, would it at least be possible to approach this ideal approximately? In the literature, there are a number of promising approaches to this more modest goal. For example, Edmonds and Pruhs [369, 370] approximate fairness in cake-cutting protocols, on the one hand, by considering "approximately" fair pieces with respect to their value to the players and, on the other hand, by merely "approximating" the positions of cuts in cut requests. Here, we will introduce yet another approach that is due to Lindner and Rothe [659]: the "degree of guaranteed envy-freeness." Their approach is based on an idea similar to the one Brams, Jones, and Klamler [195] proposed for a special variant of the Divide & Conquer protocol (see the BJK Divide & Conquer protocol in Figure 8.27 on page 556). One difference is that, to reduce envy, Brams, Jones, and Klamler [195] in addition allow players to swap their portions right before they get them assigned (see Footnote 16 on page 557 and Footnote 27 on page 601), whereas Lindner and Rothe [659] do not allow trading.

The degree of guaranteed envy-freeness quantifies how much envy-freeness the players are guaranteed to enjoy even in the worst case when executing a (finite bounded) cake-cutting protocol. This notion is based on the number of envy relations—or, envy-free relations—that may exist among the players with respect to some division of the cake. To discuss this in more detail, let us first introduce some notation. Recall that a binary relation on a set M is a subset $R \subseteq M \times M$.

Definition 8.13 (envy relation and envy-free relation). Let $X = \bigcup_{i=1}^{n} X_i$ be a given division of the cake for a set $P = \{p_1, p_2, \ldots, p_n\}$ of players, where v_i is player p_i's valuation function and X_i is p_i's portion. *Envy relations* (denoted by $\Vdash$) and *envy-free relations* (denoted by $\nVdash$) for this division are binary relations on P:

- p_i envies p_j ($p_i \Vdash p_j$), $1 \leq i, j \leq n$, $i \neq j$, if $v_i(X_i) < v_i(X_j)$.
- We write $p_i \nVdash p_j$ to express that p_j is not envied by p_i.

It is easy to see that envy relations are irreflexive, and envy-free relations are reflexive, as no player can envy herself. The trivial envy-free relation $p_i \nVdash p_i$ will always be disregarded in the analysis of envy(-free) relations. Moreover, neither envy nor envy-free relations are transitive: From Edgar preferring Belle's portion to his own, and Belle preferring Anna's portion to her own (i.e., Edgar $\Vdash$ Belle and Belle $\Vdash$ Anna), one cannot conclude that Edgar also prefers Anna's portion to his own. Of course, Anna's portion might by far be less valuable to Edgar than his own (thus, Edgar $\nVdash$ Anna is well possible). Similarly, it is possible that Belle is not envied by Edgar and does not envy Anna (i.e., Edgar $\nVdash$ Belle and Belle $\nVdash$ Anna), but Edgar nonetheless envies Anna (thus, Edgar $\Vdash$ Anna is well possible).

Now we define the degree of guaranteed envy-freeness for proportional cake-cutting protocols.

Definition 8.14 (degree of guaranteed envy-freeness). For $n \geq 1$, the *degree of guaranteed envy-freeness* (for short, DGEF) of a proportional cake-cutting protocol defined as the maximum number of envy-free relations that exist in every division obtained by this protocol (provided that all players follow the rules and strategies of the protocol), i.e., the DGEF (expressed as a function of n) is the number of envy-free relations that can be guaranteed even in the worst case.

The degree of guaranteed envy-freeness is restricted to proportional cake-cutting protocols, since otherwise this notion might make a misleading impression of fairness. Consider, for example, the following nonproportional protocol for $n \geq 1$ players, the "Anna protocol." One of the players involved is always named Anna, and there are $n - 1$ other players. There is just a single rule: "Anna gets all the cake."

Is the Anna protocol fair? Hardly.[26] The DGEF of the Anna protocol, however, can easily be determined to be $n - 1 + (n - 1)(n - 2) = n^2 - 2n + 1$, since Anna certainly will not envy any of the other $n - 1$ players, and they certainly will not envy anyone except Anna. This value of a DGEF is already very close to the best possible DGEF of $n(n - 1) = n^2 - n$, which holds for any envy-free protocol.

---

[26] At least not with $n > 1$ players. Except perhaps from Anna's point of view.

The DGEF of a proportional protocol is defined via the *number* of guaranteed envy-free relations. That is, there may exist envy-free relations between certain players for some valuation functions, yet envy-free relations between entirely different players for other valuation functions. What matters is not to identify specific envy-free relations, but only *how many* of them are guaranteed to exist in any case.

But, what is actually meant by a *guaranteed* envy-free relation? First, consider another, weaker (or more general) notion: A *case-enforced envy-free relation* is an envy-free relation that exists only for the case of certain suitable valuation functions of the players. For example, looking at the Last Diminisher protocol (see Figure 8.14 on page 535), this means that an envy-free relation from the player dropping out first (call him p_1) to the player who chooses in the final round when Cut & Choose is being executed (call her p_n) is merely case-enforced (provided it exists at all). Depending on p_1's valuation function, it might be that he either envies p_n ($p_1 \Vdash p_n$) or he doesn't ($p_1 \nVdash p_n$). By contrast, a *guaranteed* envy-free relation exists in any case—independent of the players' valuation functions. A guaranteed envy-free relation thus is a special case of case-enforced envy-free relation: It is enforced in every— even in the worst—case. For example, again looking at the Last Diminisher protocol with p_1 dropping out in the first round and p_n lasting until the final round, there exists a guaranteed envy-free relation from p_n to p_1 (i.e., $p_n \nVdash p_1$ is true for all valuation functions of the players). That is, it is guaranteed in any case that the player dropping out last (here p_n) will not envy the player dropping out first (here p_1).

The reason for this difference between a merely case-enforced envy-free relation (such as $p_1 \nVdash p_n$, in case it holds) and a guaranteed envy-free relation (such as $p_n \nVdash p_1$) is that p_n "rubber-stamped" p_1's portion when evaluating it in the first round as not being super-proportional to her, and therefore did not trim it. On the other hand, p_1 is no longer in the game when p_n is assigned her portion (assuming there are $n \geq 3$ players), so he might envy p_n for her share. As we will see later on, such asymmetries among players occur not only with the Last Diminisher protocol, but with all other proportional cake-cutting protocols from Section 8.4.2 as well, causing different DGEF values specific to these protocols.

In addition to explaining the difference between case-enforced and guaranteed envy-free relations, the above example also illustrates that there exist both two-way and one-way envy-free relations. (Of course, the same is true for envy relations.[27])

---

[27] If after executing a cake-cutting protocol, there is a cycle of one-way envy relations (see Footnote 16 on page 557 for an example; a two-way envy relation, which expresses mutual envy by two players, can be viewed as a special case of such a cycle), then these can be resolved by all pairs of adjacent players in the cycle swapping their portions in sequence. This would increase the number of envy-free relations afterwards. However, such a trade commonly is not considered to be part of a cake-cutting protocol, but

8.4 Cake-Cutting Protocols

A two-way envy-free relation exists when two players are mutually not envious ($p_i \not\Vdash p_j$ and $p_j \not\Vdash p_i$). A one-way envy-free relation means, however, that one of the players does not envy the other ($p_i \not\Vdash p_j$), but is envied by this other player ($p_j \Vdash p_i$). Among each pair of distinct players, there can exist no more than two envy-free relations. Thus, in total there are no more than $n(n-1)$ envy-free relations for n players, and whenever all $n(n-1)$ possible envy-free relations are guaranteed, then no player will envy any of the other players, independent of the players' valuation functions. Hence, as already mentioned above, every envy-free cake-cutting protocol for n players has a DGEF of at most $n(n-1)$.

Since we consider the DGEF only for proportional cake-cutting protocols, there is also a lower bound of n guaranteed envy-free relations: In a proportional protocol, every player does not envy at least one other player, as one can easily verify.

Proposition 8.6. *Letting $d(n)$ denote the DGEF of a proportional cake-cutting protocol with $n \geq 2$ players, we have $n \leq d(n) \leq n(n-1)$.*

Thus, the DGEF—just like a yardstick—describes a way of measuring the extent of fairness in proportional cake-cutting protocols. To allow practicability, it deviates somewhat from the ideal concept of envy-freeness for the benefit of finite boundedness. The DGEF of a specific proportional protocol is the number of guaranteed envy-free relations; depending on the protocol, it may be quite obvious how to determine the DGEF, or it may require a rather involved argumentation (see the work of Lindner and Rothe [659]).

Table 8.10 DGEF of selected finite bounded proportional cake-cutting protocols

Protocol	DGEF
Last Diminisher	$2 + n(n-1)/2$
Lone Chooser	n
Lone Divider	$2n - 2$
Cut-your-own-Piece	n
Divide & Conquer	$n \cdot \lfloor \log n \rfloor + 2n - 2^{\lfloor \log n \rfloor + 1}$
Parallelized Last Diminisher [659]	$\lceil n^2/2 \rceil + 1$

Table 8.10 shows the results of analyzing selected proportional cake-cutting protocols most of which have been presented in Section 8.4.2. The "parallelized Last Diminisher protocol" mentioned in Table 8.10 was proposed by Lindner and Rothe [659]; its DGEF is better than that of the original Last Diminisher protocol (and than that of other proportional protocols) due to a

---

rather to be a subsequent improvement routine. That is why we will disregard all envy-free relations resulting from such a trade when analyzing the DGEF of a protocol.

suitable "parallelization": Instead of cutting one proportional piece from the cake and passing it around for the other players to evaluate and possibly trim it, *two* proportional pieces are cut, passed around, evaluated, and possibly trimmed by all remaining players in parallel in each round; the protocol also describes a way of assigning the pieces to the players in a way that increases the DGEF compared to the original Last Diminisher protocol. Further results and proofs regarding the notion of degree of guaranteed envy-freeness can be found in their paper [659].

The results in Table 8.10 can—hopefully—be extended and further improved. The notion of degree of guaranteed envy-freeness may provide some incentive to develop new cake-cutting protocols with increased fairness, indicated by a DGEF greater than those in the table.

8.4.9 Overview of Some Cake-Cutting Protocols

Table 8.11 gives an overview of the most important cake-cutting protocols mentioned in this chapter, and their properties. As already noted in Section 8.3.2, we see that none of the cake-cutting protocols considered here is Pareto-optimal. For the benefit of fairness, they seemingly have to abstain from efficiency in the sense of Pareto optimality.

On the other hand, various ways of relaxing fairness requirements have been proposed. For example, the degree of guaranteed envy-freeness [659], dealt with in Section 8.4.8, abstains from the ideal of envy-freeness for the benefit of finite boundedness of the protocol.

Another way to increase the number of envy-free relations (equivalently, to decrease the number of envy relations) for finite bounded protocols is to allow the players to trade their portions after they have been assigned to them by the protocol. This proposal has been made, for instance, by Feldman and Kirman [447] and also by Brams, Jones, and Klamler [195].

Despite the many important insights that have been made in the area of cake-cutting during the past decades—often by extremely original ideas and arguments—and some of which have been presented in this chapter, many questions remain unanswered. For example, as mentioned in Section 8.4.7 starting on page 577, some open questions concern the minimal number of cuts required by finite protocols (see, e.g., the nonboldfaced—and thus possibly not yet optimal—entries in Tables 8.5 and 8.9 on pages 589 and 598).

Many very interesting results from the area of cake-cutting could not be presented here, simply for reasons of limited space. For example, we presented neither the finite (but unbounded) protocol by Brams and Taylor [204] that guarantees an envy-free division among any number of players, nor the celebrated finite bounded, envy-free cake-cutting protocol for any number of players by Aziz and Mackenzie [60, 61, 62], nor the result by Robertson and Webb [813] (see also [814]) that no finite protocol can guarantee an *exact*

8.4 Cake-Cutting Protocols

Table 8.11 Some cake-cutting protocols and their properties. Key: FB = finite bounded, FU = finite unbounded, MK = moving-knife, PR = proportional, EF = envy-free, EX = exact, and e = exercise.
The "no[7]" entry for Austin's protocol refers to the fact that, as stated in Theorem 8.1, even though this protocol is not immune to manipulation in the sense of Definition 8.6, it does satisfy the weaker notion (8.2) stated in Footnote 7 on page 526.
The "yes[13]" entry for the Last Diminisher protocol refers to Footnote 13 on page 544.

Protocol name [reference]	Number of players	Runtime	Fairness	Pareto-optimal?	Immune to manip.?	Contiguous?	Page reference
Cut & Choose	$n = 2$	FB	EF	no	yes	yes	p. 529
Austin [36]	$n = 2$	MK	EX	no	no[7]	no	p. 532
Last Diminisher (see [889])	$n \geq 1$	FB	PR	no	yes	yes[13]	p. 535
Dubins–Spanier [360]	$n \geq 1$	MK	PR	no	yes	yes	p. 541
Lone Chooser [451]	$n \geq 1$	FB	PR	no	yes	no	p. 542
Lone Divider [625]	$n \geq 1$	FB	PR	no	yes	no	p. 545
Cut Your Own Piece [891]	$n \geq 1$	FB	PR	no	yes	yes	p. 548
Divide & Conquer [416]	$n \geq 1$	FB	PR	no	yes	yes	p. 551
Selfridge–Conway (see, e.g., [482])	$n = 3$	FB	EF	no	yes	no	p. 565
Stromquist [894]	$n = 3$	MK	EF	no	e	yes	p. 570
Brams–Taylor–Zwicker [206]	$n = 4$	MK	EF	no	e	no	p. 572
Brams–Taylor [203]	$n \geq 1$	FU	EF	no	e	no	p. 572

division of the cake among two players,[28] and we refer to these two books and also to the book chapter by Procaccia [789] for further details. Still, we do hope that this chapter has conveyed some of the mathematical beauty and plain elegance of the solutions to some eminently challenging problems that arise when several players are going to fairly divide a cake.

---

[28] This result sharply contrasts with Theorem 8.1 on page 533 regarding Austin's exact moving-knife protocol from Figure 8.12.

Chapter 9
Fair Division of Indivisible Goods

Jérôme Lang · Jörg Rothe

9.1 Introduction

Chapters 4 through 7 focused on collective decision-making problems (in particular, voting) with the default assumption that all agents are concerned with the outcome as a whole, and therefore, all agents are expected to have, and to express, preferences over all alternatives. We now consider another subfield of collective decision-making where every agent will be only concerned by a part of the outcome: There are a set of resources, or goods, to be allocated to the agents, each agent is only concerned with the share she will get, and as a consequence, will only express her preferences over sets of resources (and not over all allocations).[1]

We now proceed by giving a few examples.

Example 9.1. The following six scenarios will be used later on (in Section 9.2.2 starting on page 608) to classify allocation problems in terms of their typical features.

1. The agents are high school teachers and the resources are courses and time slots. The teachers express their preferences over combinations of courses and time slots. For instance, Professor A. has a slight preference for teaching on mornings, but above all prefers to have consecutive slots, that is, she prefers (14–15 and 15–16) to (9–10 and 11–12). In addition, she prefers to avoid teaching class 1, and if she has to, she does not want to teach them more than one slot a

---

[1] There are, of course, some exceptions to this rule. If George hates Helena, he will hope she gets the worst, and would even prefer to get something he likes a little bit less, provided that it worsens her share! However, the assumption that the agents care only about their own shares, called "no externality," makes sense in most cases. See Section 9.10.4 for fair division with externalities.

day. Once the agents have reported their preferences, the allocation decision will be made centrally, by the high school administration.

2. (a) The agents are George and Helena, who are engaged in a divorce settlement process. They remain good friends and their divorce is not conflictual; therefore, they decide to do without a lawyer, and decide by themselves that Helena gets the books and George the bookshelves, etc.
(b) Unlike George and Helena, John and Kindra are unable to negotiate alone, and need to involve a lawyer, who helps them deciding that the children's custody will be shared equally between them, and that, in addition, Kindra gets the house, while John gets the cat plus some monetary compensation from Kindra.

3. The agents are two countries (say, France and Germany) that have jointly bought a very expensive Earth observation satellite. Every day, each country's responsible committee expresses its preferences over the photos it wants to be made; now, there are some physical constraints on the satellite that restrict the set of photos that can be made on a single day, which needs a process to decide in a fair way which photos will be made. This may be complicated by the fact that France paid for two thirds of the satellite while Germany paid only for one third, which leads to different entitlements on the number of photos (see [649]).

4. The agents are employees in a company and the resources are offices.

5. The agents are two children (Anna and David) that have to form two sport teams. Resources are players. David chooses first one member of his team, then Anna one, then again David, then Anna, etc.

6. The agents are people who are interested in buying a house. There is only one resource: the house, sold by means of an auctioneer.

Other examples include the allocation of radio spectrum for wireless communication or of airport landing slots, and even choosing a set of representatives in a committee—as discussed in Chapter 6—can in some sense be viewed as a resource allocation problem [873]. The reader may want to consult the survey by Chevaleyre et al. [284], which gives more detailed examples.

In all these examples, *except one*, we see that the resources are indivisible and nonshareable. The only exception is ... the children in scenario (2b) of Example 9.1, who can be shared, for example, fifty-fifty between John and Kindra, *timewise*.[2] This chapter will focus on the allocation of such indivisible, nonshareable goods; the allocation of *divisible* goods (*cake-cutting*) has been

---

[2] One could perhaps argue that the resource is the time spent with the children, rather than the children themselves.

9.2 Definition and Classification of Allocation Problems

Fig. 9.1 George and Helena want to divide nine objects

discussed extensively in Chapter 8. Also, in this chapter we will focus on *fair, equitable* methods for resource allocation and will not devote space to auctions and related mechanisms where fairness is not the main focus of interest. For more background on resource allocation of indivisible goods, we point the reader to the surveys by Chevaleyre et al. [284], Nguyen, Roos, and Rothe [746, 750], as well as to the book chapters by Bouveret, Chevaleyre, and Maudet [179] and Sandholm [831], the books by Brams and Taylor [204], Moulin [717], Shoham and Leyton-Brown [864], and Wooldridge [950], and to the book edited by Cramton, Shoham, and Steinberg [316]. Finally, a very nice survey of hot topics in fair division of indivisible goods, together with a list of open problems, is due to Amanatidis et al. [19].

9.2 Definition and Classification of Allocation Problems

In this section, we first define allocation problems and then classify them according to a number of properties.

9.2.1 Allocation Problems

Throughout this chapter, we consider a set $R = \{r_1, r_2, \ldots, r_m\}$ of *indivisible resources* (also called *goods*, *objects*, or *items*) and a set $A = \{1, 2, \ldots, n\}$ of agents (of course, agents may also have names instead of numbers). Sometimes, there can be one divisible item called *money*; sometimes, some items can exist in multiple quantities and can therefore be divided, but in only a discrete way (unlike the divisibility defined in the context of cake-cutting, see Section 8.2).

An *allocation* is a function $\pi : A \to 2^R$. We often write π_i instead of $\pi(i)$ for the bundle of items (i.e., the subset of R) allocated to agent $i \in A$, and we write $\pi = (\pi_1, \pi_2, \ldots, \pi_n)$ for the corresponding allocation. For better readability, we occasionally write $\pi = (\pi_1|\pi_2|\cdots|\pi_n)$ and omit parentheses and commas; e.g., $(\{a,d,e\},\{b,f\},\{c,g\})$ will then be written as $(ade|bf|cg)$.

Unless specified otherwise, we require every item to be allocated, i.e., $\bigcup_{i=1}^{n} \pi_i = R$. Such allocations are said to be *complete*. Also, we require items (almost always) to be *nonshareable*, i.e., $\pi_i \cap \pi_j = \emptyset$ for each $i, j \in A$ with $i \neq j$. Sometimes, though, we do allow *shareable* items where allocations may contain overlapping bundles of items allocated to distinct agents.

For example, a data set collected (illegally or not) by the *National Security Agency* may be viewed to be neither a set of *indivisible* items (because one can cull certain special data from it, say, the movement profile of some particular person), nor a set of *nonshareable* items (because several secret services— such as the British MI5, the Israeli Mossad, the US-American CIA, and the German BND) may share and exploit the same data. By contrast, the art objects auctioned off at *Sotheby's* or *Christie's* are usually both indivisible and nonshareable (at least most bidders hope they are).

For example, if George and Helena want to divide the nine objects shown in Figure 9.1 among each other, an allocation that assigns the car and the buddha to George and the seven garden gnomes to Helena would be represented as $\pi = (\pi_{\text{George}}, \pi_{\text{Helena}})$ with

$$\pi_{\text{George}} = \left\{ \raisebox{-2pt}{\includegraphics[height=1em]{car}}, \raisebox{-2pt}{\includegraphics[height=1em]{buddha}} \right\};$$

$$\pi_{\text{Helena}} = \left\{ \raisebox{-2pt}{\includegraphics[height=1em]{g1}}, \raisebox{-2pt}{\includegraphics[height=1em]{g2}}, \raisebox{-2pt}{\includegraphics[height=1em]{g3}}, \raisebox{-2pt}{\includegraphics[height=1em]{g4}}, \raisebox{-2pt}{\includegraphics[height=1em]{g5}}, \raisebox{-2pt}{\includegraphics[height=1em]{g6}}, \raisebox{-2pt}{\includegraphics[height=1em]{g7}} \right\}.$$

9.2.2 Classification of Allocation Problems

In this section, we informally describe various crucial—often dichotomous— features of allocation problems, of the agents' preferences, and of allocation mechanisms that will allow us to classify allocation problems. These features will be illustrated using the scenarios from Example 9.1 on pages 605 and 606.

9.2.2.1 Allocating Single Objects versus Bundles of Objects

In some resource allocation problems, the set of objects is not structured and they are allocated independently of how other objects are allocated. For example, in a *single-item auction* on eBay, objects are typically auctioned off one by one. However, imagine that these single objects are 20 shoes: It often

9.2 Definition and Classification of Allocation Problems

makes a difference for a bidder whether she gets just a left shoe or a pair of shoes (more precisely, a pair of a left and a right shoe of the same brand, size, and color). For a pair of matching shoes, she might be willing to pay a *hell of a lot more* money than for a single left shoe, since the latter would be almost useless for her.[3] That is why an auctioneer might hope to make a higher profit by bundling a pair of matching shoes than when auctioning the two single shoes individually. On the other hand, it might also make sense for him to offer a discount on a big bundle of identical items.

As opposed to single-item allocation problems, in a *multiple-item* (or, *combinatorial*) *allocation problem* the set of items is structured, i.e., bundles of objects are considered and the agents express their preferences (in the specific case of *combinatorial auctions*, the bidders make their bids) for such bundles. The value of a bundle can be different from the sum of the values of the objects contained in it. This is somewhat comparable with the fact that in a cooperative game (see Chapter 3) the gain of the players in a coalition may be different from the sum of the gains of the single players contained in it (see, e.g., Figure 3.1 on page 142). For example, the total gain of a coalition of players in a superadditive game can be higher than the sum of the gains of the participating single players. Single-item auctions have been informally treated in Section 1.3.3 starting on page 12, and we have explained there why combinatorial auctions, a vast field closely related to multiagent resource allocation, will not be further discussed in this chapter.

Among our scenarios from Example 9.1 above, (4) and (6) deal with single objects, and the others can be seen either as a single-item or a multiple-item allocation problem, depending on the context.

9.2.2.2 Centralized versus Decentralized Mechanisms

Finding an optimal or a satisfactory allocation—or even evaluating the quality of a given allocation—requires the agents to express, in one way or another, their preferences over (possibly, bundles of) objects.

The process that consists in querying the agents about their preferences (possibly in an incremental and interactive way) is called *preference elicitation*. Here comes a serious problem. There are an exponential number of bundles, which leads to the following dilemma: Should we ask the agents to report their preferences in an explicit way, by ranking all bundles, or listing them all with a utility value; or should we elicit only a small part of their preferences, but then

---

[3] Except, perhaps, in case it is Joschka Fischer's left sneaker that he was wearing during the swearing into office as the Hessian minister of the environment in 1985—for some a day of historic dimension and the shoe thus a historic relic. Or except, say, in case it is Oliver Bierhoff's left soccer shoe that he used to score the very first *golden goal* in the history of soccer tournaments, which enabled Germany to reach the final against Czechia and then to win the UEFA European Championship of 1996—again, a historic event for some, and a historic shoe.

risk to not be able to determine an optimal allocation? Again, these issues parallel the problem of compactly representing cooperative games where the number of coalitions is exponential in the number of players as discussed, e.g., in Sections 3.4 and 3.6.

An important dichotomy for allocation methods is based on the way preference elicitation and the determination of an allocation are made:

- Some methods are *centralized*: There is a central authority that elicits the agents' preferences, and then determines the output allocation.
- Some are *decentralized*, or *distributed*: There is no central authority, and the agents themselves compute the allocation, revealing their preferences by certain specific (inter)actions.

Example 9.2. To give a very simple example, suppose that two siblings (Belle and Edgar) have to share four candies: one chocolate candy (c), one lemon candy (ℓ), one strawberry candy (s), and one pepper candy (p).

A centralized method could be the following: Belle and Edgar's mother asks them to whisper into her ear in which order they like the candies; Belle whispers that she prefers c to ℓ, ℓ to s, and s to p, and Edgar that he prefers c to s, s to ℓ, and ℓ to p. Once she knows this, the mother decides that it is fair enough to give c and p to Belle, and s and ℓ to Edgar.

A decentralized method could be the following: Belle and Edgar decide that they will pick candies in alternation. After tossing a coin, Edgar is the one to start: He picks c, then Belle picks ℓ, then Edgar s, and finally Belle takes the leftover p.

In both cases, the agents can express only some part of their preferences. In the "mother-centralized" method, this is so because Edgar could not express that even if he likes strawberry better than lemon in general, this preference is reversed when he has a chocolate candy already (just because lemon and chocolate fit so well together—and if this is not convincing enough, replace "lemon" by "mint" and suppose these are ice cream scoops rather than candies). In the decentralized picking sequences method, they could express only partial preferences because Belle did not even have a chance to say that she prefers chocolate above everything else: Edgar had already picked it when it was her first turn to pick. We see here that the first method needs more elicitation work (that is, more communication); but perhaps it also guarantees a fairer outcome. Indeed, as we will see later on, there is generally a trade-off between the quantity of communication and the quality of the solution obtained.

9.2 Definition and Classification of Allocation Problems

Among our scenarios from Example 9.1, (1), (2b), (4), and (6) deal with centralized allocation, and (2a) and (5) with decentralized allocation. In scenario (3), it is not specified whether the allocation will be determined in a centralized or decentralized way—both are possible.

Let us focus on the centralized allocation case, which is the focus of most of the literature (and, as a side-effect, of our chapter). The notion of allocation mechanism is defined differently depending on whether we are in a centralized or distributed setting. An *allocation procedure*, or *mechanism*, assigns the objects to the agents. As noted above, it can be centralized or decentralized.

In a *decentralized allocation procedure*, the agents follow the single steps of a protocol to negotiate the allocation of objects among each other. The cake-cutting protocols in Section 8.4 are examples of decentralized allocation procedures, and more will follow later on in Section 9.9.

In a *centralized allocation procedure*, however, there is a central authority in charge of allocating the objects to the agents, based on the agents' individual valuations of the objects. *Auctions* are typical examples of centralized allocation procedures. In their simplest form, goods are allocated (sold) one by one, and the central authority is the *auctioneer* who can accept a bid using his auction hammer and thus assigns (sells) the corresponding object to this bidder. When agents (also called buyers or bidders) have preferential dependencies, it becomes suboptimal (both for the buyers and the auctioneer) to allocate the objects one by one. By contrast, in a *combinatorial auction*, the bidders make their bids for *bundles* of objects. This leads to several difficulties, both in terms of communication (in the worst case, bidders may have to specify the amount of money they are willing to pay for every subset of objects) and computation (the optimal allocations are generally hard to compute). The field of combinatorial auctions is a vast research area (arguably one of the most important ones in economics and computation); they have given rise to numerous research articles, and even a full book [316]. However, as already explained at the end of Section 9.2.2.1 (as well as in Section 1.3.3), we will not expose this field in this chapter.

The common point of all centralized protocols for computing fair allocations is that they are composed of two stages: a first one where the agents report their preferences (in some representation language, with some possible domain restrictions) to a central authority, and a second stage where the central authority computes an optimal allocation according to some fairness and efficiency criteria.

Two difficulties with these protocols are (a) the communication complexity of the first stage (how many information bits should the agents send to the central authority?) and (b) the complexity of computing the output of the second stage. The first difficulty will be adressed in detail in Section 9.3, and the second one in Sections 9.4, 9.5, and 9.6.

9.2.2.3 Ordinal versus Cardinal Preferences

Another important dichotomy deals with the nature of the preferences expressed: Are the agents allowed to associate *numerical values* with different objects or bundles, or are they only allowed to *rank* them? In Example 9.2, it probably would make little sense to ask Belle and Edgar to give numbers to the candies. However, let us consider the following example: Helena and John are taking part in an auction. The objects that are being auctioned off are a 1955 bottle of Château-Margaux, and a box of candies. Helena is willing to pay up to $ 500 for the bottle and $ 5 for the candy box; John's values are, respectively, $ 400 and $ 2. They secretly tell this the auctioneer, who is the central authority here, and who can then decide that Helena will get both objects (and will have to pay a certain amount of money which depends on the kind of auction mechanism used; see Section 1.3.3 starting on page 12). Here, it is not very helpful to know that both agents prefer the bottle to the box; moreover, due to the possibility of monetary transfers it is wise to ask the agents to report their preferences using quantitative information.[4] In such a case we say that the preferences are *cardinal*, whereas in the first case (see Example 9.2) we say that they are *ordinal*.

Among our scenarios from Example 9.1, (1), (4), and (5) deal with ordinal preferences, whereas (2a) and (3) most likely, and (2b) and (6) definitely deal with cardinal preferences. Note that the latter two cases, (2b) and (6), involve monetary transfers.

9.2.2.4 Fairness versus Efficiency

In some contexts, what counts above all is to be fair and equitable to the agents. For example, while fairness is an important issue in the case of Belle and Edgar dividing candies in Example 9.2 on page 610, this seems to be much less the case for Helena and John at their auction described in the previous subsection. Regarding our scenarios in Example 9.1 on page 605, in (1), for instance, the high school administration should avoid Professor A. to complain that Professor B. got the classes and teaching slots he wanted. Fairness is even more important for (2a), (2b), and (4). Of course, it is not always possible to be completely fair: In scenario (4) of Example 9.1, it can happen that everyone wants the office on the 6th floor with a view on the Eiffel Tower and no one wants the office on the 2nd floor with a view on the highway, but the company must assign the latter to someone and it cannot assign the former to everyone! In scenario (3), the allocation should be fair as well, but it should also take into account the fact that one country invested twice as much

---

[4] This need for quantitative preferences goes far beyond auctions, and is meaningful even if no money is involved in the problem. Quantitative preferences allow for conveying much more information than ordinal preferences, such as preference intensities and explicit trade-offs.

money as the other one. In scenario (5), fairness is what will hopefully allow the two teams to be of roughly equal strength and lead to a more interesting game. On the other hand, scenario (6) has nothing to do with fairness: The house should be sold to the agent who values it most (moneywise); what counts there is the point of view of the auctioneer, who does not care about fairness or equity but about *Pareto efficiency* (see Definition 9.2 on page 622) and *maximizing social welfare* (see Section 9.6) and *maximizing revenue* (see Section 1.3.3 starting on page 12).

9.3 Preference Elicitation and Compact Representation

We now focus on the elicitation and expression of preferences in more detail.

In centralized mechanisms, agents are required to express their preferences about all possible bundles of objects. As we have said already, there are 2^m such bundles for m objects, which makes the method infeasible in practice if the number of objects exceeds a few units: Who would be willing to spend hours ranking sets of candies? A radical way to get rid of the problem consists in asking the agents to rank *single* objects only—as we did above for the allocation of candies in Example 9.2. This, however, leads to ignoring *preferential dependencies* between objects. Intuitively, a preferential dependency is what happened with the "chocolate-and-lemon" example above (see the paragraph right after Example 9.2): An agent may prefer an object r_i to an object r_j if the rest of her bundle is A, but r_j to r_i if it is B. Consider, as another example, an auction with the following objects: a left shoe, a (matching) right shoe, a lemonade, and a beer. It is perfectly reasonable to expect from an agent to be willing to pay, say, $\$5$ for the left shoe only, $\$5$ for the right shoe only, but $\$50$ for both shoes; and likewise, $\$3$ for a beer, $\$2$ for a lemonade but only $\$3.50$ for a beer and a lemonade.

More formally, we define the following notions. Throughout this chapter, $\succeq$ will be a *weak order* (i.e., $\succeq$ is a transitive and total—thus, in particular, a reflexive—relation; recall these notions from page 199 in Section 3.6) and will be used to denote *weak preference relations*: $X \succeq Y$ means that an agent likes the bundle X of goods at least as much as the bundle Y. We write $X \succ Y$ if $X \succeq Y$ but not $Y \succeq X$, and $X \sim Y$ if $X \succeq Y$ and $Y \succeq X$. While $X \succ Y$ means that X *is preferred to* Y, $X \sim Y$ expresses *indifference between* X *and* Y.

Definition 9.1 (weakly separable, separable, additively separable). Given a set R of objects, and a preference relation $\succeq$ on 2^R, we say that $\succeq$ is

- *weakly separable* if for all $r_i, r_j \in R$ and all $C, D \subseteq R$ such that $\{r_i, r_j\} \cap (C \cup D) = \emptyset$, we have $C \cup \{r_i\} \succeq C \cup \{r_j\}$ if and only if $D \cup \{r_i\} \succeq D \cup \{r_j\}$,
- *separable* if for all $C, D, X, Y \subseteq R$ such that $(C \cup D) \cap (X \cup Y) = \emptyset$, we have $C \cup X \succeq C \cup Y$ if and only if $D \cup X \succeq D \cup Y$, and

- *additively separable* if there exists a function $s: R \to \mathbb{R}$ such that for any $X, Y \subseteq R$, we have $X \succeq Y$ if and only if $\sum_{r \in X} s(r) \geq \sum_{r \in Y} s(r)$.

It is readily seen that additive separability implies separability, which in turn implies weak separability. However, the converse implications do not hold [598]: Separability is stronger than weak separability for $m \geq 4$ objects, and additive separability is stronger than separability for $m \geq 6$ objects.

If preferences are expressed cardinally, for each agent $i \in A$, a *utility function* $u_i : 2^R \to \mathbb{R}$ gives i's utility for each bundle $X \subseteq R$ of resources, independently of the other agents' utilities. This is referred to as *utility without externalities*. We say that a utility function u is

- *additively independent* if for each bundle $X \subseteq R$, $u(X) = \sum_{r \in X} u(\{r\})$;
- *submodular* if for any two bundles $S, T \subseteq R$, we have $u(S \cup T) + u(S \cap T) \leq u(S) + u(T)$; and
- *subadditive* if for any two disjoint bundles $S, T \subseteq R$, we have $u(S \cup T) \leq u(S) + u(T)$.

A utility function u induces a preference relation, by

$$X \succeq_u Y \iff u(X) \geq u(Y).$$

Clearly, if u is additively independent then $\succeq_u$ is additively separable.[5] Additively separable ordinal preferences can be easily expressed by giving m numbers $s(r)$, one for each $r \in R$; when we assume, for instance, that $s(r) \in \mathcal{O}(2^m)$ for each r, this needs only $\mathcal{O}(m^2)$ communication bits. When $\succeq$ is a weakly separable preference relation, it makes sense to ask the agent to rank single resources, that is, to provide the restriction of $\succeq$ to singletons. This requires very little communication, more precisely, $\mathcal{O}(m \cdot \log m)$; for additively separable preferences, this requires even less communication than specifying utilities, but it is also much less informative. Indeed, if Belle says $c \succ \ell \succ s \succ p$, separability implies, for instance, that

$$\{c, \ell\} \succ \{c, s\} \succ \{\ell, s\} \succ \{\ell, p\} \succ \{s, p\},$$

but does not allow to say which of $\{c, p\}$ and $\{\ell, s\}$ she prefers.

The discussion above leads to a dilemma: Should we

1. assume that preferences are additively separable, and then enjoy a cheap communication complexity, or
2. assume that they are fully general, and expose ourselves to a prohibitive communication complexity?

Obviously, both solutions are highly undesirable: (1) is overly restrictive (many real-world preferences are *not* separable, as we have seen in our previous examples), and (2) is overly costly. A middle way consists in using a

---

[5] The converse implication does not hold: If $R = \{a, b\}$ and $u(a) = u(b) = 1$, $u(\emptyset) = 0$, and $u(\{a, b\}) = 2.1$, then $\succeq_u$ is additively separable even though u is not additively independent.

9.3 Preference Elicitation and Compact Representation

compact representation language. Such languages are tailored to expressing preferences over combinatorial domains (here, sets of goods) while using a reasonable amount of space only. There are two types of such languages:[6] Those that rely on a domain restriction (separability being an extreme case) and have a guaranteed gain on the amount of communication needed; and those that are fully expressive, that is, that do not make any restriction, and which, in the worst case, will need an exponential representation space, but that in many cases take much less space. Another distinction has to do with the type of preferences that they aim at expressing: *ordinal* (preference relations) or *cardinal* (utility functions). We give here a short overview of such compact preference representation languages in the context of resource allocation. We start with ordinal languages.

9.3.1 Ordinal Preference Languages

Ordinal preference representation languages aim at expressing preference relations over 2^R compactly.

9.3.1.1 Singleton Ranking Form

When preferences are weakly separable, it makes sense to ask each agent to specify her preference relation over single objects, which is the simplest compact preference language that can be designed. Knowing an agent's preferences over single objects, plus weak separability, allows us to derive more comparisons between sets of objects. Now, an important assumption, often made in resource allocation, is *free disposal*: Every agent can dispose, for free, of the resources she does not want. This implies that preferences are *monotonic*: $X \supset Y$ implies $X \succeq Y$. Sometimes, we even assume *strict monotonicity*: $X \supset Y$ implies $X \succ Y$. With the assumption of strict monotonicity, we know that the most preferred bundle is R and the least preferred bundle is $\emptyset$. If, moreover, preferences are weakly separable, and if we know the agent's preference relation over singletons, then we know more than that: In Example 9.2 on page 610, recall that Belle's preference over single candies is $c \succ \ell \succ s \succ p$; then, we know that Belle's preference relation over sets of candies is a completion of the partial order over 2^R shown in Figure 9.2.

More generally, if we have m goods and a ranking over singletons (say, without loss of generality, $r_1 \succeq r_2 \succeq \cdots \succeq r_m$), the *responsive (or monotonic and separable) extension* of $\succeq$ on 2^R is the partial order defined as follows: For all $S, T \subseteq R$, $S \succeq T$ if and only if there exists an injective mapping σ from T to S such that for every $t \in T$, we have $\sigma(t) \succeq t$. For a detailed discussion

---
[6] Similar considerations have been made in the context of cooperative game theory; see Section 3.4 starting on page 171.

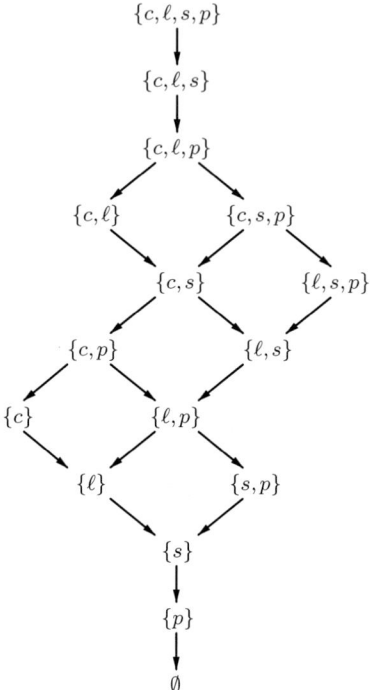

Fig. 9.2 Partial order over 2^R, where $R = \{c, \ell, s, p\}$, for Example 9.2

and references, see the survey by Barberà, Bossert, and Pattanaik [88] and see also Section 9.7 starting on page 649.

However, this cheap elicitation complexity comes not only with a domain restriction (weakly separable preferences) but also with a loss of information, since knowing only an agent's preferences over single objects does not allow us to compare each two pairs of bundles of objects. For instance, in the example above (see also Figure 9.2), $\{\ell\}$ and $\{s,p\}$ are incomparable.

Kannai and Peleg [593] were probably the first to study the problem of extending preferences over objects to preferences over sets of objects axiomatically. See the survey by Barberà, Bossert, and Pattanaik [88] [88], and also the work of Geist and Endriss [495] who apply SAT solving techniques to automatic verification in this area.

9.3.1.2 Conditional Importance Form

For the agent to be able to express preferences between bundles of objects that cannot be derived from preferences between single objects, we may extent the language in the following way: A *conditional preference statement* is a statement of the form

9.3 Preference Elicitation and Compact Representation

$$S^+, S^- : S_1 \triangleright S_2,$$

expressing that if this agent has the objects in S^+ and none of the objects in S^-, then he prefers to have the objects in S_1 rather than the objects in S_2, *ceteris paribus*. Formally, for any pair of bundles $X,Y \subseteq R$ satisfying (a) $S^+ \subseteq X \cap Y$, (b) $S^- \cap (X \cup Y) = \emptyset$, (c) $S_1 \subseteq X$, (d) $S_2 \subseteq Y$, and (e) $X \setminus S_1 = Y \setminus S_2$, we have $X \succ Y$. For instance, if $R = \{a,b,c,d,e,f,g,h\}$, the conditional importance statement $\{a,b\}, \{c\} : \{d\} \triangleright \{e,f\}$ allows us to derive

$$\{a,b,d\} \succ \{a,b,e,f\}, \qquad \{a,b,d,g\} \succ \{a,b,e,f,g\},$$
$$\{a,b,d,h\} \succ \{a,b,e,f,h\}, \qquad \{a,b,d,g,h\} \succ \{a,b,e,f,g,h\}.$$

We refer to the work of Bouveret, Endriss, and Lang [180] for more details about the language of conditional preference statements, called *conditional importance networks*. It should be noted that deciding whether $X \succ Y$ can be derived from a given set of conditional preference statements is, unfortunately, PSPACE-complete [180].

Conditional preference statements come with no domain restriction; any preference relation on 2^R can be expressed by a set of conditional preference statements. This full generality comes not only with a high complexity, but also with the fact that in the worst case, specifying a preference relation by a set of conditional preference statements can be exponentially large.

9.3.2 Cardinal Preference Languages

We are now concerned with expressing utility functions over 2^R.

9.3.2.1 The Bundle Form

In the *bundle form*, all bundles $B \subseteq R$ with a nonzero utility are simply listed, i.e., this list contains the pair $(B, u_i(B))$ for each agent $i \in A$, whenever $u_i(B) \neq 0$. The bundle form is fully expressive, since every utility function can be represented in this way. However, the representation size can be exponential in the number m of resources.

Example 9.3. The example with the shoes at the beginning of Section 9.2.2 on page 608 can be expressed as follows in the bundle form:

$$A = \{1,2,3\}; \quad R = \{S_1^\ell, S_1^r, S_2^\ell, S_2^r, \ldots, S_{10}^\ell, S_{10}^r\}; \quad U = \{u_1, u_2, u_3\},$$

where S_j^ℓ denotes the left and S_j^r denotes the right shoe of the jth pair, $1 \leq j \leq 10$, and the utility functions in U are represented according to the chosen method. For example, assuming $u_i(\emptyset) = 0$ for each $i \in \{1,2,3\}$, the bundle form may require to list up to $3(2^{20} - 1) = 3,145,725$ pairs $(B, u_i(B))$ with $\emptyset \neq B \subseteq R$, $i \in \{1,2,3\}$.

This example will be continued later on.

9.3.2.2 The k-Additive Form

For a fixed positive integer k, the *k-additive form* defines the utility of each agent i for every bundle $B \subseteq R$ of resources by

$$u_i^k(B) = \sum_{T \subseteq B,\ \|T\| \leq k} \alpha_i^T,$$

where α_i^T is a unique coefficient for each bundle $T \subseteq B$ with $\|T\| \leq k$. This coefficient expresses the "synergetic" value of agent i possessing all the resources in T. The k-additive form is more compact than the bundle form, provided that k is chosen small enough, but only for sufficiently large k is it fully expressive. Originally, k-additive functions were proposed by Grabisch [514]; they have also been considered and used for multiagent resource allocation and, more specifically, in the field of combinatorial auctions by, e.g., Conitzer, Sandholm, and Santi [311].

Continuing Example 9.3, the difference between the bundle and the k-additive form is now shown for concrete utility functions of the three agents.

Example 9.4 (continuing Example 9.3). Suppose agent 1 has the following utility function:

$$u_1(B) = \begin{cases} 10 \cdot p + s & \text{if } B \text{ contains a total of } p < 10 \text{ matching} \\ & \text{pairs and in addition } s \text{ single shoes} \\ 80 & \text{if } B = R. \end{cases} \quad (9.1)$$

In particular, this agent has a utility of (a) $u_1(B_0) = 1$ for each bundle $B_0 = \{S_j^x\}$ with only a single shoe; (b) $u_1(B_1) = 10$ for each bundle $B_1 = \{S_i^\ell, S_i^r\}$ with exactly one pair of matching shoes; (c) $u_1(B_2) = 2$ for each bundle $B_2 = \{S_i^x, S_j^y\}$, $x,y \in \{\ell,r\}$ and $i \neq j$, with exactly two shoes that do not match; (d) $u_1(B_3) = 10 \cdot 2 + 3 = 23$ for the bundle $B_3 = \{S_1^\ell, S_1^r, S_3^\ell, S_3^r, S_4^\ell, S_5^r, S_9^\ell\}$; and (e) $u_1(R) = 80$ (and not $10 \cdot 10 = 100$)

9.3 Preference Elicitation and Compact Representation

for the bundle R that contains all ten pairs of shoes, as agent 1 expects to get some discount in this case.

In the bundle form, we would have to list a total of $2^{20} - 1 = 1{,}048{,}575$ pairs $(B, u_1(B))$ with $B \neq \emptyset$ only for this agent.

In the 2-additive form, however, it is enough to give the coefficients α_1^T for all bundles $T \subseteq R$ with $\|T\| \leq 2$:

$$\alpha_1^T = \begin{cases} 0 & \text{if } T = \emptyset \\ 1 & \text{if } \|T\| = 1 \\ 0 & \text{if } T = \{S_i^x, S_j^y\} \text{ for } x,y \in \{\ell,r\} \text{ and } 1 \leq i,j \leq 10 \text{ with } i \neq j \\ 8 & \text{if } T = \{S_i^\ell, S_i^r\} \text{ for some } i,\ 1 \leq i \leq 10. \end{cases}$$

In particular, it is easy to check that in the 2-additive form, we have (a) $u_1^2(B_0) = 1$ if B_0 is a bundle containing a single shoe only; (b) $u_1^2(B_1) = 10$ if B_1 contains exactly one pair of matching shoes; (c) $u_1^2(B_2) = 2$ if B_2 contains two shoes that do not match; and (d) $u_1^2(B_3) = 23$ for $B_3 = \{S_1^\ell, S_1^r, S_3^\ell, S_3^r, S_4^\ell, S_5^r, S_9^\ell\}$. In more detail, the utility of, say, bundle B_3 is calculated as follows in the 2-additive form:

$$u_1^2(B_3) = \sum_{T \subseteq B_3,\ \|T\| \leq 2} \alpha_1^T = \alpha_1^\emptyset + \sum_{T \subseteq B_3,\ \|T\|=1} \alpha_1^T + \sum_{T \subseteq B_3,\ \|T\|=2} \alpha_1^T$$

$$= 0 + 7 + \sum_{T \subseteq B_3,\ T=\{S_i^x, S_j^y\},\ i \neq j} \alpha_1^T + \sum_{T \subseteq B_3,\ T=\{S_i^\ell, S_i^r\}} \alpha_1^T$$

$$= 0 + 7 + 0 + 8 + 8 = 23.$$

Note that this coincides with the true utility $u_1(B_3) = 23$ from (9.1), and the same applies to the other examples above, (a)–(c). (e) However, for the bundle R of all shoes, we obtain the utility value $u_1^2(R) = 100$ in the 2-additive form, which does *not* coincide with the true utility $u_1(R) = 80$. This shows that if k is chosen too small, the k-additive form is not fully expressive.

In this example, agent 1's utility function from (9.1) can be correctly expressed only for $k = 20$ in the k-additive form. The disadvantage that u_1^2 differs from u_1 for one bundle has to be weighed against the advantage that merely

$$\sum_{i=0}^{2} \binom{20}{i} = 1 + 20 + 190 = 211$$

coefficients α_1^T are needed for $T \subseteq R$ with $\|T\| \leq 2$.

For agent 2, we assume that he has utility for only one pair of shoes, $\{S_3^\ell, S_3^r\}$, because that is the only one in his size. Therefore, his utility function u_2 in the bundle form would look like, say, $(B,100)$ for every bundle B that contains $\{S_3^\ell, S_3^r\}$, i.e., for this one pair of matching shoes, he'd be willing to pay the reasonable price of 100 (and doesn't care if he gets other shoes in addition), but all other bundles of shoes, which don't contain $\{S_3^\ell, S_3^r\}$, are completely worthless to him. The 2-additive form with the coefficients $\alpha_2^T = 100$ if $T = \{S_3^\ell, S_3^r\}$, and $\alpha_2^T = 0$ otherwise for all $T \subseteq R$ with $\|T\| \le 2$ is enough to fully represent this utility function u_2.

Agent 3, finally, gives a different rational value from the interval $[0, 100]$ to each bundle $B \subseteq R$, by first enumerating all bundles (say, $B^0 = \emptyset$, $B^1 = \{S_1^\ell\}$, $B^2 = \{S_1^r\}$, $B^3 = \{S_2^\ell\}$, ..., $B^{1,048,575} = R$) and then setting her utility for the ith bundle to $u_3(B^i) = \frac{100 \cdot i}{1,048,575}$.

We leave it as an exercise for the reader to determine the coefficients of u_3 in the 2-additive form.

Since both the bundle form and, for sufficiently large k, the k-additive form are fully expressive, and thus can represent any utility function, it is clear that one form can be transformed into the other, and vice versa. Which one is better? Well, it depends. Chevaleyre et al. [286] give two examples of utility functions such that

- one (namely, u defined by $u(B) = 1 \iff \|B\| = 1$, and $u(B) = 0$ otherwise, for all $B \subseteq R$) can be compactly represented in the bundle form, but requires exponentially many coefficients in the k-additive form, while
- the other one (namely, u' defined by $u'(B) = \|B\|$ for all $B \subseteq R$) has a compact k-additive representation, but requires exponential size in the bundle form.

9.3.2.3 The GAI Form

The *generalized additivity form*, proposed by Bacchus and Grove [71] and Gonzales and Perny [508], is a generalization of the k-additive form: Instead of expressing u as the additive decomposition of subutilities bearing each on k variables, we express u as the sum of subutilities u_{S_j} for a number of subsets $S_j \subseteq R$.

Formally, let $\mathcal{S} = \{S_1, \ldots, S_q\} \subseteq 2^R$. We say that u is *generalized additively independent* with respect to $\mathcal{S}$ ($\mathcal{S}$-GAI) if $u = \sum_j u_{S_j}$, where for each j, u_{S_j} is a utility function over 2^{S_j}. Clearly, k-additivity corresponds to generalized additive independence with respect to $\{S \subseteq R \mid \|S\| = k\}$.

Example 9.5. Let $R = \{a, b, c, d\}$ and $\mathcal{S} = \{\{a, b\}, \{b, c\}, \{d\}\}$ with

9.3 Preference Elicitation and Compact Representation

$u_{\{a,b\}}(\{a,b\}) = 10; \quad u_{\{a,b\}}(\{a\}) = u_{\{a,b\}}(\{b\}) = u_{\{a,b\}}(\emptyset) = 0;$
$u_{\{b,c\}}(\{b,c\}) = u_{\{b,c\}}(\{b\}) = u_{\{b,c\}}(\{c\}) = 6; \quad u_{\{b,c\}}(\emptyset) = 0;$
$u_{\{d\}}(\{d\}) = 4; \quad u_{\{d\}}(\emptyset) = 0$

be given. Then we have

$u(\{a,b,c,d\}) = u(\{a,b,d\}) = 20; \quad u(\{a,b,c\}) = u(\{a,b\}) = 16;$
$u(\{a,c,d\}) = u(\{b,c,d\}) = u(\{b,d\}) = u(\{c,d\}) = 10;$
$u(\{b,c\}) = u(\{b\}) = u(\{c\}) = u(\{a,c\}) = 6; \quad u(\{a,d\}) = u(\{d\}) = 4;$
$u(\{a\}) = u(\emptyset) = 0.$

Note that u is 2-additive; however, the representation as an $\mathcal{S}$-GAI function is more compact than its 2-additive representation.

9.3.2.4 Weighted Logical Formulas

Going further, one can use the expressive power of propositional logic to build even more succinct representation languages. The *weighted logic language* consists in specifying a utility function by a set of pairs

$$\Phi = \{(\varphi_i, w_i) \mid 1 \leq i \leq q\},$$

where for each i, φ_i is a propositional formula over the propositional language whose set of propositional symbols is R, and w_i is a real number. Then, for each $S \subseteq R$, we have

$$u_\Phi(S) = \sum_{i: S \models \varphi_i} w_i,$$

where $S \models \varphi_i$ is to be understood as the satisfaction of φ_i by the interpretation that gives the value true to each $s \in S$ and the value false to each $r \in R \setminus S$.

For example, the utility function of Example 9.5 can be expressed as u_Φ, with $\Phi = \{(a \wedge b, 10), (b \vee c, 6), (d, 4)\}$. The weighted logical form is even more compact than the GAI form, and is also fully expressive. For references on the weighted logical form, see for instance [636, 287, 566, 182, 920, 919].

There are also other representation languages, such as *straight line programs*; see, e.g., [364, 743, 744] and the survey by Chevaleyre et al. [284]. Also, a number of *bidding languages* have been developed specifically for auctions (see, e.g., the book chapter by Nisan [754]). For a review of languages for succint preference representation, see also the survey chapter by Kaci, Lang, and Perny [581].

9.4 Pareto Efficiency and Envy-Freeness

In order to choose a "best" allocation, or a subset of admissible allocations, or a ranking of allocations according to their quality, we need to define criteria.

The most two important criteria, by far, are *envy-freeness* and *Pareto efficiency*. The search for allocations satisfying both, and the impossibility—in general domains—to guarantee both, structures the field according to the means aiming to achieve them as much as possible.

9.4.1 Pareto Efficiency

Informally, an allocation is *Pareto-efficient* if it cannot be improved to another allocation which is at least as good for every agent and strictly better for at least one agent (see also Definition 2.3 on page 49 and Definition 8.5 on page 522 for the same term in other contexts—game theory and cake-cutting). The terms *Pareto efficiency* and *Pareto optimality* are used synonymously.

Definition 9.2 (Pareto dominance and Pareto efficiency). Let A be a set of agents and R a set of resources, let $\pi = (\pi_1, \ldots, \pi_n)$ and $\pi' = (\pi'_1, \ldots, \pi'_n)$ be two allocations for A and R, and let $\succeq_i$ be the preference relation of agent $i \in A$ on 2^R. We say that

- π *Pareto-dominates* π' if (a) for every $i \in A$, we have $\pi_i \succeq_i \pi'_i$, and (b) for some $j \in A$, we have $\pi_j \succ_j \pi'_j$;
- π *strongly Pareto-dominates* π' if for every $i \in A$, we have $\pi_i \succ_i \pi'_i$.

π is *Pareto-efficient* (or, *Pareto-optimal*; *PO*, for short) if no allocation π' Pareto-dominates π, and π is *weakly Pareto-efficient* if no allocation π' strongly Pareto-dominates π.

Pareto efficiency is clearly an *ordinal* criterion in the sense that we only need to know the agents' preferences over bundles of goods to decide if an allocation is Pareto-efficient or not.

We will sometimes use a simpler and (almost) weaker form of efficiency: *completeness*. Recall that an allocation π is said to be *complete* just if it allocates every good, that is, $\bigcup_i \pi(i) = R$. As soon as every good is valued positively by some agent (that is, for each $g \in R$ we have $\{g\} \succ_i \emptyset$ for some i), Pareto efficiency implies completeness.

A stronger notion of efficiency, considered by Barman, Krishnamurthy, and Vaish [93] and Murhekar and Garg [720], is *fractional* Pareto efficiency: An allocation is *fractionally Pareto-optimal* (*fPO*, for short) if no *fractional* allocation exists that makes any of the agents better off without making someone else worse off. A *fractional allocation* distributes, for each good, a unit weight among agents. Agent i receiving a share s_i^j of good r_j can be

9.4 Pareto Efficiency and Envy-Freeness

interpreted as the probability of receiving r_j (see Section 9.11.3 starting on page 678 for randomized fair division), or as the fraction of r_j assigned to i (considering r_j as divisible into units, such as pieces of land).

9.4.2 Envy-Freeness

Informally, an allocation is *envy-free* if no agent prefers the share of another agent to her own (see also Definition 8.3 on page 518 for the same notion in the context of cake-cutting). This notion was introduced first by Foley [462].

Definition 9.3 (envy-freeness). Let A be a set of agents and R a set of resources, and let $\succeq_i$ be the preference relation of agent $i \in A$ on 2^R. An allocation $\pi = (\pi_1, \ldots, \pi_n)$ is *envy-free* (*EF*, for short) if for all $i, j \in A$, we have $\pi_i \succeq_i \pi_j$.

Envy-freeness is an ordinal criterion, as it requires only that an agent be able to compare her bundle with another agent's bundle. Envy-freeness and Pareto efficiency can both be seen as stability criteria: If one of them is not satisfied, then at least one agent has a strong objection against the allocation.

Ideally, we should try to output allocations that satisfy both. Unfortunately, while each of them is easy to satisfy in isolation, there are resource allocation problems for which there exists no allocation that is both Pareto-efficient and envy-free. This can be seen from the very simple example with two objects, a and b, and two agents who both prefer a to b (note that this example shows that there may even exist no allocation that is both envy-free and complete). The basic impossibility of fair division of indivisible goods is therefore:

Theorem 9.1 (Fundamental impossibility result). *No deterministic fair division mechanism guarantees both Pareto efficiency and envy-freeness.*

9.4.3 Envy-Free and Pareto-Efficient Allocations: Existence and Computation

Bouveret and Lang [182] assume the preference relations of the agents to be *dichotomous*, i.e., they express a partition into acceptable and inacceptable shares. More formally, this means that for each agent i, there exists a subset $D_i \subseteq 2^R$ such that for all bundles $X, Y \subseteq R$, $X \succeq_i Y$ if and only if $X \in D_i$ and $Y \notin D_i$. While this restriction imposes a certain loss of expressivity, it does allow to represent preferences over 2^R compactly via boolean formulas whose variables correspond to goods. Using this logical representation, Bouveret and Lang [182] express envy-freeness and Pareto efficiency and show that the problem of deciding whether there exists a Pareto-efficient, envy-free allocation for a given profile of monotonic, dichotomous preferences over a set of goods (represented via boolean formulas) is complete for the complexity class Σ_2^p, the second level of the polynomial hierarchy (recall the definition from Section 1.5.3.2 starting on page 35). If all agents happen to have identical monotonic, dichotomous preferences, the complexity of this problem is lowered to NP-completeness.

Further, de Keijzer et al. [599] show that deciding whether or not there exists a Pareto-efficient, envy-free allocation is Σ_2^p-complete for additive utility functions as well.

The parameterized complexity of finding a Pareto-efficient and envy-free allocation under dichotomous or additive preferences, for three natural parameters (number of agents, of goods, and of utility levels) is studied by Bliem, Bredereck, and Niedermeier [157]. The parameterized complexity of finding efficient and envy-free allocations, with multi-unit items, is studied by Bredereck et al. [234] while Bredereck et al. [233] show how they can be computed efficiently with integer linear programming.

The tension between Pareto efficiency and envy-freeness entails paradoxes, which have been first analyzed by Brams, Edelman, and Fishburn [189]. Dickerson et al. [341] provide a probabilistic study of this tension, by computing the probability, for various distributions over the agents' preferences (and assuming agents' valuations for items to be independently drawn from a common distribution), that a complete and envy-free allocation exists. They show that an envy-free allocation is likely to exist for $m = \Omega(n \log n)$ but not for $m = n + o(n)$, where m is the number of resources and n the number of agents. This result is generalized by Bai and Gölz [79], who assume that agents' utilities are drawn independently from a distribution specific to the agent, and by Bai et al. [78], who start from a worst-case utility profile, for which no envy-free allocation exists, which they very slightly perturb at random; they show then that in the perturbed profile, envy-free allocations exist with high probability and can be found efficiently. Manurangsi and Suksompong [679] go even further and study the phase transition from nonexistence to existence for such allocations. They show that, somewhat surprisingly, the transition is governed by the divisibility relation between m and n; in particular, if m is divisible by n, an envy-free allocation exists with high probability as long as $m \geq 2n$, while in the general case, an envy-free allocation is unlikely to exist even when $m = \Theta(\frac{n \log n}{\log \log n})$.

9.4.4 Relaxing Pareto-Efficiency: Partial Assignments

The fundamental impossibility result implies that in order to find an allocation, we have to sacrifice determinism, envy-freeness, or Pareto efficiency. Randomized allocations, which sacrifice determinism, will be discussed in Section 9.11.3 starting on page 678. Relaxing envy-freeness will be discussed in Section 9.5 starting on page 625. Relaxing Pareto efficiency is less common; still, a reasonable way of relaxing it is to search for small subsets of goods to remain unallocated; unallocated goods can be considered to be deleted or donated. The goal is to minimize the number of items to be deleted so as to obtain an allocation satisfying some fairness notion. This path has been followed by Caragiannis, Gravin, and Huang [250], Chaudhury et al. [279] (both discussed in Section 9.5.3 starting on page 628), Dorn, de Haan, and Schlotter [353] (also discussed in Section 9.5.4 starting on page 629), as well as by Boehmer et al. [160]. This methodology can be seen as a form of "control by deleting items." Other forms of control in fair division have been considered by Aziz, Schlotter, and Walsh [68].

9.5 Relaxations of Envy-Freeness

We have seen that it is not always possible to find an allocation that is both Pareto-efficient and envy-free. While the early literature (until 2015) had focused on making envy a relative notion and defining degrees of envy, most of the recent literature went along a different way: defining various *relaxations* of envy-freeness, with the hope of finding some that are guaranteed to be satisfied by some (Pareto-)efficient allocation.

9.5.1 Degrees of Envy

Because it is not always possible to ensure envy-freeness while preserving Pareto efficiency, some authors have proposed to relax envy-freeness, by defining *degrees of envy* (cf. the different notion of "degree of envy-freeness" introduced in Section 8.4.8 in the context of cake-cutting). Several such degrees have been proposed by several authors [322, 660, 289, 285, 747]. Most of them are defined in three steps:

1. One first defines a *local degree of envy*, for each pair of agents: $E(i,j,\pi)$ is the degree to which agent i envies agent j in allocation π;
2. these degrees are then aggregated into degrees of envy for each agent, $E(i,\pi)$;
3. finally, these degrees are aggregated into a *global degree of envy*, $E(\pi)$.

    Common definitions for the local degree of envy $E(i,j,\pi)$ are:

- $E(i,j,\pi) = 1$ if $\pi_j \succ_i \pi_i$, and $E(i,j,\pi) = 0$ otherwise;
- $E(i,j,\pi) = \max(0, u_i(\pi_j) - u_i(\pi_i))$; and
- $E(i,j,\pi) = \max\left(1, \frac{u_i(\pi_j)}{u_i(\pi_i)}\right)$.

Common choices for defining $E(i,\pi)$ are

- $E(i,\pi) = \max_{j \neq i} E(i,j,\pi)$ and
- $E(i,\pi) = \sum_{j \neq i} E(i,j,\pi)$.

Finally, common choices for defining the global degree of envy $E(\pi)$ are:

- $E(\pi) = \max_i E(i,\pi)$ and
- $E(\pi) = \sum_i E(i,\pi)$.

For instance, the choice of the first definition for $E(i,j,\pi)$ together with the choices of max for $E(i,\pi)$ and $\sum$ for $E(\pi)$ lead to defining $E(\pi)$ as the number of agents i who envy at least one other agent in π; the choice of the first definition for $E(i,j,\pi)$ together with the choice of $\sum$ both for $E(i,\pi)$ and $E(\pi)$ lead to defining $E(\pi)$ as the number of pairs of agents (i,j) such that i envies j in π; and so on.

Shams et al. [855] aggregate degrees on envy using *ordered weighted average* (*OWA*) operators, generalizing both the sum of envies and the maximum of envies. Such cardinal notions of envy allow to define approximations of envy-free mechanisms, where the degree of envy is guaranteed to be bounded. For instance, with the choice (max, min), an allocation π is α-*envy-free* (α-*EF*, for short) if for all agents i and j, it holds that $u_i(\pi_i) \geq \alpha u_i(\pi_j)$. Results on approximating envy-freeness until 2016 are surveyed by Markakis [682].

9.5.2 Max-Min Fair Share and Min-Max Fair Share

Budish [241] defined the *max-min fair share* of an agent i with utility function u_i as the maximum, over all allocations, of the utility of the worst share that i gets according to his own utility function, whereas his *min-max fair share* [186] is defined as the minimum, over all allocations, of the utility of the best share that i gets according to his own utility function. An allocation π satisfies the max-min (respectively, min-max) fair share criterion if each agent gets a share that he values at least as much as his max-min (respectively, min-max) fair share. Formally:

Definition 9.4 (max-min fair share and min-max fair share). Let A be a set of agents, R be a set of resources, $\pi = (\pi_1, \ldots, \pi_n)$ be an allocation for A and R, and let u_i be agent i's utility function over 2^R. Agent i's *max-min fair share* is defined by

$$\textit{max-min-FS}(i) = \max_\pi \min_j u_i(\pi_j)$$

and i's *min-max fair share* is defined by

$$\textit{min-max-FS}(i) = \min_\pi \max_j u_i(\pi_j).$$

9.5 Relaxations of Envy-Freeness

We say π satisfies the *max-min* (respectively, *min-max*) *fair share criterion* if for all $i \in A$, we have $u_i(\pi_i) \geq$ *max-min-FS(i)* (respectively, $u_i(\pi_i) \geq$ *min-max-FS(i)*). We abbreviate the max-min fair share criterion by MMS.

Although we formulated these criteria using utility values, it can in fact be seen that they are purely ordinal: π satisfies the max-min fair share criterion if and only if for all $i \in A$ and for all π', there exists some $j \in A$ such that $\pi_i \succeq_i \pi'_j$, and π satisfies the min-max fair share criterion if and only if for all $i \in A$, there exists some π' such that for all $j \in A$, we have $\pi_i \succeq_i \pi'_j$.

As most of the work on the max-min fair share criterion focus on additive valuations, in the paragraphs below, unless stated otherwise, we assume that agents' preferences are specified by additive valuations. Under this assumption, Bouveret and Lemaître [186] show that deciding the existence of an allocation satisfying the max-min fair share property (recall Definition 9.4 on page 626) is NP-complete. That is, all agents are *guaranteed* to receive at least their max-min (respectively, min-max) fair share in such an allocation $\pi = (\pi_1, \pi_2, \cdots, \pi_n)$, i.e., each agent i is assigned a bundle π_i with $u_i(\pi_i) \geq$ *max-min-FS(i)* $= \max_\pi \min_j u_i(\pi_j)$ (respectively, $u_i(\pi_i) \geq$ *min-max-FS(i)* $= \min_\pi \max_j u_i(\pi_j)$).

For some time it was an open problem whether there exist instances with additive preferences for which there is no complete allocation satisfying the max-min fair share criterion. Extensive generation of random instances did not lead to finding even a single one [186]. Yet, it was answered in the negative by Kurokawa, Procaccia, and Wang [630]; their counterexample is built by a very clever ("Sudoku-like," as they call it) construction.

From then on, the research focused on designing algorithms finding a max-min fair share with high probability and approximations of max-min fair share [630]: An allocation π is α-*MMS* if for any agent i, $u_i(\pi_i) \geq \alpha \cdot$ *max-min-FS(i)*. Note that such a notion is actually a *relaxation of a relaxation of envy-freeness*. A series of approximation algorithms with improving bounds have been published: Kurokawa, Procaccia, and Wang [630] design a $\frac{2}{3}$-approximation algorithm that runs in polynomial time (only for a constant number of agents, though); Amanatidis et al. [22] provide a $(\frac{2}{3} - \epsilon)$-approximation algorithm that runs in polynomial time for any number of agents (and goods), which was improved to a polynomial-time $(\frac{3}{4} - \epsilon)$-approximation algorithm by Barman and Krishnamurthy [91] for any $\epsilon > 0$. Ghodsi et al. [496, 497] prove the existence of a $\frac{3}{4}$-MMS allocation and give a polynomial-time $(\frac{3}{4} - \epsilon)$-approximation algorithm for any $\epsilon > 0$. Garg and Taki [491] improve the bounds further with a polynomial-time $\frac{3}{4}$-approximation algorithm and the proof of existence of a $(\frac{3}{4} + \frac{1}{12n})$-MMS allocation, which is the best result to date.

Hosseini, Searns, and Segal-Halevi [560] consider the ℓ-*out-of-d max-min fair share* (ℓ-*out-of-d MMS*), which is defined as the value an agent is guaranteed to receive by partitioning the goods into d bundles and then choosing the ℓ least preferred ones. This notion obviously generalizes MMS, which is the same as 1-out-of-n MMS (for n agents). However, since MMS is very sensitive to small perturbations of the agents' cardinal utilities, Hosseini, Searns, and Segal-Halevi [560] focus on agents who *ordinally* rank all bundles,[7] yielding a more robust notion. In particular, they show that for each $\ell \geq 1$, there always exists an ℓ-out-of-$\lfloor (\ell + 1/2)n \rfloor$ MMS allocation; for the special case of $\ell = 1$, they provide a polynomial-time algorithm that computes a 1-out-of-$\lceil 3n/2 \rceil$ MMS allocation; and for each $\ell > 1$, they design another polynomial-time algorithm that computes an ℓ-out-of-$\lceil (\ell + 1/2)n + \mathcal{O}(n^{2/3}) \rceil$ MMS allocation.

Again, all the above results hold for *additive valuations*; additional results exist for nonadditive valuations [91, 498, 274].

9.5.3 Envy-Freeness up to One Good and up to Any Good

An allocation satisfies *envy-freeness up to one good* (*EF1*, for short) [660] if for any pair of agents (i, j), if i envies j, the envy can be eliminated by removing a single good from the bundle of agent j (naturally, to achieve this goal, it would make most sense to remove a good from j's bundle that i values most). EF1 is an ordinal criterion.

Again, as most results bear on preferences represented by additive valuations, unless we say otherwise we will assume this is the case in the rest of this section.

EF1 is easy to obtain in isolation, but its combination with Pareto efficiency is not easy to obtain; it is, however, guaranteed to exist [660], even in conjunction with Pareto efficiency: Maximum Nash welfare allocations are both PO and EF1 [255] (see Section 9.6.3 starting on page 642 for the definition of Nash social welfare). However, since computing an exact maximum Nash welfare allocation is hard (again, see Section 9.6.3), this does not give an efficient way of computing an PO+EF1 allocation, and whether a PO+EF1 allocation can be computed in polynomial time is an open problem. Barman, Krishnamurthy, and Vaish [93] give a pseudo-polynomial algorithm for computing PO+EF1 allocations, and also prove that there always exists an allocation that is both EF1 and *fractionally* PO.

An allocation satisfies *envy-freeness up to any good* (*EFX*) [255] if for any pair of agents (i, j), if i envies j, the envy can be eliminated by removing *any* good from the bundle of agent j (or equivalently, by removing a good i values least among those allocated to j). EFX is again an ordinal criterion. Of course, EFX implies EF1. For additive valuations, it is not known whether a complete EFX allocation always exists; this question is considered as today's most enigmatic question about the allocation of indivisible goods [255]. Al-

---

[7] See Section 9.7 starting on page 649 for some work in centralized fair division with ordinal preferences.

9.5 Relaxations of Envy-Freeness

though the main question remains open, such EFX allocations are known to exist for three agents [276], or when the agents' valuations are identical [783] or binary [329, 94] or bi-valued [18]. EFX allocations for additive valuations are also known to exist for four agents provided one item may be left unallocated [133].

Approximation variants of EFX have been considered, which we denote by α-EFX as before: Plaut and Roughgarden [783] show that there always exists a 0.5-EFX allocation; this bound of 0.5 is improved to 0.618 by Amanatidis, Markakis, and Ntokos [23].

Akrami, Rezvan, and Seddighin [7] consider a relaxation of EFX, called EFkX, which requires that no agent envies another agent after the removal of any k goods, and design an algorithm that finds a complete EF2X allocation for restricted additive valuations, where every good has a nonnegative value, and every agent is interested in only some of the goods.

Another direction is to look for *partial* EFX allocations: An EFX allocation is guaranteed to exist if we allow to leave at most $n-1$ goods unallocated [279]. Reducing the number of unallocated goods is a challenging question: Berger et al. [132] address the case of four agents, and Chaudhury et al. [277] show that for all $\epsilon > 0$, there exists an $(1-\epsilon)$-EFX allocation with a sublinear number of unallocated goods. Caragiannis, Gravin, and Huang [250] (already mentioned in Section 9.4.4 on page 624) show how to allocate a large subset of the items while guaranteeing EFX and a good approximation of maximum Nash social welfare.

For random instances, EFX allocations exist with high probability [681]. Suksompong [899] establishes tight lower bounds on the number of allocations satisfying EF1 (respectively, EFX) in the case of two agents.

The related notion of *envy-freeness up to a random good* (*EFR*, for short), due to Farhadi et al. [445], is weaker than EFX, yet stronger than EF1.

9.5.4 Proportional Fair Share

The *(proportional) fair share* of an agent is the ratio of his value for the whole set of objects by the number of agents [889] (see also Definition 8.2 on page 514 for the same notion in the context of cake-cutting). An allocation satisfies the proportional fair share criterion if every agent has at least his fair share. Formally:

Definition 9.5. Let A be a set of agents, R be a set of resources, $\pi = (\pi_1, \ldots, \pi_n)$ be an allocation for A and R, and let u_i be agent i's utility function over 2^R. Agent i's *(proportional) fair share* is defined by

$$FS(i) = \frac{u_i(R)}{n}.$$

We say π satisfies the *(proportional) fair share criterion* (*PFS criterion*, for short) if for all $i \in A$, we have $u_i(\pi_i) \geq FS(i)$.

Unlike the min-max and max-min fair share criteria, this notion needs numerical utilities (and thus indeed is a *cardinal* criterion), but it does not need any interpersonal comparison of utilities.

As shown by Bouveret and Lemaître [186], assuming additive valuations of the agents, envy-freeness implies the min-max fair share criterion, which in turn implies the fair share criterion, which in turn implies the max-min fair share criterion. Interestingly, for k-additive utility functions with $k \geq 2$, Heinen et al. [532, 533] show that there exist allocations that satisfy the fair share criterion but violate the max-min fair share criterion. In particular, even for two agents having symmetric and submodular utility functions,[8] there cannot exist a $\left(\frac{1}{2}+\epsilon\right)$-max-min fair share allocation for any $\epsilon > 0$; on the other hand, there is a polynomial-time approximation algorithm that gives to all agents a share of at least $1/4$ of their max-min share [532].

Example 9.6. Let $n = 2$, $R = \{a,b,c,d\}$, and let the agents' additive utility functions, u_1 and u_2, be defined by their utilities for single objects:

	a	b	c	d
u_1	10	5	7	0
u_2	9	4	8	1

The preference relations induced by u_1 and u_2 are the following:

- Agent 1: $\{a,b,c\} \succ_1 \{a,c\} \succ_1 \{a,b\} \succ_1 \{b,c\} \succ_1 \{a\} \succ_1 \{c\} \succ_1 \{b\} \succ_1 \emptyset$, and for all $X \subseteq \{a,b,c\}$, $X \cup \{d\} \sim_1 X$.
- Agent 2: $\{a,b,c,d\} \succ_2 \{a,b,c\} \succ_2 \{a,c,d\} \succ_2 \{a,c\} \succ_2 \{a,b,d\} \succ_2 \{a,b\} \sim_2 \{b,c,d\} \succ_2 \{b,c\} \succ_2 \{a,d\} \succ_2 \{a\} \sim_2 \{c,d\} \succ_2 \{c\} \succ_2 \{b,d\} \succ_2 \{b\} \succ_2 \{d\} \succ_2 \emptyset$.

We have

$$\textit{max-min-FS}(1) = 10, \quad \textit{min-max-FS}(1) = 12, \quad FS(1) = 11,$$
$$\textit{max-min-FS}(2) = 10, \quad \textit{min-max-FS}(2) = 12, \quad FS(2) = 11.$$

Exactly three allocations satisfy the max-min fair share criterion: $(a|bcd)$, $(bc|ad)$, and $(ad|bc)$. However, no allocation satisfies the PFS criterion, and *a fortiori*, none satisfies the min-max fair share criterion, and none is envy-free. Finally, all allocations that give d to agent 2 (and only those) are Pareto-efficient.

---

[8] A utility function u is said to be *symmetric* if $u(B) = u(\overline{B})$ for any bundle B of goods.

9.5 Relaxations of Envy-Freeness

Similarly as for envy-freeness, proportionality can be relaxed into *proportionality up to one good* (*PROP1*). This notion has been initially defined in the context of allocation of public goods [303]: Reformulated in the private goods setting, an allocation satisfies *PROP1* if for any agent, her utility becomes at least $1/n$ of her utility for all items if at most one item is added to her bundle, where n is the number of agents. A polynomial-time algorithm for finding a PO+PROP1 allocation has been given by Barman and Krishnamurthy [92]. *Proportionality up to any good* (*PROPX*) can be defined in a similar way as EFX; it is not known either whether it is guaranteed to exist for additive valuations. An intermediate notion is studied by Baklanov et al. [80]: *Proportionality up to the maximin item* (*PROPm*): Given an allocation π, the *maximin item for agent i* is the maximum, over all agents j, of an object in j's share i values the least. Somewhat surprisingly, the existence of a PROPm allocation is guaranteed for any number of agents with additive valuations, and it can be computed in polynomial time [81].

The probability of existence of a proportionally fair allocation is studied by Suksompong [896] who shows that for independent, uniformly drawn additive valuations, this probability tends to one in these two situations: when m grows but remains a multiple of n; when m grows asymptotically faster than n (where, recall, we have m goods and n agents).

Demko and Hill [337] show that deciding whether there exists a proportional allocation of indivisible goods is NP-complete if the agents have additive utilities. This result has been strengthened by Markakis and Psomas [683], who show that for any constant $c \geq 1$, it is also NP-complete to decide whether there is an allocation where each of the n agents receives a bundle worth at least $1/cn$.

Finally, Dorn, de Haan, and Schlotter [353] look for a minimum subset of goods to delete (or leave unallocated) so that a proportional allocation in the remaining instance is guaranteed to exist.

9.5.5 Pairwise Max-Min Fair Share and Group Max-Min Fair Share

Caragiannis et al. [255] define that an allocation $\pi = (\pi_1, \ldots, \pi_n)$ satisfies *pairwise max-min fair share* (*PMMS*) if for any pair of agents (i, j),

$$u_i(\pi_i) \geq \max_{B \subseteq \pi_i \cup \pi_j} \min(u_i(B), u_i((\pi_i \cup \pi_j) \setminus B)).$$

While PMMS considers only redistributions among pairs of agents, and MMS considers redistributions among the grand coalition, *group max-min fair share* (*GMMS*) [89, 304] considers redistributions among any possible group, and thus is stronger than both MMS and PMMS. GMMS and PMMS are ordinal criteria. To illustrate these definitions, we take this example, with additive valuations, due to Amanatidis, Birmpas, and Markakis [21]:

	a b c d e
Anna	3 1 1 1 4
Belle	4 3 3 1 4
Chris	3 2 1 3 4

Note that

- $(e|bc|ad)$ is EF, whereas
- $(a|be|cd)$ is not EF (as both Anna and Chris envy Belle); it is not EFX (as even after removing b from Belle's bundle, Anna still envies him); and not PMMS either because

$$u_{\text{Anna}}(\{a\}) = 3 < 4 = \min(4,4)$$
$$= \min(u_{\text{Anna}}(\{a,b\}), u_{\text{Anna}}(\{a,b,e\} \smallsetminus \{a,b\}));$$

and *a fortiori* not GMMS; however, it is EF1 (as both Anna's and Chris's envy can be eliminated by removing e from Belle's bundle).

For additive valuations, it is not known whether a PMMS allocation always exists. However, it is known that a GMMS (and thus a PMMS and an MMS) allocation exists when the number of goods does not exceed the number of agents by more than two [23]. We refer to the survey by Amanatidis et al. [19] for more details about these and some of the other notions discussed in Section 9.5 on relaxations of envy-freeness.

9.5.6 Epistemic Envy-Freeness

While an allocation is envy-free if each agent does not envy anyone (which presupposes they each know everyone's share), *epistemic envy-freeness* (*EEF*) holds if each agent considers it possible that she does not envy anyone: For each agent i, there is an allocation π^i such that $\pi^i_i = \pi_i$ and for each j, agent i does not prefer π^i_j to π^i_i [41]. This notion presupposes that agents know their own share, and the set of all items, but they do not know how the items that they did not receive are allocated to other agents. Epistemic envy-freeness means that for each agent i, the items allocated to other agents could be allocated to them in a way that i does not envy any of them. Of course, envy-freeness implies epistemic envy-freeness. Here comes an example (with additive valuations) that shows that the converse is not true:

9.5 Relaxations of Envy-Freeness

	a	b	c	d
Anna	**10**	8	8	1
Belle	10	**8**	**8**	1
Chris	1	8	8	**10**

The allocation $\pi = (a|bc|d)$ is not EF because Anna envies Belle (and Chris envies Belle, too); however, it is EEF: Anna considers it possible that one of Belle or Chris gets $\{b\}$ and the other one gets $\{c,d\}$. Similarly, Chris considers it possible that one of Anna and Belle gets $\{a,b\}$ and the other one $\{c\}$.

EEF also implies min-max fair share, but the converse is not true, as the next example shows:

	a	b	c	d	e	f
Anna	9	9	9	16	15	15
Belle	8	8	8	20	**13**	**13**
Chris	8	8	8	**24**	12	12

The min-max fair share values of Anna, Belle, and Chris are 25, 26, and 24, respectively; the allocation $(abc|ef|d)$ is min-max fair share. However, it can be verified that there is no EEF allocation. Therefore, epistemic envy-freeness is strictly stronger than min-max fair share.

9.5.7 Envy-Freeness up to k Hidden Goods

Hosseini et al. [561] define that an allocation $\pi = (\pi_1, \ldots, \pi_n)$ satisfies *envy-freeness up to k hidden goods* (HEFk) if there is a subset of items $S \subseteq R$ with $\|S\| = k$ such that for all i, j, $u_i(\pi_i) \geq u_i(\pi_j \setminus S)$. Clearly, HEF0 = EF, and HEF1 implies EF1. Let us look at the following example:

	a	b	c	d	e	f
Anna	9	9	9	28	13	13
Belle	8	8	8	31	**15**	**15**
Chris	8	8	8	**25**	12	12

The allocation $(abc|ef|d)$ satisfies HEF1: Envy-freeness can be restored by hiding d.

9.5.8 Objective Envy-Freeness

A radically different style of relaxing envy-freeness is defined by Shams et al. [854]. Instead of considering envy as a purely subjective notion where an agent is the only one to decide whether she prefers the share of another agent to hers, they consider it a partially objective notion: An agent i will be considered to envy another agent j not only if i alone thinks so, but if in addition a sufficient number of other agents also prefer j's share to i's share.

9.5.9 Discussion

For additive valuations, apart from EF1, PROPm, and *a fortiori* PROP1, or unless one confines them to certain approximation variants like $\frac{3}{4}$-MMS or 0.618-EFX, the considered relaxations of envy-freeness are not guaranteed to exist, or at least, not *known* to be guaranteed to exist. For GMMS, PMMS, EFR, and EFX, it is not known whether they are guaranteed to exist or not (unless, for GMMS and PMMS, one is confined to a special case like having at most two goods more than agents). In Figure 9.3, we depict all considered relaxations of envy-freeness together with their implication relations.

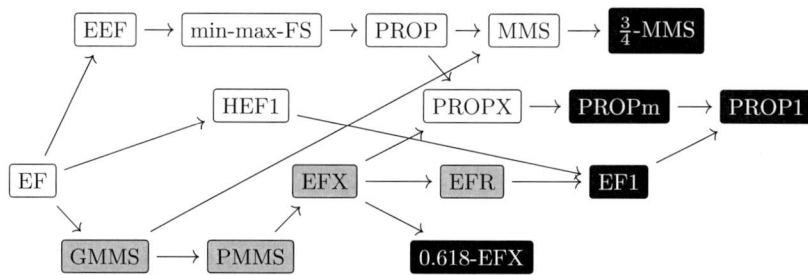

Fig. 9.3 Implications between various relaxations of envy-freeness. In white boxes, relaxations are displayed that are not guaranteed to exist for additive valuations; in black boxes, relaxations that are guaranteed to exist for additive valuations; and in gray boxes, relaxations for which the question remains open.

For various of these notions for which allocations with that property are not known to be guaranteed to exist, approximation variants have been studied; this is in particular the case for EF, MMS, and EFX. All these approximation variants are of course cardinal notions, but they do not require interpersonal utility comparison.

Fairness notions defined via a minimum fairness value associated with each agent, such as the max-min fair share or the proportional fair share, raise an intriguing questions: If agents themselves are asked to report this value, is there a way of ensuring that they will tell the truth? This question is explored by Babaioff and Feige [70].

The vast majority of the recent work in fair division of indivisible goods assumes that agents have *additive valuations*. The prominence of this assumption may be due to fact that it allows for a succinct, simple input and yet allows to ask incredibly many questions of deep complexity. This is to be compared with auctions where nonadditive valuations are considered much more often.

Still, some papers consider nonadditive valuations. Benabbou et al. [129] consider a subclass of submodular valuations (matroid-rank functions) and prove that when agents have such valuations, the allocations maximizing Nash welfare and the leximin allocations satisfy EF1. An EF1 allocation is more generally guaranteed to be exist for subadditive valuations [278, 445]; Chaudhury, Garg, and Mehta [278] also focus on EFX for which they find good approximations. Max-min fair share for subadditive valuations has been studied by Ghodsi et al. [498]. Plaut and Roughgarden [783] show that in the case of arbitrary but *identical* preferences for all agents (for which an EFX allocation is guaranteed to exist), leximin-optimality does not imply EFX, but that a refinement of it (which they call leximin++) does. On the other hand, fair division with *Leontief* utilities (which aim at representing complementary goods, as opposed to submodular utilities) has been studied by Yin et al. [967] with a focus on medical resource allocation to hospitals.

Of course, things may become simpler if on top of additivity some additional *restrictions* are made on valuation functions. Several classes of such restrictions have been considered, such as: *binary* valuations (see, e.g., [527]); *lexicographic* valuations (see, e.g., [558, 562, 866]); few agents or few distinct utility values for goods (see, e.g., [157, 18, 489]) or few agent or resource types (see, e.g., [751, 748]); single-minded valuations where an agent wants a specific bundle and nothing else (see, e.g., [222]).

A domain restriction not of additive valuations but of submodular valuations, namely, valuations that are *rank functions of matroids*, is studied by Barman and Verma [95] and leads to interesting positive results; they complement their results by considering also domain extensions, such as weighted-rank valuations, which generalize matroid-rank functions, and show that then max-min fair allocations are no longer guaranteed to exist.

9.6 Maximizing Social Welfare: The Santa Claus Problem

When preferences are represented quantitatively by utility functions, a systematic way of computing an optimal allocation consists in maximizing the social welfare of allocations, computed as a function of the utilities of the agents. Formally, $SW(\pi) = \star_{i=1}^{n} u_i(\pi_i)$, where $\star$ is a symmetric, nondecreasing mapping from $\mathbb{R}^n$ to $\mathbb{R}$. The three most common choices for $\star$ are $+$, $\min$, and product, leading to *utilitarian, egalitarian*, and *Nash (product)* social welfare. For simplicity, we sometimes write $u_i(\pi)$ instead of $u_i(\pi_i)$ to refer to agent i's utility for her own share π_i in allocation $\pi = (\pi_1, \ldots, \pi_n)$.

9.6.1 Pure Egalitarian Social Welfare

We start with pure egalitarian social welfare, providing first its definition and then surveying results about computing allocations that maximize egalitarian social welfare. Later, in Section 9.6.2, we will consider a refined notion of egalitarianism, so-called leximin social welfare, before we turn to the Nash product in Section 9.6.3 and utilitarianism in Section 9.6.4.

9.6.1.1 Definition

This social welfare measure maximizes the minimum of the agents' utilities; therefore, an allocation with maximum egalitarian social welfare is sometimes referred to as a *max-min allocation*. The problem of finding such a solution to the allocation problem is also known as the *Santa Claus problem* (see, e.g., the work of Bansal and Sviridenko [85]): Santa Claus wants to give Christmas presents to the children such that the most unhappy child is as happy as possible (see Figure 9.4). What Santa is looking for is a max-min allocation, i.e., he wants to maximize egalitarian social welfare.

Fig. 9.4 Santa Claus wants to make the most unhappy child as happy as possible

Egalitarian social welfare, formally defined by

$$sw_e(\pi) = \min\{u_j(\pi) \,|\, j \in A\},$$

describes the utility realized by an agent who is worst off in allocation π.

For example, as a real-world scenario where this may be the right social welfare measure to use, Nguyen et al. [744, 816] consider a disaster area (hit

9.6 Maximizing Social Welfare: The Santa Claus Problem

by an earthquake or a tsunami or alike) and the task of distributing humanitarian aid items—food, medical aid, tents, blankets, and so on—among the needy population. Of course, it is most important to make sure that every survivor stays alive, so it makes sense to first serve those who are worst off. This, however, is best captured by egalitarian social welfare.

Example 9.7 (Example 9.6 continued). Consider again the resource allocation scenario of Example 9.6. The three max-min allocations are $\pi = (a|bcd)$, $\pi' = (bc|ad)$, and $\pi'' = (ad|bc)$ with

$$\min(u_1(\pi_1), u_2(\pi_2)) = \min(10, 13) = 10,$$
$$\min(u_1(\pi'_1), u_2(\pi'_2)) = \min(12, 10) = 10, \text{ and}$$
$$\min(u_1(\pi''_1), u_2(\pi''_2)) = \min(10, 12) = 10$$

whereas the allocation maximizing utilitarian social welfare is $\pi'' = (ab|cd)$, with $u_1(\pi''_1) + u_2(\pi''_2) = 15 + 9 = 24$. Note that if utilities are transferable, then the latter allocation followed by a transfer of three utility units from agent 1 to agent 2 gives final utilities of 12 and 12 to both agents.

For axiomatic studies of the egalitarian solution see, in particular, the work of Thomson [912] and Sprumont [884].

9.6.1.2 Computation

As a decision problem this can be formalized as follows:

EGALITARIAN-SOCIAL-WELFARE-OPTIMIZATION (ESWO)
Given: A set A of agents, a set R of goods, the agents' utility functions over 2^R (represented by one of the cardinal preference languages described in Section 9.3.2 starting on page 617), and a positive integer K.
Question: Does there exist an allocation π such that $sw_e(\pi) \geq K$?

More precisely, the Santa Claus problem refers to the special case of this problem where the agents have additive utilities (i.e., they are represented in the 1-additive form, recall Section 9.3.2.2 starting on page 618). This problem (and, more generally, ESWO with utilities represented in the k-additive form, for each $k \geq 1$) is known to be NP-complete, as shown explicitly by Bouveret et al. [176, 187] (this result also follows implicitly from the work of Lipton et al. [660]). Solving a conjecture by Chevaleyre et al. [284] in the affirmative, NP-completeness of ESWO has been shown also for the bundle form (recall Section 9.3.2.1 starting on page 617) by Roos and Rothe [816] (see also the journal version by Nguyen et al. [744]), and the same result for utilities represented by straight line programs is due to Nguyen et al. [743, 744].

EXACT-EGALITARIAN-SOCIAL-WELFARE-OPTIMIZATION (XESWO), the exact variant of ESWO, asks, for the same input, whether $\max_\pi sw_e(\pi) = K$. Roos and Rothe [816] (see also [744]) show that XESWO is complete for DP, the class of differences of NP sets (recall the definition from Section 1.5.3.3 on page 37), both for the bundle form and for k-additive utility functions, for each $k \geq 2$. See Table 9.1 on page 647 for an overview.

Since the decision problem ESWO is NP-hard, it makes sense to consider efficient approximation algorithms for its optimization variant:

MAXIMUM-EGALITARIAN-SOCIAL-WELFARE (MAX-ESW)

Given: A set A of agents, a set R of goods, and the agents' utility functions over 2^R (represented by one of the cardinal preference languages described in Section 9.3.2 starting on page 617).

Output: $\max\{sw_e(\pi) \,|\, \pi$ is an allocation of the goods in R to the agents in $A\}$.

Much of this work has focused on the Santa Claus problem, i.e., on the approximability of MAX-ESW with utilities in the 1-additive form. For example, Bezáková and Dani [144] have applied both matching techniques and rounding techniques for linear programming relaxation to show that this problem can be approximated in a factor of $1/(m-n+1)$, where m is the number of goods and n the number of agents. On the other hand, they also provide the currently best inapproximability result for this problem: MAX-ESW with 1-additive utilities cannot be approximated in polynomial time within a factor of $\alpha > 1/2$, unless P = NP. An approximation algorithm due to Asadpour and Saberi [34, 35] achieves a performance guarantee of $\Omega(1/\sqrt{n}\log^3 n)$, and the currently best approximation result in terms of the number m of goods is due to Chakrabarty, Chuzhoy, and Khanna [264] who proposed an $\mathcal{O}(1/m^\varepsilon)$-approximation algorithm running in time $m^{\mathcal{O}(1/\varepsilon)}$, for each $\varepsilon \in \Omega(\log\log m/\log m)$; note that their technique also provides an approximation algorithm with respect to the number of agents. Annamalai, Kalaitzis, and Svensson [27] are the first to design a polynomial approximation algorithm for the Santa Claus problem with a constant approximation factor. Chan, Tang, and Wu [272] obtain better results for the restricted problem where the value of an item for an agent is either 1 or a small value ϵ.

Also restricted variants of the Santa Claus problem have been investigated in terms of their approximability. For example, Golovin [507] studies the *"Big Goods/Small Goods"* variant of the Santa Claus problem where, to express their utilities, the agents may choose among three values only: "Small" goods can have utilities 0 or 1, while "big" goods can have utilities 0 or $z > 1$. Golovin [507] shows that this problem can be efficiently approximated within a factor of $1/\sqrt{n}$ by applying min-cut and network flow techniques (see, for example, Section 3.4.2.1 starting on page 176 and Section 3.4.2.2 starting on page 176 for other applications of such techniques in cooperative game theory). For a more general variant of this Big Goods/Small Goods model,

9.6 Maximizing Social Welfare: The Santa Claus Problem

Khot and Ponnuswami [611] provide an (α/n)-approximation algorithm running in time $m^{\mathcal{O}(1)} n^{\mathcal{O}(\alpha)}$. Note that there is a trade-off between the running time and the performance guarantee of their algorithm. Bansal and Srividenko [85] consider another model where each agent j is restricted to choose either a value z_j or zero as utility for good r_j. They provide a polynomial-time $\mathcal{O}(\log \log \log n / \log \log n)$-approximation for this restriction of the problem.

Bateni, Charikar, and Guruswami [104] study the effect of bounding the "degree" of the goods: Every good is required to have a nonzero value for at most D agents. They show that the case of $D = 3$ is essentially equivalent to the general case (i.e., for unbounded degree), and they provide a ($1/4$)-approximation algorithm for $D = 2$.

For yet another restriction of MAX-ESW with 1-additive utilities, namely for the same number of agents and goods ($m = n$) and assuming, as is common, that the empty bundle has zero utility (which implies that every agent must be assigned at least one good to have an egalitarian social welfare distinct from zero),[9] Golovin [507] shows that this problem is solvable in polynomial time. The same result holds for MAX-ESW with k-additive utilities and the same restriction, for each $k \geq 1$. Moreover, building on the work of Deuermeyer, Friesen, and Langston [340] and Csirik, Kellerer, and Woeginger [317], Woeginger [941] provides a polynomial-time approximation scheme (recall the definition from Section 1.5.2.4 starting on page 33) for the problem of maximizing the minimum completion time of jobs to be scheduled on parallel identical machines, which is another formulation of the problem MAX-ESW when all agents have identical 1-additive utilities.

In general, for any $k \geq 2$, MAX-ESW with k-additive utilities cannot be approximated in polynomial time within any factor, unless P = NP. This has been observed by Nguyen, Roos, and Rothe [746] who also survey the above-mentioned approximability results and provide more details on the techniques applied. They also survey known approximation results for MAX-ESW in the bundle form, including results by Golovin [507], Goemans et al. [503], Chekuri, Vondrák, and Zenklusen [280], and Khot and Ponnuswami [611] in a variety of models and restrictions, and we refer to their survey for a detailed overview. Regarding inapproximability of MAX-ESW for the bundle form, Nguyen, Roos, and Rothe [746] show that this problem cannot be approximated in polynomial time within any factor, unless P = NP, even if every agent can only choose among two values, zero and one, for each good.

Ferraioli, Gourvès, and Monnot [450] consider the restriction of the Santa Claus problem where each agent must receive exactly k objects (which they call a *regular* allocation problem). They show that the problem is strongly NP-

---

[9] Deviating from this common normalization assumption, Lange, Nguyen, and Rothe [642] study the complexity of problems in fair division of indivisible goods settings where agents can have *nonzero* utility for the empty bundle. In particular, they identify cases of maximizing egalitarian or Nash product social welfare where dropping this normalization assumption exacts a price in terms of a raised computational complexity.

hard as soon as $k \geq 2$ (it is obviously in P for $k = 1$, since it then corresponds to an optimal matching problem), but becomes polynomial-time solvable if the utility u_i^j given by agent i to object r_j can take only two possible values. For an overview, see Table 9.2 on page 648.

The practical computation of optimal max-min allocations via mixed integer linear programming was addressed by Lesca and Perny [652]. Additionally, they go beyond the max-min fair share criterion and propose a general family of criteria, based on ordered weighted average operators that span from maximum egalitarian to maximum utilitarian social welfare. Another approach that aims at finding a trade-off between egalitarianism and utilitarianism, based on infinite Lorenz dominance, has been proposed by Golden and Perny [505].

Demko and Hill [337] show that, when the n agents have additive utilities that sum up to 1, there is a nonincreasing function $V_n(\alpha)$ from $(0,1]$ to $(0,1]$ such that if the maximum value given to an object by an agent is at most $\alpha \in (0,1]$, there always exists an allocation giving at least $V_n(\alpha)$ to each agent. For instance, let $n = 2$. If $\alpha = 1/3$, then the worst case occurs with three objects, each equally valued by each of the two agents: Then one agent receives two objects and the other one only one; thus the least happy agent has utility $1/3$. If $1/3 \leq \alpha \leq 1/2$, this bound of $1/3$ does not change: It suffices to give to one agent an object o that she values at least $1/3$ (such an agent and an object must exist, because $\alpha \geq 1/3$) and all other objects to the other agent, which she values at least $1/3$ because she values o at most $2/3$. If $\alpha = 1/4$, we can do better, since $V_n(1/4) = 2/5$. Of course, when α tends to 0, $V_n(\alpha)$ tends to 1. Markakis and Psomas [683] give a polynomial-time algorithm for finding an allocation with such a guarantee, and the lower bound $V_n(\cdot)$ was refined by a tighter lower bound due to Gourvès, Monnot, and Tlilane [512].

Instead of simply minimizing egalitarian social welfare, one can view the allocation problem as an efficiency-fairness tradeoff and express it as a bi-criteria optimization problem. Choosing max-min fairness and utilitarian efficiency as the two criteria, Nguyen and Rothe [751, 748] (see also their related work [749] on bi-criteria approximation for load balancing on unrelated machines with costs) show that when there are only a few item types or a few agent types, one can construct an approximate Pareto set (with respect to these two criteria) in time polynomial in the input size.

9.6.2 Maximum Leximin Social Welfare

While egalitarian social welfare aims at guaranteeing fairness by maximizing the utility of the least happy agent, it does so in an extreme way, to the detriment of efficiency. Leximin social welfare is a refinement of pure egalitarian social welfare: While the latter only pays attention to the utility of the least happy agent, the former allows to break ties between two allocations

9.6 Maximizing Social Welfare: The Santa Claus Problem

maximizing the utility of the least happy agent by paying attention to the second least happy agent, and in case there are still ties, then to the third least happy agent, and so on.

Definition 9.6 (leximin domination and leximin optimality). Let $\vec{u}(\pi) = (u_1(\pi_1), \ldots, u_n(\pi_n))$ be the vector of utilities obtained by the agents for some allocation π. Let $\vec{u}^\uparrow(\pi)$ be the vector obtained by reordering the values of $\vec{u}(\pi)$ nondecreasingly: Thus, $u_1^\uparrow(\pi)$ is the utility of the least happy agent, $u_2^\uparrow(\pi)$ is the utility of the second least happy agent, etc. We say that π *leximin-dominates* π', denoted by $\pi \succ_{leximin} \pi'$, if for some $k \leq n$,

1. for all $j < k$, $u_j^\uparrow(\pi) = u_j^\uparrow(\pi')$ and
2. $u_k^\uparrow(\pi) > u_k^\uparrow(\pi')$.

Finally, π is *leximin-optimal* if there is no π' such that $\pi' \succ_{leximin} \pi$.

Example 9.8. Let $n = 4$, $R = \{a, b, c, d\}$, and let the agents' additive utility functions, $u_1, \ldots, u_4$, be defined by their utilities for single objects:

	a	b	c	d
u_1	10	2	2	5
u_2	5	10	2	2
u_3	6	5	10	4
u_4	4	4	8	3

There are two allocations maximizing egalitarian social welfare: $\pi = (d|b|a|c)$ and $\pi' = (d|a|b|c)$. Note that π' is Pareto-dominated by π, since both agent 2 and agent 3 benefit from swapping a and b.

Now, we have $\vec{u}(\pi) = (5, 10, 6, 8)$, $\vec{u}^\uparrow(\pi) = (5, 6, 8, 10)$, and $\vec{u}(\pi') = \vec{u}^\uparrow(\pi') = (5, 5, 5, 8)$. Therefore, $\pi \succ_{leximin} \pi'$ and π is the only leximin-optimal allocation.

Note that the allocation maximizing utilitarian social welfare is $(ad|b|c|\emptyset)$.

Although their definitions make use of numerical utility functions, egalitarian (pure and leximin) social welfare criteria can be defined on an arbitrary scale, provided that this scale is common to all agents (in other terms, it requires interpersonal comparison of preferences). On the other hand, the utilitarian social welfare criterion does need numbers, since utilities are summed.

The leximin mechanism has been studied in detail for random fair division under dichotomous preferences [163] (see also Section 9.11.3 starting on page 678) and applied to the fair allocation of unused classrooms in public schools to charter schools [629].

The problem of computing leximin-optimal solutions (recall Definition 9.6 on page 641) has been studied in depth by Bouveret and Lemaître [185], who describe five specific algorithms using constraint programming and compare them both theoretically and experimentally. For further work on resource allocation by a leximin approach, we refer to the survey by Luss [674].

Related to maximum egalitarian and leximin social welfare is the notion of *equitability* [513, 468, 720] (see also Section 8.3.1.7 starting on page 520 for this notion in the related domain of cake-cutting). An allocation is *(perfectly) equitable* if all agents get the same utility. While this notion is very strong, and usually hard to fulfill, it makes more sense to study some of its relaxations, such as *equitability up to one good* or *up to any good*, or approximations thereof. Freeman et al. [468] show that leximin optimality implies equitability up to any good and Pareto optimality, and Murhekar and Garg [720] show that for additive preferences with positive valuations there always exists an allocation that is equitable up to one good and fractionally Pareto-efficient. Schneckenburger, Dorn, and Endriss [840] also aim at minimizing inequality between agent utilities and make use of the well-known Atkinson index to measure inequality.

9.6.3 Maximum Nash Social Welfare

Social welfare by the Nash product (NSW, for short), defined by

$$sw_N(\pi) = \prod_{j \in A} u_j(\pi),$$

can be seen as some kind of compromise between egalitarian and utilitarian social welfare.[10] This notion was introduced by Nash [728] (see also, e.g., the work of Kaneko and Nakamura [592]). Note that maximizing the Nash social welfare is equivalent to maximizing utilitarian social welfare after the agents' utilities have been replaced by their logarithms.

We define the corresponding problems for the Nash product, NASH-PRODUCT-SOCIAL-WELFARE-OPTIMIZATION (NPSWO) and its exact variant XNPSWO, by replacing $sw_e(\pi)$ in the definition of ESWO and XESWO on page 637 by $sw_N(\pi) = \prod_{j \in A} u_j(\pi)$.

This social welfare notion enjoys a lot of interesting properties. For example, unlike egalitarian (but just like utilitarian) social welfare, sw_N is monotonic in the sense that increasing any agent's utility (all else being equal) increases social welfare by the Nash product. Moreover, sw_N is "fairer" than utilitarian social welfare in the sense that the more "balanced" the single agents' utilities are in an allocation, the higher is their Nash product: If the total utility is distributed among the agents in an allocation π so that

---

[10] For the Nash product, we require all agents' utilities to be nonnegative.

9.6 Maximizing Social Welfare: The Santa Claus Problem

they all realize the same utility, then $sw_N(\pi)$ is maximal. And assuming all utility functions to map to nonnegative values only (as postulated in Footnote 10), $sw_N(\pi)$ reaches the minimum value of zero if and only if some agent does not realize any utility at all in allocation π; and in this case, we have $sw_N(\pi) = sw_e(\pi)$.

The following example compares the utilitarian, egalitarian, and Nash social welfare for the shoe auction from Section 9.2.2 starting on page 608 and Section 9.3.2 starting on page 617.

Example 9.9 (Examples 9.3 and 9.4 continued). Recall the set $A = \{1,2,3\}$ of agents and the set $R = \{S_1^\ell, S_1^r, S_2^\ell, S_2^r, \ldots, S_{10}^\ell, S_{10}^r\}$ of resources from Example 9.3 on page 617, and the utility functions of the three agents defined in Example 9.4 on page 618. Consider the allocation π that assigns

- bundle $B_1 = \{S_4^\ell, S_4^r, \ldots, S_{10}^\ell, S_{10}^r\}$ to agent 1,
- bundle $B_2 = \{S_1^r, S_2^\ell, S_2^r, S_3^\ell, S_3^r\}$ to agent 2, and
- bundle $B_3 = \{S_1^\ell\}$ to agent 3.

Then we have the following individual utilities:

$$u_1(\pi) = 70, \quad u_2(\pi) = 100, \quad \text{and} \quad u_3(\pi) = \frac{100}{1,048,575}.$$

Hence, we have the following social welfare values:

$$sw_e(\pi) = \min\{u_1(\pi), u_2(\pi), u_3(\pi)\} = \frac{100}{1,048,575} = 0,0000953675225902,$$
$$sw_u(\pi) = u_1(\pi) + u_2(\pi) + u_3(\pi) = 170.0000953675225902, \text{ and}$$
$$sw_N(\pi) = u_1(\pi) \cdot u_2(\pi) \cdot u_3(\pi) = 0.6675726581312733948.$$

The following example shows that maximizing social welfare according to different measures can lead to different optimal allocations.

Example 9.10 (Example 9.8 continued). Consider the same resource allocation scenario as in Example 9.8 on page 641. The allocation maximizing Nash social welfare is $(a|b|d|c)$, with a value of 3200. Recall from Example 9.8 that the allocation maximizing utilitarian social welfare is $(ad|b|c|\emptyset)$, with a value of 35, and that the allocations maximizing egalitarian social welfare are $\pi = (d|b|a|c)$ and $\pi' = (d|a|b|c)$, both with a value of 5.

Moulin [717] presents further properties of Nash social welfare. In particular, the ordering of allocations induced by the Nash social welfare measure is the only one that simultaneously satisfies the following three properties:

- *independence of unconcerned agents* (when comparing two utility vectors, agents who realize the same utility in both vectors can be ignored);
- *independence of common scale* (multiplying all utilities by a positive constant does not change the induced social welfare ordering);
- *independence of individual scale of utilities* (multiplying the utility function of a single agent does not change the induced social welfare ordering).

Caragiannis et al. [255] show that maximizing Nash social welfare enjoys many fairness and efficiency properties for the allocation of indivisible goods: It implies Pareto efficiency, EF1, and provides a good approximation to maxmin fair share. Also, while the exact computation of the maximum Nash social welfare allocation is hard, it can efficiently be approximated, and scales well on real data. This paper is seen by many as a turning point in the literature of fair division of indivisible goods, and several authors now consider maximum Nash welfare as "the ultimate solution for allocating indivisible goods."

For the Nash product, NP-completeness of NPSWO is due, independently, to Ramezani and Endriss [797] and Roos and Rothe [816] (see also the journal version by Nguyen et al. [744]) for the bundle form, and to Roos and Rothe [816] (again, see also [744]) for the k-additive form, for each $k \geq 1$. The exact variants of both problems, XUSWO and XNPSWO, have been shown to be DP-complete by Roos and Rothe [816] (again, see also Nguyen et al. [744]), both for the bundle form and for the k-additive form, where $k \geq 2$ in the case of XUSWO and $k \geq 3$ in the case of XNPSWO.

Nguyen, Roos, and Rothe [746] survey the extensive body of research on the approximability of the optimization variant MAX-USW of the NP-hard decision problem USWO. Most of this work is concerned with the bundle form to represent the agents' utilities and, in addition to the general MAX-USW problem, has focused on approximability of its special variants for restricted classes of utility functions, such as submodular, subadditive, and fractionally subadditive utility functions,[11] and we refer to their survey for a detailed overview. To the best of our knowledge, the only results about MAX-USW in the k-additive form are due to Chevaleyre et al. [286], who show that MAX-USW is in P if $k = 1$, and to Nguyen, Roos, and Rothe [746], who observe that a reduction provided by Chevaleyre et al. [286] can be employed to show that MAX-USW for k-additive utilities with $k \geq 2$ cannot be approximated in polynomial time within any factor better than $\alpha = 21/22$, unless P = NP.

---

[11] Recall that a utility function u is *submodular* if for any two bundles $S, T \subseteq R$, we have $u(S \cup T) + u(S \cap T) \leq u(S) + u(T)$, and that u is *subadditive* if for any two disjoint bundles $S, T \subseteq R$, we have $u(S \cup T) \leq u(S) + u(T)$. Further, u is *fractionally subadditive* if $u(S) \leq \sum_k \alpha_k u(T_k)$ whenever $0 \leq \alpha_k \leq 1$ and $\sum_{k: r_j \in T_k} \alpha_k \geq 1$ for each good $r_j \in S$.

9.6 Maximizing Social Welfare: The Santa Claus Problem

Nguyen and Rothe [747] also study the problem of maximizing *social welfare by the average Nash product*, which is defined by taking the nth root of the Nash product: $\left(\prod_{j \in A} u_j(\pi)\right)^{1/n}$. Assuming (sub)additive utility functions, they give a greedy $\frac{1}{m-n+1}$-approximation algorithm for this problem that is based on dynamic programming, and whenever all agents have the same additive utility function, they present a polynomial-time approximation scheme for it. The proof of the latter result applies a rounding technique and integer programming that is largely inspired by the work of Alon et al. [14].

Regarding the Nash product, approximation results on the optimization variant MAX-NPSW of the NP-hard decision problem NPSWO are thin on the ground. Inspired by the work of Woeginger [941] on the problem of maximizing the minimum completion time of jobs to be scheduled on parallel identical machines mentioned in Section 9.6.1.2, Nguyen, Roos, and Rothe [746] provide a polynomial-time approximation scheme (again, recall the definition from Section 1.5.2.4 starting on page 33) for MAX-NPSW for two agents having the same 1-additive utility function (where the empty bundle is assumed to have zero utility). They also show that MAX-NPSW with k-additive utilities, $k \geq 1$, can be solved exactly in polynomial time whenever there are the same number of agents and goods ($m = n$) and the empty bundle has zero utility. On the other hand, they observe that, unless P = NP, MAX-NPSW for k-additive utilities cannot be approximated in polynomial time within any factor better than $\alpha = {}^{21}/_{22}$ if $k = 2$, and cannot be approximated in polynomial time within any factor at all if $k \geq 3$, and the latter result also applies to the bundle form.

Maximizing the Nash social welfare has attracted a lot of attention in these last years. Even when restricted to additive valuations, not only is the problem of deciding whether there exists an allocation whose Nash social welfare exceeds a given threshold NP-hard [744], the maximization variant of the problem is also APX-hard, i.e., hard to approximate [646]. On the positive side, still for additive valuations, there are efficient constant-factor approximation algorithms: in chronological order, the constant factors achieved are 2.89 [297, 298], 2.71 [24], 2 [296], and 1.45 [93].

These approximation algorithms focus exclusively on maximizing Nash social welfare, but may lose the other fairness or efficiency guarantees offered by a max NSW allocation, namely PO, EF1 (and therefore Prop1), and approximate MMS. This issue was first addressed by McGlaughlin and Garg [696], which give a 2-approximation of max NSW that furthermore satisfies Prop1, $\frac{1}{n}$-MMS, and PO (while the 1.45-approximation due to Barman, Krishnamurthy, and Vaish [93] also satisfies $(1+\epsilon)$-EF1 but not Prop1).

Going more general than additive valuations: Anari et al. [25] show that a 2-approximation exists for separable, piecewise-linear concave utilities; for budget-additive valuations (representing an important class of submodular functions), Garg, Hoefer, and Mehlhorn [485] give a $2.404 + \epsilon$-approximation of max NSW in time polynomial in the input size and $\frac{1}{\epsilon}$, for all $\epsilon > 0$; Chaudhury, Garg, and Mehta [278] consider subadditive valuations and give a polynomial-time algorithm that outputs an $O(n)$ approximation of max NSW while guaranteeing EF1 and $\frac{1}{2}$-EFX. Seddighin and Seddighin [847] study max-min fair share fairness when agents have subadditive valuations.

Going less general, Barman, Krishnamurthy, and Vaish [94] give a 1.061-approximation algorithm when agents have identical additive valuations, and they show that an exact solution can be found in polynomial time for binary valuations. Also, Akrami et al. [6] study the restriction where the value of each agent for each good is one of two integer values, p or q, and they design an algorithm with approximation ratio of at most 1.0345.

Brânzei, Gkatzelis, and Mehta [221] consider agents who report preferences strategically, and focus on the Nash implementability of an approximate max NSW allocation, for various classes of valuation functions.

Wu, Li, and Gan [951] focus on maximizing NSW under budget-feasibility constraints: Each item has a cost that will be incurred to the agent it is allocated to, and each agent has a budget constraint on the total cost of items she receives; they show that a budget-feasible allocation maximizing NSW satisfies $\frac{1}{4}$-EF1.

Caragiannis, Gravin, and Huang [250] show that for every instance with additive valuations, there is an EFX allocation of a subset of items with a Nash welfare that is at least half of the maximum possible Nash welfare for the original set of items; they prove that this bound is the best possible.

Garg, Kulkarni, and Kulkarni [487] give approximations of maximum NSW for asymmetric agents (one for additive valuations, which further satisfies EF1; another one for submodular valuations). A very general class of valuations (so-called *Rado valuations*) has been studied by Garg, Husic, and Végh [486].

9.6.4 Utilitarian Social Welfare

Utilitarian social welfare, defined by

$$sw_u(\pi) = \sum_{j \in A} u_j(\pi),$$

is defined as the sum of the utilities all agents realize in allocation π. This criterion makes perfect sense in settings where utilities are transferable (cf. the notion of transferable utility games from Section 3.1.1 starting on page 141): Then, equality between agents can be recovered by a post-allocation utility transfer; see Section 9.8.

Utilitarianism is also the standard optimization criterion in settings where a central authority seeks to maximize overall revenue: For example, the auc-

9.6 Maximizing Social Welfare: The Santa Claus Problem

Table 9.1 Complexity of decision problems for (exact) social welfare optimization

Problem	Bundle form	Reference	k-additive form	Reference
USWO	NP-complete	[286]	NP-complete, $k \geq 2$	[286]
ESWO	NP-complete	[816, 744]	NP-complete, $k \geq 1$	[176, 187, 660]
NPSWO	NP-complete	[797, 816, 744]	NP-complete, $k \geq 1$	[744]
XUSWO	DP-complete	[816, 744]	DP-complete, $k \geq 2$	[816, 744]
XESWO	DP-complete	[816, 744]	DP-complete, $k \geq 2$	[816, 744]
XNPSWO	DP-complete	[744]	DP-complete, $k \geq 3$	[744]

tioneer in a combinatorial auction (see [316]) does not care so much about what utility can be realized by which individual agent. For this purpose—and for capturing the average utility of all agents as well—sw_u is an appropriate social welfare measure. It does not reflect, though, how utility is distributed among the agents. For example, it might be the case that all goods are assigned to just a single agent under π, so this agent alone would realize the entire utility of $sw_u(\pi)$, whereas all other agents come away empty-handed. In other words, allocations with a high utilitarian social welfare can still be quite unfair. Egalitarian, leximin, and Nash social welfare aim at correcting this.

Replacing $sw_e(\pi)$ in the definition of ESWO and XESWO on page 637 by $sw_u(\pi) = \sum_{j \in A} u_j(\pi)$, we define the corresponding problems for utilitarian social welfare, UTILITARIAN-SOCIAL-WELFARE-OPTIMIZATION (USWO) and its exact variant XUSWO.

For utilitarian social welfare, Chevaleyre et al. [286] have shown that USWO is NP-complete for both the bundle form and the k-additive form, for each $k \geq 2$, and Dunne, Wooldridge, and Laurence [364] have obtained the same result when the agents' utilities are represented by straight line programs.

Utilitarian social welfare is an efficiency notion, and as such it makes sense to consider it in conjunction with a fairness notion or with other efficiency notions. For instance, one may want to output the maximum social welfare allocation among those that are Pareto-efficient [152], or among those that satisfy EF1 or PROP1 [57].

9.6.5 Summary

Table 9.1 summarizes the results on the complexity of the decision problems modeling (exact) social welfare optimization.

Table 9.2 gives an overview of some of the known (in)approximability results for social welfare optimization problems in the k-additive form discussed above, essentially those surveyed by Nguyen, Roos, and Rothe [746].

Table 9.2 Summary of some (in)approximability of social welfare optimization problems for the k-additive form (adapted from [746]). The approximation results shown here assume that the empty bundle always has value zero for all agents (recall Footnote 9 on page 639 about work by Lange, Nguyen, and Rothe [642] dropping this normalization assumption). Key: m is the number of resources and n is the number of agents.

Problem	Approximability	Reference
	$\frac{1}{m-n+1}$	[144]
	$\frac{1}{n}$	[507]
Max-ESW (1-additive)	NP-hard in any factor $\alpha > \frac{1}{2}$	[144, 507]
	$\Omega\left(\frac{1}{\sqrt{n}\log^3 n}\right)$	[34, 35]
	$\frac{1}{m^\varepsilon}$ (for any $\varepsilon = \Omega\left(\frac{\log\log m}{\log m}\right)$)	[264]
Max-ESW (1-additive) in *Big Goods/Small Goods* model	$\frac{1}{\sqrt{n}}$	[507]
Max-ESW (1-additive) $u_i(r) = u_j(r)$ for all i,j	PTAS	[941]
Max-ESW (1-additive) $u_i(r_j) \in \{0,1,z\}$ for all i,j	$\frac{\alpha}{n}$ (for any $\alpha \leq \frac{n}{2}$)	[611]
Max-ESW (k-additive, $k \geq 1$) $m = n$	P	[507]
Max-ESW (1-additive) $u_i(r_j) \in \{0,z_j\}$ for all i,j	$\mathcal{O}\left(\frac{\log\log\log n}{\log\log n}\right)$	[85]
Max-ESW (k-additive, $k \geq 2$)	NP-hard in any factor	[746]
Max-USW (1-additive)	P	[286]
Max-USW (k-additive, $k \geq 2$) even for two agents	NP-hard in any factor $\alpha > \frac{21}{22}$	[746], based on [286]
Max-NPSW (k-additive, $k \geq 1$) $m = n$	P	[746]
Max-NPSW (1-additive) two agents, $u_1(r) = u_2(r)$	PTAS	[746]
Max-NPSW (2-additive)	NP-hard in any factor $\alpha > \frac{21}{22}$	[746]
Max-NPSW (k-additive, $k \geq 3$)	NP-hard in any factor	[746]

9.7 Centralized Fair Division with Ordinal Preferences

We assume that each agent i has a preference relation $\succ_i$ over 2^R. However, the communication dilemma, which we have already discussed in Section 9.3 starting on page 613, can be restated here as follows: Either i specifies $\succ_i$ *in extenso*, which needs $\Omega(2^m)$ communication bits and makes the mechanism practically infeasible, or we make a domain restriction on the domain of admissible preferences. The most common restriction is *separability* (together with *monotonicity*); under this assumption, it makes sense for i to specify her preferences over *single objects only*; let $\rhd_i$ be this preference relation over R. However, as we have already noticed, this comes with a loss of information: For instance, if agent 1 ranks objects a, b, and c in this order, we cannot deduce from separability and monotonicity whether she prefers $\{a\}$ to $\{b,c\}$ or *vice versa*.

A first solution to this problem is to reason from the (incomplete) preference order $\succ_i$ induced from $\rhd_i$ by the separability and monotonicity assumptions (see Section 9.3 and, in particular, Definition 9.1 on page 613). This is the way followed by, e.g., Brams, Edelman, and Fishburn [190], Brams and King [200], Bouveret, Endriss, and Lang [181], and Aziz et al. [55]. However, because the induced preference relations $\succ_i$ are incomplete, one can generally not say whether or not a given allocation satisfies an ordinal property such as Pareto efficiency or envy-freeness, and we have to redefine these notions in a "modal" way: Following Brams, Edelman, and Fishburn [190] but using the terminology of Bouveret, Endriss, and Lang [181] (cf. also Section 4.3.2 starting on page 297 for a similar approach in the context of voting), we say that an allocation is *possibly envy-free* if it is envy-free for some set of complete preferences that are consistent with the known incomplete preferences, and it is *necessarily envy-free* if it is envy-free under all possible completions. We define *possible* and *necessary Pareto efficiency* analogously.

Example 9.11 (adapted from [181]). Four kids, Anna (also abbreviated by A), Belle (B), Chris (C), and David (D), are competing for six candies, namely a(nise), b(anana), c(hocolate), d(urian), e(ggplant), and f(ig). When asked for their preferences about the single candies, they say the following:

Anna: $a \rhd_A b \rhd_A c \rhd_A d \rhd_A e \rhd_A f$
Belle: $a \rhd_B d \rhd_B b \rhd_B c \rhd_B e \rhd_B f$
Chris: $b \rhd_C a \rhd_C c \rhd_C d \rhd_C f \rhd_C e$
David: $b \rhd_D a \rhd_D c \rhd_D e \rhd_D f \rhd_D d$

Consider the allocation π that gives a to Anna, d and f to Belle, b to Chris, and c and e to David. π is possibly envy-free and possibly

Pareto-efficient. For instance, Anna is possibly not envious of Belle because there is a completion $\succ_A$ of $\rhd_A$ satisfying $\{a\} \succ_A \{d,f\}$. Because there is also a completion $\succ'_A$ of $\rhd_A$ satisfying $\{d,f\} \succ'_A \{a\}$, π is not necessarily envy-free, and in fact there is no complete and necessarily envy-free allocation at all.

Also, if f were unavailable, there would not be any complete possibly envy-free allocation, for the following reasons. First, if some agent receives nothing, she necessarily envies all those who receive something; therefore, without loss of generality, one agent receives two objects and three agents receive one object. Assume that both Anna and Belle receive only one object; then one of them does not receive a and necessarily envies the agent who receives a; the case where both Chris and David receive only one object is similar.

Finally, if Anna and David are punished and so are forbidden to eat candies, then giving $\{a,d,e\}$ to Belle and $\{b,c,f\}$ to Chris is necessarily envy-free! To see why: Belle necessarily prefers $\{a,d,e\}$ to $\{b,c,f\}$, because she prefers a to b, d to c, and e to f; and Chris necessarily prefers $\{b,c,f\}$ to $\{a,d,e\}$, because he prefers b to a, c to d, and f to e.

Computing allocations that are (possibly/necessarily) envy-free and (possibly/necessarily) Pareto-efficient is not always easy. The computation of such allocations has first been considered by Bouveret, Endriss, and Lang [181], and then by Aziz et al. [55]. While possibly envy-free allocations (possibly combined with possible or necessary Pareto efficiency) are easy to compute, this is not the case for necessarily envy-free allocations: Deciding whether there exists a complete necessarily envy-free allocation, or whether there exists a complete possibly Pareto-efficient, necessarily envy-free allocation, or whether there exists a complete necessarily Pareto-efficient, necessarily envy-free allocation, are all NP-complete problems (but become easy if there are only two agents). *Reallocation* (with respect to an initial allocation) under ordinal preferences has been studied by Aziz et al. [40]; it is also discussed in Section 9.10.7 starting on page 674.

Brams, Kilgour, and Klamler [198] have a closer look at the case of two agents, for which they obtain better results: Their algorithm, composed of two stages (a greedy allocation of the "noncontested" objects, followed by a division of the "contested pile"—see Section 9.9 starting on page 659 for more details about contested-pile-based methods), outputs a (possibly incomplete) necessarily envy-free allocation assigning a maximal number of objects. Unfortunately, their algorithm does not scale up to more than two agents.

Using the responsive extension to induce comparisons between subsets from comparisons between single goods has the drawback that it is "very incomplete," that is, it has very many complete extensions. As a consequence, the "possible" variants of fairness notions tend to be too weak, and the "necessary" variants too strong. It can therefore be often the case that we have, for instance, no necessarily envy-free allocations and many possibly envy-free allocations.

There are at least two ways to cope with this. The first way, advocated by Segal-Halevi, Hassidim, and Aziz [848], consists in assuming that the utility difference between goods ranked consecutively decreases with their position in the ranking; for instance, if a, b, c, and d are ranked in positions 1, 2, 3, and 4, then the utility difference between a and b is larger than the utility difference between c and d, which allows us to derive that $\{a,d\}$ is preferred to $\{b,c\}$, which the responsive extension cannot do.

A second way is to define complete extensions of the $\rhd_i$'s. A way to resolve, initiated by Brams and King [200] and further developed by Baumeister et al. [109], consists in "cardinalizing" $\rhd_i$: Let $\vec{s} = (s_1, s_2, \ldots, s_m)$ be a scoring vector with $s_1 \geq s_2 \geq \cdots \geq s_m \geq 0$,[12] and define the utility $u_i(o)$ of an object o for an agent i as s_j if this agent ranks o in position j, and then define u_i on sets $A \subseteq R$ of objects as, e.g., $u_i(A) = \sum_{o \in A} u_i(o)$ or $u_i(A) = \min_{o \in A} u_i(o)$. These are called *scoring allocation correspondences* or *scoring allocation rules*. Coming back to Example 9.11, and choosing the Borda-like[13] scoring vector $\vec{s} = (6,5,4,3,2,1)$, under the allocation π given in this example, the bundles that Anna, Belle, Chris, and David get give them a utility of, respectively, 6, 6, 6, and 7. This approach transforms the problem into a centralized fair division problem with cardinal utilities without money.

Baumeister et al. [109] also study the axiomatic properties—such as separability and monotonicity—of this approach of "cardinalizing" ordinal preferences. Regarding separability, Baumeister et al. [109] show that some common scoring allocation rules (i.e., such rules for common choices of an aggregation function $\star$, such as $+$ and min, and of a scoring vector s) in fact fail to be separable. These results led them to *"conjecture that (perhaps under mild conditions on s and $\star$), no positional scoring allocation rule is separable"* [109, p. 636]. This conjecture, however, was then refuted by Kuckuck and Rothe [622] who showed that (1) the family of sequential allocation rules—an elicitation-free protocol for allocating indivisible goods based on picking sequences [617], which are to be introduced in Section 9.9.2 starting on page 663—in fact *is* separable for each coherent family of picking sequences, and (2) every sequential allocation rule can be expressed as a scoring allocation rule for a suitable choice of scoring vector and social welfare ordering.

Regarding monotonicity, Baumeister et al. [109] specifically introduce two types: *object monotonicity*, which means that adding new objects makes no agent worse off, and *duplication monotonicity*, which informally speaking means that duplicating an agent by introducing another one with the same

---

[12] Cf. the notion of scoring protocols in voting, see Section 4.1.1 starting on page 235.

[13] Recall from Section 4.1.1 starting on page 235 that in voting the Borda scoring vector is usually defined by $(m-1, m-2, \ldots, 1, 0)$. While this shifting has no impact for scoring voting rules (see [535]), it does so for scoring allocation rules (see [109]).

preferences will not make their joint share worse than if they were alone in the allocation process. Again, both types of monotonicity come in a *possible* and a *necessary* variant. As mentioned by Kuckuck and Rothe [623], violating duplication monotonicity is akin to the twin paradox in voting [715] (see Section 4.2.7 starting on page 280); alternatively, duplication monotonicity also resembles control by adding voters [102] (see Section 4.3.4.3 starting on page 333); and, finally, it is also related to false-name manipulation in weighted voting games [39, 805] where agents may benefit from cheating by splitting their weights and posing as more than one agent (see Section 3.5.2.2 starting on page 185). Answering many of the questions left open by Baumeister et al. [109], Kuckuck and Rothe [623] show that while necessary duplication monotonicity fails to hold for many common scoring vectors together with egalitarian social welfare, possible duplication monotonicity is satisfied by each scoring allocation rule with a strictly decreasing scoring vector together with any of the prominent social welfare orderings $\succsim^p$, which interpolate continuously between utilitarian social welfare (expressed as $\succsim^1$) and egalitarian social welfare (expressed as $\succsim^{-\infty}$), with the parameter p, $-\infty \leq p \leq 1$, specifying the tradeoff between fairness and efficiency in allocating items.

In addition, Baumeister et al. [109] explore the computational aspects of scoring allocation rules by showing, in particular, that determining whether a given allocation is optimal for the Borda-like scoring vector and the min aggregation operator (i.e., egalitarian social welfare) is coNP-complete, while checking whether there *exists* such an optimal allocation is NP-complete. They obtain similar complexity results for a number of other scoring vectors and for other aggregation operators, such as leximin. In the same model, Darmann and Schauer [329] show that when maximizing Nash product social welfare, Borda-like scoring is NP-complete, yet approval scoring is efficiently solvable. Relatedly, Nguyen, Baumeister, and Rothe [742] characterize strategy-proofness of scoring allocation correspondences for indivisible goods.

Brams, Kilgour, and Klamler [197] look for allocations satisfying the proportional fair share criterion (see Section 9.5.4 starting on page 629). They show that a proportional allocation exists if and only if there is an allocation in which each agent receives one of her *minimal bundles*, that is, a subset of objects with the property that the subtraction of any object would make it less than a proportional share. Their allocation mechanism, called *proportional algorithm* (*PR*), outputs a proportional allocation when there exists one, and else outputs an allocation that is proportional for a maximum number of agents (plus an additional property which we omit here). Although the presentation of PR involves cardinal utilities, these are used for making the exposition easier and they are actually not needed: PR can be defined from purely ordinal information, which, in that case, consists of each agent ranking her minimal bundles. Note that this information may be exponentially large (which may justify the use of a cardinal, additive representation).

The specific case of fair division with ordinal preferences and only *two agents* has received special attention by Ramaekers [796], who axiomatically characterizes the rule that outputs an allocation minimizing the maximum, over both agents, of the rank of their allocated bundle in their preference relation. Earlier work by Brams and Fishburn [193] also proves some positive results regarding the two-agent case.

9.8 Fair Division with Money, and Related Issues

Sometimes the possibility to allow for some money transfers between agents, or from the central authority to the agents, makes it easier to find a fair allocation. The specific property of money that helps here is that it is divisible. More generally, allowing one item to be divisible helps in a similar way, and this is the way the Adjusted Winner procedure works, which will be presented in Section 9.8.2 starting on page 655. The even more general setting of fair division with a mix of divisible and indivisible goods has been addressed by Bei et al. [125, 126].

9.8.1 Money

Sometimes, in addition to the indivisible goods, we also have some amount M, an infinitely divisible resource called *money*, and we assume that for each agent i and each bundle $S \subseteq R$, there is a valuation $v_i(S)$ such that i is indifferent between receiving either S and no money or receiving $\emptyset$ and $v_i(S)$ monetary units. Consider the following example with two agents, 1 and 2, four indivisible objects, a, b, c, and d, some amount M of money, and assume that the valuation functions v_i are additive and defined as follows:

	a	b	c	d
Agent 1	10	5	7	0
Agent 2	9	4	8	1

If $M = 0$ then we obtain a problem equivalent to the one in Example 9.6 on page 630, in which we recall (from Example 9.7 on page 637) that the allocations maximizing egalitarian social welfare are $(a|bcd)$, $(bc|ad)$, and $(ad|bc)$. We also recall from Example 9.6 on page 630 that none of them is envy-free: In allocations $(a|bcd)$ and $(ad|bc)$, agent 1 envies agent 2; in allocation $(bc|ad)$, agent 2 envies agent 1. What is the minimal amount M of money needed to ensure that there exists a complete envy-free allocation? We see that if $M \geq 2$ then the allocation in which agent 1 receives a together with $\frac{M+2}{2}$ monetary units and agent 2 receives bcd together with $\frac{M-2}{2}$ monetary units, which we denote by $(a, \frac{M+2}{2}|bcd, \frac{M-2}{2})$, is envy-free. On the other hand, if $M < 2$, there is still no envy-free allocation. If we have enough money then we can obtain an envy-free allocation maximizing utilitarian social welfare: Here, we must have $M \geq 4$ and the desired allocation is $(ab, \frac{M-4}{2}|cd, \frac{M+4}{2})$. Finally, if we have enough money, we can do even better and ensure that the allocation is perfectly equitable, in the sense that all agents have exactly the same utility: This is the case here with $M \geq 6$ and the allocation $(ab, \frac{M-6}{2}|cd, \frac{M+6}{2})$.

We now give a brief overview of selected results.

Maskin [687] shows that when the agents' preferences are monotonic (that is, all objects are desirable), and there is the same number of objects as agents, then there is always an amount of money M allowing for envy-freeness and Pareto efficiency. Alkan, Demange, and Gale [12] also consider the case where some objects may be undesirable (so-called *chores*) and the amounts of money can be negative (e.g., costs to be shared by the agents). They deal with economies where the agents get a single good only, and formulate the computation of the minimal amount allowing for envy-freeness and Pareto efficiency as a linear programming problem (see also the earlier work of Alkan [11]). They notice that increasing the amount of money cannot make agents less happy, but that adding objects may make an agent less happy, as in the following example taken from the paper of Alkan, Demange, and Gale [12, page 1033]:

Example 9.12. There are three agents and there is one a(pple) that agents 1 and 2 value to be worth 6 units of money and agent 3 values it at 1, and there are $M = 12$ units of money in total. Assigning a to agent 3 would not be Pareto-efficient. The Pareto-efficient and envy-free allocations are $(a|6|6)$ and $(6|a|6)$. Now, assume that we add two more apples, a' and a''. Then—under the constraint that no agent should receive more than one object—the Pareto-efficient and envy-free allocation is now $(a, 4|a', 4|a'', 4)$, which makes agent 3 worse off.

Beviá [143] studies an extension of the framework of Alkan, Demange, and Gale [12] where the agents can receive more than one object and where side payments are allowed. Notice the difference between having money as a resource (as discussed at the beginning of this section) and allowing side payments (i.e., monetary transfers) between agents: With side payments, an agent may have to give away some money. Beviá [143] shows that in this setting, assuming the agents have reservation prices associated with any bundle of goods (that is, for each agent i and each bundle $A \subseteq R$ of goods, there is a value $v_i(A)$ representing the amount of money that i is willing to sacrifice in order to receive the goods in A), a Pareto-efficient and envy-free allocation always exists.

9.8 Fair Division with Money, and Related Issues

Other studies on fair division of indivisible goods with money include the work of Svensson [903], Quinzii [795], Tadenuma and Thomson [905, 906], and other more recent papers, such as those of Klijn [613], Fujinaka and Sakai [475], and Haake, Raith, and Su [521]; see also the survey by Thomson [913]. Aziz [38] gives a sufficient condition and an algorithm to achieve envy-freeness and equitability with monetary transfers.

Another way of using money to achieve envy-freeness and/or equity is to use *subsidy*: A third party is providing a small amount of money that can be allocated along with the allocated goods. The key question then is to search for the minimal subsidy required to guarantee, for instance, envy-freeness. A polynomial-time algorithm for finding a minimal subsidy needed for achieving envy-freeness, under additive utilities, is given by Halpern and Shah [528], which make a conjecture about the worst-case sufficient subsidy for achieving envy-freeness under additive utilities; this conjecture has then be proven by Brustle et al. [240]. Caragiannis and Ioannidis [252] study approximation algorithms for minimzing the amount of money necessary to obtain envy-freeness. Barman et al. [90] study envy-freeness via subsidy for agents with restricted valuations, with the marginal value for any good being either zero or one. Goko et al. [504] design *truthful* mechanisms with limited subsidy. Finally, Aziz [38] presents a sufficient condition and an algorithm to achieve envy-freeness and equitability when monetary transfers are allowed; for the case of additive utilities, he presents a characterization of allocations that can simultaneously be made equitable and envy-free via payments.

9.8.2 One Divisible Good: The Adjusted Winner Procedure

Do you still remember the "battle of the sexes" (described in Section 2.1.2 on page 52)? Meanwhile, a few years have passed by, but George and Helena are still fighting their battle.

George and Helena are celebrating their seventh anniversary this year. As always on this very special day, Helena is enjoying her concert ...

"Goal!!!", yells George, jumping up from his seat and throwing his arms up in the air. As always on this very special day, he is enjoying a fun match in the stadium ...

When they get home late in the evening—by this time pretty ill-humoredly—they both give mouth to their thoughts and say at once, *"I want a divorce!"*

The seven-year itch! However sad their divorce may be for the two, they do want to do it consensually and fairly. For most of the things they need to

Table 9.3 Adjusted winner procedure: George's and Helena's valuations of the objects

Number	Object	George	Helena
1	car	**10**	9
1	buddha	13	**28**
7	garden gnome	**77**	63
	Total	100	100

divide, it is clear whom they belong to, as it is clear who brought them into their marriage. There is disagreement only over some of the things they have purchased together during their marriage (recall Figure 9.1 on page 607):

-  : the *car*,

- : a precious, massive *buddha statue* made of gold, which they once had bought from a gang of children (recall the story from Section 3.1.2.2 and Figure 3.2 on page 146), and

-  : seven *garden gnomes*,

  most precious collector's items that stand on the lakeside property that George and Helena—together with Felix—inherited a while ago from uncle Ian (recall Section 8.4.2.5 and Figure 8.20 on page 548).

How can George and Helena divide these objects among each other so as to achieve a divorce settlement as fair as possible?

To this end, Brams and Taylor [204] developed the following method, known as the *adjusted winner procedure*—a.k.a. the *divorce formula* (in Germany, for instance, where it is called the *"Scheidungsformel"*). Note that although this procedure is described as a distributed protocol by Brams and Taylor [204] (and also by others, e.g., by Bouveret, Chevaleyre, and Maudet [179]), in fact nothing prevents us from viewing it as a centralized protocol, since the two agents have to report their preferences at once.

The *adjusted winner procedure* works as follows:

1. First, George and Helena both give their personal valuation of each of the objects to be divided, where they each may distribute a total of 100 points. Table 9.3 shows their individual valuations. For example, George values each of the garden gnomes to be worth 11 points, while for Helena each gnome is worth only 9 points—just as many as the car for her, to which George, however, assigns one point more. On the other hand, Helena takes more pleasure in the buddha than George—she gives 28, he only 13 points.

9.8 Fair Division with Money, and Related Issues

2. Now, every object goes to whoever values it higher (indicated by a boldface number in Table 9.3—a possible tie can be broken arbitrarily). Thus, George receives the car and all seven garden gnomes, while Helena has to make do with the buddha for the time being. Of course, she thinks that is unfair. After all, she realizes only a total of 28 of her points, whereas George rakes in 87 of his points.
3. That is why the current winner (George in our example) is "adjusted" in the next step, i.e., George has to give away as many of his objects to Helena until a balance is achieved. To this end, we calculate the ratio of the valuations of George and Helena for each object such that this ratio is greater than or equal to one, and then they are ordered by their size: At the head of the list we have all objects whose ratio is equal to one (if there exist any objects George and Helena value the same), then we have the objects with the next-larger ratio, etc., where ties among objects with the same ratio can be broken arbitrarily. In our example, we obtain the order of objects given in Table 9.4.

Table 9.4 Adjusted winner procedure: sorting the objects

Object	🚗	💂	💂	💂	💂	💂	💂	💂	🧘
Ratio	$\frac{10}{9}$	$\frac{11}{9}$	$\frac{11}{9}$	$\frac{11}{9}$	$\frac{11}{9}$	$\frac{11}{9}$	$\frac{11}{9}$	$\frac{11}{9}$	$\frac{28}{13}$

Now, if George cedes the first object in this list, the car, to Helena, she will be somewhat less cheesed off, but still not happy because—although she does realize $28 + 9 = 37$ of her points now—George is still much better off with $87 - 10 = 77$ of his points. However, if he then gives two of the garden gnomes to her, too, their scores are leveled with $37 + 2 \cdot 9 = 55 = 77 - 2 \cdot 11$ points. Figure 9.5 shows the resulting division.

Note that the two (ex)partners' scores in the produced allocation of objects do not always tally as nicely as in our example. Thus there would be a problem if all the objects were indivisible. Therefore, one of the objects is required to be divisible and can thus be split, which is why the adjusted winner procedure is described in the current section about fair division with cardinal preferences and *money*.[14]

Suppose, for example, that the collection of garden gnomes has a value of 77 points for George and of 63 points for Helena *only as a whole*. Then the division in Figure 9.5 would not be possible. One would merely be able to determine the *ratio* of this one object comprising seven garden

---

[14] In fact, this is not really a problem of "fair division of indivisible goods with money" but rather a problem of "fair division of indivisible goods with one divisible good," especially given the argument, exposed twice in this section, that if the good that the procedure asks to be divded cannot be divided in practice, it will be sold.

gnomes that would have to go to George and Helena, respectively, so as to establish a perfectly equitable divorce settlement:

$$77\alpha = 9 + 63(1-\alpha) + 28, \tag{9.2}$$

where we have the values of the objects currently (i.e., after George ceded the car to Helena) owned by George on the left-hand side of (9.2) and those owned by Helena on the right-hand side of (9.2). Solving this equation for α, one obtains with $\alpha = 5/7$ the ratio of the gnome collection George is entitled to receive, and the corresponding ratio of $1-\alpha = 2/7$ Helena is entitled to receive. Substituting these values in (9.2), we would end up with exactly a score of 55 points for both.

Of course, if the garden gnome collection indeed is "indivisible" for George and Helena, they will have to sell it to be able to split the money resulting from this sale among each other according to these ratios.

(a) George's share

(b) Helena's share

Fig. 9.5 Division of nine objects by the adjusted winner procedure

This method has a number of advantages. For example, the division it produces has the property that:

1. it is *Pareto-optimal* according to Definition 9.2 on page 622, i.e., every other allocation of objects that makes one of the two better off would make the other one worse off at the same time,
2. it is *envy-free* according to Definition 9.3 on page 623, i.e., none of them would like to swap their own bundle of objects with the other one's bundle, and
3. it is *perfectly equitable* as defined on page 642, i.e., both George and Helena assign the same value to their own share of the objects.

For an informal description of the arguments as to why these properties hold, and for a detailed discussion, we refer to the book by Brams and Taylor [204, Section 4.3]. They also elaborate on the point that George and Helena may not announce their honest values, and they analyze strategic and game-theoretic concepts such as Nash equilibrium (see Definition 2.4 on page 50 and Definition 2.5 on page 63) for the adjusted winner procedure.

Dall'Aglio and Mosca [323] design a more efficient algorithm for the adjusted winner procedure, with linear running time, which proceeds by allocating a large subset of objects directly without having to rank them. They also study a procedure for solving the max-min allocation problem for two agents, that mimics a branch-and-bound algorithm and makes a repeated use of the adjusted winner procedure.

Aziz et al. [49] give several characterizations of the adjusted winner procedure, and study its strategic aspects.

One potential problem with the adjusted winner procedure is that the output generally requires one of the goods to be split, *and we don't know beforehand which one will be split*. According to Karp, Kazachkov, and Procaccia [594], it could be, for instance, that the fair division of the goods in a divorce settlement end up dividing a flat. In this case, what the agents would probably do in practice is to sell the good and split the money resulting from the sale. This has lead Karp, Kazachkov, and Procaccia [594] to consider fair division of indivisible, *sellable* goods. They find out that the results about the price of envy-freeness (due to Caragiannis et al. [253]) become much less negative with sellable goods.

9.9 Decentralized Allocation Protocols

We now review a few classes of approaches that are partially or fully decentralized. We say that an approach is *fully decentralized* if no central authority is needed at all: The agents execute the allocation protocol themselves. At the other extreme, recall that *centralized* approaches proceed in two successive, noninteracting stages: a first one where the agents report their preferences to the central authority, followed by a second one where the central authority computes the allocation. All approaches inbetween, where the agents interact in a nontrivial way with the help of a central authority, are *partially decentralized*. Thus, centralized approaches can be seen as a particular case of a partially decentralized protocol.

There are at least two good reasons for using decentralized allocation procedures: (a) It may alleviate a lot the communication burden, and (b) it avoids (partially or totally) to ask the agents to reveal their preferences (at least in a direct, explicit way).

As fas as (a) is concerned, one may want to look for a cheapest protocol, for computing an allocation satisfying a given property (if any exists), defined as one for which the worst-case total amount of communication to be sent by the agents is minimum. This is precisely the aim of *communication complexity*: Given a property P of allocations, and a class of valuations C, the communication complexity for P and C is the cost of a cheapest protocol for computing an allocation satisfying P (or detect that there is none), where the worst case is taken over all instances made up of valuations from C. Of course, this cost depends on the number m of items and the number n of agents. Plaut and Roughgarden [784] were the first to consider the problem in a systematic way. As for the choice of P, they focus on envy-freeness, proportionality, and approximations thereof; as for C, they consider submodular, subadditive, and unrestricted valuations. They show that when $n \geq 3$, as soon as we move away from additive valuations, the communication complexity is exponential, for all properties considered. However, for two agents, the situation is surprisingly complex and diverse. In particular, Plaut and Roughgarden [783] show that the communication complexity of computing a leximin-optimal allocation is exponential, too, even for two agents with identical (but arbitrary) valuations; they also explore the communication complexity of finding an EFX allocation. The communication complexity for finding a Pareto-efficient and envy-free allocation is investigated by Cole and Tao [299].

Now, specific classes of decentralized protocols for fair division of indivisible goods have been studied. The criteria for evaluating such a protocol are:

- The possible restrictions on the number of agents.
- The communication complexity of the protocol.
- When the protocol involves some elicitation process, the nature (cardinal; ordinal with or without interpersonal comparability) of the preferences that the agents need to report, and possibly the domain restrictions that have to be made for the protocol to be implementable.
- The quality of the returned solution.
- The vulnerability to strategic behavior.

We now review four classes of protocols, and try to evaluate them along these criteria (except the last one, which we will not discuss in detail, but see Section 9.11.1 starting on page 675 for some pointers to the literature).

9.9.1 The Descending Demand Protocols

The original version of the *descending demand protocol* (*DDP*), due to Herreiner and Puppe [553], is conceptually very simple: The agents speak in turns, according to a predefined order, and communicate, at each step k, their kth best subset of objects; the protocol stops when we reach a stage k at which there exist $\pi_1, \ldots, \pi_n$ such that

9.9 Decentralized Allocation Protocols

(a) $(\pi_1, \ldots, \pi_n)$ is a complete allocation, that is, $\bigcup_{i=1}^{n} \pi_i = R$ and $\pi_i \cap \pi_j = \emptyset$ for all $i, j \in A$ with $i \neq j$, and
(b) for all $i \in A$, $rank(\pi_i, \succ_i) \leq k$, where $rank(\pi_i, \succ_i)$ is the rank of bundle π_i in the preference relation $\succ_i$ of agent i.

Then the protocol outputs the allocation $\pi = (\pi_1, \ldots, \pi_n)$.

Example 9.13. George and Helena want to allocate an a(pple), a b(anana), a piece of c(ookie), and a d(onut). They apply the descending demand protocol.

1. They communicate their most preferred bundles, which are, for both of them, *abcd*. (This is obvious if they have monotonic preferences; note that the descending demand protocol works even if preferences are not necessarily monotonic.) The stopping criterion is not met.
2. Their second most preferred bundles are *bcd* for George (for whom the heavier the food, the better) and *abc* for Helena. The stopping criterion is not met.
3. Their 3rd most preferred bundles are *acd* (George) and *abd* (Helena). The stopping criterion is not met.
4. Their 4th most preferred bundles are *cd* (George, who prefers the cookie and the donut to only one of them plus two pieces of fruit) and *bcd* (Helena). The stopping criterion is not met.
5. Their 5th most preferred bundles are *abc* (George) and *acd* (Helena, who prefers bundles with more items to bundles with fewer items). The stopping criterion is not met.
6. Their 6th most preferred bundles are *abd* (George) and *bd* (Helena, who prefers a well-balanced bundle composed of a fruit and a pastry to two fruits or two pastries). The stopping criterion is not met.
7. Their 7th most preferred bundles are *bc* (George) and *ac* (Helena). The stopping criterion is not met.
8. Their 8th most preferred bundles are *bd* (George) and *bc* (Helena). The stopping criterion is met: We have found an allocation such that both agents get one of their first eight bundles, namely, (*bd*|*ac*).

George's preferences from Example 9.13 are shown in Figure 9.6(a) and Helena's in Figure 9.6(b). The allocation (*bd*|*ac*) where the stopping criterion is met is marked by horizontal lines.

The descending demand protocol is partially centralized, because the preferences are reported to the central authority which (using a computer program) can tell when the stopping criterion is met and then computes the resulting allocation; but it can also be made fully decentralized, by requiring that the agents themselves take the responsibility to store the information

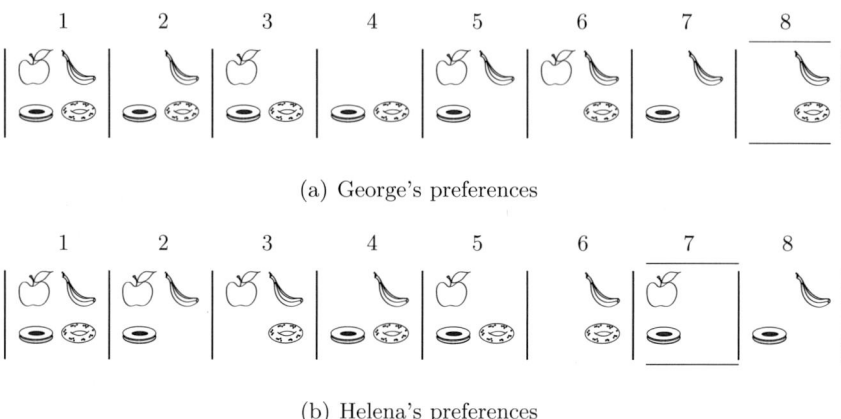

Fig. 9.6 Descending Demand Protocol: Preferences over bundles in Example 9.13

and compute the allocation. Moreover, the descending demand protocol enjoys some nice properties:

- It outputs a *balanced* allocation (that is, an allocation maximizing egalitarian social welfare) if we assume that the utility of a bundle for any agent is an agent-independent, nonincreasing function of the rank of the bundle in the preference relation. Moreover, the returned allocation is Pareto-optimal.
- It does not require any restriction on the agents' preferences.
- It uses only ordinal preference information.

On the other hand, the resulting allocation may not be envy-free. In our example, Helena envies George for his share: She prefers the banana-donut combination, which was given to George, to the apple-cookie one that she got. To cope with this, Herreiner and Puppe [553] define a *modified descending demand protocol (MDDP)* that aims at returning envy-free allocations. It does so perfectly in the two-agent case: It then outputs an envy-free allocation, provided that there exists one, and this resulting allocation is also the envy-free allocation with maximal egalitarian social welfare (in the sense explained above). With more than two agents, however, the protocol is sound but not complete: When it outputs an allocation, then it is envy-free, but it may fail to give an output, even when an envy-free allocation exists.

The communication complexity of the descending demand protocol can be huge: As we see in our example, the agents may have to specify about half of their preference relation (that is, an exponential amount of communication) before the stopping criterion is met. Even worse, this case is not going to be a pathological case, but, if the agents' preferences are similar enough, this is going to be the typical case. This is the price to pay for dealing with restriction-free preferences. A further problem with DDP is the fact that the evaluation of the stopping criterion, and therefore the computation of the resulting allocation, is hard. In conclusion, DDP is a nice procotol if the number of items is small enough.

9.9.2 The Picking Sequences Protocols

Even if they do not envy each other, George and Helena are not very happy with the descending demand protocol because the time they spent on reporting their preferences made them lose their appetite. They consult with each other and come up with a brilliant idea: Why do we need this long and tedious procedure with this central computer while we can do much simpler by taking single items in turn? They have just reinvented the *strict alternation picking sequences protocol* (also called *round-robin picking sequences protocol*): Next time they have to divide a set of items, they will start by tossing a coin; if that's heads, Helena starts, if that's tails, George starts. Assume Helena starts. She picks any item she wants; then George takes one (any one he wants, except the one already taken by Helena); then Helena again; then George again; etc., until all objects have been allocated. We denote this sequence by HGHG···.

Is this so simple? For greedy George, yes, as his preferences are separable: Whatever else he has in his bundle, he prefers the cookie to the donut, the donut to the banana, and the banana to the apple. Assume he picks first, then he picks the cookie. For Helena this is less simple, as it's not obvious to her what her most preferred item is because her preferences are *not* separable: The strict alternation policy will give her exactly two items; and if she has the apple in her bundle, then she prefers the cookie to the donut, but this is the other way round if she has the banana. After George has picked the cookie, she might think of picking the banana (because $\{b,d\}$ is her most preferred bundle with two items) ... but then George will pick the donut, leaving the apple to Helena, who now realizes that she has the apple and the banana, her least preferred bundle with two items.[15] The conclusion is that the notion of sincere picking strategy is not well-defined for agents whose preferences are nonseparable.

However, if both agents have weakly separable preferences, then the notion of sincere picking strategy is well-defined, and, assuming that the agents behave sincerely (which they are likely to do, for instance, if they don't know anything about the other agents' preferences), we can predict from their preferences which objects they will pick. Take Example 9.13 again and assume that, instead of sharing with Helena, George has to share the objects with Georgina, who has the very same preferences as George. Then, assuming

---

[15] Note that if she chooses the donut instead of the banana, then she gets a better share in the end.

George starts, he will pick the cookie, then Georgina the donut, then George the banana, and Georgina the apple.

Is this fair? This is not what Georgina thinks: George has got his first and third choice while Georgina has got her second and fourth choice only. Could we have done better? Yes! Instead of strict alternation, we could have chosen *balanced alternation*: One agent (say George) starts picking one object, then Georgina picks *two* objects, and George takes the last remaining object. In this case, George gets the cookie and the apple, and Georgina the donut and the banana.

Balanced alternation was proposed first by Brams and Taylor [205]. For two agents (G and H) and eight objects, for instance, balanced alternation would be the sequence GHHGHGGH (George picks one, Helena two, George one, Helena one, George two, and Helena one). But balanced alternation also has limitations: What if we have more than two agents? What if the number of objects is not a multiple of the number of agents? (Take a moment to consider the case of two agents and five objects.) A solution, advocated by Bouveret and Lang [183], consists in defining criteria allowing to compute *optimal picking sequences*:

(a) a *scoring vector* $(s_1, \ldots, s_m)$ *defining utilities for single objects from the ranks of the objects in a preference relation* (and thus assuming implicitly that the agents have additive preferences): For instance, the Borda-like scoring vector (recall Footnote 13 on page 651) gives utilities $m, m-1, \ldots, 1$ to the objects ranked first, second, ..., last;
(b) an *aggregation function for defining social welfare from individual utilities* (with the two obvious choices being + and min, that is, utilitarianism and egalitarianism);
(c) a *prior probability distribution over preference profiles* (with two extreme choices being full independence—uniform distribution over all profiles—and full correlation—all agents have identical preferences).

For a given number of agents and a given number of objects, a scoring vector, a social welfare aggregation function, and a prior distribution allow to define optimal picking sequences as those maximizing expected social welfare. For instance, for three agents (named 1, 2, 3), eight objects, Borda-like scoring, and full independence, the optimal sequence is 11332232 for egalitarian social welfare and, respectively, 12312312 for utilitarian social welfare (see [183] and Sylvain Bouveret's web page on sequential resource allocation: https://recherche.noiraudes.net/en/sequences.php). The computation of optimal sequences was further investigated by Kalinowski, Narodytska, and Walsh [590], who proved that the optimal picking sequence for utilitarian social welfare and full independence can be computed in polynomial time, and that if, in addition, there are two agents, then a strictly alternating policy (such as 12121212) is optimal.

As to the strategic aspects of picking sequences, there are two lines of work. First, the characterization and computation of subgame-perfect Nash equilibria (recall this notion from Section 2.3.2.3 on page 101) for the picking sequences game is initiated by Kohler and Chandrasekaran [617] and studied further by Kalinowski et al. [591] and by Aziz, Goldberg, and Walsh [56]. Second, the manipulation of picking sequences has been given increasing attention for the last few years, for instance by Bouveret and Lang [184], who study single-agent and coalitional manipulation of the picking sequences protocol, and later on by Tominaga, Todo, and Yokoo [917], who focus on the two-agent case for random picking sequences, and Aziz et al. [42], who investigate the complexity of manipulating sequential allocation procedures. The parameterized complexity of computing an optimal manipulation has been studied by Flammini and Gilbert [459] and by Xiao and Ling [964].

The study of picking sequences in *weighted* fair division problems (where agents may have different entitlements) has been initiated by Chakraborty, Schmidt-Kraepelin, and Suksompong [266].

Gourvès, Lesca, and Wilczynski [510] show that picking sequences can be used for introducing novel fairness criteria, and study their connections with known criteria.

Payan and Zick [773] use a modification of strict alternation to assign *reviewers to papers*.

To sum up, the picking sequences protocols are very simple, fully distributed, elicitation-free protocols with cheap communication complexity and no "online" computational complexity (what we mean here is that computing an optimal picking sequence for a given choice of criteria can be hard, but this is done offline). On the other hand, they need weakly separable preferences to work well, and they offer no strict guarantee of fairness or envy-freeness; only Pareto efficiency is guaranteed.

9.9.3 Contested Pile-Based Protocols: Undercut

This class mainly consists of one protocol, proposed by Brams, Kilgour, and Klamler [196]: the *undercut protocol*. It is only partially distributed, and works only for two agents with monotonic, weakly separable preferences. Roughly speaking, the undercut protocol can be thought of as a discrete analogue of the well-known Cut & Choose protocol in cake-cutting (recall Figure 8.8 on page 529), except that the "cutter" here does not offer a choice between two portions she values equally (which may be impossible for indivisible goods) but instead proposes a division into two portions, one for herself and the other for the "chooser," who has now the choice between either accepting this split (and thus her portion) or undercutting the cutter's portion by taking a slightly less valuable subset of it for herself. In fact, both agents will make such proposals and a central authority picks one of them. The option of undercutting will give an incentive to both agents to not ask too much for themselves, but instead to propose a split where their own portion is preferrable, yet in a certain sense minimal.

In more detail, the undercut protocol consists in two successive phases.

In the first phase, the two agents (1 and 2) communicate to the central authority their best objects; if they name different objects, each of them gets the object they named; else the object goes into the contested pile; the process is iterated until all objects have been named or placed into the contested pile. At the end of the process, if the contested pile is empty, the protocol ends; else we enter the second phase.

In the second phase, both agents communicate to the central authority, in secret, their *minimal bundles* from the *contested pile* (denoted CP), defined as follows: For $i \in \{1,2\}$, let $>_i$ be the monotonic and separable extension (see Section 9.3.1.1 starting on page 615) induced from agent i's preference relation over single objects, and let $\succeq_i$ be i's full preference relation; then $S \subseteq CP$ is a *minimal bundle* for i if and only if

(a) $S \succeq_i CP \setminus S$ and
(b) for every $T <_i S$, $CP \setminus T \succ_i T$.

Let MB_1 and MB_2 be the sets of minimal bundles of agents 1 and 2. The rest of the procedure is as follows:

1. If $MB_1 \neq MB_2$, then each agent sends to the central authority, in secret, a ranking of her minimal bundles; the central authority chooses 1's most preferred minimal bundle, call it S_1, among those that are not in MB_2 (or vice versa). Then 2 has the choice between

   a. accepting $CP \setminus S_1$, or
   b. else to *undercut* 1's proposal by taking the most preferred bundle, according to $>_2$, among those that are $>_2$-dominated by S_1.

2. If $MB_1 = MB_2$, then the central authority will check whether there exists some $S \in MB_1$ such that $S \sim_1 CP \setminus S$ (and, equivalently, $S \sim_2 CP \setminus S$);

   a. if such an S exists, 1 receives S and 2 receives $CP \setminus S$ (or vice versa);
   b. else, a minimal bundle S_1 of agent 1 is chosen randomly and agent 2 has to choose between accepting $CP \setminus S_1$, or undercutting S_1.

In Cases 1 and 2.a, the allocation *resulting from the second phase* is envy-free. The global allocation is also envy-free, provided that the agents' preferences are separable. Only in Case 2.b can it fail to be envy-free (and even though there may exist an envy-free allocation).

Example 9.14. Anna (also abbreviated by A) and Belle (B) have to split six objects: an a(pple), a b(anana), a c(ookie), and a d(onut) as before in Example 9.13 on page 661, plus an e(ggplant) and a f(lower). In the first phase, they rank the single objects:

Anna: $e \succ_A a \succ_A b \succ_A f \succ_A c \succ_A d$
Belle: $f \succ_B e \succ_B a \succ_B b \succ_B c \succ_B d$

Anna names the eggplant, Belle the flower, and they both take the object they named. Then both name the apple, which thus goes to the contested pile. So do the banana, the cookie, and the donut. At the end of the first phase, the contested pile is $CP = \{a, b, c, d\}$.

Assume that Anna's and Belle's preferences over bundles (which we don't ask them to report) are

Anna: $abcd \succ_A abc \succ_A abd \succ_A acd \succ_A bcd \succ_A ab \succ_A ac \succ_A bc \succ_A$
$ad \succ_A bd \succ_A cd \succ_A a \succ_A b \succ_A c \succ_A d \succ_A \emptyset$

Belle: $abcd \succ_B abc \succ_B abd \succ_B ab \succ_B acd \succ_B ac \succ_B bcd \succ_B ad \succ_B$
$a \succ_B bc \succ_B bd \succ_B cd \succ_B b \succ_B c \succ_B d \succ_B \emptyset$

Anna's set of minimal bundles is $MB_A = \{bc\}$, and Belle's is $MB_B = \{bcd, ad\}$. The central authority will, for instance, take Belle's most preferred minimal bundle, bcd, and propose it to Anna, who can either choose its complement a, or undercut bcd into bc; since she prefers bc to a, she will rather choose bc. (One can check that starting from Anna's unique minimal bundle bc leads to the same allocation, which is not always the case.) Finally, the resulting allocation is $(bce|adf)$. Note that it is envy-free (which we knew already, since we were in Case 1).

To sum up, the undercut protocol is a partially distributed protocol that works only for two agents with weakly separable preferences. It offers no strict guarantee of envy-freeness, but envy-freeness is obtained in most cases, provided that the agents' preferences are separable. The communication complexity can be exponential in the worst-case.[16] The strategic behavior of contested-pile protocols has been studied by Vetschera and Kilgour [923].

9.9.4 Protocols Based on Local Exchanges

This last class of decentralized protocols contains fully distributed protocols only, but makes an additional assumption: *There is a prior allocation* (see also Section 9.10.7 starting on page 674). This allocation can be thought of as a default allocation, a random allocation, or an initial allocation from which the agents will evolve. Then the agents may contract local exchanges,

---

[16] However, it has been shown by Brams, Kilgour, and Klamler [196] that assuming the uniform distribution over preference profiles, the size of the contested pile remains small in expectation, which makes the communication complexity low.

respecting some rationality criteria. Depending on the *number of agents and objects involved in an exchange*, the nature of the *rationality criteria* for defining admissible exchanges, the *possibility (or not) of side payments*, and *possible restrictions on the preferences*, the process may or may not converge towards a fair allocation, for some given fairness criteria. The study of this class of protocols has been initiated by Sandholm [830] and then pursued by Endriss and Maudet [399], Endriss et al. [400], and Chevaleyre, Endriss, and Maudet [290].

Chevaleyre et al. [285, 291] propose a distributed negotiation framework that, under certain conditions, allows the agents to reach Pareto-efficient, envy-free allocations (or, if these conditions are not met, at least allocations with "minimal" envy). For example, they show that if all utility functions are supermodular,[17] then a Pareto-efficient, envy-free allocation can always be reached in their negotiation framework, and they establish general convergence theorems that show which conditions are sufficient for a sequence of negotiation deals to reach a Pareto-efficient, envy-free allocation.

Chevaleyre, Endriss, and Maudet [289] introduce another distributed negotiation framework that implements a "negotiation topology" on graphs whose vertices correspond to the agents, and two vertices are connected by an edge if and only if the corresponding agents can negotiate a deal to exchange goods. In this setting, an agent is said to envy another agent if she would prefer to swap places with him in that graph. A convergence theorem is proven, stating under which conditions a *"graph-envy-free"* allocation can be reached by a sequence of negotiation deals. In particular, the agents' utility functions are required to be supermodular for the convergence theorem to hold. Chevaleyre, Endriss, and Maudet [289] also prove NP-completeness of the problem of deciding whether or not a given negotiation state admits any deal that is beneficial to each agent involved and at the same time reduces envy in the society of agents.

Social networks—e.g., implemented via *negotiation topologies* by Chevaleyre, Endriss, and Maudet [289] as described above—are a simple way to express whether two agents can trade objects with each other or to model hierarchies over the agents. Gourvès, Lesca, and Wilczynski [509] model a housing market where trades are only allowed between neighbors in a social network. They study the complexity of finding a given allocation or of checking whether a designated object can be obtained by rational trades (i.e., trades no agent is disadvantaged by). Some of their open questions about the complexity of the latter problem, called REACHABLE-OBJECT, were later answered by Bentert et al. [131]: It is polynomial-time solvable if the social network is a path or if all agents have preference lists of length at most three, yet it is NP-hard on cliques and generalized caterpillars, even if all agents have preference lists of length at most four.

---

[17] A utility function u is said to be *supermodular* if for any two bundles $S, T \subseteq R$, we have $u(S \cup T) + u(S \cap T) \geq u(S) + u(T)$.

Bredereck, Kaczmarczyk, and Niedermeier [237] study the complexity of some problems related to graph-envy-free allocations in directed social networks. In particular, they show that it is NP-complete to check whether there is a graph-envy-free allocation that maximizes utilitarian social welfare in acyclic social networks. Their results are complemented by Lange and Rothe [643] who prove that the problem of maximizing social welfare for graph-envy-free allocations is NP-complete for other social welfare measures as well, such as egalitarian and Nash social welfare. Aziz et al. [41] study epistemic envy-freeness (recall this notion from Section 9.5.6 starting on page 632) in social networks. Bredereck et al. [235] suggest to improve an existing allocation by allowing resources being shared with social network neighbors of the resource owners, and to this end they compute an optimal sharing between neighbors. Unlike the papers mentioned above that each are concerned with social networks whose vertices are agents, Bouveret et al. [177] consider social networks whose vertices are goods instead (see also Section 9.10.2.1 on page 670).

9.10 Extensions and Variants

The basic setting that we considered so far is fair division of indivisible goods (plus possibly money or allowing one good to be divided), either centralized or decentralized, without any specific constraints on the feasible allocations, where agents are considered equally and without an initial endowment, have preferences without externalities, and have full knowledge of the final allocation. In this section, we review extensions and variants of the basic setting that relax these assumptions: dividing bads instead of goods; constraints over feasible allocations; agents having epistemic access to only part of the final allocation; fair division among groups of agents; online fair division; preferences with externalities; unequal agents and private endowments. The review of the corresponding work will be very brief, though.

9.10.1 Fair Division of Bads

Sometimes, what has to be allocated is not goods but *bads*, such as chores, tasks (a paradigmatic example being reviewing papers), or nuisances, which bring negative utilities to agents. Research on dividing bads is much more recent and less developed than that on dividing goods, which partly explains why so many papers on this topic have been published in the last five years.

While it seems at first glance that dividing goods and dividing bads are symmetric problems and that mechanisms for fairly dividing goods can be easily adapted to dividing bads, it turns our that this is not the case at all, as explained by Bogomolnaia et al. [165, 166] and Brânzei and Sandomirskiy [223]. Still, some notions initially designed for goods can be adapted to bads, such as max-min fair share [66, 559], approximate max-min fair share [59, 563], EF1 [145], equity [469], and EFX and PMMS [563]. Garg, Murhekar, and Qin [490] and Ebadian, Peters, and Shah [368] independently showed that when agents have bivalued valuations, an allocation of chores satisfying EF1 and PO is guaranteed to exist and can be computed in polynomial time. For connections between fairness and efficiency criteria, see the work of Sun, Chen, and Doan [900].

The joint allocation of goods and chores—often referred to as a *mixed manna*—has also received a considerable attention in the last years [165, 51, 67, 488, 627, 484, 626, 275].

9.10.2 Constrained Fair Division

Sometimes allocations are subject to various kinds of feasibility constraints.

9.10.2.1 Connected Fair Division

While until now in this chapter the items (goods or bads) were not structured, here they form a topological structure: $G = (R, E)$ is an undirected graph, called the *item graph*, whose vertices are the items. A subset X of R is *connected* if it induces a connected subgraph of G. An allocation π is *connected* if for each agent i, $\pi(i)$ is connected. It is assumed that agents are only interested in receiving a connected bundle of items. Examples of domains where this is relevant are when the items have a temporal or spatial structure: For instance, if we wish to allocate time slots to agents, it may be required that these time slots be contiguous; likewise, when allocating spatial resources such as offices to research departments, it may be required that the set of offices allocated to a single department form a single connected component.

Connected fair division has been initiated by Bouveret et al. [177], who show that special classes of graphs lead to positive results: For instance, if the graph is acyclic then an allocation satisfying max-min fair share is guaranteed to exist and can be found in polynomial time. Suksompong [898] considers the specific class of paths and gives several approximation results for envy-freeness and proportionality. The existence and computation of max-min fair share allocations for cycles has been studied by Lonc and Truszczynski [665]. The complexity of computing max-min fair share connected allocations has been studied in further detail by Greco and Scarcello [519]. Igarashi and Peters [567] study the problem of finding connected Pareto-optimal allocations (where Pareto optimality is defined with respect to the set of connected allocations, namely, a connected allocation is *Pareto-optimal* if it is not Pareto-dominated by another connected allocation). EF1 for connected allocations is considered by Bilò et al. [146] and Oh, Procaccia, and Suksompong [759]. Caragiannis, Micha, and Shah [256] show that the existence of a fair allocation of a graph (for some new notions of fairness that they define) can be guaranteed by leaving $n-1$ nodes unallocated, where n is the number of agents. The *price of connectivity*, defined as the worst-case gap between the max-min fair share of an agent computed from all connected allocations and the one that would be obtained without connectivity constraints, has been studied by Bei et al. [124]. The parameterized complexity of connected fair division has been studied by Deligkas et al. [336]. Finally, Bouveret, Cechlárová, and Lesca [178] initiated the study of connected fair division of bads.

9.10.2.2 Multi-Type Fair Division and Matroid Constraints

A *categorized domain* is a partition of the items into several *categories* or *types*. Allocations in categorized domains are constrained by a minimal or maximal number of items of each type that each agent should receive. (In a more specific setting, Ferraioli, Gourvès, and Monnot [450] do not consider multiple types but require that each agent should receive the same number of items.) Multi-type fair division originates from the book by Moulin [716]. It has been further studied by Mackin and Xia [675] (who assume that each agent must get at least one item of each type) and Biswas and Barman [153] (who assume that for each type, each agent must get a specified maximum number of items, which depends on the type). Sikdar, Adali, and Xia [865] consider a multi-type version of housing markets, to which they generalize the "Top Trading Cycle" (TTC) mechanism. Sikdar et al. [866] design sequential mechanisms for multi-type allocation. Wang et al. [932] study multi-type randomized mechanisms and propose a multi-type version of the probabilistic serial and random priority mechanisms, where agents express partial preferences.

Multi-type fair division is a particular case of fair division with *matroid feasibility constraints*. In this more general setting, constraints on the set of feasible bundles that an agent can get are represented by matroids: Partition matroids correspond to multi-type fair division (possibly with different bounds for different agents). Fair division with more general matroid constraints, allowing for agents to have different types, or for items to have a more complex structure such as hierarchies of types (*laminar* matroids), is studied by Biswas and Barman [154] and Dror, Feldman, and Segal-Halevi [358]. As Gourvès, Monnot, and Tlilane [513, 511] point out, the requirement is that the *union* of bundles received by the agents satisfies the matroid constraints.

9.10.3 Partial Knowledge

We now briefly discuss scenarios with only partial knowledge, either on the side of the agents or on the side of the central authority.

9.10.3.1 When Agents Have Partial Knowledge of the Allocation

Envy-freeness is not only a strong notion in the sense that it is difficult to guarantee an envy-free allocation, but it is also informationally demanding, in the sense that it presupposes that agents have full knowledge of the allocation. One way of relaxing this consists in assuming that they know only part of the allocation. For example, as mentioned in Section 9.9.4 starting on page 667, Chevaleyre, Endriss, and Maudet [291] and Bredereck, Kaczmarczyk, and Niedermeier [237] assume that the agents are members of a social network, leading to a more local (and more plausible) notion of envy-freeness that merely requires the agents to not envy any of their neighbors. The parameterized complexity of finding an envy-free allocation with respect to a social network has been studied by Eiben et al. [372]. Aziz et al. [41] also make use of a social network but define a different relaxation of envy-freeness with respect to a social network: Agents know the share of their neighbors and ask themselves if the entire allocation can be possibly envy-free; this notion collapses to epistemic envy-freeness (see Section 9.5.6 starting on page 632) with an empty network (and of course to envy-freeness with a completely connected network).

Chan et al. [271] define *maximin-aware fairness*, which is similar to epistemic envy-freeness in the sense that agents know only their own share; an allocation is *maximin-aware fair* if each agent i knows that there is at least one agent that i does not envy (even if i cannot identify this agent). Hosseini et al. [561] define yet another fairness notion where the agents can withhold (or hide) some of the goods in their bundle and reveal only the remaining goods to the other agents; they show that in practice, withholding only a small number of goods overall is enough to obtain envy-freeness in this sense.

9.10.3.2 When the Central Authority Has Partial Knowledge of the Agents' Preferences

Arunachaleswaran, Barman, and Rathi [32] propose a model where the central authority knows the valuations of all agents except one; with this incomplete information it is still possible to design a protocol (letting the nth agent choose among several bundles) that outputs an EF1 allocation.

Amanatidis, Birmpas, and Markakis [20] show that if agents have private additive valuations but only their ordinal preferences are elicited by the central authority, it is not possible to guarantee more than a $H(n)$-approximation of MMS, where $H(n) = 1 + \frac{1}{2} + \cdots + \frac{1}{n}$ is the nth harmonic number. Halpern and Shah [529] go even further and assume that only a ranking of their k most preferred goods is elicited for each of them; they determine the value of k needed to achieve EF1 and approximate MMS.

Another context where the central authority has partial knowledge of the agents' preferences is fair division with ordinal preferences, reviewed in Section 9.7 starting on page 649.

9.10.4 Fair Division among Groups of Agents and Externalities

The fairness properties we considered so far apply to single agents: In particular, envy-freeness says that no single agent envies the share of another agent. In some settings, agents are clustered in groups that share some specific characteristics, such as families, research departments, ethnic groups, or even different instances of the same individual on different days. In this case, it may be more reasonable to apply fairness criteria to groups rather than single agents: The single agents in a group feel solidary or empathic enough with each other (or, even, can transfer utility within the group) so that the dissatisfaction of some of its members can be compensated by the satisfaction of some other members. There are two orthogonal research directions.

In the first one, agents still get their own share each but fairness properties are evaluated groupwise. Different groupwise generalizations of envy-freeness have been defined and studied by Todo et al. [916], Manurangsi and Suksompong [678, 680], Segal-Halevi and Suksompong [849], Aleksandrov and Walsh [9], Kyropoulou, Suksompong, and Voudouris [633], and Benabbou et al. [128]. Note that these notions suppose that the groups are pre-existing. By contrast, notions such as *groupwise fair share guarantee*, which consider all possible groups, lead to relaxations of envy-freeness and as such are evoked in Section 9.5 starting on page 625.

In the second research direction, goods are allocated to *groups of agents* and shared by the members of a group. This is an intermediate setting between classical resource allocation (where all agents get their own share of the goods) and public good selection (where all selected goods are shared among all agents). Suksompong [897] generalizes MMS to such a setting and Segal-Halevi and Suksompong [849] introduce the concept of *democratic fairness*, which aims to satisfy a certain fraction of the agents in each group. Somewhat related to this second research line is *fair division with externalities* [846]: Agents are partitioned in a network and care about the allocation of their neighbors, so the share allocated to an agent may affect the utilities of other agents.[18] For another work on fair division with externalities, see [464].

9.10.5 Sharing Goods

A way of coping with the incompatibility of efficiency and fairness consists in allowing some goods to be *shared* among agents, rather than assigned to a single agent. This direction was initiated by Airiau and Endriss [4]. Sandomirskiy and Segal-Halevy [833] determine the smallest number of objects that must be shared among two or more agents to attain a fair (proportional or envy-free) and efficient allocation. Bredereck et al. [235] (already cited in Section 9.9.4 starting on page 667) restrict sharing to neighbors in a social network.

Skowron, Faliszewski, and Slinko [873] model the multiwinner voting rules of Monroe [713] and Chamberlin and Courant [270] (see Section 6.2.2 starting on page 407) as resource allocation rules where the alternatives are shareable resources and each agent receives exactly one resource.

9.10.6 Online Fair Division

Online fair division[19]—or how to dynamically allocate indivisible goods that arrive online, one after the other—has been addressed by Aleksandrov et al. [8, 9], Benade et al. [130], He et al. [531], Bogomolnaia, Moulin, and Sandomirskiy [164], Zeng and Psomas [970], and Gkatzelis, Psomas, and Tan [502]. For an overview of the very young field of online fair division and for more details, see the survey by Aleksandrov and Walsh [10].

9.10.7 Private Endowments and Asymmetric Agents

A variant of the resource allocation problem considered in this chapter is obtained when each agent is initially endowed with a set of indivisible goods. Under this assumption, not all allocations are admissible: An agent is not willing to participate in an allocation mechanism if she considers her new

---

[18] This is also somewhat related to the notion of altruistic hedonic games [605] described in Section 3.6 starting on page 198.

[19] This is closely related to online cake cutting studied by Walsh [930], and remotely related to online manipulation [543], online control [545, 544, 740], and online bribery [546] in sequential elections, as discussed in Sections 4.3.3.10, 4.3.4.11, and 4.3.5.5 starting on pages 321, 354, and 363, respectively.

share worse than her initial share. The *individual rationality* property states exactly this: An allocation π is *individually rational* if no agent i strictly prefers her initial endowment σ_i to π_i. In problems with initial endowments, we do not look for allocations computed "from scratch" but rather for *exchanges* (cf. Section 9.9.4 starting on page 667). For a few key references about fair division with private endowments, see the work of Sönmez [882], Pápai [769], and Sonoda et al. [883].

Finding a Pareto improvement over an initial allocation can be seen as a *reallocation* problem. For a computational study of reallocation for additive cardinal utilities or responsive ordinal preferences, see the work of Aziz et al. [40].

Sometimes, agents are *asymmetric* in the sense that some deserve a higher share than others (cf. Section 8.4.4 starting on page 562). For instance, party lists that merge after a first election round may divide the seats according to their relative first-round score. This leads to an extension of fair division of indivisible goods when agents are *weighted*: See the work of Aziz et al. [52, 64], Chakraborty et al. [265, 266], Farhadi et al. [444], and Chakraborty, Segal-Halevi, and Suksompong [267]; in particular, the case of chores for asymmetric agents has been handled by Aziz, Chan, and Li [52].

9.11 Further Issues

There are a number of issues that we chose not to discuss at great length. We briefly evoke them here, together with a few standard references.

9.11.1 Strategic Agents

In this chapter, we have considered a number of fair division mechanisms requiring agents to report their preferences either to a central authority, or to the other agents in a decentralized way. In both cases, mechanisms are prone to strategic manipulation, just like for voting rules (see Section 4.2.5.1 starting on page 274; see also Sections 4.3.3–4.3.5 on manipulation, control, and bribery attacks in voting) or judgement aggregation procedures (see Sections 7.3.3–7.3.5 for manipulation, bribery, and control attacks on JA procedure).

A major difference with voting, and a reason why the Gibbard–Satterthwaite theorem (see Theorem 4.2 on page 275) does not apply here, is that in fair division the preferences of the agents over allocations depend only on their own share, that is, on only a part of the outcome. Let us define *serial dictatorship* as an allocation rule that, at each step, assigns to a designated agent (the "current dictator") her best possible set of objects composed of objects that have not yet been assigned, and possibly subject to some constraints such as cardinality bounds on the number of objects (when there is no constraint, the current dictator takes her preferred bundle composed of the remaining objects; she may take all of them, or leave some that she does not want). Serial dictatorships can be defined in a centralized setting (shares are assigned by the central authority) or in a distributed setting (agents choose their share when it is their turn to do so).

Clearly, serial dictatorships are strategy-proof. It turns out to be difficult to find other allocation rules that are strategy-proof and satisfy some minimal set of properties. If we require only Pareto efficiency and neutrality,[20] and if there are $n \geq 3$ agents, then there are strategy-proof allocation rules that are not serial dictatorships, such as this one: Suppose that each agent is entitled to get only one object, and that $n = m$; then agent 1 gets her most preferred object, and for each $j \geq 2$, agent j gets agent 1's jth preferred object. This rule has the undesirable property that agent 1 influences the way the remaining goods are assigned to the other agents "more than needed," that is, even when it does not change her share.

Let us say that an allocation rule F is *nonbossy* if for all profiles $P = (P_1, P_2, \ldots, P_n)$ consisting of preference relations over 2^R, and letting $\pi = (\pi_1, \ldots, \pi_n) = F(P)$ be the outcome of F on input P and letting $\pi' = (\pi'_1, \ldots, \pi'_n) = F(P'_1, P_{-1})$ be the outcome of F given the profile resulting from P when replacing P_1 by any preference relation P'_1 (all else being equal), then $\pi_1 = \pi'_1$ implies that for all $i \geq 2$, $\pi_i = \pi'_i$. Svensson [904] shows that if $n = m$, then each strategy-proof, nonbossy, and neutral mechanism is a serial dictatorship. When agents can be assigned more than one object, Pápai [768] shows that mechanisms that are nonbossy, strategy-proof, and Pareto-optimal are *sequential dictatorships*, which differ from serial dictatorships in that the identity of dictator i may depend on the preferences reported by dictators 1 to $i-1$. Ehlers and Klaus [371] prove a similar result in the case where the agents have separable preferences, and so does Hatfield [530] in the case where each agent desires to have a fixed number of objects: Strategy-proofness, Pareto optimality, and nonbossiness combined again implies sequential dictatorship, whereas requiring neutrality on top of that leads to serial dictatorships.

Much work has been done to identify the particular conditions under which certain allocation rules guarantee strategy-proofness. For example, as hinted at earlier, Nguyen, Baumeister, and Rothe [742] provide a necessary and sufficient condition for utilitarian scoring allocation correspondences to satisfy strategy-proofness in the sense of Kelly [601] and Gärdenfors [481]: There are at most two goods to allocate or fewer goods than twice the number of occurrences of the greatest value in the scoring vector. In particular, the

---

[20] *Neutrality* is the property that the allocation mechanism treats all objects equally—this is similar to neutrality in voting, replacing objects by alternatives, see Section 4.2.6.2 on page 277.

Borda-like scoring vector $(m, m-1, \ldots, 1)$ for $m > 2$ does not yield a strategy-proof utilitarian scoring allocation correspondence in this sense.

Truthful mechanisms for max-min fair share allocations have been studied by Amanatidis, Birmpas, and Markakis [20] and truthful mechanisms for binary valuations by Halpern et al. [527] and Babaioff, Ezra, and Feige [69]. A complete characterization of truthful mechanisms for allocating goods to two agents with additive valuations is given by Amanatidis et al. [17]. Barman and Verma [96] consider strategic agents whose valuations are submodular functions with binary marginals. Walsh [931] surveys the computational questions related to strategic behavior in the fair division of indivisible goods.

9.11.2 Matching

A very specific kind of resource allocation is *one-sided bipartite matching*: Each agent should be assigned to *one* object, such as a house.[21] In most cases, preferences are assumed to be ordinal. Although a one-sided bipartite matching problem can be viewed as a fair division problem (with indivisible resources), the fact that agents receive a single object has strong implications on the relevance of some of the criteria we have considered in this chapter. For instance, envy-freeness, even if it makes sense, is very hard to satisfy, since it requires each agent to receive her most preferred object among those that are assigned. The proportional fair share and max-min fair share criteria make even less sense. However, Pareto efficiency remains a crucial criterion.

Since the celebrated paper by Gale and Shapley[22] [479] most work so far has focused on the computation of matchings that are stable, that is, that are resistant to possible deviations by individual agents or groups of agents. There

---

[21] There are also *two-sided bipartite matching* problems where not only agents have preferences over objects but "objects" also have preferences over agents. For example, agents are junior doctors (called "medical residents") and "objects" are hospital posts; the doctors have preferences over the hospitals and the hospitals over the doctors; in the USA, a bipartite matching mechanism is used on a nationwide scale to assign junior doctors to hospital posts. As another classical example, suppose the agents are students and the "objects" are universities; both the students may have preferences over the universities—aiming to choose a top university—and the universities may have preferences over the students—aiming to attract the more talented ones. And as a last example of a two-sided bipartite matching problem, suppose that men have preferences over women and women have preferences over men, and we are looking for arranging *stable marriages* among them. There also exist *nonbipartite* matching problems, such as roommate assignment problems, where agents and resources coincide. Nonbipartite matching problems can be considered a restriction of hedonic games (see Section 3.6 starting on page 198) where coalitions cannot have more than two players (see Section 3.6 starting on page 198 for pointers to the literature on hedonic games).

[22] As mentioned in Chapter 3 already, Lloyd S. Shapley received the 2012 *Nobel Prize in Economics*, jointly with Alvin E. Roth, for the theory of stable allocations and the practice of market design.

is also a version of one-sided bipartite matching where the agents have initial endowments (such as the house exchange problem, where the agents have an initial house and may want to swap houses with other agents), which is the matching version of fair division with private endowments to be discussed briefly in the following section. For an overview of matching with preferences and its computational aspects, see the book chapter by Klaus, Manlove, and Rossi [612] and the book by Manlove [677].

We conclude this section by mentioning a work by Freeman, Micha, and Shah [466] that introduces a new model for two-sided matching which borrows fairness notions from the fair division literature such as envy-freeness up to one good and the max-min share guarantee.

9.11.3 Randomized Fair Division

Throughout this chapter we focused on deterministic allocations. Sometimes it makes sense to define mechanisms that output a random allocation that assigns each agent a probability distribution over objects or sets of objects. Two prominent randomized assignment rules are random serial dictatorship and the probabilistic serial rule.

According to the *random serial dictatorship* (*RSD*) rule, a sequence of agents is selected randomly, and then each agent, when it is her turn, takes her most preferred share from among the remaining goods, subject to some constraints (such as "one good per agent"), that is, given the agents' preferences over the objects, RSD returns a random allocation, along with the probabilities with which the objects are allocated to the agents. More specifically, RSD picks a permutation uniformly at random and lets the agents in the permutation serially choose their most preferred object among those not yet allocated. RSD is the rule that is known to be anonymous, strategy-proof, and *ex post* efficient (which means, that it randomizes over Pareto-efficient allocations). The computational aspects of random serial dictatorship have been studied (also in the context of voting) by Aziz, Brandt, and Brill [45] and, independently, by Sabán and Sethuraman [828], who show that the problem of computing the RSD probabilities is #P-complete (where #P is the "counting version" of NP that has been introduced by Valiant [921]). Aziz and Mestre [63] establish fixed-parameter tractability results for computing the RSD probabilities for parameters such as the number of agents (or of agent types) and the number of objects; see also the survey by Aziz et al. [46] for an overview.

A randomized allocation should not be confused with a fractional allocation. The outcome of the *probabilistic serial rule* is a fractional allocation: For instance, if the input is $(a \succ b \succ c,\ a \succ c \succ b,\ b \succ c \succ a)$, the outcome of the probabilistic serial rule is $\left(\frac{1}{2}a + \frac{1}{4}b + \frac{1}{4}c,\ \frac{1}{2}a + \frac{1}{2}c,\ \frac{3}{4}b + \frac{1}{4}c\right)$, to be primarily interpreted in a world of divisible goods: Agent 1 gets one half of

9.11 Further Issues

good a, etc. Now, the theorem due to von Neumann [735] and Birkhoff [151] says that such a fractional allocation can be decomposed as a probability distribution over deterministic one-to-one allocations: In our example, such a decomposition is $\frac{1}{2}(a,c,b) + \frac{1}{4}(c,a,b) + \frac{1}{4}(b,a,c)$. We observe that this lottery over deterministic allocations has good fairness properties *ex ante*, but will inevitably have envy *ex post*, since two agents have a as their most preferred item.

In general, it is much easier to obtain strong fairness guarantees *ex ante* than *ex post*. In the more general context of many-to-one fair division of indivisible goods, this question of how to conciliate a strong *ex-ante* fairness property, such as envy-freeness, with a milder *ex-post* fairness property, such as EF1 or approximate MMS, is highly desirable. Aleksandrov et al. [8] analyze the *ex-ante* and *ex-post* fairness guarantees for binary utilities. Assuming more general additive valuations, Freeman, Shah, and Vaish [467] show that *ex-ante* envy-freeness can be achieved in combination with *ex-post* envy-freeness up to one good, and that such a randomized allocation is computable in polynomial time, using a recursive generalization of the probabilistic serial rule; a simpler proof of the same result is given by Aziz [37]. Babaioff, Ezra, and Feige [69] show that *ex-ante* envy-freeness is compatible with a constant fraction of the MMS *ex post*. Caragiannis, Kanellopoulos, and Kyropoulou [254] introduce *interim envy-freeness*, a property that lies between *ex-ante* and *ex-post* envy-freeness: A randomized allocation is *interim envy-free* if all agents, after seeing the realization of *their own share* but not the realization of the other agents' shares, prefer their own share to the expected value of every other agent's share. The underlying epistemic notion is closely related to epistemic envy-freeness, discussed in Section 9.5.6 starting on page 632.

Randomized fair division should also not be confused either with deterministic fair division of "risky" goods, that is, goods whose quality (and its repercussion on the agents' utilities) is defined by a probability distribution; see the work of Lumet, Bouveret, and Lemaître [673] for a computational study, and see the references therein for pointers to the literature.

Kawase and Sumita [597] study the problem of finding randomized allocations that maximize the expected *ex-ante* egalitarian social welfare, and design polynomial-time approximation algorithms for submodular utilities. They also consider the computation of *ex-ante* envy-free allocations.

9.11.4 Fair Allocation of Public Goods

We conclude this chapter by mentioning a research area which, although not dealing with the allocation of goods *to agents*, is highly related to fair division of indivisible goods: the fair selection of public goods. Public goods can be members of an elected committee, common projects to be funded by a municipality, or issues in a multiple-referenda election. The output here is common to all agents—there is nothing like an agent's share. There is an abundant research on this topic, which is closely related to *participatory budgeting* and *multiwinner elections* (surveyed in Chapter 6 of this book); however, we mention this line of research here as well because it makes a considerable use of notions that directly stem from the literature on fair division of indivisible goods. Since agents do not have their own share, some criteria, notably envy-freeness, become meaningless in this setting, but some others, such as max-min fair share or proportional fair share, can be generalized to public decision-making in a meaningful way [303, 417].

An intermediate setting between classical allocation (where all agents get their own share) and public good selection (where all selected goods are shared among all agents) is when goods are allocated to *groups of agents* and shared by the members of the group, as discussed in Section 9.10.4 starting on page 673.

References

1. Aarts, E., Lenstra, J. (eds.): Local Search in Combinatorial Optimization. Princeton University Press (2003)
2. Adams, D.: The Hitchhiker's Guide to the Galaxy. Pan Books (1979)
3. Ahuja, R., Magnanti, T., Orlin, J.: Network Flows: Theory, Algorithms, and Applications. Prentice-Hall (1993)
4. Airiau, S., Endriss, U.: Multiagent resource allocation with sharable items: Simple protocols and Nash equilibria. In: Proceedings of the 9th International Conference on Autonomous Agents and Multiagent Systems, pp. 167–174. IFAAMAS (2010)
5. Akca, N.: Auktionen zur nationalen Reallokation von Treibhausgas-Emissionsrechten und Treibhausgas-Emissionsgutschriften auf Unternehmensebene: Ein spieltheoretischer nicht-kooperativer Modellierungs- und Lösungsansatz für das Reallokationsproblem. Gabler Edition Wissenschaft (2009). Dissertation an der Universität Duisburg-Essen, Germany.
6. Akrami, H., Chaudhury, B., Hoefer, M., Mehlhorn, K., Schmalhofer, M., Shahkarami, G., Varricchio, G., Vermande, Q., van Wijland, E.: Maximizing Nash social welfare in 2-value instances. In: Proceedings of the 36th AAAI Conference on Artificial Intelligence, pp. 4760–4767. AAAI Press (2022)
7. Akrami, H., Rezvan, R., Seddighin, M.: An EF2X allocation protocol for restricted additive valuations. In: Proceedings of the 31st International Joint Conference on Artificial Intelligence, pp. 17–23. ijcai.org (2022)
8. Aleksandrov, M., Aziz, H., Gaspers, S., Walsh, T.: Online fair division: Analysing a food bank problem. In: Proceedings of the 24th International Joint Conference on Artificial Intelligence, pp. 2540–2546. AAAI Press/IJCAI (2015)
9. Aleksandrov, M., Walsh, T.: Group envy freeness and group Pareto efficiency in fair division with indivisible items. In: Proceedings of the 41st German Conference on AI, pp. 57–72. Springer-Verlag *Lecture Notes in Computer Science #11117* (2018)
10. Aleksandrov, M., Walsh, T.: Group envy freeness and group Pareto efficiency in fair division with indivisible items. Tech. Rep. arXiv:2006.15893 [cs.GT], ACM Computing Research Repository (CoRR) (2020)
11. Alkan, A.: Existence and computation of matching equilibria. European Journal of Political Economy **5**, 285–296 (1989)
12. Alkan, A., Demange, G., Gale, D.: Fair allocation of indivisible goods and criteria of justice. Econometrica **59**(4), 1023–1039 (1991)
13. Allis, L.: Searching for solutions in games and artificial intelligence. Ph.D. thesis, University of Limburg, The Netherlands (1994)
14. Alon, N., Azar, Y., Woeginger, G., Yadid, T.: Approximation schemes for scheduling on parallel machines. Journal of Scheduling **1**(1), 55–66 (1998)

15. Alon, N., Bredereck, R., Chen, J., Kratsch, S., Niedermeier, R., Woeginger, G.: How to put through your agenda in collective binary decisions. In: Proceedings of the 3rd International Conference on Algorithmic Decision Theory, pp. 30–44. Springer-Verlag *Lecture Notes in Artificial Intelligence #8176* (2013)
16. Alon, N., Falik, D., Meir, R., Tennenholtz, M.: Bundling attacks in judgment aggregation. In: Proceedings of the 27th AAAI Conference on Artificial Intelligence, pp. 39–45. AAAI Press (2013)
17. Amanatidis, G., Birmpas, G., Christodoulou, G., Markakis, E.: Truthful allocation mechanisms without payments: Characterization and implications on fairness. In: Proceedings of the 18th ACM Conference on Economics and Computation, pp. 545–562. ACM Press (2017)
18. Amanatidis, G., Birmpas, G., Filos-Ratsikas, A., Hollender, A., Voudouris, A.: Maximum Nash welfare and other stories about EFX. Theoretical Computer Science **863**, 69–85 (2021)
19. Amanatidis, G., Birmpas, G., Filos-Ratsikas, A., Voudouris, A.: Fair division of indivisible goods: A survey. In: Proceedings of the 31st International Joint Conference on Artificial Intelligence, pp. 5385–5393. ijcai.org (2022)
20. Amanatidis, G., Birmpas, G., Markakis, E.: On truthful mechanisms for maximin share allocations. In: Proceedings of the 25th International Joint Conference on Artificial Intelligence, pp. 31–37. AAAI Press/IJCAI (2016)
21. Amanatidis, G., Birmpas, G., Markakis, E.: Comparing approximate relaxations of envy-freeness. In: Proceedings of the 27th International Joint Conference on Artificial Intelligence, pp. 42–48. ijcai.org (2018)
22. Amanatidis, G., Markakis, E., Nikzad, A., Saberi, A.: Approximation algorithms for computing maximin share allocations. ACM Transactions on Algorithms **13**(4), 52:1–52:28 (2017)
23. Amanatidis, G., Markakis, E., Ntokos, A.: Multiple birds with one stone: Beating 1/2 for EFX and GMMS via envy cycle elimination. Theoretical Computer Science **841**, 94–109 (2020)
24. Anari, N., Gharan, S., Saberi, A., Singh, M.: Nash social welfare, matrix permanent, and stable polynomials. In: Proceedings of the 8th Innovations in Theoretical Computer Science Conference, *LIPIcs*, vol. 67, pp. 36:1–36:12. Schloss Dagstuhl – Leibniz-Zentrum für Informatik (2017)
25. Anari, N., Mai, T., Gharan, S., Vazirani, V.: Nash social welfare for indivisible items under separable, piecewise-linear concave utilities. In: Proceedings of the 29th Annual ACM-SIAM Symposium on Discrete Algorithms, pp. 2274–2290. Society for Industrial and Applied Mathematics (2018)
26. Anderson, A., Shoham, Y., Altman, A.: Internal implementation. In: Proceedings of the 9th International Conference on Autonomous Agents and Multiagent Systems, pp. 191–198. IFAAMAS (2010)
27. Annamalai, C., Kalaitzis, C., Svensson, O.: Combinatorial algorithm for restricted max-min fair allocation. ACM Transactions on Algorithms **13**(3), 13:1–13:28 (2017)
28. Arora, S., Lund, C.: Hardness of approximations. In: D. Hochbaum (ed.) Approximation Algorithms for NP-Hard Problems, chap. 10, pp. 399–446. PWS Publishing Company (1996)
29. Arrow, K.: Social Choice and Individual Values. John Wiley and Sons (1951 (revised edition 1963))
30. Arrow, K., Sen, A., Suzumura, K. (eds.): Handbook of Social Choice and Welfare, vol. 1. North-Holland (2002)
31. Arrow, K., Sen, A., Suzumura, K. (eds.): Handbook of Social Choice and Welfare, vol. 2. North-Holland (2011)
32. Arunachaleswaran, E., Barman, S., Rathi, N.: Fair division with a secretive agent. In: Proceedings of the 33rd AAAI Conference on Artificial Intelligence, pp. 1732–1739. AAAI Press (2019)

References

33. Arzi, O., Aumann, Y., Dombb, Y.: Throw one's cake – and eat it too. In: Proceedings of the 4th International Symposium on Algorithmic Game Theory, pp. 69–80. Springer-Verlag *Lecture Notes in Computer Science #6982* (2011)
34. Asadpour, A., Saberi, A.: An approximation algorithm for max-min fair allocation of indivisible goods. In: Proceedings of the 39th ACM Symposium on Theory of Computing, pp. 114–121. ACM Press (2007)
35. Asadpour, A., Saberi, A.: An approximation algorithm for max-min fair allocation of indivisible goods. SIAM Journal on Computing **39**(7), 2970–2989 (2010)
36. Austin, A.: Sharing a cake. Mathematical Gazette **66**(437), 212–215 (1982)
37. Aziz, H.: Simultaneously achieving ex-ante and ex-post fairness. In: Proceedings of the 16th International Workshop on Internet & Network Economics, pp. 341–355. Springer-Verlag *Lecture Notes in Computer Science #12495* (2020)
38. Aziz, H.: Achieving envy-freeness and equitability with monetary transfers. In: Proceedings of the 35th AAAI Conference on Artificial Intelligence, pp. 5102–5109. AAAI Press (2021)
39. Aziz, H., Bachrach, Y., Elkind, E., Paterson, M.: False-name manipulations in weighted voting games. Journal of Artificial Intelligence Research **40**, 57–93 (2011)
40. Aziz, H., Biró, P., Lang, J., Lesca, J., Monnot, J.: Efficient reallocation under additive and responsive preferences. Theoretical Computer Science **790**, 1–15 (2019)
41. Aziz, H., Bouveret, S., Caragiannis, I., Giagkousi, I., Lang, J.: Knowledge, fairness, and social constraints. In: Proceedings of the 32nd AAAI Conference on Artificial Intelligence, pp. 4638–4645. AAAI Press (2018)
42. Aziz, H., Bouveret, S., Lang, J., Mackenzie, S.: Complexity of manipulating sequential allocation. In: Proceedings of the 31st AAAI Conference on Artificial Intelligence, pp. 328–334. AAAI Press (2017)
43. Aziz, H., Brandl, F.: Existence of stability in hedonic coalition formation games. In: Proceedings of the 11th International Conference on Autonomous Agents and Multiagent Systems, pp. 763–770. IFAAMAS (2012)
44. Aziz, H., Brandl, F., Brandt, F., Harrenstein, P., Olsen, M., Peters, D.: Fractional hedonic games. ACM Transactions on Economics and Computation **7**(2), 1–29 (2019)
45. Aziz, H., Brandt, F., Brill, M.: The computational complexity of random serial dictatorship. Economics Letters **121**(3), 341–345 (2013)
46. Aziz, H., Brandt, F., Brill, M., Mestre, J.: Computational aspects of random serial dictatorship. SIGecom Exchanges **13**(2), 26–30 (2014)
47. Aziz, H., Brandt, F., Harrenstein, P.: Pareto optimality in coalition formation. Games and Economic Behavior **82**, 562–581 (2013)
48. Aziz, H., Brandt, F., Seedig, H.: Computing desirable partitions in additively separable hedonic games. Artificial Intelligence **195**, 316–334 (2013)
49. Aziz, H., Brânzei, S., Filos-Ratsikas, A., Frederiksen, S.: The adjusted winner procedure: Characterizations and equilibria. In: Proceedings of the 31st International Joint Conference on Artificial Intelligence, pp. 454–460. ijcai.org (2022)
50. Aziz, H., Brill, M., Conitzer, V., Elkind, E., Freeman, R., Walsh, T.: Justified representation in approval-based committee voting. Social Choice and Welfare **48**(2), 461–485 (2017)
51. Aziz, H., Caragiannis, I., Igarashi, A., Walsh, T.: Fair allocation of indivisible goods and chores. In: Proceedings of the 28th International Joint Conference on Artificial Intelligence, pp. 53–59. ijcai.org (2019)
52. Aziz, H., Chan, H., Li, B.: Maxmin share fair allocation of indivisible chores to asymmetric agents. In: Proceedings of the 18th International Conference on Autonomous Agents and Multiagent Systems, pp. 1787–1789. IFAAMAS (2019)
53. Aziz, H., Elkind, E., Huang, S., Lackner, M., Sánchez-Fernández, L., Skowron, P.: On the complexity of extended and proportional justified representation. In: Proceedings of the 32nd AAAI Conference on Artificial Intelligence, pp. 902–909. AAAI Press (2018)

54. Aziz, H., Gaspers, S., Gudmundsson, J., Mackenzie, S., Mattei, N., Walsh, T.: Computational aspects of multi-winner approval voting. In: Proceedings of the 14th International Conference on Autonomous Agents and Multiagent Systems, pp. 107–115. IFAAMAS (2015)
55. Aziz, H., Gaspers, S., Mackenzie, S., Walsh, T.: Fair assignment of indivisible objects under ordinal preferences. In: Proceedings of the 13th International Conference on Autonomous Agents and Multiagent Systems, pp. 1305–1312. IFAAMAS (2014)
56. Aziz, H., Goldberg, P., Walsh, T.: Equilibria in sequential allocation. In: Proceedings of the 5th International Conference on Algorithmic Decision Theory, pp. 270–283. Springer-Verlag *Lecture Notes in Artificial Intelligence #10576* (2017)
57. Aziz, H., Huang, X., Mattei, N., Segal-Halevi, E.: Computing welfare-maximizing fair allocations of indivisible goods. Tech. Rep. arXiv:2012.03979v2 [cs.GT], ACM Computing Research Repository (CoRR) (2021)
58. Aziz, H., Lee, B.: The expanding approvals rule: Improving proportional representation and monotonicity. Social Choice and Welfare **54**(1), 1–45 (2020)
59. Aziz, H., Li, B., Wu, X.: Strategyproof and approximately maxmin fair share allocation of chores. In: Proceedings of the 28th International Joint Conference on Artificial Intelligence, pp. 60–66. ijcai.org (2019)
60. Aziz, H., Mackenzie, S.: A discrete and bounded envy-free cake cutting protocol for any number of agents. In: Proceedings of the 57th IEEE Symposium on Foundations of Computer Science, pp. 416–427. IEEE Computer Society Press (2016)
61. Aziz, H., Mackenzie, S.: A discrete and bounded envy-free cake cutting protocol for four agents. In: Proceedings of the 48th Annual ACM Symposium on Theory of Computing, pp. 454–464. ACM Press (2016)
62. Aziz, H., Mackenzie, S.: A bounded and envy-free cake cutting algorithm. Communications of the ACM **63**(4), 119–126 (2020)
63. Aziz, H., Mestre, J.: Parametrized algorithms for random serial dictatorship. Mathematical Social Sciences **72**, 1–6 (2014)
64. Aziz, H., Moulin, H., Sandomirskiy, F.: A polynomial-time algorithm for computing a Pareto optimal and almost proportional allocation. Operations Research Letters **48**(5), 573–578 (2020)
65. Aziz, H., Paterson, M.: False name manipulations in weighted voting games: Splitting, merging and annexation. In: Proceedings of the 8th International Conference on Autonomous Agents and Multiagent Systems, pp. 409–416. IFAAMAS (2009)
66. Aziz, H., Rauchecker, G., Schryen, G., Walsh, T.: Algorithms for max-min share fair allocation of indivisible chores. In: Proceedings of the 31st AAAI Conference on Artificial Intelligence, pp. 335–341. AAAI Press (2017)
67. Aziz, H., Rey, S.: Almost group envy-free allocation of indivisible goods and chores. In: Proceedings of the 29th International Joint Conference on Artificial Intelligence, pp. 39–45. ijcai.org (2020)
68. Aziz, H., Schlotter, I., Walsh, T.: Control of fair division. In: Proceedings of the 25th International Joint Conference on Artificial Intelligence, pp. 67–73. AAAI Press/IJCAI (2016)
69. Babaioff, M., Ezra, T., Feige, U.: Fair and truthful mechanisms for dichotomous valuations. In: Proceedings of the 35th AAAI Conference on Artificial Intelligence, pp. 5119–5126. AAAI Press (2021)
70. Babaioff, M., Feige, U.: Fair shares: Feasibility, domination and incentives. In: Proceedings of the 23rd ACM Conference on Economics and Computation, p. 435. ACM Press (2022)
71. Bacchus, F., Grove, A.: Graphical models for preference and utility. In: Proceedings of the 11th Annual Conference on Uncertainty in Artificial Intelligence, pp. 3–10. Morgan Kaufmann (1995)
72. Bachmann, P.: Analytische Zahlentheorie, vol. 2. Teubner (1894)

73. Bachrach, Y., Elkind, E.: Divide and conquer: False-name manipulations in weighted voting games. In: Proceedings of the 7th International Conference on Autonomous Agents and Multiagent Systems, pp. 975–982. IFAAMAS (2008)
74. Bachrach, Y., Elkind, E., Meir, R., Pasechnik, D., Zuckerman, M., Rothe, J., Rosenschein, J.: The cost of stability in coalitional games. In: Proceedings of the 2nd International Symposium on Algorithmic Game Theory, pp. 122–134. Springer-Verlag *Lecture Notes in Computer Science #5814* (2009)
75. Bachrach, Y., Meir, R., Zuckerman, M., Rothe, J., Rosenschein, J.: The cost of stability in weighted voting games (extended abstract). In: Proceedings of the 8th International Conference on Autonomous Agents and Multiagent Systems, pp. 1289–1290. IFAAMAS (2009)
76. Bachrach, Y., Porat, E.: Path disruption games. In: Proceedings of the 9th International Conference on Autonomous Agents and Multiagent Systems, pp. 1123–1130. IFAAMAS (2010)
77. Bachrach, Y., Rosenschein, J.: Power in threshold network flow games. Journal of Autonomous Agents and Multi-Agent Systems **18**(1), 106–132 (2009)
78. Bai, Y., Feige, U., P-Gölz, Procaccia, A.: Fair allocations for smoothed utilities. In: Proceedings of the 23rd ACM Conference on Economics and Computation, pp. 436–465. ACM Press (2022)
79. Bai, Y., Gölz, P.: Envy-free and Pareto-Optimal allocations for agents with asymmetric random valuations. In: Proceedings of the 31st International Joint Conference on Artificial Intelligence, pp. 53–59. ijcai.org (2022)
80. Baklanov, A., Garimidi, P., Gkatzelis, V., Schoepflin, D.: Achieving proportionality up to the maximin item with indivisible goods. In: Proceedings of the 35th AAAI Conference on Artificial Intelligence, pp. 5143–5150. AAAI Press (2021)
81. Baklanov, A., Garimidi, P., Gkatzelis, V., Schoepflin, D.: PROPm allocations of indivisible goods to multiple agents. In: Proceedings of the 30th International Joint Conference on Artificial Intelligence, pp. 24–30. ijcai.org (2021)
82. Baldwin, J.: The technique of the Nanson preferential majority system of election. Transactions and Proceedings of the Royal Society of Victoria **39**, 42–52 (1926)
83. Ballester, C.: NP-completeness in hedonic games. Games and Economic Behavior **49**(1), 1–30 (2004)
84. Banerjee, S., Konishi, H., Sönmez, T.: Core in a simple coalition formation game. Social Choice and Welfare **18**(1), 135–153 (2001)
85. Bansal, N., Sviridenko, M.: The Santa Claus problem. In: Proceedings of the 38th ACM Symposium on Theory of Computing, pp. 31–40. ACM Press (2006)
86. Banzhaf III, J.: Weighted voting doesn't work: A mathematical analysis. Rutgers Law Review **19**, 317–343 (1965)
87. Barbanel, J.: Super envy-free cake division and independence of measures. Journal of Mathematical Analysis and Applications **197**(1), 54–60 (1996)
88. Barberà, S., Bossert, W., Pattanaik, P.: Ranking sets of objects. In: S. Barberà, P. Hammond, C. Seidl (eds.) Handbook of Utility Theory, vol. 2: Extensions, pp. 893–977. Kluwer Academic Publisher (2004)
89. Barman, S., Biswas, A., Krishnamurthy, S., Narahari, Y.: Groupwise maximin fair allocation of indivisible goods. In: Proceedings of the 32nd AAAI Conference on Artificial Intelligence, pp. 917–924. AAAI Press (2018)
90. Barman, S., Krishna, A., Narahari, Y., Sadhukhan, S.: Achieving envy-freeness with limited subsidies under dichotomous valuations. In: Proceedings of the 31st International Joint Conference on Artificial Intelligence, pp. 60–66. ijcai.org (2022)
91. Barman, S., Krishnamurthy, S.: Approximation algorithms for maximin fair division. In: Proceedings of the 18th ACM Conference on Economics and Computation, pp. 647–664. ACM Press (2017)
92. Barman, S., Krishnamurthy, S.: On the proximity of markets with integral equilibria. In: Proceedings of the 33rd AAAI Conference on Artificial Intelligence, pp. 1748–1755. AAAI Press (2019)

93. Barman, S., Krishnamurthy, S., Vaish, R.: Finding fair and efficient allocations. In: Proceedings of the 19th ACM Conference on Economics and Computation, pp. 557–574. ACM Press (2018)
94. Barman, S., Krishnamurthy, S., Vaish, R.: Greedy algorithms for maximizing Nash social welfare. In: Proceedings of the 17th International Conference on Autonomous Agents and Multiagent Systems, pp. 7–13. IFAAMAS (2018)
95. Barman, S., Verma, P.: Existence and computation of maximin fair allocations under matroid-rank valuations. In: Proceedings of the 20th International Conference on Autonomous Agents and Multiagent Systems, pp. 169–177. IFAAMAS (2021)
96. Barman, S., Verma, P.: Truthful and fair mechanisms for matroid-rank valuations. In: Proceedings of the 36th AAAI Conference on Artificial Intelligence, pp. 4801–4808. AAAI Press (2022)
97. Barrot, N., Gourvès, L., Lang, J., Monnot, J., Ries, B.: Possible winners in approval voting. In: Proceedings of the 3rd International Conference on Algorithmic Decision Theory, pp. 57–70. Springer-Verlag *Lecture Notes in Artificial Intelligence #8176* (2013)
98. Bartholdi III, J., Trick, M.: Stable matching with preferences derived from a psychological model. Operations Research Letters **5**(4), 165–169 (1986)
99. Bartholdi III, J., Orlin, J.: Single transferable vote resists strategic voting. Social Choice and Welfare **8**(4), 341–354 (1991)
100. Bartholdi III, J., Tovey, C., Trick, M.: The computational difficulty of manipulating an election. Social Choice and Welfare **6**(3), 227–241 (1989)
101. Bartholdi III, J., Tovey, C., Trick, M.: Voting schemes for which it can be difficult to tell who won the election. Social Choice and Welfare **6**(2), 157–165 (1989)
102. Bartholdi III, J., Tovey, C., Trick, M.: How hard is it to control an election? Mathematical and Computer Modelling **16**(8/9), 27–40 (1992)
103. Basu, K., Weibull, J.: Strategy subsets closed under rational behavior. Economics Letters **36**, 141–146 (1991)
104. Bateni, M., Charikar, M., Guruswami, V.: MaxMin allocation via degree lower-bounded arborescences. In: Proceedings of the 41st ACM Symposium on Theory of Computing, pp. 543–552. ACM Press (2009)
105. Battigalli, P.: Dynamic consistency and imperfect recall. Games and Economic Behavior **20**(1), 31–50 (1997)
106. Baumeister, D., Boes, L.: Distortion in attribute approval committee elections (extended abstract). In: Proceedings of the 22nd International Conference on Autonomous Agents and Multiagent Systems, pp. 2649–2651. IFAAMAS (2023)
107. Baumeister, D., Boes, L., Laußmann, C., Rey, S.: Bounded approval ballots: Balancing expressiveness and simplicity for multiwinner elections. In: Proceedings of the 22nd International Conference on Autonomous Agents and Multiagent Systems, pp. 1400–1408. IFAAMAS (2023)
108. Baumeister, D., Böhnlein, T., Rey, L., Schaudt, O., Selker, A.: Minisum and minimax committee election rules for general preference types. In: Proceedings of the 22nd European Conference on Artificial Intelligence, pp. 1656–1657. IOS Press (2016)
109. Baumeister, D., Bouveret, S., Lang, J., Nguyen, N., Nguyen, T., Rothe, J., Saffidine, A.: Positional scoring-based allocation of indivisible goods. Journal of Autonomous Agents and Multi-Agent Systems **31**(3), 628–655 (2017)
110. Baumeister, D., Brandt, F., Fischer, F., Hoffmann, J., Rothe, J.: The complexity of computing minimal unidirectional covering sets. Theory of Computing Systems **53**(3), 467–502 (2013)
111. Baumeister, D., Erdélyi, G., Erdélyi, O., Rothe, J.: Control in judgment aggregation. In: Proceedings of the 6th European Starting AI Researcher Symposium, pp. 23–34. IOS Press (2012)

References

112. Baumeister, D., Erdélyi, G., Erdélyi, O., Rothe, J.: Computational aspects of manipulation and control in judgment aggregation. In: Proceedings of the 3rd International Conference on Algorithmic Decision Theory, pp. 71–85. Springer-Verlag *Lecture Notes in Artificial Intelligence #8176* (2013). An extended version appears in the proceedings of the *5th International Workshop on Computational Social Choice*, June 2014.
113. Baumeister, D., Erdélyi, G., Erdélyi, O., Rothe, J.: Complexity of manipulation and bribery in judgment aggregation for uniform premise-based quota rules. Mathematical Social Sciences **76**, 19–30 (2015)
114. Baumeister, D., Erdélyi, G., Erdélyi, O., Rothe, J., Selker, A.: Complexity of control in judgment aggregation for uniform premise-based quota rules. Journal of Computer and System Sciences **112**, 13–33 (2020)
115. Baumeister, D., Erdélyi, G., Hemaspaandra, E., Hemaspaandra, L., Rothe, J.: Computational aspects of approval voting. In: J. Laslier, R. Sanver (eds.) Handbook on Approval Voting, chap. 10, pp. 199–251. Springer (2010)
116. Baumeister, D., Erdélyi, G., Rothe, J.: How hard is it to bribe the judges? A study of the complexity of bribery in judgment aggregation. In: Proceedings of the 2nd International Conference on Algorithmic Decision Theory, pp. 1–15. Springer-Verlag *Lecture Notes in Artificial Intelligence #6992* (2011)
117. Baumeister, D., Faliszewski, P., Lang, J., Rothe, J.: Campaigns for lazy voters: Truncated ballots. In: Proceedings of the 11th International Conference on Autonomous Agents and Multiagent Systems, pp. 577–584. IFAAMAS (2012)
118. Baumeister, D., Hogrebe, T.: On the complexity of predicting election outcomes and estimating their robustness. In: Proceedings of the 18th European Conference on Multi-Agent Systems, pp. 228–244. Springer-Verlag *Lecture Notes in Artificial Intelligence #12802* (2021)
119. Baumeister, D., Neveling, M., Roos, M., Rothe, J., Schend, L., Weishaupt, R., Xia, L.: The possible winner with uncertain weights problem. Journal of Computer and System Sciences **138**, 103 464 (2023)
120. Baumeister, D., Roos, M., Rothe, J.: Computational complexity of two variants of the possible winner problem. In: Proceedings of the 10th International Conference on Autonomous Agents and Multiagent Systems, pp. 853–860. IFAAMAS (2011)
121. Baumeister, D., Roos, M., Rothe, J., Schend, L., Xia, L.: The possible winner problem with uncertain weights. In: Proceedings of the 20th European Conference on Artificial Intelligence, pp. 133–138. IOS Press (2012)
122. Baumeister, D., Rothe, J.: Taking the final step to a full dichotomy of the possible winner problem in pure scoring rules. Information Processing Letters **112**(5), 186–190 (2012)
123. Baumeister, D., Rothe, J., Selker, A.: Strategic behavior in judgment aggregation. In: U. Endriss (ed.) Trends in Computational Social Choice, chap. 8, pp. 145–168. AI Access Foundation (2017)
124. Bei, X., Igarashi, A., Lu, X., Suksompong, W.: The price of connectivity in fair division. In: Proceedings of the 35th AAAI Conference on Artificial Intelligence, pp. 5151–5158. AAAI Press (2021)
125. Bei, X., Li, Z., Liu, J., Liu, S., Lu, X.: Fair division of mixed divisible and indivisible goods. In: Proceedings of the 34th AAAI Conference on Artificial Intelligence, pp. 1814–1821. AAAI Press (2020)
126. Bei, X., Liu, S., Lu, X., Wang, H.: Maximin fairness with mixed divisible and indivisible goods. In: Proceedings of the 35th AAAI Conference on Artificial Intelligence, pp. 5167–5175. AAAI Press (2021)
127. Beigel, R., Hemachandra, L., Wechsung, G.: Probabilistic polynomial time is closed under parity reductions. Information Processing Letters **37**(2), 91–94 (1991)
128. Benabbou, N., Chakraborty, M., Elkind, E., Zick, Y.: Fairness towards groups of agents in the allocation of indivisible items. In: Proceedings of the 28th International Joint Conference on Artificial Intelligence, pp. 95–101. ijcai.org (2019)

129. Benabbou, N., Chakraborty, M., Igarashi, A., Zick, Y.: Finding fair and efficient allocations when valuations don't add up. In: Proceedings of the 13th International Symposium on Algorithmic Game Theory, pp. 32–46. Springer-Verlag *Lecture Notes in Computer Science #12283* (2020)
130. Benade, G., Kazachkov, A., Procaccia, A., Psomas, A.: How to make envy vanish over time. In: Proceedings of the 19th ACM Conference on Economics and Computation, pp. 593–610. ACM Press (2018)
131. Bentert, M., Chen, J., Froese, V., Woeginger, G.: Good things come to those who swap objects on paths. Tech. Rep. arXiv:1905.04219 [cs.DS], ACM Computing Research Repository (CoRR) (2019)
132. Berger, B., Cohen, A., Feldman, M., Fiat, A.: (Almost full) EFX exists for four agents (and beyond). Tech. Rep. arXiv:cs/0608057v2 [cs.GT], ACM Computing Research Repository (CoRR) (2021)
133. Berger, B., Cohen, A., Feldman, M., Fiat, A.: Almost full EFX exists for four agents. In: Proceedings of the 36th AAAI Conference on Artificial Intelligence, pp. 4826–4833. AAAI Press (2022)
134. Berman, P., Fujito, T.: On approximation properties of the independent set problem for low degree graphs. Theory of Computing Systems **32**(2), 115–132 (1999)
135. Betzler, N., Bredereck, R., Chen, J., Niedermeier, R.: Studies in computational aspects of voting — a parameterized complexity perspective. In: H. Bodlaender, R. Downey, F. Fomin, D. Marx (eds.) The Multivariate Algorithmic Revolution and Beyond. Essays Dedicated to Michael R. Fellows on the Occasion of His 60th Birthday, pp. 318–363. Springer-Verlag *Lecture Notes in Computer Science #7370* (2012)
136. Betzler, N., Bredereck, R., Niedermeier, R.: Partial kernelization for rank aggregation: Theory and experiments. In: V. Conitzer, J. Rothe (eds.) Proceedings of the 3rd International Workshop on Computational Social Choice, pp. 31–42. Heinrich-Heine-Universität Düsseldorf, Düsseldorf, Germany (2010)
137. Betzler, N., Dorn, B.: Towards a dichotomy for the possible winner problem in elections based on scoring rules. Journal of Computer and System Sciences **76**(8), 812–836 (2010)
138. Betzler, N., Fellows, M., Guo, J., Niedermeier, R., Rosamond, F.: Fixed-parameter algorithms for Kemeny rankings. Theoretical Computer Science **410**(45), 4554–4570 (2009)
139. Betzler, N., Guo, J., Niedermeier, R.: Parameterized computational complexity of Dodgson and Young elections. Information and Computation **208**(2), 165–177 (2010)
140. Betzler, N., Niedermeier, R., Woeginger, G.: Unweighted coalitional manipulation under the Borda rule is NP-hard. In: Proceedings of the 22nd International Joint Conference on Artificial Intelligence, pp. 55–60. AAAI Press/IJCAI (2011)
141. Betzler, N., Slinko, A., Uhlmann, J.: On the computation of fully proportional representation. Journal of Artificial Intelligence Research **47**, 475–519 (2013)
142. Betzler, N., Uhlmann, J.: Parameterized complexity of candidate control in elections and related digraph problems. Theoretical Computer Science **410**(52), 5425–5442 (2009)
143. Beviá, C.: Fair allocation in a general model with indivisible goods. Review of Economic Design **3**(3), 195–213 (1998)
144. Bezáková, I., Dani, V.: Allocating indivisible goods. SIGecom Exchanges **5**(3), 11–18 (2005)
145. Bhaskar, U., Sricharan, A., Vaish, R.: On approximate envy-freeness for indivisible chores and mixed resources. In: Proceedings of Approximation, Randomization and Combinatorial Optimization. Algorithms and Techniques (APPROX/RANDOM 2021), *Leibniz International Proceedings in Informatics (LIPIcs)*, vol. 207, pp. 1:1–1:23. Schloss Dagstuhl – Leibniz-Zentrum für Informatik (2021)

146. Bilò, V., Caragiannis, I., Flammini, M., Igarashi, A., Monaco, G., Peters, D., Vinci, C., Zwicker, W.: Almost envy-free allocations with connected bundles. Games and Economic Behavior **131**, 197–221 (2022)
147. Bilò, V., Fanelli, A., Flammini, M., Monaco, G., Moscardelli, L.: Nash stable outcomes in fractional hedonic games: Existence, efficiency and computation. Journal of Artificial Intelligence Research **62**, 315–371 (2018)
148. Bilò, V., Monaco, G., Moscardelli, L.: Hedonic games with fixed-size coalitions. In: Proceedings of the 36th AAAI Conference on Artificial Intelligence, pp. 9287–9295. AAAI Press (2022)
149. Binkele-Raible, D., Erdélyi, G., Fernau, H., Goldsmith, J., Mattei, N., Rothe, J.: The complexity of probabilistic lobbying. Discrete Optimization **11**(1), 1–21 (2014)
150. Binmore, K.: Playing for Real: A Text on Game Theory. Oxford University Press (2007)
151. Birkhoff, G.: Three observations on linear algebra. Universidad Nacional de Tucuma, Revista A **5**, 147–151 (1946)
152. Biró, P., Gudmundsson, J.: Complexity of finding Pareto-efficient allocations of highest welfare. European Journal of Operational Research **291**(2), 614–628 (2021)
153. Biswas, A., Barman, S.: Fair division under cardinality constraints. In: Proceedings of the 27th International Joint Conference on Artificial Intelligence, pp. 91–97. ijcai.org (2018)
154. Biswas, A., Barman, S.: Matroid constrained fair allocation problem. In: Proceedings of the 33rd AAAI Conference on Artificial Intelligence, pp. 9921–9922. AAAI Press (2019)
155. Black, D.: On the rationale of group decision-making. Journal of Political Economy **56**(1), 23–34 (1948)
156. Black, D.: The Theory of Committees and Elections. Cambridge University Press (1958)
157. Bliem, B., Bredereck, R., Niedermeier, R.: Complexity of efficient and envy-free resource allocation: Few agents. In: Proceedings of the 25th International Joint Conference on Artificial Intelligence, pp. 102–108. AAAI Press/IJCAI (2016)
158. Bodlaender, H., Thilikos, D., Yamazaki, K.: It is hard to know when greedy is good for finding independent sets. Information Processing Letters **61**, 101–106 (1997)
159. Boehmer, N., Bredereck, R., Faliszewski, P., Niedermeier, R.: Winner robustness via swap- and shift-bribery: Parameterized counting complexity and experiments. In: Proceedings of the 30th International Joint Conference on Artificial Intelligence, pp. 66–72. ijcai.org (2021)
160. Boehmer, N., Bredereck, R., Heeger, K., Knop, D., Luo, J.: Multivariate algorithmics for eliminating envy by donating goods. In: Proceedings of the 21st International Conference on Autonomous Agents and Multiagent Systems, pp. 127–135. IFAAMAS (2022)
161. Boehmer, N., Bullinger, M., Kerkmann, A.: Causes of stability in dynamic coalition formation. In: Proceedings of the 37th AAAI Conference on Artificial Intelligence, pp. 5499–5506. AAAI Press (2023)
162. Bogomolnaia, A., Jackson, M.: The stability of hedonic coalition structures. Games and Economic Behavior **38**(2), 201–230 (2002)
163. Bogomolnaia, A., Moulin, H.: Random matching under dichotomous preferences. Econometrica **72**(1), 257–279 (2004)
164. Bogomolnaia, A., Moulin, H., Sandomirskiy, F.: On the fair division of a random object. Tech. Rep. arXiv:1903.10361v4 [cs GT], ACM Computing Research Repository (CoRR) (2021)
165. Bogomolnaia, A., Moulin, H., Sandomirskiy, F., Yanovskaya, E.: Competitive division of a mixed manna. Econometrica **85**(6), 1847–1871 (2017)
166. Bogomolnaia, A., Moulin, H., Sandomirskiy, F., Yanovskaya, E.: Dividing bads under additive utilities. Social Choice and Welfare **52**(3), 395–417 (2019)

167. Booth, K.: Isomorphism testing for graphs, semigroups, and finite automata are polynomially equivalent problems. SIAM Journal on Computing **7**(3), 273–279 (1978)
168. Booth, K., Lueker, G.: Testing for the consecutive ones property, interval graphs, and graph planarity using PQ-tree algorithms. Journal of Computer and System Sciences **13**(3), 335–379 (1976)
169. Borda, J.: Mémoire sur les élections au scrutin. Histoire de L'Académie Royale des Sciences, Paris (1781). English translation appears in [518].
170. Borel, É.: La théorie du jeu et les équations intégrales à noyau symétrique gauche. In: Comptes rendus hebdomadaires des séances de l'Académie des sciences, vol. 173. 1304–1308 (1921)
171. Borel, É.: Traité du calcul des probabilités et ses applications. In: Applications aux jeux des Hazard, vol. IV, fascicule 2. Gautier-Villars, Paris (1938)
172. Borodin, A., El-Yaniv, R.: Online Computation and Competitive Analysis. Cambridge University Press (1998)
173. Botan, S.: Manipulability of Thiele methods on party-list profiles. In: Proceedings of the 20th International Conference on Autonomous Agents and Multiagent Systems, pp. 223–231. IFAAMAS (2021)
174. Boutilier, C., Rosenschein, J.: Incomplete information and communication in voting. In: F. Brandt, V. Conitzer, U. Endriss, J. Lang, A. Procaccia (eds.) Handbook of Computational Social Choice, chap. 10, pp. 223–257. Cambridge University Press (2016)
175. Bouton, C.: Nim, a game with a complete mathematical theory. Annals of Mathematics **3**(1–4), 35–39 (1901–1902)
176. Bouveret, S.: Fair allocation of indivisible items: Modeling, computational complexity and algorithmics. Ph.D. thesis, Institut Supérieur de l'Aéronautique et de l'Espace, Toulouse, France (2007)
177. Bouveret, S., Cechlárová, K., Elkind, E., Igarashi, A., Peters, D.: Fair division of a graph. In: Proceedings of the 26th International Joint Conference on Artificial Intelligence, pp. 135–141. ijcai.org (2017)
178. Bouveret, S., Cechlárová, K., Lesca, J.: Chore division on a graph. Journal of Autonomous Agents and Multi-Agent Systems **33**(5), 540–563 (2019)
179. Bouveret, S., Chevaleyre, Y., Maudet, N.: Fair allocation of indivisible goods. In: F. Brandt, V. Conitzer, U. Endriss, J. Lang, A. Procaccia (eds.) Handbook of Computational Social Choice, chap. 12, pp. 284–310. Cambridge University Press (2016)
180. Bouveret, S., Endriss, U., Lang, J.: Conditional importance networks: A graphical language for representing ordinal, monotonic preferences over sets of goods. In: Proceedings of the 21st International Joint Conference on Artificial Intelligence, pp. 67–72. IJCAI (2009)
181. Bouveret, S., Endriss, U., Lang, J.: Fair division under ordinal preferences: Computing envy-free allocations of indivisible goods. In: Proceedings of the 19th European Conference on Artificial Intelligence, pp. 387–392. IOS Press (2010)
182. Bouveret, S., Lang, J.: Efficiency and envy-freeness in fair division of indivisible goods: Logical representation and complexity. Journal of Artificial Intelligence Research **32**, 525–564 (2008)
183. Bouveret, S., Lang, J.: A general elicitation-free protocol for allocating indivisible goods. In: Proceedings of the 22nd International Joint Conference on Artificial Intelligence, pp. 73–78. AAAI Press/IJCAI (2011)
184. Bouveret, S., Lang, J.: Manipulating picking sequences. In: Proceedings of the 21st European Conference on Artificial Intelligence, *Frontiers in Artificial Intelligence and Applications*, vol. 263, pp. 141–146. IOS Press (2014)
185. Bouveret, S., Lemaître, M.: Computing leximin-optimal solutions in constraint networks. Artificial Intelligence **173**(2), 343–364 (2009)

186. Bouveret, S., Lemaître, M.: Characterizing conflicts in fair division of indivisible goods using a scale of criteria. Journal of Autonomous Agents and Multi-Agent Systems **30**(2), 259–290 (2016)
187. Bouveret, S., Lemaître, M., Fargier, H., Lang, J.: Allocation of indivisible goods: A general model and some complexity results (extended abstract). In: Proceedings of the 4th International Joint Conference on Autonomous Agents and Multiagent Systems, pp. 1309–1310. ACM Press (2005)
188. Bovet, D., Crescenzi, P.: Introduction to the Theory of Complexity. Prentice Hall (1993)
189. Brams, S., Edelman, P., Fishburn, P.: Paradoxes of fair division. Journal of Philosophy **98**(6), 300–314 (2001)
190. Brams, S., Edelman, P., Fishburn, P.: Fair division of indivisible items. Theory and Decision **55**(2), 147–180 (2003)
191. Brams, S., Fishburn, P.: Approval voting. American Political Science Review **72**(3), 831–847 (1978)
192. Brams, S., Fishburn, P.: Paradoxes of preferential voting. Mathematics Magazine **56**(4), 207–216 (1983)
193. Brams, S., Fishburn, P.: Fair division of indivisible items between two people with identical preferences: Envy-freeness, Pareto-optimality, and equity. Social Choice and Welfare **17**(2), 247–267 (2002)
194. Brams, S., Jones, M., Klamler, C.: Better ways to cut a cake. Notices of the AMS **53**(11), 1314–1321 (2006)
195. Brams, S., Jones, M., Klamler, C.: Divide-and-Conquer: A proportional, minimal-envy cake-cutting algorithm. SIAM Review **53**(2), 291–307 (2011)
196. Brams, S., Kilgour, M., Klamler, C.: The undercut procedure: An algorithm for the envy-free division of indivisible items. Social Choice and Welfare **39**(2–3), 615–631 (2012)
197. Brams, S., Kilgour, M., Klamler, C.: An algorithm for the proportional division of indivisible items. Tech. Rep. MPRA paper No. 56587, Munich Personal RePEc Archive (2014)
198. Brams, S., Kilgour, M., Klamler, C.: Two-person fair division of indivisible items: An efficient, envy-free algorithm. Notices of the AMS **61**(2), 130–141 (2014)
199. Brams, S., Kilgour, M., Sanver, R.: A minimax procedure for negotiating multilateral treaties. In: M. Wiberg (ed.) Reasoned Choices: Essays in Honor of Hannu Nurmi. Finnish Political Science Association (2004)
200. Brams, S., King, D.: Efficient fair division: Help the worst off or avoid envy? Rationality and Society **17**(4), 387–421 (2005)
201. Brams, S., Sanver, R.: Critical strategies under approval voting: Who gets ruled in and ruled out. Electoral Studies **25**(2), 287–305 (2006)
202. Brams, S., Sanver, R.: Voting systems that combine approval and preference. In: S. Brams, W. Gehrlein, F. Roberts (eds.) The Mathematics of Preference, Choice, and Order: Essays in Honor of Peter C. Fishburn, pp. 215–237. Springer (2009)
203. Brams, S., Taylor, A.: An envy-free cake division protocol. The American Mathematical Monthly **102**(1), 9–18 (1995)
204. Brams, S., Taylor, A.: Fair Division: From Cake-Cutting to Dispute Resolution. Cambridge University Press (1996)
205. Brams, S., Taylor, A.: The Win-Win Solution. Guaranteeing Fair Shares to Everybody. W. W. Norton & Company (2000)
206. Brams, S., Taylor, A., Zwicker, W.: A moving-knife solution to the four person envy-free cake-division problem. Proceedings of the American Mathematical Society **125**(2), 547–554 (1997)
207. Brandes, U., Laußmann, C., Rothe, J.: Voting for centrality (extended abstract). In: Proceedings of the 21st International Conference on Autonomous Agents and Multiagent Systems, pp. 1554–1556. IFAAMAS (2022)

208. Brandt, F.: Some remarks on Dodgson's voting rule. Mathematical Logic Quarterly **55**(4), 460–463 (2009)
209. Brandt, F., Brill, M., Harrenstein, P.: Tournament solutions. In: F. Brandt, V. Conitzer, U. Endriss, J. Lang, A. Procaccia (eds.) Handbook of Computational Social Choice, chap. 3, pp. 57–84. Cambridge University Press (2016)
210. Brandt, F., Brill, M., Hemaspaandra, E., Hemaspaandra, L.: Bypassing combinatorial protections: Polynomial-time algorithms for single-peaked electorates. In: Proceedings of the 24th AAAI Conference on Artificial Intelligence, pp. 715–722. AAAI Press (2010)
211. Brandt, F., Bullinger, M.: Finding and recognizing popular coalition structures. Journal of Artificial Intelligence Research **74**, 569–626 (2022)
212. Brandt, F., Bullinger, M., Tappe, L.: Single-agent dynamics in additively separable hedonic games. In: Proceedings of the 36th AAAI Conference on Artificial Intelligence, pp. 4867–4874. AAAI Press (2022)
213. Brandt, F., Bullinger, M., Wilczynski, A.: Reaching individually stable coalition structures. ACM Transactions on Economics and Computation **11**(1–2), 4:1–4:65 (2023)
214. Brandt, F., Conitzer, V., Endriss, U.: Computational social choice. In: G. Weiß (ed.) Multiagent Systems, second edn., pp. 213–283. MIT Press (2013)
215. Brandt, F., Conitzer, V., Endriss, U., Lang, J., Procaccia, A. (eds.): Handbook of Computational Social Choice. Cambridge University Press (2016)
216. Brandt, F., Faliszewski, P. (eds.): Proceedings of the 4th International Workshop on Computational Social Choice. AGH University of Science and Technology, Kraków, Poland (2012)
217. Brandt, F., Fischer, F.: Computing the minimal covering set. Mathematical Social Sciences **56**(2), 254–268 (2008)
218. Brandt, F., Fischer, F., Holzer, M.: Symmetries and the complexity of pure Nash equilibrium. Journal of Computer and System Sciences **75**(3), 163–177 (2009)
219. Brandt, F., Fischer, F., Holzer, M.: Equilibria of graphical games with symmetries. Theoretical Computer Science **412**(8–10), 675–685 (2011)
220. Brandt, F., Geist, C.: Finding strategyproof social choice functions via SAT solving. In: Proceedings of the 13th International Conference on Autonomous Agents and Multiagent Systems, pp. 1193–1200. IFAAMAS (2014)
221. Brânzei, S., Gkatzelis, V., Mehta, R.: Nash social welfare approximation for strategic agents. In: Proceedings of the 18th ACM Conference on Economics and Computation, pp. 611–628. ACM Press (2017)
222. Brânzei, S., Lv, Y., Mehta, R.: To give or not to give: Fair division for single minded valuations. In: Proceedings of the 25th International Joint Conference on Artificial Intelligence, pp. 123–129. AAAI Press/IJCAI (2016)
223. Brânzei, S., Sandomirskiy, F.: Algorithms for competitive division of chores. Tech. Rep. arXiv:1907.01766 [cs.GT], ACM Computing Research Repository (CoRR) (2019)
224. Bredereck, R., Chen, J., Faliszewski, P., Guo, J., Niedermeier, R., Woeginger, G.: Parameterized algorithmics for computational social choice: Nine research challenges. Tsinghua Science and Technology **19**(4), 358–373 (2014)
225. Bredereck, R., Chen, J., Faliszewski, P., Nichterlein, A., Niedermeier, R.: Prices matter for the parameterized complexity of shift bribery. In: Proceedings of the 28th AAAI Conference on Artificial Intelligence, pp. 1398–1404. AAAI Press (2014)
226. Bredereck, R., Chen, J., Hartung, S., Kratsch, S., Niedermeier, R., Suchý, O.: A multivariate complexity analysis of lobbying in multiple referenda. In: Proceedings of the 26th AAAI Conference on Artificial Intelligence, pp. 1292–1298. AAAI Press (2012)
227. Bredereck, R., Chen, J., Woeginger, G.: Are there any nicely structured preference profiles nearby? In: Proceedings of the 23rd International Joint Conference on Artificial Intelligence, pp. 62–68. AAAI Press/IJCAI (2013)

228. Bredereck, R., Faliszewski, P., Furdyna, M., Kaczmarczyk, A., Lackner, M.: Strategic campaign management in apportionment elections. In: Proceedings of the 29th International Joint Conference on Artificial Intelligence, pp. 103–109. ijcai.org (2020)
229. Bredereck, R., Faliszewski, P., Kaczmarczyk, A., Niedermeier, R., Skowron, P., Talmon, N.: Robustness among multiwinner voting rules. Artificial Intelligence **290**, 103 403 (2021)
230. Bredereck, R., Faliszewski, P., Niedermeier, R., Talmon, N.: Large-scale election campaigns: Combinatorial shift bribery. In: Proceedings of the 14th International Conference on Autonomous Agents and Multiagent Systems, pp. 67–75. IFAAMAS (2015)
231. Bredereck, R., Faliszewski, P., Niedermeier, R., Talmon, N.: Complexity of shift bribery in committee elections. In: Proceedings of the 30th AAAI Conference on Artificial Intelligence, pp. 2452–2458. AAAI Press (2016)
232. Bredereck, R., Faliszewski, P., Niedermeier, R., Talmon, N.: Complexity of shift bribery in committee elections. ACM Transactions on Computation Theory **13**(3), 20:1–20:25 (2021)
233. Bredereck, R., Figiel, A., Kaczmarczyk, A., Knop, D., Niedermeier, R.: High-multiplicity fair allocation made more practical. In: Proceedings of the 20th International Conference on Autonomous Agents and Multiagent Systems, pp. 260–268. IFAAMAS (2021)
234. Bredereck, R., Kaczmarczyk, A., Knop, D., Niedermeier, R.: High-multiplicity fair allocation: Lenstra empowered by n-fold integer programming. In: Proceedings of the 20th ACM Conference on Economics and Computation, pp. 505–523. ACM Press (2019)
235. Bredereck, R., Kaczmarczyk, A., Luo, J., Niedermeier, R., Sachse, F.: On improving resource allocations by sharing. In: Proceedings of the 36th AAAI Conference on Artificial Intelligence, pp. 4875–4883. AAAI Press (2022)
236. Bredereck, R., Kaczmarczyk, A., Niedermeier, R.: On coalitional manipulation for multiwinner elections: Shortlisting. Journal of Autonomous Agents and Multi-Agent Systems **35**, 38 (2021)
237. Bredereck, R., Kaczmarczyk, A., Niedermeier, R.: Envy-free allocations respecting social networks. Artificial Intelligence **305**, 103 664 (2022)
238. Brelsford, E., Faliszewski, P., Hemaspaandra, E., Schnoor, H., Schnoor, I.: Approximability of manipulating elections. In: Proceedings of the 23rd AAAI Conference on Artificial Intelligence, pp. 44–49. AAAI Press (2008)
239. Brueggemann, T., Kern, W.: An improved deterministic local search algorithm for 3-SAT. Theoretical Computer Science **329**(1-3), 303–313 (2004)
240. Brustle, J., Dippel, J., Narayan, V., Suzuki, M., Vetta, A.: One dollar each eliminates envy. In: Proceedings of the 21st ACM Conference on Economics and Computation, pp. 23–39. ACM Press (2020)
241. Budish, E.: The combinatorial assignment problem: Approximate competitive equilibrium from equal incomes. Journal of Political Economy **119**(6), 1061–1103 (2011)
242. Bullinger, M.: Pareto-optimality in cardinal hedonic games. In: Proceedings of the 19th International Conference on Autonomous Agents and Multiagent Systems, pp. 213–221. IFAAMAS (2020)
243. Bullinger, M.: Boundaries to single-agent stability in additively separable hedonic games. In: Proceedings of the 47th International Symposium on Mathematical Foundations of Computer Science, *LIPIcs*, vol. 241, pp. 26:1–26:15. Schloss Dagstuhl – Leibniz-Zentrum für Informatik (2022)
244. Bullinger, M., Kober, S.: Loyalty in cardinal hedonic games. In: Proceedings of the 30th International Joint Conference on Artificial Intelligence, pp. 66–72. ijcai.org (2021)

245. Bullinger, M., Romen, R.: Online coalition formation under random arrival or coalition dissolution. In: Proceedings of the 31st Annual European Symposium on Algorithms, *LIPIcs*, vol. 274, pp. 27:1–27:18. Schloss Dagstuhl – Leibniz-Zentrum für Informatik (2023)
246. Bullinger, M., Suksompong, W.: Topological distance games. In: Proceedings of the 37th AAAI Conference on Artificial Intelligence, pp. 5549–5556. AAAI Press (2023)
247. Byrka, J., Skowron, P., Sornat, K.: Proportional approval voting, harmonic k-median, and negative association. In: Proceedings of the 45th International Colloquium on Automata, Languages, and Programming, *Leibniz International Proceedings in Informatics (LIPIcs)*, vol. 107, pp. 26:1–26:14. Schloss Dagstuhl – Leibniz-Zentrum für Informatik (2018)
248. Cai, J., Gundermann, T., Hartmanis, J., Hemachandra, L., Sewelson, V., Wagner, K., Wechsung, G.: The boolean hierarchy I: Structural properties. SIAM Journal on Computing **17**(6), 1232–1252 (1988)
249. Cai, J., Gundermann, T., Hartmanis, J., Hemachandra, L., Sewelson, V., Wagner, K., Wechsung, G.: The boolean hierarchy II: Applications. SIAM Journal on Computing **18**(1), 95–111 (1989)
250. Caragiannis, I., Gravin, N., Huang, X.: Envy-freeness up to any item with high Nash welfare: The virtue of donating items. In: Proceedings of the 20th ACM Conference on Economics and Computation, pp. 527–545. ACM Press (2019)
251. Caragiannis, I., Hemaspaandra, E., Hemaspaandra, L.: Dodgson's rule and Young's rule. In: F. Brandt, V. Conitzer, U. Endriss, J. Lang, A. Procaccia (eds.) Handbook of Computational Social Choice, chap. 5. Cambridge University Press (2016). 103–126
252. Caragiannis, I., Ioannidis, S.: Computing envy-freeable allocations with limited subsidies. Tech. Rep. arXiv:2002.02789 [cs.GT], ACM Computing Research Repository (CoRR) (2020)
253. Caragiannis, I., Kaklamanis, C., Kanellopoulos, P., Kyropoulou, M.: The efficiency of fair division. Theory of Computing Systems **50**(4), 589–610 (2012)
254. Caragiannis, I., Kanellopoulos, P., Kyropoulou, M.: On interim envy-free allocation lotteries. In: Proceedings of the 22nd ACM Conference on Economics and Computation, pp. 264–284. ACM Press (2021)
255. Caragiannis, I., Kurokawa, D., Moulin, H., Procaccia, A., Shah, N., Wang, J.: The unreasonable fairness of maximum Nash welfare. ACM Transactions on Economics and Computation **7**(3), 12:1–12:32 (2019)
256. Caragiannis, I., Micha, E., Shah, N.: A little charity guarantees fair connected graph partitioning. In: Proceedings of the 36th AAAI Conference on Artificial Intelligence, pp. 4908–4916. AAAI Press (2022)
257. Carleton, B., Chavrimootoo, M., Hemaspaandra, L., Narváez, D.: Search versus search for collapsing electoral control types (extended abstract). In: Proceedings of the 22nd International Conference on Autonomous Agents and Multiagent Systems, pp. 2682–2684. IFAAMAS (2023)
258. Carleton, B., Chavrimootoo, M., Hemaspaandra, L., Narváez, D., Taliancich, C., Welles, H.: Separating and collapsing electoral control types. In: Proceedings of the 22nd International Conference on Autonomous Agents and Multiagent Systems, pp. 1743–1751. IFAAMAS (2023)
259. Cary, D.: Estimating the margin of victory for instant-runoff voting. In: Website Proceedings of the Electronic Voting Technology Workshop/Workshop on Trustworthy Elections (2011)
260. Caskurlu, B., Kizilkaya, F., Ozen, B.: Hedonic expertise games. Annals of Mathematics and Artificial Intelligence **92**(3) (2024). URL https://doi.org/10.1007/s10472-023-09900-y. To appear.
261. Cechlárová, K., Hajduková, J.: Computational complexity of stable partitions with B-preferences. International Journal of Game Theory **31**(3), 353–364 (2003)

262. Cechlárová, K., Hajduková, J.: Stable partitions with $\mathcal{W}$-preferences. Discrete Applied Mathematics **138**(3), 333–347 (2004)
263. Cechlárová, K., Romero-Medina, A.: Stability in coalition formation games. International Journal of Game Theory **29**(4), 487–494 (2001)
264. Chakrabarty, D., Chuzhoy, J., Khanna, S.: On allocating goods to maximize fairness. In: Proceedings of the 50th IEEE Symposium on Foundations of Computer Science, pp. 107–116. IEEE Computer Society Press (2009)
265. Chakraborty, M., Igarashi, A., Suksompong, W., Zick, Y.: Weighted envy-freeness in indivisible item allocation. In: Proceedings of the 19th International Conference on Autonomous Agents and Multiagent Systems, pp. 231–239. IFAAMAS (2020)
266. Chakraborty, M., Schmidt-Kraepelin, U., Suksompong, W.: Picking sequences and monotonicity in weighted fair division. In: Proceedings of the 30th International Joint Conference on Artificial Intelligence, pp. 82–88. ijcai.org (2021)
267. Chakraborty, M., Segal-Halevi, E., Suksompong, W.: Weighted fairness notions for indivisible items revisited. In: Proceedings of the 36th AAAI Conference on Artificial Intelligence, pp. 4949–4956. AAAI Press (2022)
268. Chalkiadakis, G., Elkind, E., Wooldridge, M.: Computational Aspects of Cooperative Game Theory. Synthesis Lectures on Artificial Intelligence and Machine Learning. Morgan and Claypool Publishers (2011)
269. Chalkiadakis, G., Wooldridge, M.: Weighted voting games. In: F. Brandt, V. Conitzer, U. Endriss, J. Lang, A. Procaccia (eds.) Handbook of Computational Social Choice, chap. 16, pp. 377–395. Cambridge University Press (2016)
270. Chamberlin, J., Courant, P.: Representative deliberations and representative decisions: Proportional representation and the Borda rule. The American Political Science Review **77**(3), 718–733 (1983)
271. Chan, H., Chen, J., Li, B., Wu, X.: Maximin-aware allocations of indivisible goods. In: Proceedings of the 18th International Conference on Autonomous Agents and Multiagent Systems, pp. 1871–1873. IFAAMAS (2019)
272. Chan, T., Tang, Z., Wu, X.: On $(1,\epsilon)$-restricted max-min fair allocation problem. Algorithmica **80**(7), 2181–2200 (2018)
273. Chandra, A., Kozen, D., Stockmeyer, L.: Alternation. Journal of the ACM **26**(1), 114–133 (1981)
274. Chaudhury, B., Cheung, Y., Garg, J., Garg, N., Hoefer, M., Mehlhorn, K.: Fair division of indivisible goods for a class of concave valuations. Journal of Artificial Intelligence Research **74**, 111–142 (2022)
275. Chaudhury, B., Garg, J., McGlaughlin, P., Mehta, R.: Competitive allocation of a mixed manna. In: Proceedings of the 32nd Annual ACM-SIAM Symposium on Discrete Algorithms, pp. 1405–1424. Society for Industrial and Applied Mathematics (2021)
276. Chaudhury, B., Garg, J., Mehlhorn, K.: EFX exists for three agents. In: Proceedings of the 21st ACM Conference on Economics and Computation, pp. 1–19. ACM Press (2020)
277. Chaudhury, B., Garg, J., Mehlhorn, K., Mehta, R., Misra, P.: Improving EFX guarantees through rainbow cycle number. In: Proceedings of the 22nd ACM Conference on Economics and Computation, pp. 310–311. ACM Press (2021)
278. Chaudhury, B., Garg, J., Mehta, R.: Fair and efficient allocations under subadditive valuations. In: Proceedings of the 35th AAAI Conference on Artificial Intelligence, pp. 5269–5276. AAAI Press (2021)
279. Chaudhury, B., Kavitha, T., Mehlhorn, K., Sgouritsa, A.: A little charity guarantees almost envy-freeness. SIAM Journal on Computing **50**(4), 1336–1358 (2021)
280. Chekuri, C., Vondrák, J., Zenklusen, R.: Dependent randomized rounding via exchange properties of combinatorial structures. In: Proceedings of the 51st IEEE Symposium on Foundations of Computer Science, pp. 575–584. IEEE Computer Society Press (2010)

281. Chen, J., Csáji, G., Roy, S., Simola, S.: Hedonic games with friends, enemies, and neutrals: Resolving open questions and fine-grained complexity. In: Proceedings of the 22nd International Conference on Autonomous Agents and Multiagent Systems, pp. 251–259. IFAAMAS (2023)
282. Chen, J., Faliszewski, P., Niedermeier, R., Talmon, N.: Elections with few voters: Candidate control can be easy. In: Proceedings of the 29th AAAI Conference on Artificial Intelligence, pp. 2045–2051. AAAI Press (2015)
283. Chen, X., Deng, X.: Settling the complexity of two-player Nash equilibrium. In: Proceedings of the 47th IEEE Symposium on Foundations of Computer Science, pp. 261–272. IEEE Computer Society Press (2006)
284. Chevaleyre, Y., Dunne, P., Endriss, U., Lang, J., Lemaître, M., Maudet, N., Padget, J., Phelps, S., Rodríguez-Aguilar, J., Sousa, P.: Issues in multiagent resource allocation. Informatica **30**(1), 3–31 (2006)
285. Chevaleyre, Y., Endriss, U., Estivie, S., Maudet, N.: Reaching envy-free states in distributed negotiation settings. In: Proceedings of the 20th International Joint Conference on Artificial Intelligence, pp. 1239–1244. IJCAI (2007)
286. Chevaleyre, Y., Endriss, U., Estivie, S., Maudet, N.: Multiagent resource allocation in k-additive domains: Preference representation and complexity. Annals of Operations Research **163**(1), 49–62 (2008)
287. Chevaleyre, Y., Endriss, U., Lang, J.: Expressive power of weighted propositional formulas for cardinal preference modeling. In: Proceedings of the 7th International Conference on Principles of Knowledge Representation and Reasoning, pp. 145–152. AAAI Press (2006)
288. Chevaleyre, Y., Endriss, U., Lang, J., Maudet, N.: A short introduction to computational social choice. In: Proceedings of the 33rd International Conference on Current Trends in Theory and Practice of Computer Science, pp. 51–69. Springer-Verlag *Lecture Notes in Computer Science #4362* (2007)
289. Chevaleyre, Y., Endriss, U., Maudet, N.: Allocating goods on a graph to eliminate envy. In: Proceedings of the 22nd AAAI Conference on Artificial Intelligence, pp. 700–705. AAAI Press (2007)
290. Chevaleyre, Y., Endriss, U., Maudet, N.: Simple negotiation schemes for agents with simple preferences: Sufficiency, necessity and maximality. Journal of Autonomous Agents and Multi-Agent Systems **20**(2), 234–259 (2010)
291. Chevaleyre, Y., Endriss, U., Maudet, N.: Distributed fair allocation of indivisible goods. Artificial Intelligence **242**, 1–22 (2017)
292. Chevaleyre, Y., Lang, J., Maudet, N., Monnot, J.: Possible winners when new candidates are added: The case of scoring rules. In: Proceedings of the 24th AAAI Conference on Artificial Intelligence, pp. 762–767. AAAI Press (2010)
293. Chevaleyre, Y., Lang, J., Maudet, N., Monnot, J., Xia, L.: New candidates welcome! Possible winners with respect to the addition of new candidates. Mathematical Social Sciences **64**(1), 74–88 (2012)
294. Christian, R., Fellows, M., Rosamond, F., Slinko, A.: On complexity of lobbying in multiple referenda. Review of Economic Design **11**(3), 217–224 (2007)
295. Church, A.: An unsolvable problem of elementary number theory. American Journal of Mathematics **58**, 345–363 (1936)
296. Cole, R., Devanur, N., Gkatzelis, V., Jain, K., Mai, T., Vazirani, V., Yazdanbod, S.: Convex program duality, Fisher markets, and Nash social welfare. In: Proceedings of the 18th ACM Conference on Economics and Computation, pp. 459–460. ACM Press (2017)
297. Cole, R., Gkatzelis, V.: Approximating the Nash social welfare with indivisible items. In: Proceedings of the 47th ACM Symposium on Theory of Computing, pp. 371–380. ACM Press (2015)
298. Cole, R., Gkatzelis, V.: Approximating the Nash social welfare with indivisible items. SIAM Journal on Computing **47**(3), 1211–1236 (2018)

References

299. Cole, R., Tao, Y.: On the existence of Pareto efficient and envy-free allocations. Journal of Economic Theory **193**, 105,207 (2021)
300. Condorcet, J.: Essai sur l'application de l'analyse à la probabilité des décisions rendues à la pluralité des voix (1785). Facsimile reprint of original published in Paris, 1972, by the Imprimerie Royale. English translation appears in [697]: I. McLean and A. Urken, *Classics of Social Choice*, University of Michigan Press, 1995, pages 91–112.
301. Conitzer, V.: Making decisions based on the preferences of multiple agents. Communications of the ACM **53**(3), 84–94 (2010)
302. Conitzer, V., Davenport, A., Kalagnanam, J.: Improved bounds for computing Kemeny rankings. In: Proceedings of the 21st National Conference on Artificial Intelligence, pp. 620–626. AAAI Press (2006)
303. Conitzer, V., Freeman, R., Shah, N.: Fair public decision making. In: Proceedings of the 18th ACM Conference on Economics and Computation, pp. 629–646. ACM Press (2017)
304. Conitzer, V., Freeman, R., Shah, N., Wortman Vaughan, J.: Group fairness for the allocation of indivisible goods. In: Proceedings of the 33rd AAAI Conference on Artificial Intelligence, pp. 1853–1860 (2019)
305. Conitzer, V., Rognlie, M., Xia, L.: Preference functions that score rankings and maximum likelihood estimation. In: Proceedings of the 21st International Joint Conference on Artificial Intelligence, pp. 109–115. IJCAI (2009)
306. Conitzer, V., Rothe, J. (eds.): Proceedings of the 3rd International Workshop on Computational Social Choice. Heinrich-Heine-Heinrich-Heine-Universität Düsseldorf, Düsseldorf, Germany (2010)
307. Conitzer, V., Sandholm, T.: Complexity results about Nash equilibria. In: Proceedings of the 18th International Joint Conference on Artificial Intelligence, pp. 765–771. Morgan Kaufmann (2003)
308. Conitzer, V., Sandholm, T.: Nonexistence of voting rules that are usually hard to manipulate. In: Proceedings of the 21st National Conference on Artificial Intelligence, pp. 627–634. AAAI Press (2006)
309. Conitzer, V., Sandholm, T.: A technique for reducing normal-form games to compute a Nash equilibrium. In: Proceedings of the 5th International Joint Conference on Autonomous Agents and Multiagent Systems, pp. 537–544. ACM Press (2006)
310. Conitzer, V., Sandholm, T., Lang, J.: When are elections with few candidates hard to manipulate? Journal of the ACM **54**(3), Article 14 (2007)
311. Conitzer, V., Sandholm, T., Santi, P.: Combinatorial auctions with k-wise dependent valuations. In: Proceedings of the 20th National Conference on Artificial Intelligence, pp. 248–254. AAAI Press (2005)
312. Conitzer, V., Walsh, T.: Barriers to manipulation in voting. In: F. Brandt, V. Conitzer, U. Endriss, J. Lang, A. Procaccia (eds.) Handbook of Computational Social Choice, chap. 6, pp. 127–145. Cambridge University Press (2016)
313. Cook, S.: The complexity of theorem-proving procedures. In: Proceedings of the 3rd ACM Symposium on Theory of Computing, pp. 151–158. ACM Press (1971)
314. Copeland, A.: A "reasonable" social welfare function. Mimeographed notes from a Seminar on Applications of Mathematics to the Social Sciences, University of Michigan (1951)
315. Cormen, T., Leiserson, C., Rivest, R., Stein, C.: Introduction to Algorithms, second edn. MIT Press and McGraw-Hill (2001)
316. Cramton, P., Shoham, Y., Steinberg, R. (eds.): Combinatorial Auctions. MIT Press (2006)
317. Csirik, J., Kellerer, H., Woeginger, G.: The exact LPT-bound for maximizing the minimum completion time. Operations Research Letters **11**(5), 281–287 (1992)
318. Culberson, J.: Sokoban is PSPACE-complete. Tech. Rep. TR 97-02, The University of Alberta (1997)

319. Custer, C.: Cake-cutting Hugo Steinhaus style: Beyond $N = 3$. Department of Mathematics, Union College, Schenectady, NY, USA (1994). Senior Thesis.
320. Cygan, M., Fomin, F., Kowalik, L., Lokshtanov, D., Marx, D., Pilipczuk, M., Pilipczuk, M., Saurabh, S.: Parameterized Algorithms. Springer (2015)
321. Cygan, M., Kowalik, L., Socala, A., Sornat, K.: Approximation and parameterized complexity of minimax approval voting. Journal of Artificial Intelligence Research **63**, 495–513 (2018)
322. Dall'Aglio, M., Hill, T.: Maximin share and minimax envy in fair-division problems. Journal of Mathematical Analysis and Applications **281**(1), 346–361 (2003)
323. Dall'Aglio, M., Mosca, R.: How to allocate hard candies fairly. Mathematical Social Sciences **54**(3), 218–237 (2007)
324. Dantsin, E., Goerdt, A., Hirsch, E., Kannan, R., Kleinberg, J., Papadimitriou, C., Raghavan, P., Schöning, U.: A deterministic $(2 - 2/(k+1))^n$ algorithm for k-SAT based on local search. Theoretical Computer Science **289**(1), 69–83 (2002)
325. Dantsin, E., Wolpert, A.: Derandomization of Schuler's algorithm for SAT. In: Proceedings of the 7th International Conference on Theory and Applications of Satisfiability Testing, pp. 80–88. Springer-Verlag *Lecture Notes in Computer Science #3542* (2004)
326. Dantzig, G., Thapa, M.: Linear Programming 1: Introduction. Springer-Verlag (1997)
327. Dantzig, G., Thapa, M.: Linear Programming 2: Theory and Extensions. Springer-Verlag (2003)
328. Darmann, A., Elkind, E., Kurz, S., Lang, J., Schauer, J., Woeginger, G.: Group activity selection problem with approval preferences. International Journal of Game Theory **47**(3), 767–796 (2018)
329. Darmann, A., Schauer, J.: Maximizing Nash product social welfare in allocating indivisible goods. European Journal of Operational Research **247**(2), 548–559 (2015)
330. Daskalakis, C., Goldberg, P., Papadimitriou, C.: The complexity of computing a Nash equilibrium. In: Proceedings of the 38th ACM Symposium on Theory of Computing, pp. 71–78. ACM Press (2006)
331. Daskalakis, C., Goldberg, P., Papadimitriou, C.: The complexity of computing a Nash equilibrium. Communications of the ACM **52**(2), 89–97 (2009)
332. Daskalakis, C., Goldberg, P., Papadimitriou, C.: The complexity of computing a Nash equilibrium. SIAM Journal on Computing **39**(1), 195–259 (2009)
333. Daskalakis, C., Papadimitriou, C.: Continuous local search. In: Proceedings of the Twenty-Second Annual ACM-SIAM Symposium on Discrete Algorithms, pp. 790–804. SIAM (2011)
334. Davies, J., Katsirelos, G., Narodytska, N., Walsh, T.: Complexity of and algorithms for Borda manipulation. In: Proceedings of the 25th AAAI Conference on Artificial Intelligence, pp. 657–662. AAAI Press (2011)
335. Dawson, C.: An algorithmic version of Kuhn's lone-divider method of fair division. Missouri Journal of Mathematical Sciences **13**(3), 172–177 (2001)
336. Deligkas, A., Eiben, E., Ganian, R., Hamm, T., Ordyniak, S.: The parameterized complexity of connected fair division. In: Proceedings of the 30th International Joint Conference on Artificial Intelligence, pp. 139–145. ijcai.org (2021)
337. Demko, S., Hill, T.: Equitable distribution of indivisible objects. Mathematical Social Sciences **16**(2), 45–158 (1988)
338. Deng, X., Papadimitriou, C.: On the complexity of cooperative solution concepts. Mathematics of Operations Research **19**(2), 257–266 (1994)
339. Desmedt, Y., Elkind, E.: Equilibria of plurality voting with abstentions. In: Proceedings of the 11th ACM Conference on Electronic Commerce, pp. 347–356. ACM Press (2010)

340. Deuermeyer, B., Friesen, D., Langston, M.: Scheduling to maximize the minimum processor finish time in a multiprocessor system. SIAM Journal on Algebraic and Discrete Methods **3**(2), 452–454 (1982)
341. Dickerson, J., Goldman, J., Karp, J., Procaccia, A., Sandholm, T.: The computational rise and fall of fairness. In: Proceedings of the 28th AAAI Conference on Artificial Intelligence, pp. 1405–1411. AAAI Press (2014)
342. Diestel, R.: Graphentheorie. Springer-Verlag (2006)
343. Dietrich, F.: A generalised model of judgment aggregation. Social Choice and Welfare **28**(4), 529–565 (2007)
344. Dietrich, F., List, C.: Arrow's theorem in judgment aggregation. Social Choice and Welfare **29**(1), 19–33 (2007)
345. Dietrich, F., List, C.: Judgment aggregation by quota rules: Majority voting generalized. Journal of Theoretical Politics **19**(4), 391–424 (2007)
346. Dietrich, F., List, C.: Strategy-proof judgment aggregation. Economics and Philosophy **23**(3), 269–300 (2007)
347. Dimitrov, D., Borm, P., Hendrickx, R., Sung, S.: Simple priorities and core stability in hedonic games. Social Choice and Welfare **26**(2), 421–433 (2006)
348. Dobzinski, S., Procaccia, A.: Frequent manipulability of elections: The case of two voters. In: Proceedings of the 4th International Workshop on Internet & Network Economics, pp. 653–664. Springer-Verlag *Lecture Notes in Computer Science #5385* (2008)
349. Dodgson, C.: A method of taking votes on more than two issues (1876). Pamphlet printed by the Clarendon Press, Oxford, and headed "not yet published" (see the discussions in [697, 156], both of which reprint this paper).
350. Doignon, J., Falmagne, J.: A polynomial time algorithm for unidimensional unfolding representations. Journal of Algorithms **16**(2), 218–233 (1994)
351. Dokow, E., Holzman, R.: Aggregation of binary evaluations. Journal of Economic Theory **145**(2), 495–511 (2010)
352. Dor, D., Zwick, U.: SOKOBAN and other motion planning problems. Computational Geometry **13**(4), 215–228 (1999)
353. Dorn, B., de Haan, R., Schlotter, I.: Obtaining a proportional allocation by deleting items. Algorithmica **83**(5), 1559–1603 (2021)
354. Dorn, B., Krüger, D.: Being caught between a rock and a hard place in an election – voter deterrence by deletion of candidates. In: Proceedings of the 39th International Conference on Current Trends in Theory and Practice of Computer Science, pp. 182–193. Springer-Verlag *Lecture Notes in Computer Science #7741* (2013)
355. Downey, R., Fellows, M.: Parameterized Complexity. Springer-Verlag (1999)
356. Drèze, J., Greenberg, J.: Hedonic coalitions: Optimality and stability. Econometrica **48**(4), 987–1003 (1980)
357. Driessen, T.: Cooperative Games, Solutions and Applications. Kluwer Academic Publishers (1988)
358. Dror, A., Feldman, M., Segal-Halevi, E.: On fair division under heterogeneous matroid constraints. In: Proceedings of the 35th AAAI Conference on Artificial Intelligence, pp. 5312–5320. AAAI Press (2021)
359. Dubey, P., Shapley, L.: Mathematical properties of the Banzhaf power index. Mathematics of Operations Research **4**(2), 99–131 (1979)
360. Dubins, L., Spanier, E.: How to cut a cake fairly. The American Mathematical Monthly **68**(1), 1–17 (1961)
361. Dudycz, S., Manurangsi, P., Marcinkowski, J., Sornat, K.: Tight approximation for proportional approval voting. In: Proceedings of the 29th International Joint Conference on Artificial Intelligence, pp. 276–282. ijcai.org (2020)
362. Duggan, J., Schwartz, T.: Strategic manipulability without resoluteness or shared beliefs: Gibbard–Satterthwaite generalized. Social Choice and Welfare **17**(1), 85–93 (2000)

363. Dung, P.: On the acceptability of arguments and its fundamental role in nonmonotonic reasoning, logic programming and n-person games. Artificial Intelligence **77**(2), 321–357 (1995)
364. Dunne, P., Wooldridge, M., Laurence, M.: The complexity of contract negotiation. Artificial Intelligence **164**(1–2), 23–46 (2005)
365. Dutta, B.: Covering sets and a new Condorcet choice correspondence. Journal of Economic Theory **44**(1), 63–80 (1988)
366. Dutta, B., Jackson, M., Le Breton, M.: Strategic candidacy and voting procedures. Econometrica **69**(4), 1013–1037 (2001)
367. Dwork, C., Kumar, R., Naor, M., Sivakumar, D.: Rank aggregation methods for the web. In: Proceedings of the 10th International World Wide Web Conference, pp. 613–622. ACM Press (2001)
368. Ebadian, S., Peters, D., Shah, N.: How to fairly allocate easy and difficult chores. In: Proceedings of the 21st International Conference on Autonomous Agents and Multiagent Systems, pp. 372–380. IFAAMAS (2022)
369. Edmonds, J., Pruhs, K.: Balanced allocations of cake. In: Proceedings of the 47th IEEE Symposium on Foundations of Computer Science, pp. 623–634. IEEE Computer Society Press (2006)
370. Edmonds, J., Pruhs, K.: Cake cutting really is not a piece of cake. ACM Transactions on Algorithms **7**(4), 51:1–51:12 (2011)
371. Ehlers, L., Klaus, B.: Coalitional strategy-proofness, resource-monotonicity, and separability for multiple assignment problems. Social Choice and Welfare **21**(2), 265–280 (2003)
372. Eiben, E., Ganian, R., Hamm, T., Ordyniak, S.: Parameterized complexity of envy-free resource allocation in social networks. In: Proceedings of the 34th AAAI Conference on Artificial Intelligence, pp. 7135–7142. AAAI Press (2020)
373. Elkind, E., Faliszewski, P.: Approximation algorithms for campaign management. In: Proceedings of the 6th International Workshop on Internet & Network Economics, pp. 473–482. Springer-Verlag *Lecture Notes in Computer Science #6484* (2010)
374. Elkind, E., Faliszewski, P., Laslier, J., Skowron, P., Slinko, A., Talmon, N.: What do multiwinner voting rules do? An experiment over the two-dimensional Euclidean domain. In: Proceedings of the 31st AAAI Conference on Artificial Intelligence, pp. 494–501. AAAI Press (2017)
375. Elkind, E., Faliszewski, P., Skowron, P., Slinko, A.: Properties of multiwinner voting rules. Social Choice and Welfare **48**(3), 599–632 (2017)
376. Elkind, E., Faliszewski, P., Slinko, A.: Swap bribery. In: Proceedings of the 2nd International Symposium on Algorithmic Game Theory, pp. 299–310. Springer-Verlag *Lecture Notes in Computer Science #5814* (2009)
377. Elkind, E., Faliszewski, P., Slinko, A.: Cloning in elections: Finding the possible winners. Journal of Artificial Intelligence Research **42**, 529–573 (2011)
378. Elkind, E., Faliszewski, P., Slinko, A.: Clone structures in voters' preferences. In: Proceedings of the 13th ACM Conference on Electronic Commerce, pp. 496–513 (2012)
379. Elkind, E., Fanelli, A., Flammini, M.: Price of Pareto optimality in hedonic games. Artificial Intelligence **288**, 103 357 (2020)
380. Elkind, E., Goldberg, L., Goldberg, P., Wooldridge, M.: Computing good Nash equilibria in graphical games. In: Proceedings of the 8th ACM Conference on Electronic Commerce, pp. 162–171. ACM Press (2007)
381. Elkind, E., Goldberg, L., Goldberg, P., Wooldridge, M.: On the computational complexity of weighted voting games. Annals of Mathematics and Artificial Intelligence **56**(2), 109–131 (2009)
382. Elkind, E., Lackner, M., Peters, D.: Preference restrictions in computational social choice: A survey. Tech. Rep. arXiv:2205.09092v1 [cs.GT], ACM Computing Research Repository (CoRR) (2022)

References

383. Elkind, E., Slinko, A.: Rationalizations of voting rules. In: F. Brandt, V. Conitzer, U. Endriss, J. Lang, A. Procaccia (eds.) Handbook of Computational Social Choice, chap. 8, pp. 169–196. Cambridge University Press (2016)
384. Elkind, E., Wooldridge, M.: Hedonic coalition nets. In: Proceedings of the 8th International Conference on Autonomous Agents and Multiagent Systems, pp. 417–424. IFAAMAS (2009)
385. Elkind, E., Xia, L. (eds.): Proceedings of the 7th International Workshop on Computational Social Choice. Rensselaer Polytechnic Institute, Troy, NY, USA (2018)
386. Emerson, P.: The original Borda count and partial voting. Social Choice and Welfare **40**(2), 353–358 (2013)
387. Endriss, U.: Judgment aggregation. In: F. Brandt, V. Conitzer, U. Endriss, J. Lang, A. Procaccia (eds.) Handbook of Computational Social Choice, chap. 17, pp. 399–426. Cambridge University Press (2016)
388. Endriss, U. (ed.): Trends in Computational Social Choice. AI Access Foundation (2017)
389. Endriss, U.: Judgment aggregation with rationality and feasibility constraints. In: Proceedings of the 17th International Conference on Autonomous Agents and Multiagent Systems, pp. 946–954. IFAAMAS (2018)
390. Endriss, U., Goldberg, P. (eds.): Proceedings of the 2nd International Workshop on Computational Social Choice. University of Liverpool, Liverpool, UK (2008)
391. Endriss, U., Grandi, U., de Haan, R., Lang, J.: Succinctness of languages for judgment aggregation. In: Proceedings of the 15th International Conference on the Principles of Knowledge Representation and Reasoning, pp. 176–186. AAAI Press (2016)
392. Endriss, U., Grandi, U., Porello, D.: Complexity of judgment aggregation: Safety of the agenda. In: Proceedings of the 9th International Conference on Autonomous Agents and Multiagent Systems, pp. 359–366. IFAAMAS (2010)
393. Endriss, U., Grandi, U., Porello, D.: Complexity of winner determination and strategic manipulation in judgment aggregation. In: V. Conitzer, J. Rothe (eds.) Proceedings of the 3rd International Workshop on Computational Social Choice, pp. 139–150. Heinrich-Heine-Universität Düsseldorf, Düsseldorf, Germany (2010)
394. Endriss, U., Grandi, U., Porello, D.: Complexity of judgment aggregation. Journal of Artificial Intelligence Research **45**, 481–514 (2012)
395. Endriss, U., de Haan, R.: Complexity of the winner determination problem in judgment aggregation: Kemeny, Slater, Tideman, Young. In: Proceedings of the 14th International Conference on Autonomous Agents and Multiagent Systems, pp. 117–125. IFAAMAS (2015)
396. Endriss, U., de Haan, R., Slavkovik, M., Lang, J.: The complexity landscape of outcome determination in judgment aggregation. Journal of Artificial Intelligence Research **69**, 687–731 (2020)
397. Endriss, U., de Haan, R., Szeider, S.: Parameterized complexity results for agenda safety in judgment aggregation. In: Proceedings of the 14th International Conference on Autonomous Agents and Multiagent Systems, pp. 127–136. IFAAMAS (2015)
398. Endriss, U., Lang, J. (eds.): Proceedings of the 1st International Workshop on Computational Social Choice. Universiteit van Amsterdam, Amsterdam, The Netherlands (2006)
399. Endriss, U., Maudet, N.: On the communication complexity of multilateral trading: Extended report. Journal of Autonomous Agents and Multi-Agent Systems **11**(1), 91–107 (2005)
400. Endriss, U., Maudet, N., Sadri, F., Toni, F.: Negotiating socially optimal allocations of resources. Journal of Artificial Intelligence Research **25**, 315–348 (2006)
401. Ephrati, E., Rosenschein, J.: The Clarke Tax as a consensus mechanism among automated agents. In: Proceedings of the 9th National Conference on Artificial Intelligence, pp. 173–178. AAAI Press (1991)

402. Ephrati, E., Rosenschein, J.: Multi-agent planning as a dynamic search for social consensus. In: Proceedings of the 13th International Joint Conference on Artificial Intelligence, pp. 423–429. Morgan Kaufmann (1993)
403. Ephrati, E., Rosenschein, J.: A heuristic technique for multi-agent planning. Annals of Mathematics and Artificial Intelligence **20**(1–4), 13–67 (1997)
404. Erdélyi, G., Fellows, M.: Parameterized control complexity in Bucklin voting and in fallback voting. In: V. Conitzer, J. Rothe (eds.) Proceedings of the 3rd International Workshop on Computational Social Choice, pp. 163–174. Heinrich-Heine-Universität Düsseldorf, Düsseldorf, Germany (2010)
405. Erdélyi, G., Fellows, M., Piras, L., Rothe, J.: Control complexity in Bucklin and fallback voting. Tech. Rep. arXiv:1103.2230 [cs.CC], ACM Computing Research Repository (CoRR) (2011). Revised August, 2012
406. Erdélyi, G., Fellows, M., Rothe, J., Schend, L.: Control complexity in Bucklin and fallback voting: A theoretical analysis. Journal of Computer and System Sciences **81**(4), 632–660 (2015)
407. Erdélyi, G., Fellows, M., Rothe, J., Schend, L.: Control complexity in Bucklin and fallback voting: An experimental analysis. Journal of Computer and System Sciences **81**(4), 661–670 (2015)
408. Erdélyi, G., Hemaspaandra, L., Rothe, J., Spakowski, H.: Frequency of correctness versus average polynomial time. Information Processing Letters **109**(16), 946–949 (2009)
409. Erdélyi, G., Hemaspaandra, L., Rothe, J., Spakowski, H.: Generalized juntas and NP-hard sets. Theoretical Computer Science **410**(38–40), 3995–4000 (2009)
410. Erdélyi, G., Lackner, M., Pfandler, A.: Computational aspects of nearly single-peaked electorates. In: Proceedings of the 27th AAAI Conference on Artificial Intelligence, pp. 283–289. AAAI Press (2013)
411. Erdélyi, G., Neveling, M., Reger, C., Rothe, J., Yang, Y., Zorn, R.: Towards completing the puzzle: Complexity of control by replacing, adding, and deleting candidates or voters. Journal of Autonomous Agents and Multi-Agent Systems **35**(2), 41:1–41:48 (2021)
412. Erdélyi, G., Nowak, M., Rothe, J.: Sincere-strategy preference-based approval voting fully resists constructive control and broadly resists destructive control. Mathematical Logic Quarterly **55**(4), 425–443 (2009)
413. Erdélyi, G., Piras, L., Rothe, J.: The complexity of voter partition in Bucklin and fallback voting: Solving three open problems. In: Proceedings of the 10th International Conference on Autonomous Agents and Multiagent Systems, pp. 837–844. IFAAMAS (2011)
414. Erdélyi, G., Rothe, J.: Control complexity in fallback voting. In: Proceedings of Computing: The 16th Australasian Theory Symposium, *Conferences in Research and Practice in Information Technology Series (CRPIT)*, vol. 32, pp. 39–48. Australian Computer Society (2010)
415. Escoffier, B., Lang, J., Öztürk, M.: Single-peaked consistency and its complexity. In: Proceedings of the 18th European Conference on Artificial Intelligence, pp. 366–370. IOS Press (2008)
416. Even, S., Paz, A.: A note on cake cutting. Discrete Applied Mathematics **7**(3), 285–296 (1984)
417. Fain, B., Munagala, K., Shah, N.: Fair allocation of indivisible public goods. In: Proceedings of the 19th ACM Conference on Economics and Computation, pp. 575–592. ACM Press (2018)
418. Faliszewski, P.: Nonuniform bribery (extended abstract). In: Proceedings of the 7th International Conference on Autonomous Agents and Multiagent Systems, pp. 1569–1572. IFAAMAS (2008)
419. Faliszewski, P., Elkind, E., Wooldridge, M.: Boolean combinations of weighted voting games. In: Proceedings of the 8th International Conference on Autonomous Agents and Multiagent Systems, pp. 185–192. IFAAMAS (2009)

420. Faliszewski, P., Gawron, G., Kusek, B.: Robustness of greedy approval rules. In: Proceedings of the 19th European Conference on Multi-Agent Systems, pp. 116–133. Springer-Verlag *Lecture Notes in Artificial Intelligence #13442* (2022)
421. Faliszewski, P., Hemaspaandra, E., Hemaspaandra, L.: The complexity of bribery in elections. In: Proceedings of the 21st National Conference on Artificial Intelligence, pp. 641–646. AAAI Press (2006)
422. Faliszewski, P., Hemaspaandra, E., Hemaspaandra, L.: How hard is bribery in elections? Journal of Artificial Intelligence Research **35**, 485–532 (2009)
423. Faliszewski, P., Hemaspaandra, E., Hemaspaandra, L.: Using complexity to protect elections. Communications of the ACM **53**(11), 74–82 (2010)
424. Faliszewski, P., Hemaspaandra, E., Hemaspaandra, L.: Multimode control attacks on elections. Journal of Artificial Intelligence Research **40**, 305–351 (2011)
425. Faliszewski, P., Hemaspaandra, E., Hemaspaandra, L.: The complexity of manipulative attacks in nearly single-peaked electorates. Artificial Intelligence **207**, 69–99 (2014)
426. Faliszewski, P., Hemaspaandra, E., Hemaspaandra, L.: Weighted electoral control. Journal of Artificial Intelligence Research **52**, 507–542 (2015)
427. Faliszewski, P., Hemaspaandra, E., Hemaspaandra, L., Rothe, J.: Llull and Copeland voting broadly resist bribery and control. In: Proceedings of the 22nd AAAI Conference on Artificial Intelligence, pp. 724–730. AAAI Press (2007)
428. Faliszewski, P., Hemaspaandra, E., Hemaspaandra, L., Rothe, J.: Copeland voting fully resists constructive control. In: Proceedings of the 4th International Conference on Algorithmic Aspects in Information and Management, pp. 165–176. Springer-Verlag *Lecture Notes in Computer Science #5034* (2008)
429. Faliszewski, P., Hemaspaandra, E., Hemaspaandra, L., Rothe, J.: Llull and Copeland voting computationally resist bribery and constructive control. Journal of Artificial Intelligence Research **35**, 275–341 (2009)
430. Faliszewski, P., Hemaspaandra, E., Hemaspaandra, L., Rothe, J.: A richer understanding of the complexity of election systems. In: S. Ravi, S. Shukla (eds.) Fundamental Problems in Computing: Essays in Honor of Professor Daniel J. Rosenkrantz, chap. 14, pp. 375–406. Springer (2009)
431. Faliszewski, P., Hemaspaandra, E., Hemaspaandra, L., Rothe, J.: The shield that never was: Societies with single-peaked preferences are more open to manipulation and control. Information and Computation **209**(2), 89–107 (2011)
432. Faliszewski, P., Hemaspaandra, E., Schnoor, H.: Copeland voting: Ties matter. In: Proceedings of the 7th International Conference on Autonomous Agents and Multiagent Systems, pp. 983–990. IFAAMAS (2008)
433. Faliszewski, P., Hemaspaandra, E., Schnoor, H.: Manipulation of Copeland elections. In: Proceedings of the 9th International Conference on Autonomous Agents and Multiagent Systems, pp. 367–374. IFAAMAS (2010)
434. Faliszewski, P., Hemaspaandra, E., Schnoor, H.: Weighted manipulation for four-candidate Llull is easy. In: Proceedings of the 20th European Conference on Artificial Intelligence, pp. 318–323. IOS Press (2012)
435. Faliszewski, P., Hemaspaandra, L.: The complexity of power-index comparison. Theoretical Computer Science **410**(1), 101–107 (2009)
436. Faliszewski, P., Lackner, M., Peters, D., Talmon, N.: Effective heuristics for committee scoring rules. In: Proceedings of the 32nd AAAI Conference on Artificial Intelligence, pp. 1023–1030 (2018)
437. Faliszewski, P., Manurangsi, P., Sornat, K.: Approximation and hardness of shift-bribery. In: Proceedings of the 33rd AAAI Conference on Artificial Intelligence, pp. 1901–1908. AAAI Press (2019)
438. Faliszewski, P., Procaccia, A.: AI's war on manipulation: Are we winning? AI Magazine **31**(4), 53–64 (2010)

439. Faliszewski, P., Reisch, Y., Rothe, J., Schend, L.: Complexity of manipulation, bribery, and campaign management in Bucklin and fallback voting. Journal of Autonomous Agents and Multi-Agent Systems **29**(6), 1091–1124 (2015)
440. Faliszewski, P., Rothe, J.: Control and bribery in voting. In: F. Brandt, V. Conitzer, U. Endriss, J. Lang, A. Procaccia (eds.) Handbook of Computational Social Choice, chap. 7, pp. 146–168. Cambridge University Press (2016)
441. Faliszewski, P., Skowron, P., Slinko, A., Talmon, N.: Multiwinner analogues of the plurality rule: Axiomatic and algorithmic views. Social Choice and Welfare **51**(3), 513–550 (2018)
442. Faliszewski, P., Skowron, P., Slinko, A., Talmon, N.: Committee scoring rules: Axiomatic characterization and hierarchy. ACM Transactions on Economics and Computation **7**(1), 3:1–3:39 (2019)
443. Faliszewski, P., Skowron, P., Talmon, N.: Bribery as a measure of candidate success: Complexity results for approval-based multiwinner rules. In: Proceedings of the 16th International Conference on Autonomous Agents and Multiagent Systems, pp. 6–14. IFAAMAS (2017)
444. Farhadi, A., Ghodsi, M., Hajiaghayi, M., Lahaie, S., Pennock, D., Seddighin, M., Seddighin, S., Yami, H.: Fair allocation of indivisible goods to asymmetric agents. Journal of Artificial Intelligence Research **64**, 1–20 (2019)
445. Farhadi, A., Hajiaghayi, M., Latifian, M., Seddighin, M., Yami, H.: Almost envy-freeness, envy-rank, and Nash social welfare matchings. In: Proceedings of the 35th AAAI Conference on Artificial Intelligence, pp. 5355–5362. AAAI Press (2021)
446. Feige, U., Tennenholtz, M.: On fair division of a homogeneous good. Games and Economic Behavior **87**, 305–321 (2014)
447. Feldman, A., Kirman, A.: Fairness and envy. The American Economic Review **64**(6), 995–1005 (1974)
448. Felsenthal, D., Machover, M.: Postulates and paradoxes of relative voting power – A critical re-appraisal. Theory and Decision **38**(2), 195–229 (1995)
449. Felsenthal, D., Machover, M.: Voting power measurement: A story of misreinvention. Social Choice and Welfare **25**(2), 485–506 (2005)
450. Ferraioli, D., Gourvès, L., Monnot, J.: On regular and approximately fair allocations of indivisible goods. In: Proceedings of the 13th International Conference on Autonomous Agents and Multiagent Systems, pp. 997–1004. IFAAMAS (2014)
451. Fink, A.: A note on the fair division problem. Mathematics Magazine **37**(5), 341–342 (1964)
452. Fischer, F., Hudry, O., Niedermeier, R.: Weighted tournament solutions. In: F. Brandt, V. Conitzer, U. Endriss, J. Lang, A. Procaccia (eds.) Handbook of Computational Social Choice, chap. 4, pp. 85–102. Cambridge University Press (2016)
453. Fishburn, P.: Even-chance lotteries in social choice theory. Theory and Decision **3**(1), 18–40 (1972)
454. Fishburn, P.: Condorcet social choice functions. SIAM Journal on Applied Mathematics **33**(3), 469–489 (1977)
455. Fitzsimmons, Z., Hemaspaandra, E.: High-multiplicity election problems. Journal of Autonomous Agents and Multi-Agent Systems **33**(4), 383–402 (2019)
456. Fitzsimmons, Z., Hemaspaandra, E.: Using weighted matching to solve 2-approval/veto control and bribery. In: Proceedings of the 26th European Conference on Artificial Intelligence, *Frontiers in Artificial Intelligence and Applications*, vol. 372, pp. 732–739. IOS Press (2023)
457. Fitzsimmons, Z., Hemaspaandra, E., Hemaspaandra, L.: Control in the presence of manipulators: Cooperative and competitive cases. In: Proceedings of the 23rd International Joint Conference on Artificial Intelligence, pp. 113–119. AAAI Press/IJCAI (2013)
458. Fitzsimmons, Z., Lackner, M.: Incomplete preferences in single-peaked electorates. Journal of Artificial Intelligence Research **67**, 797–833 (2020)

459. Flammini, M., Gilbert, H.: Parameterized complexity of manipulating sequential allocation. In: Proceedings of the 24th European Conference on Artificial Intelligence, *Frontiers in Artificial Intelligence and Applications*, vol. 325, pp. 99–106. IOS Press (2020)
460. Flammini, M., Monaco, G., Moscardelli, L., Shalom, M., Zaks, S.: On the online coalition structure generation problem. Journal of Artificial Intelligence Research **72**, 1215–1250 (2021)
461. Flum, J., Grohe, M.: Parameterized Complexity Theory. EATCS Texts in Theoretical Computer Science. Springer-Verlag (2006)
462. Foley, D.: Resource allocation and the public sector. Yale Economic Essays **7**(1), 45–98 (1967)
463. Fortnow, L., Lipton, R., van Melkebeek, D., Viglas, A.: Time-space lower bounds for satisfiability. Journal of the ACM **52**(6), 835–865 (2005)
464. Fotakis, D., Gourvès, L., Kasouridis, S., Pagourtzis, A.: Object allocation and positive graph externalities. In: Proceedings of the 24th European Conference on Artificial Intelligence, *Frontiers in Artificial Intelligence and Applications*, vol. 325, pp. 107–114. IOS Press (2020)
465. Fréchet, M.: Émile Borel, initiator of the theory of psychological games and its application. Econometrica **21**(1), 95–96 (1953)
466. Freeman, R., Micha, E., Shah, N.: Two-sided matching meets fair division. In: Proceedings of the 30th International Joint Conference on Artificial Intelligence, pp. 203–209. ijcai.org (2021)
467. Freeman, R., Shah, N., Vaish, R.: Best of both worlds: Ex-Ante and ex-post fairness in resource allocation. In: Proceedings of the 21st ACM Conference on Economics and Computation, pp. 21–22. ACM Press (2020)
468. Freeman, R., Sikdar, S., Vaish, R., Xia, L.: Equitable allocations of indivisible goods. In: Proceedings of the 28th International Joint Conference on Artificial Intelligence, pp. 280–286. ijcai.org (2019)
469. Freeman, R., Sikdar, S., Vaish, R., Xia, L.: Equitable allocations of indivisible chores. In: Proceedings of the 19th International Conference on Autonomous Agents and Multiagent Systems, pp. 384–392. IFAAMAS (2020)
470. Freixas, J., Marciniak, D.: On the notion of dimension and codimension of simple games. In: Proceedings of the 3rd International Conference on Contributions to Game Theory and Management, pp. 67–81. St. Petersburg University, St. Petersburg, Russia (2010)
471. Freixas, J., Puente, M.: A note about games-composition dimension. Discrete Applied Mathematics **113**, 265–273 (2001)
472. Freixas, J., Puente, M.: Dimension of complete simple games with minimum. European Journal of Operational Research **188**(2), 555–568 (2008)
473. Friedgut, E., Kalai, G., Nisan, N.: Elections can be manipulated often. In: Proceedings of the 49th IEEE Symposium on Foundations of Computer Science, pp. 243–249. IEEE Computer Society Press (2008)
474. Friedgut, E., Keller, N., Kalai, G., Nisan, N.: A quantitative version of the Gibbard–Satterthwaite theorem for three alternatives. SIAM Journal on Computing **40**(3), 934–952 (2011)
475. Fujinaka, Y., Sakai, T.: Maskin monotonicity in economies with indivisible goods and money. Economics Letters **94**(2), 253–258 (2007)
476. Fulkerson, D., Gross, G.: Incidence matrices and interval graphs. Pacific Journal of Mathematics **15**(5), 835–855 (1965)
477. Gailmard, S., Patty, J., Penn, E.: Arrow's theorem on single-peaked domains. In: E. Aragonès, C. Beviá, H. Llavador, N. Schofield (eds.) The Political Economy of Democracy, pp. 335–342. Fundación BBVA (2009)
478. Gairing, M., Savani, R.: Computing stable outcomes in symmetric additively separable hedonic games. Mathematics of Operations Research **44**(3), 1101–1121 (2019)

479. Gale, D., Shapley, L.: College admissions and the stability of marriage. The American Mathematical Monthly **69**(1), 9–14 (1962)
480. Galeotti, A., Goyal, S., Jackson, M., Vega-Redondo, F., Yariv, L.: Network games. The Review of Economic Studies **77**(1), 218–244 (2010)
481. Gärdenfors, P.: Manipulation of social choice functions. Journal of Economic Theory **13**(2), 217–228 (1976)
482. Gardner, M.: Aha! Insight. W. H. Freeman and Company (1978)
483. Garey, M., Johnson, D.: Computers and Intractability: A Guide to the Theory of NP-Completeness. W. H. Freeman and Company (1979)
484. Garg, J., Hoefer, M., McGlaughlin, P., Schmalhofer, M.: When dividing mixed manna is easier than dividing goods: Competitive equilibria with a constant number of chores. In: Proceedings of the 14th International Symposium on Algorithmic Game Theory, pp. 329–344. Springer-Verlag *Lecture Notes in Computer Science #12885* (2021)
485. Garg, J., Hoefer, M., Mehlhorn, K.: Approximating the Nash social welfare with budget-additive valuations. In: Proceedings of the 29th Annual ACM-SIAM Symposium on Discrete Algorithms, pp. 2326–2340. Society for Industrial and Applied Mathematics (2018)
486. Garg, J., Husic, E., Végh, L.: Approximating Nash social welfare under Rado valuations. SIGecom Exchanges **19**(1), 45–51 (2021)
487. Garg, J., Kulkarni, P., Kulkarni, R.: Approximating Nash social welfare under submodular valuations through (un)matchings. In: Proceedings of the 31st Annual ACM-SIAM Symposium on Discrete Algorithms, pp. 2673–2687. Society for Industrial and Applied Mathematics (2020)
488. Garg, J., McGlaughlin, P.: Computing competitive equilibria with mixed manna. In: Proceedings of the 19th International Conference on Autonomous Agents and Multiagent Systems, pp. 420–428. IFAAMAS (2020)
489. Garg, J., Murhekar, A.: Computing fair and efficient allocations with few utility values. In: Proceedings of the 14th International Symposium on Algorithmic Game Theory, pp. 345–359. Springer-Verlag *Lecture Notes in Computer Science #12885* (2021)
490. Garg, J., Murhekar, A., Qin, J.: Fair and efficient allocations of chores under bivalued preferences. In: Proceedings of the 36th AAAI Conference on Artificial Intelligence, pp. 5043–5050. AAAI Press (2022)
491. Garg, J., Taki, S.: An improved approximation algorithm for maximin shares. Artificial Intelligence **300**, 103 547 (2021)
492. Gasarch, W.: The P =? NP poll. SIGACT News **33**(2), 34–47 (2002)
493. Gaspers, S., Kalinowski, T., Narodytska, N., Walsh, T.: Coalitional manipulation for Schulze's rule. In: Proceedings of the 12th International Conference on Autonomous Agents and Multiagent Systems, pp. 431–438. IFAAMAS (2013)
494. Gawron, G., Faliszewski, P.: Robustness of approval-based multiwinner voting rules. In: Proceedings of the 6th International Conference on Algorithmic Decision Theory, pp. 17–31. Springer-Verlag *Lecture Notes in Artificial Intelligence #11834* (2019)
495. Geist, C., Endriss, U.: Automated search for impossibility theorems in social choice theory: Ranking sets of objects. Journal of Artificial Intelligence Research **40**, 143–174 (2011)
496. Ghodsi, M., Hajiaghayi, M., Seddighin, M., Seddighin, S., Yami, H.: Fair allocation of indivisible goods: Improvements and generalizations. In: Proceedings of the 19th ACM Conference on Economics and Computation, pp. 539–556. ACM Press (2018)
497. Ghodsi, M., Hajiaghayi, M., Seddighin, M., Seddighin, S., Yami, H.: Fair allocation of indivisible goods: Improvement. Mathematics of Operations Research **46**(3), 1038–1053 (2021)

498. Ghodsi, M., Hajiaghayi, M., Seddighin, M., Seddighin, S., Yami, H.: Fair allocation of indivisible goods: Beyond additive valuations. Artificial Intelligence **303**, 103 633 (2022)
499. Ghosh, S., Mundhe, M., Hernandez, K., Sen, S.: Voting for movies: The anatomy of recommender systems. In: Proceedings of the 3rd Annual Conference on Autonomous Agents, pp. 434–435. ACM Press (1999)
500. Gibbard, A.: Manipulation of voting schemes: A general result. Econometrica **41**(4), 587–601 (1973)
501. Gill, J.: Computational complexity of probabilistic Turing machines. SIAM Journal on Computing **6**(4), 675–695 (1977)
502. Gkatzelis, V., Psomas, A., Tan, X.: Fair and efficient online allocations with normalized valuations. In: Proceedings of the 35th AAAI Conference on Artificial Intelligence, pp. 5440–5447. AAAI Press (2021)
503. Goemans, M., Harvey, N., Iwata, S., Mirrokni, V.: Approximating submodular functions everywhere. In: Proceedings of the 20th ACM-SIAM Symposium on Discrete Algorithms, pp. 535–544. Society for Industrial and Applied Mathematics (2009)
504. Goko, H., Igarashi, A., Kawase, Y., Makino, K., Sumita, H., Tamura, A., Yokoi, Y., Yokoo, M.: Fair and truthful mechanism with limited subsidy. In: Proceedings of the 21st International Conference on Autonomous Agents and Multiagent Systems, pp. 534–542. IFAAMAS (2022)
505. Golden, B., Perny, P.: Infinite order Lorenz dominance for fair multiagent optimization. In: Proceedings of the 9th International Conference on Autonomous Agents and Multiagent Systems, pp. 383–390. IFAAMAS (2010)
506. Goldreich, O.: Notes on Levin's theory of average-case complexity. Tech. Rep. TR97-058, Electronic Colloquium on Computational Complexity (1997)
507. Golovin, D.: Max-min fair allocation of indivisible goods. Tech. Rep. CMU-CS-05-144, School of Computer Science, Carnegie Mellon University (2005)
508. Gonzales, C., Perny, P.: GAI networks for utility elicitation. In: Proceedings of the 9th International Conference on Principles of Knowledge Representation and Reasoning, pp. 224–234. AAAI Press (2004)
509. Gourvès, L., Lesca, J., Wilczynski, A.: Object allocation via swaps along a social network. In: Proceedings of the 26th International Joint Conference on Artificial Intelligence, pp. 213–219. ijcai.org (2017)
510. Gourvès, L., Lesca, J., Wilczynski, A.: On fairness via picking sequences in allocation of indivisible goods. In: Proceedings of the 7th International Conference on Algorithmic Decision Theory, pp. 258–272. Springer-Verlag *Lecture Notes in Artificial Intelligence #13023* (2021)
511. Gourvès, L., Monnot, J.: On maximin share allocations in matroids. Theoretical Computer Science **754**, 50–64 (2019)
512. Gourvès, L., Monnot, J., Tlilane, L.: A matroid approach to the worst case allocation of indivisible goods. In: Proceedings of the 23rd International Joint Conference on Artificial Intelligence, pp. 136–142. AAAI Press/IJCAI (2013)
513. Gourvès, L., Monnot, J., Tlilane, L.: Near fairness in matroids. In: Proceedings of the 21st European Conference on Artificial Intelligence, *Frontiers in Artificial Intelligence and Applications*, vol. 263, pp. 393–398. IOS Press (2014)
514. Grabisch, M.: Alternative representations of discrete fuzzy measures for decision making. International Journal of Uncertainty, Fuzziness and Knowledge-Based Systems **5**(5), 587–608 (1997)
515. Grandi, U.: Binary aggregation with integrity constraints. Ph.D. thesis, University of Amsterdam (2012)
516. Grandi, U., Endriss, U.: Lifting integrity constraints in binary aggregation. Artificial Intelligence **199–200**, 45–66 (2013)
517. Grandi, U., Rosenschein, J. (eds.): Proceedings of the 6th International Workshop on Computational Social Choice. University of Toulouse, Toulouse, France (2016)

518. de Grazia, A.: Mathematical deviation of an election system. Isis **44**(1–2), 41–51 (1953)
519. Greco, G., Scarcello, F.: The complexity of computing maximin share allocations on graphs. In: Proceedings of the 34th AAAI Conference on Artificial Intelligence, pp. 2006–2013. AAAI Press (2020)
520. Gurski, F., Rothe, I., Rothe, J., Wanke, E.: Exakte Algorithmen für schwere Graphenprobleme. eXamen.Press. Springer-Verlag (2010)
521. Haake, C., Raith, M., Su, F.: Bidding for envy-freeness: A procedural approach to n-player fair-division problems. Social Choice and Welfare **19**(4), 723–749 (2002)
522. de Haan, R.: Complexity results for manipulation, bribery and control of the Kemeny judgment aggregation procedure. In: Proceedings of the 16th International Conference on Autonomous Agents and Multiagent Systems, pp. 1151–1159. IFAAMAS (2017)
523. de Haan, R.: Hunting for tractable languages for judgment aggregation. In: Proceedings of the 16th International Conference on the Principles of Knowledge Representation and Reasoning, pp. 194–203. AAAI Press (2018)
524. de Haan, R., Slavkovik, M.: Complexity results for aggregating judgments using scoring or distance-based procedures. In: Proceedings of the 16th International Conference on Autonomous Agents and Multiagent Systems, pp. 952–961. IFAAMAS (2017)
525. Hačijan, L.: A polynomial algorithm in linear programming. Soviet Mathematics Doklady **20**(1), 191–194 (1979)
526. Hägele, G., Pukelsheim, F.: The electoral writings of Ramon Llull. Studia Lulliana **41**(97), 3–38 (2001)
527. Halpern, D., Procaccia, A., Psomas, A., Shah, N.: Fair division with binary valuations: One rule to rule them all. In: Proceedings of the 16th International Workshop on Internet & Network Economics, pp. 370–383. Springer-Verlag *Lecture Notes in Computer Science #12495* (2020)
528. Halpern, D., Shah, N.: Fair division with subsidy. In: Proceedings of the 12th International Symposium on Algorithmic Game Theory, pp. 374–389. Springer-Verlag *Lecture Notes in Computer Science #11801* (2019)
529. Halpern, D., Shah, N.: Fair and efficient resource allocation with partial information. In: Proceedings of the 30th International Joint Conference on Artificial Intelligence, pp. 224–230. ijcai.org (2021)
530. Hatfield, J.: Strategy-proof, efficient, and nonbossy quota allocations. Social Choice and Welfare **33**(3), 505–515 (2009)
531. He, J., Procaccia, A., Psomas, A., Zeng, D.: Achieving a fairer future by changing the past. In: Proceedings of the 28th International Joint Conference on Artificial Intelligence, pp. 343–349. ijcai.org (2019)
532. Heinen, T., Nguyen, N., Nguyen, T., Rothe, J.: Approximation and complexity of the optimization and existence problems for maximin share, proportional share, and minimax share allocation of indivisible goods. Journal of Autonomous Agents and Multi-Agent Systems **32**(6), 741–778 (2018)
533. Heinen, T., Nguyen, N., Rothe, J.: Fairness and rank-weighted utilitarianism in resource allocation. In: Proceedings of the 4th International Conference on Algorithmic Decision Theory, pp. 521–536. Springer-Verlag *Lecture Notes in Artificial Intelligence #9346* (2015)
534. Hemachandra, L.: The strong exponential hierarchy collapses. Journal of Computer and System Sciences **39**(3), 299–322 (1989)
535. Hemaspaandra, E., Hemaspaandra, L.: Dichotomy for voting systems. Journal of Computer and System Sciences **73**(1), 73–83 (2007)
536. Hemaspaandra, E., Hemaspaandra, L., Menton, C.: Search versus decision for election manipulation problems. In: Proceedings of the 30th Annual Symposium on Theoretical Aspects of Computer Science, *LIPIcs*, vol. 20, pp. 377–388. Schloss Dagstuhl – Leibniz-Zentrum für Informatik (2013)

537. Hemaspaandra, E., Hemaspaandra, L., Rothe, J.: Exact analysis of Dodgson elections: Lewis Carroll's 1876 voting system is complete for parallel access to NP. Journal of the ACM **44**(6), 806–825 (1997)
538. Hemaspaandra, E., Hemaspaandra, L., Rothe, J.: Raising NP lower bounds to parallel NP lower bounds. SIGACT News **28**(2), 2–13 (1997)
539. Hemaspaandra, E., Hemaspaandra, L., Rothe, J.: Anyone but him: The complexity of precluding an alternative. In: Proceedings of the 20th National Conference on Artificial Intelligence, pp. 95–101. AAAI Press (2005)
540. Hemaspaandra, E., Hemaspaandra, L., Rothe, J.: Hybrid elections broaden complexity-theoretic resistance to control. Tech. Rep. arXiv:cs/0608057v2 [cs.GT], ACM Computing Research Repository (CoRR) (2006). Revised, September 2008. Conference version appeared in the *Proceedings of the 20th International Joint Conference on Artificial Intelligence (IJCAI 2007)*.
541. Hemaspaandra, E., Hemaspaandra, L., Rothe, J.: Anyone but him: The complexity of precluding an alternative. Artificial Intelligence **171**(5–6), 255–285 (2007)
542. Hemaspaandra, E., Hemaspaandra, L., Rothe, J.: Hybrid elections broaden complexity-theoretic resistance to control. Mathematical Logic Quarterly **55**(4), 397–424 (2009)
543. Hemaspaandra, E., Hemaspaandra, L., Rothe, J.: The complexity of online manipulation of sequential elections. Journal of Computer and System Sciences **80**(4), 697–710 (2014)
544. Hemaspaandra, E., Hemaspaandra, L., Rothe, J.: The complexity of controlling candidate-sequential elections. Theoretical Computer Science **678**, 14–21 (2017)
545. Hemaspaandra, E., Hemaspaandra, L., Rothe, J.: The complexity of online voter control in sequential elections. Journal of Autonomous Agents and Multi-Agent Systems **31**(5), 1055–1076 (2017)
546. Hemaspaandra, E., Hemaspaandra, L., Rothe, J.: The complexity of online bribery in sequential elections. Journal of Computer and System Sciences **127**, 66–90 (2022)
547. Hemaspaandra, E., Rothe, J.: Recognizing when greed can approximate maximum independent sets is complete for parallel access to NP. Information Processing Letters **65**(3), 151–156 (1998)
548. Hemaspaandra, E., Rothe, J., Spakowski, H.: Recognizing when heuristics can approximate minimum vertex covers is complete for parallel access to NP. R.A.I.R.O. Theoretical Informatics and Applications **40**(1), 75–91 (2006)
549. Hemaspaandra, E., Schnoor, H.: Dichotomy for pure scoring rules under manipulative electoral actions. In: Proceedings of the 22nd European Conference on Artificial Intelligence, pp. 1071–1079. IOS Press (2016)
550. Hemaspaandra, E., Spakowski, H., Vogel, J.: The complexity of Kemeny elections. Theoretical Computer Science **349**(3), 382–391 (2005)
551. Hemaspaandra, L., Ogihara, M.: The Complexity Theory Companion. EATCS Texts in Theoretical Computer Science. Springer-Verlag (2002)
552. Hemaspaandra, L., Williams, R.: An atypical survey of typical-case heuristic algorithms. SIGACT News **43**(4), 71–89 (2012)
553. Herreiner, D., Puppe, C.: A simple procedure for finding equitable allocations of indivisible goods. Social Choice and Welfare **19**(2), 415–430 (2002)
554. Holcombe, R.: Absence of envy does not imply fairness. Southern Economic Journal **63**(3), 797–802 (1997)
555. Homan, C., Hemaspaandra, L.: Guarantees for the success frequency of an algorithm for finding Dodgson-election winners. Journal of Heuristics **15**(4), 403–423 (2009)
556. Homer, S., Selman, A.: Computability and Complexity Theory. Texts in Computer Science. Springer-Verlag (2001)
557. Hopcroft, J., Motwani, R., Ullman, J.: Introduction to Automata Theory, Languages, and Computation, second edn. Addison-Wesley (2001)

558. Hosseini, H., Larson, K.: Multiple assignment problems under lexicographic preferences. In: Proceedings of the 18th International Conference on Autonomous Agents and Multiagent Systems, pp. 837–845. IFAAMAS (2019)
559. Hosseini, H., Searns, A., Segal-Halevi, E.: Ordinal maximin share approximation for chores. In: Proceedings of the 21st International Conference on Autonomous Agents and Multiagent Systems, pp. 597–605. IFAAMAS (2022)
560. Hosseini, H., Searns, A., Segal-Halevi, E.: Ordinal maximin share approximation for goods. Journal of Artificial Intelligence Research **74**, 353–391 (2022)
561. Hosseini, H., Sikdar, S., Vaish, R., Wang, H., Xia, L.: Fair division through information withholding. In: Proceedings of the 34th AAAI Conference on Artificial Intelligence, pp. 2014–2021. AAAI Press (2020)
562. Hosseini, H., Sikdar, S., Vaish, R., Xia, L.: Fair and efficient allocations under lexicographic preferences. In: Proceedings of the 35th AAAI Conference on Artificial Intelligence, pp. 5472–5480. AAAI Press (2021)
563. Huang, X., Lu, P.: An algorithmic framework for approximating maximin share allocation of chores. In: Proceedings of the 22nd ACM Conference on Economics and Computation, pp. 630–631. ACM Press (2021)
564. Huang, Z., Tröbst, T.: Online matching. In: F. Echenique, N. Immorlica, V. Vazirani (eds.) Online and Matching-Based Market Design. Cambridge University Press (2023)
565. Hudry, O.: On the complexity of Slater's problems. European Journal of Operational Research **203**(1), 216–221 (2010)
566. Ieong, S., Shoham, Y.: Marginal contribution nets: A compact representation scheme for coalitional games. In: Proceedings of the 6th ACM Conference on Electronic Commerce, pp. 193–202. ACM Press (2005)
567. Igarashi, A., Peters, D.: Pareto-optimal allocation of indivisible goods with connectivity constraints. In: Proceedings of the 33rd AAAI Conference on Artificial Intelligence, pp. 2045–2052. AAAI Press (2019)
568. Isaksson, M., Kindler, G., Mossel, E.: The geometry of manipulation: A quantitative proof of the Gibbard–Satterthwaite theorem. In: Proceedings of the 51st IEEE Symposium on Foundations of Computer Science, pp. 319–328. IEEE Computer Society Press (2010)
569. Isaksson, M., Kindler, G., Mossel, E.: The geometry of manipulation – A quantitative proof of the Gibbard–Satterthwaite theorem. Combinatorica **32**(2), 221–250 (2012)
570. Jackson, M.: Social and Economic Networks. Princeton University Press (2008)
571. Jackson, M., van den Nouweland, A.: Strongly stable networks. Games and Economic Behavior **51**(2), 420–444 (2005)
572. Jackson, M., Srivastava, S.: On the relation between Nash equilibria and undominated strategies for two person, finite games. Economics Letters **45**(3), 315–318 (1994)
573. Jackson, M., Watts, A.: Social games: Matching and the play of finitely repeated games. Games and Economic Behavior **70**(1), 170–191 (2010)
574. Jackson, M., Zenou, Y.: Games on networks. In: P. Young, S. Zamir (eds.) Handbook on Game Theory, vol. 4. Elsevier Science (2014)
575. Jain, K., Vohra, R.: On stability of the core. Technical Report, Northwestern University (2006)
576. Jain, M., Korzhyk, D., Vaněk, O., Conitzer, V., Pěchouček, M., Tambe, M.: A double oracle algorithm for zero-sum security games on graphs. In: Proceedings of the 10th International Conference on Autonomous Agents and Multiagent Systems, pp. 327–334. IFAAMAS (2011)
577. Janeczko, L., Faliszewski, P.: Ties in multiwinner approval voting. In: Proceedings of the 32nd International Joint Conference on Artificial Intelligence, pp. 2765–2773. ijcai.org (2023)

References

578. Johnson, D.: The NP-completeness column: An ongoing guide. Journal of Algorithms **2**(4), 393–405 (1981). First column in a series of columns on NP-completeness appearing in the same journal
579. Johnson, D., Papadimitriou, C., Yannakakis, M.: How easy is local search? Journal of Computer and System Sciences **37**(1), 79–100 (1988)
580. Jones, N., Lien, Y., Laaser, W.: New problems complete for nondeterministic log space. Mathematical Systems Theory **10**(1), 1–17 (1976)
581. Kaci, S., Lang, J., Perny, P.: Compact representation of preferences. In: P. Marquis, O. Papini, H. Prade (eds.) A Guided Tour of Artificial Intelligence Research, vol. 1, Knowledge Representation, Reasoning and Learning, chap. 7. Springer (2020)
582. Kaczmarczyk, A., Faliszewski, P.: Algorithms for destructive shift bribery. Journal of Autonomous Agents and Multi-Agent Systems **33**(3), 275–297 (2019)
583. Kaczmarek, J., Rothe, J.: Manipulation in communication structures of graph-restricted weighted voting games. In: Proceedings of the 7th International Conference on Algorithmic Decision Theory, pp. 194–208. Springer-Verlag *Lecture Notes in Artificial Intelligence #13023* (2021)
584. Kaczmarek, J., Rothe, J.: Controlling weighted voting games by deleting or adding players with or without changing the quota. Annals of Mathematics and Artificial Intelligence **92**(3) (2024). URL https://doi.org/10.1007/s10472-023-09874-x. To appear.
585. Kaczmarek, J., Rothe, J.: NP^{PP}-completeness of control by adding players to change the Penrose–Banzhaf power index in weighted voting games (extended abstract). In: Proceedings of the 23rd International Conference on Autonomous Agents and Multiagent Systems. IFAAMAS (2024). To appear.
586. Kaczmarek, J., Rothe, J., Talmon, N.: Complexity of control by adding or deleting edges in graph-restricted weighted voting games. In: Proceedings of the 26th European Conference on Artificial Intelligence, *Frontiers in Artificial Intelligence and Applications*, vol. 372, pp. 1190–1197. IOS Press (2023)
587. Kakutani, S.: A generalization of Brouwer's fixed point theorem. Duke Mathematical Journal **8**(3), 457–459 (1941)
588. Kalai, E., Zemel, E.: Generalized network problems yielding totally balanced games. Operations Research **30**(5), 998–1008 (1982)
589. Kalai, E., Zemel, E.: Totally balanced games and games of flow. Mathematics of Operations Research **7**(3), 476–478 (1982)
590. Kalinowski, T., Narodytska, N., Walsh, T.: A social welfare optimal sequential allocation procedure. In: Proceedings of the 23rd International Joint Conference on Artificial Intelligence, pp. 227–233. AAAI Press/IJCAI (2013)
591. Kalinowski, T., Narodytska, N., Walsh, T., Xia, L.: Strategic behavior when allocating indivisible goods sequentially. In: Proceedings of the 27th AAAI Conference on Artificial Intelligence, pp. 452–458. AAAI Press (2013)
592. Kaneko, M., Nakamura, K.: The Nash social welfare function. Econometrica **47**(2), 423–436 (1979)
593. Kannai, Y., Peleg, B.: A note on the extension of an order on a set to the power set. Journal of Economic Theory **32**(1), 172–175 (1984)
594. Karp, J., Kazachkov, A., Procaccia, A.: Envy-free division of sellable goods. In: Proceedings of the 28th AAAI Conference on Artificial Intelligence, pp. 728–734. AAAI Press (2014)
595. Karp, R.: Reducibility among combinatorial problems. In: R. Miller, J. Thatcher (eds.) Complexity of Computer Computations, pp. 85–103. Plenum Press (1972)
596. Karp, R., Vazirani, U., Vazirani, V.: An optimal algorithm for on-line bipartite matching. In: Proceedings of the 22nd ACM Symposium on Theory of Computing, pp. 352–358. ACM Press (1990)
597. Kawase, Y., Sumita, H.: On the max-min fair stochastic allocation of indivisible goods. In: Proceedings of the 34th AAAI Conference on Artificial Intelligence, pp. 2070–2078. AAAI Press (2020)

598. Keeney, R., Raiffa, H.: Decisions with Multiple Objectives: Preferences and Value Tradeoffs. John Wiley and Sons (1976)
599. de Keijzer, B., Bouveret, S., Klos, T., Zhang, Y.: On the complexity of efficiency and envy-freeness in fair division of indivisible goods with additive preferences. In: Proceedings of the 1st International Conference on Algorithmic Decision Theory, pp. 98–110. Springer-Verlag *Lecture Notes in Artificial Intelligence #5783* (2009)
600. Kellerer, H., Pferschy, U., Pisinger, D.: Knapsack Problems. Springer-Verlag (2004)
601. Kelly, J.: Strategy-proofness and social choice functions without single-valuedness. Econometrica **45**(2), 439–446 (1977)
602. Kemeny, J.: Mathematics without numbers. Dædalus **88**(4), 571–591 (1959)
603. Kerkmann, A., Cramer, S., Rothe, J.: Altruism in coalition formation games. Annals of Mathematics and Artificial Intelligence (2023). URL https://doi.org/10.1007/s10472-023-09881-y. To appear.
604. Kerkmann, A., Lang, J., Rey, A., Rothe, J., Schadrack, H., Schend, L.: Hedonic games with ordinal preferences and thresholds. Journal of Artificial Intelligence Research **67**, 705–756 (2020)
605. Kerkmann, A., Nguyen, N., Rey, A., Rey, L., Rothe, J., Schend, L., Wiechers, A.: Altruistic hedonic games. Journal of Artificial Intelligence Research **75**, 129–169 (2022)
606. Kerkmann, A., Nguyen, N., Rothe, J.: Local fairness in hedonic games via individual threshold coalitions. Theoretical Computer Science **877**, 1–17 (2021)
607. Kerkmann, A., Rothe, J.: Altruism in coalition formation games. In: Proceedings of the 29th International Joint Conference on Artificial Intelligence, pp. 347–353. ijcai.org (2020)
608. Kerkmann, A., Rothe, J.: Popularity and strict popularity in altruistic hedonic games and minimum-based altruistic hedonic games (extended abstract). In: Proceedings of the 21st International Conference on Autonomous Agents and Multiagent Systems, pp. 1657–1659. IFAAMAS (2022)
609. Kern, P., Neugebauer, D., Rothe, J., Schilling, R., Stoyan, D., Weishaupt, R.: A closer look at the cake-cutting foundations through the lens of measure theory. In: R. Meir, W. Zwicker (eds.) Proceedings of the 8th International Workshop on Computational Social Choice. Nonarchival proceedings, Technion, Haifa, Israel (2021)
610. Kern, P., Neugebauer, D., Rothe, J., Schilling, R., Stoyan, D., Weishaupt, R.: Cutting a cake is not always a 'piece of cake': A closer look at the foundations of cake-cutting through the lens of measure theory. Tech. Rep. arXiv:2111.05402v2 [cs.GT], ACM Computing Research Repository (CoRR) (2023)
611. Khot, S., Ponnuswami, A.: Approximation algorithms for the max-min allocation problem. In: Proceedings of Approximation, Randomization and Combinatorial Optimization. Algorithms and Techniques. 10th International Workshop APPROX 2007 and 11th International Workshop RANDOM 2007, pp. 204–217. Springer-Verlag *Lecture Notes in Computer Science #4627* (2007)
612. Klaus, B., Manlove, D., Rossi, F.: Matching under preferences. In: F. Brandt, V. Conitzer, U. Endriss, J. Lang, A. Procaccia (eds.) Handbook of Computational Social Choice, chap. 14, pp. 333–355. Cambridge University Press (2016)
613. Klijn, F.: An algorithm for envy-free allocations in an economy with indivisible objects and money. Social Choice and Welfare **17**(2), 201–215 (2000)
614. Knop, D., Koutecký, M., Mnich, M.: Voting and bribing in single-exponential time. In: Proceedings of the 34th Annual Symposium on Theoretical Aspects of Computer Science, *LIPIcs*, vol. 66, article 46, pp. 1–14. Leibniz-Zentrum für Informatik, Schloss Dagstuhl, Germany (2017)
615. Kober, S., Weltge, S.: Improved lower bound on the dimension of the EU Council's voting rules. Optimization Letters **15**, 1293–1302 (2021)
616. Köbler, J., Schöning, U., Wagner, K.: The difference and truth-table hierarchies for NP. R.A.I.R.O. Informatique théorique et Applications **21**(4), 419–435 (1987)

617. Kohler, D., Chandrasekaran, R.: A class of sequential games. Operations Research **19**(2), 270–277 (1971)
618. Konczak, K., Lang, J.: Voting procedures with incomplete preferences. In: Proceedings of the Multidisciplinary IJCAI-05 Workshop on Advances in Preference Handling, pp. 124–129 (2005)
619. Könemann, J.: Gewinnstrategie für ein Streichholzspiel. In: B. Vöcking, H. Alt, M. Dietzfelbinger, R. Reischuk, C. Scheideler, H. Vollmer, D. Wagner (eds.) Taschenbuch der Algorithmen, chap. 26, pp. 267–273. Springer-Verlag (2008)
620. Kornhauser, L., Sager, L.: Unpacking the court. Yale Law Journal **96**(1), 82–117 (1986)
621. Koutsoupias, E., Papadimitriou, C.: Worst-case equilibria. Computer Science Review **3**(2), 65–69 (2009)
622. Kuckuck, B., Rothe, J.: Sequential allocation rules are separable: Refuting a conjecture on scoring-based allocation of indivisible goods. In: Proceedings of the 17th International Conference on Autonomous Agents and Multiagent Systems, pp. 650–658. IFAAMAS (2018)
623. Kuckuck, B., Rothe, J.: Duplication monotonicity in the allocation of indivisible goods. AI Communications **32**(4), 253–270 (2019)
624. Kuhlisch, W., Roos, M., Rothe, J., Rudolph, J., Scheuermann, B., Stoyan, D.: A statistical approach to calibrating the scores of biased reviewers of scientific papers. Metrika **79**(1), 37–57 (2016)
625. Kuhn, H.: On games of fair division. In: M. Shubik (ed.) Essays in Mathematical Economics in Honor of Oskar Morgenstern, pp. 29–37. Princeton University Press (1967)
626. Kulkarni, R., Mehta, R., Taki, S.: Indivisible mixed manna: On the computability of MMS+PO allocations. In: Proceedings of the 22nd ACM Conference on Economics and Computation, pp. 683–684. ACM Press (2021)
627. Kulkarni, R., Mehta, R., Taki, S.: On the PTAS for maximin shares in an indivisible mixed manna. In: Proceedings of the 35th AAAI Conference on Artificial Intelligence, pp. 5523–5530. AAAI Press (2021)
628. Kullmann, O.: New methods for 3-SAT decision and worst-case analysis. Theoretical Computer Science **223**(1–2), 1–72 (1999)
629. Kurokawa, D., Procaccia, A., Shah, N.: Leximin allocations in the real world. ACM Transactions on Economics and Computation **6**(3–4), 11:1–11:24 (2018)
630. Kurokawa, D., Procaccia, A., Wang, J.: Fair enough: Guaranteeing approximate maximin shares. Journal of the ACM **65**(2), 8:1–8:27 (2018)
631. Kurz, S., Napel, S.: Dimension of the Lisbon voting rules in the EU Council: A challenge and new world record. Optimization Letters **10**(6), 1245–1256 (2016)
632. Kusek, B., Bredereck, R., Faliszewski, P., Kaczmarczyk, A., Knop, D.: Bribery can get harder in structured multiwinner approval election. In: Proceedings of the 22nd International Conference on Autonomous Agents and Multiagent Systems, pp. 1725–1733. IFAAMAS (2023)
633. Kyropoulou, M., Suksompong, W., Voudouris, A.: Almost envy-freeness in group resource allocation. In: Proceedings of the 28th International Joint Conference on Artificial Intelligence, pp. 400–406. ijcai.org (2019)
634. Labunski, R.: James Madison and the Struggle for the Bill of Rights. Oxford University Press (2006)
635. Ladner, R., Lynch, N., Selman, A.: A comparison of polynomial time reducibilities. Theoretical Computer Science **1**(2), 103–124 (1975)
636. Lafage, C., Lang, J.: Logical representation of preferences for group decision making. In: Proceedings of the 7th International Conference on Principles of Knowledge Representation and Reasoning, pp. 457–468. AAAI Press (2000)
637. Lampis, M.: Determining a Slater winner is complete for parallel access to NP. In: Proceedings of the 39th Annual Symposium on Theoretical Aspects of Computer

Science, *LIPIcs*, vol. 219, pp. 45:1–45:14. Schloss Dagstuhl – Leibniz-Zentrum für Informatik (2022)
638. Landau, E.: Handbuch der Lehre von der Verteilung der Primzahlen. Teubner (1909)
639. Lang, J., Pigozzi, G., Slavkovik, M., van der Torre, L., Vesic, S.: A partial taxonomy of judgment aggregation rules and their properties. Social Choice and Welfare **48**(2), 327–356 (2017)
640. Lang, J., Pini, M., Rossi, F., Salvagnin, D., Venable, B., Walsh, T.: Winner determination in voting trees with incomplete preferences and weighted votes. Journal of Autonomous Agents and Multi-Agent Systems **25**(1), 130–157 (2012)
641. Lang, J., Xia, L.: Voting in combinatorial domains. In: F. Brandt, V. Conitzer, U. Endriss, J. Lang, A. Procaccia (eds.) Handbook of Computational Social Choice, chap. 9, pp. 197–222. Cambridge University Press (2016)
642. Lange, P., Nguyen, N., Rothe, J.: The price to pay for forgoing normalization in fair division of indivisible goods. Annals of Mathematics and Artificial Intelligence **88**(7), 817–832 (2020)
643. Lange, P., Rothe, J.: Optimizing social welfare in social networks. In: Proceedings of the 6th International Conference on Algorithmic Decision Theory, pp. 81–96. Springer-Verlag *Lecture Notes in Artificial Intelligence #11834* (2019)
644. Laslier, J., Sanver, R. (eds.): Handbook on Approval Voting. Springer (2010)
645. Laußmann, C., Rothe, J., Seeger, T.: Apportionment with thresholds: Strategic campaigns are easy in the top-choice but hard in the second-chance mode. In: Proceedings of the 49th International Conference on Current Trends in Theory and Practice of Computer Science, pp. 355–368. Springer-Verlag *Lecture Notes in Computer Science #14519* (2024)
646. Lee, E.: APX-hardness of maximizing Nash social welfare with indivisible items. Information Processing Letters **122**, 17–20 (2017)
647. LeGrand, R.: Analysis of the Minimax Procedure. Tech. Rep. WUCSE-2004-67, Department of Computer Science and Engineering, Washington University, St. Louis, Missouri (2004)
648. LeGrand, R., Markakis, E., Mehta, A.: Some results on approximating the minimax solution in approval voting. In: Proceedings of the 6th International Conference on Autonomous Agents and Multiagent Systems, p. 198. IFAAMAS (2007)
649. Lemaître, M., Verfaillie, G., Bataille, N.: Exploiting a common property resource under a fairness constraint: A case study. In: Proceedings of the 16th International Joint Conference on Artificial Intelligence, pp. 206–211. Morgan Kaufmann (1999)
650. Lemke, C., Howson Jr., J.: Equilibrium points of bimatrix games. SIAM Journal on Applied Mathematics **12**(2), 413–423 (1964)
651. Lenstra Jr., H.: Integer programming with a fixed number of variables. Mathematics of Operations Research **8**(4), 538–548 (1983)
652. Lesca, J., Perny, P.: LP solvable models for multiagent fair allocation problems. In: Proceedings of the 19th European Conference on Artificial Intelligence, pp. 393–398. IOS Press (2010)
653. Levenglick, A.: Fair and reasonable election systems. Behavioral Science **20**(1), 34–46 (1975)
654. Levin, L.: Average case complete problems. SIAM Journal on Computing **15**(1), 285–286 (1986)
655. Lichtenstein, D., Sipser, M.: GO is polynomial-space hard. Journal of the ACM **27**(2), 393–401 (1980)
656. Lin, A.: The complexity of manipulating k-approval elections. In: Proceedings of the 3rd International Conference on Agents and Artificial Intelligence, pp. 212–218. SciTePress (2011)
657. Lin, A.: Solving hard problems in election systems. Ph.D. thesis, Rochester Institute of Technology, Rochester, NY, USA (2012)

658. Lindner, C., Rothe, J.: Fixed-parameter tractability and parameterized complexity, applied to problems from computational social choice. In: A. Holder (ed.) Mathematical Programming Glossary. INFORMS Computing Society (2008)
659. Lindner, C., Rothe, J.: Degrees of guaranteed envy-freeness in finite bounded cake-cutting protocols. In: Proceedings of the 5th International Workshop on Internet & Network Economics, pp. 149–159. Springer-Verlag *Lecture Notes in Computer Science #5929* (2009)
660. Lipton, R., Markakis, E., Mossel, E., Saberi, A.: On approximately fair allocations of indivisible goods. In: Proceedings of the 5th ACM Conference on Electronic Commerce, pp. 125–131. ACM Press (2004)
661. List, C.: The discursive dilemma and public reason. Ethics **116**(2), 362–402 (2006)
662. List, C.: The theory of judgment aggregation: An introductory review. Synthese **187**(1), 179–207 (2012)
663. List, C., Pettit, P.: Aggregating sets of judgments: An impossibility result. Economics and Philosophy **18**(1), 89–110 (2002)
664. List, C., Puppe, C.: Judgment aggregation: A survey. In: P. Anand, P. Pattanaik, C. Puppe (eds.) The Handbook of Rational and Social Choice, chap. 19. Oxford University Press (2009)
665. Lonc, Z., Truszczynski, M.: Maximin share allocations on cycles. In: Proceedings of the 27th International Joint Conference on Artificial Intelligence, pp. 410–416. ijcai.org (2018)
666. Loreggia, A.: Iterative voting and multi-mode control in preference aggregation. Intelligenza Artificiale **8**(1), 39–51 (2014)
667. Loreggia, A.: Iterative voting, control and sentiment analysis. Ph.D. thesis, University of Padova (2016)
668. Loreggia, A., Narodytska, N., Rossi, F., Venable, B., Walsh, T.: Controlling elections by replacing candidates: Theoretical and experimental results. In: Proceedings of the 8th Multidisciplinary Workshop on Advances in Preference Handling, pp. 61–66 (2014)
669. Loreggia, A., Narodytska, N., Rossi, F., Venable, B., Walsh, T.: Controlling elections by replacing candidates or votes (extended abstract). In: Proceedings of the 14th International Conference on Autonomous Agents and Multiagent Systems, pp. 1737–1738. IFAAMAS (2015)
670. Lu, T., Boutilier, C.: Budgeted social choice: A framework for multiple recommendations in consensus decision making. In: V. Conitzer, J. Rothe (eds.) Proceedings of the 3rd International Workshop on Computational Social Choice, pp. 55–66. Heinrich-Heine-Universität Düsseldorf, Düsseldorf, Germany (2010)
671. Lucas, W.: The proof that a game may not have a solution. Transactions of the AMS **136**, 219–229 (1969)
672. Lucas, W.: Von Neumann-Morgenstern stable sets. In: Handbook of Game Theory, vol. 1. Elsevier (1992)
673. Lumet, C., Bouveret, S., Lemaître, M.: Fair division of indivisible goods under risk. In: Proceedings of the 20th European Conference on Artificial Intelligence, pp. 564–569. IOS Press (2012)
674. Luss, H.: On equitable resource allocation problems: A lexicographic minimax approach. Operations Research **47**(3), 361–378 (1999)
675. Mackin, E., Xia, L.: Allocating indivisible items in categorized domains. In: Proceedings of the 25th International Joint Conference on Artificial Intelligence, pp. 359–365. AAAI Press/IJCAI (2016)
676. Magrino, T., Rivest, R., Shen, E., Wagner, D.: Computing the margin of victory in IRV elections. In: Website Proceedings of the Electronic Voting Technology Workshop/Workshop on Trustworthy Elections (2011)
677. Manlove, D.: Algorithmics of Matching Under Preferences. World Scientific Publishing (2013)

678. Manurangsi, P., Suksompong, W.: Asymptotic existence of fair divisions for groups. Mathematical Social Sciences **89**, 100–108 (2017)
679. Manurangsi, P., Suksompong, W.: When do envy-free allocations exist? SIAM Journal on Discrete Mathematics **34**(3), 1505–1521 (2020)
680. Manurangsi, P., Suksompong, W.: Almost envy-freeness for groups: Improved bounds via discrepancy theory. In: Proceedings of the 30th International Joint Conference on Artificial Intelligence, pp. 335–341. ijcai.org (2021)
681. Manurangsi, P., Suksompong, W.: Closing gaps in asymptotic fair division. SIAM Journal on Discrete Mathematics **35**(2), 668–706 (2021)
682. Markakis, E.: Approximation algorithms and hardness results for fair division with indivisible goods. In: U. Endriss (ed.) Trends in Computational Social Choice, chap. 12, pp. 231–247. AI Access Foundation (2017)
683. Markakis, E., Psomas, C.: On worst-case allocations in the presence of indivisible goods. In: Proceedings of the 7th International Workshop on Internet & Network Economics, pp. 278–289. Springer-Verlag *Lecture Notes in Computer Science #7090* (2011)
684. Marple, A., Rey, A., Rothe, J.: Bribery in multiple-adversary path-disruption games is hard for the second level of the polynomial hierarchy (extended abstract). In: Proceedings of the 13th International Conference on Autonomous Agents and Multiagent Systems, pp. 1375–1376. IFAAMAS (2014)
685. Marple, A., Shoham, Y.: Equilibria in finite games with imperfect recall. Tech. Rep. SSRN-id2188646, Social Science Research Network (SSRN) (2012). Available at http://dx.doi.org/10.2139/ssrn.2188646
686. Maschler, M., Peleg, B., Shapley, L.: Geometric properties of the kernel, nucleolus, and related solution concepts. Mathematics of Operations Research **4**(4), 303–338 (1979)
687. Maskin, E.: On the fair allocation of indivisible goods. In: G. Feiwel (ed.) Arrow and the Foundations of the Theory of Economic Policy (essays in honor of Kenneth Arrow), pp. 341–349. MacMillan Publishing Company (1987)
688. Matsui, Y., Matsui, T.: NP-completeness for calculating power indices of weighted majority games. Theoretical Computer Science **263**(1–2), 305–310 (2001)
689. Maushagen, C., Neveling, M., Rothe, J., Selker, A.: Complexity of shift bribery for iterative voting rules. Annals of Mathematics and Artificial Intelligence **90**(10), 1017–1054 (2022)
690. Maushagen, C., Niclaus, D., Nüsken, P., Rothe, J., Seeger, T.: Toward completing the picture of control in Schulze and ranked pairs elections. In: Proceedings of the *18th International Symposium on Artificial Intelligence and Mathematics* (2024). URL https://isaim2024.cs.ou.edu/papers/ISAIM2024_Maushagen_etal.pdf. Nonarchival website proceedings.
691. Maushagen, C., Rothe, J.: Complexity of control by partitioning veto and maximin elections and of control by adding candidates to plurality elections. In: Proceedings of the 22nd European Conference on Artificial Intelligence, pp. 277–285. IOS Press (2016)
692. Maushagen, C., Rothe, J.: Complexity of control by partition of voters and of voter groups in veto and other scoring protocols. In: Proceedings of the 16th International Conference on Autonomous Agents and Multiagent Systems, pp. 615–623. IFAAMAS (2017)
693. Maushagen, C., Rothe, J.: Complexity of control by partitioning veto elections and of control by adding candidates to plurality elections. Annals of Mathematics and Artificial Intelligence **82**(4), 219–244 (2018)
694. Maushagen, C., Rothe, J.: The last voting rule is home: Complexity of control by partition of candidates or voters in maximin elections. In: Proceedings of the 24th European Conference on Artificial Intelligence, *Frontiers in Artificial Intelligence and Applications*, vol. 325, pp. 163–170. IOS Press (2020)

695. McCabe-Dansted, J., Pritchard, G., Slinko, A.: Approximability of Dodgson's rule. Social Choice and Welfare **31**(2), 311–330 (2008)
696. McGlaughlin, P., Garg, J.: Improving Nash social welfare approximations. Journal of Artificial Intelligence Research **68**, 225–245 (2020)
697. McLean, I., Urken, A.: Classics of Social Choice. University of Michigan Press (1995)
698. Mehrizi, M., D'Angelo, G.: Multi-winner election control via social influence: Hardness and algorithms for restricted cases. Algorithms **13**, 251 (2020)
699. Meir, R., Bachrach, Y., Rosenschein, J.: Minimal subsidies in expense sharing games. In: Proceedings of the 3rd International Symposium on Algorithmic Game Theory, pp. 347–358. Springer-Verlag *Lecture Notes in Computer Science #6386* (2010)
700. Meir, R., Procaccia, A., Rosenschein, J., Zohar, A.: Complexity of strategic behavior in multi-winner elections. Journal of Artificial Intelligence Research **33**, 149–178 (2008)
701. Meir, R., Zwicker, W. (eds.): Proceedings of the 8th International Workshop on Computational Social Choice. Technion – Israel Institute of Technology, Haifa, Israel (2021)
702. Menon, V., Larson, K.: Computational aspects of strategic behaviour in elections with top-truncated ballots. Journal of Autonomous Agents and Multi-Agent Systems **31**(6), 1506–1547 (2017)
703. Menton, C.: Normalized range voting broadly resists control. Theory of Computing Systems **53**(4), 507–531 (2013)
704. Menton, C., Singh, P.: Manipulation and control complexity of Schulze voting. Tech. Rep. arXiv:1206.2111v1 [cs.GT], ACM Computing Research Repository (CoRR) (2012)
705. Menton, C., Singh, P.: Control complexity of Schulze voting. In: Proceedings of the 23rd International Joint Conference on Artificial Intelligence, pp. 286–292. AAAI Press/IJCAI (2013)
706. Meyer, A., Stockmeyer, L.: The equivalence problem for regular expressions with squaring requires exponential space. In: Proceedings of the 13th IEEE Symposium on Switching and Automata Theory, pp. 125–129. IEEE Computer Society Press (1972)
707. Miller, G.: The magical number seven, plus or minus two: Some limits on our capacity for processing information. Psychological Review **63**(2), 81–97 (1956)
708. Miller, M., Osherson, D.: Methods for distance-based judgment aggregation. Social Choice and Welfare **32**(4), 575–601 (2009)
709. Mirrlees, J.: An exploration in the theory of optimum income taxation. The Review of Economic Studies **38**(2), 175–208 (1971)
710. Monderer, D., Tennenholtz, M.: K-implementation. Journal of Artificial Intelligence Research **21**, 37–62 (2004)
711. Mongin, P.: The doctrinal paradox, the discursive dilemma, and logical aggregation theory. Theory and Decision **73**(3), 315–355 (2012)
712. Monien, B., Speckenmeyer, E.: Solving satisfiability in less than 2^n steps. Discrete Applied Mathematics **10**(3), 287–295 (1985)
713. Monroe, B.: Fully proportional representation. The American Political Science Review **89**(4), 925–940 (1995)
714. Mossel, E., Rácz, M.: A quantitative Gibbard–Satterthwaite theorem without neutrality. In: Proceedings of the 44th ACM Symposium on Theory of Computing, pp. 1041–1060. ACM Press (2012)
715. Moulin, H.: Condorcet's principle implies the no show paradox. Journal of Economic Theory **45**(1), 53–64 (1988)
716. Moulin, H.: Cooperative Microeconomics. Princeton University Press (1995)
717. Moulin, H.: Fair Division and Collective Welfare. MIT Press (2004)

718. Muller, E., Satterthwaite, M.: The equivalence of strong positive association and strategy-proofness. Journal of Economic Theory **14**, 412–418 (1977)
719. Munagala, K., Shen, Z., Wang, K.: Optimal algorithms for multiwinner elections and the Chamberlin–Courant rule. In: Proceedings of the 22nd ACM Conference on Economics and Computation, pp. 697–717. ACM Press (2021)
720. Murhekar, A., Garg, J.: On fair and efficient allocations of indivisible goods. In: Proceedings of the 35th AAAI Conference on Artificial Intelligence, pp. 5595–5602. AAAI Press (2021)
721. Muroga, S.: Threshold Logic and its Applications. John Wiley and Sons (1971)
722. Myerson, R.: Graphs and cooperation in games. Mathematics of Operations Research **2**(3), 225–229 (1977)
723. Nagel, R.: Unraveling in guessing games: An experimental study. The American Economic Review **85**(5), 1313–1326 (1995)
724. Nanson, E.: Methods of election. Transactions and Proceedings of the Royal Society of Victoria **19**, 197–240 (1882)
725. Napel, S., Nohn, A., Alonso-Meijide, J.: Monotonicity of power in weighted voting games with restricted communication. Mathematical Social Sciences **64**(3), 247–257 (2012)
726. Narodytska, N., Walsh, T., Xia, L.: Manipulation of Nanson's and Baldwin's rules. In: Proceedings of the 25th AAAI Conference on Artificial Intelligence, pp. 713–718. AAAI Press (2011)
727. Nasar, S.: A Beautiful Mind: A Biography of John Forbes Nash, Jr., Winner of the Nobel Prize in Economics, 1994. Simon & Schuster (1998)
728. Nash, J.: The bargaining problem. Econometrica **18**(2), 155–162 (1950)
729. Nash, J.: Equilibrium points in n-person games. Proceedings of the National Academy of Sciences **36**(1), 48–49 (1950)
730. Nash, J.: Non-cooperative games. Annals of Mathematics **54**(2), 286–295 (1951)
731. Nehring, K., Puppe, C.: The structure of strategy-proof social choice. Part II: Nondictatorship, anonymity, and neutrality (2005). Unpublished manuscript.
732. Nehring, K., Puppe, C.: The structure of strategy-proof social choice. Part I: General characterization and possibility results on median space. Journal of Economic Theory **135**(1), 269–305 (2007)
733. Nemhauser, G., Wolsey, L., Fisher, M.: An analysis of approximations for maximizing submodular set functions. Mathematical Programming **14**(1), 265–294 (1978)
734. von Neumann, J.: Zur Theorie der Gesellschaftsspiele. Mathematische Annalen **100**(1), 295–320 (1928)
735. von Neumann, J.: A certain zero-sum game equivalent to the optimal assignment problem. In: W. Kuhn, A. Tucker (eds.) Contributions to the Theory of Games, vol. 2, pp. 5-12. Princeton University Press (1953)
736. von Neumann, J., Morgenstern, O.: Theory of Games and Economic Behavior. Princeton University Press (1944)
737. Neveling, M., Rothe, J.: Closing the gap of control complexity in Borda elections: Solving ten open cases. In: Proceedings of the 18th Italian Conference on Theoretical Computer Science, vol. 1949, pp. 138–149. CEUR-WS.org (2017)
738. Neveling, M., Rothe, J.: Solving seven open problems of offline and online control in Borda elections. In: Proceedings of the 31st AAAI Conference on Artificial Intelligence, pp. 3029–3035. AAAI Press (2017)
739. Neveling, M., Rothe, J.: The complexity of cloning candidates in multiwinner elections. In: Proceedings of the 19th International Conference on Autonomous Agents and Multiagent Systems, pp. 922–930. IFAAMAS (2020)
740. Neveling, M., Rothe, J.: Control complexity in Borda elections: Solving all open cases of offline control and some cases of online control. Artificial Intelligence **298**, 103 508 (2021)

References

741. Neveling, M., Rothe, J., Weishaupt, R.: The possible winner problem with uncertain weights revisited. In: Proceedings of the 23rd International Symposium on Fundamentals of Computation Theory, pp. 399–412. Springer-Verlag *Lecture Notes in Computer Science #12867* (2021)
742. Nguyen, N., Baumeister, D., Rothe, J.: Strategy-proofness of scoring allocation correspondences for indivisible goods. Social Choice and Welfare **50**(1), 101–122 (2018)
743. Nguyen, N., Nguyen, T., Roos, M., Rothe, J.: Complexity and approximability of egalitarian and Nash product social welfare optimization in multiagent resource allocation. In: Proceedings of the 6th European Starting AI Researcher Symposium, pp. 204–215. IOS Press (2012)
744. Nguyen, N., Nguyen, T., Roos, M., Rothe, J.: Computational complexity and approximability of social welfare optimization in multiagent resource allocation. Journal of Autonomous Agents and Multi-Agent Systems **28**(2), 256–289 (2014)
745. Nguyen, N., Rey, A., Rey, L., Rothe, J., Schend, L.: Altruistic hedonic games. In: Proceedings of the 15th International Conference on Autonomous Agents and Multiagent Systems, pp. 251–259. IFAAMAS (2016)
746. Nguyen, T., Roos, M., Rothe, J.: A survey of approximability and inapproximability results for social welfare optimization in multiagent resource allocation. Annals of Mathematics and Artificial Intelligence **68**(1–3), 65–90 (2013)
747. Nguyen, T., Rothe, J.: Minimizing envy and maximizing average Nash social welfare in the allocation of indivisible goods. Discrete Applied Mathematics **179**, 54–68 (2014)
748. Nguyen, T., Rothe, J.: Approximate Pareto set for fair and efficient allocation: Few agent types or few resource types. In: Proceedings of the 29th International Joint Conference on Artificial Intelligence, pp. 290–296. ijcai.org (2020)
749. Nguyen, T., Rothe, J.: Improved bi-criteria approximation schemes for load balancing on unrelated machines with cost constraints. Theoretical Computer Science **858**, 35–48 (2021)
750. Nguyen, T., Rothe, J.: Complexity results and exact algorithms for fair division of indivisible items: A survey. In: Proceedings of the 32nd International Joint Conference on Artificial Intelligence, pp. 6732–6740. ijcai.org (2023)
751. Nguyen, T., Rothe, J.: Fair and efficient allocation with few agent types, few item types, or small value levels. Artificial Intelligence **314**, 103 820 (2023)
752. Nicolo, A., Yu, Y.: Strategic divide and choose. Games and Economic Behavior **64**(1), 268–289 (2008)
753. Niedermeier, R.: Invitation to Fixed-Parameter Algorithms. Oxford University Press (2006)
754. Nisan, N.: Bidding languages for combinatorial auctions. In: P. Cramton, Y. Shoham, R. Steinberg (eds.) Combinatorial Auctions. MIT Press (2006)
755. Nisan, N., Roughgarden, T., Tardos, É., Vazirani, V. (eds.): Algorithmic Game Theory. Cambridge University Press (2007)
756. Norden, L., Burstein, A., Hall, J., Chen, M.: Post-Election Audits: Restoring Trust in Elections. Brennan Center for Justice at the New York University School of Law and the Samuelson Law, Technology & Public Policy Clinic at the University of California, Berkeley School of Law (Boalt Hall) (2007)
757. Obraztsova, S., Elkind, E., Hazon, N.: Ties matter: Complexity of voting manipulation revisited. In: Proceedings of the 22nd International Joint Conference on Artificial Intelligence, pp. 2698–2703. AAAI Press/IJCAI (2011)
758. Ogiwara, M., Watanabe, O.: On polynomial-time bounded truth-table reducibility of NP sets to sparse sets. SIAM Journal on Computing **20**(3), 471–483 (1991)
759. Oh, H., Procaccia, A., Suksompong, W.: Fairly allocating many goods with few queries. SIAM Journal on Discrete Mathematics **35**(2), 788–813 (2021)

760. Ohta, K., Barrot, N., Ismaili, A., Sakurai, Y., Yokoo, M.: Core stability in hedonic games among friends and enemies: Impact of neutrals. In: Proceedings of the 31st AAAI Conference on Artificial Intelligence, pp. 359–365. AAAI Press (2017)
761. Olsen, M.: On defining and computing communities. In: Proceedings of Computing: The 18th Australasian Theory Symposium, *Conferences in Research and Practice in Information Technology (CRPIT)*, vol. 128, pp. 97–102 (2012)
762. Osborne, M., Rubinstein, A.: A Course in Game Theory. MIT Press (1994)
763. Papadimitriou, C.: On the complexity of the parity argument and other inefficient proofs of existence. Journal of Computer and System Sciences **48**(3), 498–532 (1994)
764. Papadimitriou, C.: Computational Complexity, second edn. Addison-Wesley (1995)
765. Papadimitriou, C.: Algorithms, games, and the internet. In: Proceedings of the 33rd ACM Symposium on Theory of Computing, pp. 749–753. ACM Press (2001)
766. Papadimitriou, C., Yannakakis, M.: The complexity of facets (and some facets of complexity). Journal of Computer and System Sciences **28**(2), 244–259 (1984)
767. Papadimitriou, C., Zachos, S.: Two remarks on the power of counting. In: Proceedings of the 6th GI Conference on Theoretical Computer Science, pp. 269–276. Springer-Verlag *Lecture Notes in Computer Science #145* (1983)
768. Pápai, S.: Strategyproof single unit award rules. Social Choice and Welfare **18**(4), 785–798 (2001)
769. Pápai, S.: Exchange in a general market with indivisible goods. Journal of Economic Theory **132**(1), 208–235 (2007)
770. Parkes, D., Procaccia, A.: Dynamic social choice with evolving preferences. In: Proceedings of the 27th AAAI Conference on Artificial Intelligence, pp. 767–773. AAAI Press (2013)
771. Parkes, D., Xia, L.: A complexity-of-strategic-behavior comparison between Schulze's rule and ranked pairs. In: Proceedings of the 26th AAAI Conference on Artificial Intelligence, pp. 1429–1435. AAAI Press (2012)
772. Pauly, M., van Hees, M.: Logical constraints on judgment aggregation. Journal of Philosophical Logic **35**(1), 569–585 (2006)
773. Payan, J., Zick, Y.: I will have order! Optimizing orders for fair reviewer assignment. In: Proceedings of the 21st International Conference on Autonomous Agents and Multiagent Systems, pp. 1711–1713. IFAAMAS (2022)
774. Peleg, B., Sudhölter, P.: Introduction to the Theory of Cooperative Games. Kluwer Academic Publishers (2003)
775. Penrose, L.: The elementary statistics of majority voting. Journal of the Royal Statistical Society **109**(1), 53–57 (1946)
776. Peters, D., Lackner, M.: Preferences single-peaked on a circle. Journal of Artificial Intelligence Research **68**, 463–502 (2020)
777. Peters, D., Pierczyński, G., Skowron, P.: Proportional participatory budgeting with additive utilities. In: Proceedings of the 35th Conference on Neural Information Processing Systems, pp. 12,726–12,737. Curran Associates, Inc. (2021)
778. Pettit, P.: Deliberative democracy and the discursive dilemma. Philosophical Issues **11**(1), 268–299 (2001)
779. Piccione, M., Rubinstein, A.: On the interpretation of decision problems with imperfect recall. Games and Economic Behavior **20**(1), 3–24 (1997)
780. Pigozzi, G.: Belief merging and the discursive dilemma: An argument-based account of paradoxes of judgment. Synthese **152**(2), 285–298 (2006)
781. Pini, M., Rossi, F., Venable, B., Walsh, T.: Incompleteness and incomparability in preference aggregation: Complexity results. Artificial Intelligence **175**(7), 1272–1289 (2011)
782. Pivato, M., Lev, O. (eds.): Proceedings of the 9th International Workshop on Computational Social Choice. Ben-Gurion University of the Negev, Beersheba, Israel (2023)

783. Plaut, B., Roughgarden, T.: Almost envy-freeness with general valuations. SIAM Journal on Discrete Mathematics **34**(2), 1039–1068 (2020)
784. Plaut, B., Roughgarden, T.: Communication complexity of discrete fair division. SIAM Journal on Computing **49**(1), 206–243 (2020)
785. Porter, R., Nudelman, E., Shoham, Y.: Simple search methods for finding a Nash equilibrium. Games and Economic Behavior **63**(2), 642–662 (2008)
786. Poundstone, W.: Gaming the Vote: Why Elections Aren't Fair (and What We Can Do about It). Hill and Wang, a division of Farrar, Straus and Giroux (2008)
787. Prasad, K., Kelly, J.: NP-completeness of some problems concerning voting games. International Journal of Game Theory **19**(1), 1–9 (1990)
788. Procaccia, A.: Thou shalt covet thy neighbor's cake. In: Proceedings of the 21st International Joint Conference on Artificial Intelligence, pp. 239–244. IJCAI (2009)
789. Procaccia, A.: Cake cutting algorithms. In: F. Brandt, V. Conitzer, U. Endriss, J. Lang, A. Procaccia (eds.) Handbook of Computational Social Choice, chap. 13, pp. 311–329. Cambridge University Press (2016)
790. Procaccia, A., Rosenschein, J.: Junta distributions and the average-case complexity of manipulating elections. Journal of Artificial Intelligence Research **28**, 157–181 (2007)
791. Procaccia, A., Rosenschein, J., Kaminka, G.: On the robustness of preference aggregation in noisy environments. In: Proceedings of the 6th International Conference on Autonomous Agents and Multiagent Systems, pp. 416–422. IFAAMAS (2007)
792. Procaccia, A., Rosenschein, J., Zohar, A.: On the complexity of achieving proportional representation. Social Choice and Welfare **30**(3), 353–362 (2008)
793. Procaccia, A., Walsh, T. (eds.): Proceedings of the 5th International Workshop on Computational Social Choice. Carnegie Mellon University, Pittsburgh, USA (2014)
794. Puppe, C., Tasnádi, A.: Optimal redistricting under geographical constraints: Why "pack and crack" does not work. Economics Letters **105**(1), 93–96 (2009)
795. Quinzii, M.: Core and competitive equilibria with indivisibilities. International Journal of Game Theory **13**(1), 41–60 (1984)
796. Ramaekers, E.: Fair allocation of indivisible goods: The two-agent case. Social Choice and Welfare **41**(2), 359–380 (2013)
797. Ramezani, S., Endriss, U.: Nash social welfare in multiagent resource allocation. In: Agent-Mediated Electronic Commerce. Designing Trading Strategies and Mechanisms for Electronic Markets, pp. 117–131. Springer-Verlag *Lecture Notes in Business Information Processing #79* (2010)
798. von Randow, G.: Das Ziegenproblem: Denken in Wahrscheinlichkeiten. Rowohlt (2004)
799. Reisch, S.: Gobang ist PSPACE-vollständig. Acta Informatica **13**(1), 59–66 (1980)
800. Reisch, Y., Rothe, J., Schend, L.: The margin of victory in Schulze, cup, and Copeland elections: Complexity of the regular and exact variants. In: Proceedings of the 7th European Starting AI Researcher Symposium, pp. 250–259. IOS Press (2014)
801. Rey, A., Rothe, J.: Complexity of merging and splitting for the probabilistic Banzhaf power index in weighted voting games. In: Proceedings of the 19th European Conference on Artificial Intelligence, pp. 1021–1022. IOS Press (2010)
802. Rey, A., Rothe, J.: Merging and splitting for power indices in weighted voting games and network flow games on hypergraphs. In: Proceedings of the 5th European Starting AI Researcher Symposium, pp. 277–289. IOS Press (2010)
803. Rey, A., Rothe, J.: Bribery in path-disruption games. In: Proceedings of the 2nd International Conference on Algorithmic Decision Theory, pp. 247–261. Springer-Verlag *Lecture Notes in Artificial Intelligence #6992* (2011)
804. Rey, A., Rothe, J.: Probabilistic path-disruption games. In: Proceedings of the 20th European Conference on Artificial Intelligence, pp. 923–924. IOS Press (2012). An extended version appears in the proceedings of the *6th European Starting AI Researcher Symposium*, IOS Press, pages 264–269, August 2012

805. Rey, A., Rothe, J.: False-name manipulation in weighted voting games is hard for probabilistic polynomial time. Journal of Artificial Intelligence Research **50**, 573–601 (2014)
806. Rey, A., Rothe, J.: Structural control in weighted voting games. The B.E. Journal on Theoretical Economics **18**(2), 1–15 (2018)
807. Rey, A., Rothe, J., Marple, A.: Path-disruption games: Bribery and a probabilistic model. Theory of Computing Systems **60**(2), 222–252 (2017)
808. Rey, A., Rothe, J., Schadrack, H., Schend, L.: Toward the complexity of the existence of wonderfully stable partitions and strictly core stable coalition structures in enemy-oriented hedonic games. Annals of Mathematics and Artificial Intelligence **77**(3), 317–333 (2016)
809. Rieck, C.: Spieltheorie: Eine Einführung, 10th edn. Christian Rieck Verlag (2010)
810. Riege, T., Rothe, J.: Completeness in the boolean hierarchy: Exact-Four-Colorability, minimal graph uncolorability, and exact domatic number problems – a survey. Journal of Universal Computer Science **12**(5), 551–578 (2006)
811. Riege, T., Rothe, J.: Improving deterministic and randomized exponential-time algorithms for the satisfiability, the colorability, and the domatic number problem. Journal of Universal Computer Science **12**(6), 725–745 (2006)
812. Roberts, K.: Voting over income tax schedules. Journal of Public Economics **8**(3), 329–340 (1977)
813. Robertson, J., Webb, W.: Near exact and envy free cake division. Ars Combinatoria **45**, 97–108 (1997)
814. Robertson, J., Webb, W.: Cake-Cutting Algorithms: Be Fair If You Can. A K Peters (1998)
815. Rogers Jr., H.: The Theory of Recursive Functions and Effective Computability. McGraw-Hill (1967)
816. Roos, M., Rothe, J.: Complexity of social welfare optimization in multiagent resource allocation. In: Proceedings of the 9th International Conference on Autonomous Agents and Multiagent Systems, pp. 641–648. IFAAMAS (2010)
817. Roos, M., Rothe, J., Scheuermann, B.: How to calibrate the scores of biased reviewers by quadratic programming. In: Proceedings of the 25th AAAI Conference on Artificial Intelligence, pp. 255–260. AAAI Press (2011)
818. Roth, A. (ed.): The Shapley Value: Essays in Honor of Lloyd S. Shapley. Cambridge University Press (1988)
819. Rothe, J.: Complexity Theory and Cryptology. An Introduction to Cryptocomplexity. EATCS Texts in Theoretical Computer Science. Springer-Verlag (2005)
820. Rothe, J.: Borda count in collective decision making: A summary of recent results. In: Proceedings of the 33rd AAAI Conference on Artificial Intelligence, pp. 9830–9836. AAAI Press (2019)
821. Rothe, J.: Thou shalt love thy neighbor as thyself when thou playest: Altruism in game theory. In: Proceedings of the 35th AAAI Conference on Artificial Intelligence, pp. 15 070–15 077. AAAI Press (2021)
822. Rothe, J., Baumeister, D., Lindner, C., Rothe, I.: Einführung in Computational Social Choice: Individuelle Strategien und kollektive Entscheidungen beim Spielen, Wählen und Teilen. Spektrum Akademischer Verlag (2011)
823. Rothe, J., Schend, L.: Control complexity in Bucklin, fallback, and plurality voting: An experimental approach. In: Proceedings of the 11th International Symposium on Experimental Algorithms, pp. 356–368. Springer-Verlag *Lecture Notes in Computer Science #7276* (2012)
824. Rothe, J., Schend, L.: Challenges to complexity shields that are supposed to protect elections against manipulation and control: A survey. Annals of Mathematics and Artificial Intelligence **68**(1–3), 161–193 (2013)
825. Rothe, J., Spakowski, H., Vogel, J.: Exact complexity of the winner problem for Young elections. Theory of Computing Systems **36**(4), 375–386 (2003)

References

826. Russel, N.: Complexity of control of Borda count elections. Master's thesis, Rochester Institute of Technology (2007)
827. Saari, D.: Which is better: the Condorcet or Borda winner? Social Choice and Welfare **27**(1), 107–129 (2006)
828. Sabán, D., Sethuraman, J.: The complexity of computing the random priority allocation matrix. In: Proceedings of the 9th International Workshop on Internet & Network Economics, p. 421. Springer-Verlag *Lecture Notes in Computer Science #8289* (2013)
829. Sánchez-Fernández, L., Elkind, E., Lackner, M., Fernández, N., Fisteus, J., Basanta Val, P., Skowron, P.: Proportional justified representation. In: Proceedings of the 31st AAAI Conference on Artificial Intelligence, pp. 670–676. AAAI Press (2017)
830. Sandholm, T.: Contract types for satisficing task allocation: I Theoretical results. In: Proceedings of the AAAI Spring Symposium: Satisficing Models, pp. 68–75. AAAI Press (1998)
831. Sandholm, T.: Distributed rational decision making. In: G. Weiß (ed.) Multiagent Systems, pp. 201–258. MIT Press (1999)
832. Sandholm, T., Gilpin, A., Conitzer, V.: Mixed-integer programming methods for finding Nash equilibria. In: Proceedings of the 20th National Conference on Artificial Intelligence, pp. 495–501. AAAI Press (2005)
833. Sandomirskiy, F., Segal-Halevi, E.: Efficient fair division with minimal sharing. Operations Research **70**(3), 1762–1782 (2022)
834. Sarwate, A., Checkoway, S., Shacham, H.: Risk-limiting audits and the margin of victory for nonplurality elections. Statistics, Politics and Policy pp. 29–64 (2012)
835. Satterthwaite, M.: Strategy-proofness and Arrow's conditions: Existence and correspondence theorems for voting procedures and social welfare functions. Journal of Economic Theory **10**(2), 187–217 (1975)
836. Savitch, W.: Relationships between nondeterministic and deterministic tape complexities. Journal of Computer and System Sciences **4**(2), 177–192 (1970)
837. Scarf, H.: The approximation of fixed points of a continuous mapping. SIAM Journal on Applied Mathematics **15**(5), 1328–1343 (1967)
838. Schiermeyer, I.: Pure literal look ahead: An $\mathcal{O}(1.497^n)$ 3-satisfiability algorithm. In: Proceedings of the Workshop on the Satisfiability Problem, pp. 127–136 (1996). Also available as Technical Report No. 96-230, Universität zu Köln, Germany
839. Schlotter, I., Faliszewski, P., Elkind, E.: Campaign management under approval-driven voting rules. In: Proceedings of the 25th AAAI Conference on Artificial Intelligence, pp. 726–731. AAAI Press (2011)
840. Schneckenburger, S., Dorn, B., Endriss, U.: The Atkinson inequality index in multiagent resource allocation. In: Proceedings of the 16th International Conference on Autonomous Agents and Multiagent Systems, pp. 272–280. IFAAMAS (2017)
841. Schöning, U.: Complete sets and closeness to complexity classes. Mathematical Systems Theory **19**(1), 29–42 (1986)
842. Schöning, U.: Algorithmics in exponential time. In: Proceedings of the 22nd Annual Symposium on Theoretical Aspects of Computer Science, pp. 36–43. Springer-Verlag *Lecture Notes in Computer Science #3404* (2005)
843. Schulze, M.: A new monotonic, clone-independent, reversal symmetric, and Condorcet-consistent single-winner election method. Social Choice and Welfare **36**(2), 267–303 (2011)
844. Schwartz, T.: On the possibility of rational policy evaluation. Theory and Decision **1**(1), 89–106 (1970)
845. Schwartz, T.: Rationality and the myth of the maximum. Noûs **6**(2), 97–117 (1972)
846. Seddighin, M., Saleh, H., Ghodsi, M.: Maximin share guarantee for goods with positive externalities. Social Choice and Welfare **56**(2), 291–324 (2021)
847. Seddighin, M., Seddighin, S.: Improved maximin guarantees for subadditive and fractionally subadditive fair allocation problem. In: Proceedings of the 36th AAAI Conference on Artificial Intelligence, pp. 5183–5190. AAAI Press (2022)

848. Segal-Halevi, E., Hassidim, A., Aziz, H.: Fair allocation with diminishing differences. Journal of Artificial Intelligence Research **67**, 471–507 (2020)
849. Segal-Halevi, E., Suksompong, W.: Democratic fair allocation of indivisible goods. Artificial Intelligence **277**, 103 167 (2019)
850. Selker, A.: Manipulative Angriffe auf Judgment-Aggregation-Prozeduren. Master's thesis, Heinrich-Heine-Universität Düsseldorf, Institut für Informatik, Düsseldorf, Germany (2014)
851. Selman, A.: A taxonomy of complexity classes of functions. Journal of Computer and System Sciences **48**(2), 357–381 (1994)
852. Selten, R.: Reexamination of the perfectness concept for equilibrium points in extensive games. International Journal of Game Theory **4**(1), 25–55 (1975)
853. Selten, R., Nagel, R.: Das Zahlenwahlspiel – Ergebnisse und Hintergrund. Spektrum der Wissenschaft **1998**(2), 16–22 (1998)
854. Shams, P., Beynier, A., Bouveret, S., Maudet, N.: Fair in the eyes of others. In: Proceedings of the 24th European Conference on Artificial Intelligence, *Frontiers in Artificial Intelligence and Applications*, vol. 325, pp. 203–210. IOS Press (2020)
855. Shams, P., Beynier, A., Bouveret, S., Maudet, N.: Minimizing and balancing envy among agents using ordered weighted average. In: Proceedings of the 7th International Conference on Algorithmic Decision Theory, pp. 289–303. Springer-Verlag *Lecture Notes in Artificial Intelligence #13023* (2021)
856. Shapley, L.: A value for n-person games. In: H. Kuhn, A. Tucker (eds.) Contributions to the Theory of Games, *Annals of Mathematics Studies 40*, vol. II. Princeton University Press (1953)
857. Shapley, L.: Simple games: An outline of the descriptive theory. Behavioral Science **7**(1), 59–66 (1962)
858. Shapley, L.: Cores of convex games. International Journal of Game Theory **1**(1), 11–26 (1971)
859. Shapley, L.: Measurement of power in political systems. In: W. Lucas (ed.) Game Theory and its Applications. American Mathematical Society (1981). Proceedings of Symposia in Applied Mathematics, volume 24
860. Shapley, L., Shubik, M.: A method of evaluating the distribution of power in a committee system. The American Political Science Review **48**(3), 787–792 (1954)
861. Shapley, L., Shubik, M.: Quasi-cores in a monetary economy with non-convex preferences. Econometrica **34**(4), 805–827 (1966)
862. Shiryaev, D., Yu, L., Elkind, E.: On elections with robust winners. In: Proceedings of the 12th International Conference on Autonomous Agents and Multiagent Systems, pp. 415–422. IFAAMAS (2013)
863. Shoham, Y.: Computer science and game theory. Communications of the ACM **51**(8), 74–79 (2008)
864. Shoham, Y., Leyton-Brown, K.: Multiagent Systems: Algorithmic, Game-Theoretic, and Logical Foundations. Cambridge University Press (2009)
865. Sikdar, S., Adali, S., Xia, L.: Mechanism design for multi-type housing markets. In: Proceedings of the 31st AAAI Conference on Artificial Intelligence, pp. 684–690. AAAI Press (2017)
866. Sikdar, S., Guo, X., Wang, H., Xia, L., Cao, Y.: Sequential mechanisms for multi-type resource allocation. In: Proceedings of the 20th International Conference on Autonomous Agents and Multiagent Systems, pp. 1209–1217. IFAAMAS (2021)
867. Simpson, P.: On defining areas of voter choice: Professor Tullock on stable voting. The Quarterly Journal of Economics **83**(3), 478–490 (1969)
868. Singh, S.: The Simpsons and their Mathematical Secrets. Bloomsbury, New York (2013)
869. Skibski, O.: Closeness centrality via the Condorcet principle. Social Networks **74**, 13–18 (2023)

870. Skibski, O., Michalak, T., Sakurai, Y., Yokoo, M.: A pseudo-polynomial algorithm for computing power indices in graph-restricted weighted voting games. In: Proceedings of the 24th International Joint Conference on Artificial Intelligence, pp. 631–637. AAAI Press/IJCAI (2015)
871. Skowron, P.: Proportionality degree of multiwinner rules. In: Proceedings of the 22nd ACM Conference on Economics and Computation, pp. 820–840. ACM Press (2021)
872. Skowron, P., Faliszewski, P., Lang, J.: Finding a collective set of items: From proportional multirepresentation to group recommendation. Artificial Intelligence **241**, 191–216 (2016)
873. Skowron, P., Faliszewski, P., Slinko, A.: Fully proportional representation as resource allocation: Approximability results. In: Proceedings of the 23rd International Joint Conference on Artificial Intelligence, pp. 353–359. AAAI Press/IJCAI (2013)
874. Slater, P.: Inconsistencies in a schedule of paired comparisons. Biometrika **48**(3–4), 303–312 (1961)
875. Slinko, A.: On asymptotic strategy-proofness of classical social choice rules. Theory and Decision **52**(4), 389–398 (2002)
876. Slinko, A.: On asymptotic strategy-proofness of the plurality and the run-off rules. Social Choice and Welfare **19**(2), 313–324 (2002)
877. Slinko, A.: How large should a coalition be to manipulate an election? Mathematical Social Sciences **47**(3), 289–293 (2004)
878. Sloth, B.: The theory of voting and equilibria in noncooperative games. Games and Economic Behavior **5**(1), 152–169 (1993)
879. Smith, J.: Aggregation of preferences with variable electorate. Econometrica **41**(6), 1027–1041 (1973)
880. Smith, J., Lim, C.: Algorithms for network interdiction and fortification games. In: A. Chinchuluun, P. Pardalos, A. Migdalas, L. Pitsoulis (eds.) Pareto Optimality, Game Theory and Equlibria, pp. 609–644. Springer (2008)
881. Sonar, C., Dey, P., Misra, N.: On the complexity of winner verification and candidate winner for multiwinner voting rules. In: Proceedings of the 29th International Joint Conference on Artificial Intelligence, pp. 89–95. ijcai.org (2020)
882. Sönmez, T.: Strategy-proofness and essentially single-valued cores. Econometrica **67**(3), 677–690 (1999)
883. Sonoda, A., Fujita, E., Todo, T., Yokoo, M.: Two case studies for trading multiple indivisible goods with indifferences. In: Proceedings of the 28th AAAI Conference on Artificial Intelligence, pp. 791–797. AAAI Press (2014)
884. Sprumont, Y.: Axiomatizing ordinal welfare egalitarianism when preferences may vary. Journal of Economic Theory **68**(1), 77–110 (1996)
885. Stark, P.: Conservative statistical post-election audits. Annals of Applied Statistics **2**(2), 435–776 (2008)
886. Stark, P.: A sharper discrepancy measure for post-election audits. Annals of Applied Statistics **2**(3), 982–985 (2008)
887. Stark, P.: Risk-limiting postelection audits: Conservative-values from common probability inequalities. IEEE Transactions on Information Forensics and Security **4**(4), 1005–1014 (2009)
888. Stark, P.: Super-simple simultaneous single-ballot risk-limiting audits. In: Website Proceedings of the Electronic Voting Technology Workshop/Workshop on Trustworthy Elections (2010)
889. Steinhaus, H.: The problem of fair division. Econometrica **16**(1), 101–104 (1948)
890. Steinhaus, H.: Sur la division pragmatique. Econometrica **17**(Supplement: Report of the Washington Meeting), 315–319 (1949)
891. Steinhaus, H.: Mathematical Snapshots, third edn. Oxford University Press (1969)
892. Stockmeyer, L.: The polynomial-time hierarchy. Theoretical Computer Science **3**(1), 1–22 (1976)

893. Storer, J.: On the complexity of chess. Journal of Computer and System Sciences **27**(1), 77–100 (1983)
894. Stromquist, W.: How to cut a cake fairly. The American Mathematical Monthly **87**(8), 640–644 (1980)
895. Stromquist, W.: Envy-free cake divisions cannot be found by finite protocols. The Electronic Journal of Combinatorics **15**, R11 (2008)
896. Suksompong, W.: Asymptotic existence of proportionally fair allocations. Mathematical Social Sciences **81**, 62–65 (2016)
897. Suksompong, W.: Approximate maximin shares for groups of agents. Mathematical Social Sciences **92**, 40–47 (2018)
898. Suksompong, W.: Fairly allocating contiguous blocks of indivisible items. Discrete Applied Mathematics **260**, 227–236 (2019)
899. Suksompong, W.: On the number of almost envy-free allocations. Discrete Applied Mathematics **284**, 606–610 (2020)
900. Sun, A., Chen, B., Doan, X.: Connections between fairness criteria and efficiency for allocating indivisible chores. In: Proceedings of the 20th International Conference on Autonomous Agents and Multiagent Systems, pp. 1281–1289. IFAAMAS (2021)
901. Sung, S., Dimitrov, D.: On core membership testing for hedonic coalition formation games. Operations Research Letters **35**(2), 155–158 (2007)
902. Sung, S., Dimitrov, D.: Computational complexity in additive hedonic games. European Journal of Operational Research **203**(3), 635–639 (2010)
903. Svensson, L.: Large indivisibles: An analysis with respect to price equilibrium and fairness. Econometrica **51**(4), 939–954 (1983)
904. Svensson, L.: Strategy-proof allocation of indivisible goods. Social Choice and Welfare **16**, 557–567 (1999)
905. Tadenuma, K., Thomson, W.: No-envy and consistency in economies with indivisible goods. Econometrica **59**(6), 1755–1767 (1991)
906. Tadenuma, K., Thomson, W.: The fair allocation of an indivisible good when monetary compensations are possible. Mathematical Social Sciences **25**(2), 117–132 (1993)
907. Taylor, A.: Mathematics and Politics. Springer-Verlag (1995)
908. Taylor, A.: Social Choice and the Mathematics of Manipulation. Cambridge University Press (2005)
909. Taylor, A., Zwicker, W.: Weighted voting, multicameral representation, and power. Games and Economic Behavior **5**(1), 170–181 (1993)
910. Taylor, A., Zwicker, W.: Simple Games: Desirability Relations, Trading, Pseudoweightings. Princeton University Press (1999)
911. Tennenholtz, M.: Transitive voting. In: Proceedings of the 13th ACM Conference on Electronic Commerce, pp. 230–231. ACM Press (2004)
912. Thomson, W.: Problems of fair division and the egalitarian solution. Journal of Economic Theory **31**(2), 211–226 (1983)
913. Thomson, W.: Introduction to the theory of fair allocation. In: F. Brandt, V. Conitzer, U. Endriss, J. Lang, A. Procaccia (eds.) Handbook of Computational Social Choice, chap. 11, pp. 261–283. Cambridge University Press (2016)
914. Tideman, N.: Independence of clones as a criterion for voting rules. Social Choice and Welfare **4**(3), 185–206 (1987)
915. Toda, S.: PP is as hard as the polynomial-time hierarchy. SIAM Journal on Computing **20**(5), 865–877 (1991)
916. Todo, T., Li, R., Hu, X., Mouri, T., Iwasaki, A., Yokoo, M.: Generalizing envy-freeness toward group of agents. In: Proceedings of the 22nd International Joint Conference on Artificial Intelligence, pp. 386–392. AAAI Press/IJCAI (2011)
917. Tominaga, Y., Todo, T., Yokoo, M.: Manipulations in two-agent sequential allocation with random sequences. In: Proceedings of the 15th International Conference on Autonomous Agents and Multiagent Systems, pp. 141–149. IFAAMAS (2016)

918. Turing, A.: On computable numbers, with an application to the Entscheidungsproblem. Proceedings of the London Mathematical Society, ser. 2 **42**, 230–265 (1936). Correction, *ibid*, vol. 43, pp. 544–546, 1937.
919. Uckelman, J., Chevaleyre, Y., Endriss, U., Lang, J.: Representing utility functions via weighted goals. Mathematical Logic Quarterly **55**(4), 341–361 (2009)
920. Uckelman, J., Endriss, U.: Compactly representing utility functions using weighted goals and the max aggregator. Artificial Intelligence **174**(15), 1222–1246 (2010)
921. Valiant, L.: The complexity of computing the permanent. Theoretical Computer Science **8**(2), 189–201 (1979)
922. Vazirani, V.: Approximation Algorithms, second edn. Springer-Verlag (2003)
923. Vetschera, R., Kilgour, M.: Strategic behavior in contested-pile methods for fair division of indivisble items. Group Decision and Negotiation **22**(2), 299–319 (2013)
924. Wagner, K.: More complicated questions about maxima and minima, and some closures of NP. Theoretical Computer Science **51**(1–2), 53–80 (1987)
925. Wagner, K.: Bounded query classes. SIAM Journal on Computing **19**(5), 833–846 (1990)
926. Wagner, K., Wechsung, G.: Computational Complexity. D. Reidel Publishing Company (1986). Distributors for the U.S.A. and Canada: Kluwer Academic Publishers.
927. Walsh, T.: Uncertainty in preference elicitation and aggregation. In: Proceedings of the 22nd AAAI Conference on Artificial Intelligence, pp. 3–8. AAAI Press (2007)
928. Walsh, T.: Where are the really hard manipulation problems? The phase transition in manipulating the veto rule. In: Proceedings of the 21st International Joint Conference on Artificial Intelligence, pp. 324–329. IJCAI (2009)
929. Walsh, T.: An empirical study of the manipulability of single transferable voting. In: Proceedings of the 19th European Conference on Artificial Intelligence, pp. 257–262. IOS Press (2010)
930. Walsh, T.: Online cake cutting. In: Proceedings of the 2nd International Conference on Algorithmic Decision Theory, pp. 292–305. Springer-Verlag *Lecture Notes in Artificial Intelligence #6992* (2011)
931. Walsh, T.: Strategic behaviour when allocating indivisible goods. In: Proceedings of the 30th AAAI Conference on Artificial Intelligence, pp. 4177–4183. AAAI Press (2016)
932. Wang, H., Sikdar, S., Guo, X., Xia, L., Cao, Y., Wang, H.: Multi-type resource allocation with partial preferences. In: Proceedings of the 34th AAAI Conference on Artificial Intelligence, pp. 2260–2267. AAAI Press (2020)
933. Wang, J.: Average-case computational complexity theory. In: L. Hemaspaandra, A. Selman (eds.) Complexity Theory Retrospective II, pp. 295–328. Springer-Verlag (1997)
934. Wang, J.: Average-case intractable NP problems. In: D. Du, K. Ko (eds.) Advances in Languages, Algorithms, and Complexity, pp. 313–378. Kluwer Academic Publishers (1997)
935. Washburn, A., Wood, K.: Two-person zero-sum games for network interdiction. Operations Research **43**(2), 243–251 (1995)
936. Webb, W.: An algorithm for super envy-free cake division. Journal of Mathematical Analysis and Applications **239**(1), 175–179 (1999)
937. Wechsung, G.: Vorlesungen zur Komplexitätstheorie, *Teubner-Texte zur Informatik*, vol. 32. Teubner (2000)
938. Wegener, I.: Komplexitätstheorie. Grenzen der Effizienz von Algorithmen. Springer-Verlag (2003)
939. Weller, D.: Fair division of a measurable space. Journal of Mathematical Economics **14**(1), 5–17 (1985)
940. Wiechers, A., Rothe, J.: Stability in minimization-based altruistic hedonic games. In: Proceedings of the 9th European Starting AI Researcher Symposium, vol. 2655, pp. 3:1–3:8. CEUR-WS.org (2020)

941. Woeginger, G.: A polynomial-time approximation scheme for maximizing the minimum machine completion time. Operations Research Letters **20**(4), 149–154 (1997)
942. Woeginger, G.: Exact algorithms for NP-hard problems. In: M. Jünger, G. Reinelt, G. Rinaldi (eds.) Combinatorical Optimization: "Eureka, you shrink!", pp. 185–207. Springer-Verlag *Lecture Notes in Computer Science #2570* (2003)
943. Woeginger, G.: Core stability in hedonic coalition formation. In: Proceedings of the 39th International Conference on Current Trends in Theory and Practice of Computer Science, pp. 33–50. Springer-Verlag *Lecture Notes in Computer Science #7741* (2013)
944. Woeginger, G.: A hardness result for core stability in additive hedonic games. Mathematical Social Sciences **65**(2), 101–104 (2013)
945. Woeginger, G., Sgall, J.: On the complexity of cake cutting. Discrete Optimization **4**(2), 213–220 (2007)
946. Woodall, D.: Dividing a cake fairly. Journal of Mathematical Analysis and Applications **78**(1), 233–247 (1980)
947. Woodall, D.: A note on the cake-division problem. Journal of Combinatorial Theory, Series A **42**(2), 300–301 (1986)
948. Woodall, D.: Properties of preferential election rules. Voting Matters **3**, 8–15 (1994)
949. Woodall, D.: Monotonicity of single-seat preferential election rules. Discrete Applied Mathematics **77**(1), 81–88 (1997)
950. Wooldridge, M.: An Introduction to MultiAgent Systems, second edn. John Wiley and Sons (2009)
951. Wu, X., Li, B., Gan, J.: Budget-feasible maximum Nash social welfare is almost envy-free. In: Proceedings of the 30th International Joint Conference on Artificial Intelligence, pp. 465–471. ijcai.org (2021)
952. Xia, L.: Computing the margin of victory for various voting rules. In: Proceedings of the 13th ACM Conference on Electronic Commerce, pp. 982–999. ACM Press (2012)
953. Xia, L., Conitzer, V.: Generalized scoring rules and the frequency of coalitional manipulability. In: Proceedings of the 9th ACM Conference on Electronic Commerce, pp. 109–118. ACM Press (2008)
954. Xia, L., Conitzer, V.: A sufficient condition for voting rules to be frequently manipulable. In: Proceedings of the 9th ACM Conference on Electronic Commerce, pp. 99–108. ACM Press (2008)
955. Xia, L., Conitzer, V.: Stackelberg voting games: Computational aspects and paradoxes. In: Proceedings of the 24th AAAI Conference on Artificial Intelligence, pp. 697–702. AAAI Press (2010)
956. Xia, L., Conitzer, V.: Strategy-proof voting rules over multi-issue domains with restricted preferences. In: Proceedings of the 6th International Workshop on Internet & Network Economics, pp. 402–414. Springer-Verlag *Lecture Notes in Computer Science #6484* (2010)
957. Xia, L., Conitzer, V.: Determining possible and necessary winners given partial orders. Journal of Artificial Intelligence Research **41**, 25–67 (2011)
958. Xia, L., Conitzer, V., Lang, J.: Aggregating preferences in multi-issue domains by using maximum likelihood estimators. In: Proceedings of the 9th International Conference on Autonomous Agents and Multiagent Systems, pp. 399–408. IFAAMAS (2010)
959. Xia, L., Conitzer, V., Lang, J.: Strategic sequential voting in multi-issue domains and multiple-election paradoxes. In: Proceedings of the 12th ACM Conference on Electronic Commerce, pp. 179–188. ACM Press (2011)
960. Xia, L., Conitzer, V., Procaccia, A.: A scheduling approach to coalitional manipulation. In: Proceedings of the 11th ACM Conference on Electronic Commerce, pp. 275–284. ACM Press (2010)

961. Xia, L., Lang, J., Conitzer, V.: Hypercubewise preference aggregation in multi-issue domains. In: Proceedings of the 22nd International Joint Conference on Artificial Intelligence, pp. 158–163. AAAI Press/IJCAI (2011)
962. Xia, L., Lang, J., Monnot, J.: Possible winners when new alternatives join: New results coming up! In: Proceedings of the 10th International Conference on Autonomous Agents and Multiagent Systems, pp. 829–836. IFAAMAS (2011)
963. Xia, L., Zuckerman, M., Procaccia, A., Conitzer, V., Rosenschein, J.: Complexity of unweighted coalitional manipulation under some common voting rules. In: Proceedings of the 21st International Joint Conference on Artificial Intelligence, pp. 348–353. IJCAI (2009)
964. Xiao, M., Ling, J.: Algorithms for manipulating sequential allocation. In: Proceedings of the 34th AAAI Conference on Artificial Intelligence, pp. 2302–2309. AAAI Press (2020)
965. Yang, Y.: On the complexity of destructive bribery in approval-based multi-winner voting. In: Proceedings of the 19th International Conference on Autonomous Agents and Multiagent Systems, pp. 1584–1592. IFAAMAS (2020)
966. Yang, Y., Wang, J.: Parameterized complexity of multiwinner determination: More effort towards fixed-parameter tractability. Journal of Autonomous Agents and Multi-Agent Systems **37**(2), 28:1–28:35 (2023)
967. Yin, S., Wang, S., Zhang, L., Kroer, C.: Dominant resource fairness with meta-types. In: Proceedings of the 30th International Joint Conference on Artificial Intelligence, pp. 486–492. ijcai.org (2021)
968. Young, P.: Extending Condorcet's rule. Journal of Economic Theory **16**(2), 335–353 (1977)
969. Zankó, V.: #P-completeness via many-one reductions. International Journal of Foundations of Computer Science **2**(1), 76–82 (1991)
970. Zeng, D., Psomas, A.: Fairness-efficiency tradeoffs in dynamic fair division. In: Proceedings of the 21st ACM Conference on Economics and Computation, pp. 911–912. ACM Press (2020)
971. Zick, Y., Skopalik, A., Elkind, E.: The Shapley value as a function of the quota in weighted voting games. In: Proceedings of the 22nd International Joint Conference on Artificial Intelligence, pp. 490–496. AAAI Press/IJCAI (2011)
972. Zuckerman, M., Faliszewski, P., Bachrach, Y., Elkind, E.: Manipulating the quota in weighted voting games. Artificial Intelligence **180–181**, 1–19 (2012)
973. Zuckerman, M., Lev, O., Rosenschein, J.: An algorithm for the coalitional manipulation problem under maximin. In: Proceedings of the 10th International Conference on Autonomous Agents and Multiagent Systems, pp. 845–852. IFAAMAS (2011)
974. Zuckerman, M., Procaccia, A., Rosenschein, J.: Algorithms for the coalitional manipulation problem. Artificial Intelligence **173**(2), 392–412 (2009)

Index

$>$, 235
$\succ$, 199, 297, 492, 613
$\succeq$, 199, 492, 613
$\sim$, 199, 492, 613
$\leq_{m}^{P}$, 30
$\leq_{\text{d-tt}}^{P}$, 311
$\neg$, 25
$\wedge$, 24
$\vee$, 24
$\Longrightarrow$, 25
$\Longleftrightarrow$, 25
$\lfloor \cdot \rfloor$, 256
$\lceil \cdot \rceil$, 479
$\| \cdot \|$, 72
$\uplus$, 479
$[\cdot]$, 179, 407
$[\cdot]_k$, 411

A Beautiful Mind, 50
Aarts, E., 130
$ACC\text{-}Score(\cdot)$, 428
Adali, S., 671
Adams, D., 22, 322
adjusted winner procedure, 655, 656
advertisement campaign game
   in normal form, 100
Aesop, 562
agenda, 472
   closure of an – under complement, 472
   closure of an – under propositional
      variables, 480
   median property of an, 486
   set of conclusions of an, 479
      closure under complement of the, 479
   set of premises of an, 479
      closure under complement of the, 479
agent
   asymmetric, 675
   independence of unconcerned –s, 644
   set of –s, 607
   strategic, 675
   utility function of an, *see* utility
      function
   weighted, 675
aggregation
   binary – with integrity constraints, 474
   judgment, *see* judgment aggregation
   preference, *see* preference aggregation
Ahuja, R., 195
Airiau, S., 674
Akca, N., 16
Akrami, H., 629, 646
Aleksandrov, M., 673, 674, 679
algorithm
   approximation, 320
   frequently self-knowingly correct, 319
algorithmics, 19
Alkan, A., 654
Allis, L., 86
allocation, 608
   α-envy-free, 626
   balanced, 662
   complete, 608, 622
   connected, 670
      Pareto-optimal, 670
   envy-free, 623
      necessarily, 649
      possibly, 649
   fractional, 622
   graph-envy-free, 668
   individually rational, 675
   max-min, 636
   of parts of a divisible good, *see* division
   Pareto-efficient, 622

necessarily, 649
possibly, 649
weakly, 622
perfectly equitable, 642
allocation mechanism, *see* allocation procedure
allocation method, *see* allocation procedure
allocation problem, 607
  combinatorial, *see* allocation problem, multiple-item
  multiple-item, 609
  regular, 639
  single-item, 608
allocation procedure, 611
  centralized, 610, 611
  decentralized, 610, 611
    fully, 659
    partially, 659
  distributed, *see* allocation procedure, decentralized
  neutral, 676
  nonbossy, 676
  strategy-proof, 675
allocation rule, *see* allocation procedure
Alon, N., 500, 504, 645
Alonso-Meijide, J., 193
alphabet, 29
alternative, 234
  set of –s, 234
alternative vote, 252
Altman, A., 158
Amanatidis, G., 607, 627, 629, 632, 635, 672, 677
Anari, N., 645
Anderson, A., 158
Annamalai, C., 638
anonymity, *see* voting system, anonymous, *see* JA procedure, anonymous
antiplurality, *see* veto
k-approval, *see* k-approval *(under K)*
approval CC, *see* multiwinner voting rule, approval Chamberlin–Courant
approval Chamberlin–Courant multiwinner rule, *see* multiwinner voting rule, approval Chamberlin–Courant
approval score, 250
approval strategy, 261
  admissible, 262
  profile of –ies, 263
  sincere, 262
approval vector, 250
approval voting, 249, 264, 342, 361

attribute, *see* multiwinner voting rule, approval voting, attribute
  expanding, 426
  minimax, *see* multiwinner voting rule, approval voting, minimax
  multiwinner, *see* multiwinner voting rule, approval voting
  proportional, *see* multiwinner voting rule, approval voting, proportional
  satisfaction, *see* multiwinner voting rule, approval voting, satisfaction
approval winner, 250
approximation algorithm, 33
  performance guarantee of an, *see* approximation factor
approximation factor, 33
approximation ratio, *see* approximation factor
argumentation theory, 160
Arora, S., 34
Arrow, K., 2, 8, 18, 235, 268, 272, 273, 279, 373, 485
  impossibility theorem of, 8, 268, 273, 279, 373, 485
Arunachaleswaran, E., 672
Arzi, O., 520
Asadpour, A., 638
$AScore(\cdot)$, 250
Atkinson index, 642
attribute approval committee election, *see* multiwinner voting rule, approval voting, attribute
attribute multiwinner approval election, *see* multiwinner voting rule, approval voting, attribute
  distortion in –s, 441
auction, 611
  all-pay
    first-price, sealed-bid, 16
  American, 15
  clock, 14
  combinatorial, 2, 12, 609, 611
  Dutch, 14
  English, 13
  expected revenue of an, 16
  first-price, 12
  multiple-item, *see* auction, combinatorial
  open-cry, 12
    ascending-price, 12, 15
    descending-price, 12
  sealed-bid, 12
    first-price, 13
    second-price, *see* auction, Vickrey

Index 733

second-price, 13
single-item, 12, 608
Vickrey, 14
winner determination in an, 12
auctioneer, 611
risk-averse, 17
Aumann, Y., 520
Austin, A., 532, 545, 565, 603
Australian Electoral Commission, 335
average-case polynomial time, 318
AvgP, 318
Azar, Y., 645
Aziz, H., 171, 187, 188, 206, 219, 220, 227, 228, 426, 428, 431, 433, 435, 449, 564, 572, 598, 602, 625, 632, 647, 649–652, 655, 659, 665, 669, 670, 672, 674, 675, 678, 679

Babaioff, M., 634, 677, 679
Bacchus, F., 620
Bachmann, P., 23
Bachrach, Y., 36, 157, 158, 160, 170, 171, 177, 178, 183, 187, 188, 192, 652
backward induction, 101
Bai, Y., 624
Baklanov, A., 631
Baldwin winner, 257
Baldwin, J., 257, 365
voting system of, 257, 264
Ballester, C., 206
ballot, 234
bounded approval, 439
generalized preference, 439
set of –s, 234
Banach, S., 535, 603
Banerjee, S., 198
Bansal, N., 636, 639
Banzhaf$(\cdot,\cdot)$, 169
Banzhaf$^*(\cdot,\cdot)$, 169
$\overline{\text{Banzhaf}}(\cdot,\cdot)$, 169
Banzhaf III, J., 168, 169
Banzhaf index, see game, cooperative, Banzhaf index of a player in a simple
normalized, see game, cooperative, normalized Banzhaf index of a player in a simple
additivity of the, 170
efficiency of the, 170
null player property of the, 170
symmetry property of the, 170
valuation property of the, 170

probabilistic, see game, cooperative, probabilistic Banzhaf index of a player in a simple
additivity of the, 170
efficiency of the, 170
null player property of the, 170
symmetry property of the, 170
valuation property of the, 170
Barbanel, J., 519
Barberà, S., 616
Barman, S., 622, 627–629, 631, 635, 645, 646, 655, 671, 672, 677
Barrot, N., 207, 302
Bartholdi III, J., 288, 289, 291, 301, 305, 308, 309, 312, 329–331, 336–338, 340–342, 357, 374, 397, 401, 463, 486, 652
Basanta Val, P., 436
Basu, K., 160
Bataille, N., 606
Bateni, M., 639
Battigalli, P., 103
battle of the sexes, 4, 52, 68
Baumeister, D., ix, xii, 2, 18, 160, 233, 301, 303–306, 344, 363, 365, 403, 426, 439, 441, 492–497, 499, 500, 503, 504, 651, 652, 676
Bayes, T., 106, 108
formula of, 106–108
theorem of, 108
Bei, X., 653, 671
Beigel, R., 38, 187
Bellman, R., 90
Benabbou, N., 635, 673
Benade, G., 674
BENEFICIAL-MERGE, see WVG-ℙI-BENEFICIAL-MERGE
BENEFICIAL-SPLIT, see WVG-ℙI-BENEFICIAL-SPLIT
Bentert, M., 668
Berger, B., 629
Berman, P., 34
Betzler, N., 245, 288, 290, 301, 309, 364, 366, 451, 459
Beviá, C., 654
Beynier, A., 626, 633
Bezáková, I., 638
Bhaskar, U., 670
Biathlon World Cup, 501
bidder
risk-loving, 17
risk-neutral, 17
bidding language, see utility function, represented by a bidding language

Bierhoff, O., 609
Big Bang Theory, The, 58
Bilò, V., 229
Bilò, V., 671
Bilò, V., 216
binary tree, 253
  balanced, 253
Binkele-Raible, D., 290, 366, 500
Binmore, K., 114, 117
Birkhoff, G., 679
Birmpas, G., 607, 629, 632, 635, 672, 677
Biró, P., 647, 650, 675
Biswas, A., 631, 671
Black winner, 258
Black, D., 9, 258, 373
  voting system of, 258, 264
Bliem, B., 624, 635
bloc rule, *see* multiwinner voting rule, bloc
BND, 608
Bobo, E., 105
Bodlaender, H., 33
Boehmer, N., 225, 363, 625
Boes, L., 439, 441
Bogomolnaia, A., 198, 203, 204, 211–213, 216, 227, 641, 670, 674
Böhnlein, T., 439
boolean circuit, 27
boolean constant, 25
boolean hierarchy, *see* NP, boolean hierarchy over
boolean operation, 24, 25
  conjunction, 24, 472
  disjunction, 24, 472
  equivalence, 25, 472
  implication, 25, 472
  negation, 25, 472
boolean variable, *see* formula, boolean, variable of a
boolean weighted voting game, *see* weighted voting game, boolean
Booth, K., 375, 397
k-Borda multiwinner rule, *see* multiwinner voting rule, k-Borda
Borda paradox, 266
Borda winner, 237
Borda, J., 237
  multiwinner voting system of, *see* multiwinner voting rule, k-Borda
  voting system of, 237, 264, 286, 302, 303, 313, 346, 361
  modified, 304
Borel, É., 1, 4, 43, 47, 110, 113
Borm, P., 204, 219, 220

Bossert, W., 616
Botan, S., 463
Boutilier, C., 288, 302, 486
Bouton, C., 88
Bouveret, S., 607, 617, 621, 623, 624, 626, 627, 630, 632, 633, 637, 642, 647, 649–652, 656, 664, 665, 669–672, 679
Bovet, D., 19
Brams, S., 18, 249, 259, 261–263, 266, 283, 284, 344, 427, 438–440, 509, 513, 520, 528, 530, 546, 556–558, 561, 562, 565, 570, 572, 588, 598, 602, 603, 607, 624, 649–653, 655, 656, 659, 664, 665, 667
  adjusted winner procedure of –
  and Taylor, *see* adjusted winner procedure
Brams–Taylor protocol, 572, 603
Brandes, U., 366
Brandl, F., 206, 219
Brandt, F., 18, 136, 160, 206, 212, 216, 219, 220, 222, 223, 227, 228, 243, 244, 246, 275, 276, 280, 281, 292, 360, 389, 393, 394, 486, 678
Brânzei, S., 635, 646, 659, 670
Bredereck, R., 288, 290, 335, 363, 365, 396, 451, 460–464, 500, 504, 624, 625, 635, 668, 669, 672, 674
Brelsford, E., 361
BRIBERY, 358
  \$, 361
  \$ ⚄⚄, 361
  ⚄⚄, 361
bribery, *see also* bribery broblem; election, bribery in an; JA procedure, bribery for –s
  constructive, 360
  destructive, 362
  extension, 365
  in multiwinner voting, 462
  negative, 360
  of single-peaked electorates, 393
  priced, 359
  shift, 364
    combinatorial, 365
    destructive, 365
    for multiwinner voting rules, 464
  strongnegative, 360
  support, 365
  swap, 363
    constructive, 363
    destructive, 363
    unweighted, 363
    weighted, 363

Index 735

unpriced, 358
unweighted, 358
weighted, 359
bribery problem, *see* BRIBERY, *see also* JA procedure, bribery for –s
  priced, *see* BRIBERY, $
    weighted, *see* BRIBERY, $ ⚖
  weighted, *see* BRIBERY, ⚖
    priced, *see* BRIBERY, $ ⚖
  with price function, *see* bribery problem, priced
Brill, M., 244, 246, 292, 360, 389, 393, 394, 428, 431, 486, 678
British House of Commons, 148
Brouwer, L., 80, 82, 84
Brueggemann, T., 28
Brustle, J., 655
Bryla, K., 58
$BScore^i(\cdot)$, 256
$BScore(\cdot)$, 256
Bucklin, 305
  simplified, 302
Bucklin score, *see* $BScore(\cdot)$
  in stage i, *see* $BScore^i(\cdot)$
Bucklin winner, 256
Bucklin, J., 256
  voting system of, 256, 264, 286, 342, 349, 361
Budish, E., 626
Bullinger, M., ix, 2, 139, 206, 212, 216, 219, 222, 223, 225–229
Bundesliga, 241
  German champion in the, 241
Bundestag, *see* Deutscher Bundestag
bundle, 608
  minimal, 652, 666
bundle form, *see* utility function, in bundle form
Burstein, A., 362
Byrka, J., 458

$\mathfrak{C} \xrightarrow{i} \mathfrak{C}'$, 200
$\mathfrak{C}(\cdot)$, 199
Cai, J., 37
cake, 507
  division of a, 507
  heterogeneous, 509
  homogeneous, 509
  normalization of a, 510
  piece of a, 508
    acceptable, 545, 593
    preferred, 593
    strictly preferred, 593
cake-cutting, x, xi, 10, 507

cake-cutting procedure, 2
cake-cutting protocol, 528
  action in a, 525
  continuous, 527
  cut in a, 579
  cut request in a, 511, 588
  dirty-work, 574
  efficiency of a, 522
  envy-free, 518
  equitable, 518
    proportionally, 520, 530
  evaluation request in a, 511, 588
  exact, 514
  fairness of a, 513
  finite, 527
  finite bounded, 527
  finite unbounded, 527
  immunity to manipulation of a, 525
    for risk-averse players, 525
  manipulability of a, 523
    for risk-averse players, 526
  marking in a, 579
  minimal number of cuts required by a, 577, 579, 587
    lower bound for the, 588
    upper bound for the, 588
  moving-knife, 527
  properties of a, 603
  proportional, 514
    degree of guaranteed envy-freeness of a, 599
  rule of a, 512
  runtime of a, 527
  strategy in a, 512
  strategy-proof, 526
  super-envy-free, 518
  super-proportional, 514
Calkins, G., 105
campaign game
  in extensive form
    game tree for the, 100
candidate, 234
  q-affordable, 443
  cloning of –s, 276, 425
    necessary, *see* cloning, necessary
    possible, *see* cloning, possible
  cost of a, 441
  score of a, 442
  set of –s, 234
  utility of a, 442
candidate monotonic, *see* monotonicity, candidate
Cao, Y., 635, 671

Caragiannis, I., 296, 486, 625, 628, 629, 631, 632, 644, 646, 655, 659, 669–672, 679
Carleton, B., 307, 330, 337, 338
Carroll, L., see Dodgson, C.
Cary, D., 362
Caskurlu, B., 441
CC, see multiwinner voting rule, Chamberlin–Courant
CCAC, 331
CCAC, 338, 340, 342, 346, 357
CCAUC, 331, 338, 340, 342, 346
CCAV, 334
CCAV, 338, 340, 342, 346, 357
CCDC, 331
CCDC, 338, 340, 342, 346, 357
CCDV, 334
CCDV, 338, 340, 342, 346, 357
CCM, 312
CCPC-TE, 333, 338, 340, 342, 346
CCPC-TP, 333, 338, 340, 342, 346
CCPV-TE, 336, 338, 340, 342, 346, 349
CCPV-TP, 336, 338, 340, 342, 346, 349
CCRC, 357
CCRPC-TE, 333, 338, 340, 342, 346
CCRPC-TP, 333, 338, 340, 342, 346
CCRV, 357
$CC\text{-}Score(\cdot)$, 408
CCWM, 312, 313
CDK-Kemeny ranking, 247
CDK-Kemeny rule, 247
Cechlárová, K., 207, 669–671
chair, see election chair, see JA chair
Chakrabarty, D., 638
Chakraborty, M., 635, 665, 673, 675
Chalkiadakis, G., xii, 172, 193
Chamberlin, J., 407, 674
Chamberlin–Courant multiwinner rule, see multiwinner voting rule, Chamberlin–Courant
Chamberlin–Courant score, see $CC\text{-}Score(\cdot)$
Chan, H., 672, 675
Chan, T., 638
Chandra, A., 35
Chandrasekaran, 651, 665
characteristic function, see game, cooperative, characteristic function of a
Charikar, M., 639
Chaudhury, B., 625, 628, 629, 635, 646, 670
Chavrimootoo, M., 307, 330, 337, 338
checkers, 84

Checkoway, S., 362
Chekuri, C., 639
Chen, B., 670
Chen, J., 207, 221, 290, 346, 365, 396, 500, 504, 668, 672
Chen, M., 362
Chen, X., 136
chess, 84
  winning strategies in, 96
Cheung, Y., 628
Chevaleyre, Y., 18, 303, 606, 607, 620, 621, 625, 637, 644, 647, 656, 668, 672
chicken game, 54, 69
chore division, 574
chores, 654
Christian, R., 366, 500
Christie's, 13, 608
Christodoulou, G., 677
Church, A., 20
  thesis of, 20
Chuzhoy, J., 638
CIA, 608
citizens' sovereignty, 274, 286
Clarke, E., 15
clone, 276
  in approval voting, 276
clone set, 425
cloning
  necessary, 425
    in the general-cost model, 425
    in the unit-cost model, 425
    in the zero-cost model, 425
  possible, 425
    in the general-cost model, 425
    in the unit-cost model, 425
    in the zero-cost model, 425
closure, see agenda, closure of an – under complement; agenda, closure of an – under propositional variables; agenda, set of conclusions of an, closure under complement of the; agenda, set of premises of an, closure under complement of the; complexity class, closure of a; $\mathcal{L}_P$, closure of – under negation
CM, 312
CNF, see formula, boolean, in conjunctive normal form
k-CNF, 26
co-winner model, see nonunique-winner model
coalition structure, see game, cooperative, coalition structure for a

Index   737

coalitional function, *see* game, coop-
    erative, coalitional function of
    a
coalitional weighted manipulation
    constructive, 311, 389
    destructive, 312
COALITIONAL-MANIPULATION, 309
Cohen, A., 629
cohesive, *see* voter, cohesive group of –s
β-cohesive, *see* voter, β-cohesive group of
    –s
ℓ-cohesive, *see* voter, ℓ-cohesive group of
    –s
Cole, R., 645, 660
Colfer, E., 322
collective decision-making, *see* decision-
    making, collective
combinatorial auction, *see* auction,
    combinatorial
committee, 404
committee election, *see* voting, multiwin-
    ner
    robustness level of a, 460
    robustness radius of a, 462
committee enlargement monotonic,
    see monotonicity, committee
    enlargement
committee position, 411
committee scoring function, *see* scoring
    function, committee
common knowledge, 111
common scale of utilities
    independence of, 644
completeness, *see* game, noncooperative,
    with complete information; JA
    procedure, complete; judgment
    set, complete; NP-completeness;
    problem, completeness of a
complexity
    average-case, 318
    parameterized, 289
    worst-case, 28, 318
complexity class, 21
    $\leq_m^P$-closure of a, *see* complexity class,
        closure of a
    $\leq_m^P$-completeness in a, *see* problem,
        completeness of a
    $\leq_m^P$-hardness for a, *see* problem,
        hardness of a
    closure of a, 29
    parameterized, 39
complexity function
    asymptotic growth of a, 23
complexity theory, xi

foundations of, 19
    parameterized, 290
computability theory, 19
computational complexity, *see* complexity
    theory
computational social choice, vii, 2, 17,
    287, 471
    international workshop on, 17
conclave, 241
conclusion-based procedure, *see* JA
    procedure, conclusion-based
conditional importance form, 616
conditional importance network, 617
conditional preference statement, 616
Condorcet criterion, 265, 286
    weak, 266
Condorcet cycle, 8, 240
Condorcet loser, 240, 266
    weak, 240
Condorcet paradox, 7
Condorcet top cycle, 240, 249
Condorcet winner, 7, 228, 239, 265
    weak, 228, 239
Condorcet, J.-A.-N. de Caritat, Marquis
    de, 7, 8, 239
    voting system of, 7, 239, 264, 302, 342,
        347, 357
Conitzer, V., 18, 136, 137, 178, 247, 290,
    301, 302, 309, 310, 312–314, 316,
    318–321, 326, 327, 364, 366, 428,
    431, 486, 618, 631, 680
conjunction, *see* boolean operation,
    conjunction
coNP, 36
consecutive ones property, 374, 375
consistency, *see* JA procedure, consistent;
    judgment set, consistent; voting
    system, consistent
    narrow-top, 410, 420
CONSTRUCT-CORE
    ISG-, *see* ISG-CONSTRUCT-CORE
    WVG-, *see* WVG-CONSTRUCT-CORE
CONSTRUCT-LEAST-CORE, *see* WVG-
    CONSTRUCT-LEAST-CORE
CONSTRUCTIVE-COALITIONAL-
    WEIGHTED-MANIPULATION,
    311
CONSTRUCTIVE-CONTROL-BY-ADDING-
    AN-UNLIMITED-NUMBER-OF-
    CANDIDATES, 331
CONSTRUCTIVE-CONTROL-BY-ADDING-
    CANDIDATES, 331
CONSTRUCTIVE-CONTROL-BY-ADDING-
    VOTERS, 334

CONSTRUCTIVE-CONTROL-BY-DELETING-CANDIDATES, 331
CONSTRUCTIVE-CONTROL-BY-DELETING-VOTERS, 334
CONSTRUCTIVE-CONTROL-BY-PARTITION-OF-CANDIDATES, 333
CONSTRUCTIVE-CONTROL-BY-PARTITION-OF-VOTERS, 336
CONSTRUCTIVE-CONTROL-BY-RUN-OFF-PARTITION-OF-CANDIDATES, 333
contested pile, 666
 minimal bundle from a, 666
contested pile-based protocol, 665
contractual deviation, see hedonic game, coalition structure for a, contractual deviation from a
contractual individual deviation, see hedonic game, coalition structure for a, contractual individual deviation from a
contradiction, 472
control, see election, control of an; electoral control; JA procedure, control for -s
 in multiwinner voting, 462
CONTROL-BY-DELETING-PLAYERS-TO-DECREASE-$\mathbb{PI}$, see WVG-CONTROL-BY-DELETING-PLAYERS-TO-DECREASE-$\mathbb{PI}$
CONTROL-BY-DELETING-PLAYERS-TO-INCREASE-$\mathbb{PI}$, see WVG-CONTROL-BY-DELETING-PLAYERS-TO-INCREASE-$\mathbb{PI}$
CONTROL-BY-DELETING-PLAYERS-TO-MAINTAIN-$\mathbb{PI}$, see WVG-CONTROL-BY-DELETING-PLAYERS-TO-MAINTAIN-$\mathbb{PI}$
convex set, 72
convexity, see game, cooperative, convex
Conway, J., 565, 567, 603
Cook, S., 30
$C^\alpha Score(\cdot)$, 241
Copeland score, 241
Copeland voting, see Copeland, A., voting system of
 second-order, 308
Copeland winner, 241
Copeland$^\alpha$ score, see $C^\alpha Score(\cdot)$
Copeland$^\alpha$ voting, 241, 264, 342, 357, 361
Copeland$^\alpha$ winner, 242
Copeland^0, 242, 286
Copeland^1, 241
Copeland$^{1/2}$, 241

Copeland$^{1/3}$, 241
Copeland, A., 241
 family of voting systems of Llull and –, 241, 264, 342, 357, 361
 voting system of, 241, 264, 302, 303, 305, 313, 342, 357, 361
$Core(\cdot)$, 150
core, see game, cooperative, core of a
core stability, see game, cooperative, core of a; hedonic game, coalition structure for a, core-stable
CORE-STABILITY-EXISTENCE, 209
CORE-STABILITY-VERIFICATION, 209
Cormen, T., 220
$CoS(\cdot)$, 159
cost$(\cdot)$, 441, 442
cost of implementation, 158
cost of stability, see game, cooperative, cost of stability of a
COST-OF-STABILITY, see WVG-COST-OF-STABILITY
Courant, P., 407, 674
CPLEX, see ILP solver, CPLEX
cracking, 335
Cramer, S., 226
Cramton, P., 12, 15, 607, 611, 647
Crescenzi, P., 19
Cry Baby, 55
Csáji, G., 207, 221
CSE, 209
Csirik, J., 639
CSV, 209
Culberson, J., 96
cup protocol, 253, 264, 302, 303, 313
 schedule of the, 254
 tree of the, 253
 winner of the, 254
CURB set, 160
Custer, C., 546, 547
Cut & Choose protocol, 529
Cut Your Own Piece protocol, 548, 603
Cygan, M., 439, 453, 456

D'Angelo, G., 463
Dall'Aglio, M., 625, 659
Dani, V., 638
Dantsin, E., 28
Dantzig, G., 123
Darmann, A., 203, 215, 221, 629, 652
Daskalakis, C., 18, 125–127, 130, 134, 136
Davenport, A., 247
Davies, J., 309, 321, 354
Dawson, C., 545
DCAC, 338, 340, 342, 346, 357

Index 739

DCAUC, 338, 340, 342, 346
DCAV, 338, 340, 342, 346, 357
DCDC, 338, 340, 342, 346, 357
DCDV, 338, 340, 342, 346, 357
DCM, 312
DCPC-TE, 338, 340, 342, 346
DCPC-TP, 338, 340, 342, 346
DCPV, 347
DCPV-TE, 338, 340, 342, 346
DCPV-TP, 338, 340, 342, 346
DCRC, 357
DCRPC-TE, 338, 340, 342, 346
DCRPC-TP, 338, 340, 342, 346
DCRV, 357
DCWM, 312, 313
DDP, 660
Dean, J., 55
decision-making
  collective, xi
degree of guaranteed envy-freeness, *see* cake-cutting protocol, proportional, degree of guaranteed envy-freeness of a
Deligkas, A., 671
Δ_2^p, 36
Δ_i^p, 36
Demange, G., 654
demaverickification, 400, 402
Demko, S., 631, 640
Deng, X., 136, 176, 184, 194, 196, 197
Depp, J., 55
descending demand protocol, 660
  modified, 662
Desmedt, Y., 326
DESTRUCTIVE-COALITIONAL-WEIGHTED-MANIPULATION, 312
Deuermeyer, B., 639
Deutscher Bundestag, 148, 335
  allocation of seats, 335
    in the 17th, 148
Devanur, N., 645
Dey, P., 451
$d_G(\cdot,\cdot)$, 162, 163, 165
DGEF, 599
dichotomy result
  for CCWM, 312, 389
    in the single-peaked case, 389
  for ONLINE-CCWM, 327
  for POSSIBLE-WINNER, 301
  for priced, weighted bribery, 361
Dickerson, J., 624
dictator, 268
dictatorship
  sequential, 676

  serial, 675
    random, 678
Diestel, R., 85
Dietrich, F., 471, 479, 486, 489, 490, 493
Dimitrov, D., 204, 216, 219–221
Dippel, J., 655
dirty-work problem, 574
dirty-work protocol, *see* cake-cutting protocol, dirty-work
discursive dilemma, 9, 470
disjoint union, *see* ⊎
disjunction, *see* boolean operation, disjunction
distance
  Euclidean, 81
  Hamming, *see* Hamming distance
distance-based procedure, *see* JA procedure, distance-based
diversity of dislike, 312
Divide & Conquer protocol, 551, 603
  BJK, 556–558
  dirty-work, 576
  minimal number of cuts required by the, 584
divide et impera, 551
divide-and-conquer principle, 585, 589
division, 508
  envy relation in a, 599
    cycle of one-way –s in a, 600
    one-way, 600
    two-way, 600
  envy-free, 518
  envy-free relation in a, 599
    case-enforced, 600
    guaranteed, 600
    one-way, 600
    two-way, 600
  equitable, 518
  exact, 514
  fair, *see* fair division
  of a cake, 508
  of Germany, 507
  of indivisible goods, *see* allocation
  proportional, 514
    minimal number of cuts required for a, 589
  simply fair, *see* division, proportional
  strongly fair, *see* division, super-proportional
  super-envy-free, 518
  super-proportional, 514
divorce formula, 656
DM, 312

DNF, *see* formula, boolean, in disjunctive normal form
Doan, X., 670
Dobzinski, S., 320
doctrinal paradox, 9, 470
Dodgson score, *see* $DScore(\cdot)$
Dodgson winner, 243
Dodgson, C., 242, 243, 291
   voting system of, 242, 243, 264, 286, 483
      homogeneous variant of the, 281
Doignon, J., 375, 397
Dokow, E., 486
Dombb, Y., 520
DOMINATING-SET
   k-DOMINATING-SET, 497
Dor, D., 96
Dorn, B., 301, 364, 366, 625, 631, 642
dove strategy, 56
Downey, R., 39, 290, 453, 456, 499
DP, 37
Drèze, J., 198, 227
Driessen, T., 145
Dror, A., 671
$DScore(\cdot)$, 243
Dubey, P., 162, 169, 170
Dubins, L., 541, 542, 557, 558, 579, 590, 603
Dudycz, S., 435, 458
Duggan, J., 275
Dung, P., 160
Dunne, P., 18, 606, 607, 621, 637, 647
Dutta, B., 160, 355
Dwork, C., 287
dynamic programming, 90

e-commerce, 12
ϵ-$Core(\cdot)$, 155
ϵ-core, *see* game, cooperative, ϵ-core of a
Ebadian, S., 670
eBay, 15, 608
EBU, 332
Edelman, P., 624, 649
Edmonds, J., 588, 590, 598
Edwards, D., 105
EEF, 632
EF, 623
   α-EF, 626
EF1, 628
efficiency, *see* game, cooperative, efficient, *see* Pareto optimality
EFkX, 629
EFR, 629
EFX, 628

α-EFX, 629
EGALITARIAN-SOCIAL-WELFARE-OPTIMIZATION, 637
Ehlers, L., 676
Eiben, E., 671, 672
EJR, *see* justified representation, extended, *see* justified representation, extended, for additive utilities
EJR-up-to-one, 443
election, 234
   apportionment, 335
      with thresholds, 335
   bribery in an, 358
   control of an, 328
      constructive, 330
      destructive, 330
   manipulation of an, 306, 310, 314, 316–318
      coalitional, *see* manipulation, coalitional weighted
      constructive, 310
      destructive, 310
      for single-peaked preferences, 386
      weighted, *see* manipulation, coalitional weighted
   multiwinner, *see* voting, multiwinner, 680
   sequential
      online bribery of a, 366
      online candidate control of a, 356
      online control of a, 356
      online manipulation of a, 321
      online voter control of a, 356
   winner of an, 234, 290
      necessary, 297, 301
      possible, 297, 298, 317
      unique, 234, 290
   with irrational voters, 365
   with weighted voters, 310
election chair, 328
electoral control, 328
   by adding an unlimited number of candidates, 331
   by adding candidates, 330
   by adding voters, 333
   by deleting candidates, 330
   by deleting voters, 333
   by partition of candidates, 332
   by partition of voters, 335
   by replacing candidates, 355
   by replacing voters, 355
   by run-off partition of candidates, 332
   constructive, 330
   destructive, 330

Index 741

immunity to, 338
multimode, 355
of single-peaked electorates, 354, 376
overview of complexity results for, 341
overview of problems of, 336
pairs of collapsing – types, 337
resistance to, 338
susceptibility to, 338
unweighted, 354
vulnerability to, 338
  certifiable, 349
weighted, 354
electoral participation, 282
electorate, *see* preference profile
  single-peaked, *see* single-peaked electorate
Elkind, E., ix, xii, 18, 137, 139, 158, 160, 171, 172, 175, 181–183, 187, 188, 192, 193, 203, 206, 215, 221, 227, 229, 286, 289, 326, 346, 358, 362–365, 375, 401, 407, 415–417, 421, 425, 428, 431, 433, 436, 486, 652, 669, 670, 673
ellipsoid method, 122
Emerson, P., 304
EMPTY-CORE
  ISG-, *see* ISG-EMPTY-CORE
  WVG-, *see* WVG-EMPTY-CORE
END-OF-THE-LINE, 131
Endriss, U., 10, 18, 474, 476, 477, 483, 486–490, 494, 495, 504, 606, 607, 616, 617, 620, 621, 625, 637, 642, 644, 647, 649, 650, 668, 672, 674
envy, *see* envy-freeness
  degree of, 625
    global, 625
    local, 625
envy-freeness, *see* allocation, envy-free; cake-cutting protocol, envy-free; cake-cutting protocol, proportional, degree of guaranteed envy-freeness of a; division, envy-free; price of envy-freeness; share, envy-free, 622
  epistemic, 632
  interim, 679
  objective, 633
  relaxations of, 625
  up to a random good, 629
  up to any good, 628
  up to k hidden goods, 633
  up to one good, 628
Ephrati, E., 288
EQUAL-SUBSETS, 126
equilibrium, 45
  subgame-perfect, 101
equitability, *see* allocation, nearly equitable; allocation, perfectly equitable; cake-cutting protocol, equitable; division, equitable; price of perfect equitability; share, equitable, *see* allocation, perfectly equitable
equitability up to any good, 642
equitability up to one good, 642
equivalence, *see* boolean operation, equivalence; formula, boolean, equivalence of –s
Erdélyi, G., x, 18, 263, 290, 319, 332, 342, 344, 345, 350, 354–357, 366, 396, 492–497, 499, 500, 503
Erdélyi, O., 492–496, 499, 500, 503
Erhard, L., 10
Escoffier, B., 375, 397
Estivie, S., 620, 625, 644, 647, 668
ESWO, 637
European Broadcasting Union, 332
Eurovision Song Contest, 237, 332
Even, S., 551, 587, 593, 594, 597, 603
event, 107
  atomic, 107
  independent –s, 108
exact cover, *see* set, exact cover of a
EXACT-COVER-BY-THREE-SETS, 350
EXACT-COVER-BY-THREE-SETS, 216
EXACT-EGALITARIAN-SOCIAL-WELFARE-OPTIMIZATION, 638
EXACT-JA-BRIBERY, 499
EXACT-JA-CONTROL-BY-ADDING-JUDGES, 503
EXACT-JA-MICROBRIBERY, 499
exactness, *see* cake-cutting protocol, exact; division, exact; JA procedure, bribery for –s, exact; JA procedure, control for –s, exact; JA procedure, manipulation for –s, exact; JA procedure, microbribery for –s, exact; margin of victory, exact variant of the – problem; share, exact
excess mandate, 335
extended justified representation, *see* justified representation, extended
Ezra, T., 677, 679

Fain, B., 680
fair division, vii, ix–xi, 2
  centralized
    with ordinal preferences, 649
  decentralized, 659

of a heterogeneous resource, 507
of a homogeneous resource, 509
online, 674
problem of, 507
with externalities, 673
with matroid feasibility constraints, 671
with money, 653
fair share
  ℓ-out-of-d max-min, 627
  max-min, 626
  min-max, 626
  proportional, 629
fair share criterion
  group max-min, 631
  groupwise, 673
  max-min, 627
  min-max, 627
  pairwise max-min, 631
  proportional, 630
fairness, 612
  democratic, 673
  maximin-aware, 672
Falik, D., 500
Faliszewski, P., x, xii, 2, 18, 43, 175, 185, 188, 192, 241, 289, 290, 292, 304, 309, 312–314, 316, 318, 330, 331, 335, 337, 342, 343, 346, 354, 355, 357, 358, 360–366, 375, 376, 386, 388, 390, 393, 394, 396, 397, 399–401, 403, 407, 412, 415–417, 420, 421, 425, 426, 435, 449, 451, 454, 459–462, 464, 486, 606, 674
fallback score, see $FScore(\cdot)$
  in stage i, see $FScore^i(\cdot)$
fallback voting, 259, 264, 305, 342, 357, 361
fallback winner, 259, 260
Falmagne, J., 375, 397
false-name manipulation, see manipulation, false-name
Fanelli, A., 227, 229
Fargier, H., 637, 647
Farhadi, A., 629, 635, 675
feasibility constraint, 476
Federal Constitutional Court of Germany, 336
Feige, U., 509, 624, 634, 677, 679
Feldman, A., 602
Feldman, M., 629, 671
Fellows, M., 39, 290, 332, 342, 344, 345, 350, 354, 366, 453, 456, 499, 500
Felsenthal, D., 170
Fernández, N., 436
Fernau, H., 290, 366, 500

Ferraioli, D., 639, 671
Fiat, A., 629
FIFA World Cup, 332
  1938 through 2002, 332
  1966, 513
  1990, 56
Figiel, A., 624
Filos-Ratsikas, A., 607, 629, 632, 635, 659
Fink, A., 542, 603
Fischer, F., 136, 160, 486
Fischer, J., 609
Fishburn, P., 243, 248, 249, 258, 278–284, 394, 427, 463, 624, 649, 653
Fisher, M., 458
Fisteus, J., 436
Fitzsimmons, Z., 305, 355–357, 375
five in a row, see Go-Moku
fixed point theorem
  Brouwer's, 80, 82
  Kakutani's, 84
fixed-parameter tractability, see problem, fixed-parameter tractable
Flammini, M., 227, 229, 665, 671
Flum, J., 39, 290, 453, 456
Foley, D., 623
Fomin, F., 453, 456
formula
  boolean, 24
    atomic proposition of a, 24
    clause of a – in CNF, 26
    complement of a, 472
    equivalence of –s, 26
    implicant of a – in DNF, 26
    in conjunctive normal form, 26
    in disjunctive normal form, 26
    proposition of a, 24
    satisfiable, 25
    set of –s, see $\mathcal{L}_P$
    variables of a, 24
  propositional, see formula, boolean
Fortnow, L., 29
Fotakis, D., 674
FP, 29
fPO, 622
FPT, see problem, fixed-parameter tractable
FPTAS, 34
Fréchet, M., 113
Frederiksen, S., 659
free disposal, 615
Freeman, R., 428, 431, 631, 642, 670, 678–680
Freixas, J., 173, 174
Friedgut, E., 319, 320

Index   743

Friesen, D., 639
Frobenius–König theorem, 545, 546
Froese, V., 668
$FS(\cdot)$, 629
$FScore^i(\cdot)$, 260
$FScore(\cdot)$, 260
Fujinaka, Y., 655
Fujita, E., 675
Fujito, T., 34
Fulkerson, D., 374, 375, 397
fully polynomial-time approximation scheme, 34
function
  continuous, 80
  factorial, 163
  monotonically increasing, 277
    strictly, 277
  polynomial-time computable, 29
  potential, 216
  social choice, *see* social choice function
  social welfare, *see* social welfare function
  utility, *see* agent, utility function of an; utility function
  valuation, *see* player, valuation function of a; valuation function
Furdyna, M., 335

γ-$Score(\cdot)$, 412
GAI, *see* utility function, additively independent, generalized
Gailmard, S., 9, 373
Gairing, M., 223
Gale, D., 654, 677
Galeotti, A., 178
game
  Bayesian, 111, 112
  combinatorial, 86
  cooperative, xi, 139
    adjusted, 159
    Banzhaf index of a player in a simple, 168, 169
    characteristic function of a, 141
    coalition structure for a, 142
    coalitional function of a, *see* game, cooperative, characteristic function of a
    convex, 145, 152
    core of a, 149, 150, 152, 181
    cost of stability of a, 156, 159, 183
    ϵ-core of a, 155, 182
    efficient, 143
    external stability of a, 160
    fair outcome of a, 162
    grand coalition of a, 142
    graph-restricted, 193
    imputation for a, 150
    individually rational, 150
    induced subgraph, *see* induced subgraph game
    internal stability in a, 160
    kernel of a, 161
    least core of a, 155, 182
    monotonic characteristic function of a, 141
    network flow, *see* network flow game
    normalized Banzhaf index of a player in a simple, 169
    normalized characteristic function of a, 141
    nucleolus of a, 161
    null player in a simple, 162
    outcome of a, 143
    path disruption, *see* path disruption game
    payoff vector for a, 143, 150
    pivotal player in a simple, 162
    power index of a player in a simple, 162
    probabilistic Banzhaf index of a player in a simple, 169
    raw Banzhaf index of a player in a simple, 169
    raw Shapley–Shubik index of a player in a simple, 163
    set of payoff vectors for a, *see* $\mathcal{PV}(\cdot)$
    Shapley value of a player in a, 165, 167
    Shapley–Shubik index of a player in a simple, 162, 164
    simple, 146
    stable set of a, 160
    strong ϵ-core of a, 155, 182
    subgame of a, 145
    superadditive, 144
    supermodular characteristic function of a, 145
    value of $-s$, 167
    veto player in a simple, 153
    with nontransferable utility, 143
    with transferable utility, 141
  hedonic, *see* hedonic game
  noncooperative, xi, 47
    in extensive form, 84
    normal form of a, 45, 47
    with complete information, 111
    with imperfect recall, 103
    with incomplete information, 103

with perfect information, 85, 112
  rule of a, 44
  run-and-chase, 208
  structure of a, 111
  weighted voting, *see* weighted voting game
game theory
  algorithmic, 1
  cooperative, ix, 43, 139
    altruism in, 226
  noncooperative, x, 43, 139
    altruism in, 226
game tree, 85
  decision vertex in a, 85
  equilibrium in a, 97
  leaf of a, 85
  root of a, 85
games against nature, 109, 112
Gan, J., 646
Ganian, R., 671, 672
Gärdenfors, P., 463, 676
Gardner, M., 565, 574, 603
Garey, M., 31, 32, 123, 182, 216, 221, 450, 456, 487
Garg, J., 622, 627–629, 635, 642, 645, 646, 670
Garg, N., 628
Garimidi, P., 631
Gasarch, W., 22
Gaspers, S., 313, 435, 449, 649, 650, 674, 679
Gawron, G., 462
Geist, C., 275, 616
generalized additivity form, *see* utility function, in generalized additivity form
GEOGRAPHY, 93
German reunification, 269
gerrymandering, 335
get-out-the-votes drive, 334
Gharan, S., 645
Ghodsi, M., 627, 628, 635, 673, 675
Ghosh, S., 288
Giagkousi, I., 632, 669, 672
Gibbard, A., 8, 275, 279, 288
Gibbard–Satterthwaite theorem, 8, 275, 288, 306, 320
  quantitative version of the, 320
Gilbert, H., 665
Gill, J., 37, 185
Gilpin, A., 137
Gkatzelis, V., 631, 645, 646, 674
GMMS, 631
go, 84

winning strategies in, 96
Go-Moku, 86
  winning strategies in, 96
Goemans, M., 639
Goerdt, A., 28
Goko, H., 655
Goldberg, L., 137, 181, 182, 192
Goldberg, P., 18, 125–127, 134, 136, 137, 181, 182, 192, 665
golden goal, 609
Golden, B., 640
Goldman, J., 624
Goldreich, O., 318
Goldsmith, J., 290, 366, 500
Golovin, D., 638, 639
Gölz, P., 624
Gonzales, C., 620
good, *see* resource
Gourvès, L., 302, 639, 640, 642, 665, 668, 671, 674
Goyal, S., 178
Grabisch, M., 618
grand coalition, *see* game, cooperative, grand coalition of a
  subsidy to the, 156
Grand Junction system, *see* Bucklin, J., voting system of
Grandi, U., 10, 18, 474, 476, 486–490, 494, 495
graph, 85
  acyclic, 85
  clique number of a vertex in a, 210
  connected, 85
  directed
    indegree of a vertex in a, 128
    outdegree of a vertex in a, 128
  edge of a, 85
  independent set of a, 32
  partition of the vertex set of a – into cliques
    wonderfully stable, 210
  s-t-cut in a, 177
  sink of a, 176, 178
  source of a, 176, 178
  strongly connected component of a, 220
  undirected
    degree of a vertex in an, 128
  vertex cover of a, 37
  vertex of a, 85
    inner, 85
    with bounded treewidth, 193
graph theory, 85
graph-restricted cooperative game, *see* game, cooperative, graph-restricted

Index 745

graph-restricted weighted voting game, *see* weighted voting game, graph-restricted
Gravin, N., 625, 629, 646
Grazer, B., 50
Greco, G., 670
Greenberg, J., 198, 227
Groening, M., 58
Grohe, M., 39, 290, 453, 456
Gross, G., 375, 397
group activity selection problem, 203
groupwise fair share guarantee, *see* fair share criterion, groupwise
Grove, A., 620
Groves, T., 15
Gudmundsson, J., 435, 449, 647
guessing numbers game, 60
Gundermann, T., 37
Guo, J., 245, 290, 365
Guo, X., 635, 671
Gurobi, *see* ILP solver, Gurobi
Gurski, F., 23, 85
Guruswami, V., 639

$H(\cdot)$, 433
Haake, C., 655
de Haan, R., x, 2, 467, 476–478, 483, 488, 489, 494, 499, 504, 625, 631
Hačijan, L., 122
  algorithm of, 123
Hägele, G., 241
Hajduková, J., 207
Hajiaghayi, M., 627–629, 635, 675
Hall's marriage theorem, 545, 546
Hall, J., 362
Hall, M., 104, 109
Hall, P., 545
Halpern, D., 635, 655, 673, 677
halting problem, 19, 20
Hamm, T., 671, 672
Hamming distance, 438, 481
  between two complete judgment sets, 481
  between two incomplete judgment sets, 490
hardness, *see* problem, hardness of a
harmonic Borda rule, *see* multiwinner voting rule, harmonic Borda
harmonic number
  nth, 673
harmonic numbers, 433
Harrenstein, P., 206, 227, 486
Harsanyi, J., 4, 50, 111, 112, 121
  transformation of, 112

Hartmanis, J., 37
Hartung, S., 500
Harvey, N., 639
Hassidim, A., 651
Hatfield, J., 676
hawk strategy, 56
hawk-dove game, 56
Hazon, N., 286
HB rule, *see* multiwinner voting rule, harmonic Borda
HD-JA-MANIPULATION, 494
He, J., 674
hedonic coalition net, 206
hedonic expertise game, 441
hedonic game, 198, 199
  additively separable, 204
    symmetric, 215, 219
  altruistic, 226
    altruistic-treatment preferences in –s, 226
    equal-treatment preferences in –s, 226
    minimum-based, 226
    selfish-first preferences in –s, 226
  anonymous, 203
    black-and-white, 204
    single-peaked, 211
  appreciation-of-friends, 204
    loyal variant of symmetric –s, 226
  aversion-to-enemies, 204
  coalition in a
    blocking, 201
    individually rational, 200
    weakly blocking, 202
  coalition structure for a, 199
    contractual deviation from a, 200
    contractual individual deviation from a, 201
    contractually individually stable, 201
    contractually Nash-stable, 201
    core-stable, 201
    individual deviation from a, 200
    individually rational, 200
    individually stable, 200
    mixed popular, 228
    Nash deviation from a, 200
    Nash-stable, 200
    Pareto-optimal, 227
    perfect, 203
    popular, 228
    single-player deviation from a, 200
    strictly popular, *see* hedonic game, coalition structure for a, strongly popular
    strictly core-stable, 201

strongly popular, 228
wonderfully stable, 210
dynamics in
  necessarily converging, 222
  necessarily converging from a starting partition, 222
  possibly converging, 222
  possibly converging from a starting partition, 222
dynamics in a, 222
enemy-oriented, *see* hedonic game, aversion-to-enemies
fractional, 206
  price of a best Nash-stable outcome in –s, 229
friend-oriented, *see* hedonic game, appreciation-of-friends
list of individually rational coalitions in a, 206
local fairness in –s, 199
modified fractional, 207
network of friends in a, 206
online, 229
outcome of a, 199
  contractual deviation from a, 200
  contractual individual deviation from a, 201
  contractually individually stable, 201
  contractually Nash-stable, 201
  core-stable, 201
  individual deviation from a, 200
  individually rational, 200
  individually stable, 200
  Nash deviation from a, 200
  Nash-stable, 200
  perfect, 203
  price of a worst Nash-stable, 229
  single-player deviation from a, 200
  strictly core-stable, 201
  wonderfully stable, 210
set of enemies in a, 204
set of friends in a, 204
set of neutral players in a, 207
singleton encoding of a, 207
  optimistic, 207
  pessimistic, 207
stability in –s, 199
with ordered characteristics, 211
Heeger, K., 625
van Hees, M., 486
  impossibility theorem of Pauly and –, 486
HEFk, 633
Heinen, T., 630

Hemachandra, L., *see* Hemaspaandra, L.
Hemaspaandra, E., x, xii, 18, 36, 37, 229, 241, 244, 246, 247, 289–294, 296, 305, 307, 309, 311–314, 316, 323–327, 330, 331, 333, 337–340, 342–347, 349, 354–358, 360–362, 365, 366, 369, 375, 376, 386, 388–390, 393, 394, 396, 397, 399–401, 463, 486, 651, 674
Hemaspaandra, L., x, xii, 18, 19, 36–38, 185, 187, 188, 192, 229, 241, 244, 246, 289–294, 296, 305, 307, 311, 312, 319, 323–327, 330, 331, 333, 337–340, 342–347, 349, 354–358, 360–362, 365, 366, 369, 375, 376, 386, 388–390, 393, 394, 396, 397, 399–401, 463, 486, 651, 674
Hendrickx, R., 204, 219, 220
Hernandez, K., 288
Herreiner, D., 660, 662
Hicks, J., 8, 273
Hill, T., 625, 631, 640
Hirsch, E., 28
Hoefer, M., 628, 645, 646, 670
Hoffmann, J., 160
Hogrebe, T., 363
Holcombe, R., 530
Hollender, A., 629, 635
Holzer, M., 136
Holzman, R., 486
Homan, C., 319
Homer, S., 19
homogeneity, *see* voting system, homogeneous
Hopcroft, J., 19
Hosseini, H., 627, 628, 633, 635, 670, 672
Howard, R., 50
Howson jr., J., 136
  algorithm of Lemke and –, 136
Hu, X., 673
Huang, S., 433
Huang, X., 625, 629, 646, 647, 670
Huang, Z., 229
Hudry, O., 291, 486
Husic, E., 646

$\mathcal{I}(\cdot)$, 150
Ieong, S., 172, 206, 621
Igarashi, A., 635, 655, 669–671, 675
IIA, *see* independence, of irrelevant alternatives
ILP, *see* integer linear programming
  mixed, 454
ILP solver

Index 747

CPLEX, 452
Gurobi, 452
implication, *see* boolean operation, implication
impossibility theorem
  for electoral control, 343
  for JA procedures, 485
imputation, *see* game, cooperative, imputation for a
  dominance relation between two –s, 160
  dominance relation between two –s via a coalition, 160
IN-CORE
  ISG-, *see* ISG-IN-CORE
  WVG-, *see* WVG-IN-CORE
IN-LEAST-CORE, *see* WVG-IN-LEAST-CORE
increasing marginal return, 145
independence, *see* JA procedure, independent
  of clones, 276, 286, 425
  of irrelevant alternatives, 272, 286
INDEPENDENT-SET, 32
  k-DEGREE-INDEPENDENT-SET, 33
individual deviation, *see* hedonic game, coalition structure for a, individual deviation from a
individual rationality, *see* game, cooperative, individually rational, *see* hedonic game, outcome of a, individually rational
individual stability, *see* hedonic game, coalition structure for a, individually stable
  contractual, *see* hedonic game, coalition structure for a, contractually individually stable
INDIVIDUAL-STABILITY-EXISTENCE, 208
INDIVIDUAL-STABILITY-VERIFICATION, 208
indiviual scale of utilities
  independence of, 644
induced subgraph game, 176
information set, 102, 103, 116
instant-runoff voting, 252
integer linear programming, 435
integrity constraint, 474
interior point method, 122, 123
internal implementation, 158
International Monetary Fund, 148
Internationale Skatordnung, *see* skat, international laws of
Ioannidis, S., 655
IS dynamics

execution of the, 222
  starting partition of an, 222
Isaksson, M., 320
ISE, 208
ISV, 208
ISG-CONSTRUCT-CORE, 194
ISG-EMPTY-CORE, 194
ISG-IN-CORE, 194
Ismaili, A., 207
item, *see* resource
  categorized domain –s, 671
  set of connected –s, 670
item graph, 670
Iwasaki, A., 673
Iwata, S., 639

JA chair, 500
JA problem, 486
JA procedure, 472
  anonymous, 484
  bribery for –s, 495
    exact, 499
  by a quota rule, 479
  by a uniform premise-based quota rule, 481
  by a uniform quota rule, 479
  by the majority rule, 478
  by the ranked agenda rule, 483
  complement-free, 473
  complete, 473
  conclusion-based, 479, 480
  consistent, 473
  control for –s, 500
    by adding judges, 500
    by bundling judges, 502
    by deleting judges, 500
    by replacing judges, 500
    exact, 503
  distance-based, 481
  Dodgson, 483
  immunity of a
    to a control type, 503
  independent, 484
  Kemeny, *see* JA procedure, distance-based
  manipulation for –s, 489
    exact, 495
  microbribery for –s, 499
    exact, 499
  monotonic, 485
  neutral, 484
  nondictatorial, 484
  premise-based, 479, 480
  rational, 473

safety of the agenda for –s, 488
strategy-proof – for Hamming-distance-respecting induced preferences, 493
strategy-proof – for induced preferences of some type
  necessarily, 493
  possibly, 493
strategy-proof – in the sense of Dietrich and List, see JA procedure, strategy-proof – for induced preferences of some type, necessarily
susceptibility of a
  to a control type, 503
systematic, 484
unanimous, 483
universal domain of a, 473
vulnerability of a
  to a control type, 503
Young, 483
JA-Bribery, 495
JA-Control-by-Adding-Judges, 500
JA-Control-by-Bundling-Judges, 503
JA-Control-by-Deleting-Judges, 500
JA-Control-by-Replacing-Judges, 500
JA-Microbribery, 499
JA-Necessary-Manipulation, 494
JA-Possible-Manipulation, 493
JA-Winner, 487
Jackson, M., vii, xii, 71, 99, 136, 177, 178, 198, 203, 204, 211–213, 216, 227, 288, 355
Jain, K., 161, 645
Jain, M., 178
Janeczko, L., 451
Johnson, D., 31, 32, 123, 130, 182, 216, 221, 450, 456, 487
Jones, M., 520, 530, 556–558, 598, 602
Jones, N., 26, 28
JR, see justified representation
judge, 471
judgment aggregation, vii, ix–xi, 9, 467
judgment aggregation procedure, see JA procedure
judgment set, 472
  collective, 472
  complement-free, 472, 474
  complete, 472, 474
  consistent, 472, 474
  desired, 489
  individual, 472
  rational, 472, 474
junta distribution, 319

justified representation, 429
  extended, 431
  for additive utilities, 443
  proportional, 436

k-additive form, see utility function, in k-additive form
k-approval, 237, 305, 357
k-Borda multiwinner rule, see multiwinner voting rule, k-Borda
k-maverick-single-peakedness, 397
k-maverick-SP, 397
k-veto, 237, 357
Kaci, S., 621
Kaczmarczyk, A., 335, 365, 451, 460–464, 624, 669, 672, 674
Kaczmarek, J., 190, 191, 193
Kaklamanis, C., 659
Kakutani, S., 84
Kalagnanam, J., 247
Kalai, E., 177
Kalai, G., 319, 320
Kalaitzis, C., 638
Kalinowski, T., 313, 664, 665
Kaminka, G., 362, 366
Kaneko, M., 642
Kanellopoulos, P., 659, 679
Kannai, Y., 616
Kannan, R., 28
Karp, J., 624, 659
Karp, R., 216, 229
Kasouridis, S., 674
Kass, S., 58
Katsirelos, G., 309, 321, 354
Kavitha, T., 625, 629
Kawase, Y., 655, 679
Kazachkov, A., 659, 674
Keeney, R., 614
de Keijzer, B., 624
Keller, N., 319
Kellerer, H., 192, 639
Kelly, J., 184, 463, 676
Kemeny consensus, 247
Kemeny score, see $KScore_{(\cdot)}(\cdot)$
Kemeny winner, 247
Kemeny, J., 246, 291, 481
  voting system of, 246, 264, 291, 481
Kerkmann, A., 199, 207, 225, 226, 228, 674
Kern, P., 509
Kern, W., 28
kernel, see game, cooperative, kernel of a
Khanna, S., 638
Khot, S., 639

Index

Kilgour, M., 438–440, 650, 652, 665, 667
Kindler, G., 320
King, D., 649, 651
Kirman, A., 602
Kizilkaya, F., 441
Klamler, C., 520, 530, 556–558, 598, 602, 650, 652, 665, 667
Klaus, B., 676, 678
Kleinberg, J., 28
Klijn, F., 655
Klos, T., 624
Knaster, B., 535, 557, 603
Knop, D., 365, 464, 624, 625
Kober, S., 175, 226
Köbler, J., 36, 292
Kohler, D., 651, 665
Konczak, K., 289, 297, 299–301, 326
Könemann, J., 90
Konishi, H., 198
Kornhauser, L., 9, 470
Korzhyk, D., 178
Koutecký, M., 365
Koutsoupias, E., 229
Kowalik, L., 439, 453, 456
Kozen, D., 35
Kratsch, S., 500, 504
Krishna, A., 655
Krishnamurthy, S., 622, 627–629, 631, 645, 646
Kroer, C., 635
$KScore_{(\cdot)}(\cdot)$, 247
Kuckuck, B., 651, 652
Kuhlisch, W., 366
Kuhn, H., 545, 603
Kulkarni, P., 646
Kulkarni, R., 646, 670
Kullmann, O., 28
Kumar, R., 287
Kurokawa, D., 627, 628, 631, 641, 644
Kurz, S., 175, 203, 215, 221
Kusek, B., 462, 464
Kyropoulou, M., 659, 673, 679

Laaser, W., 26, 28
Labunski, R., 335
Lackner, M., 335, 375, 396, 433, 436, 459
Ladner, R., 311
Lafage, C., 621
Lahaie, S., 675
Lampis, M., 291
Landau, E., 23
Lang, J., x, xii, 2, 18, 203, 207, 215, 221, 228, 289, 297, 299–304, 309, 310, 312–314, 316, 326, 327, 365, 366, 375, 397, 426, 435, 449, 476, 477, 482, 483, 486, 488, 605–607, 617, 621, 623, 632, 637, 647, 649–652, 664, 665, 669, 672, 675
Lange, P., 639, 648, 669
Langston, M., 639
Larson, K., 304, 426, 635
Laslier, J., 287, 421
Last Diminisher protocol, 535, 603
  dirty-work, 574
  minimal number of cuts required by the, 582
  modified, 582
    minimal number of cuts required by the, 582
Latifian, M., 629, 635
Laurence, M., 621, 647
Laußmann, C., 335, 366, 439
LCP, 136
Le Breton, M., 355
least core, see game, cooperative, least core of a
  value of the, 156
LEAST-CORE, see WVG-LEAST-CORE
Lee, B., 426
Lee, E., 645
LeGrand, R., 439
Leiserson, C., 220
Lemaître, M., 18, 606, 607, 621, 626, 627, 630, 637, 642, 647, 679
Lemke, C., 136
  algorithm of – and Howson, 136
Lenstra jr., H., 123
Lenstra, J., 130
Lesca, J., 640, 650, 665, 668, 671, 675
Lev, O., 18, 320
Levenglick, A., 246
Levin, L., 318
leximin domination, 641
leximin optimality, 641
leximin++, 635
Leyton-Brown, K., 12, 15, 136, 288, 607
Li, B., 646, 670, 672, 675
Li, R., 673
Li, Z., 653
Lichtenstein, D., 96
Lien, Y., 26, 28
Lim, C., 178
Lin, A., 357, 361, 401
Lindner, C., ix, x, 290, 507, 548, 598, 601, 602
linear complementary problem, 136
linear program, 122, 123
  integer, 123

linear programming, 122, 281
Lineker, G., 56
Ling, J., 665
lion's share, 563
Lipton, R., 29, 625, 628, 637, 647
List, C., 471, 479, 485, 486, 488–490, 493, 504
   impossibility theorem of – and Pettit, 485
literal, 26
Liu, J., 653
Liu, S., 653
lizard, 58
Llull, R., 241
   family of voting systems of – and Copeland, 241, 264, 342, 357, 361
   voting system of, 241, 264, 342, 357, 361
local search, 130
   continuous, 130
LOCAL-OPTIMUM, 130
logic
   boolean, see logic, propositional
   propositional, 24
      satisfiability problem of, 24
Lokshtanov, D., 453, 456
Lonc, Z., 670
lone chooser, 542
Lone Chooser protocol, 542, 603
   minimal number of cuts required by the, 583
lone divider, 545
Lone Divider protocol, 545, 603
Loreggia, A., 346, 355, 357
Lower House of the German Parliament, see Deutscher Bundestag
$\mathcal{L}_P$, 471
   closure of – under negation, 472
Lu, P., 670
Lu, T., 288
Lu, X., 653, 671
Lucas, W., 160
Lueker, G., 375
Lumet, C., 679
Lund, C., 34
Luo, J., 625, 669, 674
Luss, H., 642
Lv, Y., 635
Lynch, N., 311

Machover, M., 170
Mackenzie, S., 435, 449, 564, 572, 598, 602, 649, 650, 665
Mackin, E., 671

Magnanti, T., 195
Magrino, T., 362
Mai, T., 645
$maj(\cdot)$, 256
majority
   absolute, 267
   relative, 267
majority criterion, 267, 286
   simple, 267
majority graph, 239
   weighted, 249
      strength of a path in the, 249
majority rule, see JA procedure, by the majority rule
majority voting, 286, 302, 303, 305, 313, 342
MAJORITY-SAT, 37
MAJORITYWISE-ACCEPTED-BALLOT, 504
Makino, K., 655
MANIPULATION, 308
manipulation, see also election, manipulation of an; JA procedure, manipulation for –s; voting system, manipulable
   coalitional unweighted
      constructive, 309
      destructive, 309
   coalitional weighted
      constructive, 310
      destructive, 310
   false-name, 38, 186
   in multiwinner voting, 462
   of single-peaked electorates, x, xi, 386
Manlove, D., 678
Manurangsi, P., 364, 435, 458, 624, 629, 673
many-one reducibility, see problem, reducibility between –s
   functional, 129
Marcinkowski, J., 435, 458
margin of victory, 362
   exact variant of the – problem, 363
marginal contribution net, 172
Markakis, E., 439, 625–629, 631, 632, 637, 640, 647, 672, 677
Marple, A., 36, 103, 178
Marx, D., 453, 456
Maschler, M., 155
Maskin, E., 654
matching, 677
   bipartite
      one-sided, 677
      two-sided, 677
   nonbipartite, 677

Index   751

online, 229
matching pennies game, 57
matrix game, 123
Matsui, T., 184
Matsui, Y., 184
Mattei, N., 290, 366, 435, 449, 500, 647
Maudet, N., 18, 303, 606, 607, 620, 621,
    625, 626, 633, 637, 644, 647, 656,
    668, 672
Maushagen, C., 342, 346, 355, 365
maverick, 372
Maverick, S., 396
maverick-single-peakedness, *see* k-
    maverick-single-peakedness
    (under K)
maverick-SP, *see* k-maverick-SP *(under K)*
Max-ESW, 638
max-flow/min-cut theorem, 177, 195
Max-Independent-Set, 33
    minimum-degree greedy heuristic for,
    33
Max-k-Degree-Independent-Set, 33
$max\text{-}min\text{-}FS(\cdot)$, 626
Max-NPSW, 645
Max-SAT, 33
Max-SAT-ASG$_=$, 37, 292
Max-Satisfying-Assignment-
    Equality, 37
Max-USW, 644
maximin item, 631
maximin score, *see* $SScore(\cdot)$
maximin voting, *see* Simpson, P., voting
    system of
Maximum-Egalitarian-Social-
    Welfare, 638
Maximum-Set-Packing-Compare, 293
McCabe-Dansted, J., 244, 319
McGlaughlin, P., 645, 670
MDDP, 662
Median-Property, 489
Mehlhorn, K., 625, 628, 629, 645, 646
Mehrizi, M., 463
Mehta, A., 439
Mehta, R., 629, 635, 646, 670
Meir, R., 18, 157, 158, 160, 183, 192, 462,
    500
van Melkebeek, D., 29
Menon, V., 304, 426
Menton, C., 250, 307, 330, 337, 338, 342,
    345, 346, 357
Merkel, A., 269, 336
Mestre, J., 678
method of equal shares, *see* multiwinner
    voting rule, method of equal shares

Meyer, A., 35, 36, 178
MI5, 608
Micha, E., 671, 678
Michalak, T., 193
Microbribery, 365
millennium problem, 22
Miller, G., 426
Miller, M., 481
MILP, *see* ILP, mixed
min-cost flow theorem, 366
$min\text{-}max\text{-}FS(\cdot)$, 626
Min-Vertex-Cover, 37
    edge deletion heuristic for, 37
    maximum-degree greedy heuristic for,
    37
minimal covering set, 160
minimax approval voting, *see* multiwinner
    voting rule, approval voting,
    minimax
minimax theorem, 123, 124
Mirrlees, J., 14, 464
Mirrokni, V., 639
Misra, N., 451
Misra, P., 629
mixed manna, 670
mixed-strategy profile
    expected gain of a, 63
MMS, 627
    α-MMS, 627
    ℓ-out-of-d MMS, 627
Mnich, M., 365
Monaco, G., 216, 229, 671
Monderer, D., 158
money, 607, 653
Mongin, P., 470
Monien, B., 28
Monnot, J., 302, 303, 639, 640, 642, 650,
    671, 675
monotonicity, *see* function, monotonically
    increasing; game, monotonic charac-
    teristic function of a; JA procedure,
    monotonic; preference, extension of
    a, monotonic; preference, monotonic;
    voting system, monotonic
    candidate, 416
    committee enlargement, 407, 415
    noncrossing, 417
    strict, *see* preference, monotonic,
    strictly
    strong, *see* voting system, monotonic,
    strongly
    top-member, 410, 420
Monroe, B., 674

Monty Hall dilemma, *see* Monty Hall problem
Monty Hall problem, 104
　intuitive solutions to the, 105
　solutions to the – using the formula of Bayes, 109
　solutions to the – using the law of total probability, 106
Morgenbesser, S., 272
Morgenstern, C., 43
Morgenstern, O., 1, 4, 43, 110, 113, 139, 146, 160
Mosca, R., 659
Moscardelli, L., 216, 229
Mossad, 608
Mossel, E., 320, 625, 628, 637, 647
Motwani, R., 19
Moulin, H., 18, 283–285, 607, 628, 631, 641, 644, 652, 670, 671, 674, 675
Mouri, T., 673
moving-knife protocol, 527
　Austin's, 532, 603
　of Brams, Taylor, and Zwicker, 572, 603
　of Dubins and Spanier, 541, 603
　Stromquist's, 570, 603
MSPC, 293
Muller, E., 279
　impossibility theorem of – and Satterthwaite, 279
multiagent resource allocation, 2, 605, 609
multimode control, *see* electoral control, multimode
multiwinner election, *see* voting, multiwinner
multiwinner voting, *see* voting, multiwinner
multiwinner voting rule, 404
　approval Chamberlin–Courant, 428
　approval voting, 427
　　attribute, 441
　　minimax, 438
　　proportional, 433
　　satisfaction, 440
　bloc, 414
　k-Borda, 405
　Chamberlin–Courant, 408
　harmonic Borda, 420
　method of equal shares, 443
　Phragmén sequential, 435
　proportional approval voting
　　greedy, 456
　　sequential, 456
　single nontransferable vote, 413
　single transferable vote, 423

w-Thiele, 451
Munagala, K., 459, 680
Mundhe, M., 288
Murhekar, A., 622, 642, 670
Myerson, R., 193

ℕ, 23
Nagel, R., 61
Nakamura, K., 642
Nanson winner, 257
Nanson, E., 257, 365
　voting system of, 257, 264
　　Baldwin's variant of the, 257
　　Schwartz's variant of the, 257
Naor, M., 287
Napel, S., 175, 193
Narahari, Y., 631, 655
Narayan, V., 655
Narodytska, N., 309, 313, 321, 346, 354, 355, 357, 664, 665
Narváez, D., 307, 330, 337, 338
Nasar, S., 50, 84
Nash deviation, *see* hedonic game, coalition structure for a, Nash deviation from a
Nash equilibrium
　Bayesian, 114
　　for risk-averse players, 116
　　for risk-loving players, 116
　　for risk-neutral players, 116
　complexity of a
　　in normal form games, 125
　　in zero-sum games, 122
　existence of a, 71, 82
　implausible, 101
　in mixed strategies, 63, 82
　in pure strategies, 50
　symmetric, 68
Nash stability, *see* hedonic game, coalition structure for a, Nash-stable
　contractual, *see* hedonic game, coalition structure for a, contractually Nash-stable
Nash, J., 4, 50, 71, 72, 82–84, 122, 124, 642
NASH-EQUILIBRIUM, 134
NASH-PRODUCT-SOCIAL-WELFARE-OPTIMIZATION, 642
NASH-STABILITY-EXISTENCE, 208
NASH-STABILITY-VERIFICATION, 208
National Security Agency, 608
negation, *see* boolean operation, negation
NEGATIVE-BRIBERY, 360
negotiation topology, 668

Index 753

Nehring, K., 486
Nemhauser, G., 458
network
  centrality index in a, 366
    based on a voting rule, 366
  closeness centrality in a, 367
  degree centrality in a, 367
network flow game, 176
  totally balanced, 177
network interdiction, 178
network science, 366
Neugebauer, D., xii, 509
von Neumann, J., 1, 4, 43, 47, 50, 84, 110, 113–115, 121–123, 139, 146, 160, 679
  simplified poker variante due to, 113
  simplification of the, 114
neutrality, see voting system, neutral, see JA procedure, neutral, see allocation procedure, neutral
Neveling, M., 36, 305, 306, 309, 346, 355–357, 365, 425, 674
Nguyen, N., xii, 199, 226, 228, 621, 630, 636, 638, 639, 644, 645, 647, 648, 651, 652, 674, 676
Nguyen, T., 18, 607, 621, 625, 635, 636, 638–640, 644, 645, 647, 648, 651, 652
Nichterlein, A., 365
Niclaus, D., 342, 346, 355
Nicolo, A., 530
Niedermeier, R., 39, 245, 288, 290, 309, 346, 363, 365, 451, 453, 456, 460–464, 486, 500, 624, 635, 669, 672, 674
Niedermeier, S., 504
Nikzad, A., 627
Nim, 88, 91
  winning strategy for, 89
nine men's morris, 84
Nisan, N., 18, 319, 320, 621
no-show paradox, 282, 283
  strong variant of the, 283
Nohn, A., 193
noncrossing monotonic, see monotonicity, noncrossing
nondecreasing marginal return, see increasing marginal return
nondeterministic logarithmic space, 26
nondictatorship, see voting system, nondictatorial, see JA procedure, nondictatorial
nonlinear complementary problem, 136
NONUNIFORM-BRIBERY, 365
nonunique-winner model, 234, 290
Norden, L., 362

normalized range voting, see range voting, normalized
van den Nouweland, A., 177
Nowak, M., 263, 342, 344
NP, 21
  boolean hierarchy over, 37
NP = coNP? question, 36
NP function
  class of total –s, 131
NP oracle, 291
NP-completeness, 30
NPSWO, 642
NRV, 250, 357
NS dynamics
  execution of the, 222
NSE, 208
NSV, 208
NSW, 642
Ntokos, A., 629, 632
nucleolus, see game, cooperative, nucleolus of a
Nudelman, E., 137
NULL, see WVG-NULL
number
  set of integer –s, 123
  set of nonnegative integer –s, 23
  set of nonnegative real –s, 141
  set of real –s, 47
Nüsken, P., 342, 346, 355

$\mathcal{O}(\cdot)$, 23
$o(\cdot)$, 23
object, see resource
Obraztsova, S., 286
OCS principle, see one-cut-suffices principle
ODD-DEGREE-VERTEX, 130
ODD-SAT, 37
$\omega_G(\cdot)$, 210
Ogihara, M., 19, 319
Ogiwara, M., see Ogihara, M.
Oh, H., 671
Ohta, K., 207
Olsen, M., 206, 207
$\Omega(\cdot)$, 23
$\omega(\cdot)$, 23
one-cut-suffices principle, 580, 582
one-way envy relation, see division, envy relation in a, one-way
online manipulation setting, 323
online voting, 334
ONLINE-CCM, 324
ONLINE-CCM$[k]$, 324
ONLINE-CCWM$[k]$, 324

ONLINE-CCWM, 324
ONLINE-CONSTRUCTIVE-COALITIONAL-MANIPULATION, 324
ONLINE-DCM[k], 325
ONLINE-DCM, 325
ONLINE-DCWM[k], 325
ONLINE-DCWM, 325
OPTIMAL-LOBBYING, 366
order
  linear, 235
  partial, 297
  preference, see preference relation
  weak, 199, 492, 613
ordered weighted average, 626
Ordyniak, S., 671, 672
Orlin, J., 195, 288, 301, 309, 312, 486
Osborne, M., 150, 161
Osherson, D., 481
OWA, 626
Ozen, B., 441
Öztürk, M., 375, 397

#P, 184, 678
P, 21
P = NP? question, 22
packing, 335
packing and cracking, 335
Padget, J., 18, 606, 607, 621, 637
Pagourtzis, A., 674
Papadimitriou, C., 18, 19, 28, 29, 37, 122, 125–127, 130, 134, 136, 176, 184, 194, 196, 197, 229, 292, 487
Pápai, S., 675, 676
paper, 58
paper-rock-scissors game, 58, 59, 67
paper-rock-scissors-lizard-spock game, 58
parallel access to NP, 36, 292
parallel-universes tie-breaking, 290
Pareto consistency, 268, 269, 286
Pareto dominance, 49, 622
  strong, 49, 622
  weak, 49
Pareto efficiency, see Pareto optimality
Pareto optimality, 48, 49, see hedonic game, coalition structure for a, Pareto-optimal, 522, 612, 622
  fractional, 622
  necessary, 649
  possible, 649
  price of , 229
  weak, 49, 622
Pareto optimum, 49
Parkes, D., 326, 342, 345, 361, 362, 364
parliamentary elections

for the German Bundestag
  1990, 269
  2009, 282
  2013, 282
for the German Reichstag
  1936, 282
  1939, 282
for the German Volkskammer
  1963, 282
  1967, 282
  1981, 282
  1986, 282
participation criterion, 282, 283, 286
participatory budgeting, 447, 680
PARTITION, 182, 316
Pasechnik, D., 158, 160, 183, 192
Paterson, M., 171, 187, 188, 652
path disruption game, 177
Pattanaik, P., 616
Patty, J., 9, 373
Pauly, M., 486
  impossibility theorem of – and van Hees, 486
PAV, see multiwinner voting rule, proportional approval voting
  greedy, see multiwinner voting rule, proportional approval voting, greedy
  sequential, see multiwinner voting rule, proportional approval voting, sequential
$PAV\text{-}Score(\cdot)$, 433
Payan, J., 665
payoff vector, see game, cooperative, payoff vector for a
Paz, A., 551, 587, 593, 594, 597, 603
$PBP(\cdot)$, 480
Pěchouček, M., 178
Peleg, B., 155, 170, 616
penalty game, 56, 66
Penn, E., 9, 373
Pennock, D., 675
Penrose, L., 168
Penrose–Banzhaf index, see game, cooperative, normalized Banzhaf index of a player in a simple
perfectness, see hedonic game, coalition structure for a, perfect
permutation, 162
  set of all –s of a set, 163
Perny, P., 620, 621, 640
Peters, D., 206, 375, 443, 445, 459, 669–671
Pettit, P., 9, 470, 471, 485, 488
  impossibility theorem of List and –, 485

Index 755

Pfandler, A., 396
Pferschy, U., 192
PFS, 630
Phelps, S., 18, 606, 607, 621, 637
Phragmén sequential rule, *see* multiwinner voting rule, Phragmén sequential
Π_0^p, 36
Π_1^p, 36
Piccione, M., 103
picking sequence, 663
　optimal, 664
picking sequences protocol, 663
　balanced alternation, 664
　round-robin, 663
　strict alternation, 663
Pierczyński, G., 443, 445
PIGEONHOLE-FUNCTION, 128
Pigozzi, G., 481–483
Π_i^p, 36
Pilipczuk, M., 453, 456
Pilipczuk, M., 453, 456
Pini, M., 302, 366
Piras, L., *see* Schend, L.
Pisinger, D., 192
Pivato, M., 18
PJR, 436
Plaut, B., 629, 635, 660
player
　action of a, 525
　beneficial merging of –s, 185
　beneficial splitting of a, 185
　coalition of –s, 139, 141
　gain of a, 47
　irrevocable advantage of a, 567
　marginal contribution of a
　　to a permutation, 163
　　to the coalition of predecessors, 152
　　to the gains of a coalition, 145, 162
　move of a, 84
　null, *see* game, cooperative, null player in a simple
　payoff of a, 525
　peak of a, 211
　pivotal, *see* game, cooperative, pivotal player in a simple
　portion of a, 508
　　contiguous, 544
　risk-averse, 63, 112, 525
　risk-loving, 63, 112
　risk-neutral, 63, 112
　share of a, 508
　strategy of a, 45, 512
　symmetric –s, 166, 215
　type of a, 111, 112

valuation function of a, 508
　additively separable, 204
　additivity of a, 509
　box representation of a, 531
　continuous, 533
　divisibility of a, 510
　nonnegativity of a, 509
　normalization of a, 509
　positivity of a, 509
　veto, *see* game, cooperative, veto player in a simple
PLS, 127, 130
plurality, 236, 264, 286, 302, 303, 305, 313, 342, 357
plurality rule, *see* plurality
plurality voting, *see* plurality
plurality winner, 236
plurality with run-off, 251, 264, 357
　winner in, 251
PMMS, 631
$P_{\|}^{NP}$, 36, 291
　search variant of, 488
P^{NP}, 36, 328
$P^{NP[1]}$, 36, 328
$P^{NP[\log]}$, 36
PO, 622
poker, 110, 113, 114
poker face, 121
political cynicism, 282
polynomial hierarchy, 35
polynomial local search, 127, 130
polynomial parity argument
　for directed graphs, 128, 131
　for graphs, 128, 130
polynomial pigeonhole principle, 127, 128
polynomial space
　deterministic, 35
　nondeterministic, 35
polynomial time
　alternating, 35
　deterministic, 21
　nondeterministic, 21
　probabilistic, 37
polynomial-time approximation scheme, 34
Ponnuswami, A., 639
Porat, E., 36, 177, 178
Porello, D., 10, 486–490, 494, 495
Porter, R., 137
portion, *see* player, portion of a
pos(·), 408, 411
POSSIBLE-WINNER, 299, 302
POSSIBLE-WINNER-NEW-ALTERNATIVES, 303

Possible-Winner-Top/Bottom/Doubly-Truncated-Ballots, 304
POSSIBLE-WINNER-UNCERTAIN-WEIGHTS, 305
POSSIBLE-WINNER-UNCERTAIN-WEIGHTS, 305
post-election audits, 362
    risk-limiting, 362
potential function technique, 216
Poundstone, W., 272
power index, *see* game, cooperative, power index of a player in a simple
    additivity of a, 167
    efficiency of a, 167
    null player property of a, 167
    symmetry of a, 167
    valuation property of a, 170
POWER-COMPARE, *see* WVG-$\mathbb{PI}$-POWER-COMPARE
Π_P, 163
PP, 37, 185
PPA, 128, 130
PPAD, 128, 131
PPP, 127, 129
PR, 652
Prasad, K., 184
preference, 234
    compact representation of –s, 613
    extension of a
        monotonic, 615
        responsive, 207, 615
        separable, 615
    generalized – ballot, 439
    induced by a judgment set
        closeness-respecting, 492
        Hamming-distance-respecting, 493
        necessary, 493
        possible, 493
        top-respecting, 492
        unrestricted, 492
        weak necessary, 493
        weak possible, 493
    monotonic, 615
        strictly, 615
    single-peaked –s, 369, 373
        in anonymous hedonic games, 212
    k-maverick-, 397
    swoon-, 398
preference aggregation, ix, xi, 233
preference elicitation, 613
preference language
    cardinal, 617
    ordinal, 615
preference list

partial, 297
    total extension of a, 298
preference profile, 199, 234
    single-peaked, 369
preference relation, 235, 492, 613
    dichotomous, 623
    linear, 235
    partial, 297
    separable, 613
        additively, 204, 614
        weakly, 613
    weak, 199, 492, 613
        between outcomes of a hedonic game, 199
preferential dependency, 613
premise-based procedure, *see* JA procedure, premise-based
price function
    $discrete, 360
    discrete, 360
    swap-bribery, 363
price of anarchy, 229
price of connectivity, 671
principle of excluded contradiction, 472
principle of the excluded third, 472
prisoners' dilemma, 4, 45, 69
    as a Bayesian game, 121
    Bayesian Nash equilibrium for the, 121
Pritchard, G., 244, 319
probabilistic polynomial time, 185
probabilistic serial rule, 678
PROBABILISTIC-LOBBYING, 366
probability, 107
    conditional, 108
    total, 107
        law of, 106, 108
probability distribution, 107
probability space
    finite, 107
problem
    completeness of a, 29
    complexity of a, 21
        lower bound on the, 19, 23, 29
        upper bound on the, 19, 22
    decidable, 19
    fixed-parameter tractable, 453
    hardness of a, 29
    parameterized, 39
    reducibility between –s, 29
    undecidable, 19
Procaccia, A., 18, 309, 318–321, 326, 354, 362, 366, 451, 462, 588, 590, 603, 624, 627, 628, 631, 635, 641, 644, 659, 671, 674, 677

Index

PROP1, 631
PROPm, 631
proportional algorithm, 652
proportional approval voting, *see* multi-winner voting rule, approval voting, proportional
proportional justified representation, *see* justified representation, proportional
proportionality, *see* cake-cutting protocol, proportional; division, proportional; fair share, proportional; fair share criterion, proportional; price of proportionality; proportional algorithm; share, proportional, *see* fair share criterion, proportional
  up to any good, 631
  up to one good, 631
  up to the max-min item, 631
proportionality degree, 432
proportionality for solid coalitions, 423
PROPX, 631
Pruhs, K., 588, 590, 598
Psomas, A., 635, 674, 677
Psomas, C., 631, 640
PSPACE, 35
PTAS, 34
Puente, M., 173, 174
Pukelsheim, F., 241
Puppe, C., 355, 486, 504, 660, 662
$\mathcal{PV}(\cdot)$, 155
PW-Top/Bottom/Doubly-TB, 304

QBF, 35
QBF_i, 35
$\overline{\mathrm{QBF}_i}$, 35
Qin, J., 670
QUANTIFIED-BOOLEAN-FORMULA, 35
Quarter protocol
  for four players, 594
  for three players, 591
Quinzii, M., 655
quota, 479
quota rule, *see* JA procedure, by a quota rule

$\mathbb{R}$, 47
$\mathbb{R}^+$, 141
Rácz, M., 320
Raghavan, P., 28
Raiffa, H., 614
Raith, M., 655
Ramaekers, E., 653
Ramezani, S., 644, 647
random variable
  discrete, 115
  expectation of a, 115
von Randow, G., 104, 105
range voting, 250, 264, 357, 361
  normalized, 250, 264, 357
ranked pairs, 248, 264, 302, 305, 361
RANKING
  Dodgson-, 293, 294
  Kemeny-, 293, 294
  Young-, 293
Rathi, N., 672
rationality constraint, 476
Rauchecker, G., 670
REACHABLE-OBJECT, 668
Rebel Without a Cause, 55
recommender system, 288
recursive function theory, 19
reducibility, *see* problem, reducibility between $-$s
  polynomial-time many-one, 29
referee, *see* judge
Reger, C., 355–357
Reisch, S., 94, 96
Reisch, Y., 312, 361–365
relation
  antisymmetric, 199
  asymmetric, 235
  connected, 235
  indifference, 199, 492, 613
  irreflexive, 235
  preference, *see* preference relation
  reflexive, 199, 235
  total, 199
  transitive, 199, 235
representation language
  compact, 615
resoluteness, *see* voting system, resolute
resource
  divisible, 507
    allocation of $-$s, 507
    fair division of $-$s, x, xi, 507
  division of a, 507
  heterogeneous, 509
  homogeneous, 509
  indivisible, 607
    allocation of $-$s, vii, ix–xi, 605
  nonshareable, 608
  shareable, 608
response strategy
  best, 50
  strictly best, 50
responsive extension principle, *see* preference, extension of a, responsive
  bipolar extension of the, 207

del Rey, L., 53
Rey, A., xii, 36, 170, 178, 187–190, 207, 210, 221, 226, 228, 652, 674
Rey, L., 226, 228, 439, 674
Rey, S., 439, 670
Rezvan, R., 629
Rieck, C., 109
Riege, T., 28, 37
Ries, B., 302
Rivest, R., 220, 362
Roberts, K., 464
Robertson, J., 509, 511, 513, 515, 528, 545, 563, 565, 570, 577, 582, 587–589, 597, 602
rock, 58
Roddenberry, G., 59
Rodríguez-Aguilar, J., 18, 606, 607, 621, 637
Rogers jr., H., 19
Rognlie, M., 290
roll-call voting game, 326
Romen, R., 229
Romero-Medina, A., 207
Roos, M., 18, 303–306, 366, 607, 621, 636–639, 644, 645, 647, 648
Rosamond, F., 290, 366, 500
Rosenschein, J., 18, 157, 158, 160, 170, 183, 192, 288, 302, 309, 319–321, 354, 362, 366, 451, 462, 486
Rösler, P., 269
Rossi, F., 302, 346, 355, 357, 366, 678
Roth, A., 4, 140, 162, 677
Rothe, I., ix, x, xii, 23, 43, 85
Rothe, J., ix, x, xii, 1, 2, 18, 19, 23, 27–29, 31, 34–37, 43, 85, 139, 158, 160, 170, 178, 183, 187–193, 199, 207, 210, 221, 226, 228, 229, 233, 241, 244–246, 263, 281, 290–294, 296, 301, 303–307, 309, 311, 312, 319, 321, 323–327, 330–333, 335, 337, 339, 340, 342–347, 349, 350, 354–358, 361–366, 369, 375, 376, 386, 388, 390, 394, 397, 399, 401, 403, 425, 426, 463, 467, 486, 487, 492–497, 499, 500, 503, 504, 507, 509, 548, 598, 601, 602, 605, 607, 621, 625, 630, 635–640, 644, 645, 647, 648, 651, 652, 669, 674, 676
Roughgarden, T., 18, 629, 635, 660
Rowling, J., 322
Roy, S., 207, 221
RSD, 678
Rubinstein, A., 103, 150, 161
Rudolph, J., 366

Russel, N., 346

Saari, D., 258
Sabán, D., 678
Saberi, A., 625, 627, 628, 637, 638, 645, 647
Sachs, R., 105
Sachse, F., 669, 674
Sadhukhan, S., 655
Sadri, T., 668
Saffidine, A., 651, 652
Sager, L., 9, 470
Sakai, T., 655
Sakurai, Y., 193, 207
Saleh, H., 673
Salvagnin, D., 302
Sánchez-Fernández, L., 433, 436
Sandholm, T., 17, 136, 137, 309, 310, 312–314, 316, 319, 607, 618, 624, 668
Sandomirskiy, F., 670, 674, 675
Santa Claus problem, 635, 636
Santi, P., 618
Sanver, R., 259, 261–263, 266, 287, 344, 438–440
Sarwate, A., 362
SAT, 24
k-SAT, 26
SAT solver, 27
SAT-UNSAT, 37
satisfaction approval voting, see multi-winner voting rule, approval voting, satisfaction
SATISFIABILITY, 24
Satterthwaite, M., 8, 275, 279, 288
impossibility theorem of Muller and –, 279
Saurabh, S., 453, 456
SAV, see multiwinner voting rule, satisfaction approval voting
Savani, R., 223
vos Savant, M., 105, 106
Savitch, W., 35
theorem of, 35
Scarcello, F., 670
Scarf, H., 136
algorithm of, 136
Schadrack, H., 207, 210, 221, 228
Schaudt, O., 439
Schauer, J., 203, 215, 221, 629, 652
Scheidungsformel, see adjusted winner procedure
Schend, L., xii, 18, 207, 210, 221, 226, 228, 290, 304–306, 309, 312, 321, 332, 342, 344, 345, 350, 354, 361–365, 674

Index 759

Scheuermann, B., 366
Schiermeyer, I., 28
Schiller, F., 43
Schilling, R., 509
Schlotter, I., 365, 625, 631
Schmalhofer, M., 646, 670
Schmidt-Kraepelin, U., 665, 675
Schneckenburger, S., 642
Schnorr, H., 309, 313, 314, 316, 346, 361
Schnorr, I., 361
Schoepflin, D., 631
Schöning, U., 27, 28, 36, 292, 319
Schryen, G., 670
Schulze winner, 249
Schulze, M., 248, 249, 345
   voting system of, 248, 249, 264, 361
Schwartz sequential dropping, 249
Schwartz winner, 248
Schwartz, T., 248, 257, 275
   voting system of, 248, 264
scissors, 58
SCORE
   Dodgson-, 294
   Kemeny-, 294
   Young-, 293
scoring allocation correspondence, *see* scoring allocation rule
scoring allocation rule, 651
   duplication-monotonic, 651
   monotonic, 651
   object-monotonic, 651
   separable, 651
   strategy-proof, 652, 676
scoring function, 308
   t-approval, 412
   Borda, 412
   committee, 411
   submodular, 456
scoring protocol, 235, 264, 361
   generalized, 320, 362
   pure, 301
   winner in a, 235
scoring rule, *see* scoring protocol
   committee, 411
      sentation-focused, 413
      separable, 413
      weakly separable, 413
scoring vector, 235
SCSE, 209
SCSV, 209
search problem, 125
Searns, A., 627, 628, 670
Seddighin, M., 627–629, 635, 646, 673, 675

Seddighin, S., 627, 628, 635, 646, 675
Seedig, H., 219, 220, 227, 228
Seeger, T., 335, 342, 346, 355
Segal-Halevi, E., 627, 628, 647, 651, 670, 671, 673–675
Selfridge, J., 565, 567, 603
Selfridge–Conway protocol, 565, 603
Selker, A., 18, 365, 439, 500, 503, 504
Selman, A., 19, 126, 311
Selten, R., 4, 50, 61, 101
Sen, A., 8, 18
Sen, S., 288
set
   bounded, 80
   cardinality of a, 72
   closed, 80
   compact, 80
   complement of a, 107
   convex, 72
   exact cover of a, 217, 350
   power set of a, 141, 473
Sethuraman, J., 678
Sewelson, V., 37
Sgall, J., 588, 590
Sgouritsa, A., 625, 629
Shacham, H., 362
Shah, N., 628, 631, 635, 641, 644, 655, 670, 671, 673, 677–680
Shahkarami, G., 646
Shalom, M., 229
Shams, P., 626, 633
Shapley$(\cdot,\cdot)$, 165
Shapley$^*(\cdot,\cdot)$, 165
Shapley value, *see* game, cooperative, Shapley value of a player in a
   additivity of the, 166, 167
   axiomatic characterization of the, 167
   efficiency of the, 167
   null player property of the, 167
      converse of the, 167
   raw, 165
   symmetry of the, 167
Shapley's saddle, 160
Shapley, L., 4, 145, 146, 152, 155, 160, 162, 165, 167, 169, 170, 677
Shapley–Shubik index, *see* game, cooperative, Shapley–Shubik index of a player in a simple
Shapley-Shubik$(\cdot,\cdot)$, 164
Shapley-Shubik$^*(\cdot,\cdot)$, 163
share, *see also* player, share of a
   envy-free, 518
   equitable, 518
   exact, 514

guaranteed, 514, 517, 518
proportional, 514
simply fair, *see* share, proportional
strongly fair, *see* share, super-proportional
super-envy-free, 518
super-proportional, 514
Shen, E., 362
Shen, Z., 459
SHIFT-BRIBERY, 364
Shiryaev, , 362, 364
Shoham, Y., vii, xii, 12, 15, 103, 122, 136, 137, 158, 172, 206, 288, 607, 611, 621, 647
Shubik, M., 155, 162
side payment, 654, 668
Σ^*, 29
Σ_0^p, 35
Σ_1^p, 35
Σ_i^p, 35
Sikdar, S., 633, 635, 642, 670–672
Simola, S., 207, 221
simple game, *see* game, cooperative, simple
  losing coalition in a, 146
    maximal, 171
  winning coalition in a, 146
    minimal, 171
simplex, 73
  face of a, 73
  standard, 73
  vertex of a, 73
simplex method, 123
simplotope, 82
Simpson score, *see* $SScore(\cdot)$
Simpson winner, 244
Simpson's paradox, 269
Simpson, B., 60
Simpson, L., 60
Simpson, P., 244, 269
  voting system of, 244, 264, 302, 303, 305, 313, 357, 361
Simpsons, The, 58, 60
sincere-strategy preference-based approval voting, *see* SP-AV
  winner of, *see* SP-AV winner
Singh, M., 645
Singh, P., 345, 346
Singh, S., 58, 60
single nontransferable vote rule, *see* multiwinner voting rule, single nontransferable vote

single transferable vote, 252, *see* multiwinner voting rule, single transferable vote
  winner of, *see* STV winner
single-peaked electorate, x, xi, 9, 369, 373
  k-maverick-, 397
  swoon-, 398
single-peaked society, *see* single-peaked electorate
single-peakedness, *see* single-peaked electorate
single-player deviation, *see* $\mathfrak{C} \xrightarrow{i} \mathfrak{C}'$; hedonic game, coalition structure for a, single-player deviation from a
singleton ranking form, 615
Sipser, M., 96
Sivakumar, D., 287
skat, 44
  international laws of, 44
Skibski, O., 193, 367
Skopalik, A., 192
Skowron, P., xi, 2, 403, 407, 412, 415–417, 420, 421, 432, 433, 435, 436, 443, 445, 449, 451, 454, 458, 460–462, 464, 606, 674
Slater, P., 291
  voting system of, 291
Slavkovik, M., 476, 477, 482, 483, 488
Slinko, A., 244, 289, 319, 320, 346, 358, 363, 364, 366, 401, 407, 412, 415–417, 420, 421, 425, 451, 454, 459, 486, 500, 606, 674
Sloth, B., 326
Smith, J., 178, 269
SNTV, *see* multiwinner voting rule, single nontransferable vote
Socala, A., 439
social choice correspondence, 234
social choice function, 234
  surjectivity of a, 274
social choice theory, 8
social game, 71
  finitely repeated, 71
social welfare
  egalitarian, 636
  Nash product, 642, 645
  utilitarian, 646
social welfare function, 234
societal axis, 373
Sokoban puzzles
  winning strategies in, 96
Sonar, C., 451
Sönmez, T., 198, 675
Sonoda, A., 675

Index 761

Sornat, K., 364, 435, 439, 458
Sotheby's, 13, 608
Sousa, P., 18, 606, 607, 621, 637
SP-AV, 261, 264, 342
SP-AV winner, 262
Spakowski, H., 37, 245–247, 281, 291–294, 296, 319, 343
Spanier, E., 541, 542, 557, 558, 579, 590, 603
Speckenmeyer, E., 28
Sperner, E., 77, 134
  lemma of, 77, 134
spock, 59
Spock, Mr., 59
Sprumont, Y., 637
Sricharan, A., 670
Srivastava, S., 136
Srividenko, M., 639
$SScore(\cdot)$, 244
stability, *see* game, cooperative, cost of stability of a; game, cooperative, external stability of a; game, cooperative, internal stability in a; hedonic game, core stability of a; hedonic game, individual contractual stability of a; hedonic game, Nash stability of a; stability of a solution,
stability of a solution, 50
stable marriage problem, 677
stable semantics, 160
stable set, *see* game, cooperative, stable set of a
Stackelberg voting game, 326
Star Trek, 59
Stark, P., 362
Stein, C., 220
Steinberg, R., 12, 15, 607, 611, 647
Steinhaus, H., 11, 508, 535, 545, 548, 557, 564, 579, 603, 629
Still, M., 105
Stockmeyer, L., 35, 36, 178
Storer, J., 96
Stoyan, D., 366, 509
straight line program, *see* utility function, represented by a straight line program
strategy
  dominant, 45, 48
  mixed, 47, 63
    fully, 64
    support of a, 64
  pure, 47
    set of profiles of –ies, 47

strictly dominant, 48
weakly dominant, 48
strategy-proofness, *see* cake-cutting protocol, strategy-proof; JA procedure, strategy-proof – for Hamming-distance-respecting induced preferences; JA procedure, strategy-proof – for induced preferences of some type; scoring allocation correspondence, strategy-proof; voting system, strategy-proof;
strict core stability, *see* hedonic game, coalition structure for a, strictly core-stable
STRICT-CORE-STABILITY-EXISTENCE, 209
STRICT-CORE-STABILITY-VERIFICATION, 209
Stromquist, W., 545, 565, 570–572, 577, 603
strong ϵ-core, *see* game, cooperative, strong ϵ-core of a
STRONGNEGATIVE-BRIBERY, 360
StrongYoung score, *see* $SYScore(\cdot)$
StrongYoung voting, 245
STV, *see* single transferable vote, 264, 286, 302, 313, 361, *see* multiwinner voting rule, single transferable vote
STV winner, 252
Su, F., 655
SUBSET-SUM, 126
subsidy, 655
successive elimination, *see* voting by successive elimination
Suchý, O., 500
Sudhölter, P., 170
Suksompong, W., 216, 624, 629, 631, 665, 670, 671, 673, 675
Sumita, H., 655, 679
Sun, A., 670
Sung, S., 204, 216, 219–221
super-envy-freeness, *see* cake-cutting protocol, super-envy-free; division, super-envy-free; share, super-envy-free
SUPER-IMPUTATION-STABILITY, *see* WVG-SUPER-IMPUTATION-STABILITY
super-proportionality, *see* cake-cutting protocol, super-proportional; division, super-proportional; share, super-proportional
superadditivity, *see* game, cooperative, superadditive

supermodularity, *see* game, cooperative, supermodular characteristic function of a
Suzuki, M., 655
Suzumura, K., 18
Svensson, L., 655, 676
Svensson, O., 638
Sviridenko, M., 636
SWAP-BRIBERY, 363
swoon-single-peakedness, 398
swoon-SP, 398
$\overline{SYScore}(\cdot)$, 245
$SYScore(\cdot)$, 245
systematicity, *see* JA procedure, systematic
Szeider, S., 489

Tadenuma, K., 655
Taki, S., 627, 670
Taliancich, C., 337, 338
Talmon, N., 193, 346, 365, 412, 415, 417, 420, 421, 451, 454, 459–462, 464
Tambe, M., 178
Tamura, A., 655
Tan, X., 674
Tang, Z., 638
Tao, Y., 660
Tappe, L., 216, 219, 223
Tardos, É, 18
Tasnádi, A., 355
TAUTOLOGY, 487
tautology, 472
Taylor, A., 18, 146, 173, 174, 272, 273, 509, 513, 528, 546, 561, 562, 565, 570, 572, 588, 602, 603, 607, 655, 656, 659, 664
  adjusted winner procedure of Brams and –, *see* adjusted winner procedure
TE, 333
Tennenholtz, M., 158, 326, 500, 509
TFNP, 131
Thapa, M., 123
theory of formal languages and automata, 19
$\Theta(\cdot)$, 23
w-Thiele rule, *see* multiwinner voting rule, w-Thiele
Thiele rule, *see* multiwinner voting rule, w-Thiele
Thilikos, D., 33
Thomson, W., 637, 655
Tic-Tac-Toe, 86
  game tree for, 87
Tideman, N., 243, 248, 276, 330, 345

voting system of, *see* ranked pairs
tie-breaking rule
  randomized, 287
tie-handling rule, 333
Tierney, J., 105
ties eliminate, 333
ties promote, 333
Tlilane, L., 640, 642, 671
Toda, S., 187
Todo, T., 665, 673, 675
Tolkien, J., 321
Tominaga, Y., 665
Toni, F., 668
top cycle, *see* Schwartz, T., voting system of, 264
top cycle winner, *see* Schwartz winner
Top Trading Cycle, 671
van der Torre, L., 482, 483
Tovey, C., 288, 289, 291, 305, 308, 329–331, 336–338, 340–342, 357, 401, 463, 486, 652
TP, 333
tree, 85
  game, *see* game tree
  leaf of a, 85
  root of a, 85
  voting, *see* voting tree
Trick, M., 288, 289, 291, 305, 308, 329–331, 336–338, 340–342, 357, 374, 397, 401, 463, 486, 652
trimming algorithm, 582
Tröbst, T., 229
Truszczynski, M., 670
truth value, 25
truth-table reducibility, 311
  disjunctive, 311
TTC, 671
Turing machine, 20
  alternating, 35
  configuration of a, 20
  deterministic, 20
  end configuration of a, 20
    accepting, 20
    rejecting, 20
  halting configuration of a, *see* Turing machine, end configuration of a
  instantaneous description of a, 20
  memory of a, *see* Turing machine, space of a
  nondeterministic, 20
    computation tree of a, 21
  probabilistic, 38
  space of a, 21
  time of a, 19

Index 763

Turing, A., 20
twin, 284
twin paradox, 285
twins-welcome criterion, 284, 285
two-player game, 46, 52
  multiple-move, 84
  one-move, 84
  properties of a, 70
  sequential, 84
two-way envy relation, *see* division, envy relation in a, two-way

Überhangmandat, *see* excess mandate
Uckelman, J., 621
UCO, 320
UEFA Champions League, 253
UEFA European Championship 1996, 609
Uhlmann, J., 290, 451, 459
Ullman, J., 19
unanimity, *see* JA procedure, unanimous; voting system, unanimous; weighted voting game, unanimous
UNANIMOUSLY-ACCEPTED-BALLOT, 504
undercut protocol, 665
uniform distribution, 107
uniform premise-based quota rule, *see* JA procedure, by a uniform premise-based quota rule
uniform quota rule, *see* JA procedure, by a uniform quota rule
Unique-WARP, 340
unique-winner model, 234, 290
United Nations Security Council, 148
UNWEIGHTED-COALITIONAL-OPTIMIZATION, 320
$UPQR_q(\cdot)$, 481
US Congress
  1789 election of the First, 335
US House of Representatives, 148, 335
US presidential election of 2000, 330, 331
USWO, 647
UTILITARIAN-SOCIAL-WELFARE-OPTIMIZATION, 647
utility function
  additively independent, 614
    generalized, 620
  in bundle form, 617
  in generalized additivity form, 620
  in k-additive form, 618
  Leontief, 635
  Rado, 646
  represented by a bidding language, 621
  represented by a straight line program, 621
  represented by a weighted logic language, 621
  subadditive, 614, 644
    fractionally, 644
  submodular, 614, 644
  supermodular, 668
  symmetric, 630
utility without externalities, 614

Vaish, R., 622, 628, 629, 633, 635, 642, 645, 646, 670, 672, 679
Valiant, L., 184, 193, 678
valuation, *see* utility function
valuation criteria, 513
valuation function, 204
  box representation of a, 530
  dynamic, 225
valuation property, *see* power index, valuation property of a
Vaněk, O., 178
Varricchio, G., 646
Vazirani, U., 229
Vazirani, V., 18, 34, 229, 645
vector
  affinely independent –s, 72
  convex combination of –s, 72
  linearly independent –s, 73
  unit, 73
vector weighted voting game, *see* weighted voting game, vector
  dimension of a, 173
Vega-Redondo, F., 178
Végh, L., 646
Venable, B., 302, 346, 355, 357, 366
Verfaillie, G., 606
Verma, P., 635, 677
Vermande, Q., 646
Vesic, S., 482, 483
veto, 236, 264, 302, 303, 305, 313, 357
k-veto, *see* k-veto *(under K)*
veto rule, *see* veto
veto voting, *see* veto
veto winner, 236
veto with run-off, 252, 357
Vetschera, R., 667
Vetta, A., 655
Vickrey, W., 14, 15
Vickrey–Clarke–Groves mechanism, 15
Viglas, A., 29
Vinci, C., 671
Virginia House of Delegates, 335
Vogel, J., 245–247, 281, 291–294, 296, 343

Vohra, R., 161
Vondrák, J., 639
vote, 234
　list of –s, 234
vote by mail, 334
voter
　β-cohesive group of –s, 442, 443
　ℓ-cohesive group of –s, 431
　cohesive group of –s, 428
　electoral behavior of a, 336
　irrational, 365
voting
　multiwinner, ix–xi, 403
voting by successive elimination, 355
voting rule, see voting system
voting system, vii, 234
　anonymous, 277, 286
　based on pairwise comparisons, 237, 264
　Condorcet-consistent, 240
　consistent, 269, 286
　convexity of a, 269
　dictatorial, 268
　homogeneous, 280, 286
　hybrid, 258, 264
　immunity of a – to a control type, 338
　manipulable, 274
　monotonic, 277, 278, 286
　　strongly, 278
　　weakly, 319
　multiwinner, see multiwinner voting rule
　monotonicity of a, 407
　neutral, 277, 286
　nondictatorial, 268
　proceeding in stages, 251, 264
　properties of –s, 286
　pure, 258
　resistance of a – to a control type, 339
　resolute, 274
　separability of a, 269
　strategy-proof, 274, 286
　susceptibility of a – to a control type, 338
　unanimous, 355
　voiced, 311, 340
　vulnerability of a – to a control type, 339
　weakCondorcet-consistent, 240
　winner determination in a, 289
voting tree, 253
Voudouris, A., 607, 629, 632, 635, 673

W[1], 456

W[2], 456
W-hierarchy, 456
w-$Score(\cdot)$, 451
Wagner, D., 362
Wagner, K., 19, 36, 37, 292, 294
Walsh, T., 18, 302, 309, 313, 318, 321, 346, 354, 355, 357, 366, 386, 428, 431, 435, 449, 486, 625, 649, 650, 664, 665, 670, 673, 674, 677, 679
Wang, H., 633, 635, 653, 671, 672
Wang, J., 318, 454, 627, 628, 631, 644
Wang, K., 459
Wang, S., 635
Wanke, E., 23, 85
WARP, 340
Washburn, A., 178
wasted votes effect, 335
Watanabe, O., 319
Waters, J., 55
Watts, A., 71
Weak Axiom of Revealed Preference, 340
　Unique, 340
Webb, W., 509, 511, 513, 515, 519, 528, 545, 563, 565, 570, 577, 582, 587–589, 597, 602
Wechsung, G., 19, 37, 38, 187
Wegener, I., 19
Weibull, J., 160
weighted logic language, see utility function, represented by a weighted logic language
weighted majority game, see weighted voting game
weighted threshold game, see weighted voting game
weighted voting game, 146, 148, 310
　boolean, 175
　complexity of power indices in –s, 184
　complexity of problems for –s, 180
　control by adding players to –s, 189, 190
　control by deleting players from –s, 189
　graph-restricted, 193
　　control by adding edges to –s, 193
　　control by deleting edges from –s, 193
　　losing coalition in a, 193
　　probabilistic Banzhaf index of a player in a, 193
　　Shapley–Shubik index of a player in a, 193
　　winning coalition in a, 193
　losing coalition in a, 146, 148
　manipulating the quota in –s, 191
　quota in a, 146

Index

unanimous, 188
vector, 173
winning coalition in a, 146, 148
Weishaupt, R., 305, 306, 509
Weller, D., 523
Welles, H., 337, 338
Weltge, S., 175
Wiechers, A., 226, 228, 674
van Wijland, E., 646
Wilczynski, A., 212, 222, 665, 668
Williams, R., 319
WINNER, 289
  Dodgson-, 293
  Kemeny-, 293
  Young-, 296
winner determination
  in Copeland elections, 290
  in Dodgson elections, 291
  in Kemeny elections, 291
  in scoring protocols, 290
  in voting systems proceeding in stages, 290
  in Young elections, 291
winner model, *see* nonunique-winner model
winner's curse, 14
winner-turns-loser paradox, 278
winning committee
  approval CC, 428
  approval Chamberlin–Courant, *see* winning committee, approval CC
  bloc, 414
  k-Borda, 405
  CC, 409
  Chamberlin–Courant, *see* winning committee, CC
  feasible, 441
  harmonic Borda, *see* winning committee, HB
  HB, 420
  method of equal shares, 443
  minimax approval voting, 438
  multiwinner approval voting, 428
  PAV, 433
  Phragmén sequential rule, 436
  proportional approval voting, *see* winning committee, PAV
  satisfaction approval voting, *see* winning committee, SAV
  SAV, 440
  single nontransferable vote, *see* winning committee, SNTV
  single transferable vote, *see* winning committee, STV

SNTV, 413
STV, 423
w-Thiele, 451
Woeginger, G., 28, 203, 207, 210, 215, 219–221, 309, 365, 396, 504, 588, 590, 639, 645, 668
Wolpert, A., 28
Wolsey, L., 458
wonderful stability, *see* hedonic game, coalition structure for a, wonderfully stable
WONDERFUL-STABILITY-EXISTENCE, 211
WONDERFUL-STABILITY-VERIFICATION, 211
Wood, K., 178
Woodall, D., 269, 283, 560, 561, 565
Wooldridge, M., v, xii, 12, 14, 15, 17, 137, 172, 175, 181, 182, 192, 193, 206, 288, 607, 621, 647
World Rock Paper Scissors Society, 60
Wortman Vaughan, J., 631
WSE, 211
WSV, 211
Wu, X., 638, 646, 670, 672
WVG-CONSTRUCT-CORE, 181
WVG-CONSTRUCT-LEAST-CORE, 182
WVG-CONTROL-BY-DELETING-PLAYERS-TO-DECREASE-$\mathbb{PI}$, 190
WVG-CONTROL-BY-DELETING-PLAYERS-TO-INCREASE-$\mathbb{PI}$, 190
WVG-CONTROL-BY-DELETING-PLAYERS-TO-MAINTAIN-$\mathbb{PI}$, 190
WVG-COST-OF-STABILITY, 183
WVG-EMPTY-CORE, 181
WVG-IN-CORE, 181
WVG-IN-LEAST-CORE, 182
WVG-LEAST-CORE, 182
WVG-NULL, 184
WVG-$\mathbb{PI}$, 184
WVG-$\mathbb{PI}$-BENEFICIAL-MERGE, 188
WVG-$\mathbb{PI}$-BENEFICIAL-SPLIT, 187
WVG-$\mathbb{PI}$-POWER-COMPARE, 185
WVG-SUPER-IMPUTATION-STABILITY, 183

X3C, 350
X3C, 216
XESWO, 638
Xia, L., 18, 290, 301–306, 309, 313, 320, 321, 326, 327, 342, 345, 361, 362, 364, 366, 486, 633, 635, 642, 665, 670–672
Xiao, M., 665
XNPSWO, 642

XUSWO, 647

Yadid, T., 645
Yahtzee, 6
Yamazaki, K., 33
Yami, H., 627–629, 635, 675
Yang, Y., 355–357, 454, 464
Yannakakis, M., 37, 130
Yanovskaya, E., 670
Yariv, L., 178
Yazdanbod, S., 645
yin and yang, 19
Yin, S., 635
Yokoi, Y., 655
Yokoo, M., 193, 207, 655, 665, 673, 675
Young score, *see* $YScore(\cdot)$
Young winner, 245
Young, P., 245, 281, 291, 296
   see also StrongYoung voting, 245
   voting system of, 245, 264, 286, 291, 483
      homogeneous variant of the, 282, 296
$\overline{YScore}(\cdot)$, 245
$YScore(\cdot)$, 245
Yu, L., 362, 364

Yu, Y., 530

$\mathbb{Z}$, 123
Zachos, S., 292
Zaks, S., 229
Zankó, V., 129
Zemel, E., 177
Zeng, D., 674
Zenklusen, R., 639
Zenou, Y., 178
zero-sum game, 116, 122, 123
   Nash equilibrium in a, 122
   Nash equilibrium in a two-player, 122
zero-sum security game on graphs, 178
Zhang, L., 635
Zhang, Y., 624
Zick, Y., 192, 635, 665, 673, 675
von Ziegesar, C., 322
Zohar, A., 451, 462
Zorn, R., 355–357
Zuckerman, M., 158, 160, 183, 192, 309, 320, 321, 354
Zwick, U., 96
Zwicker, W., 18, 146, 173, 174, 572, 603, 671

SPRINGER NATURE

GPSR Compliance

The European Union's (EU) General Product Safety Regulation (GPSR) is a set of rules that requires consumer products to be safe and our obligations to ensure this.

If you have any concerns about our products, you can contact us on ProductSafety@springernature.com

In case Publisher is established outside the EU, the EU authorized representative is:

Springer Nature Customer Service Center GmbH
Europaplatz 3
69115 Heidelberg, Germany

The manufacturer's authorised representative in the EU is Springer Nature Customer Service Centre GmbH, Europaplatz 3, 69115 Heidelberg, Germany. If you have any concerns regarding our products, please contact ProductSafety@springernature.com

Printed and bound by CPI Group (UK) Ltd, Croydon, CR0 4YY

25/03/2026

02078199-0001